MECHANISMS OF RESISTANCE TO PLANT DISEASES

T0215638

Mechanisms of Resistance to Plant Diseases

Edited by

A.J. Slusarenko

RWTH,
Aachen, Germany

R.S.S. Fraser

Society for General Microbiology,
Reading, United Kingdom,
Honorary Visiting Professor,
University of Manchester, U.K.

and

L.C. van Loon

Utrecht University,
The Netherlands

KLUWER ACADEMIC PUBLISHERS
DORDRECHT / BOSTON / LONDON

A C.I.P. Catalogue record for this book is available from the Library of Congress.

ISBN 1-4020-0399-4
Transferred to Digital Print 2001

Published by Kluwer Academic Publishers,
P.O. Box 17, 3300 AA Dordrecht, The Netherlands.

Sold and distributed in North, Central and South America
by Kluwer Academic Publishers,
101 Philip Drive, Norwell, MA 02061, U.S.A.

In all other countries, sold and distributed
by Kluwer Academic Publishers,
P.O. Box 322, 3300 AH Dordrecht, The Netherlands.

Printed on acid-free paper

Cover illustration:
Part B, the dispersal of pigment throughout a dead cell,
of Figure 22 from the article by J.W. Mansfield (p.325-370).
Provided by Ralph Nicholson.

Table of Contents

1 CASE STUDIES

   1 Resistance to Tobacco Mosaic Virus in Tobacco Plants   1
      R.S.S. Fraser

   2 Black Rot of Crucifers   21
      Anne M. Alvarez

   3 The *Cladosporium fulvum*-Tomato Interaction   53
      A Model System to Study Gene-For-Gene Relationships
      Pierre J.G.M. De Wit

   4 The Barley/*Blumeria* (Syn. *Erysiphe*) *graminis* Interaction   77
      Hans Thordal-Christensen, Per L. Gregersen & David B. Collinge

2 GENETICS OF DISEASE RESISTANCE   101
Basic Concepts and Application in Resistance Breeding
Beat Keller, Catherine Feuillet & Monika Messmer

3 RESISTANCE IN POPULATIONS   161
J. Frantzen

4 RESISTANCE GENES AND THE PERCEPTION AND   189
TRANSDUCTION OF ELICITOR SIGNALS IN HOST-
PATHOGEN INTERACTIONS
Thomas Boller & Noel T. Keen

5 STRUCTURAL ASPECTS OF DEFENSE   231
Bruno Moerschbacher & Kurt Mendgen

6 THE HYPERSENSITIVE RESPONSE   279
Thorsten Jabs & Alan J. Slusarenko

7 ANTIMICROBIAL COMPOUNDS AND RESISTANCE   325
The Role of Phytoalexins and Phytoanticipins
J.W. Mansfield

8 **INDUCED AND PREFORMED ANTIMICROBIAL PROTEINS**   371
W.F. Broekaert, F.R.G. Terras & B.P.A. Cammue

9 **SPECIAL ASPECTS OF RESISTANCE TO VIRUSES**   479
R.S.S. Fraser

10 **SYSTEMIC INDUCED RESISTANCE**   521
L.C. Van Loon

11 **TRANSGENIC APPROACHES TO CONTROL EPIDEMIC**   575
SPREAD OF DISEASES
Ben J.C. Cornelissen & André Schram

SUBJECT INDEX   601
SPECIES INDEX   615

CHAPTER 1

CASE STUDIES

RESISTANCE TO TOBACCO MOSAIC VIRUS IN TOBACCO PLANTS

R.S.S. FRASER

Society for General Microbiology, Marlborough House, Basingstoke Road, Spencers Wood, Reading RG7 1AE, UK (r.fraser@sgm.ac.uk)

Summary

Tobacco mosaic virus (TMV) was first studied scientifically as a plant pathogen 100 years ago. A form of hypersensitive resistance transferred from the wild species *Nicotiana glutinosa* to cultivated tobacco was shown 60 years ago to be due to a single dominant gene, named *N*. This causes the virus to be localised to necrotic lesions which form around each site of infection: the hypersensitive response (HR). *N*-gene resistance has proved extremely durable: only one virulent (resistance-breaking) TMV isolate has been reported to date. Another resistance gene, *N'*, thought to be allelomorphic with *N*, causes a hypersensitive reaction to avirulent isolates of TMV, but numerous virulent isolates are also known. These do not induce necrosis but spread systemically and cause normal mosaic symptoms. The single known example of virulence against *N* has been mapped on TMV RNA to the replicase gene, whereas virulence against *N'* in different TMV isolates has been mapped to a number of locations, all within the coat protein gene. The *N* gene has been isolated and sequenced: it shows structural and possibly functional features in common with certain other genes for resistance to bacterial and fungal pathogens, and to other genes with known functions in control of development or response to hormones in animals. These similarities give some clues about how the *N*-gene product might be involved in TMV recognition and in signalling the cascade of resistance and other responses which follows. The actual mechanism which inhibits TMV spread or multiplication after resistance is induced is not yet fully clear, but may involve an inhibition of multiplication or blocking of cell-to-cell spread of the infection front.

Abbreviations

HR   hypersensitive response
TMV   tobacco mosaic virus
PR   pathogenesis-related protein

A. Slusarenko, R.S.S. Fraser, and K. van Loon (eds), Mechanisms of Resistance to Plant Diseases. 1-19.

I.   Introduction

A.   HISTORY

It is highly fitting that the case study of virus resistance chosen for this book is not only one of the best understood examples of a mechanism of resistance to a plant-virus, but also the subject of a notable centenary in 1998, the year of writing. Tobacco mosaic virus (TMV) had been identified in the 1880-1890 period (Matthews, 1991) as a sap-transmissible disease-causing agent that could pass though bacterial filters, but it was not until 1898 that the phenomenon became fully accepted (Beijerinck, 1898). This is now generally recognised as the birth of the study of plant virology. Forty years later, Holmes (1938) demonstrated that resistance to TMV, transferred from the wild species *Nicotiana glutinosa* to the cultivated tobacco *N. tabacum* by Clausen and Goodspeed (1925), was controlled by a single dominant gene which he named *N*. These publications may be regarded as the start of the scientific study of resistance to plant viruses and of its exploitation in practical crop protection. The *N* gene has also played a central role in the development of the classical technology of plant virus studies, such as infectivity assay (Holmes, 1929; Kleczkowski, 1950), and was the first gene for resistance to a plant virus to be isolated and sequenced (Dinesh-Kumar *et al.*, 1995).

B.   THE VIRUS AND ITS HOSTS

1.   *TMV as a model virus*

Before describing the phenomenon of TMV resistance, and the components of plant-virus interactions involved, it may be useful to describe TMV in more general terms as a structure and pathogen, especially for the reader not familiar with plant virology. In the context of this book, there are massive differences between viruses on the one hand, and bacteria and fungi on the other, both in terms of genetic and structural complexity, and in mechanisms of pathogenesis. TMV is taken in this section as a model virus which exemplifies a number of features of the viral 'lifestyle', an understanding of which is an essential foundation for studying the resistance mechanisms and other types of plant-virus interactions described here and in chapter 9. A more detailed description of the diversity of viruses as plant pathogens is given in the introduction to chapter 9. It must be stressed here that plant viruses have evolved many different ways of solving their 'lifestyle issues'; TMV is one well-understood example, but should not be considered in any way typical.

TMV infects host plants through natural or experimentally-caused wounds: its only biotic transmission agents are man or animals which may transfer it casually between infected and healthy plants by surface contact. This contrasts with the highly-specific interactions other viruses have with particular vectors such as leaf-feeding insects (see chapter 9).

In susceptible tobacco plants, TMV causes the characteristic light-green/dark-green mosaic pattern from which it derives its name, together with stunting and distortion of those leaves which become infected at an early stage of development (Fig. 1). The virus can accumulate to concentrations as high as 5-10 mg g^{-1} fresh weight of leaf (Fraser, 1987).

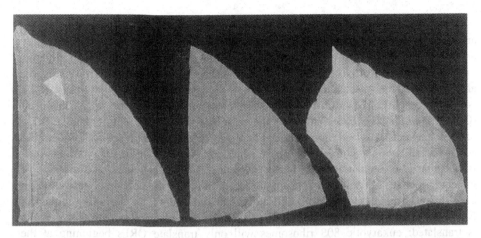

Figure 1. Different types of response to TMV infection in tobacco leaves. Centre: uninoculated, healthy leaf. Right: leaf of a susceptible variety showing the characteristic mosaic of light green/dark green tissue, after systemic spread of the virus. Left: inoculated leaf of a resistant variety carrying the *N* gene, showing the local lesions (arrowed) which form around each site of infection (HR).

TMV is a rigid, rod-shaped particle, 300 nm long and 18 nm in diameter, with a central hollow core 4 nm in diameter (Fig. 2). The rod contains a single RNA molecule of molecular mass 2.1 mDa, assembled with about 2100 coat protein molecules of molecular mass 17.6 kDa. In the particle structure, the RNA forms a helix, embedded within a helix of assembled coat protein molecules. Good illustrations of particle architecture are to be found in Matthews (1991), and excellent three-dimensional rotating depictions in Sgro (1995).

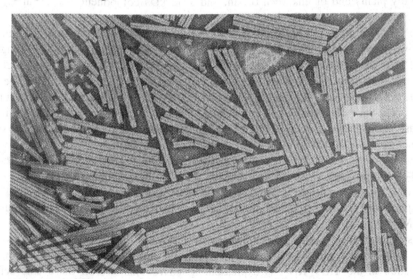

Figure 2. Electron micrograph of TMV particles. TMV in sap from an infected tobacco plant was negatively stained with methylamine tungstate. The hollow core of the particle is clearly visible in short lengths of broken particle seen end on. The scale bar indicates 50 nm. Photomicrograph by courtesy of Colin Clay.

The TMV RNA contains the information required to specify virus multiplication and pathogenesis, although both processes clearly involve interactions with the host and participation of host-coded components. TMV RNA from the virus particle is messenger sense (positive), in that it can be translated into protein in vitro or in vivo, although the actual mechanism of expression of all the proteins encoded by the RNA is rather complex. TMV RNA was the first plant virus RNA to be fully sequenced (Goelet *et al.*, 1982). Examination of the sequence for open reading frames (ORFs: regions between start and stop codons coding for a recognisable sequence of amino acids) has allowed the compilation of the genetic map shown in Fig. 3. From an AUG start codon after an untranslated leader sequence of 69 nucleotides, two proteins of molecular masses 126 and 183 kDa are produced, the larger by occasional read-through of a leaky termination codon (UAG, amber) at the end of the 126 kDa sequence. No other significant polypeptides are produced when the full length TMV RNA is translated: eukaryotic 80S ribosomes will only translate ORFs beginning at the 5'-proximal start codon.

The 126 and 183 kDa proteins are components of the TMV replicase (Osman and Buck, 1996). By comparisons with the amino acid sequences of replicase proteins from other putatively related single-stranded RNA plant and animal viruses forming what is known as the Sindbisvirus supergroup, three functional domains have been postulated. D1 is a methyltransferase activity thought to be involved in capping of mRNA with 7-methylguanosine triphosphate, which enhances translation. The full length TMV RNA and sub-genomic coat protein mRNA (see below) are capped; the other sub-genomic mRNAs are not. D2 is a helicase, and D3 is an RNA polymerase. The purified, functional replicase also contains a plant-coded protein (56 kDa) which is required for activity (see chapter 9), two components (54 and 50 kDa) not found in healthy plants and of unknown origin, and a 32 kDa component also occurring in healthy plants (Osman and Buck, 1997).

The TMV replicase is involved in the synthesis of a complementary full length (negative sense) RNA (Fig. 3). From this, three 3'-co-terminal sub-genomic messenger-sense RNAs are produced: all are detectable in infected plants. These contain at or close to their 5' ends the ORFs for the 54, 30 and 17.6 kDa proteins respectively. The 30 and 17.6 kDa proteins have been detected in infected plants but the 54 kDa protein has not: it is not yet established whether it is the 54 kDa component found in the purified replicase (Osman and Buck, 1997). The 17.6 kDa protein is the virus coat protein; the function of the 30 kDa protein will be described later.

The negative sense full length TMV RNA also functions as the template for synthesis of full-length progeny virus RNA molecules (Fig. 3). These then assemble with coat protein to form the progeny virus particles. The process of assembly requires no other participating macromolecular components, and can indeed be carried out in vitro (Matthews, 1991).

This description of the molecular biology of TMV multiplication covers the events within an individual infected cell, based largely on an understanding of how the genetic information in the TMV RNA is expressed. It leaves unanswered the events before and after that process: how the infection commences after the virus particle penetrates the wound, and how infection spreads through the plant from the initially infected cell.

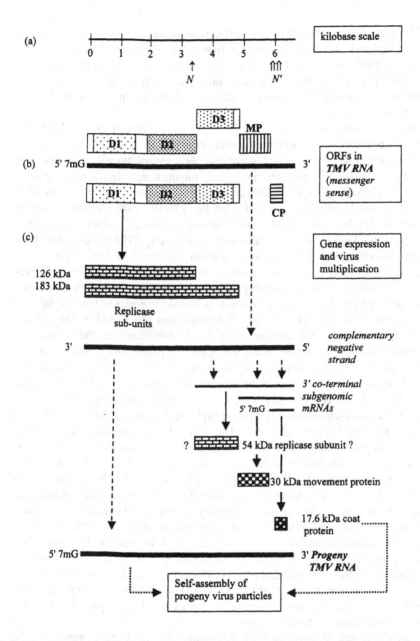

Figure 3. Organization and expression of the TMV genome. (a) Scale bar. (b) The various possible open reading frames (ORFs) on the genomic RNA are shown, together with putative functional domains in the replicase, as described in the text and in Lewandowski and Dawson (1995). (c) Replication and expression of the genome. Transcription to RNA copies is shown by broken arrows; translation to protein by solid arrows, and assembly of progeny virus particles by dotted arrows. The 54 kDa replicase subunit protein has not yet been shown to exist in infected plants, although its sub-genomic messenger RNA has been detected. This protein may equate to a 54 kDa protein found in the purified replicase (Osman and Buck, 1996, 1997) but not yet fully characterised. The kilobase scale (a) also shows the location of the determinants of virulence/avirulence against the *N* and *N'* genes (Padgett and Beachy, 1993; Taraporewala and Culver, 1996, 1997).

The TMV particle is extremely robust: infectivity is retained for decades in dead infected plant material, soil and groundwater. Infectivity will survive heating to 80°C, and treatment of particles with high levels of ribonuclease, or mutagens such as HNO_2 and ultraviolet light (Matthews, 1991). How then does such a 'survivor' expose its RNA to initiate infection? The answer is that it exploits the molecular biology of the multiplication process.

The coat protein subunits covering the 80 or so nucleotides at the 5' end of the genomic RNA appear to be comparatively loosely bound to the RNA and the other coat proteins in the particle. It is thought that in a region of the cytoplasm with comparatively high pH (8.0) and low Ca^{2+} concentration, these subunits may disassemble, leaving the 5' end of the TMV RNA exposed. This allows a ribosome to attach to the first AUG initiation codon, and commence translation of the 126/183 kDa replicase proteins. As it moves along the TMV RNA, the ribosome displaces further coat protein subunits, a process known as co-translational disassembly. The process has been demonstrated in vitro, and TMV particles partially stripped of coat proteins and expressing replicase proteins have been found in vivo (Shaw et al., 1986; Wilson et al., 1990). Co-translational disassembly effectively exposes the first three quarters of the genome from the 5' end, but translation is unable to proceed beyond the stop codon at the end of the 183 kDa protein. It appears that the 3' end of the genome is uncoated by removal of the subunits in a 3'→5' direction (i.e. the opposite to co-translational disassembly) by the newly expressed replicase, and concomitant with the synthesis of the negative strand viral RNA (Wu and Shaw, 1997). Presumably the nucleotide sequence at the 3' end of the genomic RNA where the replicase attaches and initiates transcription is accessible in the 5'→3' partially stripped particle.

TMV particles are much too large to pass through the plasmodesmata which provide cytoplasmic continuity between adjacent plant cells (Lucas and Gilbertson, 1994). Experiments with micro-injected dyes indicate that the size exclusion limit is of the order of 1.5-2.0 nm molecular diameter, equivalent to a molecular mass of around 1 kDa (Terry and Robards, 1987). The TMV 30 kDa protein has been shown to be tightly bound to the cell wall fraction and associated with plasmodesmata; it increases the plasmodesmatal size exclusion limit markedly to about 5-9 nm, although this is still not enough to allow the passage of intact virions (Wolf et al., 1989). It appears instead that the infectious entity that moves from cell to cell is the viral RNA. The 30 kDa protein has a single-stranded RNA-binding function, which opens up the free-folded TMV RNA (mean diameter 10 nm) to an extended, thinner, transferable form with a diameter around 2.5 nm (Lartey et al., 1997). An association of the movement protein-TMV RNA complex with elements of the cytoskeleton may also facilitate delivery of the complexes to the plasmodesmata.

Cell-to-cell movement of infection does not require the TMV coat protein, as coat protein-less mutants move with the same efficiency as the wild type (Dawson et al., 1988). Coat protein is, however, required for long-distance transport of infection in the phloem: it is possible that the movement of the infectious entity from mesophyll cell to sieve tube element is a different process from that involved in local cell-to-cell spread.

2.   *Host Range*

TMV is a member of the tobamovirus genus which contains 12 members infecting diverse plant groups, including tobacco, tomato, cucumber, orchids, frangipani and pepper (*Capsicum*) (Lewandowski and Dawson, 1995; Brunt *et al.*, 1996). Although tobamoviruses from different host groups tend to have a high degree of sequence similarity, with a minimum in the range 60-80%, the evidence is that in nature each is adapted to its particular host species or related group of species, with a comparatively narrow host range (Bald, 1960). In contrast, TMV and certain other tobamoviruses have been shown to have very wide *experimental* host ranges in laboratory tests of hundreds of species in numerous families (Horvath, 1978).

TMV causes systemic mosaic on numerous *Nicotiana* species, including *N. sylvestris, N. tomentosa* and *N. tomentosiformis.* An *N*-gene type of HR is also found in numerous species, including *N. glutinosa, N. rustica, N. gosseii, N. suaveolens* and *N. repanda* (Valleau, 1952). This may suggest that the association of TMV with the genus *Nictotiana* is one of long standing, and that resistance to the virus evolved at an early stage of speciation (Holmes, 1951). An alternative theory (Valleau, 1952), that the virus first spread from the wild host plantago to cultivated tobacco crops which had evolved susceptibility, appears untenable.

II.   The Genetics and Phenotype of Resistance

A.   THE *N* GENE

In *N. glutinosa* and *N. tabacum* cultivars containing the *N* gene, TMV does not spread systemically, but is localised to small areas of several hundred infected cells around each point of infection. After a few days, the infected cells become necrotic — the hypersensitive response (HR). Virus multiplication ceases (Fig. 4), although infectious virus can still be isolated from the lesions (Siegel, 1960). As described elsewhere in this book, HR is a common resistance response to invading bacteria, fungi and viruses of many species or types, although for viruses in particular there are numerous other types of resistance response (chapter 9).

The *N* gene was transferred from *N. glutinosa* to *N. tabacum* via an interspecific synthetic hybrid, *N. digluta* (Clausen and Goodspeed, 1925). It was thought that there had been a substitution of the entire *N. glutinosa* chromosome carrying the *N* gene (chromosome Hg) for the *N. tabacum* chromosome H, and that the *N. glutinosa* chromosome has been physiologically stable within the *N. tabacum* genetic background. This was thought to explain the ease with which the resistance gene was introgressed into commercial cultivars. Later, Gerstel (1948) did observe some exchange of segments between the H and Hg chromosomes. A useful review of the complex literature on the genetic and breeding history of the *N* gene is given by Dunigan *et al.* (1987).

Since its introduction, *N*-gene resistance has been widely incorporated into commercial tobacco cultivars; cigar-smoking readers may have noticed its presence as betrayed by the occasional lesion on the cigar outer leaf. However, TMV still causes severe crop losses in important tobacco-growing areas such as North and South Carolina, USA. This is because it has not yet been possible to breed flue-cured tobacco cultivars carrying the *N*-gene with the same quality and yield characteristics as susceptible cultivars (Barnett, 1995), despite the early discovery of the gene and elucidation of the inheritance of resistance. The problem may stem from other genes affecting yield and quality in the persisting Hg chromosomes or segments of it. Growers clearly prefer to risk losing a portion of their crop each year to TMV, in return for the economic benefits of higher quality and hoped for higher yield of the susceptible cultivars.

The *N* gene, where used in tobacco cultivars, has been remarkably durable, in that the resistance has not been overcome by TMV types occurring in tobacco cultivation. A single resistance-breaking (virulent) strain has been isolated from pepper (Csillery *et al.*, 1983). This was initially characterised as an isolate of the related tobamovirus tomato mosaic virus, but later as another related tobamovirus, *Solanum dulcamare* yellow fleck virus (Sanfaçon *et al.*, 1993).

Resistance conferred by the *N* gene to avirulent isolates is temperature sensitive, breaking down at 28-30°C to allow systemic spread of the virus (Takahashi, 1975; De Laat and Van Loon, 1983), but this does not appear to have detracted from the usefulness of the gene in practical crop protection.

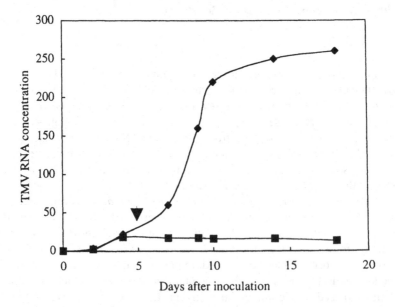

Figure 4. The effects of the hypersensitive response on multiplication of TMV. Tobacco cv. White Burley, which contains the *N'* resistance gene, was inoculated with virulent (♦) or avirulent (■) isolates of TMV. Virus concentration was measured by polyacrylamide gel electrophoresis of extracted nucleic acids, and TMV RNA concentration expressed as µg g $^{-1}$ fresh weight. Necrotic lesions began to appear on leaves inoculated with the avirulent isolate at the point shown by the arrow.

B.   THE *N'* AND *EN* GENES

These genes are of interest for cytogenetic reasons and for what they tell us about plant-virus interactions, rather than because of any practical value in plant breeding. The cultivated species *N. tabacum* does not occur naturally in the wild, but is an amphidiploid derived from the wild species *N. sylvestris* and *N. tomentosiformis* and presumably selected by an early plant breeder on the American continent. *N. sylvestris* is systemically infected by many isolates of TMV, but forms the typical local lesions of HR when inoculated with other (avirulent) isolates (Weber, 1951). This response is controlled by a dominant gene, named *N'*, thought to be allelomorphic with *N* (Valleau, 1952; Dunigan *et al.*, 1987). A gene from *N. sylvestris* and certain *N. tabacum* cultivars with similar ability to discriminate TMV strains, but showing incomplete dominance, was named n^s by Weber (1951), who suggested that it might be synonymous with *N'*.

N. tomentosiformis plants are systemically infected by (susceptible to) all isolates of TMV. *N. tabacum* cultivars may exhibit HR to avirulent isolates of TMV if they have inherited the N' gene, or are infected systemically by all isolates of the virus if they do not have *N'* from their natural parentage and have not had *N* transferred by the plant breeder. In the former case, it must be assumed that tobacco cultivars showing systemic infection by all strains of TMV have somehow lost the *N'* genes on the chromosome pairs derived from the *N. sylvestris* parent, or that *N'* has become suppressed or masked in some way.

N. tomentosiformis plants, and *N. tabacum* cultivars susceptible to all isolates of TMV, are referred to as containing the *n* gene. This may be either an inactive allelic form of *N* and *N'*, or may be a null allele of them (Fraser 1986; Dunigan *et al.*, 1987). The latter explanation assumes that when a resistance gene is transferred from a wild species via an introgressed chromosome or chromosomal segment, there may be no corresponding DNA sequence in the susceptible parent genome. This question is difficult to resolve by classical cytogenetic methods, but the functional status or otherwise of the putative *n* allele may be approachable now that the *N*-gene and related sequences can be studied directly (Dinesh-Kumar *et al.*, 1995).

Melchers *et al.* (1966) found a spontaneous mutation of the *nn* variety Samsun which gave a necrotic reaction to the same isolates of TMV as *N'*-containing varieties, and may have represented a back mutation of *n* to *N'*, or a release from a masking effect. They named this gene *EN*.

Genetically, the tobacco-TMV interaction displays a limited form of the gene-for-gene interaction between resistance/susceptibility in the host and virulence/avirulence in the pathogen, which is set out in Table 1. The TMV-*N'* gene interaction has an interesting historical significance in the development of genetic knowledge: the ability to create local-lesion-forming isolates by nitrous acid mutagenesis of systemic mosaic-inducing strains provided some of the first experimental evidence that mutation resulted from a single base alteration (Gierer and Mundry, 1958).

More complex gene-for-gene interactions between plants and viruses are discussed in chapter 9.

Table 1.   The gene-for-gene interaction between TMV and tobacco

	TMV isolate	
Host resistance	Virulent	Avirulent
N	systemic mosaic	local lesion
	(very rare)	hypersensitive response
N'	systemic mosaic	local lesion
	(common)	hypersensitive response
n	systemic mosaic	systemic mosaic[1]

[1] In a virus isolate, virulence/avirulence is assigned in the light of interaction with a specific resistance gene. Strictly speaking, the term avirulence has no meaning in the context of a host with no known resistance gene against that virus, but this is a matter of semantics, and does not compromise the integrity of the gene-for-gene interaction.

III.   The Biochemistry of the Resistance Response

A.   RECOGNITION

The widely supported model for HR, for bacterial and fungal pathogens as well as viruses, is that a 'recognition event' occurs between some product of the resistance gene, and an avirulence determinant in the pathogen. This recognition then induces a series of responses, possibly by complex signalling pathways, which give effect to the resistance mechanism(s) that block the pathogenic process, and a cascade of associated changes. The challenges have been to identify the critical recognition event and the molecular participants in it, the nature of the resistance mechanism(s) induced, and those changes which are secondary. The comparatively simple gene-for-gene interaction between tobacco and TMV has lent itself to analysis of some of these questions.

1.   Mapping of the Determinants of Virulence and Avirulence on TMV

TMV is a comparatively simple plant virus. The genetic map (Fig. 3) shows that almost the entire genome is taken up by the replicase, movement protein and coat protein functions, essential to the full replicative cycle in the susceptible plant. There is no 'spare capacity' that could be devoted purely and solely to determinants of virulence or avirulence, if these determinants operated at the protein level. Even if the viral determinant in the recognition event were to operate at the RNA level, or as a product of a virus-coded protein, this would have to be in the overall context of another function for that gene in the pathogenicity of the virus.

The ability to make cDNA copies of plant viruses with RNA genomes, and the fact that these copies can themselves be infectious (Weber et al., 1992) or can be used to produce infectious RNA transcripts (Meshi *et al.*, 1986), revolutionised the approach to genetic mapping and analysis of virulence. Knowledge of sequence data and appropriate restriction enzymes have been used to make artificial recombinants between virulent and avirulent isolates, which could then be tested for biological activity in resistant and susceptible plants. This allowed mapping of determinants first to particular functional regions of the viral genome, and ultimately with the aid of site-directed mutagenesis to individual nucleotide residues and consequent changes in single amino acids in the derived proteins. Intriguingly, given the proposed allelomorphic nature of *N* and *N'*, the determinants of virulence/avirulence mapped to different viral functions. For *N'* the determinant is in the coat protein gene (Culver and Dawson, 1989; Pfitzner and Pfitzner, 1992). Later work demonstrated that a number of amino acid residues at non-contiguous positions in the coat protein sequence are essential for triggering the HR response, but in the three-dimensional folded structure of the protein these come together to form a surface which was proposed as the binding site with the host receptor (Taraporewala and Culver, 1996; 1997). For *N,* the rare example of a virulent isolate allowed mapping of the determinant to the 126 kDa replicase gene (Ikeda *et al.*, 1993; Padgett and Beachy, 1993) (Fig. 3).

Examples of virulence/avirulence determinants for other resistance genes which map to other viral functions are given in chapter 9.

2. *Isolation of the N Gene*

Early attempts to isolate the *N* gene utilised differential hybridization procedures to compare messenger RNA or translated protein populations from HR-expressing and non-expressing hosts, and while several additional proteins were found to be expressed during HR, none could be specifically associated with *N*-gene activity (Smart *et al.*, 1987, Dunigan *et al.*, 1987).

Later work involved the use of transposon tagging to disable *N*-gene function, and selection procedures to detect mutant plants with the tagged *N* gene (Whitham *et al.*, 1994; Dinesh-Kumar *et al.*, 1995). The maize transposon *Ac* integrates itself into the plant genome and can disable gene function at the point of integration. Tobacco plants with the *NN* genotype were treated with *Ac*, and crossed with *nn* plants. The *Nn* progeny seedlings were then inoculated with TMV and maintained at 30°C, to allow systemic multiplication of the virus. The seedlings were then shifted to 21°C, when those with a functional *N* gene suffered lethal systemic necrosis, whereas those with the *N* gene disabled by transposon insertion survived. One unstable mutant line, giving rise to sectored progeny, was shown to contain an *Ac* element insertion in a large open reading frame. This was used to recover a homologous full length gene from a genomic library of *N. glutinosa*. This gene was shown to be the *N* gene by transformation into susceptible tobacco, which subsequently exhibited HR when inoculated with TMV. Further confirmation that this single gene was enough to switch on the HR response after infection came from the demonstration that it could also confer HR in transformed tomato plants (Whitham *et al.*, 1996). This also demonstrated that all the genetic elements downstream from the initial recognition event between the *N* gene product and TMV, and necessary for the full expression of HR, are present in tomato as well as in tobacco.

3.   Sequence of the N Gene and Similarities to Other Genes

Sequence analysis of the *N* gene cDNA and genomic clones indicated that the gene contains five exons which could be spliced to give a protein with 1144 amino acids and a molecular mass of 131 kDa. An alternative splicing route might give rise to a truncated protein of 75 kDa, containing the N-terminal region and an additional C-terminal region of 36 amino acids (Whitham *et al.*, 1994, Dinesh-Kumar *et al.*, 1995) (Fig. 5). Full-length and truncated forms have been reported for other receptor proteins such as mammalian cytokine and growth factor receptors, although the possible significance of the two forms in vivo and in regulation of *N*-gene activity is not known.

Structures of pathogen resistance genes in plants

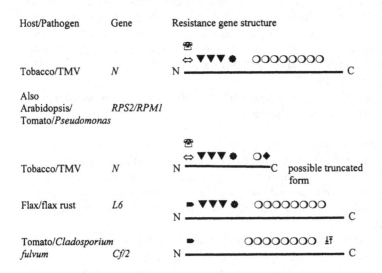

Figure 5.   Structural features of the *N* gene for TMV resistance, in full length and possible truncated form, based on data from Whitham *et al.* (1994), and comparison with other known genes for resistance to bacterial and fungal pathogens. (OOOO) leucine-rich repeats; (▼) nucleotide binding sites; (☎) cytoplasmic signalling domains which may affect nuclear transcription factors; (●) regions of conserved nucleotide sequence; (◆) additional sequence of 30 amino acid residues in the putative truncated form of the *N*-gene protein; (⇔) leucine zipper or Toll-like domain; (➡) signal peptide for membrane transport; (⌶) transmembrane domain. Based on data in Bent (1996) and Gebhardt (1997).

Comparisons of the amino acid sequence of the *N* gene with sequences of other genes of known activities revealed a number of intriguing similarities, which may provide clues to the mechanism of gene action. Firstly, the *N* gene shows some close structural similarities to a number of other plant genes for disease resistance, including the *RPS2* and *RPM1* genes for resistance to *Pseudomonas syringae* pathovars in *Arabidopsis* (Bent *et al.*, 1994; Grant *et al.*, 1995); *Prf* for resistance to *P. syringae* in tomato (Salmeron *et al.*, 1994), and to a lesser extent, *L6* for resistance to *Melampsora lini* in flax (Lawrence *et al.*, 1995). There are therefore common structural elements in genes for resistance to quite distinct types of pathogen: viral, bacterial and fungal. As shown in Fig. 5, the common elements include those referred to as cytoplasmic elements, three nucleotide binding sites, and a region of leucine-rich repeats. The possible functional significance of these common structural regions will be considered in the next section. It is important to point out, however, that the product of the *N* gene is predicted to be cytoplasmic, on the basis of its sequence, whereas several of the genes for resistance to fungal and bacterial pathogens contain trans-membrane domains. This difference may reflect the cytoplasmic location of TMV multiplication, and the possible involvement of membranes in the recognition of and response to microbial pathogens.

Although there is strong similarity between these genes in terms of functional regions, this is not reflected in strong conservation of nucleic acid sequence, apart from two highly conserved regions (Gebhardt, 1997). Other plant resistance genes against bacterial and fungal pathogens are also known which do not share some or all of these functional regions (Bent, 1996, Gebhardt, 1997).

B.   SIGNAL TRANSDUCTION

The main structural motifs described above in *N* and certain other resistance genes also occur in a number of other proteins whose function is more clearly understood, for example in the activity of mammalian hormones, or control of insect development. There are helpful indicators here of how the resistance genes may operate against pathogens, although no single clear-cut mechanism emerges. Rather, the implication is that the resistance response is likely to be multifaceted, and composed of pathogen-specific and non-specific elements.

Leucine-rich repeats have been identified in genes in a number of organisms and associated with diverse functions; they are thought to mediate protein-protein interactions (Kobe and Deisenhofer, 1995), and have been proposed as the site of pathogen-specific recognition through interaction with the avirulence gene product (Bent, 1996; Jones and Jones, 1997). The leucine-rich regions are imperfect repeats of units approximately 25 amino acids long: pathogen-specificity might be determined by the other amino acids present. It is likely that the leucine and other hydrophobic residues form the inside of a 'fist'-like structure with the other, intervening amino acids exposed (Bent, 1996). An alternative model is that the leucine-rich repeats might be involved in the interaction with other proteins involved in later stages of signal transduction and activation of the various components of the resistance response (Dixon *et al.*, 1996).

The nucleotide binding sites appear to be essential for the HR-inducing function, as mutations in these sites may eliminate function, although the mechanism of action is still unclear (Bent, 1996).

The large N-terminal domain of the *N* gene bears similarity to the so-called cytoplasmic signalling domains in the *Drosophila* Toll protein, which is a receptor involved in the establishment of polarity during insect development, and in the mammalian interleukin-1 receptor involved in response to cytokine IL-1 (Whitham *et al.*, 1994). Both are thought to activate transcription factors which cause changes in gene expression. It has been suggested that this region of the *N* gene might similarly be involved in the mechanism of activation of genes involved in the HR.

A further proposal by analogy with the animal pathways is that this region of
N might be involved in the stimulated production of activated oxygen species.
The 'oxidative burst' is one of the earliest detectable events in the onset of HR
(Doke and Ohashi, 1988; see chaper 6). It may be involved in redox regulation of
transcription factors affecting gene expression, as well as potentially in other disease
resistance and signalling mechanisms (Tenhaken *et al.*, 1995; Hammond-Kosack and
Jones, 1996). Specifically, it is likely that the oxidative burst is involved in the
induction of salicylic acid synthesis, an important signalling molecule for further
downstream changes (Klessig and Malamy, 1994; Leon *et al.*, 1995). The development
of HR in *N* gene tobacco plants inoculated with TMV was inhibited when the plants
were exposed to low oxygen pressure, but virus multiplication was not inhibited by
low oxygen (Mittler *et al.*, 1996).

C.   HOW DOES *N*-GENE RESISTANCE TO TMV WORK?

During the hypersensitive response to TMV — and indeed to numerous other pathogens
in many plant species — a series of changes occurs in the host as part of programmed
cell death in the lesion, in surrounding non-necrotic tissue, and systemically. The
changes include accumulation of phytoalexins, synthesis of proteins not normally
present in healthy plants, such as the pathogenesis-related (PR) proteins (see chapters
9 and 10), increased activities of certain enzymes such as phenylalanine ammonia
lyase and chalcone synthase, alterations in plant cell wall structure, and changes in
concentrations of plant growth regulators. The early literature is reviewed in Fraser
(1985; 1987) and more recent developments in Hammond-Kosack and Jones (1996).
It does seem that many of the observed changes, although induced by TMV infection
and occurring as part of the local lesion response, are not directly involved in virus
resistance. Thus there is convincing evidence that some PR proteins are involved in
resistance to fungal and bacterial pathogens (Zhu *et al.*, 1994) but no firm evidence
that they have any direct role in virus resistance (Fraser, 1982), although contrary
statements do appear in the literature (e.g. Chivasa *et al.*, 1997).
    The early pattern of multiplication of TMV in resistant plants, such as shown in
Fig. 4, and the fact that HR resistance does not operate when isolated protoplasts are
infected (Otsuki *et al.*, 1972) suggest that the restriction of virus multiplication and
spread requires induction of the host mechanism; it is not constitutive in the initially
infected cells. The fact that virus particles can be detected outside the necrotic region
(Da Graca and Martin 1976; Konate *et al.*, 1983) suggests that cell necrosis in the core
of the lesion is not the active defence mechanism, but a secondary response.
    This leaves two likely mechanisms. The tissue around the focus of infected cells
may have induced in it some inhibitor of virus multiplication, which increasingly
restricts multiplication. Alternatively or additionally, the spread of virus from cell to
cell might be inhibited.
    Possible inhibitors of virus replication (IVR) have been isolated from TMV-infected
N-gene tobacco plants (Gera *et al.*, 1990, 1993), and these might be involved in
the resistance response. The extent of inhibition of virus multiplication in the assays
used appeared too low to account for the complete effectiveness of *N* in stopping

infection, and non-specific inhibitory effects on plant metabolism were not excluded. Another group of putative antiviral factors (AVF) was reported by Sela *et al.* (1987) and extensive comparisons were drawn with the human interferon system, but the relationship to *N*-gene resistance *in vivo* is not clear. Some of the plant proteins purified using monoclonal antibodies to human interferon have now been shown to have no sequence similarity to interferon (Edelbaum *et al.*, 1990), and to correspond to the TMV-induced PR proteins β-1,3-glucanase and another PR of unknown function (Edelbaum *et al.*, 1991).

Inhibition of cell-to-cell spread appears an attractive possible explanation of localizing resistance, and for other examples of mechanisms of resistance to plant viruses the evidence that it is the main mode of action is quite compelling (chapter 9). What is known is that creating the conditions required for cell-to-cell spread of infection is an active process, involving a virus-coded movement protein and modification of the plasmodesmata (Lucas and Gilbertson, 1994). In the case of the HR response to TMV infection of tobacco, it may be notable that the amount of 30 kDa movement protein in the cell wall fraction decreases sharply as virus localization commences and necrosis appears (Moser *et al.*, 1988).

It is also possible that restriction of virus spread involves the creation of physical barriers by alterations in wall structure or plasmodesmatal properties. There are several reports of increased lignification and deposition of callose at the lesion edge (reviewed by Fraser, 1985; Hammond-Kosack and Jones, 1996). Interestingly, *N*-gene tobacco plants transformed to express an antisense construct of the basic tobacco class 1 β-1,3-glucanase, and as a result showing reduced glucanase activity, were shown to have decreased spread of TMV when inoculated (Beffa *et al.*, 1996). This was attributed to an increased deposition of callose on the plasmodesmatal pores, which may have hindered cell-to-cell spread of the virus. While this is an attractive possible explanation of enhanced localisation, it is difficult to reconcile with the *increased* glucanase activity stimulated by necrotic infection in the form of PR proteins, and the smaller lesions formed on inoculation of leaves with acquired systemic resistance (chapters 9 and 10). Clearly, there are many intriguing clues to how *N*-gene resistance to TMV may operate: the elucidation of the mechanism, or the relative importance of different components of the mechanism, is eagerly awaited.

References

Bald JG (1960) Forms of tobacco mosaic virus. Nature 188: 645—647

Barnett OW (1995) Plant virus disease — economic aspects. In: Webster RG and Granoff A (eds) Encyclopedia of Virology, CD-ROM Edition. Academic Press, London

Beffa RS, Hofer RM, Thomas M and Meins F (1996) Decreased susceptibility to viral disease of β-1,3-glucanase-deficient plants generated by antisense transformation. Plant Cell 8: 1001—1011

Beijerinck MW (1898) Over een contagium vivum fluidum als oorzaak van de vlekziekte der tabaksbladen. Versl Gewone Vergad Afd Wis-Natuurkd Kon Akad Wetensch Amsterdam 7: 229—235

Bent AF (1996) Plant disease resistance genes: function meets structure. Plant Cell 8: 1757—1771

Bent AF, Kunkel BN, Dahlbeck D, Brown KL, Schmidt RL, Giraudat J, Leung JL and Staskawicz BJ

(1994) *RPS2* of *Arabidopsis thaliana*: a leucine-rich repeat class of plant disease resistance genes. Science 265: 1856—1860

Brunt AA, Crabtree K, Dallwitz MJ, Gibbs AJ and Watson L (1996) Viruses of Plants. Descriptions and Lists from the VIDE Database. CAB International, Wallingford

Chivasa S, Murphy AM, Naylor M and Carr JP (1997) Salicylic acid interferes with tobacco mosaic virus replication via a novel salicylhydroxamic acid-sensitive mechanism. Plant Cell 9: 547—557

Clausen RE and Goodspeed TH (1925) Interspecific hybridization in *Nicotiana*. II. A tetraploid *glutinosa-tabacum* hybrid, an experimental verification of Winge's hypothesis. Genetics 10: 278—284

Csillery G, Tobias I and Rusko J (1983) A new pepper strain of tomato mosaic virus. Acta Phytopathol Acad Sci Hung 18: 195—200

Culver JN and Dawson WO (1989) Point mutations in the coat protein gene of tobacco mosaic virus induce hypersensitivity in *Nicotiana sylvestris*. Mol Plant-Microbe Interact 2: 209—213

Da Graca JV and Martin MM (1976) An electron microscope study of hypersensitive tobacco infected with tobacco mosaic virus at 32°C. Physiol Plant Pathol 8: 215—219

Dawson WO, Bubrick P and Grantham GL (1988) Modification of the tobacco mosaic virus coat protein gene affecting replication, movement and symptomatology. Phytopathology 78: 783—789

De Laat AMM and Van Loon LC (1983) Effects of temperature, light and leaf age on ethylene production and symptom expression in virus-infected tobacco leaves. Physiol Plant Pathol 22: 275—283

Dinesh-Kumar SP, Whitham S, Choi D, Hehl R, Corr C and Baker B (1995) Transposon tagging of tobacco mosaic virus resistance gene *N*: its possible role in the TMV-*N*-mediated signal transduction pathway. Proc Natl Acad Sci USA 92: 4175—4180

Dixon MS, Jones DA, Keddie JS, Thomas CM, Harrison K and Jones JDG (1996) The tomato *Cf-2* disease resistance locus comprises two functional genes encoding leucine-rich repeat proteins. Cell 84: 451—459

Doke N and Ohashi Y (1988) Involvement of an O_2^- generating system in the induction of necrotic lesions on tobacco leaves infected with tobacco mosaic virus. Physiol Mol Plant Pathol 32: 163—175

Dunigan DD, Golemboski DB and Zaitlin M (1987) Analysis of the *N* gene of *Nicotiana*. In: Evered D and Harnett S (eds) Plant Resistance to Viruses (Ciba Foundation Symposium 133), pp 120—135. John Wiley and Sons, Chichester

Edelbaum O, Ilan N, Grafi G, Sher N, Stram Y, Novick D, Tal N, Sela I and Rubinstein M (1990) Two antiviral proteins from tobacco: purification and characterization by monoclonal antibodies to human interferon. Proc Natl Acad Sci USA 87: 588—592

Edelbaum O, Sher N, Rubinstein M, Novick D, Tal N, Moyer M, Ward E, Ryals J and Sela I (1991) Two antiviral proteins, gp35 and gp22, correspond to β-1,3-glucanase and an isoform of PR-5. Plant Mol Biol 17: 171—173

Fraser RSS (1982) Are 'pathogenesis-related' proteins involved in acquired systemic resistance of tobacco plants to tobacco mosaic virus? J Gen Virol 58: 305—313

Fraser RSS (1985) Mechanisms involved in genetically controlled resistance and virulence: virus diseases. In: Fraser RSS (ed) Mechanisms of Resistance to Plant Diseases, pp 143—196. Martinus Nijhoff/Dr W Junk, Dordrecht

Fraser RSS (1986) Genes for resistance to plant viruses. Crit Rev Plant Sci 3: 257—294

Fraser RSS (1987) Biochemistry of Virus-Infected Plants. Research Studies Press, Letchworth/John Wiley and Sons, New York and Chichester

Gebhardt C (1997) Plant genes for pathogen resistance — variation on a theme. Trends Plant Sci 2: 243—244

Gera A, Loebenstein G, Saloman R and Frank A (1990) An inhibitor of virus replication (IVR) from protoplasts of a hypersensitive tobacco cultivar infected with tobacco mosaic virus, is associated with a 23K protein species. Phytopathology 80: 78—81

Gera A, Tam Y, Teverovsky E and Loebenstein G (1993) Enhanced tobacco mosaic virus production and suppressed synthesis of the inhibitor of virus replication in protoplasts and plants of local lesion responding cultivars exposed to 35°C. Physiol Mol Plant Pathol 43: 299—306

Gerstel DU (1948) Transfer of the mosaic-resistance factor between H chromosomes of *Nicotiana glutinosa* and *N. tabacum*. J Agric Res 76: 219—223

Gierer A and Mundry KW (1958) Production of mutants of tobacco mosaic virus by chemical alteration of its ribonucleic acid *in vitro*. Nature 182: 1457—1458

Goelet P, Lomonossoff GP, Butler PJG, Akam ME, Gait MJ and Karn J (1982) Nucleotide sequence of tobacco mosaic virus RNA. Proc Natl Acad Sci USA 79: 5818—5822

Grant MR, Godiard L, Straube E, Ashfield T, Lewald J, Sattler A, Innes RW and Dangl JL (1995) Structure of the *Arabidopsis RPM1* gene enabling dual specificity disease resistance. Science 269: 843—846

Hammond-Kosack KE and Jones DGJ (1996) Resistance gene-dependent plant defence responses. Plant Cell 8: 1773—1791

Holmes FO (1929) Local lesions in tobacco mosaic. Bot Gaz (Chicago) 87: 39—55

Holmes FO (1938) Inheritance of resistance to tobacco mosaic virus in tobacco. Phytopathology 28: 553—561

Holmes FO (1951) Indications of a New-World origin of tobacco mosaic virus. Phytopathology 41: 341—349

Horvath J (1978) New artificial hosts and non-hosts of plant viruses and their role in the identification and separation of viruses. IV. Tobamovirus group: tobacco mosaic virus and tomato mosaic virus. Acta Phytopathol Acad Sci Hung 13: 57—73

Ikeda R, Watanabe E, Watanabe Y and Okada Y (1993) Nucleotide sequence of tobamovirus Ob which can spread systemically in *N* gene tobacco. J Gen Virol 73: 1939—1944

Jones DA and Jones JDG (1997) The roles of leucine-rich repeat proteins in plant defences. Adv Bot Res Adv Plant Pathol 24: 89—167

Kleczkowski A (1950) Interpreting relationships between concentrations of plant viruses and numbers of local lesions. J Gen Microbiol 4: 53—69

Klessig DF and Malamy J (1994) The salicylic acid signal in plants. Plant Mol Biol 26: 1439—1458

Kobe B and Deisenhofer J (1995) A structural basis of the interactions between leucine-rich repeats and protein ligands. Nature 374: 183-186

Konate G, Kopp M and Fritig B (1983) Studies on TMV multiplication in systemically and hypersensitively reacting tobacco varieties by means of radiochemical and immunoenzymatic methods. Agronomie 3: 95

Lartey L, Ghoshroy S, Sheng J and Citovsky V (1997) Transport through plasmodesmata and nuclear pores: cell-to-cell movement of plant viruses and nuclear import of *Agrobacterium* T-DNA. In: McCrae MA, Saunders JR, Smyth CJ and Stow ND (eds) Molecular Aspects of Host-Pathogen Interactions, pp 253—280. Society for General Microbiology Symposium Series volume 55. Cambridge University Press, Cambridge

Lawrence GJ, Finnegan EJ, Ayliffe MA and Ellis JG (1995) The *L6* gene for flax rust resistance is related to the Arabidopsis bacterial resistance gene *RPS2* and the tobacco viral resistance gene *N*. Plant Cell 7: 1195—1206

Leon J, Lawton MA and Raskin I (1995) Hydrogen peroxide stimulates salicylic acid biosynthesis in tobacco. Plant Physiol 108: 1673—1678

Lewandowski DJ and Dawson WO (1995) Tobamoviruses. In: Webster RG and Granoff A (eds) Encyclopedia of Virology, CD-ROM Edition. Academic Press, London

Lucas WJ and Gilbertson RL (1994) Plasmodesmata in relation to viral movement within leaf tissues. Annu Rev Phytopathol 32: 387—411

Matthews REF (1991) Plant Virology (Third Edition). Academic Press, San Diego.

Melchers G, Jockusch H and Sengbusch PV (1966) A tobacco mutant with a dominant allele for hypersensitivity against some TMV strains. Phytopathol Z 55: 86—88

Meshi T, Ishikawa M, Motoyoshi F, Semba K and Okada Y (1986) In vitro transcription of infectious RNAs from full-length cDNAs of tobacco mosaic virus. Proc Natl Acad Sci USA 83: 5043—5047

Mittler R, Shulaev V, Seskar M and Lam E (1996) Inhibition of programmed cell death in tobacco plants during a pathogen-induced hypersensitive response at low oxygen pressure. Plant Cell 8: 1991—2001

Moser O, Gagey MJ, Godefroy-Colburn T, Stussi-Garaud C, Ellwart-Tschurtz M and Nitschko H (1988) The fate of the transport protein of tobacco mosaic virus in systemic and hypersensitive tobacco hosts. J Gen Virol 69: 1367—1378

Osman TAM and Buck KW (1996) Complete replication in vitro of tobacco mosaic virus RNA by a template-dependent membrane-bound RNA polymerase. J Virol 70: 6227—6234

Osman TAM and Buck KW (1997) The tobacco mosaic virus RNA polymerase complex contains a plant protein related to the RNA-binding subunit of yeast eIF-3. J Virol 71: 6057—6082

Otsuki Y, Shimomura T and Takebe I (1972) Tobacco mosaic virus multiplication and expression of the *N* gene in necrotic responding tobacco varieties. Virology 50: 45—50

Padgett HS and Beachy RN (1993) Analysis of a tobacco mosaic virus strain capable of overcoming *N* gene-mediated resistance. Plant Cell 5: 577—586

Pfitzner UM and Pfitzner AJ (1992) Expression of a viral avirulence gene in transgenic plants is sufficient to induce the hypersensitive defense reaction. Mol Plant-Microbe Interact 6: 318—321

Salmeron JM, Barker SJ, Carland FM, Mehta AY and Staskawicz BJ (1994) Tomato mutants altered in bacterial disease resistance provide evidence for a new locus controlling pathogen recognition. Plant Cell 6: 511—520

Sanfaçon H, Cohen JV, Elder M, Rochon DM and French CJ (1993) Characterization of *Solanum dulcamara* yellow fleck-Ob, a tobamovirus that overcomes the *N* resistance gene. Phytopathology 83: 400—404

Sela I, Grafi G, Sher N, Edelbaum O, Yagev H and Gerassi E (1987) Resistance systems related to the *N* gene and their comparison with interferon. In: Evered D and Harnett S (eds) Plant Resistance to Viruses (Ciba Foundation Symposium 133), pp 109—119. John Wiley and Sons, Chichester

Sgro J-Y (1995) Special section on virus visualization. In: Webster RG and Granoff A (eds) Encyclopedia of Virology, CD-ROM version. Academic Press, London

Shaw JG, Plaskitt KA and Wilson TMA (1986) Evidence that tobacco mosaic virus particles disassemble co-translationally in vivo. Virology 148: 326—336

Siegel A (1960) Studies on the induction of tobacco mosaic virus mutants with nitrous acid. Virology 11: 156—167

Smart TE, Dunigan DD and Zaitlin M (1987) *In vitro* translation products of mRNAs derived from TMV-infected tobacco exhibiting a hypersensitive response. Virology 158: 461—464

Takahashi T (1975) Studies on viral pathogenesis in plant hosts: VIII. Systemic virus invasion and localization of infection in 'Samsun-NN' plants resulting from tobacco mosaic virus infection. Phytopathol Z 84: 75—87

Taraporewala ZF and Culver JN (1996) Identification of an elicitor active site within the three-dimensional structure of the tobacco mosaic virus tobamovirus coat protein. Plant Cell 6: 169—178

Taraporewala ZF and Culver JN (1997) Structural and functional conservation of the tobamovirus coat protein elicitor active site. Mol Plant-Microbe Interact 10: 597—604

Tenhaken R, Levine A, Brisson LF, Dixon RA and Lamb C (1995) Function of the oxidative burst in hypersensitive disease resistance. Proc Natl Acad Sci USA 92: 4158—4163

Terry BR and Robards AW (1987) Hydrodynamic radius alone governs the mobility of molecules through plasmodesmata. Planta 171: 145—157

Valleau WD (1952) The evolution of susceptibility to tobacco mosaic virus in *Nicotiana* and the origin of the tobacco mosaic virus. Phytopathology 42: 40—42

Weber H, Haeckel P and Pfitzner AJP (1992) A cDNA clone of tobacco mosaic virus is infectious in plants. J Virol 66: 3909—3912

Weber PVV (1951) Inheritance of a necrotic-lesion reaction to a mild strain of tobacco mosaic virus. Phytopathology 41: 593—609

Whitham S, Dinesh-Kumar SP, Choi D, Hehl R, Corr C and Baker B (1994) The product of the tobacco mosaic virus resistance gene *N*: similarity to Toll and the interleukin-1 receptor. Cell 78: 1101—1115

Whitham S, McCormick S and Baker B (1996) The *N* gene of tobacco confers resistance to tobacco mosaic virus in transgenic tomato. Proc Natl Acad Sci USA 93: 8776—8781

Wilson TMA, Plaskitt KA, Watts JW, Osbourn JK and Watkins PAC (1990) Signals and structures involved in early interactions between plants and viruses or pseudoviruses. In: Fraser RSS (ed) Recognition and Response in Plant-Virus Interactions, pp 123—145. Springer Verlag, Heidelberg

Wolf S, Deom CM, Beachy RN and Lucas WJ (1989) Movement protein of tobacco mosaic virus modifies plasmodesmatal size exclusion limit. Science 246: 377—379

Wu X and Shaw JG (1997) Evidence that a viral replicase protein is involved in the disassembly of tobacco mosaic virus particles in vivo. Virology 239: 426—434

Zhu Q, Maher EA, Masoud S, Dixon RA and Lamb CJ (1994) Enhanced protection against fungal attack by constitutive co-expression of chitinase and glucanase genes in transgenic tomato. Bio/Technology 12: 807—812

BLACK ROT OF CRUCIFERS

ANNE M. ALVAREZ

University of Hawaii,
Department of Plant Pathology
3190 Maile Way, Honolulu, HI 96822 (alvarez@hawaii.edu)

"In July, 1889, cabbage in the vicinity of Lexington was badly affected with a rot, which bore marks of being caused by the bacteria in the tissue. In some gardens two-thirds of the heads were affected, and of these more than half were completely invaded and rendered worthless... The invaded leaves became brown and watery at first; later, to become black as the decay had reached an advanced stage. The heads...gave forth a peculiarly noxious odor such as cabbage alone among vegetables is capable of producing. This final rotting was doubtless ordinary decomposition brought about by septic bacteria... it seems to me we have here a well-marked disease of cabbage... attributable, perhaps, to several causes working together, and at least encouraged by the vital activities of bacteria... it is only during periods of high temperature and excessive rainfall that the organisms are able to invade and break down the tissues of plants." - and thus proceeds the first description of black rot as observed in Lexington, Kentucky (Garman, 1890).

Summary

In just over 100 years, the focus on host-pathogen interactions in the black rot disease has shifted gradually from basic aspects of the disease cycle to enzyme production and gene regulation at the molecular level. The wealth of information provided through a long history of research makes it an interesting case study. Yet, one first asks whether any one pathosystem can provide a well-rounded view of plant-microbe interactions. Although far from complete, a thorough examination of a single bacterial disease provides a framework for deciphering the language of host-pathogen signalling mechanisms. To understand this system in depth, we take a broad overview of the disease, the pathogen and its natural variability. Then beginning with initial inoculum and epiphytic colonization we consider the steps in breaching a series of physical and chemical barriers, responding to the ionic environment, and appropriating nutrients within plant

A. Slusarenko, R.S.S. Fraser, and L.C. van Loon (eds), Mechanisms of Resistance to Plant Diseases, 21-52.
© 2000 *Kluwer Academic Publishers. Printed in the Netherlands.*

tissues. What attracts the pathogen to the hydathodes at leaf margins? And once having penetrated natural openings, what factors contribute to its movement through the epithem and into the xylem? What enables the black rot pathogen to invade xylem elements whereas closely related leaf spot pathogens are restricted to mesophyll tissues of the same hosts? What triggers the host response? What causes a delayed host response to a virulent pathogen in the compatible interaction but a rapid defense response in an incompatible interaction? What genes are responsible and how are they regulated? Are these genes interchangeable among pathogens? How have these questions been approached experimentally, and what is now known about them? What are the next steps? A historical perspective is interspersed in this chapter because it provides significant information leading to our current understanding of black rot as a model disease and deepens our comprehension of host-pathogen interactions. Finally, brief comparisons with several other diseases caused by bacteria in the genus *Xanthomonas* is made in the attempt to find appropriate areas for generalization.

I.   Introduction

A.   THE DISEASE, THE PATHOGEN, AND THE ENVIRONMENT

Black rot of crucifers is characterized by blackened vascular tissues and foliar marginal V-shaped chlorotic to necrotic lesions (Stewart and Harding, 1903; Smith, 1911; Cook *et al.*, 1952b; Sutton and Williams, 1970). As the disease progresses, parenchyma cells surrounding vessels in the main stem also turn black, and the plant becomes wilted, stunted, and finally rots; hence the name, black rot. Following the early descriptions, the disease is now found throughout the world and is still considered to be one of the most destructive diseases of crucifers (Garman, 1890; Russell, 1898; Chupp and Sherf, 1960; Anon, 1978; Williams, 1980; Onsando, 1992).

Black rot is caused by the Gram-negative bacterium, *X. campestris* pv. *campestris* (Xcc), a pathogen that infects a wide range of plants in the crucifer family Brassicaceae. Pathogenicity on Swedish turnip or rutabaga (*Brassica campestris* L.) was first demonstrated in 1895 by Pammel, who named the bacterium *Bacillus campestris* (Pammel, 1895). The fataher of modern plant bacteriology, Erwin F. Smith, made many contributions to the understanding of disease etiology and expanded the host range to cabbage, cauliflower, kale, rape, radish, and black mustard (Smith, 1897a, b). The pathogen is now known to affect most members of the family and model studies have been undertaken on the small, readily manipulated crucifer, *Arabidopsis thaliana* (L.) Heynh (Meyerowitz, 1987; Simpson and Johnson, 1990; Lummerzheim *et al.*, 1993, 1995; Parker *et al.*, 1993; Buell and Somerville, 1995).

B.   PATHOGEN VARIABILITY AND POPULATION STRUCTURE

Variations in virulence among strains of the bacterial pathogen and differential susceptibility of cruciferous hosts were reported by Smith as early as 1911. A high degree of variability has been repeatedly observed among Xcc strains that produce black

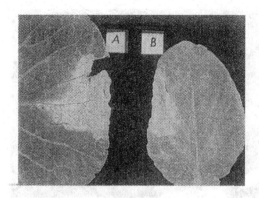

Figure 1.   Comparison of typical foliar symptom characteristic of black rot (A) with leaf blight (B), a more severe form of the disease. Note the absence of black veins in the rounded area of necrosis in the leaf showing blight. (Reprinted from Alvarez *et al.*, 1987).

rot symptoms (Fig. 1A). Highly aggressive strains produce extensive foliar necrosis, resembling blight, prior to appearance of black veins (Fig. 1B).

History of the name changes of the black rot pathogen
    Bacillus campestris Pammel (1895)
    Pseudomonas campestris (Pammel) Smith 1897b
    Bacterium campestris (Pammel) Chester 1897
    Phytomonas campestris (Pammel) Bergey, 1923
    Xanthomonas campestris (Pammel) Dowson 1939
    Xanthomonas campestris pv. *campestris* (Dowson) Dye *et al.* 1980

Type strain:
    NCPPB 528: National Collection of Plant Pathogenic Bacteria, Harpenden, England
    = ATCC 33913: American Type Culture Collection, Bethesda, MD, USA

Strains of Xcc have been grouped according to their phage types, serological reactivity patterns, outer membrane proteins, lipopolysaccharide patterns, genetic fingerprints delineated by RFLP analysis and rep-PCR (Liew and Alvarez, 1981a, b, Alvarez *et al.*, 1985, 1994; Theevachi and Schaad, 1986; Vauterin *et al.*, 1991; Ojanen *et al.*, 1993; Louws *et al.*, 1994). Typical black rot strains are characterized by reactivity patterns to a panel of six Xcc-specific monoclonal antibodies and can thereby be distinguished from weakly virulent or avirulent strains, which react to only one or none of these antibodies (Alvarez *et al.*, 1994). A leaf spot disease of crucifers (Fig. 2) is caused by a closely related pathovar, *X. campestris* pv. *armoraciae* (McCulloch 1929) Dye 1980. The latter pathogen was indistinguishable from yet another crucifer pathogen, *X. campestris* pv. *raphani* (White 1930) Dye 1980 by phage type, serotype, and host range, and it was concluded that the designation *X. campestris* pv. *armoraciae* should take precedence because of earlier publication (Machmud, 1982).

Figure 2. Leaf spots on cabbage leaves caused by *Xanthomonas campestris* pv. *armoraciae*. Lesions resulting from stomatal invasion are distributed throughout the lamina rather than being localized at leaf margins (Reprinted from Alvarez *et al.*, 1987).

Based on a comprehensive DNA-DNA hybridization study (Vauterin *et al.*, 1995) the epithet *X. campestris* pv. *campestris* proposed by Dye *et al.* (1980) was retained for the black rot pathogen. As type species of the genus, *Xanthomonas campestris* (Pammel 1895) Dowson 1939, was amended to include only pathovars obtained from crucifers (i.e., *aberrans, armoraciae, barbareae, campestris, incanae,* and *raphani*), and separate pathovar designations have been retained for the two leaf spot pathogens.

II. Overview of the disease cycle

A. ORIGINS OF INOCULUM, DISPERSAL, AND SURVIVAL

An overview of the disease cycle, starting with infected seed as the initial inoculum, is shown in Figure 3. Seed was clearly established as the primary source of inoculum, eventhough infected seeds often appear healthy (Harding *et al.*, 1904; Cook *et al.*, 1952a). During seed germination the pathogen invades vascular tissues of the epicotyl, infects foliage and is released through guttation droplets at leaf margins. Inoculum is dispersed to other plants by watersplash, wind-driven rain and aerosols (Williams, 1980; Kuan *et al.*, 1986). Xcc survives in plant debris in or on soil as long as the host tissue is not decomposed, but there is little evidence that the pathogen survives in soil in the absence of plant debris for longer than six weeks (Schaad and White, 1974; Alvarez and Cho, 1978). The disease is particularly difficult to manage in warm, moist, low elevation tropics where pressure on land use is high, and crucifers are frequently grown without sufficient rotation (Alvarez and Cho, 1978; Onsando, 1992). Xcc survived for nearly two years in cabbage residues in soil (Schaad and White, 1974). Thus, 2- to 3- year crop rotations, along with weed control and pathogen-free seed are recommended for disease control (Schaad and Alvarez, 1993).

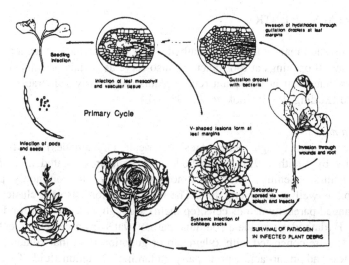

Figure 3.   Disease cycle for black rot of cabbage caused by *Xanthomonas campestris* pv. *campestris*. (Reprinted from Alvarez *et al.*, 1987).

B.   EPIPHYTIC GROWTH PRECEDING INFECTION

Infection of plants by pathogenic xanthomonads is most likely preceded by build-up of epiphytic populations after bacterial cells are deposited on plant surfaces (Rudolph *et al.*, 1994). As in the case of other xanthomonads, Xcc apparently colonizes plant surfaces in aggregates, with subsequent invasion through hydathodes (Dane and Shaw, 1996). Attachment of other xanthomonads is associated with fibrillar material (Bashan and Okon, 1986), or by fimbriae or pili in *X. hyacinthi* (previously *X. campestris* pv. *hyacinthi*) (Romantschuk *et al.*, 1994), but evidence for fibrillar attachment is lacking for Xcc. The yellow pigment, xanthomonadin, characteristic of xanthomonads (Starr, 1981), protects bacteria against photobiological damage (Jenkins and Starr, 1982) and is important in epiphytic survival of Xcc (Poplawsky and Chun, 1995, 1997, 1998; Chun *et al.*, 1997)

Colonization of the leaf margins was visually documented by Dane and Shaw (1993, 1996) using a bioluminescent strain of Xcc transformed with the *lux* operon. Colonization results in additional inoculum for further spread of the disease even prior to appearance of symptoms (Mochizuki and Alvarez, 1992; Arias *et al.*, 1996; Dane and Shaw, 1996; Shigaki, 1996). Populations of Xcc on leaf surfaces remained at relatively high levels (>100 cfu/cm^2) for more than 40 days after inoculation without symptom development (Arias *et al.*, 1996; Dane and Shaw, 1996).

C.   PORTALS OF ENTRY

Leaves are invaded mainly through hydathodes and wounds, and occasionally through stomates. Natural openings at root junctions also serve as portals of entry for Xcc, and root wounds that occur at transplant render plants particularly vulnerable to infection (Stewart and Harding, 1903; Cook *et al.*, 1952a, b).

1.   *Hydathodes*

Bacterial penetration through "water pores" (later called hydathodes) was demonstrated histologically by Smith (1911). Hydathodes are structures located at leaf margins that form natural openings into the atmosphere. They are commonly present in *Brassica* sp., as well as other crucifers. The hydathode consists of a cluster of small, loosely arranged parenchymatous cells (epithem), through which xylem fluid is exuded (cf. Fig. 3). Hydathodes provide a continual path of water from vessels to leaf margins, and they often are congested with xylem sap or guttation fluid that contains minerals, carbohydrates, and amino acids, principally glutamine. Guttation fluid of cabbage has a pH of approximately 7.0, which is optimal for bacterial growth. Ruissen and Gielink (1993) found a positive correlation between the amount of guttation fluid emitted from cabbage cultivars and development of V-shaped zones of chlorosis and necrosis, characteristic of black rot.

The pathogen shows chemotaxis towards guttation fluid exuded from hydathodes (Kamoun and Kado, 1990b). Chemotaxis is assumed to play a role only during the transitional infection phase in the life cycle of xanthomonads (Rudolph *et al.*, 1994). Cells probably lose motility as they epiphytically colonize leaf surfaces, but shift to the motile phase with chemotactic movement toward the hydathodes when free water becomes available. Chemotactic cells become non-motile when they reach guttating leaf margins of radish seedlings (Kamoun and Kado, 1990b).

2.   *Stomates*

Stomatal infections by Xcc are occasionally observed in seed beds but rarely in the field. Infection through stomates requires special conditions that reduce hydrophobicity associated with cuticular wax on epidermal cells (Cook *et al.*, 1952b). Water congestion is needed to form a continuous pathway for bacterial entry into substomatal chambers. When the bacteria occasionally gain entry via stomates they colonize the intercellular spaces of the leaf mesophyll but rarely enter the vascular system.

Symptoms resulting from stomatal invasion by the black rot pathogen, Xcc, may resemble the leaf spot disease of crucifers caused by *X. campestris* pv. *armoraciae*, which enters primarily through stomates or wounds, and causes water-soaked leaf spots and localized necrosis. Occasionally, hydathodes are infected by *X. campestris* pv. *armoraciae*, but rather than entering the vessels opening into the hydathodes, the bacteria colonize the mesophyll and parenchyma cells surrounding the veins. In a seedling bioassay, where the petiole from one cotyledon is excised from the stem and inoculum is placed onto exposed vascular tissue, *X. campestris* pv. *armoraciae* invades parenchyma cells causing a stem canker followed by stem collapse within four to five days after inoculation (Alvarez *et al.*, 1994). In the same bioassay Xcc strains travel up the xylem into the emerging primary leaf, where they produce V-shaped lesions at leaf margins.

3. *Roots*

Based on field observations, roots can be considered probable sites of infection, especially when high levels of the pathogen survive in infected plant debris (Stewart and Harding, 1903; Cook *et al.*, 1952b). Marginal V-shaped lesions were observed on cabbage leaves following wound inoculation of roots (Cook *et al.*, 1952b). The progress of the pathogen through vascular tissues from roots to leaves was traced using a bioluminescent reporter strain (McElhaney, 1991). Seedlings transplanted into soil containing infected plant debris were extensively invaded prior to symptom development (Fig. 4), but eventually chlorotic V-shaped lesions appeared on the leaves.

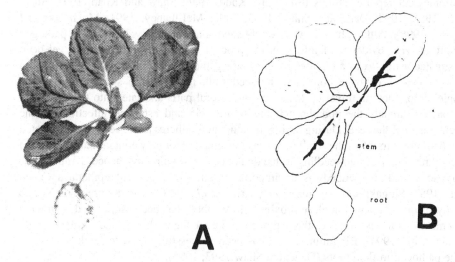

Figure 4. Infection of cabbage plants through roots two weeks after transplanting healthy seedlings into soil infested with cabbage debris infected with a bioluminescent strain of *Xanthomonas campestris* pv. *campestris* containing the lux reporter gene. A sensitive (x-ray) film was overlaid onto the symptomless leaves and exposed for 12 h. **A**. A normal photograph of the uprooted cabbage seedling. **B**. Leaf blot (autophotograph) with outline of leaf margins, stem and roots traced onto the film. The black streaks show the vessels of the cabbage leaf that were invaded by the pathogen. (Courtesy of R. McElhaney, University of Hawaii).

4. *Wounds*

Mechanical injury occurring during transplanting and other cultural practices may create wounds that favor bacterial invasion. Vascular tissue is colonized rapidly when natural defenses are breached by wounds and nutrients are released into the infection court. Hence, avoidance of wounds was recognized in very early descriptions of black rot disease as an important measure for disease control (Harding *et al.*, 1904).

D.   LATENT INFECTIONS

A bacterial infection is considered latent when the pathogen resides within host tissue prior to symptom development; the length of the latent phase may vary depending on the host-pathogen-environmental factors. Attention to the latent phase of bacterial diseases is essential for a clear understanding of the histological, biochemical, and molecular aspects of host-pathogen interactions. In black rot disease bacteria do not always multiply to sufficient numbers in vascular tissues to produce symptoms of water stress and chlorosis, but the presence of bacteria within host tissues can affect the host at the chemical and molecular level.

The extent of bacterial colonization in asymptomatic infections following inoculation of Xcc through hydathodes, wounds, and roots has been determined by tracking bioluminescent reporter strains (Shaw and Kado, 1986; Shaw and Khan, 1993; Shaw et al., 1988, 1992; Dane and Shaw, 1993, 1996; McElhaney, 1991; Mochizuki and Alvarez, 1996). Although it is well established that seed may harbor the pathogen without showing evidence of infection, the process of seed infection has not yet been investigated using lux as a reporter gene to trace pathogen invasion. Cook et al. (1952a) observed that as cabbage bolts and the seed stalk forms in the second year of the bienniel crop, bacteria travel up the xylem into floral parts and later into the seed pod. The bacteria invade the pod through the suture vein and establish infections in the funicular area of the seed. During drying, the funiculus adheres to the seed. Contaminated seed then serves to infect the seedling during germination, or may contaminate other seed during threshing. Infected seedlings often do not develop symptoms; hence, symptomless epiphytic spread in seedbeds is an important source of black rot epidemics (Yuen et al., 1987; Shigaki and Alvarez, 1993; Arias et al., 1996). Latent spread of Xcc to healthy seedlings planted in close proximity in seedbeds has been established by tracing movement of various strains of the pathogen in the seedbed (Shigaki and Alvarez, 1993; Shigaki et al., 1994). Bioluminescence has been successfully used to track movement of the pathogen in field plots (Dane and Shaw 1993, 1996).

E.   SYMPTOM EXPRESSION

Vein blackening is the first visible symptom of disease following inoculation (Sutton and Williams, 1970; Wallis et al., 1973). As bacteria progress from hydathodes through smaller veins into the midrib of the leaf, the vessels are occluded with polysaccharides of plant and bacterial origin, resulting in water stress and chlorosis. Infected tissues later become necrotic, and a light tan lesion resembling parchment progressively develops inwards from the smaller veins toward the midrib, producing the V-shaped areas of necrosis bordered by areas of chlorosis. Although the V-shaped lesions are commonly associated with invasion through hydathodes, they also occur when vessels are occluded following root invasion (Harding et al., 1904; Cook et al., 1952b).

In addition to the typical V-shaped lesions of black rot, symptoms may also appear as extended areas of necrosis resembling blight (Fig 1B). These necrotic patches are predominantly associated with "blight" strains, a subpopulation within the pathovar Xcc. Blight strains generally spread more rapidly and were more aggressive than typical black rot strains (Shigaki and Alvarez, 1993; Shigaki *et al.*, 1994). Blight symptoms can be reproduced in the greenhouse following either wound- or spray-inoculation of non-wounded leaves. The blight symptom is the result of rapid cellular death, apparently occurring as the pathogen induces water deficit in the interveinal tissues. The affected tissue becomes very dark and necrotic with little or no chlorosis, and an abrupt boundary between diseased and healthy tissue is clearly delineated, whereas, typical black rot V-shaped lesions are characterized by a network of black veins, chlorosis and a light tan area of necrosis (Fig. 1A). The distinction between blight and black rot is best observed on maturing leaves, 3 to 6 inches long. On seedlings, blight is not clearly differentiated from the severe necrosis caused by a highly virulent black rot strain. Intermediate reactions between blight and black rot were also observed in pathogenicity tests on black-rot resistant cabbage cultivars (Hunter *et al.*, 1987). The capacity to produce blight thus represents a subtle distinction among strains in the Xcc population.

Like many other xanthomonads, the pathogen affecting crucifers is heterogeneous, consisting of a spectrum of strains with varying capacities to cause severe necrosis or blight. Blight strains can be distinguished from typical black rot strains by surface antigens as well as RFLP patterns (Alvarez, *et al.*, 1994). Of 131 Xcc strains isolated from crucifer seeds, only 10 reacted with a monoclonal antibody, MAb A11, and all of these caused blight, whereas the majority of the Xcc strains reacted with two or more antibodies in a panel of six Xcc-specific monoclonal antibodies (Alvarez, *et al.*, 1994). The nature of the epitopes is not understood, but they are associated with cell surface components.

III. The compatible interaction

A compatible bacterial-plant interaction involves bacterial multiplication, avoidance of defense-related gene expression by the plant, and production of pathogenicity determinants by the pathogen. The effects of pathogenicity are assessed by bacterial growth *in planta* and finally by symptom expression. In early studies of Sutton and Williams (1970) and Wallis *et al.* (1973) bacterial production of extracellular polysaccharides and plant cell wall-degrading enzymes were considered important factors in pathogenesis.

Host-pathogen interactions are best investigated by using intact plants, natural means of infection, and environmental conditions that most closely approximate conditions prevalent during plant growth (Shaw and Kado, 1988). Wound inoculation techniques bypass some of the early host responses to Xcc infections in the incompatible interaction between pathogen and host (Shaw and Kado, 1988; Daniels, 1989) and can result in high bacterial population levels in resistant hosts (Williams *et al.*, 1972; Dane and Shaw, 1993). Breaching the host barriers by artificial means may thus produce artifacts in disease expression and lead to erroneous conclusions (Shaw and Kado, 1988; Robeson *et al.*, 1989; Daniels, 1989). A hydathode inoculation technique described by Robeson *et al.* (1989) simulates the natural process of infection. Bacteria are introduced into guttation droplets, which are later taken up by the leaf, giving bacteria access to the vascular system in the absence of wounds. Other inoculation methods may be needed when screening large numbers of genetically modified bacterial clones, and the advantages of several different methods have been reviewed (Daniels and Leach, 1993). Interpretation of results regarding the role of specific factors in host-pathogen interactions is largely dependent on the inoculation methodology and environmental conditions during incubation.

A. HISTOPATHOLOGY, PHYSIOLOGY AND BIOCHEMISTRY OF THE INFECTION PROCESS

Water deficit in affected tissues was associated with black rot in early studies (Harding *et al.*, 1904; Smith, 1911), but the host-pathogen interactions leading to the related symptoms (vein blackening, chlorosis, necrosis and eventual production of V-shaped lesions) were debated. Toxin production rather than vascular plugging alone was postulated as the cause of chlorosis (Cook *et al.*, 1952b). In order to determine the cause of vascular plugging and chlorosis the sequence of events leading to symptom production was investigated in several detailed studies using histochemical stains and/or light and transmission electron microscopy (Sutton and Williams, 1970; Wallis *et al.*, 1973; Bretschneider *et al.*, 1989).

Sutton and Williams (1970) compared multiplication of two strains that differed in their capacity to produce "mucopolysaccharides", later referred to as exo- or extracellular polysaccharides (EPS) (see section III. A. 1). A weakly virulent strain failed to produce EPS and caused no symptoms, whereas a highly virulent strain produced EPS and caused V-shaped zones of chlorosis, necrosis and black veins. By light microscopy, lysigenous cavities in degraded parenchyma and compacted masses of bacteria in vessels were observed. Bacterial numbers reached a maximum in occluded vessels just prior to symptom expression but were significantly lower in blackened areas of vessels, implicating a vascular host defense response that killed the bacteria. In many cruciferous hosts oxidation and polymerization of phenolic compounds results in production of darkened patches, thought to be melanin (Sutton and Williams, 1970). Histochemical stains were used to differentiate the EPS, other carbohydrates, pectin and "melanin". The melanin was histochemically localized in parenchyma cells surrounding vessels, but was not observed in the lumen of xylem elements. Vascular plugging was attributed to EPS and masses of living and dead bacterial cells.

Wallis *et al.* (1973) used transmission electron microscopy to show disruption of primary cell walls, loosened spiral thickenings and shredding of cellulose in secondary cell walls. In thin sections of vascular tissues they observed clear zones in xylem walls (interpreted as hemicellulose), reticulate material (interpreted as disorganized cellulose and lignin microfibrils) and fibrillar materials (interpreted as host cell wall material degraded by bacterial enzymes). Vascular plugging, observed after vein blackening had already occurred, was attributed primarily to degradation of plant cell walls with some contribution of the EPS produced by bacteria. In studies of both Sutton and Williams (1970) and Wallis *et al.* (1973) vascular plugging ultimately resulted in water stress and production of typical V-shaped lesions, starting at the site of vascular occlusion. The early suggestion that chlorosis is due to toxin formation (Cook *et al.*, 1952b) was not substantiated by either group.

Bretschneider *et al.* (1989) undertook further histological studies, incorporating several changes in methodology to remove possible artifacts. Plants were inoculated by introducing the pathogen into guttation droplets without wounding the tissues, in contrast to earlier studies in which leaf margins were notched (Sutton and Williams, 1970) or wounded at sites in tertiary veins (Wallis, *et al.*, 1973). Four to six-week old plants were used to compare susceptible (Golden Acre) and resistant (Early Fuji) cabbage cultivars to avoid possible complications associated with the lack of juvenile resistance (Williams, 1980). In compatible host-pathogen interactions, bacteria filled xylem elements but the intercellular spaces and surrounding mesophyll tissue were not heavily invaded. Plant cell walls and spiral thickenings of xylem elements were severely degraded. Bacteria within xylem elements were embedded in a fine-structured fibrilar matrix and were surrounded by an electron-lucent halo thought to be lysed bacterial cells and/or partially degraded plugging material. The histopathology was similar in the incompatible reaction (discussed later), except that the xylem was more severely disrupted and surrounding parenchyma cells showed greater plasmolysis, cell shrinkage, and chloroplast deformation. The plugging material in both compatible and incompatible interactions appeared to consist both of reticulate matter and a fibrillar matrix of undetermined composition, leaving origin and the nature of the plugging material open to further debate.

1.   *Extracellular polysaccharide (EPS)*
The EPS produced by most xanthomonads is xanthan, a cellulosic (1-4) β-D-glucose polymer with trisaccharide side chains consisting of two mannose and one glucuronic acid residues) (for review: Sutherland, 1993). As described above, EPS-deficient mutants of Xcc caused reduced symptom expression when inoculated into veins of mature plants (Sutton and Williams, 1970).

Two types of Xcc strains are often observed in culture. Larger colonies produce EPS, are nonchemotactic and virulent, whereas smaller, swarmer-type colonies do not produce EPS, are chemotactic and avirulent (Kamoun and Kado, 1990b). In culture, the small types are maintained under nutritional stress and often appear when cultures are revived from water storage. The population can switch reversibly *in planta*. Mutational analysis indicated that EPS is a virulence factor and directly related to the presence of *hrp* genes (Kamoun and Kado, 1990b) (see section III.B.1). The presence of EPS in virulent colonies was demonstrated by using cetyl trimethylammonium bromide (CTAB) to precipitate the EPS, followed by color development in the anthrone assay. The involvement of EPS in pathogenicity thus corroborated early study of Sutton and Williams (1970).

Extracellular polysaccharides also have been implicated in host-pathogen interactions in other *Xanthomonas*-induced plant diseases as well as other bacterial diseases (for review: Denny, 1995). In the *Xanthomonas*-pepper pathosystem xanthan encapsulates *X. campestris* pv. *vesicatoria* (= *X. axonopodis* pv. *vesicatoria*; Vauterin *et al.*, 1995) cells in compatible and incompatible interactions with pepper within a few hours of inoculation (Bonas *et al.*, 1991; Brown *et al.*, 1993). Similar observations were made on *Arabidopsis thaliana* leaves inoculated with Xcc. These and other studies provide additional evidence that xanthan may have a role in the early phases of disease development.

2.   Lipopolysaccharide (LPS) and the bacterial cell envelope

LPS consists of three main regions, an inner lipid region (Lipid A), a polysaccharide core, and an outer region (the O-specific side chain or O-antigen). In *Salmonella minnesota* the core is an oligosaccharide consisting of two molecules of 2-keto-3-deoxyoctulosonate (KDO) (Osbourn, 1963). The lipid region and the polysaccharide core are highly phosphorylated, while the O-antigen is largely composed of repeating units of short branched oligosaccharides that are readily detached from the bacterial outer membrane. Fractionation of bacterial cells in hot-phenol separates proteins and other phenol-soluble cell components from the totality of the polysaccharide (LPS) and nucleic acid (Westphall and Jann, 1965). The water-soluble LPS from Xcc contains galactose, 3-amino-3,6-dideoxy-D-hexose, glucose, mannose, rhamnose, galacturonic acid, phosphate and KDO (Volk, 1966; Ojanen, *et al.*, 1993). The bands forming ladder-like patterns in silver-stained polyacrylamide gels represent size-fractionated repeating units of the O-antigen. (Tsai and Frasch, 1981). The structure of the core oligosaccharide of *Xanthomonas* LPS differs significantly from that of the lipid A-(KDO)$_2$ of enteric bacteria, as it lacks heptose (Volk, 1966). Using a modification of the procedure described by of Westphall and Jann (1965), Hickman and Ashwell (1966) recovered two components from phenol-soluble LPS that they crystallized and identified as 3-acetamido-3,6-dideoxy-D-galactose and D-rhamnose. The recovery of LPS compounds in different solvent fractions is related to their partitioning in the solvent system used.

Outer membrane proteins and components of LPS have a major role in determining the surface characteristics of the bacterial cell and are largely responsible for the antigenic properties of different bacterial strains. The repeated observation that pathovar-specific LPS moieties are recognized as distinct surface antigens by diagnostically useful monoclonal and polyclonal antibodies is a further reason to suspect that LPS molecules may be associated with pathovar-host specificity (Benedict *et al.*, 1990; Alvarez *et al.*, 1991; Ojanen *et al.*, 1993; Kingsley *et al.*, 1993; Gabriel *et al.*, 1994). Lipopolysaccharides on bacterial cell surfaces may trigger alterations in plant responses to pathogens, and hence may have a governing role in initial stages of the host-pathogen interaction (Rudolph *et al.*, 1994; Newman *et al.*, 1994, 1995).

The role of LPS in host-pathogen interactions was investigated in turnip (*B. campestris*, syn. = *Brassica rapa*) using wild-type Xcc strains and LPS-deficient mutants (Newman *et al.*, 1995). The intact lipid A-lipopolysaccharide core molecule induced defense-related gene expression associated with alteration of plant cell walls and elicitation of plant defense (Newman *et al.*, 1995). Defects in the biosynthesis of the core oligosaccharide component of LPS were associated with decreased multiplication of mutants in turnip, *Arabidopsis* and a nonhost, *Datura*. Thus, a pathogenicity locus of Xcc was associated with LPS biosynthesis (Dow *et al.*, 1995; Newman *et al.*, 1995). Lipid A and core oligosaccharide components of bacterial LPS also have a role in prevention of the hypersensitive response in pepper (Newman *et al.*, 1997; described in section IV.B).

In a disease of cassava caused by a related pathogen, *X. campestris* pv. *manihotis* (= *X. axonopodis* pv. *manihotis*; Vauterin *et al.*, 1995), LPS was associated with shredded plant cell walls during pathogenesis (Boher *et al.*, 1997). LPS- and EPS-specific monoclonal antibodies were used to localize bacterial outer membrane components within leaf tissues 7 days after pathogen inoculation. Xanthan was a component of the fibrillar matrix filling the intercellular spaces of the leaf mesophyl and attached to the outer surface of bacteria, but LPS was not detected in the fibrillar matrix. [The fibrillar matrix observed by transmission electronmicroscopy resembles the fibrillar matrix observed by Wallis *et al.* (1973) and Bretschneider *et al.* (1989) in cabbage vessels infected with Xcc]. LPS was associated with degraded mesophyll cell walls. The LPS-specific monoclonal antibody reacted with periodate-sensitive, and heat-, protease-, and lysozyme-resistant molecules, and the ladder-like pattern observed on Western blots suggested that the specific antigen was a component of the O-antigen of the LPS. Detection of this antigen in the apoplast of mesophyll cells suggested a possible contact between LPS of *X. campestris* pv. *manihotis* and the host plasma membrane (Boher *et al.*, 1997). Virulent strains not reacting with the LPS-specific antibody in vitro became positive during infection, suggesting that biosynthesis of LPS occurs during pathogenesis.

B.   GENETICS OF THE HOST-PATHOGEN INTERACTION

Gene-for-gene interactions between Xcc and cruciferous hosts have not been clearly established. Although there is an indication that various crucifers are differentially affected by strains of Xcc (Kamoun *et al.*, 1992a), specific gene-for gene relationships involving pathogen races and host cultivars are obscure in crops, such as cabbage, which have multigenic resistance. Genetic studies of the host-pathogen interaction have instead involved a search for major pathogenicity genes.

Mutants of Xcc that either failed to produce a pathogenic response or showed altered pathogenicity following stem-inoculation of three-day old seedlings were isolated by Daniels *et al.* (1984a, b), and genes encoding proteases and endoglucanases were identified and cloned. In older plants, however, growth rates of wild-type and mutant strains were similar, as was symptom production, indicating that proteases and endoglucanases probably do not play a role in early stages of black rot infection. Subsequently, other strategies and other inoculation methods were used to search for and clone genes involved in pathogenicity. Osbourn *et al.* (1990a) sequenced an open reading frame associated with reduction in pathogenicity. Nevertheless, pathogenicity mutants continued to produce EPS and plant cell wall- degrading enzymes, which were characteristic of wild-type strains and thought to be involved in pathogenesis. Thus, the search continued for genes important in early stages of pathogenesis, involving, i) *hrp* genes; ii) other genes for EPS biosynthesis; iii) structural genes for production of extracellular enzymes; iv) enzyme secretion systems; and v) regulatory genes (Daniels *et al.*, 1994).

1.   *hrp genes*
Hrp stands for "<u>H</u>ypersensitive <u>R</u>eaction and <u>P</u>athogenicity" and involves elicitation of a hypersensitive reaction (HR) in nonhost plants and pathogenicity in compatible hosts. Mutagenesis of *hrp* genes causes a loss in ability to incite both HR and pathogenicity; that is, both compatible and incompatible bacterial plant interactions are affected (Dow and Daniels, 1994). The apparent dual role of *hrp* genes suggests that plants have

developed resistance mechanisms based on recognition of bacterial determinants that control pathogenicity (Arlat *et al.*, 1991).

In 1990, Kamoun and Kado (1990 a, b) described a *hrpX* locus in Xcc required for pathogenesis on crucifers and HR on non-host plants. Mutations in *hrpX* caused loss of pathogenicity in host plants, a loss in HR on non-host plants, but a gain in ability to cause HR in the host plants (Kamoun *et al.*, 1992a, b). The *hrpX* locus is present in virulent wild-type strains of Xcc that produce EPS and are nonchemotactic. Mutation of Xcc to produce *hrp⁻* mutants resulted in failure to produce EPS or symptoms on crucifers. Co-inoculation of an avirulent, *hrp⁻* mutant strain JS111 with the virulent wildtype (*hrp⁺* EPS⁺) strain 2D520 resulted in exocellular complementation *in planta* by the *hrp⁺* strain after 1 to 3 days, resulting in delayed symptom production. These observations suggested that the *hrpX* locus controls functions involved in evading the defense system of the host.

HrpX is found in several pathovars of *X. campestris* and in *X. oryzae* pv. *oryzae* (Kamdar *et al.*, 1993). The nucleotide sequence in Xcc (*hrpXc*) shows 84% identity with *hrpXo* from *X. oryzae* pv. *oryzae* (Oku *et al.*, 1995). There is a 48.7% identity in nucleotide sequence between *hrpX* and the *hrpB* gene of *Ralstonia (Burkholderia* or *Pseudomonas solanacearum) solanacearum* and a 35.8% amino acid sequence identity (45.96 similarity) between the predicted protein products of these genes (Genin *et al.*, 1992; Oku, *et al.*, 1995). Sequence similarity to other regulatory proteins, including AraC that regulates the arabinose operon of *Escherichia coli* and VirF of *Yersinia enterocolitica* indicate that the *hrpXc* gene may produce a transient protein involved in gene regulation. A putative DNA- binding domain (helix-turn-helix) present in the carboxyl terminal half of the HrpX protein is highly conserved among HrpB, AraC and VirF. The VirF protein of *Y. enterocolitica* is a transcriptional activator of the *Yersinia* virulence regulon (Cornelis *et al.*, 1989); and HrpXc by extension, may also function as a transcriptional activator of virulence genes of Xcc. Thus, *hrpX* likely encodes a protein that regulates genes involved in pathogenicity, suppression of host defense mechanisms and recognition by non-host plants (Oku *et al.*, 1998).

A *hrpX*-counterpart gene, *hrpXv*, was found in *X. campestris pv. vesicatoria* (Wengelnik and Bonas, 1996). The predicted gene products, HrpXc, HrpXo and HrpXv, are nearly identical in amino acid sequence (Oku *et al.*, 1995; Wengelnik and Bonas, 1996). *hrpX* was recently shown to be widely conserved in xanthomonads (Oku *et al.*, 1998). Primers were selected from the sequence of the 1.4 Kb fragment of *hrpXo* and used to amplify genomic DNA from sixteen *X. campestris* pathovars. Each PCR-amplified DNA fragment was then probed with an internal DNA segment of *hrpXo* (Oku *et al.*, 1998). The presence of the *hrpXo* fragment was verified by Southern hybridization analysis, and all tested strains from the sixteen pathovars hybridized to the 1.4 kb PCR-amplified DNA.

Although highly conserved among *Xanthomonas* species, *hrpX* is not part of the larger 25 kb *hrp* cluster described for *X. campestris* pv. *vesicatoria* (Bonas *et al.*, 1991), which is homologous to the 19 kb DNA region of the *R. solanacearum* cluster (Boucher *et al.*, 1987). The conserved *hrp* genes found in *Erwinia, Pseudomonas*, and *Xanthomonas* appear to be related to pathogenicity genes of *Yersinia* and *Shigella* involved in type III secretion of proteinaceous virulence factors (Van Gijsegem *et al.*, 1993). Many *X. campestris* pathovars, including Xcc contain DNA which hybridizes to the large *hrp* cluster of *R. solanacearum* (Arlat *et al.*, 1991). Mutagenesis of the corresponding regions in Xcc resulted in strains defective in both pathogenicity and HR induction, but *hrp⁻* mutants produced wild-type levels of extracellular enzyme activities and EPS (Arlat *et al.*, 1991). Thus, there are two sets of pathogenicity genes, the large *hrp* cluster and the smaller 1.4 kb unlinked *hrpXc* locus associated both with pathogenicity and EPS production (Kamoun and Kado, 1990a, b). In *R. solanacearum*, the HrpB protein is a positive regulator of the large hrp gene cluster (Van Gijsegem *et al.*, 1994). Similarly, the HrpXc homologue, HrpXv (also with sequence similarity to HrpB), is a positive regulator of five *hrp* loci (*hrpB* to *hrpF*) of the large *hrp* cluster of *X. campestris* pv. *vesicatoria* (Wengelnik and Bonas, 1996). Yet another *hrp* gene, *hrp G*, functions at the top of the *hrp* gene regulatory cascade of *X. campestris* pv. *vesicatoria*, positively regulating the *hrp A* and hrpXv genes (Wengelnik *et al.*, 1996).

The induction of *hrp* genes probably helps the bacterium to become established in the plant and exploit its host more efficiently as the pathogen responds to the physical and nutritional changes in the microenvironment and shifts from epiphytic to endophytic growth (Arlat *et al.*, 1991). Starvation appears to be a major factor in stimulating *hrp*-gene expression in Xcc (Dow and Daniels, 1994). Transcription of *hrp* genes of Xcc is repressed in rich media, as are *hrp* genes of *R. solanacearum* (Arlat *et al.*, 1991). Peptone, yeast extract, and casamino acids strongly repressed the expression of *hrp* genes, whereas sucrose, glutamate, and glycerol gave the highest levels of *hrp* gene expression (Arlat *et al.*, 1991). Sucrose and methionine were needed for efficient induction of *hrp* genes, and a medium low in phosphate favored induction (Fenselau and Bonas, 1995). McElhaney, *et al.* (1998) found that high levels of ammonium sulphate, ammonium nitrate or potassium nitrate applied to roots of cabbage seedlings repressed both colonization of Xcc in the vascular system and symptom expression in six-week-old plants, whereas nitrogen-deficient plants developed symptoms in the usual 7-10 days after inoculation. Collectively, these observations emphasize that the bacterial response to a changing nutritional environment is critical in the successful establishment of infection in compatible hosts.

The "phenotypic switch" in Xcc (Kamoun and Kado, 1990b) from a nonchemotactic, mucoid strain to a chemotactic and nonmucoid derivative was also observed for Xcc by Martinez-Salazar *et al.* (1993), leading them to question whether the phenotypic switch is *recA*-dependent. The RecA protein is involved in genetic rearrangements in other pathogens and was isolated and characterized for Xcc (Lee *et al.*, 1996; Martinez *et al.*, 1997). Although some *recA* mutants showed decreased virulence in cabbage, *the recA* mutation was not related to a genetic rearrangement that affects chemotaxis and xanthan production, and decreased virulence of *recA* mutants was attributed to an effect independent of the switch to chemotaxis (Martinez *et al.*, 1997). Using a different approach, Poplawsky and Chun (1995, 1997) demonstrated that regulation of EPS synthesis in Xcc is determined by diffusible factors, one of which also regulates xanthomonadin production (Poplawsky *et al.*, 1998). (See section III.B.5).

2. *Other genes for EPS biosynthesis*
In addition to its association with the *hrpX* locus, EPS biosynthesis has been associated with other gene clusters. Mutations which result in loss of EPS production have been mapped to at least four regions of the chromosome (Barrere *et al.*, 1986; Harding *et*

al., 1987; Hötte *et al.*, 1990; Köplin *et al.*, 1992; Lin *et al.*, 1995; Wei *et al.*, 1996). Thorne *et al.* (1987) found three unlinked clusters of genes affecting xanthan synthesis. EPS biosynthesis genes with no sequence similarity to *hrpX* also may suppress the host defence response or avoid triggering those responses. These include *pigB* and *rpfF* (discussed in section III.B.5).

3.   *Structural genes for production of extracellular enzymes*
Virulent strains of Xcc produce extracellular enzymes including polygalacturonase (PG), pectin methyl esterase (PME), polygalacturonate lyase (PGL), protease, endoglucanase, lipase and amylase, but not all of these enzymes appear to be involved in early stages of infection. No reduction in virulence was observed in mutants failing to produce pectolytic enzymes, and these enzymes apparently were activated only at later stages of infection (Dow *et al.*, 1989).

Protease.   Proteolytic enzymes (proteases) with different specificities are thought to be important in pathogenicity (Dow *et al.*, 1990; Tang *et al.*, 1987, 1990, 1991). Protease mutants, which failed to produce serine metalloprotease, were as virulent as wild-type strains when inoculated into turnip seedlings but were less virulent when inoculated into margins of mature turnip leaves (Tang *et al.*, 1987). In later studies, no determinative relationship was found between the pattern of protease gene expression and the mode of pathogenesis in early stages of vascular or mesophyllic infection (Dow *et al.*, 1993). Protease activity is associated with permeability changes and dissolution of plant membranes in leaf-spotting diseases, but in the case of black rot, no role for protease was found in early stages of disease development.

Endoglucanase.   Cellulolytic degradation is caused by multiple enzymes of two major types: Eng (endo-1,4 -β-glucanase and Exg (cellobiohydrolases). The *engXca* gene encoding the major endoglucanase of Xcc was sequenced, and a comparison of the complete deduced amino acid sequence revealed a region with a high degree of homology with the deduced sequences of endoglucanase- and exoglucanase-encoding genes of other bacteria and fungi (Gough *et al.*, 1990). The endoglucanase produced by Xcc has limited ability to degrade crystalline cellulose; hence, Xcc is not truly cellulolytic (Gough *et al.*, 1990). Endoglucanase mutants and wild-type strains colonized plants at similar rates (Gough *et al.*, 1988), but pleiotrophic effects of mutation, including loss of EPS production were later associated with reduced virulence (Barber *et al.*, 1997). The role of these genes in pathogenicity is currently under investigation (See section III.B.5).

4.   *Enzyme secretion systems*
Secretion mutants which produce enzymes but fail to export them into the medium are severely reduced in virulence (Dums *et al.*, 1991; Daniels *et al.*, 1994). The battery of enzymes such as, protease, endoglucanase (cellulase), PGL, lipase and amylase, which are involved in degradation of plant membranes, cell walls, and other structures, are all exported by a signal sequence-dependent secretion mechanism encoded by a cluster of at least 12 genes (Dow *et al.*, 1993; Dow and Daniels, 1994). Export-defective

mutants were nonpathogenic or showed reduced symptom expression, but their growth rates in leaf lamina were similar to wild-type strains. Thus, the enzymes appeared to be necessary for disease development, but not at early stages when multiplication of the pathogen is critical to establishing infection (Dow and Daniels, 1994).

5. *Regulatory genes*

Gene regulation is critical for determining rapid responses to the abrupt changes in the physico-chemical environment as plant tissues are invaded. Coping with these changes requires bacterial sensors of the osmotic potential, H^+ ion concentration and nutrients during early stages of colonization and infection. Bacterial sensor/response mechanisms result in chemotaxis toward nutrients, production of signal proteins, protection against active oxygen species, induction of plant degradative enzymes, as well as other pathogencity factors needed for establishment of infection.

Mounting evidence indicates that gene regulation plays a direct role in pathogenesis of Xcc. Regulatory mutants which fail to produce xanthan and extracellular enzymes are severely reduced in virulence and in the ability to multiply in plants, and as such, they provided the first evidence for global pleiotropic gene regulation in Xcc (Daniels *et al.*, 1983; 1984a, b) Eight genes, designated *rpf A—H* (for regulation of pathogenicity factors) were all involved in synthesis of extracellular enzymes and xanthan (Tang *et al.*, 1991; Dow and Daniels, 1994). Mutation in any one gene in the cluster resulted in reduced xanthan synthesis and enzyme production. Of particular interest are *rpfC*, which encodes a two-component regulator with sensor and regulator domains in the same protein, and *rpfG*, a two-component regulatory protein with a receptor domain. Such gene pairs are reminiscent of histidine kinase two-component regulatory systems.

Signal-transduction pathways that translate environmental cues, such as chemicals, pH, temperature and others into adaptive responses are perhaps best understood in prokaryotes. They use a pair of proteins (sensor/regulator) in signalling which are called two component regulators. One of the components, the sensor, has a transmembrane domain with a portion of the molecule in the periplasm, which senses the environmental signals, and another portion in the cytoplasm. Upon receiving the signal this protein autophosphorylates a histidine residue that is present at the conserved C-terminal end of the cytoplasmic portion of the sensor. The regulator protein, which has a conserved N-terminal end, is in the cytoplasm. The phosphate group from the histidinyl residue of the sensor is then transferred to an aspartyl residue in the N-terminal end of the regulator. This activates the regulator which modulates the expression of the target genes.

Although gene regulation in pathogenesis is indisputably complex, regulatory pathways are progressively being unraveled (Daniels *et al.*, 1984a, b, 1989, 1994; Dow and Daniels, 1994; Dow *et al.*, 1989, 1990, 1993, 1995). The *rpfA—H* gene cluster is involved in positive regulation, whereas another locus, *rpfN*, is required for binding of a protein to promoter region, resulting in negative regulation. Mutation of *rpfN* causes overproduction of enzymes and xanthan. Whether or not the positive and negative regulatory systems interact is still unknown (Dow and Daniels, 1994).

Low-molecular-weight diffusible factors were independently described as positive regulators of gene functions (Barber *et al.*, 1996, 1997; Poplawsky and Chun, 1997). The *rpfF* and *rpfB* genes generate a diffusible signal factor, DSF, that positively regulates endoglucanase (EngXca) and xanthan production (Barber *et al.*, 1997). Likewise, the *pigB* transcriptional unit was implicated in production of a diffusible factor, DF, that positively regulates pigment (xanthomonadin) and xanthan production. Production of these diffusible signals contribute to pathogenesis (DSF) and survival (DF). Cross-feeding experiments revealed that, whereas both DSF and DF regulate production of xanthan, they represent two separate intercellular signalling systems (Poplawsky *et al.*, 1998). DF is tentatively characterized as a butyrolactone (Chun *et al.*, 1997), somewhat different from the N-acyl homoserine lactones associated with "quorum sensing" and autoinduction mechanisms that contribute to virulence in other bacteria (Fuqua *et al.*, 1994).

The role of receptors and signal transduction in the crucifer defense reactions to Xcc is not well understood. Nevertheless, much insight has been gained through studies of gene regulation as related to pathogenesis in Xcc and other xanthomonads (Dow and Daniels, 1994; Tang *et al.*, 1996; Barber *et al.*, 1997; Poplawsky *et al.*, 1998).

6.   *Other virulence factors*

Chen *et al.* (1994) recovered a 5.4 kb DNA fragment from the Xcc type strain (528^T) that conferred the ability to cause blight symptoms, but not systemic movement, to a strain of *X. campestris* pv. *armoraciae* and thereby changed a mild, mesophyllic pathogen into a more aggressive pathogen. Shigaki (1996) found that all blight-inducing strains of Xcc hybridized with this 5.4 kb DNA fragment and reacted with MAb A11, specific for a LPS antigen. In contrast, typical black rot strains neither reacted with MAb A11, nor hybridized with the DNA fragment. Nevertheless, the requirement of this fragment for elicitation of blight symptoms was not established using mutational analysis (Shigaki, 1996). The strong correlation between the capacity to elicit blight symptoms and reactivity to MAb A11 may be explained by linkage disequilibrium, that is, the lack of recombination between the pathogenicity genes and the epitope modifying genes (Gabriel, personal communication). Utilizing this linkage, the epitope may be used as a marker to localize closely-linked pathogenicity genes.

IV.   The incompatible interaction

The interaction of a pathogen with a non-host or resistant host, referred to as an incompatible interaction, is often manifested in a rapid, localized necrosis that limits the spread of the invading microorganism. Permeability changes in plant cell membranes, medium alkalinization, production of active oxygen species (AOS) and irreversible membrane damage resulting in cell collapse and necrosis are plant responses to pathogens in an incompatible host-pathogen interaction. A bacterial defense against AOS involving superoxide dismutase (SOD) was demonstrated for Xcc (Smith *et al.*, 1996) (see section IV.C) Other host defense mechanisms may involve structural components and alterations in extracellular matrix proteins (Section IV.D).

A. STRUCTURAL ASPECTS OF DEFENSE

Host resistance in crucifers appears to operate at the hydathode region (Williams *et al.*, 1972). Apparently, internal, not external, colonization of the hydathode region is required for expression of host resistance because growth and survival of epiphytic populations was independent of host genotype (Dane and Shaw, 1993). Cuticular waxes on epidermal cells and surrounding stomatal openings prevent entry of Xcc via stomates under usual field conditions. The intriguing difference between a vascular pathogen such as, Xcc, and the closely related pathovars, *X. campestris* pv. *armoraciae* and *X. campestris* pv. *raphani*, which invade primarily through stomates, is repeated in other pathogen pairs, such as, *X. oryzae* pv. *oryzae* (vascular) and *X. oryzae* pv. *oryzicola* (stomatal). The differences in the host-pathogen interactions involve the anatomy of the structures invaded, as well as the biochemical responses to the respective pathovars.

Cells of Xcc placed directly into guttation fluids of unwounded hydathodes first colonized the vascular elements of the compatible cabbage cultivar, Golden Acre, whereas they colonized intercellular spaces of mesophyll and parenchyma cells of the incompatible cabbage cultivar, Early Fuji (Bretschneider *et al.*, 1989). In contrast, when *X. campestris* pv. *armoraciae* cells were placed into guttation fluids of either cultivar, the bacteria failed to invade the vascular elements but rather colonized intercellular spaces of the parenchyma and mesophyll, causing angular water-soaked lesions that later become necrotic. Plant defense mechanisms thus are tissue-, as well as pathovar-specific. Hydathode colonization by Xcc and different pathogenicity-mutants has been compared to patterns of colonization of *X. campestris* pv. *armoraciae*, using recently developed methodology developed for the *X. campestris/Arabidopsis* pathosystem (Hugovieux *et al.*, 1998). This methodology has potential for characterizing key genetic differences between these two closely related pathovars.

B. THE HYPERSENSITIVE RESPONSE

The dual role of *hrp* genes in the general HR/pathogenicity interaction was outlined in Section III.B.1. With the exception of *hrp* mutants, all Xcc pathogenicity mutants provoke normal HR on non-host plants such as tobacco and pepper (Dow and Daniels, 1994). The HR requires several hours to become established and is localized to the site of infection (for review, Bonas, 1994; Dangle *et al.*, 1996). Cabbage cultivar Early Fuji showed HR following inoculation with Xcc, in contrast to Golden Acre, which developed black veins and V-shaped zones of chlorosis when inoculated with the same strain (Robeson *et al.*, 1989; Bretschneider *et al.*, 1989). Foliar HR, achieved by introducing Xcc into non-wounded hydathodes through guttation droplets, was described as small (<3 mm) necrotic lesions, which remained localized in tissues surrounding hydathodes (Robeson *et al.*, 1989). A vascular HR reaction on resistant host plants was observed as browning of vessels at the site of inoculation, and *hrp* mutants failed to induce vascular browning (Kamoun *et al.*, 1992a).

Pepper leaves undergo HR when infiltrated with virulent Xcc, but this response can be prevented by prior infiltration of leaf panels with LPS or LPS-protein complexes (Newman *et al.*, 1997). This LPS-induced resistance response, referred to as localized induced resistance (LIR), was used to assay the ability of several forms of LPS to trigger changes in HR. Lipid A and core oligosaccharide components of bacterial LPS from both virulent and avirulent pathogens prevented HR when leaf panels of pepper were subsequently infiltrated with live cells of Xcc. The source of the LPS was not important; any complete LPS prevented HR. Hence, a plant response to LPS appears to play an important role in both compatible and incompatible plant-microbe interactions although the biochemical pathways involved are still unclear (Newman et al., 1995, 1997).

C.   THE PATHOGEN RESPONSE TO OTHER ASPECTS OF HOST DEFENSE

Irreversible membrane damage and cell collapse associated with HR are the result of several factors including enzymatic and free radical-dependent mechanisms. Increased production of AOS, resulting in an "oxidative burst" occurs during early stages of plant defense, and antioxidant enzymes produced by invading bacteria play a role in protecting them (Smith et al., 1996; Loprasert et al., 1996). Peroxidase and superoxide dismutase (SOD) are produced by Xcc and were thought to play a role as virulence factors during bacterial-plant interactions (Smith et al., 1996).

Other plant responses related to resistance/susceptibility may involve phytoalexin induction by elicitors, such as glycans, chitin, proteins, and/or glycoproteins, but have not been widely studied in the case of black rot. A gene with unknown function identified by Osbourn et al. (1990a) decreased growth of Xcc in plants and was thought to be important in early phases of infection. A mutant bacterial strain triggered an earlier accumulation of transcripts for β-1,3- glucanase (BGL), a defense-related gene in Brassica. BGL induction was studied by Newman et al. (1995) who monitored the accumulation of the transcript in compatible (Xcc) and incompatible (X. campestris pv. armoraciae) reactions. The intact lipid A-oligosaccharide core LPS induced accumulation of the transcript but the O-antigen alone was ineffective. They suggested that LPS may be released during growth in planta and trigger the host defense response. In contrast, bacterial EPS, which is thought to mask the presence of bacteria, may suppress induction of BGL, and hence, host defense. LPS from nonpathogens did not induce transcript accumulation in cabbage (Newman et al., 1995).

D.   EXTRACELLULAR MATRIX GLYCOPROTEINS
AND THE HOST RESPONSE

Proteins, glycoproteins, and polysaccharides in the extracellular matrix of plants are important in plant development and appear to have a role in defense responses to invasion by plant pathogens and symbiotic organisms (Davies et al., 1997a). Glycoproteins expressed in 3- to 4-week-old Brassica seedlings were localized by comparing toluidine-blue stained transverse sections of healthy petioles and tissue prints probed with monoclonal antibodies that recognized distinct extracellular matrix glycoproteins (Davies et al., 1997a). Localization was further confirmed by Western blot analysis of extracts from the tissues dissected from petioles as well as by electronmicroscopy using immuno-gold to localize the antigens. One glycoprotein, gp120 (Mr 120,000) was localized in vascular tissues and intercellular spaces, predominantly within the intercellular space at the three-way junctions between cells of the phloem and sclerid fibers. A smaller glycoprotein, gp45 (Mr 45,000), was observed only in the pith. Both glycoproteins have epitopes in common with known hydroxyproline-rich glycoproteins (HRGPs), although amino acid analysis showed that these molecules were rich in proline and lysine but poor in hydroxyproline. As novel glycoproteins, gp120 and gp45 are probably unrelated to extensin-like proteins and unlikely to have lectin-like binding domains similar to those found in solanaceous plants and French bean (Davies *et al.*, 1997a).

The potential role of extracellular matrix glycoproteins in plant defense was investigated by following their induction in *Brassica* petioles following wounding or inoculation with virulent and avirulent strains of *X. campestris* (Davies *et al.*, 1997b). Two new antigens appeared following wounding; gp160 appeared as a smear in Western blots, and gpS remained in the stacking gel. The latter antigen is recognized by monoclonal antibody JIM11 that recognizes an epitope carried on HRGPs produced in response to wounding. Presence of this antigen was revealed in tissue prints probed with JIM11 24 hours following inoculation. Antigen induction was pronounced in *Brassica* petioles following inoculation of *X. campestris* pv. *raphani* and pv. *armoraciae* strains that produced a strong petiole splitting response, indicative of an incompatible reaction (Fig. 5). Antigen induction was associated

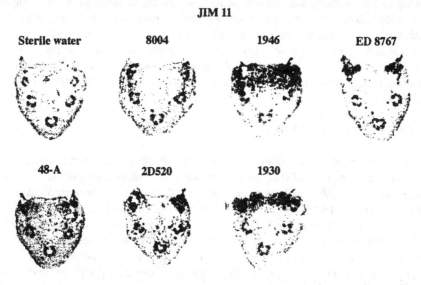

Figure 5. Tissue prints of transverse sections of *Brassica* petioles 24 h after inoculation with different bacteria, probed with monoclonal antibody JIM11 that reacts to a plant extracellular matrix glycoprotein having several components characterized as extensin-like hydroxy-proline-rich glycoproteins (HRGPs). Antigen localization was revealed by reaction with a secondary antibody conjugated to alkaline phosphatase, which reacts with the substrate BCIP (5-bromo-4-chloroindoxyl phosphate) and NBT (nitroblue tetrazolium), producing a dark stain during immunodevelopment. Bacteria (and the nature of the interaction) are *Xanthomonas campestris* pv. *campestris* strains 8004 and 2D520 (both compatible), *Xanthomonas campestris* pv. *armoraciae* strains 48-A (compatible) and 1930 (incompatible), and *Xanthomonas campestris* pv. *raphani* strain 1946 (incompatible). Inoculation with an *Escherichia coli* strain ED8767 revealed antigen induction around vascular tissues similar to the compatible interaction, whereas in incompatible interactions pith tissue is darkly stained. Control inoculation of petioles (with sterile water) had no effect on the pattern of antigens on tissue printing. (Reprinted by permission from Davies *et al.*, 1997b).

with pith tissue near the site of inoculation, whereas in compatible interactions antigen induction was restricted to phloem and sclerid fibers of the vascular bundles. The host response to *E. coli*, a nonpathogen, was similar to the low level of antigen induction in the compatible interaction (Fig. 5), but the host showed no response to infiltration with sterile water. The induced glycoprotein components were characterized as extensin-like HRGPs induced at high levels by wounding and at lower levels by pathogens and saprophytes. Analysis of the plant response using wild-type virulent strains and avirulent mutants for pathogenicity genes (*hrp* gene cluster, *hrpX* gene, extracellular enzymes, core oligosaccharide and O-antigen of LPS, and extracellular polysaccharides) led to the conclusion that in this pathosystem, these genes do not suppress plant defense responses that involve HRGP induction (Davies *et al.*, 1997). Further studies are needed to unravel the complex interactions and to clarify the differences between specific plant response to pathogens from general plant defense reactions.

V.   Is Xcc and black rot of crucifers a representative pathosystem?

A single bacterial pathosystem rarely provides all the richness of detail needed for a working model of host-pathogen interactions, so we fill in the gaps by drawing on data from other host-pathogen interactions. However, generalizations from one system to another are not always appropriate, because inherent differences in all pathosystems account for specificity. Nevertheless, comparisons with other pathosystems in terms of pathogenicity factors, *hrp* genes, gene regulation, and elicitors of plant defense mechanisms indicate that many commonalities are present (Van Gijsegem *et al.*, 1993, 1994, 1995).

A.   PATHOGENICITY GENES

Pathogenicity gene, *pthA*, belonging to an avirulence/pathogenicity gene family from *Xanthomonas*, has been established for *X. citri* where it confers host-specific virulence (Swarup *et al.*, 1991, 1992; Gabriel *et al.*, 1994) (= *X. axonopodis* pv. *citri* Vauterin *et al.*, 1995). Loss of *pthA* in *X. citri* renders it avirulent. Placement of the gene into another pathogen, *X. campestris* pv. *phaseoli* (= *X. axonopodis* pv. *phaseoli* Vauterin *et al.*, 1995) makes the latter avirulent on its homologous host, bean. No homologous sequences to *pthA* were found in Xcc. On the other hand, the pathogenicity/hypersensitivity gene, *hrpX*, of Xcc is present in *X. citri*, and the homologue, *hrpXct*, appears to function as in Xcc (T. Oku, unpublished).

B.   AVIRULENCE GENES

The avirulence gene cluster, *avrBs2*, complementary to the resistance gene *Bs2* in pepper exists in *X. campestris* pv. *vesicatoria*. The cluster is related to the gene-for-gene interaction established for races of this pathogen (for review: Bonas, 1994; Leach and White, 1996). The *avrBs2* cluster is found in other *X. campestris* pathovars, including Xcc, but it is not always related to an analogous gene-for-gene relationship in these pathosystems. Avirulence genes also may have a role in pathogenicity, making the pathogen more aggressive on its compatible host (Kearney and Staskawicz, 1990). Advances with other *Xanthomonas* pathosystems (Gabriel *et al.*, 1994; Yang and Gabriel, 1995a, b; Yang *et al.*, 1996) have provided insights on pathovar specificity that eventually may help to understand the *Xanthomonas*/crucifer pathosystem.

C.   BACTERIAL EPS AND THE HOST RESPONSE

Virulent strains of *R. solanacearum* and Xcc, both of which produce EPS, fail to trigger the natural host defense response. Mutations in *hrpB* and *hrpX* loci result in HR or incompatibility on host plants, as well as failure to produce EPS (Genin *et al.*, 1992; Kamoun *et al.*, 1992a, b). In *X. campestris* pv. *manihotis* EPS was associated with plant cell walls during ingress of bacteria into cassava vascular elements, and cell wall alterations indicated that this may be related to host defense (Boher *et al.*, 1997). Such observations implicate gene regulation as well as EPS in a common mechanism that involves masking a host defense response (for review see Denny, 1995).

VI.   Conclusions

Many genes are activated during pathogenesis, and their respective gene products are known in several cases. Reactions of the gene products with plants are not clearly understood, but they probably are secreted as those in *Yersinia* and other human and animal pathogens (Kado, 1994).

Susceptible cabbage cultivars can be infected at all stages of the crop cycle but are most vulnerable in the seedling stage when leaf surfaces are wet. The hydrophilic environment of the leaf clearly changes as the leaf dries. Irrigation water and rain leach substances from leaf surfaces, and as moisture evaporates, solute concentrations in the surface water film increase. The bacteria must respond quickly to these changes in order to colonize the leaf surface. The yellow pigment, xanthomonadin, likely protects cells from deleterious UV-irradiation, enabling bacteria to survive epiphytically, and mutants for pigment production show reduced ability to survive (Poplawsky and Chun, 1997, 1998). Bacteria show chemotaxis toward guttation fluids and invasion is commonly through the hydathodes. Once inside the tissues, the bacteria multiply on nutrients provided by vascular fluids and proceed into the major veins. As the bacteria deplete nutrients in the xylem and approach stationary phase, additional nutrients must be supplied from the plant. Nutrients are supplied by tissue degradation with subsequent release of polysaccharides needed for cellular protection and bacterial growth.

Pathogenicity factors probably are produced in response to changes in the environment as the bacteria adapt to the internal components of the plant (Dow and Daniels, 1994). Diffusible factors are involved in multiple signalling systems that up-regulate specific sets of genes as the bacteria reach stationary phase. Pathogenicity factors include EPS and extracellular enzymes, whose syntheses are regulated by complex mechanisms. EPS production may be independently regulated by different pathways, because mutations in three different classes of genes (*rpf*F, *hrp*Xc, *pig*B) result in EPS-deficient mutants (Barber *et al.*, 1997; Kamoun and Kado, 1990a, b; Poplawsky *et al.*, 1997). Starvation stimulates *hrp*-gene expression and may result in greater symptom development in compatible hosts (Dow and Daniels, 1994). Reciprocally, high nitrogen levels in the vascular tissue may repress activity of *hrp* genes and shift the host-pathogen interaction in favor of the plant, which becomes more resistant to colonization (McElhaney *et al.*, 1998). *Hrp* genes do not appear to interact with the *rpf* regulon, however, indicating that different signaling mechanisms are in place.

The complexities of the host-pathogen interactions of Xcc with crucifers are currently being deciphered and integrated into a general model with detailed studies of gene regulation. Thus, a knowledge of this bacterial pathosystem, albeit presently incomplete, provides us with the essential details from which some meaningful generalizations can later be derived for similar bacterial pathogens that invade foliage and vascular tissues.

Acknowledgements

I wish to thank M.J. Daniels, C.I. Kado, J.E. Leach, S.S. Patil, and T. Oku for providing unpublished data and insightful comments.

References

Alvarez AM and Cho JJ (1978) Black rot of cabbage in Hawaii: Inoculum source and disease incidence. Phytopathology 68:1456—1459

Alvarez AM, Benedict AA and Mizumoto CY (1985) Identification of xanthomonads and grouping of strains of *Xanthomonas campestris* pv. *campestris* with monoclonal antibodies. Phytopathology 75:722—728

Alvarez AM, Cho JJ and Hori TM (1987) Black rot of cabbage in Hawaii. Hawaii Agricultural Experiment Station, Research Series 051: 1—20.

Alvarez AM, Benedict AA, Mizumoto CY, Pollard LW and Civerolo EL (1991) Analysis of *Xanthomonas campestris* pv. *citri* and *Xanthomonas campestris* pv. *citrumelo* with monoclonal antibodies. Phytopathology 81:857—865

Alvarez AM, Benedict AA, Mizumoto CY, Hunter JE and Gabriel DW (1994) Serological, pathological, and genetic diversity among strains of *Xanthomonas campestris* infecting crucifers. Phytopathology 84:1449—1457

Anon (1978) Distribution maps of plant diseases No 136, 4th ed., Commonwealth Mycological Institute, Kew, Surry, England

Arias RS, Nelson SC and Alvarez AM (1996) MPN—microfluorplate technique to study epiphytic populations of a bioluminescent *Xanthomonas campestris* pv. *campestris*. Phytopathology 86:S35—36

Arlat M, Gough CL, Barber CE, Boucher C and Daniels MJ (1991) *Xanthomonas campestris* contains a cluster of *hrp* genes related to the larger *hrp* cluster *of Pseudomonas solanacearum*. Mol Plant-Microbe Interact 4:593—601

Barber CE, Wilson TJG, Slater H, Dow JM and Daniels MJ (1996) Some novel factors required for pathogenicity of *Xanthomonas campestris* pv. *campestris*. In: Stacey G, Mullin B and Gresshoff PM (eds) Biology of Plant-Microbe Interactions, pp 209—212. American Phytopathological Society, St Paul MN

Barber CE, Tang J-L, Feng J-X, Pan MQ, Wilson TJG, Slater H, Dow JM, Williams P and Daniels MJ (1997) A novel regulatory system required for pathogenicity of *Xanthomonas campestris* is mediated by a small diffusible signal molecule. Mol Microbiol 24:555—566

Barrere GC, Barber CE and Daniels MJ (1986) Molecular cloning of genes involved in the production of the extracellular polysaccharide xanthan by *Xanthomonas campestris* pv. *campestris*. Int J Biol Macromol 8:372—374

Bashan Y and Okon Y (1986) Internal and external infections of fruits and seeds of peppers by *Xanthomonas campestris* pv. *vesicatoria*. Can J Bot 64:2865—2871

Benedict AA, Alvarez AM and Pollard LW (1990) Pathovar-specific antigens of *Xanthomonas campestris* pv. *begoniae* and *Xanthomonas campestris* pv. *pellargoni* detected with monoclonal antibodies. Appl Environ Microbiol 56:572—574

Bergey DH (1923) Bergeyís Manual of Determinative Bacteriology, pp 176—177. Williams and Wilkins Co, Baltimore, MD

Boher B, Nicole M, Potin M and Geiger JP (1997). Extracellular polysaccharides from *Xanthomonas axonopodis* pv. *manihotis* interact with cassava cell walls during pathogenesis. Mol Plant-Microbe Interact 10:803—811.

Bonas U (1994) *hrp* genes of phytopathogenic bacteria. In: Dangle JL (ed) Current Topics in Microbiology and Immunology: Bacterial Pathogenesis of Plants and Animals. Vol 192, pp 79—98. Springer, Berlin, Heidelberg

Bonas U, Schulte R, Fenselau S, Minsavage GV, Staskawicz BJ and Stall RE (1991) Isolation of a gene cluster from *Xanthomonas campestris* pv. *vesicatoria* that determines pathogenicity and the hypersensitive response on pepper and tomato. Mol Plant-Microbe Interact 4:81—88

Boucher CA, Van Gijsegem F, Barberis PA, Arlat M and Zischek C (1987) *Pseudomonas solanacearum* genes controlling both pathogenicity on tomato and hypersensitivity on tobacco are clustered. J Bacteriol 169:5626—5632

Bretschneider KE, Gonella, MP and Robeson DJ (1989) A comparative light and electron microscopical study of compatible and incompatible interactions between *Xanthomonas campestris* pv. *campestris* and cabbage (*Brassica oleracea*). Physiol Mol Plant Pathol 34:285—297

Brown I, Mansfield J, Irlam I, Conrads-Strauch J and Bonas U (1993) Ultrastructure of interactions between *Xanthomonas campestris* pv. vesicatoria and pepper, including immunocytochemical localization of extracellular polysaccharides and the AvrBs3 protein. Mol Plant-Microbe Interact 6:376—386

Buell CR and Somerville SC (1995) Expression of defense-related and putative signaling genes during tolerant and susceptible interactions of *Arabidopsis* with *Xanthomonas campestris* pv. *campestris* Mol Plant-Microbe Interact 8:435—443

Chen J, Roberts PD and Gabriel DW (1994) Effects of a virulence locus from *Xanthomonas campestris* 528[T] on pathovar status and ability to elicit blight symptoms on crucifers. Phytopathology 84:1458—1465

Chester FD (1897) A preliminary arrangement of the species of the genus *Bacterium*. Annual Report, Delaware Agricultural Experiment Station 9:110—117

Chun W, Cui J and Poplawsky AR (1997) Purification, characterization and biological role of a pheromone produced by *Xanthomonas campestris* pv. *campestris*. Physiol Mol Plant Pathol 51:1—14

Chupp C and Sherf AF (1960) Vegetable Diseases and Their Control. Ronald Press, New York

Cook AA, Larson RH and Walker JC (1952a) Relation of the black rot pathogen to cabbage seed. Phytopathology 42:316—320

Cook AA, Walker JC and Larson RH (1952b) Studies on the disease cycle of black rot of crucifers. Phytopathology 42:162—167

Cornelis G, Sluiters C, Lambert de Rouvroit C and Michiels T (1989) Homology between *VirF*, the transcriptional activator of the *Yersinia* virulence regulon, and *AraC*, the *Escherichia coli* arabinose operon regulator. J Bacteriol 171:254—262

Dane F and Shaw JJ (1993) Growth of bioluminescent *Xanthomonas campestris* pv. *campestris* in and on susceptible and resistant host plants. Mol Plant-Microbe Interact 6:786—789

Dane F and Shaw JJ (1996) Survival and persistence of bioluminescent *Xanthomonas campestris* pv. *campestris* on host and non-host plants in the field environment. J Appl Bacteriol 80:73—80

Dangl JL, Dietrich RA, Richberg MH (1996) Death don't have no mercy: Cell death programs in plant-microbe interactions. Plant Cell 8:1793—1807

Daniels MJ (1989) Pathogenicity of *Xanthomonas* and related bacteria towards plants In: Hopwood DA and Chater KF (eds) Genetics of Bacterial Diversity, pp 353—371. Academic Press, London

Daniels MJ and Leach JE (1993) Genetics of *Xanthomonas*. In: Swings JG and Civerolo EL (eds) *Xanthomonas*. pp 301—339. Chapman and Hall, London

Daniels MJ, Turner PC, Barber CE, Cleary WG and Reed G (1983) Towards the genetical analysis of pathogenicity of *Xanthomonas campestris*. In: Pühler A (ed) Molecular Genetics of the Bacteria-Plant Interaction, pp 340—344. Springer-Verlag, Berlin

Daniels MJ, Barber CE, Turner PC, Cleary WG and Sawczyc MK (1984a) Isolation of mutants of *Xanthomonas campestris* pv. *campestris* showing altered pathogenicity. J Gen Microbiol 130:2447—2455

Daniels MJ, Barber CE, Turner PC, Sawczyc MK, Byrde RJW and Fielding AH (1984b) Cloning of genes involved in pathogenicity of *Xanthomonas campestris* pv. *campestris* using the broad host range cosmid pLAFR1. EMBO J 3:3323—3328

Daniels MJ, Dow JM, Wilson TJG, Soby SD, Tang JL, Han B and Liddle SA (1994) Regulation of gene expression in bacterial pathogens. In: Bowles DJ, Gilmartin PM, Knox JP and Lunt GG (eds) Molecular Botany: Signals and the Environment, Biochem. Soc. Symp. 60:231—240

Davies HA, Daniels MJ and Dow JM (1997a) Induction of extracellular matrix glycoproteins in *Brassica* petioles by wounding and in response to *Xanthomonas campestris*. Mol Plant-Microbe Interact 7:812—820

Davies HA, Findlay K, Daniels J and Dow J (1997b) A novel proline-rich glycoprotein associated with the extracellular matrix of vascular bundles of *Brassica* petioles. Planta 202:28—35

Denny T (1995) Involvement of bacterial exopolysaccharides in plant pathogenesis. Annu Rev Phytopathol 33:173—197

Dow JM and Daniels MJ (1994) Pathogenicity determinants and global regulation of pathogenicity of *Xanthomonas campestris* pv. *campestris*. In: Dangle JL (ed) Current Topics in Microbiology and Immunology: Bacterial Pathogenesis of Plants and Animals. Vol 192, pp 29—41. Springer, Berlin, Heidelberg

Dow JM, Milligan DE, Jamieson L, Barber CE and Daniels MJ (1989) Molecular cloning of a polygalacturonate lyase gene from *Xanthomonas campestris* pv. *campestris* and role of the gene product in pathogenicity. Physiol Mol Plant Pathol 35:113—120

Dow JM, Clarke BR, Milligan DE, Tang JL and Daniels MJ (1990) Extracellular proteases from *Xanthomonas campestris* pv. *campestris*, the black rot pathogen. Appl Environ Microbiol 56:2994—2998

Dow JM, Fan MJ, Newman M-A and Daniels MJ (1993) Differential expression of conserved protease genes in crucifer-attacking pathovars of *Xanthomonas campestris*. Appl Environ Microbiol 59: 3996—4003

Dow JM, Osbourn AE, Wilson TJ and Daniels MJ (1995) A locus determining pathogenicity of *Xanthomonas campestris* is involved in lipopolysaccharide biosynthesis. Mol Plant-Microbe Interact 8:768—777

Dowson WJ (1939) On the systematic position and generic names of the Gram-negative bacterial plant pathogens. Zentralbl f Bakt Parasit u Infekt, Abt II 100:177—193

Dums F, Dow JM and Daniels MJ (1991) Structural characterization of protein secretion genes of the bacterial phytopathogen *Xanthomonas campestris* pathovar *campestris*: relatedness to secretion systems of other gram-negative bacteria. Mol Gen Genet 229:357—364

Dye DW, Bradbury JF, Goto M, Hayward AC, Lelliott RA and Schroth MN (1980) International standards for naming pathovars of phytopathogenic bacteria and a list of pathovar names and pathotype strains. Rev Plant Pathol 59:153—168

Falke JJ, Bass RB, Butler SL, Chervitz SA, Danielson MA (1997) The two-component signaling pathway of bacterial chemotaxis: a molecular view of signal transduction by receptors, kinases, and adaption enzymes. Annu Rev Cell Dev Biol 13:457—512

Fenselau S and Bonas U (1995) Sequence and expression analysis of the *hrpB* pathogenicity operon of *Xanthomonas campestris* pv. *vesicatoria* which encodes eight proteins with similarity to components of the Hrp, Ysc, Spa, and Fli secretion systems. Mol Plant-Microbe Interact 8:845—854

Fuqua WC, Winans SC, Greenberg EP (1994) Quorum sensing in bacteria: the LuxR/LuxL family of cell density responsive transcriptional regulators. J Bacteriol 176:269—275

Gabriel DW, Kingsley MT, Yang Y, Chen J and Roberts P (1994) Host-specific virulence genes of *Xanthomonas*. In: Kado CI and Crosa JH (eds) Molecular Mechanisms of Bacterial Virulence, pp 141—158. Kluwer, Dordrecht, the Netherlands

Garman H (1890) A bacterial disease of cabbage. Third Annual Report for Year 1990, Kentucky Agricultural Experiment Station 3:43—46

Genin S, Gough CL, Zischek C, and Boucher CA (1992) Evidence that the hrpB encodes a positive regulator of *hrp* genes from *Pseudomonas solanacearum*. Mol Microbiol 6: 3065—3076

Gough CL, Dow JM, Barber CE and Daniels MJ (1988) Cloning of two endoglucanase genes in *Xanthomonas campestris* pv. *campestris*: analysis of the role of the major endoglucanase in pathogenesis. Mol Plant-Microbe Interact 1:275—281

Gough CL, Dow JM, Keen J, Henrissat B and Daniels MJ (1990) Nucleotide sequence of the *engXCA* gene encoding the major endoglucanase of *Xanthomonas campestris* pv. *campestris*. Gene 89:53—59

Harding HA, Stewart FC and Prucha MJ (1904) Vitality of the cabbage black rot germ on cabbage seed. New York Agricultural Experiment Station Bulletin 251:178—194

Harding NE, Clearly JM, Cabanas DK, Rosen IG and Kang K (1987) Genetic and physical analysis of a cluster of genes essential for xanthan gum biosynthesis in *Xanthomonas campestris* pv. *campestris*. J Bacteriol 169:2854—2861

Hickman J and Ashwell G (1966) Isolation of a bacterial liposaccharide from *Xanthomonas campestris* containing 3-acetamido-3,6-dideoxy-D-galactose and D-rhamnose. J Biol Chem 241:1424—1428

Hötte B, Rath-Arnold I, Pühler A, and Simon R (1990) Cloning and analysis of a 35.3-kilobase DNA region involved in exopolysaccharide production by *Xanthomonas campestris* pv. *campestris*. J Bacteriol 172:5165—5172

Hugouvieux V, Barber CE and Daniels MJ (1998) Entry of *Xanthomonas campestris* pv. *campestris* into hydathodes of *Arabidopsis thaliana* leaves: A system system for studying early infection events in bacterial pathogenesis. Mol Plant-Microbe Interact 11:537—543

Hunter JE, Dickson MH and Ludwig JW (1987) Source of resistance to black rot of cabbage expressed in seedlings and adult plants. Plant Disease 71:263—266

Jenkins CL and Starr MP (1982) The brominated aryl-polyene (xanthomonadin) pigments of *Xanthomonas campestris* pv. *juglandis* protect against photobiological damage. Curr Microbiol 7:323—326

Kado CI (1994) Anti-host-defense systems are elaborated by plant pathogenic bacteria. In: Kado CI and Crosa JH (eds) Molecular Mechanisms of Bacterial Virulence, pp 581—591. Kluwer, Dordrecht, the Netherlands

Kamdar HV, Kamoun S and Kado CI (1993) Restoration of pathogenicity of avirulent *Xanthomonas oryzae* pv. *oryzae* and *X. campestris* pathovars by reciprocal complementation with the *hrpXo* and *hrpXc* genes and identification of *HrpX* function by sequence analysis. J Bacteriol 175:2017—2025

Kamoun S and Kado CI (1990a) A plant-inducible gene of *Xanthomonas campestris* pv. *campestris* encodes an exocellular component required for growth in the host and hypersensitivity on nonhosts. J Bacteriol 172:5165—5172

Kamoun S and Kado CI (1990b) Phenotypic switching affecting chemotaxis, xanthan production, and virulence in *Xanthomonas campestris*. Appl Environ Microbiol 56:3855—3860

Kamoun S, Kamdar HV, Tola E and Kado CI (1992a) Incompatible interactions between crucifers and *Xanthomonas campestris* involve a vascular hypersensitive response: role of the *hrpX* locus. Mol Plant-Microbe Interact 5:22—33

Kamoun S, Tola E, Kamdar H and Kado CI (1992b) Rapid generation of directed and unmarked deletions in *Xanthomonas*. Mol Microbiol 6:809—816

Kearney B and Staskawicz BJ (1990) Widespread distribution and fitness contribution of the *Xanthomonas campestris* avirulence gene avrBs2. Nature 346:385—386

Kingsley MT, Gabriel DW, Marlow GC and Roberts PD (1993) The *opsX* locus of *Xanthomonas campestris* affects host range and biosynthesis of lipopolysaccharide and extracellular polysaccharide. J Bacteriol 175:5839—5850

Köplin R, Arnold W, Hötte R, Simon R, Wang G and Pühler A (1992) Genetics of xanthan production in *Xanthomonas campestris* pv. *campestris*: the *xanA* and *xanB* genes are involved in UDP-glucose and GDP-mannose biosynthesis. J Bacteriol 174:191—199

Kuan TL, Minsavage GV and Schaad NW (1986) Aerial dissemination of *Xanthomonas campestris* pv. *campestris* from crucifer weeds. Plant Disease 70:409—413

Leach JE and White FF (1996) Bacterial avirulence genes. Annu Rev of Phytopathol 34:153—179

Leach JE, Guo A, Reimers P, Choi SH, Hopkins CM and White, FF (1994) In: Kado CI and Crosa JH (eds) Molecular Mechanisms of Bacterial Virulence, pp 551—560. Kluwer, Dordrecht, the Netherlands

Lee TC, Lin NT and Tseng YH (1996) Isolation and characterization of the *recA* gene of *Xanthomonas campestris* pv. *campestris*. Biochem Biophys Res Commun 221:459—465

Liew KW and Alvarez AM (1981a) Biological and morphological characterization of *Xanthomonas campestris* bacteriophages. Phytopathology 71:269—273

Liew KW and Alvarez AM (1981b) Phage typing and lysotype distribution of *Xanthomonas campestris*. Phytopathology 71:274—276

Lin CS, Lin NT, Yang BY, Weng SF and Tseng YH (1995) Nucleotide sequence and expression of UDP-glucose dehydrogenase gene required for the synthesis of xanthan gum in *Xanthomonas campestris* Biochem Biophys Res Commun 207:223—230

Loprasert S, Vattanviboon P, Praituan W, Chamnongpol S and Mongkolsuk S (1996) Regulation of the oxidative stress protective enzymes, catalase and superoxide dismutase in *Xanthomonas*. Gene 179:33—37

Louws FJ, Fulbright DW, Stephens CT and De Bruijn FJ (1994) Specific genomic fingerprints of phytopathogenic *Xanthomonas* and *Pseudomonas* pathovars and strains generated with repetitive sequences and PCR. Appl and Environ Microbiol 60:2286—2295

Lummerzheim M, De Oliveira D, Castresana C, Miguens FC, Louzada E, Roby D, Van Montagu M and Timmerman B (1993) Identification of compatible and incompatible interactions between *Arabidopsis thaliana* and *Xanthomonas campestris* pv. *campestris* and characterization of the hypersensitive response. Mol Plant-Microbe Interact 5:532—544

Lummerzheim M, Sandroni M, Castresana C, De Oliveira D, Van Montagu M, Roby D and Timmerman

B (1995) Comparative microscopic and enzymatic characterization of the leaf necrosis induced in *Arabidopsis thaliana* by lead nitrate and by *Xanthomonas campestris* pv. campestris after foliar spray. Plant Cell Environ 18:499—509

McCulloch L (1929) A bacterial spot of horseradish caused by *Bacterium campestre* var armoraciae, n. var. J Agric Res 38:269—287

McElhaney R (1991) Bacterial bioluminescence: a tool to study host-pathogen interactions between *Brassica oleracea* and the bacterial phytopathogen *Xanthomonas campestris* pv. campestris in black rot of cabbage. PhD Thesis University of Hawaii, Honolulu, HI, 102 pp

McElhaney R, Alvarez AM and Kado CI (1998) Nitrogen limits *Xanthomonas campestris* pv. *campestris* invasion of the host xylem. Physiol Mol Plant Pathol 52:15—24

Machmud M (1982) *Xanthomonas campestris* pv. *armoraciae*, the causal agent of Xanthomonas leaf spot of crucifers. PhD Thesis Louisiana State University, Baton Rouge, LA, 99 pp

Martínez-Salazar JM, Palacios AN, Sánchez R, Caro AD and Soberón-Chávez G (1993) Genetic stability and xanthan gum production in *Xanthomonas campestris* pv. *campestris* NRRL B1459. Mol Microbiol 8:1053—1061

Martínez S, Martínez-Salazar JM, Camas A, Sánchez R and Soberón-Chávez G (1997) Evaluation of the role of protein in plant virulence with recA mutants of *Xanthomonas campestris* pv. *campestris*. Mol Plant-Microbe Interact 10:911—916

Meyerowitz EM (1987) *Arabidopsis thaliana*. Annu Rev Genet 21:93—111

Mochizuki GT and Alvarez AM (1992) Immunodetection of *Xanthomonas campestris* pv. *campestris* in the guttation fluids of latently infected cabbage seedlings. Phytopathology 82:1154

Mochizuki GT and Alvarez AM (1996) A bioluminescent *Xanthomonas campestris* pv. *campestris* used to monitor black rot infections in cabbage seedlings treated with Fosetyl-Al. Plant Disease 80:758—762

Newman M-A, Conrads-Strauch J, Scofield G and Daniels MJ (1994) Defense-related gene induction in *Brassica campestris* in response to defined mutants of *Xanthomonas campestris* with altered pathogenicity. Mol Plant-Microbe Interact 7:553—563

Newman M-A, Daniels MJ Dow JM (1995) Lipopolysaccharide from *Xanthomonas campestris* induces defense-related gene expression in *Brassica campestris*. Mol Plant-Microbe Interact 8:778—780

Newman M-A, Daniels MJ and Dow JM (1997) The activity of lipid A and core components of bacterial lipopolysaccharides in the prevention of the hypersensitive response in pepper. Mol Plant-Microbe Interact 7:926—928

Ojanen T, Helander IM, Haahtela K, Korhonen TK and Laakso T (1993) Outer membrane proteins and lipopolysaccharides in pathovars of *Xanthomonas campestris*. Appl Environ Microbiol 59:4143—4151

Oku T, Alvarez AM and Kado CI (1995) Conservation of the hypersensitivity-pathogenicity regulatory gene *hrpX* of *Xanthomonas campestris* and *X. oryzae*. DNA Sequence 5:245—249

Oku T, Wakasaki Y, Adachi N, Kado CI, Tsuchiya K and Hibi T (1998) Pathogenicity, non-host hypersensitivity, and host defense non-permissibility regulatory gene hrpX is highly conserved in *Xanthomonas* pathovars. J Phytopathology (in press)

Onsando JM (1992) Black rot of crucifers In: Chaube HS, Kumar J, Mukhopadhyay AN and Singh US (eds) Plant Diseases of International Importance, Vol II: Diseases of Vegetable and Oil Seed Crops, pp 243—252. Prentice Hall, Englewood Cliffs, New Jersey

Osbourn AE, Clarke BR and Daniels MJ (1990a) Identification and DNA sequence of a pathogenicity gene of *Xanthomonas campestris* pv. *campestris*. Mol Plant-Microbe Interact 3:280—285

Osbourn AE, Clarke BR, Stevens BJH and Daniels MJ (1990b) Use of oligonucleotide probes to identify members of two-component regulatory systems in *Xanthomonas campestris* pathovar *campestris*. Mol

Gen Genet 222:145—151

Osbourn MJ (1963) Studies on the gram-negative cell wall. I. Evidence for the role of 2-keto-3-deoxyoctonate in the lipopolysaccharide of *Salmonella typhimurium*. Proc Natl Acad Sci USA 50:499—506

Pammel LH (1895) Bacteriosis of rutabaga (*Bacillus campestris* n.sp.) Iowa Agricultural Experiment Statation Bulletin No 27:130—134

Parker JE, Barber C, Fan M-J, Daniels MJ (1993) Interaction of *Xanthomonas campestris* with *Arabidopsis thaliana*: Characterization of a gene from *X. c.* pv. *raphani* that confers avirulence to most *A. thaliana* accessions. Mol Plant-Microbe Interact 6:216—224

Poplawsky AR and Chun W (1995) A *Xanthomonas campestris pv. campestris* mutant negative for production of a diffusible signal is also impaired in epiphyic development. Phytopathology 85:1148

Poplawsky AR and Chun W (1997) *pigB* determines a diffusible factor needed for extracellular polysaccharide slime and xanthomonadin production in *Xanthomonas campestris* pv. *campestris*. J Bacteriol 179:439—444

Poplawsky AR and Chun W (1998) *Xanthomonas campestris* pv. *campestris* requires a functional *pigB* for epiphytic survival and host infection. Mol Plant-Microbe Interact 11:466—475

Poplawsky AR, Kawalek MD and Schaad NW (1993) A xanthomonadin-encoding gene cluster for the identification of pathovars of *Xanthomonas campestris*. Mol Plant- Microbe Interact 6:545—552

Poplawsky AR, Chun W, Slater H, Daniels MJ and Dow JM (1998) Synthesis of extracellular polysaccharide, extracellular enzymes and xanthomonadin in *Xanthomonas campestris*: Evidence for the involvement of two intercellular regulatory signals. Mol Plant-Microbe Interact 11:68—70

Robeson DJ, Bretschneider KE and Gonella MP (1989) A hydathode inoculation technique for the simulation of natural black rot infection of cabbage by *Xanthomonas campestris* pv. *campestris*. Ann Appl Biol 115:455—459

Romantschuk M, Roine E, Ojanen T, Van Doorn J, Louhelainen J, Nurmiaho-Lassila E-L and Haahtela K (1994) Fimbria (pilus) mediated attachment of *Pseudomonas syringae*, *Erwinia rhapontici* and *Xanthomonas campestris* to plant surfaces. In: Kado CI and Crosa JH (eds) Molecular Mechanisms of Bacterial Virulence, pp 67—77. Kluwer, Dordrecht, the Netherlands

Rudolph KWE, Gross M, Ebrahim-Nesbat F, Nöllenburg M, Zomorodian A, Wydra K, Neugebauer M, Hettwer U, El-Shouny W, Sonnenberg B and Klement Z (1994) The role of extracellular polysaccharides as virulence factors for phytopathogenic pseudomonads and xanthomonads. In: Kado CI and Crosa JH (eds) Molecular Mechanisms of Bacterial Virulence, pp 357—378. Kluwer, Dordrecht, the Netherlands

Ruissen MA and Gielink AJ (1993) The development of black rot in cabbage as a result of difference in guttation between cultivars. In: Proc. 8th Int Conf on Plant Pathogenic Bacteria. Versailles, France

Russell HL (1898) A bacterial rot of cabbage and allied plants. Wisconsin Agricultural Experiment Station Bulletin 65:130—135

Schaad NW and Alvarez AM (1993) *Xanthomonas campestris* pv. *campestris*: cause of black rot of crucifers. In: Swings JG, Civerolo EL (eds) *Xanthomonas*, pp 51—55. Chapman and Hall, London

Schaad NW and White WC (1974) A qualitative method for detecting *Xanthomonas campestris* in crucifer seed. Phytopathology 65:1034—1036

Shaw JJ and Kado CI (1986) Development of a *Vibrio* bioluminescent gene-set to monitor phytopathogenic bacteria during the ongoing disease process in a non-disruptive manner. Bio/Technology 4:560—564

Shaw JJ and Kado CI (1988) Whole plant wound inoculation for consistent reproduction of black rot of crucifers. Phytopathology 78:981—986

Shaw JJ and Khan I (1993) Efficient transposon mutagenesis of *Xanthomonas campestris* pathovar *campestris* by high voltage electroporation. Biotechniques 14:556—557

Shaw JJ, Settles L and Kado CI (1988) Transposon Tn*4431* mutagenesis of *Xanthomonas campestris* pv.

campestris: characterization of a nonpathogenic mutant and cloning of a locus for pathogenicity. Mol Plant-Microbe Interact 1:39—45

Shaw JJ, Dane F, Geiger D and Kloepper JW (1992) Use of bioluminescence for detection of genetically engineered microorganisms released into the environment. Appl Environ Microbiol 58:267—273

Shigaki T (1996) Differential epidemiological fitness among strains of *Xanthomonas campestris* pv. *campestris* and the genetics of pathogenicity. PhD Thesis University of Hawaii, Honolulu, HI, 86 pp

Shigaki T and Alvarez AM (1993). Differential epidemiological fitness observed for two strains of *Xanthomonas campestris* pv. *campestris*. Phytopathology 83:1405

Shigaki T, Nelson SC and Alvarez AM (1994) Greater epidemiological fitness of *Xanthomonas campestris* pv. *campestris* blight strains in the seedbed. Phytopathology 84:1112

Simpson RB and Johnson LJ (1990) *Arabidopsis thaliana* as a host for *Xanthomonas campestris* pv. *campestris*. Mol Plant-Microbe Interact 3:233—237

Smith EF (1897a) *Pseudomonas campestris* (Pammel), the cause of a brown-rot in cruciferous plants. Zentralbl f Bakt Parasit u Infekt, Abt II, 3:284—291

Smith EF (1897b) *Pseudomonas campestris* (Pammel), the cause of a brown-rot in cruciferous plants. Zentralbl f Bakt Parasit u Infekt, Abt II, 3:478—486

Smith EF (1911) Bacteria in Relation to Plant Diseases, Vol II, pp 300—334. Carnegie Inst, Washington DC

Smith SG, Wilson TJG, Dow JM and Daniels MJ (1996) A gene for superoxide dismutase from *Xanthomonas campestris* pv. *campestris* and its expression during bacterial-plant interactions. Mol Plant-Microbe Interact 7:584—593

Starr MP (1981) The genus *Xanthomonas*. In: Starr MP, Stolp H, Truper HG, Balows A and Schlegel HG (eds) The Prokaryotes Vol I. pp 742—763. Springer, Berlin, Germany

Stewart FC and Harding HA 1903 Combating the black rot of cabbage by the removal of affected leaves. New York State Agricultural Experiment Station Bulletin 232:43—65

Sutherland IW (1993) Xanthan. In: Swings JG and Civerolo EL (eds) *Xanthomonas*. pp 364—388. Chapman and Hall. London

Sutton MD and Williams PH (1970) Relation of xylem plugging to black rot lesion development in cabbage. Can J Bot 48:391—401

Swarup S, De Feyter R, Brlansky RH and Gabriel DW (1991) A pathogenicity locus from *Xanthomonas citri* enables strains from several pathovars of *X. campestris* to elicit cankerlike lesions on citrus. Phytopathology 81:802—809

Swarup S, Yang Y, Kingsley MT and Gabriel DW (1992) A *Xanthomonas citri* pathogenicity gene, *pthA*, pleiotropically encodes gratuitous avirulence on nonhosts. Mol Plant-Microbe Interact 5:204—213

Tang J-L, Gough CL, Barber CE, Dow JM and Daniels MJ (1987) Molecular cloning of protease gene(s) from *Xanthomonas campestris* pathovar *campestris*: expression in *Escherichia coli* and role in pathogenicity Mol Gen Genet 210:443—448

Tang J-L, Gough CL and Daniels MJ (1990) Cloning of genes involved in negative regulation of production of extracelular enzymes and polysaccharide of *Xanthomonas campestris* pathovar *campestris*. Mol Gen Genet 222:157—160

Tang J-L, Liu Y-N, Barber CE, Dow JM, Wootton JC and Daniels MJ (1991) Genetic and molecular analysis of a cluster of *rpf* genes involved in positive regulation of synthesis of extracellular enzymes and polysaccharide in *Xanthomonas campestris* pathovar *campestris*. Mol Gen Genet 266:409—417

Tang J-L, Feng J-X, Li Q-Q, Wen H-X, Zhou D-L, Wilson TJG, Dow JM, Ma Q-S and Daniels MJ (1996) Cloning and characterization of the *rpfC* Gene of *Xanthomonas oryzae* pv. *oryzae*: Involvement in exopolysaccharide production and virulence to rice. Mol Plant-Microbe Interact 7:664—666

Theevachi N and Schaad NW (1986) Immunological characterization of a subspecies-specific antigenic determinant of a membrane protein extract of *Xanthomonas campestris pv. campestris*. Phytopathology 76:148—153

Thorne L, Tansey L and Pollock TJ (1987) Clustering of mutations blocking synthesis of xanthan gum by *Xanthomonas campestris*. J Bacteriol 169:3593—3600

Tsai C-M and Frasch CE (1982) A sensitive silver stain for detecting lipopolysaccharides in polyacrylamide gels. Anal Biochem 119:115—119

Van Gijsegem F, Genin S and Boucher CA (1993) Conservation of secretion pathways for pathogenicity determinants of plant and animal bacteria. Trends Microbiol 1:175—180

Van Gijsegem F, Arlat M, Genin S, Gough CL, Zischek C, Barberis PA and Boucher C (1994) Genes governing the secretion of factors involved in host-bacteria interactions are conserved among animal and plant pathogenic bacteria. In: CI Kado and JH Crosa (eds) Molecular Mechanisms of Bacterial Virlence, pp 625—642. Kluwer, Dordrecht, the Netherlands

Van Gijsegem F, Gough C, Zischek C, Niqueux E, Arlat M, Genin S, Barberis P, German S, Castello P and Boucher C (1995) The *hrp* gene locus of *Pseudomonas solanacearum*, which controls the production of a type III secretion system, encodes eight proteins related to components of the bacterial flagellar biogenesis complex. Mol Microbiol 15:1095—1114

Vauterin L, Swings J, Kersters K (1991) Grouping of *Xanthomonas campestris* pathovars by SDS-PAGE of proteins. J Gen Microbiol 137:1677—1687

Vauterin L, Hoste B, Kersters K and Swings J (1995) Reclassification of *Xanthomonas*. Int J Syst Bacteriol 45:472—489

Volk WA (1966) Cell wall lipopolysaccharides from *Xanthomonas* species. J Bacteriol 91:39—42

Wallis FM, Rijkenberg FHJ, Joubert JJ and Martin WM (1973) Ultrastructural histopathology of cabbage leaves infected with *Xanthomonas campestris*. Physiol Plant Pathol 3:371—378

Wei CL, Lin NT, Weng SF and Tseng YH 1996 The gene encoding UDP-glucose pyrophosphorylase is required for the synthesis of xanthan gum in *Xanthomonas campestris* pv. *campestris*. Biochem Biophys Res Commun 226:607—612

Wengelnik K, Van den Ackerveken G and Bonas U (1996) HrpG, a key *hrp* regulatory protein of *Xanthomonas campestris* pv. *vesicatoria* is homologous to two-component response regulators. Mol Plant-Microbe Interact 8:704—712

Wengelnik K and Bonas U (1996) HrpXv, an AraC-type regulator, activates expression of five of the six loci in the *hrp* cluster of *Xanthomonas campestris* pv. *vesicatoria*. J. Bacteriol 178:3462—3469

Westphal O and Jann K (1965) Bacterial lipopolysaccharides: Extraction with phenol-water and further applications of the procedure. Methods Carbohydrate Chem 5:83—91

Williams PH (1980) Black Rot: A continuing threat to world crucifers. Plant Dis 64:736—742

Williams PH, Staub T and Sutton JC (1972) Inheritance of resistance in cabbage to black rot. Phytopathology 62:247—252

Yang Y and Gabriel DW (1995a) Intragenic recombination of a single plant pathogen gene provides a mechanism for the evolution of new host specificities. J Bacteriol 177: 4963—4968

Yang Y and Gabriel DW (1995b) *Xanthomonas* avirulence/pathogenicity gene family encodes functional plant nuclear targeting signals. Mol Plant-Microbe Interact 8:627—631

Yang YO, Yuan QP and Gabriel DW (1996) Watersoaking function (s) of XcmH1005 are redundantly encoded by members of the xanthomonas *avr/pth* gene family. Mol Plant-Microbe Interact 9:105—113

Yuen GY, Alvarez AM, Benedict AA and Trotter KJ (1987) Use of monoclonal antibodies to monitor the dissemination of *Xanthomonas campestris* pv. *campestris*. Phytopathology 77:366—370

THE *CLADOSPORIUM FULVUM*-TOMATO INTERACTION

A MODEL SYSTEM TO STUDY GENE-FOR-GENE RELATIONSHIPS

PIERRE J.G.M. DE WIT

Department of Phytopathology, Wageningen Agricultural University
P.O. Box 8025, 6700 EE Wageningen, The Netherlands
(pierre.dewit@medew.fyto.wau.nl)

Summary

The pathosystem *Cladosporium fulvum*-tomato has become a model system in molecular plant pathology as both fungus and host plant are amenable to study by molecular methods. Many resistance genes (*Cf*) are available in near-isogenic lines of tomato, while many races of *C. fulvum* exist that can overcome one or more *Cf* genes, giving rise to a gene-for-gene relationship. *C. fulvum* is a biotrophic pathogen that penetrates tomato leaves through stomata and colonizes the intercellular space around leaf mesophyll cells, where it stays extracellular during its entire life cycle. In compatible interactions the fungus produces various extracellular proteins (ECPs), some of which represent crucial virulence factors. In incompatible interactions fungal growth is arrested very soon after penetration. Resistance is associated with deposition of callose, a hypersensitive response (HR), electrolyte leakage and accumulation of phytoalexins and pathogenesis-related proteins.

HR-based resistance is induced in plants with *Cf* genes, by matching race-specific elicitors secreted by *C. fulvum* immediately after stomatal penetration. The race-specific elicitors AVR4 and AVR9 have been characterized and their encoding genes *Avr4* and *Avr9* have been cloned. Both elicitors are cystine-rich peptides of which AVR9 belongs to the family of cystine-knotted peptides. The resistance genes *Cf-2*, *Cf-4*, *Cf-5* and *Cf-9* have been cloned. They encode proteins that belong to a superfamily of leucine-rich repeat (LRR) proteins. The major part of the Cf proteins contains 25-38 LRRs which is extracellular, while a short C-terminal domain is cytoplasmic.

Two major players in the gene-for-gene system, i.e. the *Avr* gene and the corresponding *Cf* gene, have been cloned, but so far there is no evidence that their products interact directly. Membranes of both *Cf-9*-plus and *Cf-9*-minus plants contain a similar high affinity binding site for the AVR9 peptide (Kd = 70 pM). To determine whether Cf-9 represents a low affinity binding site, binding studies of AVR9 to in vitro-produced Cf-9 protein are required.

A. Slusarenko, R.S.S. Fraser, and L.C. van Loon (eds), Mechanisms of Resistance to Plant Diseases, 53-75.
© 2000 *Kluwer Academic Publishers. Printed in the Netherlands.*

treatment with matching AVR elicitors. These responses include an HR and an oxidative burst, as well as activation of enzymes such as H^+-ATPase and NADPH oxidase.

Tomato lines have been found that respond with an HR to the virulence factor ECP2, which indicates that it can also function as an avirulence factor. The resistance gene that recognizes ECP2 has been designated *Cf-ECP2* and most probably represents a durable resistance gene.

Abbreviations

AF	apoplastic fluid
Avr	avirulence gene encoding race-specific elicitor
Ecp	gene encoding extracellular protein
Cf	resistance gene against *Cladosporium fulvum*
GUS	β-glucuronidase
HR	hypersensitive response
LRR	leucine-rich repeat
NMR	nuclear magnetic resonance
PR	pathogenesis-related
PVX	potato virus X
TNF	tumor necrosis factor
TNFR	tumor necrosis factor receptor

I.   Introduction

Cladosporium fulvum Cooke (syn. *Fulvia fulva*) causes leaf mould of tomato.

C. fulvum is a Deuteromycete that was first described in 1883 (Cooke, 1883). The origin of *C. fulvum* is most probably South America, the gene centre of tomato. The fungus is a biotrophic pathogen that occurs world-wide and causes disease only on the genus *Lycopersicon*. Tomato (*Lycopersicon esculentum* Mill.) is susceptible to the fungus, but other species of the genus that occur in South America are often resistant. The fungus penetrates tomato leaves through stomata and colonization is confined to the intercellular space of tomato leaves where no specialized feeding structures such as haustoria are formed. The fungus causes pale-yellow spots on the upper surface of the leaves. At the lower side of the leaves beneath the spots, conidiophores emerge through stomata a few weeks after infection. The conidiophores produce numerous conidia that can re-infect tomato plants. The major damage caused by this disease is through the disfunctioning of stomata and loss of leaves once the fungus starts to sporulate leading to yield reduction (Butler and Jones, 1949).

During the 1930s losses in tomato crops caused by *C. fulvum* were high. At that time no resistance genes were present in commercially grown cultivars. Since, many *Cf*-resistance genes originating from related wild species of tomato such as *L. chilense, L. hirsutum, L. peruvianum,* and *L. pimpinellifolium,* have been transferred to cultivated tomato in breeding programmes (Langford, 1937; Kerr and Bailey, 1994; Stevens and

Rick, 1988). Several of these resistance genes (e.g. *Cf-2*, *Cf-3*, *Cf-4*, *Cf-5*, and *Cf-9*) are available in near-isogenic lines of tomato cultivar MoneyMaker (Boukema, 1981). Presently, tomato crops grown in glasshouses and outdoors generally carry one or more *Cf* genes. However, after introduction of commercial cultivars carrying *Cf* genes, new races of *C. fulvum* appeared which could often overcome introgressed monogenic resistances. Many races of *C. fulvum* have been described (Hubbeling, 1978; Laterrot, 1986, Lindhout *et al.*, 1989).

The availability of many near-isogenic lines of tomato that carry *Cf* genes and which respond differentially (resistant or susceptible) to the various races of *C. fulvum* has made the *C. fulvum*-tomato interaction a model system in molecular plant pathology to study gene-for-gene relationships (Day, 1957; De Wit, 1992; Oliver, 1992). During the last two decades the physiology, biochemistry and molecular biology of the *C. fulvum*-tomato interaction have been studied extensively by several research groups (Higgins, 1982; De Wit, 1992, 1995; Oliver, 1992; Hammond-Kosack and Jones, 1996). Basic questions on pathogenicity, virulence, avirulence, resistance and specificity of resistance and accompanying defence responses have been actively pursued and will be covered in this case study (Box 1).

Box 1. Pathogenicity, Virulence and Avirulence Genes of Fungi.

A pathogenicity gene refers to a gene of a pathogen that is absolutely required to cause disease on its host. A virulence gene refers to a gene with a quantitative effect on disease development. A virulence gene encodes a virulence factor that can have major or minor effects on disease development. An avirulence gene is defined as a gene that encodes a product that is recognized by a resistant plant that carries the matching resistance gene. Recognition is usually followed by a hypersensitive response (HR) which keeps the fungus localized to the primary infection site. Loss of avirulence can be due to complete loss of the avirulence gene or to mutations in the avirulence gene which result in production of a protein that no longer induces a resistance response in the cultivar with the matching resistance gene. In gene-for-gene systems virulence refers to loss of avirulence. In this context use of the term virulence gene should be avoided.

II.   The Infection Process

A.   THE COMPATIBLE INTERACTION

The first detailed microscopic studies were performed by Bond (1938). The infection process was initially studied by light microscopy. Later, more detailed studies were performed using transmission and scanning electron microscopy (Lazarovits and Higgins, 1976a; De Wit, 1977). The infection process was followed after inoculation of near-isogenic lines of susceptible cultivars of tomato with conidia of virulent races of *C. fulvum*. Conidia germinate on the lower surface of the leaf at high relative humidity and form germ tubes, the so-called "runner hyphae"(2-3 μm in diameter). A runner hypha grows until its encounters an open stoma, which is subsequently penetrated. Once inside the leaf, the diameter of the hyphae enlarges at least two-fold. The fungus colonizes the intercellular space of leaves of susceptible genotypes without visible induction of defence responses. Hyphae grow in close contact with mesophyll cells. It is not known which factors are crucial to enable and sustain growth of the fungus inside the intercellular space. Extracellular growth of the fungus allows isolation of apoplastic fluid (AF) from infected leaves that contains compounds originating from the plant-fungus interfase. AF, which is obtained by *in vacuo* infiltration of infected tomato leaves with water, followed by low speed centrifugation (De Wit and Spikman, 1982), contains compounds that are constitutively produced by the plant or fungus in addition to compounds that are produced as a result of the interaction between host plant and fungus (De Wit *et al.*, 1986; Joosten *et al.*, 1990).

1.   *Uptake of Nutrients*
In the intercellular space a large interface is present where communication between plant and fungus, and uptake of nutrients occurs. Sucrose is a possible carbon source for the fungus. The level of sucrose in the apoplast decreases significantly during the course of colonization of the leaf (Joosten *et al.*, 1990). Sucrose is the main product of photosynthesis and is loaded into the phloem via the apoplast, probably by specific sucrose carriers (Dickinson *et al.*, 1991). Invertases that are secreted by *C. fulvum* convert apoplastic sucrose into glucose and fructose, which are subsequently taken up and metabolized by the fungus. Many fungi, including *C. fulvum*, are able to convert monosaccharides into mannitol, a polyol that can be metabolized or stored by the fungus, but which cannot be metabolized by the host (Joosten *et al.* 1990 and references in there). Mannitol levels increase significantly during infection and can be considered as a marker for fungal growth (Joosten *et al.*, 1990; Noeldner *et al.*, 1994). Nitrogen-containing compounds and other nutrients that are imported into the leaves via the xylem, are probably directly or indirectly taken up by the fungus from the tissue surrounding the vascular bundles.

2.   *Virulence Factors*
The fungus does not seem to require classical pathogenicity and virulence factors such as cell wall-degrading enzymes and toxins to colonize its host (Agrios, 1997). In order to search for putative proteinaceous pathogenicity or virulence factors that enable *C. fulvum* to colonize and reproduce in the leaves of tomato, AF isolated from compatible interactions between tomato and *C. fulvum*, has been analyzed. In this way various compatible interaction-specific proteins of *C. fulvum* were identified (De Wit *et al.*, 1986). Two of these extracellular proteins, ECP1 and ECP2, have been purified (Joosten and De Wit, 1988; Wubben *et al.*, 1994b) and the encoding genes, *Ecp*1 and *Ecp*2, isolated (Van den Ackerveken *et al.* 1993b). ECP1 and ECP2, predominantly

produced by *C. fulvum in planta*, are associated with the matrix between fungal hyphae and host cell walls (Wubben *et al.*, 1994b). Whether these proteins are essential for the adherence of *C. fulvum* to the host cell wall is not known. In order to investigate whether *Ecp*1 and *Ecp*2 are required for pathogenicity or virulence of *C. fulvum*, both genes were disrupted by homologous recombination. The *Ecp*2 gene appeared not to be essential for pathogenicity or virulence of *C. fulvum* on young tomato seedlings (Marmeisse *et al.*, 1994). However, more recent experiments have shown that on mature tomato plants the *Ecp*2 gene is essential for full virulence (Laugé *et al.*, 1997). In the same study it was also shown that the *Ecp*1 gene encodes a virulence factor. In both cases gene replacement had a significant effect on sporulation of *C. fulvum* on the host (Laugé *et al.*, 1997). On mature plants the *Ecp*1-minus strain of *C. fulvum* sporulated significantly less as compared to the wild type, while the *Ecp*2-minus strain hardly sporulated. It is not known whether *Ecp*1 and *Ecp*2 play a role in uptake or metabolism of nutrients. The decrease in virulence of both disruptants was associated with strong accumulation of PR proteins. This suggests that both *Ecp*1 and *Ecp*2 play a role in pathogenesis possibly by suppressing active plant defence responses during colonization. In addition a double disruptant which lacks both *Ecp*1 and *Ecp*2, showed a phenotype very similar to that of the *Ecp*2-minus strain.

The sequence of the *Ecp*1 and *Ecp*2 genes did not provide any clues to putative functions of the encoded genes (Van den Ackerveken *et al.*, 1993b). However, the spacing of the cysteine residues of ECP1 shows similarity to the cysteine spacing in the family of the tumor necrosis factor receptors (TNFRs; Bazan *et al.*, 1993). Combined with these observations, an interesting parallel can be drawn with mammalian systems. Several mammalian viruses have been reported to produce extracellular suppressors of host defence responses (Gooding, 1992). They act on cytokines, protein mediators of the immune system, of which a number are produced by host cells upon infection, including the tumor necrosis factors-α and -β (TNF-α and TNF-β). The extracellular suppressors share structural homologies with the host cell membrane-bound receptor that senses the cytokine signal. Due to this feature they act as analogues of receptors and trap the cytokines before reaching their cellular target (Alcami and Smith, 1992; Spriggs *et al.*, 1992). In this way, they interfere with cytokine function and disturb the establishment of host defence responses. Results of experiments carried out with viruses which are deficient in these suppressor proteins show striking similarities to the results obtained with tomato plants inoculated with the ECP-deficient strains of *C. fulvum*. Increased host defence responses were associated with lower pathogenic abilities of the deficient viral strain (Mossman *et al.*, 1996). The T2 suppressor protein of the Shope Fibroma virus, shares the same type of structural homology with the TNFR family as ECP1, and was shown to inhibit competitively the binding of tumor necrosis factor to its cell surface receptor. Testing for the existence of an homologous situation in the interaction between *C. fulvum* and tomato becomes very attractive since the recent report on the cloning of a putative plant receptor that shares structural homology with the TNFR family (Becraft *et al.*, 1996).

B.   THE INCOMPATIBLE INTERACTION

In incompatible interactions growth of *C. fulvum* is arrested soon after penetration of the leaf of a resistant genotype (Lazarovits and Higgins, 1976a; De Wit, 1977). Fungal growth is restricted to the area of a few mesophyll cells, while the cells contacted by hyphae or in advance of the growing hyphae often die. Fairly accurate measurements of fungal biomass were achieved by inoculating tomato with a transformant of *C. fulvum* constitutively expressing β-glucuronidase (GUS) (Oliver *et al.*, 1993). By inoculating near-isogenic lines of tomato homozygous or heterozygous for different *Cf* genes with race 4 of *C. fulvum*, expressing GUS, it was found that the different *Cf* genes confer distinct abilities to restrict *C. fulvum* infection (Hammond-Kosack and Jones, 1994). However, one should keep in mind that the effectiveness of a given *Cf* gene is also dependent on the amount and stability of the race-specific elicitor produced by the matching *Avr* gene. In addition, it was found that all *Cf* genes show incomplete dominance. Significantly less growth of *C. fulvum* occurred in plants that are homozygous, compared to plants that are heterozygous for a given *Cf* gene.

1.   *Callose Deposition and Cell Wall Appositions*

One of the early defence responses observed in incompatible interactions involves deposition of callose and cell wall thickening. Callose deposition is predominantly associated with incompatible interactions, but under suboptimal conditions for disease development it also occurs in compatible interactions (Lazarovits and Higgins, 1976a; De Wit, 1977). Callose is deposited between the plant cell wall and plasma membrane. Cell wall thickening usually ocurrs at the interface between intercellular hyphae and mesophyll cells. Thickened cell walls contain polyphenolic compounds (Lazarovits and Higgins, 1976a). When conidia of virulent or avirulent races of *C. fulvum* are injected into the intercellular space of tomato leaves the conidia and the primary germ tube invoke strong cell wall appositions containing predominantly callose. Secondary germ tubes ceased to invoke callose deposition and cell wall appositions in susceptible plants, but continued to do so in resistant plants (Higgins, 1982). Non-specific elicitors which are present in the cell walls of conidia and primary germ tubes possibly induce those responses in both susceptible and resistant plants. However, enzymes of plant origin present in the apoplast might degrade these non-specific elicitors, explaining why secondary hyphae of virulent races no longer induce these responses in susceptible plants while avirulent races continue to induce these responses. An additional plant response associated with incompatible interactions is the enlargement of host cells that are in contact with or slightly in front of the advancing hyphae (Lazarovits and Higgins, 1976a; Hammond-Kosack and Jones, 1994).

2.   *Phytoalexins*

In tomato leaves inoculated with *C. fulvum*, accumulation of phytoalexins occurs in incompatible interactions (De Wit and Flach, 1979; De Wit and Kodde, 1981a). The phytoalexins detected in leaves are not sesquiterpenes, the common phytoalexins detected in solanaceous plants, but the polyacetylene phytoalexins, falcarinol and falcarindiol (De Wit and Kodde, 1981a). The latter two phytoalexins accumulate quicker and usually to a higher level in incompatible interactions compared to compatible ones (De Wit and Flach, 1979). Pericarp tissue of green tomato fruits responds with accumulation of the sequiterpene rishitin, after treatment with conidia of virulent or avirulent races of *C. fulvum*, or non-specific elicitors isolated from culture filtrates or cell walls of virulent or avirulent races (De Wit and Roseboom, 1980; De Wit and Kodde, 1981b). However, accumulation of rishitin in green tomato fruit tissue of genotypes carrying different *Cf* genes after treatment with those elicitors, is similar in compatible and inompatible interactions. Thus, the accumulation of polyacetylene phytoalexins in leaves is race-specific, but the accumulation of rishitin in tomato fruits is not (De Wit and Flach, 1979; De Wit and Roseboom, 1980; De Wit and Kodde, 1981a,b)

3.   The Hypersensitive Response
In incompatible interactions, after penetrating stomata, hyphae that are in contact with mesophyll cells induce necrosis which is considered a hypersensitive response (HR) (Lazarovits and Higgins, 1976a; De Wit 1977). In incompatible interactions, involving different *Cf*-genes, necrosis was sometimes observed rather late (Hammond-Kosack and Jones, 1994). AF isolated from leaves that are fully colonized by virulent races of *C. fulvum*, carrying different genes for avirulence, induced HR after injection in *Cf* lines that contain a matching resistance gene. HR induced by race-specific elicitors present in these AFs, will be discussed later. HR that occurs in the natural plant-pathogen interaction is very difficult to quantify. As will be discussed later, the outcome of a particular race-genotype interaction depends on the stability and activity of both the race specific elicitor and the receptor. Plants that are homozygous for a particular *Cf* gene respond with HR to a two-fold lower race-specific elicitor concentration as compared with with heterozygous plants. This indicates that the receptor concentration might be a limiting factor in inducing HR (Hammond-Kosack and Jones, 1994).

4.   Pathogenesis-Related Protein Accumulation
Early accumulation of several host-encoded pathogenesis-related (PR) proteins in the apoplast is characteristic for incompatible interactions (De Wit and Van der Meer, 1986; De Wit *et al.*, 1986; Joosten and De Wit, 1989). Biochemical characterization revealed that many of these proteins are 1,3-β-glucanases and chitinases which are hydrolytic enzymes potentially able to degrade hyphal walls that contain 1,3-β-glucans and chitin (Joosten and De Wit, 1989; Bol *et al.*, 1990). The cDNAs encoding the various basic and acidic 1,3-β-glucanases and chitinases have been cloned (Van Kan *et al.*, 1992; Danhash *et al.*, 1993) and the expression of the genes (Van Kan *et al.*, 1992; Danhash *et al.*, 1993; Wubben *et al.*, 1994a) and localization of the transcripts and the encoded proteins have been studied (Wubben *et al.*, 1992; Wubben *et al.*, 1994b). Although the early accumulation of hydrolytic enzymes in the incompatible interaction coincides with the expression of HR and arrest of fungal growth, it is not clear whether the induced PR proteins play a decisive role in resistance of tomato against *C. fulvum* (Joosten *et al.*, 1995).

III.   Non-Specific and Race-specific Elicitors of *Cladosporium fulvum*

A.   NON-SPECIFIC ELICITORS

The search for race-specific elicitors in the 1970s was mainly carried out with
C. fulvum grown *in vitro*. Virulent and avirulent races were grown in stationary or shake
cultures and culture filtrates and cell wall fractions were analysed for the presence of
race-specific elicitors based on different bio-assays. Assays were based on electrolyte
leakage, necrosis and callose deposition (Van Dijkman and Kaars Sijpesteijn, 1973;
Dow and Callow, 1979a,b; Lazarovits and Higgins, 1979; Lazarovits *et al.*, 1979) or
phytoalexin accumulation (De Wit and Roseboom, 1980; De Wit and Kodde, 1981a,b).
In all cases elicitor activity was observed, but this activity was never race-specific.
The elicitors appeared to be glycoproteins of different sizes containing glucose,
galactose and mannose. They were called proteoglucogalactomannans (De Wit and
Kodde, 1981b) and occur in culture filtrates as well as cell walls. Their presence and
significance during growth of *C. fulvum in planta* is unclear. If produced during the
infection process the glycoproteins are probably quickly degraded by enzymes of plant
origin present in the apoplast. It was found that AF obtained from healthy plants could
inactivate the necrosis-inducing activity of these glycoprotein elicitors (Higgins, 1982;
Peever and Higgins, 1989a). So far, the biological relevance of non-specific elicitors
of *C. fulvum* can be questioned as it is unknown whether they occur in infected plants.
If they play a role, it is probably only during the late stages of infection when tomato
leaves become necrotic.

B.   RACE-SPECIFIC ELICITORS AND THEIR ENCODING GENES

1.   *Isolation of Race-Specific Elicitors and Cloning of their Encoding Genes*

A major breakthrough in research on the *C. fulvum*-tomato interaction was the discovery
of race-specific elicitors, the inducers of HR present in AF of *C. fulvum*-colonized
tomato leaves (De Wit and Spikman, 1982; De Wit *et al.*, 1984, 1985; Higgins and
De Wit, 1985). Injection of such AF into healthy leaves of near-isogenic lines containing
different *Cf* genes, resulted in the differential induction of an HR. Proteinaceous
compounds which induced HR on resistant tomato cultivars were correlated with
the presence of avirulence genes in the races of *C. fulvum* used for inoculations.
The first race-specific peptide elicitor purified from AF of *C. fulvum*-infected tomato
leaves was a peptide of 28 amino acids (Scholtens-Toma and De Wit, 1988; De Wit, 1992).
The purified peptide elicitor specifically induced HR on tomato genotypes that carried
the matching *Cf-9* resistance gene. Races that are virulent on tomato genotype *Cf-9*
do not produce the elicitor (Van Kan *et al.*, 1991). In a similar way, the race-specific
elicitor AVR4, which induces HR on Cf4 tomato genotype, has been isolated (Joosten
et al., 1994). From both proteinaceous elicitors the amino acid sequence has been
determined and by reverse genetics the encoding avirulence (*Avr*) genes *Avr9* and *Avr4*,
have been cloned (Van Kan *et al.*, 1991; Van den Ackerveken *et al.*, 1992; Joosten *et
al.*, 1994). The cloned genes revealed that both elicitors present in *C. fulvum*-infected
tomato plants are post-translationally processed. The *Avr9* gene encodes a protein of

63 amino acids including a signal sequence of 23 amino acids. In culture filtrates of transformants of *C. fulvum* that constitutively produce the AVR9 elicitor, predominantly N-terminally processed peptides of 32, 33 or 34 amino acids are present (Van den Ackerveken *et al.*, 1993a). When these latter peptides were incubated with AF from healthy tomato leaves, they were further processed into the mature elicitor of 28 amino acids, indicating that plant proteases are required for the final processing (Van den Ackerveken *et al.*, 1993a). The *Avr4* gene encodes a protein of 135 amino acids, including a signal peptide of 18 amino acids (Joosten *et al.*, 1994). Here, the mature peptide elicitor is processed at both the N- and C-terminus leaving a mature peptide elicitor of 86 amino acids (Joosten *et al.*, 1997). Although there is clear evidence for the presence of proteinaceous elicitors in AF of infected leaves that induce HR on *Cf-2* and *Cf-5* genotypes of tomato, the matching AVR2 and AVR5 elicitors have not yet been characterized.

2.   Regulation of Avirulence Genes Avr4 and Avr9

In vitro-grown *C. fulvum* hardly produces AVR4 or AVR9 elicitors. RNA gel blot analysis indicated that expression of the *Avr4* and *Avr9* genes of *C. fulvum* is specifically induced *in planta*. Accumulation of mRNAs encoding the race-specific elicitors correlates with an increase in fungal biomass in tomato leaves during pathogenesis in compatible interactions (Van Kan *et al.*, 1991; Joosten *et al.*, 1994). Studies on the expression of the *Avr4* and *Avr9* genes *in planta*, using transformants of *C. fulvum* carrying *Avr* promoter-GUS fusions, showed activation of the promoter in hyphae immediately after penetration of stomata, with expression levels particularly high in mycelia growing in the vicinity of the vascular tissue (Van den Ackerveken *et al.*, 1994, Joosten *et al.*, 1997).

The promoter of the *Avr4* gene does not contain motifs homologous to sequences known for binding of regulatory proteins. So far, the *Avr4* gene can hardly be induced under different growth conditions *in vitro*. In contrast, the *Avr9* gene could be induced under limiting concentrations of nitrogen when grown in liquid medium (Van den Ackerveken *et al.*, 1994, P.J.G.M De Wit, unpublished; S. Snoeijers, Wageningen Agricultural University, personal communication). Analysis of the *Avr9* promoter sequence revealed six copies of the hexanucleotide TAGATA (Van den Ackerveken *et al.*, 1994), which has been identified as the recognition site of the NIT2 protein, a transcription factor which positively regulates gene expression under nitrogen-limiting conditions in *Neurospora crassa* and many other filamentous fungi (Marzluf, 1996). Therefore, the expression of the *Avr9* gene is possibly regulated in a similar way, by a *C. fulvum*-homologue of the NIT2 protein. Indeed deleting a number of the TAGATA sequences abolished the induction of the *Avr9* gene under conditions of low nitrogen concentrations *in vitro* (P.J.G.M. De Wit, unpublished; S. Snoeijers, Wageningen Agricultural University, personal communication). Two of the six TAGATA sequences are essential for *Avr9* regulation under nitrogen-limiting conditions.

3.   Avr Alleles in races of C. fulvum Virulent on Tomato Genotypes Cf4 and Cf9

Genomic DNA gel blot analysis, using *Avr4* cDNA as a probe, did not reveal any differences between races of *C. fulvum* avirulent or virulent on tomato genotype Cf4.

All races contain a homologous, single copy gene, not displaying any restriction fragment length polymorphism (Joosten *et al.*, 1994). Although none of the virulent races produce biologically active AVR4 elicitor, they produce transcripts that hybridize to an *Avr4* cDNA probe, proving that those races contain alternative alleles of *Avr4* (Joosten *et al.*, 1994,1997). Sequencing these alleles, showed single base pair mutations in the open reading frame (ORF) encoding the mature AVR4 protein, resulting in single amino acid changes in the AVR4 elicitor (Fig. 1B). In one case a frame shift mutation was observed leaving only 13 amino acids at the N-terminus of the AVR4 peptide intact (Joosten *et al.*, 1994; Joosten *et al.*,1997). The single amino acid changes in the AVR4 protein cause the peptide to become unstable and probably more sensitive to proteolytic degradation which would prevent it reaching a matching receptor in sufficiently high concentration. Expression of the mutant alleles in the PVX expression system indeed revealed that most still encode an active elicitor molecule which, however, has a much shorter active half life than the peptide elicitor produced by the functional avirulence allele (Joosten *et al.*, 1997).

Genomic DNA gel blot analysis of races of *C. fulvum* virulent on tomato genotype Cf9 revealed that in all these races the *Avr9* gene is absent (Fig. 1A). Thus, there is always a strict correlation between virulence of races on Cf9 genotypes of tomato and absence of the *Avr9* gene (Van Kan *et al.*, 1991).

4.   Structure and Function of AVR4 and AVR9 Peptides

Avirulence genes *Avr4* and *Avr9* both encode relatively small globular peptide elicitors which contain 8 and 6 cysteine residues, respectively. The structure of the AVR9 peptide has been extensively studied by ^{1}H-NMR (Vervoort *et al.*, 1997). All cysteines form disulfide bridges, which are required for elicitor activity. The AVR9 peptide consists of three anti-parallel (-strands forming a rigid region of (-sheet. AVR9 is a member of the family of cystine-knotted peptides, in which the 6 cysteine residues form a typical cystine knot found in several small proteins such as proteinase inhibitors, ion channel blockers and growth factors (Pallaghy *et al.*, 1994). However, structural homology most probably does not represent functional homology.

In contrast to the *Avr9* gene, which is absent in all races of *C. fulvum* virulent on tomato genotype Cf9, the presence of mutated *Avr4* alleles in virulent races of *C. fulvum* might suggest an essential role for its product in virulence (Fig. 1). However, one isolate with a frame-shift mutation in the ORF encoding an AVR4 homologue of only 13 amino acids showed normal virulence, indicating that the AVR4 protein is probably dispensable (Joosten *et al.*, 1997).

The observation that transformants of *C. fulvum* in which the *Avr9* gene was replaced by a selection marker, did not show impaired virulence on non *Cf9*-containing tomato genotypes, and that in nature avoidance of *Cf-9*-specific resistance is achieved by complete deletion of the *Avr9* gene, suggests that the *Avr9* gene is dispensable for growth and virulence of *C. fulvum* (Marmeisse *et al.*, 1993). However, the *Cf-9* resistance gene, which is present in tomato breeding lines since 1979, still provides good protection toward *C. fulvum* in commercial tomato crops, indicating that loss of the *Avr9* gene is not sufficient to overcome the *Cf-9* locus. It was found that isogenic strains of *C. fulvum* in which the *Avr9* gene had been replaced by a selection marker are

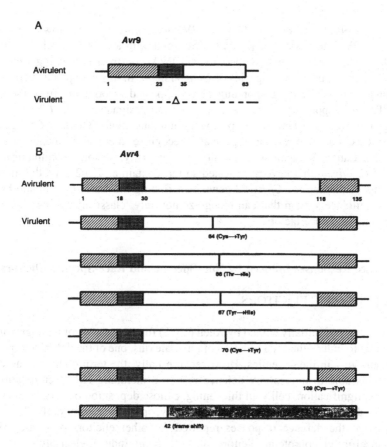

Figure 1. The structure of peptides encoded by avirulent and virulent alleles of avirulence genes *Avr9* and *Avr4*. A. Structure of the peptide encoded by avirulence gene *Avr9*. Single hatched box: signal peptide; double hatched box: stretch of amino acids removed by proteolytic cleavage; white box: mature elicitor peptide; dotted line: deletion of avirulence gene *Avr9* in virulent isolates. B. Structure of the peptide encoded by avirulence gene *Avr4*. Single hatched box: signal peptide; double hatched box: stretch of amino acids removed by proteolytic cleavage; white box: mature elicitor peptide; shaded box peptide encoded due to frame shift mutation; the various amino acid exchanges in the peptides ecoded by the virulent alleles are indicated.

only weakly virulent on *Cf-9* genotypes of tomato (Laugé *et al.*, 1998). This suggests that *Cf-9* genotypes of tomato contain functional *Cf-9* homologs which recognize an elicitor other than AVR9. Infection of *Cf-9* genotypes with *Avr9*-minus strains of *C. fulvum* was always associated with strong accumulation of PR proteins (Laugé *et al.*, 1998). most probably homologues of the *Cf-9* gene provide this protection as will be discussed later.

5. Virulence Factors with Avirulence Properties

As discussed before, *C. fulvum* secretes avirulence as well as virulence factors into the intercellular space while colonizing tomato leaves (Van den Ackerveken *et al.*, 1992;

Joosten *et al.*, 1994; Laugé *et al.*, 1997). The PVX expression system allows, within the host range of PVX, to search for plants that respond with an HR to virulence factors such as the *Ecp2* gene product. A recombinant PVX construct expressing *Ecp2* was inoculated onto various lines of species of *Lycopersicon*. In this way plants were found that responded with HR. All responding plants appeared to originate from the same ancestor. The corresponding resistance gene has been designated *Cf-ECP2* (R. Laugé, Wageningen Agricultural University, personal communication). The *Cf-ECP2* gene is anticipated to be a durable resistance gene as it recognizes a crucial virulence factor of *C. fulvum* (R. Laugé, Wageningen Agricultural University, personal communication). All races of *C. fulvum* that have been tested so far, contain an *Ecp2* gene that induces HR on plants containing the *Cf-ECP2* gene. This finding illustrates that tomato has an efficient surveillance system that can recognize not only 'classical' avirulence factors but also crucial virulence factors.

IV.   Defence Responses Induced by Non-Specific and Race-Specific Elicitors

A.   NON-SPECIFIC ELICITORS

The term elicitor was initially coined by Keen (1975) for compounds of pathogen origin which are able to induce the accumulation of phytoalexins, one of the defence responses studied intensively in the seventies. However, presently, the term elicitor is used for all compounds able to induce any defence response including HR, accumulation of phytoalexins, lignification, cell wall thickening, callose deposition or various enzymes involved in defence responses. Non-specific glycoprotein elicitors of *C. fulvum* induce most of the defence responses reported for other elicitors of biotic origin. The non-specific glycoprotein elicitors of *C. fulvum* induce electrolyte leakage, callose deposition, necrosis and accumulation of phytoalexins as discussed before (Van Dijkman and Kaars Sijpesteijn, 1973; Dow and Callow, 1979a,b; Lazarovits and Higgins, 1979; Lazarovits *et al.*, 1979; De Wit and Roseboom, 1980; De Wit and Kodde, 1981a).

B.   RACE-SPECIFIC ELICITORS

Peever and Higgins (1989b) injected a crude preparation of race-specific elicitor AVR9 in leaves of the tomato cultivar Sonatine, heterozygous for *Cf-9*, and observed a specific, elicitor-dependent, increase in electrolyte leakage. This leakage coincided with an increase in lipoxygenase activity. Similar results were obtained by others (Hammond-Kosack *et al.*, 1996; May *et al.*, 1996), who used cotyledons of 14-day-old tomato seedlings. Upon injection of AF containing the AVR9 elicitor in cotyledons of Cf9 plants, increased levels of total and oxidized glutathione were observed followed by an oxidative burst, ethylene production and loss of membrane integrity. From 12 hours onwards significant increase in free salicylic acid occurred. Inhibition of salicylic acid accumulation in Cf2 and Cf9 genotypes of tomato, transgenic for *nahG*, did not inhibit resistance toward *C. fulvum* (Hammond-Kosack and Jones, 1996).

Vera-Estrella *et al.* (1992, 1993) studied biochemical responses in cell cultures of tomato after elicitor treatment. Cell cultures, initiated from callus obtained from tomato genotypes Cf4 and Cf5, retained the specificity of the intact plants from which they originated. Within 10 minutes after treatment with AF containing matching elicitors, a marked extracellular oxidative burst was observed. In addition, the cells showed increased lipid peroxidation followed by increases in extracellular peroxidases and phenolic compounds. Cell cultures of tomato genotype Cf5 treated with AF containing AVR5 elicitor showed also a quick increase in H^+-ATPase activity causing acidification of the extracellular medium (Vera-Estrella *et al.*, 1994a,b). Activation of H^+-ATPase in cells and plasma membranes occurred through reversible dephosphorylation (Xing *et al.*, 1996) The observed acidification of the culture medium contrasts with responses to non-specific elicitors (Vera-Estrella *et al.*, 1993), which induce a rapid alkalinization of the extracellular medium. Enriched plasma membrane fractions obtained from Cf5 cells also responded to a crude preparation of AVR5 elicitor with a significant increase in NADH oxidase and cytochrome C reductase. Inhibition studies indicated that both phosphatases and G-proteins are involved in these responses (Vera-Estrella *et al.*, 1994a,b; Xing *et al.*, 1996). It was also found that plasma membrane NADPH oxidase was activated in cells after treatment with AVR5 elicitor preparations, while at the same time translocation of components of this enzyme to the plasma membrane was observed (Xing *et al.*, 1997).

In contrast to cell cultures derived from Cf4 and Cf5 genotypes of tomato, cell cultures and calli of Cf9 genotype treated with AVR9 elicitor did not show *Cf-9*-dependent defence responses such as generation of an oxidative burst and a pH shift (Honée *et al.*, 1998). This differs from leaves and shoots of tomato genotype Cf9 treated with AVR9 elicitor which gave strong *Cf-9*-dependent defence responses (Hammond *et al.*, 1996; Honée *et al.*, 1998). This suggests that AVR9-Cf-9 mediated defence is is not active in undifferentiated tissue such as callus or cells. However, cell cultures of *Cf-9* transgenic tobacco do respond with an oxidative burst and extracellular alkalinisation (C.F. De Jong, Wageningen Agricultural University, personal communication). Injection of AVR4 and AVR9 elicitors of *C. fulvum* into leaves of Cf4 or Cf9 tomato genotypes, respectively, induced expression of PR protein encoding genes (Wubben *et al.*, 1996). mRNAs encoding acidic 1,3-β-glucanase and acidic chitinase were strongly induced. In Cf0 genotypes the PR mRNAs were hardly induced. The induction pattern of basic chitinase and basic 1,3-β-glucanase was less clear. In leaves of heterozygous Cf2 and Cf9 plants treated with AF containing AVR2 and AVR9 elicitor, respectively, Ashfield *et al.* (1994) found levels of induction of acidic and basic 1,3-β-glucanases comparable to those found by Wubben *et al.* (1996).

V.   Resistance Genes in Tomato against *Cladosporium fulvum*.

Four genes for resistance against *C. fulvum* (*Cf* genes) have been mapped at two complex loci. The *Cf-9* and *Cf-4* genes have been mapped on the short arm of chromosome 1 (Jones *et al.*, 1993), and the *Cf-2* and *Cf-5* genes are located on the short arm of chromosome 6 (Dickinson *et al.*, 1993; Jones *et al.*, 1993). The *Cf-2, Cf-4, Cf-5* and *Cf-9* genes have been cloned (Jones *et al.*, 1994; Dixon *et al.*, 1996; Hammond-Kosack and Jones, 1997; Thomas *et al.*, 1997). A schematic representation of the four *Cf-* genes is shown in Figure 2.

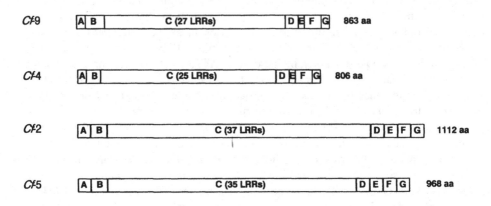

Figure 2. Schematic representation of the proteins (number of amino acids is indicated) encoded by the resistance genes *Cf-9*, *Cf-4*, *Cf-2* and *Cf-5*, respectively. See text for further details.

A	Signal peptide
B	Cysteine-rich region
C	Leucine-rich repeat (LRR) region; the number of LRRs is indicated
D	Region without specific properties
E	Acidic region
F	Transmembrane region
G	Intracellular basic region

The *Cf-9* gene encodes a 863-amino acid membrane-anchored, predominantly extracytoplasmic glycoprotein containing 27 leucine-rich repeats (LRRs) with an average length of 24 amino acids. These motifs occur in many plant resistance genes (Jones and Jones, 1996; Hammond-Kosack and Jones, 1997). In the Cf-9 protein seven domains (A to G) have been designated. The N-terminal domain A of 23 amino acids is consistent with a signal peptide; domain B of 68 amino acids is cysteine-rich; domain C contains 27 imperfect LRRs; domain D contains 28 amino acids; domain E contains 18 amino acids and is very acidic; domain F contains 37 amino acids and is the presumed trans-membrane domain; C-terminal domain G contains 21 amino acids, is very basic and concludes with the amino acids KKRY. Twenty-two potential N-glycosylation sites are distributed between domains B, C and D. Thus the *Cf-9* gene encodes a LRR protein of which the major part (A-F) is extracellular and the C-terminal part (G) is cytoplasmic. The LRR domain C in the Cf-9 protein is interrupted by a short region originally designated as LRR 24, which has only minimal LRR homology. This domain, now designated C2 is also present in the *Cf-2*, *Cf-4*, and *Cf-5* genes (Jones and Jones, 1996; Hammond-Kosack and Jones, 1997). As a result domain C has now been divided in domains C1 (the N-terminal LRRs), C2 (with minor LRR consensus) and domain C3 (the C-terminal LRRs) (Jones and Jones, 1996; Hammond-Kosack and Jones, 1997).

The LRRs match the extracytoplasmatic LRR consensus LxxLxxLxxLxLxxNxLxGxIPxx (Jones and Jones, 1996). LRR regions might be involved in various types of protein-protein interaction (Kobe and Deisenhofer, 1993) as will be discussed later.

The *Cf-4* gene is very homologous to the *Cf-9* gene. The proteins have >91% identical amino acids (Thomas *et al.*, 1997). The *Cf-4* gene encodes an 806-amino acid protein with 25 LRRs. In *Cf-4* two complete LRRs are deleted relative to the *Cf-9* gene. DNA sequence analysis suggests that *Cf-4* and *Cf-9* are derived from a common gene. The amino acids that distinguish *Cf-4* from *Cf-9* are located at the N-terminal half of the protein. The C-terminal halves of both genes are almost identical.

The *Cf-2* locus contains two functional genes that each independently confer resistance to races of *C. fulvum* carrying the *Avr2* gene (Dixon *et al.*, 1996). Each gene encodes a nearly identical 1112-amino acid protein (three amino acids are different) that is structurally very similar to the Cf-4 and Cf-9 proteins. The Cf-2 protein possesses 37 LRRs. The LRRs of Cf-2 are nearly all 24 amino acids in length and are also interrupted by a short C2 domain which divides the LRRs in a C-terminal block of 33 LRRs and a N-terminal block of 4 LRRs. The highest homology between the Cf-2, and the Cf-4 and Cf-9 proteins resides in the C-terminal part (Hammond-Kosack and Jones, 1997; Thomas *et al.*, 1997).

The *Cf-5* gene is closely linked to the *Cf-2* gene and encodes a 968-amino acid protein that is very similar to the protein encoded by *Cf-2* and contains 31 LRRs (Hammond-Kosack and Jones, 1997). The Cf-5 and Cf-2 proteins differ by an exact deletion of six LRRs in Cf-5. The C-terminal part of Cf-2 and Cf-5 is also very conserved.

A.   LRR MOTIFS IN CF PROTEINS AND THEIR POTENTIAL FUNCTION

The most simple interpretation of Flor's gene-for-gene hypothesis (Flor, 1942, 1971) would be that products of avirulence genes and resistance genes interact and activate downstream signaling events eventually leading to resistance. The best candidates for binding of AVR elicitors would be the LRRs present in the Cf proteins. The N-terminal parts of the four Cf proteins are variable and could determine specificity in recognizing race-specific elicitors, while the C-terminal parts are conserved and could function as an activation domain potentially interacting with other proteins present in the plasma membrane. For the Cf-4 and Cf-9 protein, the most appropriate ligands would be the AVR4 and AVR9 elicitors, respectively. However, proof for this hypothesis has not yet been found (Kooman-Gersmann *et al.*, 1996, 1997, 1998; De Wit 1997). Kooman-Gersmann *et al.* (1996) found that the AVR9 elicitor molecule binds to plasma membranes of Cf0 and Cf9 genotypes equally well (Kd = 70 pM). In order to get more insight in the region of the AVR9 molecule which interacts with the high affinity binding site, *in vitro* mutagenesis of *Avr9* was performed in the PVX expression system to obtain AVR9 mutants with altered biological activities. All amino acid residues present in AVR9 were exchanged one by one with alanine (alanine scan). In this way AVR9 peptides were obtained with higher, equal or lower HR-inducing activitiy compared to wild type AVR9 (Kooman-Gersmann *et al.*, 1997). Mutants of AVR9 with decreased HR-inducing activity on Cf genotypes of tomato also showed lower binding affinity to plasma membranes (Kooman-Gersmann *et al.*, 1998). However, binding was decreased to plasma membranes of both Cf9 and Cf0 genotypes of tomato. It is therefore unlikely that the LRRs of the Cf-9 protein bind the AVR9 peptide directly. In case AVR9 binds to the Cf-9 protein, the affinity is too low to detect with the presently available techniques (Kooman-Gersmann et al., 1998). Binding of AVR9 to in vitro produced Cf-9 protein is required to detect such a low affinity binding site. LRR domains of Cf proteins could also facilitate multiple interactions with other proteins which are involved in the signal transduction pathway as has been found for other LRR-containing proteins (Jones and Jones, 1997).

Similar studies are now being carried out with the AVR4 peptide. Preliminar mutational analysis of the Avr4 gene in the PVX expression system showed that all eight cysteine residues present in the AVR4 elicitor are involved in disulfide bridges (M.H.A.J. Joosten, Wageningen Agricultural University, personal communication). For the future, binding studies with the AVR4 elicitor to membranes of Cf4 and Cf0 genotypes and in vitro produced Cf-4 protein are planned.

B.   HOMOLOGUES OF CF-RESISTANCE GENES IN TOMATO

When the Cf genes are used as a probe to a Southern blot of isogenic lines of tomato, many hybridising bands can be observed (Jones et al., 1994; Dixon et al., 1996; Hammond-Kosack and Jones, 1997; Thomas et al., 1997; Parniske et al., 1997). All Cf genes appear to be members of a multigene family. The Cf-2 locus contains two nearly identical genes (Dixon et al., 1996). Both genes are functional, but it is possible that they represent different specificities. The Cf-4/Cf-9 locus contains ten homologues (Thomas et al., 1997; Parniske et al., 1997) including the functional Cf-4 and Cf-9 genes. Some of the homologues give (partial) resistance to races of C. fulvum for which the matching avirulence factors have not yet been identified. The Cf-9 cluster contains two Cf-9 homologues that confer resistance independent of AVR9 activation, while the Cf-4 cluster contains one Cf-4 homologue conferring resistance that is independent of AVR4 activation (Takken et al., 1998; Joosten, Wageningen Agricutural University, communication). The functional Cf-4 homologue is called Cf-4E (Takken et al., 1998).

VI.   Exploitation of Avirulence and Resistance Genes

Cloned avirulence and resistance genes are valuable tools to study the molecular mechanisms that determine race-specific resistance, yielding results that can be applied in molecular resistance breeding. Combining an avirulence gene and the complementary resistance gene in host plants under the control of a pathogen-inducible promoter could possibly give new horizons to molecular resistance breeding (two-component gene cassette; De Wit, 1992, 1995). Activation of the gene cassette by pathogens should occur both quickly and locally, resulting in the production of a race-specific elicitor that interacts directly or indirectly with the resistance gene product. As a result of this interaction, localized HR will be induced which will prevent further spread of an invading pathogen that is inhibited by an HR. With this system it should be possible, to use a highly specific resistance gene-avirulence gene combination (for example

Cf-9/Avr9 or *Cf-4/Avr4*) to obtain plants resistant against a wide variety of pathogens. Preliminary results with tomato plants containing the two-component gene cassette look promising (G. Honée, Wageningen Agricultural University, personal communication). For applications in plants other than tomato, the gene cassette containing both the *Cf* and *Avr* gene should be transferred to those plants. The *Cf-9* and *Avr9* gene are active in both potato and tobacco (Honée,Wageningen Agrilrural University, personal communication).

VII. Conclusion

Molecular studies on the interaction between *C. fulvum* and tomato have been very successful in recent years. This success was based on profound biological knowledge of this gene-for-gene system, generated by both plant pathologists and plant breeders in the past. Near-isogenic lines of tomato carrying different *Cf* genes and many races of *C. fulvum* able to overcome those genes, were instrumental to this success. An additional important factor is the strict apoplastic growth of this biotrophic fungus which allowed the study of the interaction *in planta* where important virulence and avirulence factors are specifically produced. Many of those factors are not, or hardly, produced when the fungus is grown in axenic culture. Four *Cf* genes and two *Avr* genes have been cloned. The cloned *Cf* genes represent major resistance genes that are members of gene families which are activated by stable AVR peptides. Many *Cf* homologues likely represent resistance genes with minor effects which are activated by AVR peptides that are probably more difficult to identify. Some activators appear crucial virulence factors, such as ECP2, recognized by the versatile surveillance system of tomato. The *Cf-ECP2* gene activated by ECP2 probably represents a durable resistance gene. The *Cf* gene products are situated mainly extracellularly with a membrane anchor and are perfectly placed to bind peptide elicitors produced by an extracellular pathogen. High affinity membrane-localized binding sites have been identified which, however, do not seem to be the products of *Cf* genes. It is possible that Cf proteins are low affinity binding sites that can only be detected in binding studies using these proteins produced in vitro. Now that the *Cf* and *Avr* genes have been cloned, one can begin to study signal transduction pathways involved in pathogenicity and plant defence. In the future the *C. fulvum*-tomato interaction will remain a very versatile biological system in these types of studies. This pathosystem is unique in that it represents a model for a strictly extracellular pathogen in contrast to many other model pathogens which grow either partially or wholly in an intracellular fashion.

Acknowledgements

The author thanks Guy Honée, Matthieu H.A.J. Joosten, Richard Laugé and Ronelle Roth for critically reading the manuscript.

References

Agrios GN (1997) Plant Pathology, Academic Press, London pp. 63—82

Alcami A and Smith GL (1992) A soluble receptor for interleukin-1(encoded by vaccinia virus: a novel mechanism of virus modulation of the host response to infection. Cell 71: 153—167

Ashfield T, Hammond-Kosack KE, Harrison K and Jones JDG (1994) *Cf* gene-dependent induction of a β-1,3-glucanase promotor in tomato plants infected with *Cladosporium fulvum*. Mol Plant Microbe Interact 7: 645—657

Bazan JF (1993) Emerging families of cytokines and receptors. Curr Biol 3: 603—606

Becraft PW, Stinard PS and McCarty DR (1996) CRINKLY4: a TNFR-like receptor kinase involved in maize epidermal differentiation. Science 273: 1406—1409

Bol JF, Linthorst, HJM and Cornelissen BJC (1990) Plant pathogenesis-related proteins induced by virus infection. Annu Rev Phytopathol 28: 113-138

Bond TET (1938) Infection experiments with *Cladosporium fulvum* Cooke and related species. Ann Appl Biol 25: 277—307

Boukema IW (1981) Races of *Cladosporium fulvum* Cke. (*Fulvia fulva*) and genes for resistance in the tomato (*Lycopersicon* Mill.). In: Philouze J (ed), Proceedings of the Meeting of the Eucarpia Tomato Working Group, Avignon, 287—292

Butler EJ and Jones SG (1949) Tomato leaf mould, *Cladosporium fulvum* Cooke. In: Buttler EJ and Jones SG (eds), Plant Pathology, MacMillan, London pp. 672-678

Cooke MC (1883) New American Fungi. Grevillea 12: 32

Danhash N, Wagemakers CAM, Van Kan JAL and De Wit PJGM (1993) Molecular characterization of four chitinase cDNAs obtained from *Cladosporium fulvum*-infected tomato. Plant Mol Biol 22: 1017—1029

Day PR (1957) Mutation to virulence in *Cladosporium fulvum*. Nature 179: 1141

De Wit PJGM (1977) A light and scanning electron-microscopic study of infection of tomato plants by virulent and avirulent races of *Cladosporium fulvum*. Neth J Plant Pathol 83: 109—122

De Wit PJGM (1992) Molecular characterization of gene-for-gene systems in plant-fungus interactions and the application of avirulence genes in control of plant pathogens. Annu Rev Phytopathol 30: 391—418

De Wit PJGM (1995) Fungal avirulence genes and plant resistance genes: unravelling the molecular basis of gene-for-gene resistance. Adv Bot Res 21: 147—185

De Wit PJGM (1997) Pathogen avirulence and plant resistance: a key role for recognition. Trends Plant Sci 2: 452—458

De Wit PJGM and Flach W (1979) Differential accumulation of phytoalexins in tomato leaves but not in fruits after inoculation with virulent and avirulent races of *Cladosporium fulvum*. Physiol Plant Pathol 15: 257—267

De Wit PJGM and Roseboom PHM (1980) Isolation, partial characterization and specificity of glycoprotein elicitors from culture filtrates, mycelium and cell walls of *Cladosporium fulvum* (syn. *Fulvia fulva*). Physiol Plant Pathol 16: 391—408

De Wit PJGM and Kodde E (1981a) Induction of polyacetylenic phytoalexins in *Lycopersicon esculentum* after inoculation with *Cladosporium fulvum* (syn. *Fulvia fulva*). Physiol Plant Pathol 18: 143—148

De Wit, PJGM and Kodde, E (1981b). Further characterization of cultivar-specificity of glycoprotein elicitors from culture filtrates and cell walls of *Cladosporium fulvum (syn. Fulvia fulva)* Physiol Mol Plant Pathol 18:297—314.

De Wit PJGM and Spikman G (1982) Evidence for the occurrence of race- and cultivar-specific elicitors of necrosis in intercellular fluids of compatible interactions between *Cladosporium fulvum* and tomato.

Physiol Plant Pathol 21: 1—11

De Wit PJGM and Van der Meer FE (1986) Accumulation of the pathogenesis-related tomato leaf protein P14 as an early indicator of incompatibility in the interaction between *Cladosporium fulvum* (syn. *Fulvia fulva*) and tomato. Physiol Mol Plant Pathol 28: 203—214

De Wit PJGM, Hofman, JE and Aarts JMMJG (1984) Origin of specific elicitors of chlorosis and necrosis occurring in intercellular fluids of compatible interactions of *Cladosporium fulvum* (syn. *Fulvia fulva*) and tomato. Physiol Plant Pathol 24: 17—23

De Wit PJGM, Hofman JE, Velthuis GCM and Kuc JA (1985) Isolation and characterization of an elicitor of necrosis isolated from intercellular fluids of compatible interactions of *Cladosporium fulvum* (syn. *Fulvia fulva*) and tomato. Plant Physiol 77: 642—647

De Wit PJGM, Buurlage MB and Hammond KE (1986) The occurrence of host, pathogen and interaction-specific proteins in the apoplast of *Cladosporium fulvum* (syn. *Fulvia fulva*) infected tomato leaves. Physiol Mol Plant Pathol 29: 159—172

Dickinson CD, Altabella T, Chripeels MJ (1991) Slow-growth phenotype of transgenic tomato expressing apoplastic invertase. Plant Physiol 95:420—425

Dickinson MJ, Jones DA and Jones JDG (1993) Close linkage between the *Cf-2/Cf-5* and *Mi* resistance loci in tomato. Mol Plant-Microbe Interact 6: 341—347

Dixon MS, Jones DA, Keddle JS, Thomas CM, Harrison K and Jones JDG (1996) The tomato *Cf-2* disease resistance locus comprises 2 functional genes encoding leucine-rich repeat proteins. Cell 84: 451—459

Dow JM and Callow JA (1979a) Partial characterization of glycopeptides from culture filtrates of *Fulvia fulva* (Cooke) Cifferi (syn. *Cladosporium fulvum*), the tomato leaf mould pathogen, J Gen Microbiol 113: 57-66

Dow JM and Callow JA (1979b) Leakage of electrolytes from isolated leaf mesophyll cells of tomato induced by glycopeptides from culture filtrates of *Fulvia fulva* (Cooke) Ciferri (syn. *Fulvia fulva*). Physiol Plant Pathol 15: 27—34

Flor HH (1942) Inheritance of pathogenicity in *Melampsora lini*. Phytopathol 32: 653—669

Flor HH (1971) Current status of the gene-for-gene concept. Annu Rev Phytopathol 9: 275—296

Gooding LR (1992) Virus proteins that counteract host immune defenses. Cell 71: 5—7

Hammond-Kosack KE and Jones JDG (1994) Incomplete dominance of tomato *Cf* genes for resistance to *Cladosporium fulvum*. Mol Plant-Microbe Interact 7: 58—70

Hammond-Kosack KE and Jones JDG (1996) Resistance gene-dependent plant defense responses. Plant Cell 8: 1773—1791

Hammond-Kosack KE and Jones JDG (1997) Plant dIsease resistance genes. Annu Rev Plant Physiol Plant Mol Biol 48: 575—607

Hammond-Kosack KE, Silverman P, Raskin I and Jones JDG (1996) Race-specific elicitors of *Cladosporium fulvum* induce changes in cell morphology and the synthesis of ethylene and salicylic acid in tomato plants carrying the corresponding *Cf* disease resistance genes. Plant Physiol 110: 1381—1394

Higgins VJ (1982) Response of tomato to leaf injection with conidia of virulent and avirulent races of *Cladosporium fulvum*. Physiol Plant Pathol 20: 145-155

Higgins VJ and De Wit PJGM (1985) Use of race- and cultivar-specific elicitors from intercellular fluids for characterizing races of *Cladosporium fulvum* and resistant tomato cultivars. Phytopathol 75: 695—699

Honée G, Buitink J, Jabs T, De Kloe J, Sijbolts F, Apotheker M, Weide R, Sijen T, Stuiver M and De Wit PJGM (1998) Induction of defence-related responses in Cf9 tomato cells by the AVR9 elicitor peptide of *Cladosporium fulvum* is developmentally regulated. Plant Physiol 117: 809—820

Hubbeling N (1978) Breakdown of resistance of the *Cf-5* gene in tomato by another new race of *Fulvia fulva*. Med Fac Landbouww Rijksuniv Gent 43: 891—894

Jones DA and Jones JDG (1996) The role of leucine-rich repeat proteins in plant defences. Adv Bot Res 24: 91—167

Jones DA, Dickinson MJ, Balint-Kurti PJ, Dixon MS and Jones JDG (1993) Two complex resistance loci revealed in tomato by classical and RFLP mapping of the *Cf-2*, *Cf-4*, *Cf-5*, and *Cf-9* genes for resistance to *Cladosporium fulvum*. Mol Plant-Microbe Interact 6: 348—357

Jones DA, Thomas CM, Hammond-Kosack KE, Balint-Kurti PJ and Jones JDG (1994) Isolation of the tomato *Cf-9* gene for resistance to *Cladosporium fulvum* by transposon tagging. Science 266: 789—793.

Joosten MHAJ, De Wit PJGM (1988) Isolation, purification and preliminary characterization of a protein specific for compatible *Cladosporium fulvum* (syn. *Fulvia fulva*)-tomato interactions. Physiol Mol Plant Pathol 33:241—253

Joosten MHAJ, De Wit PJGM (1989) Identification of several pathogenesis-related proteins in tomato leaves inoculated with *Cladosporium fulvum* (syn. *Fulvia fulva*) as 1,3-β-glucanases and chitinases. Plant Physiol 89: 945—951

Joosten MHAJ, Hendrickx LJM and De Wit PJGM (1990) Carbohydrate composition of apoplastic fluids isolated from tomato leaves inoculated with virulent or avirulent races of *Cladosporium fulvum* (syn. *Fulvia fulva*) Neth J Plant Pathol 96: 103—112

Joosten MHAJ, Cozijnsen AJ and De Wit PJGM (1994) Host resistance to a fungal tomato pathogen lost by a single base-pair change in an avirulence gene. Nature 367: 384—387

Joosten MHAJ, Verbakel M, Nettekoven ME, Van Leeuwen J, Van der Vossen RTM and De Wit PJGM (1995) The phytopathogenic fungus *Cladosporium fulvum* is not sensitive to the chitinase and 1,3-β-glucanase defence proteins of its host tomato. Physiol and Mol Plant Pathol 46: 45—59

Joosten MHAJ, Vogelsang R, Cozijnsen TJ, Verberne MC and De Wit PJGM (1997) The biotrophic fungus *Cladosporium fulvum* circumvents *Cf-4*-mediated resistance by producing instable AVR4 elicitors. Plant Cell 9: 1—13

Keen NT (1975) Specific elicitors of plant phytoalexin production: determinants of race-specificity in pathogens. Science 187: 74—75

Kerr EA and Bailey DL (1964) Resistance to *Cladosporium fulvum* Cke obtained from wild species of tomato. Can J Bot 42: 1541—1554

Kobe B and Deisenhofer J (1993) Crystal structure of porcine ribonuclease inhibitor, a protein with leucine-rich repeats. Nature 366: 751—756

Kooman-Gersmann M, Honée G, Bonnema G and De Wit PJGM (1996) A high-affinity binding site for the AVR9 peptide elicitor of *Cladosporium fulvum* is present on plasma membranes of tomato and other solanaceous plants. Plant Cell 8: 929—938

Kooman-Gersmann M, Vogelsang R, Hoogendijk ECM and De Wit PJGM (1997) Assignment of amino acid residues of the AVR9 peptide of *Cladosporium fulvum* that determine elicitor activity. Mol Plant-Microbe Interact 10: 821—829

Kooman-Gersmann M, Vogelsang R, Vossen P, Van den Hooven H, Mahé E, Honée G and De Wit PJGM (1997) Correlation between binding affinity and necrosis-inducing activity of mutant AVR9 elicitors. Plant Physiol: 117: 609—618

Langford AN (1937) The parasitism of *Cladosporium fulvum* and the genetics of resistance to it. Can J Res 15: 108—128

Laterrot H (1986) Race 2.5.9, a new race of *Cladosporium fulvum* (*Fulvia fulva*) and sources of resistance in tomato. Neth J Plant Pathol 92: 305—307

Laugé R, Joosten MHAJ, Van den Ackerveken GFJM, Van den Broek HWJ and De Wit PJGM (1997) The in planta-produced extracellular proteins ECP1 and ECP2 of *Cladosporium fulvum* are virulence factors. MPMI 10: 725—734

Laugé R, Dmitriev AP, Joosten MHAJ and De Wit PJGM (1998) Additional resistance gene(s) against *Cladosporium fulvum* present on the *Cf-9* introgression segment are associated with strong PR protein accumulation. Mol Plant-Microbe Interact 11: 301—308

Lazarovits G and Higgins VJ (1976a) Histological comparison of *Cladosporium fulvum* race 1 on immune, resistant, and susceptible tomato varieties. Can J Bot 54: 224—234

Lazarovits G and Higgins VJ (1976b) Ultrastructure of susceptible, resistant, and immune reactions of tomato to races of *Cladosporium fulvum*. Can J of Bot 54: 235—247

Lazarovits G and Higgins VJ (1979) Biological activity and specificity of a toxin produced by *Cladosporium fulvum*. Phytopathol 69: 1056—1061

Lazarovits G, Bhullar BS, Sugiyama HJ and Higgins VJ (1979) Purification and partial characterization of a glycoprotein toxin produced by *Cladosporium fulvum*. Phytopathol 69: 1062—1068

Lindhout P, Korta W, Cislik M, Vos I and Gerlagh T (1989) Further identification of races of *Cladosporium fulvum (Fulvia fulva)* on tomato originating from the Netherlands, France and Poland. Neth J Plant Pathol 95: 143—148

Marmeisse R, Van den Ackerveken GFJM, Goosen T, De Wit PJGM and Van den Broek HWJ (1993) Disruption of the avirulence gene *avr9* in two races of the tomato pathogen *Cladosporium fulvum* causes virulence on tomato genotypes with the complementary resistance gene *Cf-9*. Mol Plant-Microbe Interact 6: 412—417

Marmeisse R, Van den Ackerveken GFJM, Goosen T, De Wit PJGM and Van den Broek HWJ (1994) The *in-planta* induced *ecp2* gene of the tomato pathogen *Cadosporium fulvum* is not essential for pathogenicity. Curr Genet 26: 245—250

May MJ, Hammond-Kosack KE and Jones JDG (1996) Involvement of reactive oxygen species, glutathione metabolism, and lipid peroxidation in the *Cf*-gene-dependent defense response of tomato cotyledons induced by race-specific elicitors of *Cladosporium fulvum*. Plant Physiol 110: 1367—1379

Marzluf GA (1996) Regulation of nitrogen metabolism in mycelium fungi. In: Brambl R and Marzluf GA (eds), The Mycota III. Biochemistry and Molecular Biology. Berlin, Heidelberg, Springer Verlag: pp 357—368

Mossman K, Nation P, Macen J, Garbutt M, Lucas A and McFadden G (1996) Myxoma virus M-T7, a secreted homolog of the interferon-g receptor, is a critical virulence factor for the development of myxomatosis in European rabbits. Virology 215: 17—30

Noeldner PK-M, Coleman MJ, Faulks R and Oliver RP (1994) Purification and characterization of mannitol dehydrogenase from the fungal tomato pathogen *Cladosporium fulvum* (syn. *Fulvia fulva*). Physiol Mol Plant Pathol 45: 281—289

Oliver RP (1992) A model system for the study of plant-fungal interactions: tomato leaf mold caused by *Cladosporium fulvum*. In: Verma DPSE (ed), Molecular Signals in Plant-Microbe Communications, CRC Press, Boca Raton, Florida 97—106

Oliver RP, Farman ML, Jones JDG and Hammond-Kosack KE (1993) Use of fungal transformants expressing β-glucoronidase activity to detect infection and measure hyphal biomass in infected plant tissue. Mol Plant Microbe Interact 6: 521—525

Pallaghy PK, Nielsen KJ, Craick DJ and Norton RS (1994) A common structural motif incorporating a cystine knot and a triple-stranded beta-sheet in toxic and inhibitory polypeptides. Protein Sci 3: 1833—1839

Parniske M, Hammond-Kosack KE, Golstein C, Thomas CM, Jones DA, Harrison K, Wulff BBH and Jones

JDG (1997) Novel disease resistance specificities result from sequence exchange between tandemly repeated genes at the *Cf-4/9* locus of tomato. Cell 91: 1—20

Peever TL and Higgins VJ (1989a) Suppression of the activity of non-specific elicitor from *Cladosporium fulvum* by intercellular fluids from tomato leaves. Physiol Mol Plant Pathol 34: 471—482

Peever TL and Higgins VJ (1989b) Electrolyte leakage, lipoxygenase, and lipid peroxidation induced in tomato leaf tissue by specific and nonspecific elicitors from *Cladosporium fulvum*. Plant Physiol 90: 867—875

Scholtens-Toma IMJ and De Wit PJGM (1988) Purification and primary structure of a necrosis-inducing peptide from the apoplastic fluids of tomato infected with *Cladosporium fulvum* (syn. *Fulvia fulva*). Physiol Mol Plant Pathol 33: 59—67

Spriggs M, Hruby DE, Maliszewski CR, Pickup DJ, Sims JE, Buller RML and VanSlyke J (1992) Vaccinia and cowpox viruses encode a novel secreted interleukin-1-binding protein. Cell 71: 145—152

Stevens MA and Rick CM (1988) Genetics and Breeding. In The Tomato Crop, Atherton JG and Rudich J eds, Chapman and Hall, London 35—109

Takken FLW, Schipper D, Nijkamp HJJ and Hille J (1998) Identification and *Ds*-tagged isolation of a new gene at the *Cf-4* locus of tomato involved in disease resistance to *Cladosporium fulvum* race 5. Plant J 14: 401—411

Thomas CM, Jones DA, Pamiska M, Harrison K, Ballint-Kurti PJ, Hatzixanthis K and Jones JDG (1997) Charactarisation of the tomato *Cf-4* gene for resistance to *Cladosporium fulvum* identified sequences which determine recognitional specificity in Cf-4 and Cf-9. Plant Cell 9: 1—12

Van den Ackerveken GFJM, Van Kan JAL and De Wit PJGM (1992) Molecular analysis of the avirulence gene *avr9* of the fungal tomato pathogen *Cladosporium fulvum* fully supports the gene-for-gene hypothesis. Plant J 2: 359—366

Van den Ackerveken GFJM, Vossen JPMJ and De Wit PJGM (1993a). The AVR9 race-specific elicitor of *Cladosporium fulvum* is processed by endogenous and plant proteases. Plant Physiol 103: 91—96

Van den Ackerveken GFJM, Van Kan JAL, Joosten MHAJ, Muisers JM, Verbakel, HM and De Wit PJGM (1993b) Characterization of two putative pathogenicity genes of the fungal tomato pathogen *Cladosporium fulvum*. Mol Plant-Microbe Interact 6: 210—215

Van den Ackerveken GFJM, Dunn RM, Cozijnsen AJ, Vossen JPMJ, Van den Broek HWJ and De Wit PJGM (1994) Nitrogen limitation induces expression of the avirulence gene *avr9* in the tomato pathogen *Cladosporium fulvum*. Mol Gen Genet 243: 277—285

Van Dijkman A and Kaars Sijpesteijn A (1973) Leakage of pre-absorbed ^{32}P from tomato leaf disks infiltrated with high molecular weight products of incompatible races of *Cladosporium fulvum*. Physiol Plant Pathol 3: 57—67

Van Kan JAL, Van den Ackerveken GFJM and De Wit PJGM (1991) Cloning and characterization of cDNA of avirulence gene *avr9* of the fungal tomato pathogen *Cladosporium fulvum*, causal agent of tomato leaf mold. Mol Plant-Microbe Interact 4: 52—59

Van Kan JAL, Joosten MHAJ, Wagemakers CAM, Van den Berg-Velthuis GCM and De Wit PJGM (1992) Differential accumulation of mRNAs encoding extracellular and intracellular PR proteins in tomato induced by virulent and avirulent races of *Cladosporium fulvum*. Plant Mol Biol 20: 513—527

Vera-Estrella R, Blumwald E and Higgins VJ (1992) Effect of specific elicitors of *Cladosporium fulvum* on tomato suspension cells. Plant Physiol 99: 1208—1215

Vera-Estrella R, Blumwald E, and Higgins VJ (1993) Non-specific glycopeptide elicitors of *Cladosporium fulvum*: Evidence for involvement of active oxygen species in elicitor-induced effects on tomato cell suspensions. Physiol Mol Plant Pathol 42: 9—12

Vera-Estrella R, Higgins VJ and Blumwald E (1994a) Plant defence response to fungal pathogens. II. G-protein-mediated changes in host plasma membrane redox reactions. Plant Physiol 106: 97—102

Vera-Estrella R, Barkla BJ, Higgins VJ and Blumwald E (1994b) Plant defence response to fungal pathogens. Activation of host-plasma membrane H+-ATPase by elicitor induced enzyme dephosphorylation. Plant Physiol 104: 209—215

Vervoort J, Van den Hooven HW, Berg A, Vossen P, Vogelsang R, Joosten, MHAJ and De Wit PJGM (1997) The race-specific elicitor AVR9 of the tomato pathogen *Cladosporium fulvum*: a cystine-knot protein. Sequence-specific ¹H NMR assignments, secondary structure and global fold of the protein. FEBS Lett 404: 153—158

Wubben JP, Eijkelboom CA and De Wit PJGM (1994a) Accumulation of pathogenesis-related proteins in the epidermis of tomato leaves infected by *Cladosporium fulvum*. Neth J of Plant Pathol 99: 231—239

Wubben JP, Joosten MHAJ and De Wit PJGM (1994b) Expression and localisation of two *in planta* induced extracellular proteins of the fungal tomato pathogen *Cladosporium fulvum*. Mol Plant-Microbe Interact 7: 516—524

Wubben JP, Joosten MHAJ, Van Kan JAL and De Wit PJGM (1992) Subcellular localization of plant chitinases and 1,3-β-glucanases in *Cladosporium fulvum* (syn. *Fulvia fulva*)-infected tomato leaves. Physiol Mol Plant Pathol 41: 23—32

Wubben JP, Lawrence CB and De Wit PJGM (1996) Differential induction of chitinase and 1,3-β-glucanase gene expression in tomato by *Cladosporium fulvum* and its race-specific elicitors. Physiol Mol Plant Pathol 48: 105—116

Xing T, Higgins VJ and Blumwald E (1996) Regulation of plant defense response to fungal pathogens: Two types of protein kinases in reversible phosphorylation of the host plasma membrane H+-ATPase. Plant Cell 8: 555—564

Xing T, Higgins VJ and Blumwald E (1997) Race-specific elicitors of *Cladosporium fulvum* promote translocation of cytosolic components of NADPH oxidase to plasma membrane of tomato cells. Plant Cell 9: 249—259

THE BARLEY/*BLUMERIA* (SYN. *ERYSIPHE*) *GRAMINIS* INTERACTION

HANS THORDAL-CHRISTENSEN[1],
PER L. GREGERSEN[2] & DAVID B. COLLINGE[2]

[1] *Plant Pathology and Biogeochemistry Department*
Risoe National Laboratory, DK-4000 Roskilde, Denmark
(hans.thordal@risoe.dk)
[2] *Plant Pathology Section, Department of Plant Biology*
The Royal Veterinary and Agricultural University, Thorvaldsensvej 40
DK-1871 Frederiksberg C, Copenhagen, Denmark

Summary

Barley leaves attacked by the powdery mildew fungus is a pathosystem well suited for studies of plant-pathogen interaction mechanisms. Nearly one hundred specific resistance genes have been identified, many of which are alleles at more or less complex loci. The corresponding avirulence genes are, on the other hand, evenly distributed in the powdery mildew fungus genome. The fungus colonizes only the leaf surface, placing haustoria in the leaf epidermal cells to acquire nutrients in an obligately biotrophic manner. The fungus exhibits a highly synchronous development and the epidermal cells respond accordingly. Papillae, which are formed early, subjacent to fungal germ tubes, arrest a considerable fraction of the attempted penetrations in all combinations of pathogen and plant genotypes (i.e. both incompatible and compatible). Papillae consist of several different components, and inhibitor studies have suggested that at least callose and phenylpropanoids are directly involved in arresting the growth of the penetration peg. In compatible interactions, the haustorial nutrient uptake is believed to be driven by fungal plasma membrane H^+-ATPase activity. This is based on the apparent lack of H^+-ATPase activity at the invaginated host plasma membrane. In incompatible interactions, the hypersensitive response serves as a back-up resistance mechanism arresting the germlings which have managed to penetrate the papilla. While only limited information is available concerning the expression of pathogenicity genes in the powdery mildew fungus, significantly more is known in relation to host response gene expression. However, the biological role of the host response genes is poorly understood; data suggest that many of them are merely part of a general stress response with no direct involvement in defense.

A. Slusarenko, R.S.S. Fraser, and L.C. van Loon (eds), Mechanisms of Resistance to Plant Diseases, 77-100.
© 2000 *Kluwer Academic Publishers. Printed in the Netherlands.*

Abbreviations

Bgh:	Blumeria graminis f.sp. hordei
CAD:	cinnamyl alcohol dehydrogenase
CHS:	chalcone synthase
HR:	hypersensitive response
OMT:	O-methyltransferase
PAL:	phenylalanine ammonia lyase
PGT:	primary germ tube
PR:	pathogenesis-related

I.   Introduction

Barley (*Hordeum vulgare*), a major cereal of the temperate zones of the world, is an interesting plant species seen from the plant pathologist's point of view. Barley is often seriously diseased, the most important disease being powdery mildew. In cooler climates, it is estimated that, in the absence of fungicides, the powdery mildew fungus would cause approx. 10% yield losses (Jørgensen *et al.*, 1988). Powdery mildew fungi are very successful pathogens on many monocot and dicot plants. Species of the grass family (*Poaceae*) are attacked by *Blumeria graminis* (syn. *Erysiphe graminis* - see box). Both barley and the even more important cereal crop plant, wheat, suffer seriously from attack by this pathogen. However, the powdery mildew fungus is strictly adapted into *formae speciales* attacking individual genera of the grass family, and therefore the barley powdery mildew fungus (*B. graminis* f.sp. *hordei*) and the wheat powdery mildew fungus (*B. graminis* f.sp. *tritici*) are distinct pathotypes. *B. graminis* f.sp. *hordei* (*Bgh*) and its interaction with the barley plant has been studied extensively for many years, and it will be considered here as a model for this group of plant pathogens. Different aspects of the biology and agronomy of this interaction have been reviewed in several papers (Jørgensen, 1988; Aist and Bushnell, 1991; Wolfe and McDermott, 1994; Kunoh, 1995; Collinge *et al.*, 1997).

The *Ascomycete* fungus *Bgh* is a biotrophic, obligate parasite. The airborne asexual spores, the conidia, are blown onto the leaf blade and leaf sheath of the barley plant. The conidium germinates with a primary germ tube (PGT) and an appressorial germ tube approximately 1-2 and 4-8 hours after inoculation, respectively. Approximately 2 hours later, papilla formation and a number of other responses occur in the barley leaf epidermal cell subjacent to the germ tubes. These local responses in the outer epidermal cell wall result in the arrest of a significant proportion of the attempted penetrations by appressorial germ tubes (e.g. Carver, 1986). In compatible interactions, successful penetrations eventually lead to the formation of colonies in which vast numbers of conidia are formed in chains and "lifted" into the air. In contrast, in incompatible interactions, penetration of papillae subsequently triggers a hypersensitive response (HR) in one or more barley cells (see Aist and Bushnell, 1991; Collinge *et al.*, 1997), which combats the infection.

The success of the powdery mildews is undoubtedly related to their unique ability to infect the host under dry conditions. While other plant pathogenic fungi are hampered by their need for 100% humidity or even free water, the powdery mildew fungi can readily colonize during the dry conditions of a sunny day. It has been suggested that the PGT is able to take up water from the leaf surface, or that the conidium carries a relatively large amount of water (see below). An ability to conserve and economize the use of water, is also important for the long-distance spread of this fungus. The optimal temperature for the development of barley powdery mildew is 20°C (Pauvert and de la Tullaye, 1977). At this temperature, the time from inoculation to the development of conidia is only 5-6 days.

Blumeria or *Erysiphe* ?

Classification: Ascomycota; Euascomycetes; Erysiphales; Erysiphaceae.

The genus *Erysiphe* was formerly split into *Blumeria* and *Erysiphe* by Golovin (1958, cited by Yarwood (1978)) and subsequently by Speer (1973) on the basis of various morphological characteristics (e.g. lobed haustorium, setose secondary mycelium and primary germ tube). Although the split appears to be generally accepted by taxonomists, it has not gained full acceptance among plant pathologists. Furthermore, for formal taxonomic reasons, both names are valid until a botanical congress legitimises the split. Thus, *Erysiphe graminis* is a synonym for *Blumeria graminis*; both names will be used validly for years to come. The anamorph (imperfect) name *Oidium graminis* is occasionally encountered.

The species *B. graminis* is divided into *formae speciales* (f.sp.) on the basis of host range. The f.sp. *hordei* cause disease on the *Hordeum* species, f.sp. *tritici* on the *Triticum* species etc. In view of the biological isolation of the *formae speciales*, it has been proposed that their status be raised to full species (e.g. *B. hordei, B. tritici.*) However, due to induced accessibility (see text), different *formae speciales* can infect one host and subsequently cross (e.g. Hiura, 1978). Thus, there are insufficient grounds for this change.

II. An Interesting Experimental System

In addition to its commercial importance as a disease, the interaction of *Bgh* with barley leaves also is an interesting experimental system for studying the mechanisms underlying plant-pathogen interactions. The restriction of the fungus to the surface of the leaf makes it easy to study the development of germinating conidia. Conidia germinate synchronously, germ tubes develop to uniform length and attempt penetration within a well defined period. The individual developmental stages of the fungus exhibit limited overlap in time, and it is therefore often possible to observe plant responses come and go in several cycles within the first 3-5 days after inoculation. Because *Bgh* grows on the leaf surface, the interaction between individual conidia and individual host cells can be studied easily, and the future will undoubtedly show that this system can be exploited for studies of communication between plant cells. Signals which originate from the individual conidium are passed on to the threatened plant cell, which in turn will pass on a signal to neighbouring epidermal and mesophyll cells. This cell to cell signalling might be a cry for help, and the reply can, for instance, be envisaged to be a triggering of the HR in the originally attacked epidermal cell.

The powdery mildew fungus is an obligate biotroph, feeding via haustoria in the leaf epidermal cells. Exactly how nutrient transfer takes place is currently not well understood. Fungal genes which are expressed during disease development are being identified, and this information will help to solve the puzzle of haustorial function. It would appear that resistance towards powdery mildew fungi, and other obligately biotrophic fungi, could be manifested as direct inhibition of haustorial function, so that the fungus litterally starves to death. However, resistance acting in this way has yet to be demonstrated.

The ease by which individual conidia and individual host cells can be studied in the barley/*Bgh* system has made it the most well studied system with respect to "papilla-based resistance". This form of resistance, which is manifested at the stage of plant cell wall penetration, is considered to be non-specific in that all fungal isolates appear to have similar penetration rate (for exceptions, see below).

More than 85 race-specific resistance genes have been described in barley towards *Bgh*. Twenty-eight of these resistance genes are alleles or tightly linked genes at the *Ml-a* locus. Another ten genes have been mapped to nine different loci, while the remainder await mapping. For a comprehensive review, see Jørgensen (1994). The well characterized *Ml-a* "locus" actually consists of several loci, and recombinations between certain of those have been obtained (Jahoor *et al.*, 1993; DeScenzo *et al.*, 1994; Mahadevappa *et al.*, 1994). The powdery mildew race-specific resistance genes result in resistance phenotypes, varying from single cell to multi-cell HR based resistances. Even among the different *Ml-a* alleles, this variation in resistance phenotypes is found (e.g. Kita *et al.*, 1981; Tosa and Shishiyama, 1984; Boyd *et al.*, 1995). Molecular markers have been identified which are closely linked to several of these resistance genes, including *Ml-a* (DeScenzo *et al.* 1994), *Ml-La* (Hilbers, 1992; Giese *et al.*, 1993) and *Ml-g* (Görg *et al.*, 1993). It is to be expected that molecular clones will be obtained for these genes within a few years.

The use of molecular markers has led to the development of a genetic map of the powdery mildew fungus (Christiansen and Giese, 1990) which demonstrates that whereas a large number of the *Ml-a* resistance genes are clustered in a complex locus (see above), the corresponding avirulence genes are scattered through-out the powdery mildew genome.

A different kind of resistance, which is conferred by several recessive (mutant) alleles ($ml-o_n$) of the *Ml-o* locus, is characterized by over-sized papillae causing resistance to penetration (Skou *et al.*, 1984). The *ml-o* resistance appear to be durable under field conditions, and it is therefore now widely used for powdery mildew control.

Many near-isogenic lines have been made by back-crossing different resistance genes into several barley cultivars. The most well-developed set of near-isogenic lines exists in the cultivar Pallas (Kølster *et al.*, 1986), into which more than twenty genes have been back-crossed. Such sets of near-isogenic lines form excellent experimental material for comparative studies of defense responses.

Two mutational studies have been conducted, where mutants with lost resistance have been obtained. Torp and Jørgensen (1986) generated a large number of susceptible mutants in the cultivar Sultan-5 carrying the $Ml-a_{12}$ resistance gene. Most of these were mutated in the $Ml-a_{12}$ resistance gene itself, while a few had lesions in other genes. Using this material, Freialdenhoven et al. (1994) have defined two genes outside the $Ml-a$ locus, which are required for the HR. A similar set of susceptible mutants has been generated in a $ml-o$ resistant barley line (Freialdenhoven et al., 1996).

III. Fungal Development

A schematic presentation of the synchronous development of $Bgh,$ and the epidermal cell responses, is shown in Fig. 1. In principle, all possible outcomes can be found in any interaction between genotypes, whether compatible or incompatible, single cell HR or multi-cell HR. What differs are the frequencies of the different outcomes, and thereby the macroscopic symptoms by which the interactions are categorized into infection types.

A.   THE PRIMARY GERM TUBE

The $Blumeria$ $graminis$-powdery mildew conidium germinates with a short PGT 1-2 hours after inoculation and an appressorial germ tube 4-8 hours after inoculation. While the appressorium develops the penetration structure (the "lobe"), the PGT itself is hardly "designed" for penetration, even though scanning electro-micrographs have demonstrated penetration pegs and pores at the PGT (Kunoh et al., 1978b). What is then the purpose of the PGT, why does the conidium not germinate directly with a differentiated appressorium? For several reasons this is an intriguing question: 1) it seems that the PGT activates the defense of the leaf, i.e. this is working against the success of the fungus (see below), and 2) only the powdery mildews causing disease on the grass family (i.e. $Blumeria$) have a PGT (Kunoh et al., 1979), which may suggest that it is really not necessary. It has been suggested that physiological contact between the PGT and the leaf is required for the conidium in order to recognize a potential host (e.g. Carver and Ingerson, 1987). In support of this model, conidia with abnormal germination can occasionally be observed on a leaf. These possess several PGT-like structures, but no appressorium. A possible reason for this might be that the PGT did not contact the leaf surface, and therefore the conidium is "lost". Similar abnormal germination is observed on artificial surfaces (Carver and Ingerson, 1987). Francis et al. (1996) provides evidence that the conidium needs to take up cutin monomers in order for the appressorial germ tube to differentiate. They suggested that fungal cutinase liberates cutin monomers from the epidermal cell surface. In addition to potential uptake of compounds which signal appressoria formation, evidence for uptake of salts (Kunoh et al., 1978a) and acridine orange (Kunoh and Ishizaki, 1981) from the leaf surface through the PGT, suggests that nutrients can be taken up as well. A very important role for the PGT has been suggested by Carver and Bushnell (1983). They present evidence that contact of the PGT to the leaf surface is essential for uptake

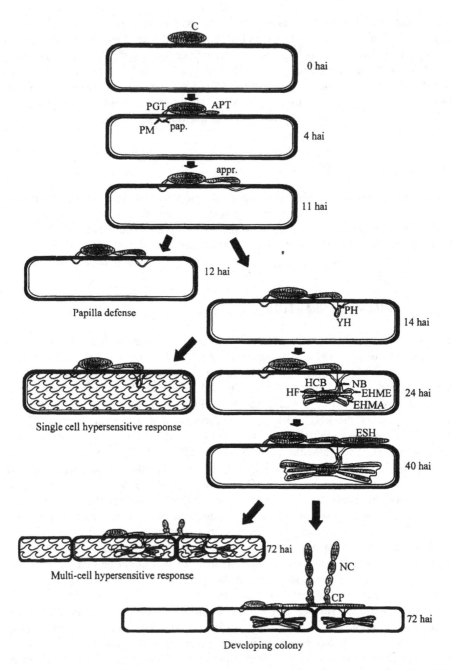

Figure 1. Development of *Blumeria graminis* f.sp. *hordei* on the barley leaf epidermis. C, conidium; PGT, primary germ tube; APT, appressorial germ tube; pap., papilla; PM, plasma membrane; appr., appressorium; PH, penetration hypha; YH, young haustorium; HCB, haustorial central body; HF, haustorial fingers; NB, neck band; EHME, extrahaustorial membrane (plant origin); EHMA, extrahaustorial matrix; ESH, elongating secondary hyphae; NC, new conidia; CP, conidiophore. Time points in hours after inoculation (hai).

of water through the PGT under conditions of low relative humidity, which possibly explains how *Blumeria graminis* can infect in a, for other fungi, hostile environment. This may justify why *Blumeria graminis* conidia germinate with a PGT in spite of all the seeming drawbacks connected with it. However, although powdery mildews attacking dicots lack a PGT, they are still capable of infecting in dry air.

A clear negative effect of the PGT is its activation of defense responses in the host. The epidermal cell responds to the PGT by forming a papilla. This is smaller, but otherwise similar to the papilla formed in response to the appressorium (see below). It is an open question whether the papilla is important for the inevitable failure of the PGT to penetrate, but the establishment of the machinery for papilla formation is undoubtedly important for the formation of papillae subjacent to the appressorium. This was illustrated nicely by Woolacott and Archer (1984), who presented data from a compatible interaction showing that it is significant for the success of a penetration attempt whether the two germ tubes encounter the same or different epidermal cells. Conidia with germ tubes encountering the same cell had a penetration success of approximately 30%, otherwise this was approximately 60%. Defense activation by the PGT might explain how induced resistance can be obtained by a temporary exposure of the leaf to powdery mildew conidia for only 2-3 hours, and that the penetration success in a compatible interaction decreases with increasing inoculum density (Cho and Smedegaard-Petersen, 1986; Thordal-Christensen and Smedegaard-Petersen, 1988a; Carver and Ingerson-Morris, 1989). Several laboratories have demonstrated that defense-related gene transcripts accumulate in "waves" of which the first coincides with the formation of papillae in response to the PGT (Brandt *et al.*, 1992; Thordal-Christensen *et al.*, 1992; Clark *et al.*, 1993, 1994, 1995; Walther-Larsen *et al.*, 1993; Boyd *et al.*, 1994a, 1994b, 1995; Gregersen *et al.*, 1997). It is therefore likely that a number of defense gene transcripts accumulate as a result of stimuli originating from the PGT.

The level to which defense is activated by the PGT does not appear to be influenced by specific resistance genes/avirulence genes. Quantification of induced resistance (Thordal-Christensen and Smedegaard-Petersen, 1988a) and defense gene transcript accumulation (Gregersen *et al.*, 1997), showed no reproducible differences in near-isogenic lines using virulent and avirulent isolates of *Bgh*.

B. THE APPRESSORIUM

The appressorial germ tube emerges approximately 4 hours after inoculation and the appressorium is fully developed at approximately 8 hours. From this time point, the first lobe starts to develop on the side of the appressorial apex. The lobe is the structure below which the penetration hypha develops. This hypha will attempt to penetrate directly through the outer epidermal cell wall of the leaf between 10 and 12 hours after inoculation. Electron micrographs of this process show that a pore is formed in the lower cell wall of the lobe, and an approximately 0.4 µm wide penetration peg develops through this pore. During the penetration of the host cell wall, the penetration hypha does not appear to have a cell wall at its tip, where a high secretory activity takes place (Edwards and Allen, 1970; Stanbridge *et al.*, 1971; McKeen and Rimmer, 1973).

This suggests that the penetration of the host cell wall is based on dissolution by fungal enzymes (see also below). In contrast, it has been suggested that the penetration through the papilla is mechanical, rather than enzymatic (Edwards and Allen, 1970). Upon successful penetration, the host cell often forms a collar-like structure surrounding the penetration hypha (e.g. Bracker, 1968; Heitefus and Ebrahim-Nesbat, 1986; Fig. 2).

In cases where the penetration fails from this first appressorial lobe, the appressorium generally forms a second lobe either on the other side of the apex or half way down the appressorium. The second lobe emerges approximately 16 hours after inoculation. The plant likewise responds to this penetration attempt by the formation of a papilla; and this second penetration attempt has a low success rate (Carver, 1986; Kunoh *et al.*, 1988a). Sometimes even a third penetration attempt is made which, however, never succeeds.

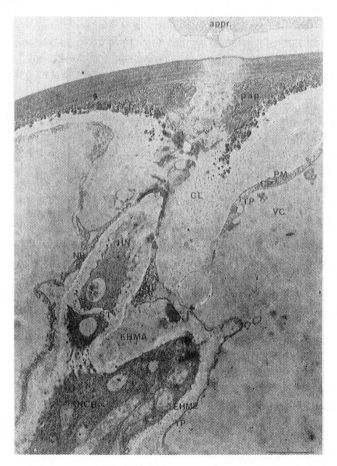

Figure 2. Electron micrograph of a cross section of a successful penetration by *Blumeria graminis* f.sp. *hordei* into a barley epidermal cell 24 hours after inoculation. appr., appressorium; pap., papilla; PM, plant plasma membrane; TP, plant tonoplast; CP, plant cytoplasm; CL, collar; VC, plant vacuole; HN, haustorial neck; ST, septum; HCB, haustorial central body; NB, neck band; EHME, extrahaustorial membrane (plant origin); EHMA, extrahaustorial matrix. Section thickness, 60 nm. Bar, 1 m. Photo provided by Drs. Fasong Zhou and Eigil de Neergaard.

C.   THE HAUSTORIUM

The successful penetration of the epidermal cell wall and the papilla allows the development of a fungal haustorium in the epidermal cell. The formation of a functional haustorium is the critical step in the establishment of the pathogen on the host, since the nutrient uptake by the fungus appears to be dependent on the haustorium. It is conceivable that haustorial function may vary in efficiency, depending on the host genotype, which could thus explain differences in partial resistance of some host plant varieties. Based on cytological observations, Kobayashi et al. (1995) suggested that the establishment of a functional haustorium requires the release of factors from the fungus which suppress the resistance response of the host plant. This is based on the well-documented phenomenon of induced susceptibility (e.g. Kunoh et al., 1986) where pre-inoculation with a virulent isolate of a biotrophic pathogen makes the infected plant susceptible to otherwise avirulent races and to non-host biotrophic pathogens. Thus, a virulent pathogenic race seems to be able to induce a so-called "state of accessibility" (Ouchi et al., 1974) in the infected host cell, where the metabolism of the host is manipulated to serve the needs of the pathogen.

1.   *The Haustorial Complex*

Strictly speaking, the haustorium is an apoplastic formation (Smith and Smith, 1990) since it does not break the host plasma membrane which, instead, is enlarged and invaginated, and surrounds the haustorium completely as an extrahaustorial membrane. The extrahaustorial membrane appears thicker and stronger than the normal host cell plasma membrane (Bracker, 1968; Fig. 2); however, the two membranes are contiguous. The extrahaustorial membrane is thus of host origin but its formation is obviously provoked by the invading pathogen. The entire infection structure inside the host cell wall is designated "the haustorial complex" and it is made up of both fungal and plant components. The haustorium of *B. graminis* develops a characteristic shape with a central body from which two sets of finger-like hyphal structures radiate. The haustorium is connected to the superficial hyphae by the extended penetration hyphae which forms a slender stalk (the haustorial neck). The neck is septate (Fig. 2) and the protoplasm of the haustorium contains the normal subcellular components and organelles of fungal cells (Bracker, 1968; Hippe-Sanwald et al., 1992). Around the haustorial neck, just inside the penetrated papilla and the collar, there is a transition zone between the extrahaustorial membrane and the normal host cell plasma membrane known as the neck band (Manners and Gay, 1977). The neck band is a complex structure in which the two membranes appear to be attached to the cell wall of the haustorial neck. The neck band seems to have two different functions (Manners and Gay, 1983): 1) It delimits the extrahaustorial membrane from the normal plasma membrane preventing the exchange of components between these two membrane domains. 2) It prevents, as a seal, diffusion of solutes from the compartment between the extrahaustorial membrane and the haustorium, called the extrahaustorial matrix, to the apoplast. Thus, influx to and efflux from the extrahaustorial matrix is controlled by passage through the extrahaustorial host membrane or through the wall and membrane of the fungal haustorium. The extrahaustorial matrix is probably of both fungal and plant origin and so it is the interphase between pathogen and host across which nutrients taken up by the fungus have to travel.

2. Haustorial Mechanisms

The obvious role of the haustorium is to function as a feeding organ for the pathogen, taking up water, minerals and metabolites from the leaf and exporting it to the superficial hyphae where it is used for mycelial growth and spore formation. A model was suggested by Spencer-Phillips and Gay (1981) in which this nutrient uptake is driven by a mechanism closely linked to the structure of the haustorial complex described above. The model assumes that the extrahaustorial membrane differs functionally from the normal plasma membrane in having no proton pump activity. The model predicts that the normal host proton pump activity drives the epidermal cell uptake of nutrients, which subsequently enters the extrahaustorial matrix due to concentration gradients. Here, it is taken up by the fungus due to a gradient created by proton pump activity in the haustorial membrane. The model was supported by cytological studies of the barley/powdery mildew system using an inhibitor and an activator of proton pump activity (Gay et al., 1987).

The preferred metabolite taken up by the haustorium appears to be glucose (Hall et al., 1992, Mendgen and Nass, 1988). Sucrose, which an earlier study (Manners and Gay, 1983) suggested to be most important, seems to be metabolized prior to uptake (Mendgen and Nass, 1988). In the haustorium and mycelium, the carbohydrates taken up are converted into other carbohydrates, especially mannitol, and lipids (Manners and Gay, 1983). Minerals needed by the fungus are potentially also taken up via the haustorium. However, very few detailed studies have been made on this aspect in B. graminis systems. One early study on wheat powdery mildew was made by Mount and Ellingboe (1969) showing that ^{32}P and ^{35}S are in fact transferred from the leaf to the superficial hyphal structures.

D.   COLONY DEVELOPMENT

Successful establishment of haustoria is succeeded by the formation of a secondary hypha from the primary appressorium around the time when the haustorial body is formed (Masri and Ellingboe, 1966; Bushnell and Gay, 1978). This first elongating secondary hypha is the starting point for the development of one fungal colony, comprising the typical macroscopic symptom of powdery mildew infected barley leaves. The hyphae in the developing colony have a tendency to grow along the epidermal cells over the anti-clinal walls. From the superficial hyphae, secondary appressoria are formed and from these, haustoria are generated in epidermal cells in a diurnal rhythm, with most haustoria formed in the middle of the dark period (Bushnell and Gay, 1978). Three to four days following the primary infection, conidiophores are formed on the hyphae, sporulation starts and spores can be spread to initiate new cycles of infections.

E.   SPECIFIC FUNGAL PROTEINS

Because the powdery mildew fungi are obligate parasites, they have not been easily amenable to physiological study. Nevertheless, progress has been made over the last few years. Enzyme activity staining of isoelectric focusing protein electrophoresis gels (zymograms) has been used to assess the level of genetic variation among *Bgh* isolates (Koch and Köhler, 1990, 1991). These studies demonstrated the presence of the following enzymes in conidia: malate dehydrogenase, phosphogluconate dehydrogenase, glucose-6-phosphate dehydrogenase, NADH-diaphorase, phosphoglucomutase, esterase, acid phosphatase, triose phosphate isomerase, glucose phosphate isomerase and superoxide dismutase. One of the first events to occur upon contact of a conidium with the surface of the barley leaf (or cellophane) is a release of liquid. Contact of this liquid with the leaf surface results in the disappearance of wax crystals from the surface (Kunoh *et al.*, 1988b) and erosion of the cuticle (Staub *et al.*, 1974; Nicholson *et al.*, 1988; Kunoh *et al.*, 1990). Esterase was released into the exudate within 15 to 30 minutes of contact (Nicholson *et al.*, 1988), and this second release could be inhibited by the translational inhibitor, cycloheximide (Kunoh *et al.*, 1990). The presence of esterase in this exudate suggests that the esterase activity observed is perhaps due to cutinase; the presence of cutinase in the infection exudate was confirmed using a specific assay by Pascholati *et al.* (1992). However, this latter study did not investigate the timing of release of the cutinase or its relationship to the individual esterase isozymes observed previously (Nicholson *et al.*, 1988). (See also Francis *et al.*, 1996.) Takahashi *et al.* (1985) used histological enzyme assays to look for hydrolytic enzymes associated with individual infection structures, and found esterase activity associated with primary and secondary germ tubes and their respective papillae. In this study, it was not possible to determine whether the enzymes were of host or pathogen origin.

Molecular studies of the powdery mildew fungus are still in their infancy. Only five gene sequences are published: two retroposon elements (Rasmussen *et al.*, 1993; Wei *et al.*, 1996), β-tubulin (Sherwood and Somerville, 1990) and two as yet unidentified gene transcripts present in germinating conidia (Justesen *et al.*, 1995).

IV. Resistance Mechanisms

A. PAPILLA FORMATION

In response to the PGT and to the appressorial lobes, the leaf forms papillae in an attempt to stop penetration. Papilla formation begins with the appearance of a cytoplasmic aggregate subjacent to these germ tubes.

1. *Cytoplasmic Aggregate*

In a normal epidermal cell, only a very thin layer of cytoplasm is found between the vacuolar tonoplast and the plasma membrane. The cytoplasm granulates locally in response to the fungal hyphae. In the case of the first appressorial lobe this happens at approximately 11 hours after inoculation and is completed within 5 to 10 min. (Bushnell and Zeyen, 1976; Aist and Israel, 1977). About half an hour later, the appressorium forms the penetration peg. Initiation of the penetration peg generally precedes initiation of the papilla by up to 1 hour, but it is not unusual that initiation of the papilla precedes initiation of the penetration peg (Aist and Israel, 1977). The cytoplasmic aggregate is characterized by high secretory activity where proteins, carbohydrates and phenols (Kita *et al.*, 1980; Russo and Bushnell, 1989) are exported across the host cell membrane in order to be built into the papilla. See also Akutsu *et al.* (1980). A recent study, using inhibitors of actin and tubulin polymerization has demonstrated that re-organisation of microfilaments and microtubules is associated with the cytoplasmic aggregation and subsequent papilla formation (Kobayashi *et al.*, 1997).

2.   *The Papilla*

The papilla is a cell wall apposition located subjacent to the fungal germ tubes on the inner surface of the outer epidermal cell wall. Papillae subjacent to the PGTs are approximately 3 μm in diameter while those formed subjacent to the appressoria are 4-5 μm in diameter. Papillae are surrounded by "haloes" which are 10-20 μm in diameter (e.g. Thordal-Christensen and Smedegaard-Petersen, 1988b). A number of compounds have been demonstrated in the papillae and haloes. These include callose, phenolics, proteins, silicon and hydrogen peroxide (Kunoh and Ishizaki, 1976; Kita *et al.*, 1980; Zeyen *et al.*, 1983; Smart *et al.*, 1986; Russo and Bushnell, 1989, Thordal-Christensen *et al.*, 1997). A "basic staining material" observed by e.g. Russo and Bushnell (1989) is potentially a guanidine radical (Wei *et al.*, 1994). The phenolic compounds in the papillae and haloes have been detected by UV-fluorescence, lacmoid, methylene blue, resorcinol and toluidine blue staining; nevertheless, the phenolic substances remain uncharacterized in detail. Several attempts have failed to demonstrate the presence of lignin using phloroglucinol-HCl (e.g. Smart *et al.*, 1986; Russo and Bushnell, 1989; Wei *et al.*, 1994). Lignin can, on the other hand, be demonstrated in wheat papilla using this stain (e.g. Wei *et al.*, 1994). Specific proteins in the papilla include hydrolytic enzymes (Takahashi *et al.*, 1985), leaf thionins (Ebrahim-Nesbat *et al.*, 1993) and peroxidase (Scott-Craig *et al.*, 1995). It has been demonstrated that proteins are covalently cross-bound in the papilla structure (Thordal-Christensen *et al.*, 1997), and the immobility of e.g. the phenolics suggests that they are covalently cross-bound as well. The papilla has a complex ultrastructure. Heitefuss and co-workers have defined several types of papilla with different organizations of e.g. layers and more or less irregular structures (see Heitefuss and Ebrahim-Nesbat, 1986). See also Hippe-Sanwald *et al.* (1992).

3.   *Penetration Success*

It is generally accepted that the resistance manifested at the early penetration stage is largely race-non-specific (see chapter by Bert Keller *et al.*) and is superimposed on resistance conferred by specific resistance genes, which is manifested through the hypersensitive response (see below). Nevertheless, the outcome of the "battle" between the penetration hypha and the papilla is dependent on a number of variable factors.

a.   *Genotypes.*   The *ml-o* resistance results in earlier and larger papillae (Skou *et al.*, 1984; Gold *et al.*, 1986) which efficiently stop the penetration attempt. It remains an open question as to which specific component makes these papillae so efficient (see below). However, *Bgh* isolates with a higher infection rate on *ml-o* resistant plants have been identified (Schwarzbach, 1979; Lyngkjær *et al.*, 1995). Data obtained by Carver (1986) in a study of varieties with different level of partial resistance suggest that the host's genetic background is important for papilla resistance.

b. *PGT-Activated Defense.* As discussed above, the PGT activates defense reactions which are manifested a few hours later against the penetration from appressoria. Resistance induced within 16 hours of exposure to *Bgh*, during which period the PGT is a major contributor to defense activation, is manifested as reduced penetration success by *Bgh*. This reduction correlates with an increased papilla diameter (Thordal-Christensen and Smedegaard-Petersen, 1988b).

c. *Host Cell Type.* There is a pronounced correlation between the size/location of the epidermal cells and their penetration. The short cells near the stomata, which are placed in rows, are more readily penetrated than the long cells between the rows of stomata (e.g. Johnson *et al.*, 1979; Thordal-Christensen and Smedegaard-Petersen, 1988b; Koga *et al.*, 1990; Görg *et al.*, 1993). A possible explanation for this might be that long cells are more likely to be encountered by PGTs, and this induces resistance to later penetration attempts. The very small stomatal subsidiary cells appear to exhibit inefficient papilla resistance, and they are susceptible even in *ml-o* resistant plants (Jørgensen and Mortensen, 1977).

4. How Do Papillae Work?

Many questions in relation to papilla-based resistance remain unanswered. For instance, where and how is the penetration peg arrested? A few published micrographs suggest that the peg stops inside the papilla (e.g. Edwards and Allen, 1970; Heitefuss and Ebrahim-Nesbat, 1986; Aist *et al.*, 1988). This is perhaps to be expected; however, the matter needs clarification. Why does the penetration peg stop growing - i.e. which papilla component is significant? Use of specific enzyme inhibitors has contributed information to this question.

Callose is a major papilla constituent and it has been suggested to be necessary for papilla efficacy, particularly in *ml-o* resistant plants (Skou *et al.*, 1984; Bayles *et al.*, 1990). Callose is synthesized from UDP-glucose by the plasma membrane bound callose synthase. In vitro callose synthase activity, measured in the microsomal fraction of barley leaf epidermal tissue, has been shown to be independent of the *ml-o* resistance gene as well as of inoculation (Pedersen, 1992), i.e. the enzyme is constitutive and its activity does not depend on resistance genotype or inoculation. The constitutive expression suggests that a regulatory mechanism for this enzyme exists. Plant callose synthase is inactive when the concentration of free Ca^{2+} in the cytoplasm remains around 0.1 µM, while half-max. activity occurs when this concentrations is around 0.8 µM (Kauss, 1992). Indeed, there is also evidence which suggests that papilla formation in *ml-o* resistant plants is Ca^{2+}-dependent (Gold *et al.*, 1986), and that inhibition of callose formation using a substrate analogue (2-deoxy-D-glucose) increases the penetration rate of *ml-o* resistant and, but to a lesser extent, in susceptible plants (Bayles *et al.*, 1990).

Phenylalanine ammonia lyase (PAL) activity increases following inoculation with *Bgh* (see below) contributing, presumably, precursors for the synthesis of phenolics. The significance of these phenolic compounds has been studied by Carver and co-workers in barley, wheat and oat. Generally, the use of specific inhibitors towards PAL and cinnamyl alcohol dehydrogenase (CAD) led to a reduction in papilla autofluorescence and at the same time increased the penetration rate (Carver *et al.*, 1992, 1994, 1996). CAD is a key enzyme for production of lignin precursors, so these data suggest that lignin or lignin-like compounds are present and important in barley papilla in spite of their inability to be detected histochemically (see above).

It appears that different components of papillae contribute to the efficacy of this cell wall apposition in preventing penetration and successful colonization.

B.   HYPERSENSITIVE RESPONSE

The HR is the defense mechanism which arrests the powdery mildew fungus in barley plants where a race-specific resistance gene is present in the host and the pathogen isolate has the corresponding avirulence gene. The HR is a back-up defense mechanism which can terminate those powdery mildew germlings which have penetrated the papillae. In certain avirulence gene/resistance gene combinations, single cell HR will be triggered rapidly after penetration, i.e. 15-24 hours after inoculation (e.g. Wright and Heale, 1988; Koga *et al.*, 1988). In other combinations, the first haustorium formed does not necessarily trigger the HR. The HR can occur in response to secondary or later penetrations. Limited fungal growth can be seen, and therefore several to many epidermal cells will undergo a HR, perhaps preceded by a HR in the subjacent mesophyll cells (e.g. Kita *et al.*, 1981; Tosa and Shishiyama, 1984). This latter multi-cell HR is often termed "necrotic lesion", as it is visible to the naked eye. "Single cell" and "multi-cell" HR describes extremes in the spectrum of resistances conferred by race-specific genes. See also Boyd *et al.* (1995). On the scale of infection types used for macroscopic evaluation of interaction phenotypes, they are rated as "0" and "2", respectively. A colonized, fully susceptible plant is rated "4".

Trypan blue staining of the HR cell can help to visualize early changes in cells undergoing the HR in this interaction. Trypan blue staining is observed from 14 hours after inoculation in single cells undergoing HR, and precedes negative vital staining, using neutral red, as well as autofluorescense (Koga *et al.*, 1988; Wright and Heale, 1988). Negative vital staining and autofluorescense were observed from 15-16 hours after inoculation. Uptake of trypan blue and urea in cells suggests membrane damage to be an early HR-related event. Whole-cell fluorescence indicates accumulation of phenolic compounds in the HR cell. Early in the development of the HR, cytoplasmic streaming halts. This is followed by cellular collapse, the final outcome of the HR, at 18-26 hours after inoculation (Bushnell, 1981; Lee-Stadelmann *et al.*, 1992). In addition to fluorescent compounds, callose, guanidine, silicon, protein and hydrogen peroxide accumulate in the HR cells (Koga *et al.*, 1988, Wei *et al.*, 1994; Thordal-Christensen *et al.*, 1997). In common with other interactions, the development of the HR in barley can be inhibited by treatment with the transcription inhibitor, cordycepin (Bushnell and Liu, 1994). Haustorial development is dramatically hampered in epidermal cells undergoing the HR. In single cell HR, they generally only develop the central body, and their ultrastructure is markedly altered, compared to haustoria in compatible interactions, and a process of haustorial degeneration occurs (Koga *et al.*, 1990; Hippe-Sanwald *et al.*, 1992).

Being dependent on living host cells, we can anticipate that the HR cell death alone can kill this pathogen, possibly unlike the situation for many necrotrophic microorganisms.

It is therefore specifically relevant in this pathosystem to know more about this cell death process and its regulation, which involves race-specific resistance genes and corresponding avirulence genes. The activation of the HR cell death pathway apparently requires presence of the haustorium inside the epidermal cell (Bushnell, 1985). Two possible scenarios can account for this: that the avirulence genes themselves are first expressed at this developmental stage, or that the avirulence gene products (i.e. specific elicitors) first come in contact with the sensory apparatus (resistance gene product *sensu lato*) after the fungus has successfully penetrated the papilla. Both may apply, but the resolution of these hypotheses awaits the characterization of the first avirulence genes from this fungus.

C.   EXPRESSION OF SPECIFIC HOST RESPONSE PROTEINS

A large number of proteins has been demonstrated to accumulate in barley leaves attacked by *Bgh*, either immunologically, by increase in enzyme activity, or indirectly in gene transcript analyses (see Collinge *et al.* 1997). These include the pathogenesis-related (PR) proteins: PR-1's (Muradov *et al.*, 1993; Bryngelsson *et al.*, 1994), β-1,3-glucanases (Jutidamrongpham *et al.*, 1991), chitinases (Kragh *et al.*, 1990, 1993), PR-4's (Gregersen *et al.*, 1997) and PR-5's (Bryngelsson and Gréen, 1989; Gregersen *et al.*, 1997). Members of the antimicrobial thionins (Bohlmann *et al.*, 1988) and lipid transfer proteins (Molina and Garcia-Olmedo, 1993) have also been found to accumulate in these leaves. So too have enzymes involved in secondary metabolism, such as PAL (Shiraishi *et al.*, 1989; Clark *et al.*, 1994; Boyd *et al.*, 1994a, 1994b, 1995), chalcone synthase (CHS) (Gregersen *et al.*, 1997; Christensen *et al.*, 1998b), flavonoid and caffeic acid *O*-methyltransferases (OMT) (Gregersen *et al.*, 1994; Christensen *et al.*, 1998a) and peroxidases (Kerby and Somerville, 1989, 1992; Thordal-Christensen *et al.*, 1992; Scott-Craig *et al.*, 1995). Other proteins potentially involved in regulation of the defense response also accumulate, e.g. the 14-3-3 protein (Brandt *et al.*, 1992), oxalate oxidases (Zhang *et al.*, 1995; Dumas *et al.*, 1995; Zhou *et al.*, 1998) and an oxalate oxidase-like protein (Wei *et al.*, 1998). Oxalate oxidases, and possibly also the oxalate oxidase-like protein, are generators of H_2O_2. H_2O_2 is a potential activator of gene expression and HR (see Mehdy, 1994). The accumulation of a chaperone located in the endoplasmic reticulum, the GRP94, is suggested to be related to the dramatic increase in protein export during the manifestation of defense response (Walther-Larsen *et al.*, 1993).

Studies of the gene transcripts for these proteins in barley leaves being attacked by *Bgh* have revealed that each gene transcript has its own unique accumulation pattern. However, many common characteristics exist among the accumulation patterns. For almost all, there are peaks at the time of papilla formation in response to the PGT and the appressorium, i.e. 4-8 hours and 12-24 hours, respectively (Davidson *et al.*, 1988; Brandt *et al.*, 1992; Thordal-Christensen *et al.*; 1992; Walther-Larsen *et al.*, 1993; Clark *et al.*, 1993, 1994, 1995; Boyd *et al.*, 1994a, 1994b, 1995; Scott-Craig *et al.*, 1995; Gregersen *et al.*, 1997; Zhou *et al.*, 1998). The 12-24-hour peak is possibly, in part, HR-correlated, but it is very significant in susceptible plants also, which suggests the relation of this peak to papilla formation subjacent to the appressorium. By comparing the expression in susceptible and HR-resistant leaves, it has been documented for a number of the gene transcripts that they also have HR-correlated expression. This is so for the gene transcripts encoding PR-1, β-1,3-glucanases, chitinase, PR-5, pe oxidase, PAL and thionin (Davidson et al., 1988; Clark et al., 1993; Boyd et al., 1994a, 1994b, 1995). These HR-correlated expressions appear 1 to 3 days after inoculation followed by a decline. The timing of this expression varies somewhat among laboratories, but seems related to the type of the HR (single or multi-cell). Similar HR-correlated gene transcript expression has been suggested for a number of the remaining response proteins mentioned above (Gregersen et al., 1997). In addition, much of the work demonstrates that these gene transcripts, which show HR-correlated expression, are also expressed during early colony development on susceptible leaves, i.e. 3 to 5 days after inoculation. In summary, most of the described gene transcripts encoding defense response protein show three distinct waves in their expression, one in response to the PGT, one in response to the appressorium and one, which is either HR-correlated in the resistant plants or appearing in response to colony formation in susceptible plants. Certain of the gene transcripts encoding defense response protein are clear exceptions from this three-wave pattern. A gene transcript has been identified which is exclusively papilla-correlated, i.e. a gene transcript encoding an oxalate oxidase-like protein (Gregersen et al., 1997; Wei et al., 1998) while 14-3-3 gene transcripts show a similar characteristic (Brandt et al., 1992). In contrast, four gene transcripts are expressed either at the time of HR in resistant leaves or at the time of colony development in susceptible leaves, i.e. the gene transcripts encoding flavonoid OMT, CHS, PR-1 and basic PR-5 (Gregersen et al., 1997).

In relation to the powdery mildew fungus, which is restricted to the leaf surface and the epidermal tissue, it is relevant to examine the location of the individual defense response components. In a study of separated epidermal and mesophyll tissues, a surprisingly low gene transcript level was observed in the epidermis for a number of the gene transcripts mentioned above (Gregersen et al., 1997). This was primarily the case for gene transcripts encoding PR-protein. These data suggest that many accumulating gene transcripts represent a general response, which is not necessarily important for the defense towards this particular pathogen. This view is supported by the fact that many of these are highly expressed in compatible interactions at the time of colony development.

V.　Conclusions

Much is now known about the interaction between barley and the powdery mildew fungus. This knowledge is largely descriptive: analyses of the timing of fungal development in relation to the timing of visible host responses, analyses of the genes in host and pathogen determining specificity, identification of gene transcripts accumulating during the defense response which encode antimicrobial proteins, and analyses of the chemical nature of components which accumulate in the key defense structures, the papillae and the HR cells. There is also an increasing body of evidence, mainly from the use of inhibitors of gene transcription, translation, specific enzymes and of cytoskeleton formation which confirm the importance of certain of these responses.

We know very little of the physiology of the fungus. We do not understand how the fungus persuades the host to allow it to establish the biotrophic relationship. We do not know how resistance works i.e. whether the possession (and timely activation) of a single component prevents the successful establishment of the parasite or whether resistance is a threshold event by which the combined action of a number of components is necessary to ensure resistance. Essentially nothing is known concerning the signal transduction pathways leading to activation of defense mechanisms. Nevertheless, the recent cloning of *Ml-o* (Büschges *et al.*, 1997) and identification of molecular markers linked to specific resistance genes, and other genes required for resistance, makes us expect substantial progress in this area within the next few years. An understanding of these biological processes will provide new inspiration for the development of new strategies for controlling the disease as well as giving an insight into a complex biological relationship.

It is expected that future studies using molecular techniques will focus on the identification of factors necessary for the establishment of different key phases of development of the fungus, for example, appressorium formation and initiation of the haustorium. Firstly, an understanding of these processes can be expected to lead to the identification of new targets for control of the disease. Secondly, remarkably little is understood of the biological processes underlying the establishment of a biotrophic relationship. Plant pathogens and symbionts (e.g. *Giberella* and *Rhizobium*) have taught us about plant growth and development (about the hormone giberellin and about *nod* factors as stimulants of organogenesis). As biotrophy depends on regulation and exploitation of the physiological processes taking place in the leaf, it is to be expected that, by understanding how the fungus regulates these processes to its own benefit, we will learn more about plants' own mechanisms for regulating their physiology.

Acknowledgements

We are grateful to our colleagues Drs. Fasong Zhou and Eigil de Neergaard for providing the electron micrograph, and to Professor Sigrun Hippe-Sanwald, Christian Albrechts University, Kiel, Germany, for discussion.

References

Aist JR and Bushnell WR (1991) Invasion of plants by powdery mildew fungi, and cellular mechanisms of the resistance. In: Cole GT and Hoch HC (eds) The Fungal Spore and Disease Initiation in Plants and Animals, pp 321—345. Plenum Press, New York

Aist JR and Israel HW (1977) Papilla formation: Timing and significance during penetration of barley coleoptiles by *Erysiphe graminis hordei*. Phytopathol 67: 455—461

Aist JR, Gold RE, Bayles CJ, Morrison GH, Chandra S and Israel HW (1988) Evidence that molecular components of papillae may be involved in *ml-o* resistance to barley powdery mildew. Physiol Mol Plant Pathol 33: 17—32

Akutsu K, Doi Y and Yora K (1980) Elementary analysis of papillae and cytoplasmic vesicles formed at the penetration site by *Erysiphe graminis* f.sp. *hordei* in epidermal cells of barley leaves. Ann Phytopathol Soc Japan 46: 667—671

Bayles CJ, Ghemawat MS and Aist JR (1990) Inhibition by 2-deoxy-D-glucose of callose formation, papilla deposition, and resistance to powdery mildew in an *ml-o* barley mutant. Physiol Mol Plant Pathol 36: 63—72

Bohlmann H, Clausen S, Behnke S, Giese H, Hiller C, Reimann-Philipp U, Schrader G, Barkholt V and Apel K (1988) Leaf-specific thionins of barley - a novel class of cell wall proteins toxic to plant-pathogic fingi and possibly involved in the defence mechanism of plants. EMBO J 7: 1559—1565

Boyd LA, Smith PH and Brown JKM (1994a) Molecular and cellular expression of quantitative resistance in barley to powdery mildew. Physiol Mol Plant Pathol 45: 47—58

Boyd LA, Smith PH, Green RM and Brown JKM (1994b) The relationship between the expression of defense-related genes and mildew development in barley. Mol Plant-Micr Interact 7: 401—410

Boyd LA, Smith PH, Foster EM and Brown JKM (1995) The effects of allelic variation at the *Mla* resistance locus in barley on the early development of *Erysiphe graminis* f.sp. *hordei* and host responses. Plant J 7: 959—968

Bracker CE (1968) Ultrastructure of the haustorial apparatus of *Erysiphe graminis* and its relationship to the epidermal cell of barley. Phytopathol 58: 12—30.

Brandt J, Thordal-Christensen H, Vad K, Gregersen PL and Collinge DB (1992) A pathogen-induced gene of barley encodes a protein showing high similarity to a protein kinase regulator. Plant J 2: 815—820

Bryngelsson T, Sommer-Knudsen J, Gregersen PL, Collinge DB, Ek B and Thordal-Christensen H (1994). Purification, characterization, and molecular cloning of basic PR-1-type pathogenesis-related proteins form barley. Mol Plant-Micr Interac 7: 267—275

Bryngelsson T and Gréen B (1989) Characterization of a pathogenesis-related protein, thaumatin-like protein isolated from barley challenged with an incompatible race of mildew. Physiol Mol Plant Pathol 35: 45—52

Büschges R, Hollricher K, Panstruga R, Simons G, Wolter M, Frijters A, van Daelen R, van der Lee T, Diergaarde P, Groenendijk J, Töpsch S, Vos P, Salamini F and Schulze-Lefert P (1997) The barley *Mlo* gene: A novel control element of plant pathogen resistance. Cell 88: 695—705

Bushnell WR (1981) Incompatibility conditioned by the *Mla* gene in powdery mildew of barley: The halt in cytoplasmic streaming. Phytopathol 71: 1062—1066

Bushnell WR (1985) Expression of race-specific incompatibility in powdery mildew: implications for mechanisms of gene action. In Sussex I, Ellingboe A, Crouch M and Malmberg R (eds) Plant Cell/Cell Interactions, pp 83—87. Cold Spring Harbor, New York

Bushnell WR and Gay J (1978) Accumulation of solutes in relation to the structure and function of haustoria in powdery mildews. In: Spencer DM (ed) The Powdery Mildews, pp 183—235. Academic Press, London

Bushnell WR and Liu Z (1994) Incompatibility conditioned by the *Mla* gene in powdery mildew of barley: timing of the effect of cordycepin on hypersensitive cell death. Physiol Molec Plant Pathol 44: 389—402

Bushnell WR and Zeyen RJ (1976) Light and electron microscope studies of cytoplasmic aggregates formed in barley cells in response to *Erysiphe graminis*. Can J Bot 54: 1647—1655

Carver TLW (1986) Histology of infection by *Erysiphe graminis* f.sp. *hordei* in spring barley lines with various levels of partial resistance. Plant Pathol 35: 232—240

Carver TLW and Bushnell WR (1983) The probable role of primary germ tubes in water uptake before infection by *Erysiphe graminis*. Physiol Plant Pathol 23: 229—240

Carver TLW and Ingerson SM (1987) Responses of *Erysiphe graminis* germlings to contact with artificial and host surfaces. Physiol Mol Plant Pathol 30: 359—372

Carver TLW and Ingerson-Morris SM (1989) Effects of inoculum density on germling development by *Erysiphe graminis* f.sp. *avenae* in relation to induced resistance of oat cells to appressorial penetration. Mycol Res 92: 18—24

Carver TLW, Zeyen RJ, Bushnell WR and Robbins MP (1994) Inhibition of phenylalanine ammonia lyase and cinnamyl alcohol dehrydrogenase increases quantitative susceptibility of barley to powdery mildew (*Erysiphe graminis* D.C.). Physiol Mol Plant Pathol 44: 261—272

Carver TLW, Zeyen RJ, Robbins MP and Dearne GA (1992) Effects of the PAL inhibitor, AOPP, on oat, barley and wheat cell responses to appropriate and inappropriate formae specials of *Erysiphe graminis* DC. Physiol Mol Plant Pathol 41: 397—409

Carver TLW, Zhang L, Zeyen RJ and Robbins MP (1996) Phenolic biosynthesis inhibitor suppress adult plant resistance to *Erysiphe graminis* in oat at 20°C and 10°C. Physiol Mol Plant Pathol 49: 121—141

Christensen AB, Gregersen PL, Olsen CE and Collinge DB (1998a) A flavonoid 7-O-methyltransferase is expressed in barley leaves in response to pathogen attack. Plant Mol Biol 36: 219—227

Christensen AB, Gregersen PL, Schröder J and Collinge DB (1998b) A chalcone synthase with unusual substrate preferences is expressed in barley leaves in response to UV light and pathogen attack. Plant Mol Biol 37: 849—857

Cho BH and Smedegaard-Petersen V (1986) Induction of resistance to *Erysiphe graminis* f.sp. *hordei* in near-isogenic barley lines. Phytopathol 76: 301—305

Christiansen SK and Giese H (1990) Genetic analysis of the obligate parastic barley powdery mildew fungus based on RFLP and virulence loci. Theor Appl Gen 79: 705—712

Clark TA, Zeyen RJ, Carver TLW, Smith AG and Bushnell WR (1995) Epidermal cell cytoplasmic events and response gene transcript accumulation during *Erysiphe graminis* attack in isogenic barley lines differing at the *Ml-o* locus. Physiol Mol Plant Pathol 46: 1—16

Clark TA, Zeyen RJ, Smith AG, Bushnell WR, Szabo LJ and Vance CP (1993) Host response gene transcript accumulation in relation to visible cytological events during *Erysiphe graminis* attack in isogenic barley lines differing at the *Ml-a* locus. Physiol Mol Plant Pathol 43: 283—298

Clark TA, Zeyen RJ, Smith AG, Carver TLW and Vance CP (1994) Phenylalanine ammonia lyase mRNA accumulation, enzyme activity and cytoplasmic responses in barley isolines, differing at *Ml-a* and *Ml-o* loci, attacked by *Erysiphe graminis* f.sp. *hordei*. Physiol Mol Plant Pathol 44: 171—185

Collinge DB, Bryngelsson T, Gregersen PL, Smedegaard-Petersen V and Thordal-Christensen H (1997) Resistance against fungal pathogens: its nature and regulation. In: Basra AS and Basra R (eds) Mechanisms of Environmental Stress Resistance in Plants, pp 335—372. Harwood Academic Publishers, Chur, Switzerland

Davidson AD, Manners JM, Simpson RS and Scott KJ (1988) Altered host gene expression in near-isogenic barley conditioned by different genes for resistance during infection by *Erysiphe graminis* f.sp. *hordei*. Physiol Mol Plant Path 32: 127—139

DeScenzo RA, Wise RP and Mahadevappa M (1994) High-resolution mapping of the *Hor1/Mla/Hor2* region on chromosome 5S in barley. Mol Plant-Micr Interact 7: 657—666

Dumas B, Freyssinet G and Pallett KE (1995) Tissue-specific expression of germin-like oxalate oxidase during development and fungal infection of barley seedlings. Plant Physiol 107: 1091—1096

Ebrahim-Nesbat F, Bohl S, Heitefuss R and Apel K (1993) Thionin in cell walls and papillae of barley in compatible and incompatible interactions with *Erysiphe graminis* f.sp. *hordei*. Physiol Mol Plant Pathol 43: 343—352

Edwards HH and Allen PJ (1970) A fine-structure of the primary infection process during infection of barley by *Erysiphe graminis* f.sp. *hordei*. Phytopathol 60: 1504—1509

Francis SA, Dewey FM and Gurr SJ (1996) The role of cutinase in germling development and infection by *Erysiphe graminis* f.sp. *hordei*. Physiol Mol Plant Pathol 49: 201—211

Freialdenhoven A, Scherag B, Hollricher K, Collinge DB, Thordal-Christensen H and Schulze-Lefert P (1994) *Nar-1* and *Nar-2*, two loci required for Mla_{12}-specified race-specific resistance to powdery mildew in barley. Plant Cell 6: 983—994

Freialdenhoven A, Peterhänsel C, Kurth J, Kreuzaler F and Schulze-Lefert P (1996) Identification of genes required for the function of non-race-specific *mlo* resistance to powdery mildew in barley. Plant Cell 8: 5—14

Gay JL, Salzberg A and Woods AM (1987) Dynamic experimental evidence for the plasma membrane ATPase domain hypothesis of haustorial transport and for ionic coupling of the haustorium of Erysiphe graminis to the host cell (*Hordeum vulgare*). New Phytol 107: 541—548

Giese H, Holm-Jensen AG, Jensen HP and Jensen J (1993) Localization of the Laevigatum powdery mildew resistance gene to barley chromosome 2 by use of RFLP markers. Theor Appl Genet 85: 897—900

Gregersen PL, Christensen AB, Sommer-Knudsen J and Collinge DB (1994) A putative *O*-methyltransferase from barley is induced by fungal pathogens and UV light. Plant Mol Biol 26: 1797—1806

Gregersen PL, Thordal-Christensen H, Förster H and Collinge DB (1997) Differential gene transcript accumulation in barley leaf epidermis and mesophyll in response to attack by *Blumeria graminis* f.sp. *hordei*. Physiol Mol Plant Pathol 51: 85—97

Gold RE, Aist JR, Hazen BE, Stolzenburg, Marshall MR and Israel HW (1986) Effect of calcium nitrate and chlortetracycline on papilla formation, *ml-o* resistance and susceptibility of barley to powdery mildew. Physiol Mol Plant Pathol 29: 115—129

Görg R, Hollricher K and Schulze-Lefert P (1993) Functional analysis and RFLP mapping of the *Mlg* resistance gene in barley. Plant J 3: 857—866

Hall JL, Aked J, Gregory AJ and Storr T (1992) Carbon metabolism and transport in a biotrophic fungal association. In: Pollock CJ, Farrar JF and Gordon AJ (eds) Carbon Partitioning Within and Between Organisms, pp 181—198. Bios Scientific Publishers, Oxford

Heitefuss R and F Ebrahim-Nesbat (1986) Ultrastructural and histochemical studies on mildew of barley (*Erysiphe graminis* DC. f.sp. *hordei* Marchal). III. Ultrastructure of different types of papillae in susceptible and adult plant resistant leaves. J Phytopathol 116: 358—373

Hippe-Sanwald S, Hermanns M and Somerville SC (1992) Ultrastructural comparison of incompatible and compatible interactions in the barley powdery mildew disease. Protoplasma 168: 27—40

Hilbers S, Fischbeck G and Jahoor A (1992) Localization of the *lavigatum* resistance gene *Ml-La* against powdery mildew in the barley genome by the use of RFLP markers. Plant Breed 109: 335—338

Hiura U (1978) Genetic basis of formae speciales in *Erysiphe graminis*. In: Spencer DM (ed) The Powdery Mildews, pp 101—128. Academic Press, London

Jahoor A, Jacobi A, Schüller CME and Fischbeck G (1993) Genetical and RFLP studies at the *Mla* locus conferring powdery mildew resistance in barley. Theor Appl Genet 85: 713—718

Johnson LEB, Bushnell WR and Zeyen RJ (1979) Binary pahways for analysis of primary infection and host response in populations of powdery mildew fungi. Can J Bot 57: 497—511

Jørgensen JH (1988) *Erysiphe graminis*, powdery mildew of cereals and grasses. Adv Plant Pathol 6: 137—157

Jørgensen JH (1994) Genetics of powdery mildew resistance in barley. Crit Rev Plant Sci 13: 97—119

Jørgensen JH and Mortensen K (1977) Primary infection by *Erysiphe graminis* f.sp. *hordei* of barley mutants with resistance genes in the *ml-o* locus. Phytopathol 67: 678—685

Jørgensen JH, Munk L and Kølster P (1988) Svampesygdom i byg koster samfundet millioner hvert år. Forskning og Samfundet 2: 10—12

Justesen A, Somerville SC, Christiansen SK and Giese H (1995) Isolation and characterization of two novel genes expressed in germinating conidia of the obligate biotroph *Erysiphe graminis* f.sp. *hordei*. Gene 170: 131—135

Jutidamrongpham W, Andersen JB, Mackinnon G, Manners JM, Simpson RS and Scott KJ (1991) Induction of β-1,3-glucanase in barley in response to infection by fungal pathogens. Mol Plant-Microb Interact 4: 234—238

Kauss H (1992) Callose and callose synthase. In: Gurr SJ, McPherson MJ and Bowles DJ (eds) Molecular Plant Pathology: A Practical Approach, Vol II, pp 1—8. IRL Press, Oxford

Kerby K and Somerville S (1989) Enhancement of specific intercellular peroxidases following inoculation of barley with *Erysiphe graminis* f.sp. *hordei*. Physiol Mol Plant Pathol 35: 323—337

Kerby K and Somerville SC (1992) Purification of an infection-related, extracellular peroxidase from barley. Plant Physiol 100: 397—402

Kita N, Toyoda H and Shishiyama J (1980) Histochemical reactions of papilla and cytoplasmic aggregate in epidermal cells of barley leaves infected by *Erysiphe graminis hordei*. Ann Phytopathol Soc Japan 46: 263—265

Kita N, Toyoda H and Shishiyama J (1981) Chronological analysis of cytological responses in powdery mildewed barley leaves. Can J Bot 59: 1761—1768

Kobayashi Y, Kobayashi I, Funaki Y, Fujimoto S, Takemoto T and Kunoh H (1997) Dynamic reorganization of microfilaments and microtubules is necessary for the expression of non-host resistance in barley coleoptile cells. Plant J 11: 525—537.

Kobayashi I, Watababe H and Kunoh H (1995) Induced accessibility and enhanced inaccessibility at the cellular level in barley coleoptiles. XIV. Evidence for elicitor(s) and suppressor(s) of host inaccessibility from *Erysiphe graminis*. Physiol Mol Plant Pathol 46: 445— 456

Koch G and Köhler W (1990) Isozyme variation and genetic distances of *Erysiphe graminis* DC formae speciales. J Phytopathol 89: 89—101

Koch G and Köhler W (1991) Isozyme variation and genetic diversity of the European barley powdery mildew population. J Phytopathol 131: 333—344

Koga H, Bushnell WR and Zeyen RJ (1990) Specificity of cell type and timing of events associated with papilla formation and the hypersensitive reaction in leaves of *Hordeum vulgare* attacked by *Erysiphe graminis* f.sp. *hordei*. Can J Bot 68: 2344—2352

Koga H, Zeyen RJ, Bushnell WR and Ahlstrand GG (1988) Hypersensitive cell death, autofluorescence, and insoluble silicon accumulation in barley leaf epidermal cells under attack by *Erysiphe graminis* f.sp. *hordei*. Physiol Mol Plant Pathol 32: 395—409

Kragh KM, Jacobsen S and Mikkelsen JD (1990) Induction, purification and characterization of barley leaf chitinase. Plant Sci 71: 55—68

Kragh KM, Jacobsen S, Mikkelsen JD and Nielsen KA (1993) Tissue specificity and induction of class I, II and III chitinase in barley (*Hordeum vulgare*). Physiol Plant 89: 490—498

Kunoh H (1995) Host-parasite specificity in powdery mildews. In: Kohmoto K, Singh US and Singh RP (eds) Pathogenesis and Host Specificity in Plant Diseases: Histopathology, Biochemistry, Genetic and Molecular Bases, Vol II: Eukaryotes, pp 239—250. Elsevier, Oxford

Kunoh H, Itoh O, Kohno M and Ishizaki H (1979) Are primary germ tubes of conidia unique to *Erysiphe graminis*? Ann Phytopathol Soc Japan 45: 675—682

Kunoh H and Ishizaki H (1976) Accumulation of chemical elements around the penetration sites of *Erysiphe graminis hordei* on barley leaf epidermis: (III) micromanipulation and X-ray microanalysis of silicon. Physiol Plant Pathol 8: 91—96

Kunoh H and Ishizaki H (1981) Cytological studies of early stages of powdery mildew in barley and wheat. VII. Reciprocal translocation of a fluorescent dye between barley coleoptile cells and conidia. Physiol Plant Patho 18: 207—211

Kunoh H, Kuroda K, Hayashimoto A and Ishizaki H (1986) Induced accessibility and enhanced resistance at the cellular level in barley coleoptiles. II. Timing and localization of induced susceptibility in a single coleoptile cell andits transfer to an adjacent cell. Can J Bot 64: 889—895

Kunoh H, Komura T, Yamaoka N and Kobayashi I (1988a) Induced accessibility and enhanced inaccessibility at the cellular level in barley coleoptiles. IV. Escape of the second lobe of the Erysiphe graminis appressorium from inaccessibility enhanced by the previous attack of the first lobe. Ann Phytopath Soc Japan 54: 577—583

Kunoh H, Nicholson RL, Yoshioka H, Yamaoka N and Kobayashi I (1990) Preparation of the infection court by Erysiphe graminis: Degradation of the host cuticle. Physiol Molec Plant Pathol 36: 397—407.

Kunoh H, Takamatsu S and Ishizaki H (1978a) Cytological studies of early stages of powdery mildew in barley and wheat. III. Distribution of residual calcium and silicon in germinated conidia of Erysiphe graminis hordei. Physiol Plant Pathol 13: 319—325

Kunoh H, Tsuzuki T and Ishizaki H (1978b) Cytological studies of early stages of powdery mildew in barley and wheat. IV. Direct ingress from superficial primary germ tubes and appressorial of Erysiphe graminis hordei on leaves of barley. Physiol Plant Pathol 13: 327—333

Kunoh H, Yamaoka N, Yoshioka H and Nicholson RL (1988b) Preparation of the infection court by Erysiphe graminis I. Contact-mediated changes in morphology of the conidium surface. Exp Mycol 12: 352—335.

Kølster P, Munk L, Stølen O and Løhde J (1986) Near-isogenic barley lines with genes for resistance to powdery mildew. Crop Sci 26: 903—907

Lee-Stadelmann OY, Curran CM and Bushnell WR (1992) Incompatibility conditioned by the Mla gene in powdery mildew of barley: change in permeability to non-electrolytes. Physiol Mol Plant Pathol 41: 165—177

Lyngkjær MF, Jensen HP and Østergård H (1995) A Japanese powdery mildew isolate with exceptionally large infection efficiency on Mlo-resistant barley. Plant Pathol 44: 786—790

Mahadevappa M, DeScenzo RA and RP Wise (1994) Recombination of alleles conferring specific resistance to powdery mildew at the Mla locus in barley. Genome 37: 460—468

Manners JM and Gay JL (1977) The morphology of haustorial complexes isolated from apple, barley, beet and vine infected with powdery mildews. Physiol Plant Pathol 11: 261—266

Manners JM and Gay JL (1983) The host-parasite interface and nutrient transfer in biotrophic parasitism. In: Callow JA (ed) Biochemical Plant Pathology, pp 163—195. John Wiley & Sons, Chichester

Masri SS and Ellingboe AH (1966) Primary infection of wheat and barley by Erysiphe graminis. Phytopathol 56: 253—378

McKeen WE and Rimmer SR (1973) Initial penetration process in powdery mildew infection of susceptible barley leaves. Phytopathol 63: 1049—1053

Mehdy MC (1994) Active oxygen species in plant defence against pathogens. Plant Physiol 105: 467—472

Mendgen K and Nass P (1988) The activity of powdery-mildew haustoria after feeding the host cells with different sugars, as measured with a potentiometric cyanine dye. Planta 174: 283—288

Molina A and García-Olmedo F (1993) Developmental and pathogen-induced expression of three barley genes encoding lipid transfer proteins. Plant J 4: 983—991

Mount MS and Ellingboe AH (1969) ^{32}P and ^{35}S transfer from susceptible wheat to Erysiphe graminis f.sp. tritici during primary infection. Phytopathol 59: 235

Muradov A, Petrasovits L, Davidson A and Scott KJ (1993) A cDNA clone for a pathogenesis-related protein

1 from barley. Plant Mol Biol 23: 439—442

Nicholson RL, Yoshioka H, Yamaoka N and Kunoh H (1988) Preparation of the infection court by *Erysiphe graminis* II. Release of esterase enzyme from conidia in response to a contact stimulus. Exp Mycol 12: 336—349

Ouchi S, Oku H, Hibino C and Akiyama I (1974) I. Induction of accessibility and enhanced resistance in leaves of barley by some races of *Erysiphe graminis*. Phytopathol Z 79: 24—34

Pascholati SF, Yoshioka H, Kunoh H and Nicholson RL (1992) Preparation of the infection court by *Erysiphe graminis* f.sp. *hordei*: cutinase is a component of the conidial exudate. Physiol Molec Plant Pathol 41: 53—59

Pauvert P and de la Tullaye B (1977) Etude des conditions de l'orge par l'oidium et la période de latence chez *Erysiphe graminis* f.sp. *hordei*. Ann Phytopathol 9: 495—501

Pedersen LH (1992) Callose Synthesis in Barley Leaves. PhD thesis. Plant Biology Section, Risø National Laboratory, Denmark and Dept. Biochem. Nutrition, Danish Technical University.

Rasmussen M, Rossen L, Giese H (1993) SINE-like properties of a highly repetitive element in the genome of the obligate parasitic fungus *Erysiphe graminis* f.sp. *hordei*. Mol Gen Genet 239: 298—303

Russo VM and Bushnell WR (1989) Responses of barley cells to puncture by microneedles and to attempted penetration by *Erysiphe graminis* f.sp. *hordei*. Can J Bot 67: 2912—2921

Schwarzbach E (1979) Response to selection for virulence against the *ml-o* based mildew resistance in barley, not fitting the gene-for-gene hypothesis. Barley Genet Newsl 9: 85—88

Scott-Craig JS, Kerby KB, Stein BD and Somerville SC (1995) Expression of an extracellular peroxidase that is induced in barley (*Hordeum vulgare*) by the powdery mildew pathogen (*Erysiphe graminis* f.sp. *hordei*). Physiol Mol Plant Pathol 47: 407—418

Sherwood JE and Somerville SC (1990) Sequence of the *Erysiphe graminis* f.sp. *hordei* gene encoding β-tubulin. Nucl Acids Res 18: 1052

Shiraishi T, Yamaoka N and Kunoh H (1989) Association between increased phenylalanine ammonia lyase activity and cinnamic acid synthesis and the induction of temporary inaccessibility caused by *Erysiphe graminis* primary germ tube penetration of the barley leaf. Physiol Mol Plant Pathol 34: 75—83

Skou JP, Jørgensen JH and Lilholt U (1984) Comparative studies of callose formation in powdery mildew compatible and incompatible barley. Phytopathol Z 109: 147—168

Smart MG, Aist JR and Israel HW (1986) Structure and function of wall appositions. 1. General histochemistry of paillae in barley coleoptiles attacked by *Erysiphe graminis* f.sp. *hordei*. Can J Bot 64: 793—801

Smith SE and Smith FA (1990) Structure and function of the interface in biotrophic symbioses as they relate to nutrient transport. New Phytol 114: 1—38

Speer EO (1973) Untersuchungen zur Morphologie und Systematik der Erysiphacen. I. Die Gattung Blumeria Golovin und ihre Typusart Erysiphe graminis DC. Sydowia 27:1—6

Spencer-Phillips PTN and JL Gay (1981) Domains of ATPase in plasma membranes and transport through infected plant cells. New Phytol 89: 393—400

Stanbridge B, Gay JL and Wood RKS (1971) Gross and fine structural changes in *Erysiphe graminis* and barley before and during infection. In: Preece TF and Dickinson CH (eds) Ecology of Leaf Surface Micro-Organisms, pp 367—379. Academic Press, London

Staub T, Dahmen H and Schwinn FJ (1974) Light- and scanning electron microscopy of cucumber and barley powdery mildew on host and nonhost plants. Phytopathol 64: 364—372

Takahashi K, Aist JR and Israel HW (1985) Distribution of hydrolytic enzymes at barley powdery mildew encounter sites: implications for resistance associated with papilla formation in a compatible system. Physiol Plant Pathol 27: 167—184

Thordal-Christensen H, Brandt J, Cho BH, Gregersen PL, Rasmussen SK, Smedegaard-Petersen V and Collinge DB (1992) cDNA cloning and characterization of two barley peroxidase transcripts induced differentially by the powdery mildew fungus *Erysiphe graminis*. Physiol Mol Plant Pathol 40: 395—409

Thordal-Christensen H and Smedegaard-Petersen V (1988a) Comparison of resistance-inducing abilities of virulent and avirulent races of *Erysiphe graminis* f.sp. *hordei* and a race of *Erysiphe graminis* f.sp. *tritici*. Plant Pathol 37: 20—27

Thordal-Christensen H and Smedegaard-Petersen V (1988b) Correlation between induced resistance and host fluorescence in barley inoculated with *Erysiphe graminis*. J Phytopathol 123: 34—46

Thordal-Christensen H, Zhang Z, Wei YD and Collinge DB (1997) Subcellular localization of H_2O_2 in plants. H_2O_2 accumulation in papillae and hypersensitive response during the barley- powdery mildew interaction. Plant J 11: 1187—1194

Torp J and Jørgensen JH (1986) Modification of barley powdery mildew resistance gene *Ml-a12* by induced mutation. Can J Genet Cytol 28: 725—731

Tosa Y and Shishiyama J (1984) Cytological aspects of events occuring after the formation of primary haustoria in barley leaves infected with powdery mildew. Can J Bot 62: 795—798

Walther-Larsen H, Brandt J, Collinge DB and Thordal-Christensen H (1993) A pathogen-induced gene of barley encodes a HSP90 homologue showing striking similarity to vertebrate forms resident in the endoplasmic reticulum. Plant Mol Biol 21: 1097—1108

Wei YD, de Neergaard E, Thordal-Christensen H, Collinge DB and Smedegaard-Petersen V (1994) Accumulation of a putative guanidine compound in relation to other early defence reactions in epidermal cells of barley and wheat exhibiting resistance to *Erysiphe graminis* f.sp. *hordei*. Physiol Mol Plant Pathol 45: 469—484

Wei YD, Collinge DB, Smedegaard-Petersen V and Thordal-Christensen H (1996) Characterization of the transcript of a new class of retroposon-type repetitive element cloned from the powdery mildew fungus, *Erysiphe graminis*. Mol Gen Genet 250: 477—482

Wei YD, Zhang Z, Andersen CH, Schmelzer E, Gregersen PL, Collinge DB, Smedegaard-Petersen V and Thordal-Christensen H (1998) An epidermis/papilla-specific oxalate oxidase-like protein in the defence response of barley attacked by the powdery mildew fungus. Plant Mol Biol 36: 101—112

Wolfe MS and McDermott JM (1994) Population genetics of plant pathogen interactions: the example of the *Erysiphe graminis - Hordeum vulgare* pathosystem. Ann Rev Phytopathol 32: 89— 113

Woolacott B and Archer SA (1984) The influence of the primary germ tube on infection of barley by *Erysiphe graminis* f.sp. *hordei*. Plant Pathol 33: 225—231

Wright AJ and Heale JB (1988) Host responses to fungal penetration in *Erysiphe graminis* f.sp. *hordei* infections in barley. Plant Pathol 37: 131—140

Yarwood CE (1978) History and taxonomy of powdery mildew. In: Spencer DM (ed) The Powdery Mildews, pp 1—37. Academic Press, London

Zeyen RJ, Carver TLW and Ahlstrand GG (1983) Relating cytoplasmic detail of powdery mildew infection to presence of insoluble silicon by sequential use of light microscopy, SEM, and X-ray microanalysis. Physiol Plant Pathol 22: 101—108

Zhang Z, Collinge DB and Thordal-Christensen H (1995) Germin-like oxalate oxidase, a H_2O_2- producing enzyme, accumulates in barley attacked by the powdery mildew fungus. Plant J 8: 139—145

Zhou F, Zhang, Z, Gregersen PL, Mikkelsen JD, de Neergaard E, Collinge DB and Thordal-Christensen (1998) Molecular characterization of the oxalate oxidase involved in the response of barley to the powdery mildew fungus. Plant Physiol 117: 33—41

CHAPTER 2

GENETICS OF DISEASE RESISTANCE

BASIC CONCEPTS AND APPLICATION IN RESISTANCE BREEDING

BEAT KELLER[1], CATHERINE FEUILLET[1] & MONIKA MESSMER[2]

[1] *Dept. of Plant Molecular Biology, Institute of Plant Biology*
University of Zürich, CH-8008 Zürich, Switzerland
(bkeller@botinst.unizh.ch)
[2] *Dept. of Quality and Resistance Breeding, Swiss Federal Research*
Station for Agroecology and Agriculture, Reckenholzstrasse 191
CH-8046 Zürich, Switzerland

Summary

Disease resistant plants are one of the prerequisites for sustainable agriculture. To understand and rationally use the naturally occurring disease resistance, its genetic basis has been investigated in great detail. These studies showed that there are two different genetic mechanisms for disease resistance: monogenic resistance is based on single genes whereas quantitative resistance depends on two or more genes. In most cases, single resistance genes confer complete resistance but are only active against certain races of the pathogen, i.e. they show a genetic interaction with genes from the pathogen. This resistance is based on an active recognition event between the product of the plant resistance gene and the product of the avirulence gene of the pathogen. Resistance genes are clustered at some loci in the genome or exist as different alleles conferring resistance towards specific pathogen races. Quantitative resistance shows no obvious genetic interaction with the pathogen and slows down the disease development by increasing latency period and other parameters related to the epidemic. Resistance breeding in crop plants depends on both types of resistance. Monogenic resistances are easy to work with but are frequently not durable. Consequently, quantitative resistance is preferred. The application of molecular markers has allowed the genetics of quantitative resistance to be determined and quantitative trait loci involved in resistance to be identified. Molecular markers have also contributed to improved breeding strategies for monogenic resistance genes in order to combine them in the "gene pyramiding" strategy for a more durable resistance. Finally, molecular markers have allowed the isolation of the first disease resistance genes. The cloning of such genes from crop plants and their wild relatives will open new possibilities for their sustainable use in breeding.

A. Slusarenko, R.S.S. Fraser, and L.C. van Loon (eds), Mechanisms of Resistance to Plant Diseases, 101-160.
© 2000 *Kluwer Academic Publishers. Printed in the Netherlands.*

I.   Introduction

The classical and molecular genetics of disease resistance in plants is one of the intellectually most challenging and practically important research topics in plant biology. The application of naturally occurring resistance in crop breeding has contributed greatly to the control of plant diseases. Recently, there has been renewed and increasing interest in genetic resistance for several reasons: in many developing countries poor farmers do not have the financial resources and the education for a safe application of pesticides whereas natural resistance is a potentially cheap and efficient way to fight diseases. It has also become clear that the use of pesticides can cause considerable environmental damage. Consequently, agricultural policies in many developing and industrialized countries have the goal to reduce the overall use of pesticides.

Originally, the genetic analysis of resistance grew out of the need to understand the basis of field observations: some plant lines were resistant to a particular disease whereas others were susceptible. In addition, resistant crops could become susceptible, even after showing good resistance in the field for several years. The genetic characterization of disease resistance in plants has been essential for the understanding of plant-pathogen interactions. It has allowed the formulation of some of the key concepts in plant pathology, thereby creating the framework for rational strategies to control plant diseases. These concepts have greatly and very successfully contributed to an efficient breeding for disease resistance in many crop plants. It is estimated that at least 75% of all important agricultural crops have an effective inherited resistance against at least one pathogen and 98% of all grain and forage crops have an inherited resistance component against one or more diseases (Schumann, 1991).

In this chapter, the basic concepts of disease resistance which resulted from genetic analysis will be described. Methodological and technical advances for the study of the genetics of resistance and their consequences for practical breeding will be discussed. Finally, we will describe in detail our current understanding of the genetics of quantitative resistance which is based on the action of several genes.

II.   The Gene-For-Gene Hypothesis for the Description of Plant-Pathogen Interactions

It was discovered early in the century that in some cases disease resistance was inherited as a monogenic trait following the laws of classical Mendelian genetics. For example, Biffen (1905) demonstrated that the resistance against yellow (stripe) rust in the wheat variety "Rivet" was due to a single, recessive gene. Recessive resistance genes are actually rare and later the more frequent dominant or semidominant resistance genes were found. However, initially there was considerable confusion about the observations. Even with genetically stable (i.e. true breeding) wheat lines, the reaction to rust differed between different locations and years. Artificial infections with spores collected in the field resulted in inconsistent data. The explanation for this variability only became clear after the discovery and characterization of defined pathogen races. The differential reaction of a variety with a resistance gene towards particular races of the pathogen suggested that there was genetic variability in the pathogen which specifically interacted with monogenic resistances in the plant. Based on reaction to the pathogen, Ausemus *et al.* (1946) described three dominant, monogenic resistances against the wheat leaf rust pathogen (*Puccinia recondita* f. sp. *tritici*) and five dominant genes against stem rust caused by *Puccinia graminis* Pers. f. sp. *tritici*. Additional studies with other pathosystems such as barley or wheat powdery mildew, flax rust, potato late blight and lettuce downy mildew revealed that dominant monogenic resistance traits in plants were quite common. Such resistance genes were the basis for Flor's pioneering analysis of resistance.

H.H. Flor was the first plant pathologist who analyzed the genetics of a resistance interaction simultaneously in both the plant and the pathogen. He studied the interaction between flax (*Linum usitatissimum*), a crop plant mainly used for fiber production, and the fungal disease flax rust caused by the Basidiomycete *Melampsora lini*. From these studies he formulated the so called gene-for-gene hypothesis as the most convincing explanation of the observed phenomena (Flor, 1942; 1955; 1971). Flor made his observations after infecting flax lines carrying different resistance genes with the progeny of crosses between different races of the rust pathogen. He used a cross of two different races of flax rust to develop segregating F_2 cultures. These cultures were then tested for the ability to multiply and grow on more than 30 different varieties of flax that had previously been selected as carrying single genes for rust reaction (Flor, 1955). The conclusion from these studies was that genetic factors of both the plant and the pathogen are required for a successful defense reaction of the plant. The specificity of a plant-pathogen interaction is determined by the interaction of an avirulence gene product encoded by a dominant gene in the pathogen and a product of the resistance gene from the plant. The basis of the plant resistance reaction is therefore a specific recognition between the two components. This recognition triggers further physiological defense reactions resulting in hypersensitive cell death and the accumulation of molecules which are toxic for the pathogen (see chapter 7; Lamb, 1994). This is also called an incompatible interaction between the plant and the pathogen. In the absence of either the resistance gene product or the avirulence gene product, there is no recognition of the pathogen by the plant. This allows the further growth of the pathogen, resulting in a compatible interaction and susceptibility. Thus, a mutation in either the avirulence or the resistance gene which results in a loss of function will result in a change from an incompatible to a compatible interaction. The presence of a resistance gene in the host plant therefore exerts a strong selective pressure for a mutation in the avirulence gene if the product of the avirulence gene is not essential for the survival of the pathogen. This selection pressure has important epidemiological consequences for the development of new pathogen races and the losses in crop production (see chapter 3 and below). Thus, resistance in the gene-for-gene interaction is race-specific whereas susceptibility is not specific. The genetic basis of specific resistance is best understood by the quadratic check (Fig. 1A) that can be used to describe the gene-for-gene interaction. In this graphical description, resistance occurs only when both a dominant *R* gene from the plant as well as the dominant avirulence gene *A* from the pathogen are present in the upper left quadrant. In all the other quadrants the interaction is compatible, resulting in susceptibility. To prove a gene-for-gene interaction, the quadratic check must be reciprocal, i.e. it must be true for at least two resistance genes in the host and two matching avirulence genes in the pathogen (Fig. 1B) (Van der Planck, 1978). If this condition is fulfilled, a gene-for-gene interaction occurs in this particular disease. A more molecular model derived from the observation of the dominant character of both the avirulence gene and the resistance gene is shown in Fig. 2. The product of the resistance gene in this model would be a receptor that actively recognizes a direct or indirect product of the avirulence gene. Only the receptor-ligand interaction (Fig. 2A) results in specific recognition indicated by the hypersensitive response and disease resistance.

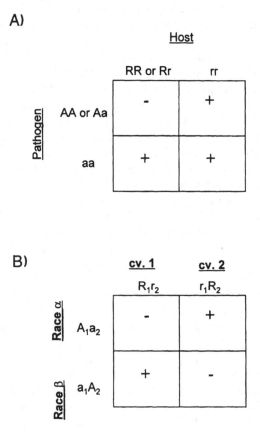

Figure 1. The gene-for-gene interaction. Quadratic check of gene combinations and the resulting different interaction types in a gene-for-gene interaction. The pathogen can grow in the compatible (+), but not in the incompatible (-) interactions. A indicates a dominant avirulence gene in the pathogen, R a dominant resistance gene in the plant.
(A): The quadratic check for a single locus in the host and in the pathogen. Only the combination of the dominant resistance and the dominant avirulence gene results in plant resistance in the upper left quadrant.
(B): Reciprocal check for two genetic loci of resistance (R_1 and R_2) in the two plant cultivars (cv. 1 and 2) and the corresponding two avirulence loci in two pathogen races (A_1 and A_2). The combination of R_1 and A_1 or R_2 and A_2 results in plant resistance. The reciprocal check defines a gene-for-gene interaction.

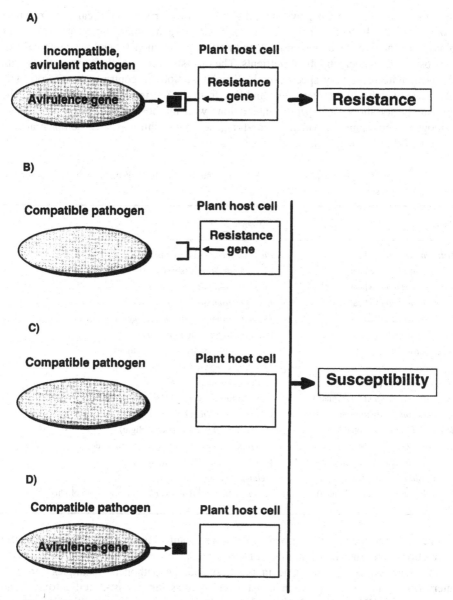

Figure 2. Molecular model of the gene-for-gene interaction (adapted from Staskawicz *et al.*, 1995). Resistance occurs only if there is a specific recognition between the resistance gene product and the product of the matching avirulence gene (A). In the absence of any recognition (B,C,D), no resistance reaction occurs and the pathogen can colonize the plant which results in a susceptible phenotype.

In an individual plant-pathogen system, there can be large number of plant resistance genes with different specificities for avirulence genes from the pathogen: more than 50 different specificities and therefore different resistance and avirulence genes have been described for example for powdery mildew in barley or stem and leaf rust in wheat (Jørgensen, 1994; McIntosh *et al.*, 1995). Gene-for-gene interactions are typical for biotrophic pathogens (such as mildew and rust, Fig. 3) which depend on living cells of the host plant for their supply of nutrients. The race-specificity of their interactions with the host indicates a very specific biological interaction. In contrast, the necrotrophic pathogens which kill the plant cells and live from the nutrients released from the cells do usually not show race-specific interactions with the host plant. Examples of host-pathogen interactions for which a gene-for-gene relationship has been demonstrated is given in Table 1, which is by no means exhaustive.

Table 1.   Incomplete list of pathosystems in crop plants for which the gene-for-gene relationship has been shown.

Plant	Pathogen
Triticum aestivum (wheat)	*Puccinia recondita* (leaf rust)
Triticum aestivum (wheat)	*Puccinia striiformis* (stripe rust)
Triticum aestivum (wheat)	*Puccinia graminis* (stem rust)
Triticum aestivum (wheat)	*Erysiphe graminis* f.sp. *tritici* (powdery mildew)
Hordeum vulgare (barley)	*Erysiphe graminis* f.sp. *hordei* (powdery mildew)
Zea mays (maize)	*Puccinia sorghi* (common rust)
Oryza sativa (rice)	*Xanthomonas oryzae* (bacterial blight)
Oryza sativa (rice)	*Pyricularia oryzae* (rice blast)
Malus sylvestris (apple)	*Venturia inaequalis* (apple scab)
Lycopersicon esculentum (tomato)	Tobacco mosaic virus (TMV)
Lycopersicon esculentum (tomato)	*Cladosporium fulvum* (leaf mold)
Solanum tuberosum (potato)	*Phytophtora infestans* (potato blight)
Solanum tuberosum (potato)	*Heterodera rostochiensis* (golden nematode)
Lactuca sativa/Serriola (lettuce)	*Bremia lactucae* (downy mildew)
Linum usitatissimum (flax)	*Melampsora lini* (flax rust)
Phaseolus vulgaris (French bean)	*Pseudomonas syringae* pv. *phaseolicola* (halo blight)

In addition to the cases described above where plant resistance is based on a specific interaction, there are also a number of cases where there is specific susceptibility to a pathogen due to a single gene in the host and the pathogen. Examples for such interactions can be found when a pathogen synthesizes a host-specific toxin. One example is the HV-toxin produced by *Helminthosporium victoriae* resulting in Victoria blight of oats (Ellingboe, 1976). There, the dominant *Vb* gene in oat is essential for sensitivity to the toxin and susceptibility to the disease. The ability of the pathogen to produce the toxin is under control of a single dominant gene in the pathogen. Only the combination of the presence of the *Vb* gene and the synthesis of the host-specific toxin results in a compatible, susceptible interaction. All the other possible gene combinations give an unspecific resistant, incompatible interaction. The selective pressure in this case is on mutational loss of an active *Vb* gene in the plant.

Figure 3. Phenotype of a typical resistance interaction conferred by a dominant resistance gene. The presence of the *Lr9* resistance gene of wheat against leaf rust results in a hypersensitive reaction after infection with an avirulent pathogen (arrows) wheareas a near-isogenic plant line without the resistance gene shows a compatible reaction and growth of the pathogen (arrowheads).

Gene-for-gene interactions were not only found between plants and biotrophic fungi but also in some instances with hemibiotrophic fungi (e.g. *Phytophtora, Colletotrichum*) nematodes, bacteria, insects and viruses. In tobacco, the resistance gene N' shows a race-specific reaction to different viral strains of the tobacco mosaic virus, i.e. some strains can spread systemically in the plant whereas the plant is resistant to other strains. This resistance trait is inherited in a Mendelian fashion as though it were conditioned by a single dominant gene. Similar race-specific resistance has been described against many bacterial pathogens such as *Pseudomonas syringae* pv. *glycinea* that causes leaf spot disease on the cultivated soybean and for which many different races are known (Fett and Sequeira, 1981). The resistance against the Hessian fly (*Mayetiola destructor*), an insect pest of wheat, was also shown to be based on a gene-for-gene relationship (Hatchett and Gallun, 1970). 25 resistance genes have been shown to be effective against the 13 reported biotypes of the Hessian fly (Patterson *et al.*, 1992). The common occurrence of gene-for-gene relationships suggests that there may be a common biological basis in the molecular recognition and signal transduction events involved in controlling resistance to diverse pathogens and pests. The recent finding that the products of resistance genes against bacterial, viral, fungal and nematode diseases have homologous domains (Bent, 1996) and can so far be grouped into only three different protein classes is a very nice confirmation of the ideas generated by classical genetics.

III.   Genetic Analysis of Race-Specific Resistance Genes

A.   DISEASE RESISTANCE GENES OCCUR IN ALLELIC SERIES AND AS GENE CLUSTERS

Classical genetic studies demonstrated that the same locus in different plant lines carried distinct alleles with different specificities to various pathogen races (Fig. 4A). In addition, resistance genes are often clustered at specific loci (Fig. 4B). In fact, it is genetically not easy to distinguish between true alleles at a resistance locus and a cluster of related genes at a particular chromosomal region. Crosses between plants with distinct resistance genes can help to clarify this question. If two genes are allelic, it will not be possible to get a chromosome with a combination of the two specificities (unless we assume the hypothetical event of an intramolecular recombination resulting in a combination of two different, specific resistances in the same gene). If the two specificities are due to two different genes, it is possible, at least theoretically if sufficient individuals in a segregating population are tested, to find recombinants. However, if the genes are physically very closely linked in a gene cluster of tandemly repeated genes, very large populations have to be built up and screened for the rare recombination events. This is often not feasible and therefore two distinct genes might be classified as alleles due to the very low recombination frequency between the two genes. Thus, alleles defined by classical genetic analysis might as well represent two closely linked genes in a molecular analysis.

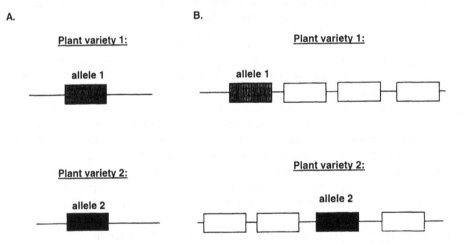

Figure 4.   Schematic model for the organization of resistance gene loci as revealed by classical genetic analysis. (A) Alleles of the same gene from different plant lines are responsible for distinct race-specific resistances against a pathogen. (B) Physically closely linked genes are arranged as tandem repeats. Each of these genes may encode a gene product with a different specificity against a pathogen.

Several interesting cases of allelic series or clusters of resistance genes have been described. The *Rp1* locus in maize confers race-specific resistance to *Puccinia sorghi*, the maize rust pathogen. The *Rp1* locus, located at the tip of chromosome 10, has at least 14 different alleles with distinct specificities called *Rp1-A* to *Rp1-N* (Saxena and Hooker, 1968). Several of these *Rp1* alleles have high natural mutation frequencies from resistance to susceptibility. Whereas normal mutation frequencies are about 10^{-6} to 10^{-7} per gene and generation, some *Rp1* alleles mutate at frequencies of 10^{-4} per generation. This high meiotic instability at the *Rp1* locus is probably due to unequal crossing-over (Sudupak *et al.*, 1992). It was also found that in all the tested cases the "alleles" at the *Rp1* locus were in fact distinct genes and recombinants with the combined specificities of the two parents could be found (Bennetzen *et al.*, 1994). Some of these progeny did not show flanking marker exchange as is expected for a normal, "symmetrical" recombination event, suggesting that genetic events such as gene conversion, intrachromosomal crossing-over or unequal sister chromatid exchange had occurred and contributed to the observed instability of the *Rp1* locus. The presence of a number of tandemly repeated copies of a very similar sequence, and the resulting unequal crossing-over events, might be the basis for the generation of new resistance specificities against the pathogen. Such a mechanism might help to compensate the potential advantage of the pathogen in its coevolution with the host plant: the pathogen with its large population size and usually fast generation time has to lose only a dominant function (i.e. the avirulence gene) to be able to grow on a previously resistant variety. In contrast, the plant must create a new dominant resistant gene to defend against such a new compatible race. Therefore, gene clusters of closely related genes might form the molecular basis for the rapid evolution of new specificities in the host plant.

In flax, 31 strain- or race-specific resistance genes have been characterized which confer resistance to different isolates of the flax rust pathogen *Melampsora lini*. They map to five distinct genetic loci, *K, L, M, N,* and *P* (Ellis *et al.*, 1988) of which the *L* and *M* locus have been particularly well studied. These two loci are also examples for two distinct strategies to evolve different specifities of resistance. Thirteen different resistance specificities map to the *L* locus. There have been many attempts to get recombinants between two different *L* locus specificities in coupling, but with no success. This suggests that, at the *L* locus, there is an allelic series of genes with different specificities. However, the genetic organization of the *M* locus is different. Seven different specific resistance genes map to this locus and recombination between different specificities was found. Obviously, the *M* genes are closely linked, tandemly repeated genes which span a genetic distance of around 0.5 centiMorgan. The relative position of four of the *M* genes was determined genetically. Flax is an ancient tetraploid species and molecular data indicate that the *L* and *M* locus are homologous, i.e. they correspond to the identical loci on the two original diploid genomes that were fused in flax (Ellis *et al.*, 1995). Obviously, the *L* and *M* loci have developed in two different ways: multiple alleles with different specificities evolved at the *L* locus whereas gene duplication, possibly followed by gene amplification through unequal crossing-over, occurred at the *M* locus. The initial duplication event might have occurred after transposon activity or by mispairing of two juxtaposed repeated sequences and subsequent non-homologous recombination. Thus, an initial duplication might have been the reason for the different evolution of the two resistance loci.

In lettuce, race-specific resistance genes are used for resistance breeding against the downy mildew pathogen *Bremia lactucae* and their genetics has been studied in detail. At least 13 different resistance genes (called *Dm* genes) have been described. These 13 genes map to four linkage groups. The group 1 gene cluster contains the genes *Dm* 1,2,3,6,14,15 and 16 which are tightly linked but show recombination and are therefore distinct genes (Farrara *et al.*, 1987). The simultaneous analysis of the genetics of avirulence in the pathogen make the lettuce-downy mildew system one of the best characterized gene-for-gene systems in plant pathology.

Another well studied resistance locus in plants is the *Mla* locus conferring resistance to powdery mildew in barley. In a recent review, 28 genes with different race-specificity were listed at the *Mla* locus (Jørgensen, 1994). These 28 genes are arranged as a large gene cluster. Recombination between some genes was detected (Wise and Ellingboe, 1985), demonstrating that not all the observed specificities result from different alleles of the same gene (thus we expect a situation as shown in Fig. 4B). It remains to be seen how many closely linked genes reside at this locus. It will be one of the most challenging research topics in the next years to analyse and study such complex loci at the molecular level. The large number of "alleles" at a single locus for different races of the same pathogen provides a unique opportunity to study the molecular basis of race-specific resistance.

B.   GENETIC EVIDENCE FOR MOLECULAR SIMILARITY AMONG DIFFERENT RESISTANCE GENES

There is some evidence from classical genetic studies that resistance genes against several diseases might be similar and thus a small number of genes would form the basis of a superfamily of resistance genes that behave according to the gene-for-gene hypothesis. In wheat, the two resistance genes *Sr15* and *Lr20,* which confer resistance to stem rust and leaf rust respectively, map to the same locus and have never shown recombination. Additionally, mutagenesis experiments showed that simultaneous changes occurred in both specificities. This is strong evidence that the two genes are identical and that a single gene can confer resistance to the two different diseases (McIntosh *et al.*, 1995). In addition, *Lr20/Sr15* is completely linked with the *Pm1* powdery mildew resistance locus. Thus, resistance genes against the three fungal wheat diseases leaf rust, stem rust and powdery mildew are either identical, or alleles or belong to the same tightly linked gene cluster.

Evidence for the genetic relatedness of resistance also came from the observation that resistance genes from different plant species recognize the same avirulence determinant in a bacterial pathogen. In several cases the transfer of isolated avirulence genes between bacterial strains of *Pseudomonas syringae* or *Xanthomonas campestris* pathovars showed that the same avirulence gene was recognized by several plant species (Michelmore, 1995). It was also shown that *Arabidopsis*, bean and soybean all have a resistance gene that recognizes the *avrRpt2* avirulence gene (Kunkel *et al.*, 1993). It is likely that the molecular basis for the resistance against *avrRpt2* is based on very similar recognition processes. This classical genetic evidence for a similarity of resistance genes will be of importance for the isolation of resistance genes by homology with known resistance genes.

C. THE HOST PLANT IN THE EVOLUTION OF GENE-FOR-GENE SYSTEMS IN THE FIELD

Here, we will only consider aspects of the evolution of gene-for-gene systems for the host plant whereas the more detailed population dynamics of the pathogen are described in chapter 3. As we have seen above, for most single resistance genes there are pathogen races that can infect a host plant carrying that particular gene. In addition, even if a race for a particular, newly used resistance gene is not found in a survey of a pathogen population, it does not mean that such an isolate does not exist. Often, a plant variety with a new, very effective resistance gene is released on the market and subsequently grown on large areas. Within a few years a new pathogen race which is compatible with (also said to be "virulent" on) the host is found and renders the gene ineffective for further use in agriculture. Thus, the cultivar did not lose the resistance gene but the resistance was no longer effective because the pathogen races had changed. The adaptation of the pathogen population towards a resistance gene can be due to a fast spread of already existing virulent races or due to the emergence of new virulent races caused by mutation, recombination or migration. Obligately biotrophic pathogens with great genetic diversity, high reproduction rate, sexual life cycles, several generations per growing season and fast dispersal by air have a very high potential to respond to the new selection pressure caused by introducing resistant host plants. The presence of an effective resistance gene will only allow the growth of virulent races. Therefore, it exerts a very strong selection pressure for virulent isolates. There are many examples for the rapid loss of effectiveness of newly introduced genes. In Australia this was particularly clear in the evolution of wheat stem rust races. After the introduction of new cultivars, virulent races developed rapidly and only the combinations of several resistance genes gave a more durable resistance (see below). In the wheat rusts, virulent races are known for almost all known resistance genes. One exception may be the *Sr26* gene that was derived from the grass *Agropyron elongatum*, a wild relative of wheat. No virulent field races for *Sr26* have been described so far (McIntosh *et al.*, 1995). As *Sr26* was introduced from a wild relative it is probably located on a large chromosomal translocation which behaves as a single Mendelian gene. In fact, this translocation could carry several resistance genes that together result in the high degree of resistance of *Sr26* and the difficulty of stem rust adapting to it.

The introduction of new resistance genes and the subsequent apparent adaptation of the pathogen by selection of virulent races is often refered to as the "boom-and-bust cycle" (Fig. 5): when a new variety with an effective resistance gene is introduced, the area on which it is grown increases rapidly resulting in the "boom". This selects the virulent race which can grow unrestrictedly, resulting in an epidemic. The farmers will then no longer grow this particular variety ("bust"), thus removing the selective advantage for the virulent race. At a first glance this "boom-and-bust" development appears like a cycle, ending where it begins. However, it should be realized that this usage of single genes in resistance breeding and agriculture is a waste of genetic resources as it makes potentially useful genes ineffective within a very short time. Thus, it becomes more and more difficult to find new sources of resistance in plant material. We will discuss in more detail below breeding strategies that allow a more efficient practical use of the monogenic resistances that fit the gene-for-gene hypothesis.

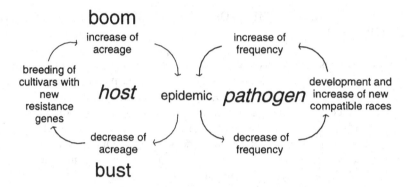

Figure 5. Boom-and-bust cycle for race-specific resistance genes in crop plants. The increase of the planted area of a new, resistant variety (boom) selects for new pathogen races which overcome the resistance gene. The variety is not resistant to these new races and no longer used in agriculture (bust) (adapted from Schumann, 1991).

It is important to remember that in the newly emerged virulent races that overcome a particular resistance gene there is no new virulence gene but rather the corresponding avirulence gene becomes mutated or inactivated, restoring the pathogenicity of the particular race on a host with a specific resistance gene. In fact, it may be better to call avirulence genes "incompatibility genes" as they cause incompatibility.

D. SUPPRESSION AND MODIFICATION OF RESISTANCE GENE ACTION

Classical genetics has revealed a number of cases where genes that otherwise fit the gene-for-gene hypothesis behave, in certain contexts, in a very unexpected way and are apparently exceptions to the rule. For example, suppressors or inhibitors of resistance gene action were found in wheat breeding during the introgression of some genes from wild relatives. Wheat, like some other important crop plants, is polyploid and problems occasionally arose when resistance in a species of lower ploidy level failed to express when it was transferred to hexaploid wheat. Wheat contains the three chromosome sets (genomes) A, B and D. Tetraploid durum wheat has the composition AABB and can be crossed with *T. tauschii*, the donor of the D-genome to produce a so called "synthetic" wheat with the composition AABBDD. When the leaf rust resistance gene *Lr23* which is present in many durum wheats was introduced into such a synthetic wheat, the resistance was suppressed and the lines were susceptible to a leaf rust isolate normally avirulent on *Lr23* genotypes. The *Lr23* gene maps about 25 cM from the centromer on chromosome 2BS. The gene which suppresses *Lr23* action is located on chromosome 2DS of the *T. tauschii* accession used for crossing, again about 25 cM from the centromer (Singh *et al.*, 1996). The very similar location of *Lr23* and its suppressor gene in the homoeologous (see Box 1) genomes suggests a direct interaction of the two genes, possibly at the level of the gene products where the suppressor protein might block the action of the resistance gene product. However, it is also possible that the two genes are similar and suppress each other by a co-suppression phenomenon as found in transgenic plants when a second copy of an already present gene is introduced, resulting in inactivation of both genes (Matzke and Matzke, 1995). The molecular cloning of resistance genes will certainly clarify the basis of suppression of resistance genes. D-genome suppressor genes in wheat are common and are a significant problem for the introduction of resistance genes into useful wheat breeding lines. In contrast to suppressor genes, complementary genes only confer resistance when both genes are present. This is the case for leaf rust resistance in wheat where both *Lr27* and *Lr31* have to be present to condition resistance. These two genes are genetically unlinked and are located on two different chromosomes (Singh and McIntosh, 1984). Dominant complementary genes were also found in oats (*Avena sativum*) for resistance to crown rust (*Puccinia coronata*). There, the crossing of two susceptible lines resulted in resistant progeny lines (Baker, 1966). Possibly, in these cases the resistance mechanism is only active when there is an interaction between the products of both resistance genes.

Box 1. **Different terms related to the concept of homology between chromosomes or genes**

Homologous genes are genes that descended from a common ancestral gene. There are two kinds of homologous genes: paralogous genes and orthologous genes.

Paralogous genes are the result of gene duplication. They have descended side by side in the same phyletic lineage and are present in the same species.

Orthologous genes are the result of speciation. They are homologous genes located either in different species or in different genomes in polyploid species.

Homologous chromosomes: chromosomes that largely or entirely contain the same linear sequence of genes and other genetic material. They may differ by structural rearrangements, such as inversions and/or duplications/deletions. In diploid genomes, one member of each pair of homologues is derived from one parent and the other from the other parent.

Homoeologous chromosomes: chromosomes that are located in different species or in different genomes in polyploid species and which originate from a common ancestral chromosome. Homoeologous chromosomes do not or poorly pair naturally.

Another class of genes, the modifying genes, do not have a direct effect on resistance itself but modify the effect of resistance genes. There are a number of examples where the expression of the resistance phenotype is modified by one or more additional genes in the plant host. In more general terms, it is frequently observed in resistance breeding that the genetic background of a variety is an essential component for the agronomically useful expression of a resistance phenotype. For example, Green *et al.* (1960) found that the two stem rust resistance genes *Sr7a* and *Sr10* were less effective after backcrossing them five times to the variety "Marquis" than in the donor parents. In some cases modifiers were clearly defined single genes as it was found for the *Sr7a* gene. There, modifiers appear to be ineffective, "overcome" resistance genes: it was found that the resistance to strain 15B-1 of stem rust which is due to the *Sr7a* resistance gene is modified by the genes *Sr9b*, *Sr10* and *Sr11*, which are all not effective against strain 15B-1 (Knott, 1989).

Modifying genes may be a subclass of genes that are involved in the resistance reaction conditioned by a particular gene that fits, the gene-for-gene hypothesis. For several resistance genes mutagenesis studies have been performed in order to find mutations which result in susceptible plants but do not map to the resistance gene. For the tomato resistance gene *Cf-9*, conditioning resistance to *Cladosporium fulvum*, two loci called *Rcr-1* and *Rcr-2* were found to be required for the expression of resistance (Hammond-Kosack *et al.*, 1994). In barley, Freialdenhoven *et al.* (1994) have also defined two loci (*Rar1* and *Rar2*) that are required for race-specific resistance conditioned by the Mla_{12} gene against barley powdery mildew. Thus, additional genes are clearly required for the expression of resistance genes. Such genes in different allelic forms might be responsible for the observed importance of the genetic background on the expression of a particular resistance phenotype.

E.   THE SYNTENY OF RESISTANCE GENE LOCALISATION IN RELATED
      SPECIES

Varying degrees in the conservation of gene order between species (synteny) have been observed. The first data reported for plants concerned the conservation of gene order between the tomato and potato genomes (Tanksley *et al.*, 1992). Subsequently, the comparative genetic analysis of cereal genomes yielded very exciting results. These studies have compared the genomes of different species of cereals such as rice and wheat (Kurata *et al.*, 1994), wheat and rye (Devos *et al.*, 1993a) barley and wheat (Devos *et al.*, 1993b) maize and sorghum (Melake Berhan *et al.*, 1993), maize and rice (Ahn and Tanksley 1993) or barley and rice (Saghai-Maroof *et al.*, 1996). The conservation of gene order as indicated by molecular markers was remarkable, even in the case of species with large differences in genome size (such as in maize and sorghum or wheat and rice). This suggests that the difference in genome size is not due to a different number of genes but rather to a different amount of highly repetitive DNA with no known function. Recently, Van Deynze *et al.* (1995a,b) reported comparative mapping data between wheat, rice, maize and oat. In the database "Graingenes" (http://wheat.pw.usda.gov/) consensus maps of several species belonging to the *Triticeae* are available. These data show that relative positions on homoeologous or orthologous (see Box 1) chromosome segments are conserved for agronomically important traits including resistance to leaf and stem rusts (Van Deynze *et al.*, 1995a). For example, the close linkage between loci containing seed storage proteins (avenins in oat, glutenins and gliadins in wheat and hordeins in barley) and resistance genes (stem rust *Pg9* and crown rust *PcX* in oat; stem rust *Sr33*, leaf rust *Lr10/Lr21* and powdery mildew *Pm3* in wheat; powdery mildew *Mla* in barley) seem to be very well conserved on the short arm of homoeologous group 1 chromosomes (Fig. 6). Thus, resistance genes in different species seem to be closely related and possibly origin from the same ancestral gene.

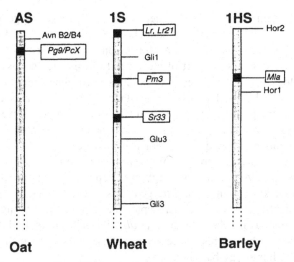

Figure 6. Schematic representation of the synteny between the end of the homologs of group 1 chromosomes in oat, wheat and barley. Loci containing genes coding for seed storage proteins (avenins (Avn), glutenins and gliadins (Glu, Gli) and hordeins (Hor) are conserved in this region as well as genes encoding resistance genes for rust (Sr, Lr, Pg, PcX) and powdery mildew (Pm, Mla) diseases.

These findings open new possibilities for gene isolation strategies using positional cloning techniques, especially in the case of those plants for which huge genome size and high amount of repetitive sequences represent a barrier in chromosome walking strategies. The conservation of gene order and composition of loci containing resistance genes will allow to use species with small genomes (rice, sorghum) and species for which tools such as YAC (yeast artificial chromosomes) and BAC (bacterial artificial chromosomes) libraries have been developed (maize, rice, barley) as genetic model plants. This should greatly help to locate more precisely and eventually isolate genes even if the locus was originally identified in species with large or even polyploid genomes. Fine scale analysis demonstrated a conserved gene organization along a small segment (1-2 cM genetically, 1 Mbp physically) of a rice chromosome if compared to the syntenic region in barley (Dunford *et al.*, 1995). The synteny between rice and barley was used by Kilian *et al.* (1995) to saturate the region of the barley *Rpg1* resistance gene in order to facilitate the map-based cloning of this gene.

The recent cloning of resistance genes against bacterial, fungal and viral diseases from different plant species showed that there are common features among sequences and conserved motifs (see chapter 4 and Jones and Jones, 1997). This opens up very exciting possibilities of gene isolation through homology-based cloning. So far, this approach has been used to successfully isolate candidate disease resistance genes in potato (Leister *et al.*, 1996) and soybean (Kanazin *et al.*, 1996) and is being tried for many other crops. In both cases the authors designed degenerate primers from two regions conserved among the nucleotide binding sites of the *N* resistance gene from tobacco, the *RPS2* gene from *Arabidopsis* and the L^6 gene from flax. They were able to amplify several classes of 500 bp DNA fragments differing in their nucleotide sequences but all containing the internal kinase 2/3 domains found in those resistance genes (see

Chapter 4). These fragments were subsequently used as probes for genetic mapping. In both species, some of the isolated sequences mapped close to known disease resistance genes. Further work using transformation is necessary to verify whether some of the cloned sequences are part of race-specific resistance genes. This example shows that homology-based strategies and the conserved gene order in the genomes of related species can be used to isolate resistance gene candidates in species for which map-based cloning strategies are very difficult to develop. However, as a note of caution it should be remembered that not all characterised resistance genes belong to the class of proteins mentioned above. For genes encoding Ser/Thr protein kinases, such as the *Pto* gene which is responsible for resistance in tomato against *Pseudomonas syringae* (Martin *et al.*, 1993b), the homology-based approach is more difficult as there is a large number of kinases in eucaryotic genomes. Nevertheless it was possible by this approach to identify a new class of receptor-like kinase segregating with the leaf rust *Lr10* resistance locus in wheat (Feuillet *et al.*, 1997). There is no doubt that in the next years a number of disease resistance genes will be isolated by using conservation in gene order and/or homology-based cloning.

IV.   Genetic Markers as Tools for the Characterization of Monogenic Resistance Genes

A.   THE DEVELOPMENT OF GENETIC MARKERS FOR MONOGENIC RESISTANCES

A genetic marker is broadly defined as a polymorphism between two plant lines which is inherited in a simple Mendelian way in the progeny of a cross between these lines. Several types of polymorphisms such as plant morphology, isozymes and DNA sequences have been used as genetic markers. The methods used to detect polymorphisms have evolved and improved in the last decades following the development of techniques in molecular biology.

In order to combine genes in a single individual (in the case of resistance genes this is called "pyramiding" resistance genes), it is necessary to determine the genotypes of plants and not just the phenotypes. The combination of complex resistances (quantitative resistances, see below) or the pyramiding of resistance genes is very difficult to achieve with classical breeding methods. By speeding up the process of selection, genetic markers represent an indispensible tool for plant improvement in general and for resistance breeding in particular. In addition, the development of high-density linkage maps composed of molecular markers is a prerequesite for resistance gene isolation through map-based cloning strategies as described in chapter 4.

Several excellent reviews on the application of genetic markers in plant breeding have been published (Paterson *et al.*, 1991; Winter and Kahl, 1995; Mohan *et al.*, 1997). Here, we want to focus on the development of markers for resistance genes and the contribution of the different marker technologies for mapping, detection and finally isolation of disease resistance genes in crop plants.

B. THE IDENTIFICATION OF GENETIC MARKERS FOR RESISTANCE GENES

The choice of the test populations for the genetic analysis of resistance represents an essential step in finding markers linked to specific resistance genes. Ideally, the test material should allow the detection of polymorphisms which are linked as closely as possible to the gene of interest. Near isogenic lines (NILs) represent some of the best material for finding polymorphisms genetically linked to the resistance gene. In this approach, a donor line with a resistance gene is crossed with a susceptible recipient parent and the resistant F_1 plants are crossed again to the recipient parent line. After several rounds of backcrossing with the susceptible line and continuous selection for resistance, the NILs should ideally have the same genetic background as the recipient parent (see Box 2). The susceptible parent and its resistant NIL should differ only by the presence of the gene of interest. Molecular differences which appear as polymorphisms between susceptible and resistant NILs should be genetically linked to the resistance gene. This can be analyzed in a segregating population by comparing the resistance phenotype and the marker genotype. Thus, it is necessary to test the markers in segregating populations derived from crosses between the resistant and susceptible NILs. This combined approach with NILs and segregating populations is illustrated in the case of tomato where most of the resistance genes studied were introgressed from wild relatives (Martin *et al.*, 1993a; Balint-Kurti *et al.*, 1994). A number of molecular markers linked to several important resistance genes in many species have been obtained through this type of strategy e.g. the *Tm-2a* viral resistance gene in tomato (Young and Tanksley, 1989), the *mlo* and *Mla* resistance loci in barley (Hinze *et al.*, 1991; Schüller *et al.*, 1992), the *Ht1* gene in maize (Bentolila *et al.*, 1991), *Pm3*, *Lr9*, *Lr24*, *Lr10*, *Lr1* in wheat (Hartl *et al.*, 1993; Schachermayr *et al.*, 1994, 1995, 1997; Feuillet *et al.*, 1995), *Xa* genes in rice (Ronald *et al.*, 1992; Yoshimura *et al.*, 1992, Zhang *et al.*, 1996), *Pg3* in oat (Penner *et al.*, 1993) and *Dm* genes in lettuce (Paran *et al.*, 1991).

Box 2. **The production of a near-isogenic line for a resistance gene by backcrossing.**

Line A is the donor line and the resistance gene has to be incorporated into the genetic background of the line B. R indicates a dominant resistance gene and the progeny are selected for its presence after each cross.

Original cross: Genotype A (homozygous RR) x Genotype B (rr)

Generation	Cross				Percentage of B genome
F1:	F1	(Rr)	x	B	50. %
F2:	F2	(Rr)	x	B	75. %
F3:	F3	(Rr)	x	B	87.5 %
F4:	F4	(Rr)	x	B	93.75 %
F5:	F5	(Rr)	x	B	96.875 %

In the absence of NILs, bulk segregant analysis (Giovannoni *et al.*, 1991; Michelmore *et al.*, 1991) can be used for the identification of genetic markers linked to a target region. In this approach, samples of a segregating population are pooled into different classes determined by their phenotype (resistant or susceptible). A marker present in only one pool but not in the other one represents a potential marker for the selected phenotypic trait. This approach has been used to characterize resistance loci such as the *plr* locus in lettuce (Kesseli *et al.*, 1993) and the cyst nematode *Ccn-D1* resistance gene in wheat (Eastwood *et al.*, 1994).

C.   CLASSES OF GENETIC MARKERS

1.   *Morphological and biochemical markers*
The first type of marker used were morphological markers for which a distinct morphological trait is determined by a single gene which is linked to the gene of interest. Such markers are principally very easy to work with and quick to score, but there are also serious limitations for their use. There are relatively few such markers in any given species and the expression of the morphological marker can be influenced by the environment. In addition, they are often expressed only at a specific growth stage or in a particular organ of the plant. Consequently, this type of marker only allows the development of reasonably good linkage maps in a few species (e.g. maize, barley, pea, tomato). Most of the morphological markers, which are often mutations, can not be used in breeding programs due to their strong and negative effects on the plant. A very limited number of morphological markers have been found which are linked to resistance genes and useful for practical breeding. Singh (1992, 1993) demonstrated a genetic linkage between leaf tip necrosis (*Ltn*, Fig. 7) and a number of resistance genes against leaf rust (*Lr34*), yellow rust (*Yr18*) and barley yellow dwarf virus (*Bdv 1*) in wheat.

Figure 7.   Leaf tip necrosis as a morphological marker for the wheat leaf rust resistance gene *Lr34*. Flag leaves of near-isogenic wheat lines with *Lr34* (left) and without *Lr34* (right) are shown. Leaf tip necrosis can be observed in the line with *Lr34*.

The second type of markers are biochemical markers which are based on differences in charged amino acids of enzymes with little or no effect on the enyzmatic activity. Such biochemically distinct isozymes are the products of different alleles of the coding gene. Maps based on isozyme markers have been established for several plant species and several isozymes have been found to be linked with disease resistance genes. They are used in many breeding programs to select efficiently for the presence of a particular resistance gene. A phosphoglucomutase is closely linked to the *Mo* gene conferring resistance to the bean yellow mosaic virus in pea (Weeden *et al.*, 1984) and an alcohol dehydrogenase isoform is a good marker for resistance to the pea enation mosaic virus (*En*) (Weeden and Provvidenti, 1987). A complete linkage was described between the endopeptidase isozyme *Ep-D1b* and an eyespot resistance gene derived from *Aegilops ventricosa* (McMillin *et al.*, 1986) as well as between *Ep-D1c* and the wheat leaf rust resistance gene *Lr19* introgressed from *Agropyron elongatum* (Winzeler *et al.*, 1995). These isozyme markers have been extensively used in improvement of crop resistance by breeding. However, similarly to morphological markers, there is only a small number of genetic loci which can be detected by these markers and their expression is often restricted to specific developmental stages or tissues (Table 2).

2. *Molecular (DNA) Markers*
The integration of DNA markers in genetic mapping projects substantially increased the number of markers available for agronomically important traits. The many advantages of DNA markers in comparison to morphological and isozyme markers (see Table 2) led to the intensive development and integration of those markers in breeding programs in recent years.

The first group of molecular markers developed were the RFLP (Restriction Fragment Length Polymorphism) markers which are based on the detection of polymorphic DNA fragments after restriction digests. Because RFLPs are mostly codominant markers (i.e. heterozygotes are easily identifiable because they combine the phenotype of the two homozygotes) and there is basically an almost unlimited number of loci, they represent very good and informative markers in linkage analysis and marker-assisted selection. The development of these markers led to a dramatic improvement in the construction of linkage maps and the elucidation of loci contributing to quantitative traits, so called quantitative trait loci (QTLs), in many crops. RFLP markers have been found to be linked with disease resistance genes in many species such as Arabidopsis, lettuce, flax, potato, tomato, maize, rice, wheat, barley etc. Despite their usefulness, the application of RFLP in marker-assisted selection remains time-consuming and expensive because of the technical difficulties and the labor intensive methods.

The recent development of markers based on the polymerase chain reaction (PCR) accelerated the development and use of molecular markers. The detection of PCR markers is faster than with the more complex technique of RFLP markers. In addition, the important steps of the reaction and also of detection can be automated, saving costs and increasing the number of samples that can be analyzed. PCR markers have been identified either by screening with arbitrary primers or with specific primers designed from known sequences.

Table 2.   Comparison of the different genetic marker systems. The usefulness of a particular marker type for map-based cloning is estimated in terms of time to find very closely linked markers and the possibility to use them as tools for gene isolation.

	Morphological	Isozyme	RFLP	RAPD	Microsatellites	AFLP
Marker inheritance	dominant	codominant, multiple alleles	codominant, multiple alleles	dominant multiple alleles	codominant	codominant, dominant
Developmental stage of detection	mostly adult plant	early stage tissue-specific expression	all tissues at early stages	all tissue at early stages (seeds)	all tissue at early stages (seeds)	all tissue at early stages (seeds)
Number of loci	limited	limited	~unlimited	~unlimited	~unlimited	~unlimited
Degree of polymorphism	low	intermediate	intermediate	high	high	very high
Synteny, comparative mapping	(yes)	no	yes	no	no	?
Analysis (time, feasibility, cost, safety)	easy in some cases	fast, easy, cheap	time consuming, no automation, expensive	fast, simple, automation, problems with reproducibility	fast, simple, automation, expensive	fast, automation, expensive
Application in breeding programs	limited	good	limited	limited	good	limited (?)
Usefulness for map-based cloning	impossible	impossible	very long	(fast if PCR screening of large insert libraries can be done)	fast if PCR screening of large insert libraries can be done. Otherwise useless as probe for further cloning (repetitive)	very fast

Arbitrary primers are used for RAPD (Random Amplified Polymorphic DNA) markers (Williams *et al.*, 1990). RAPD markers were the first PCR-based markers described. They are based on the amplification of DNA fragments between short (usually 10 bases) arbitrary oligonucleotide primers and allow the detection of polymorphisms resulting from insertions, deletions and single base changes that alter the binding sequence. RAPD markers may contain repetitive sequences within the amplified fragment and therefore may detect genomic sequences which are inaccessible to RFLP analysis. RAPDs have been used as markers for resistance genes in several species with different degree of success. In sugar beet (Uphoff and Wricke, 1992) and tomato (Klein-Lankhorst *et al.*, 1991), RAPD markers were found which are closely linked to nematode resistance genes or to a resistance gene against *Pseudomonas* in tomato (Martin *et al.*, 1991). In wheat, Schachermayr *et al.* (1994, 1995) detected RAPD markers linked to the leaf rust resistance genes *Lr9* and *Lr24*.

RAPD markers have not become widely used in marker-assisted selection because they are difficult to reproduce between different laboratories and even between different PCR thermal cyclers. In addition, they are usually dominant markers and therefore do not distinguish between heterozygotes and homozygotes. In some cases, these disadvantages have been overcome by using the RAPD fragment as a probe to highlight an RFLP on Southern (genomic DNA) blots to obtain codominant markers (Eastwood *et al.*, 1994; Martin *et al.*, 1993a). Another alternative to RAPD markers are the sequence-tagged-site (STS) or sequence characterized amplified region (SCAR) markers where PCR primers are designed from mapped, low copy number sequences (from RFLP probes or RAPD fragments). This allows the specific amplification and identification of a unique sequence at a known location in the genome. Such markers have been successfully developed among others in lettuce for *Dm* genes for resistance against downy mildew (Paran and Michelmore, 1993), in bean for the *Are* gene against anthracnose (Adam-Blondon *et al.*, 1994), in wheat for leaf rust (Schachermayr *et al.*, 1994, 1995, 1997; Feuillet *et al.*, 1995) and cereal cyst nematode (Williams *et al.*, 1996a) resistance genes and for powdery mildew resistance loci in cereals (Mohler and Jahoor, 1996).

Very recently, Zabeau and Vos (1993) developed a new technique which combines the reliability of the RFLP method with the power of the PCR technique. The amplified fragment length polymorphism (AFLP, Vos *et al.*, 1995) technique leads to the amplification of a random subset of restriction fragments without prior knowledge of nucleotide sequence. AFLP has been adapted to study complex genomes such as those of higher plants. To date, only few reports concerning AFLP markers linked to disease resistance genes have been published. However, the work of Thomas *et al.* (1995a) on AFLP markers linked to the *Cf9* resistance gene in tomato demonstrates the usefulness of AFLP analysis for positional cloning strategies. Indeed, the authors were able to detect AFLP markers located on the opposite sides of the resistance gene separated by only 15.5 kb. The development of the AFLP technology addresses one of the most limiting factors in map-based cloning strategies, i.e. the ability to rapidly test thousands of loci for polymorphisms and identify markers very closely linked to the resistance gene. For the same reasons, AFLP will probably become the technique of choice to map genetic loci contributing to more complex, quantitative traits (QTLs, see below).

In Fig. 8, different types of molecular markers are shown for some disease resistance genes. It is difficult to assign a relative advantage for one type of molecular marker in comparison to the others. Powell *et al.* (1996) recently performed an interesting comparison of the RFLP, RAPD, SSR (simple sequence repeats or microsatellites) and AFLP systems and concluded that each type of marker has different and complementary properties. It is likely that the integration of all different type of markers in genetic linkage maps will be the solution for generating highly saturated maps. The combination of RFLPs and RAPDs on tomato, maize and rice linkage maps has already enabled the detection of a number of markers tightly linked to disease resistance genes. This has represented the primary step towards the cloning of some of these genes. Indeed, the first cloned plant disease resistance gene (*HM1* in maize) was isolated by a combination of transposon mutagenesis and RFLP mapping (Johal and Briggs, 1992). Similarly, two race-specific resistance genes were cloned through map-based cloning in tomato (*Pto*, Martin *et al.*, 1993b) and rice (*Xa21*, Song *et al.*, 1995). The integration of new classes of promising marker systems such as AFLP and SSR (microsatellites, Röder *et al.*, 1995) should enable the cloning of other resistance genes from more complex genomes. Better linkage maps are also necessary for QTL analysis and the identification of genes involved in genetically complex quantitative resistance. In addition, the generation of marker systems which are easier to handle and to automate opens up interesting opportunities for application in marker-assisted breeding.

D.   IMPACT OF THE ISOLATION OF RACE-SPECIFIC RESISTANCE GENES
      FOR BREEDING

The isolation of disease resistance genes (see chapter 4) will have a big impact on resistance breeding and open up new opportunities for the use of these genes. Wild relatives of crop plants have been frequently used as a source of new resistance genes which were then introgressed into desirable varieties. The cloning of resistance genes should speed up and improve the process of gene introduction into elite lines with marker-assisted breeding. In the case of the *Xa21* resistance gene against bacterial blight of rice, Williams *et al.* (1996b) developed six markers from the resistance locus which will be useful in breeding programs to introgress *Xa21* into a number of cultivars. When the transformation procedures are routinely established for agronomically important crops, it will be possible to transfer specifically the gene conferring the resistance without associated DNA. Then, the introduction of defined resistance genes will also allow the question of linkage drag to be addressed, i.e. to distinguish the agronomical effect of the resistance gene itself from the associated genes which are generally cotransferred during classical selection processes.

A wide spectrum of resistance genes is found in the wild relatives of crop plants (see below). However, only a small percentage of the resistance potential in wild relatives has been used until now. This is often due to the fact that the introgressions from wild relatives frequently comprise large chromosomal segments. Such large fragments can carry undesirable traits which do not allow the use of the resistance genes from wild relatives in commercial varieties. The molecular cloning of these genes would solve this problem.

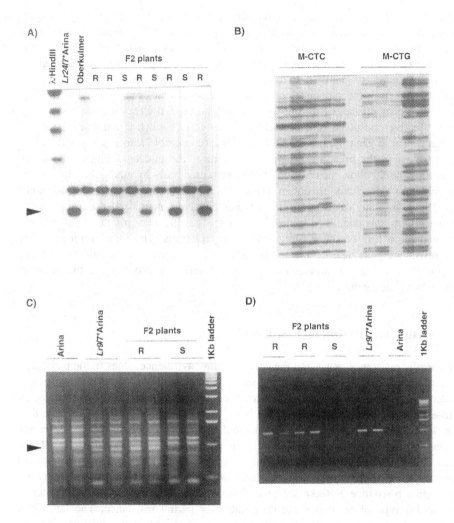

Figure 8. Different types of DNA markers:

(A) RFLP marker for the wheat leaf rust resistance gene *Lr24*. Southern blot hybridization pattern of *Hind*III digested genomic DNA from the resistant parent *Lr24*/Arina, the susceptible parent Oberkulmer and a segregating F_2 population. The arrowhead indicates the fragment which segregates with the *Lr24* resistance gene.

(B) AFLP analysis of three spring wheat and two winter wheat varieties. Template DNAs were digested with *Mse*I and *Eco*RI. The amplification step was performed with primers containing the 3'-variable extension CTC or CTG, respectively.

(C) RAPD marker for the wheat leaf rust resistance gene *Lr9*. Amplification of a polymorphic DNA fragment produced by the random primer OPR-15 on genomic DNA. The arrowhead indicates the position of the polymorphic band segregating with the *Lr9* resistance gene in a F_2 progeny derived from the cross between the susceptible parent Arina and the resistant near isogenic line *Lr9*/7*Arina.

(D) STS marker for the wheat leaf rust resistance gene *Lr9*. Amplification of a unique and specific DNA fragment with the primers J13/1+2 designed from a RAPD marker linked to the *Lr9* resistance gene. Amplification was performed on genomic DNA from the susceptible parent Arina and the resistant NIL *Lr9*/7*Arina as well as F_2 plants derived from their cross.

(The phenotype of the F_2 plants is indicated as follows: R= resistant, S=susceptible)

Durable resistance obtained by pyramiding several resistance genes by classical breeding is a very long and costly process. The use of molecular markers can improve the efficiency of the selection process (by tracking the resistance gene as well as the genetic background of the recurrent parent) but the cloning of the resistance genes themselves should also greatly help to achieve broad spectrum resistance. Indeed, DNA probes derived from the sequences of cloned resistance genes represent ideal markers to track the genes efficiently during the selection process. It is also possible that once several effective resistance genes are cloned, they can be transformed as a cassette to crop plants. These gene pyramids should confer more durable resistance and would be easy to follow in the breeding program. In addition, the understanding of resistance mechanisms should allow the creation of new resistance genes by manipulating domains responsible for specificity to obtain resistance against a broad range of pathogens (Bent, 1996). In many cases pyramiding strategies have not succeeded because of the negative effects the combined genes have in the genetic background of elite lines, often leading to loss of other agronomically useful traits. Thus, the introduction of single resistance gene against a broad spectrum of pathogen races should avoid these problems by limiting the number of steps in the selection process and the negative effect of associated genes.

V.   Vertical and Horizontal Resistance

Race-specific resistance genes and the corresponding avirulence genes in the pathogens belong to the genetically most intensively studied systems. These monogenic resistances are also of great importance for resistance breeding in many crop plants and, as discussed above, there are considerable efforts to find markers for these genes and to clone them. However, it has to be emphasized that this type of resistance represents only one form of resistance present in the gene pool of plants. Race-specific resistance is also called vertical resistance in contrast to the genetically (and physiologically) different type of non race-specific resistance which is often referred to as horizontal or polygenic resistance as there are several genes involved (see Box 3). Other terms used for this type of resistance are quantitative or partial resistance. The latter is a particularly useful term as it is purely descriptive and does not imply any genetic knowledge which is usually not present. The genes involved in this type of resistance are often called minor genes (in contrast to the major genes in vertical resistance). As this type of resistance is sometimes only expressed in the adult stage, such resistance is also described as adult plant resistance. There is a lot of discussion about the terminology of vertical and horizontal resistance. Here, we follow the terminology suggested by Robinson (1969). Horizontal resistance is characterized by the absence of genetic interactions between the host genotype and the pathogen genotype (in contrast to the race-specific genetic interaction in vertical resistance in gene-for-gene relationships). Oligogenicity (2-6 genes) is a characteristic property of horizontal resistance. If a resistance is based on several genes it is probably not race-specific (some exceptions are described below). A large part of the genetic understanding of such complex resistances comes from new studies using molecular markers to analyze the role of different genomic regions in the inheritance of a characteristic polygenic disease resistance. The methods and the results of such studies as well as the application of these results for resistance breeding will now be described.

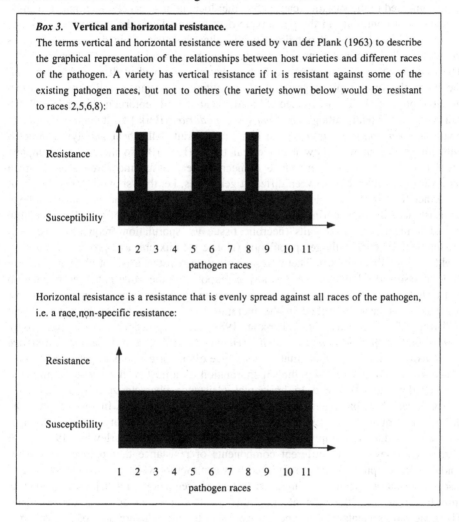

Box 3. **Vertical and horizontal resistance.**

The terms vertical and horizontal resistance were used by van der Plank (1963) to describe the graphical representation of the relationships between host varieties and different races of the pathogen. A variety has vertical resistance if it is resistant against some of the existing pathogen races, but not to others (the variety shown below would be resistant to races 2,5,6,8):

Horizontal resistance is a resistance that is evenly spread against all races of the pathogen, i.e. a race non-specific resistance:

VI. Quantitative Resistance

A. ASSESSMENT OF QUANTITATIVE RESISTANCE

In the case of qualitative or monogenic resistance fitting the gene-for-gene concept, host plants can be classified as resistant or susceptible. In contrast, quantitative resistance is characterized by a continuous distribution in the level of resistance. The degree of resistance is determined by the potential of the host plant to slow down the growth and/or the multiplication of the pathogen (Parlevliet, 1979). The continuous distribution in the expression of quantitative resistance results from the interaction of the plant genotype, the pathogen population and environmental effects. The effect of a single gene compared to the variation caused by other loci and non-genetic factors is too small to cause a discontinuity in the phenotypic distribution. Thus, the presence or absence of an individual gene will not dramatically alter the phenotypic distribution, and is therefore called a minor gene, i.e. a gene having a "minor" effect (Falconer, 1981). The more genes and environmental factors that are involved in disease development, the more the phenotypic expression of quantitative resistance will show a normal distribution (Fig. 9). The resistance of bread wheat against septoria nodorum leaf blotch caused by the fungal pathogen *Stagonospora nodorum* (Berk.) Castellani & Germano (syn. *Septoria nodorum*) is a typical quantitative trait. All wheat genotypes infected with this pathogen will show necrotic leaf tissue (Fig. 10), so there is no complete resistance. However, the percentage of necrotic leaf tissue measured after a given time varies considerably between different genotypes, i.e. they show different levels of resistance. Due to the complex interactions between the host and the pathogen, the level of resistance has to be defined in terms of developmental stage of the plant (seedling vs. adult plant), assessed traits (necrotic tissue vs. sporulation frequency), assessed organs (leaf vs. ear), pathogen population (isolate vs. mixture) and growing conditions (field vs. growth chamber). Therefore, quantitative resistance is not absolute but can only be assessed relative to other genotypes exposed to the same pathogen population and tested under the same conditions. For various host-pathogen systems quantitative resistance has been described in the literature e.g., maize/northern corn leaf blight (Ullstrup, 1977), cereals/rust (Wilcoxon, 1981), barley/powdery mildew (Jørgensen, 1994), barley/*Rhynchosporium secalis* (Habgood, 1974), and potato/*Phytophthora infestans* (Wastie, 1991). Quantitative resistance often corresponds to partial resistance which is defined as reduced pathogen sporulation on a host with susceptible infection type (Parlevliet, 1979) and to horizontal resistance characterized by the absence of significant host-pathogen interaction (Van der Plank, 1968). In this chapter, the definition of quantitative resistance is based solely on the phenotypic distribution of the trait and does not imply anything about race specificity. Parlevliet (1979) gave a good overview of the different components of resistance that reduce the rate of epidemic development. He distinguished between factors which reduce the amount of disease present at the start of infection (infection frequency) and delay the epidemic, from factors which reduce the infection rate and slow down the epidemic development. The major components of resistance which affect the reproduction rate of the pathogen are the reduction of infection frequency, lengthening of latent period and decrease of spore production. In situations with one reproductive cycle of the pathogen during disease development (e.g. smut in cereals) resistance is the sum of these components whereas in the case of a large number of reproductive cycles (e.g. rusts in cereals) the latent period is of major importance (Parlevliet, 1979). The correlation between the single components varies in different host/pathogen systems and at different developmental stages. Both factors delaying the epidemic and/or reducing the rate of epidemic development can be race-specific as well as non race-specific (Parlevliet, 1979) and can be under monogenic or polygenic control (Nelson, 1978; Parlevliet, 1979).

Box 4. **Terminology of plant disease and disease resistance**

infection: the entry of an organism or virus into a host and the establishment of a permanent or temporary parasitic relationship

resistance: the ability of an organism to withstand or oppose the operation of or to lessen or overcome the effects of an injurious or pathogenic factor

    race non-specific: resistance to all races of a pathogen

    race specific: resistance to some races of a pahtogen, but not to others

    complete resistance: multiplication of the pathogen is totally prevented, no spore production

    incomplete resistance: refers to all resistances that allow some spore production

        partial resistance: a form of incomplete resistance, in which spore production is reduced even though the host plants are susceptible to infection (susceptible infection type)

susceptibility: the inability of an organism to defend itself or to overcome the effects of invasion by a pathogenic organism or virus

tolerance: ability of the host to endure the presence and multiplication of the pathogen, can be expressed by less severe disease symptoms and/or limited yield reduction. Severity of symptoms and the amounts of damage to the host genotypes should be compared at equal amounts of the pathogen at the same stage of host development.

disease incidence: number of infected plants

disease severity: area or amount of plant tissue affected by disease

area under disease progress curve: disease severity measured several times during epidemic development

infection frequency: proportion of successful infections — spores that result in sporulating lesions

infection rate: multiplication rate of the pathogen on the host, increasse in disease severity

spore production: number of spores produced per lesion, per affected area or per time period

incubation period: time period between infection and first visible symptoms

latent period: time period from infection to spore production

infectious period: time period over which the diseased tissue sporulates

monocyclic: one reproductive cycle of the pathogen during reproductive cycle of the host

polycyclic: large number of reproductive cycles of the pathogen during reproductive cycle of the host

---

B.   BIOMETRIC APPROACH FOR QUANTITATIVE RESISTANCE TRAITS

Many efforts were made to investigate the genetic basis of quantitative resistance (for review see Geiger and Heun, 1989). Analogous to other quantitative traits such as yield or plant height, quantitative genetic theory can be applied to estimate genetic effects in controlled crosses (Falconer, 1981; Mather and Jinks, 1971). As described by Sprague (1983), quantitative genetics is an attempt to deal with the phenotypic expression

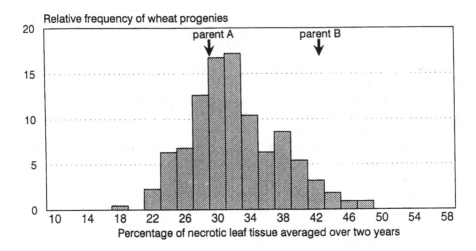

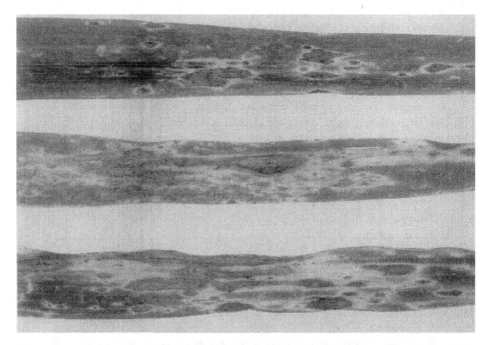

Figure 9.   Phenotypic distribution of wheat progeny (homozygous, recombinant inbred lines) derived from a cross between two parents with different levels of resistance (indicated by the arrows) against septoria nodorum leaf blotch caused by the pathogen *Stagonospora nodorum* (Berk.) Castellani & German. The damage caused by the disease was estimated as percentage of necrotic leaf tissue.

Figure 10.   Symptoms of septoria nodorum leaf blotch in wheat caused by the pathogen *Stagonospora nodorum* (Berk.) Castellani & German. *S. nodorum* is a necrotrophic pathogen causing lens-shaped chlorotic and necrotic lesions on the leaves. The three flag leaves show different percentages of necrotic leaf tissue.

resulting from the joint contribution of all genes involved.' The phenotypic value is divided into the genotypic value (complex of all genes involved) and the deviation caused by environmental effects, genotype-environmental interaction and experimental error. Accordingly, the phenotypic variation of a population observed in replicated trials over several environments can be split into the amount of variation caused by the respective effects. The broad sense heritability (h^2) is defined as the ratio between genotypic variance and phenotypic variance (Hallauer and Miranda Fo, 1981). Only the heritable part, i.e., genetic variance, can be exploited in breeding. In contrast to other quantitative traits, the expression of quantitative resistance not only depends on the genotype of the host and environmental effects but also on epidemiological factors like the genetic composition of the pathogen population, host-pathogen interaction and pathogen-environmental interaction which has an influence on the time of infection and infection pressure (Fig. 11). Since the control of the pathogen population in the field is difficult, these effects are often neglected in quantitative genetic studies and can increase the environmental effects as well as the genotype-environmental interaction.

Components of the phenotypic variance

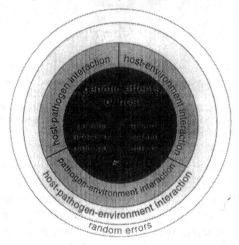

Figure 11. Composition of the phenotypic variance observed for the level of quantitative resistance on the host plant after infection with a pathogen.

The genotypic value can be further divided into different genetic effects (Mather and Jinks, 1971): the additive effect is defined as the average effect of an allele substitution at a given locus (half the difference between homozygous parents). The dominance effect is caused by allelic interaction (i.e. the heterozygous genotype deviates from the mean of the homozygous parents) whereas epistatic effects are caused by the interaction of two or more loci. Thus, the genetic variance across all loci can be partitioned into additive, dominance and epistatic variance components (Fig.12). The ratio of additive to dominance variance is important for efficient breeding strategies. While for hybrid cultivars and vegetatively propagated cultivars all variance components can be used, the breeder can only rely on additive effects for homozygous cultivars of self-pollinating species. For example, if there is a high proportion of dominance variance that covers the additive variance, selection in early generations, where there is still a high proportion of heterozygosity, is not very effective for inbred line development.

Components of the genetic variance of the host plant

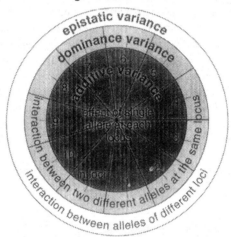

Figure 12. Composition of genotypic variance of the host plant for the level of resistance after infection with the pathogen. The sectors of the circle indicate the contribution of each locus involved in the expression of quantitative resistance. Additive variance results from the additive effects of single alleles at different genetic loci. Dominance variance is due to the interaction of different alleles present at the same locus in heterozygous plants, whereas epistatic variance is due to the interaction of alleles between different loci.

The heritability of quantitative resistance in many host/pathogen systems is generally high, indicating that genetic effects are more important for the difference in resistance than environmental effects. Thus, the ranking of cultivars based on the expressed resistance level does not change dramatically under different growing conditions. Biometric analysis of resistance against fungal diseases in small grain cereals (e.g. leaf rust in wheat, powdery mildew in rye) revealed that the additive variance was the predominant source of genetic variation for resistance for both self-pollinating and cross-pollinating species against obligate and facultative pathogens (for review see Geiger and Heun, 1989). The cumulative dominance effects across all loci were of minor importance and heterosis could be observed for resistance or susceptibility depending on the crosses for the same host/pathogen system, e.g. rye/*Erysiphe graminis* and wheat/*Stagonospora nodorum*. Therefore, heterozygous plants are not necessarily more resistant than homozygous host plants. Transgressive segregation in both directions, i.e., some progeny which are more or less resistant than either parent (as shown in Fig. 9), was also observed for resistance in wheat against leaf rust, stem rust and powdery mildew indicating that both parents, the one with the high resistance level as well as the one with the low resistance level, contributed positive and negative alleles to the progeny.

It is speculated that only a limited range of biochemical processes is involved in disease resistance. As a consequence less genes might be responsible for the expression of resistance than for other agronomic traits such as stress tolerance or yield, which are the result of many different physiological processes and therefore truely polygenic. For many pathosystems two to ten effective factors were found to be responsible for the expression of quantitative resistance which seems to be a lower limit of the number of genes involved (Geiger and Heun, 1989).

C.   QTL APPROACH FOR QUANTITATIVE RESISTANCE TRAITS

Although biometric approaches have had a great impact on breeding strategies, they give no information about the molecular basis and function of any single gene (Sprague, 1983). With the employment of genetic markers, it is now possible to dissect the genotype further into single genetic loci, so called quantitative trait loci (QTLs; Geldermann, 1975). It is assumed that the inheritance of single genes or QTLs involved in the expression of quantitative traits is not fundamentally different from those of monogenic resistances, i.e. QTLs follow a Mendelian segregation (Thoday, 1961). The cumulative effect of all QTLs involved corresponds to the genotypic value. In order to identify an individual QTL involved in the expression of a quantitative trait, one needs a method to follow the segregation of such genes. With different staining methods individual chromosomes can be identified by their banding pattern. These cytological methods allowed the development of translocation stocks (Jenkins *et al.*, 1957), aneuploids or chromosome substitution lines (Nicholson *et al.*, 1993), which were used to assign many monogenic resistance genes (McIntosh, 1988) and molecular markers (Chao *et al.*, 1989) physically to particular chromosomes. Chromosomes or chromosome arms that have an influence on the expression of the quantitative resistance traits such as resistance of wheat against septoria nodorum blotch (Nicholson *et al.*, 1993) and resistance of maize against Northern corn leaf blight (Jenkins *et al.*,1957; Jenkins and Robert, 1961) were identified this way. In contrast to the quantitative trait being investigated, the Mendelian segregation of linked genetic markers (see definition above) can be followed easily in defined crossing experiments. Linkage between seed coat color, a morphological marker, and seed size, a quantitative trait in *Phaseolus vulgaris* was first described by Sax as early as 1923. Thoday (1961) propagated the employment of genetic markers as a technique to identify individual genes involved in the expression of quantitative traits. The association of quantitative traits with single genetic markers (morphological markers or isozymes) has been demonstrated in several species, but precise localization suffered form the low number of markers available (for review see Paterson *et al.*, 1991). If the whole genome is covered with genetic markers, it should be possible to find closely linked markers for each QTL involved in the expression of the trait of interest. With the recent development of large numbers of molecular markers, it is now possible to construct genetic maps for any organism. Genetic maps, well saturated with molecular markers, are now available for almost all important crop plants: maize, sorghum, rice, wheat, barley, potato, soybean, sunflower (http://probe.nalusda.gov:8300/cgi-bin/query). These maps are a prerequisite to localize QTLs for the trait of interest, estimate their individual effects and their interactions.

***Box 5.* Basic concept of QTL analysis**

In order to dissect quantitative resistance into Mendelian factors of inheritance, a population segregating for the resistance trait has to be studied from a cross between well defined parental lines. This can be an F_2 population when the resistance trait can be assessed reliably on the F_2 progeny, otherwise replicated tests have to be performed using homozygous progeny, vegetative clones (Soller and Beckmann, 1990) or recurrent backcrosses (Tanksley and Nelson, 1996). For each individual or progeny the resistance phenotype as well as the marker genotype for a large number of molecular markers has to be assessed (see Fig. 13 for an example). At each marker locus it is determined whether this individual is homozygous for parent A (AA) or parent B (BB) or whether it is heterozygous (AB). Genetic maps are obtained by looking for cosegregation of such genetic markers in the segregating population. All markers that are linked with each other belong to one linkage group which corresponds to one chromosome. The degree of linkage determines their linear alignment on the chromosome. In contrast to the physical map which is given by the absolute length in µm of a chromosome at the metaphase or the size in kbp, the genetic length is given in centiMorgans (cM). One cM corresponds to an average of 1% crossover between two markers during one meiotic phase in the life cycle. On the basis of such genetic maps it is now possible to pin down the location of a QTL on the chromosome relative to the marker positions and to estimate the genetic effect of this QTL. Without going into statistical details which are very well explained elsewhere (Lander and Botstein, 1989; Paterson *et al.*, 1991; Haley and Knott, 1992; Lee, 1995) the basic concept of QTL analysis is explained here very briefly. Based on information of an individual marker locus, the population can be divided into the three marker classes: AA, BB or AB, respectively. If the three classes show a significant difference in the phenotypic expression of resistance (Fig. 14) it is assumed that this marker locus is linked with a QTL for resistance. In interval mapping, the information of two or more marker loci is used simultaneously. Thus, by looking at recombination frequencies between the trait and the two markers, it is possible to estimate the most likely position of the QTL within the marker interval. A measure for the significance of the presence of a QTL in the investigated marker interval is the LOD score. The LOD ($\log_{10}$ of the odds ratio) is defined as the logarithm (base 10) of the ratio of the likelihood that the phenotypic data of the segregating population have arisen assuming that there is a QTL present (i.e. linkage between marker and QTL) versus the likelihood that the data have arisen assuming the absence of a QTL (i.e. independent segregation of marker and QTL). The LOD score indicates the probability of a QTL which determines the trait in question being linked to a marker (Lander and Botstein, 1989). The LOD threshold depends on the genome size and the marker density. Typically LOD thresholds between 2 and 3 are required to limit the rate of false positives to 5%. The map position with the highest LOD score is assumed to be the most likely position of the QTL. In most studies a QTL can be placed within 10 to 20 cM with an acceptable degree of certainty (Lee, 1995). Thus, interval mapping results in a less precise localization of QTLs than the mapping of monogenic traits. For each QTL the following genetic effects are estimated: the additive effect, which is calculated as half the difference between the homozygous class AA and BB at the QTL

Box 5. continued

locus, and indicates the average effect if one allele of parent A is substituted by one allele of parent B. The dominance effect is calculated as the deviation from the heterozygous class AB from the average of the homozygous classes AA and BB. If a QTL is totally linked with a marker locus, the additive and dominance effects can be directly calculated taking the phenotypic mean of the marker classes (Fig. 14). In some studies epistatic effects between QTL are calculated as well as QTL x environment interactions. The amount of phenotypic variation that can be explained by a single QTL is expressed as the coefficient of determination (R^2). The accuracy of the QTL analysis depends on the heritability of the trait, type of population and progeny, population size, coverage of the genome and marker density (van Ooijen, 1992).

D. CHARACTERIZATION OF LOCI INVOLVED IN QUANTITATIVE RESISTANCE

In recent years several QTL studies have been undertaken to dissect the quantitative resistance of major crop plants against various pathogens and pests (for review see Young, 1996). Results of these studies contributed towards a better understanding of the genetic basis of resistance. Besides the identification and localization of QTLs involved in the expression of quantitative resistance traits, new insight was gained concerning the race specificity of individual loci, gene-for-gene interactions and the common genetic basis of different components of quantitative resistance. The cosegregation of QTLs for resistance with major resistance genes or other genes with known function might give first hints of the possible function of genes contributing to quantitative resistance.

1.   *Number of Loci Involved in Quantitative Resistance and Their Genetic Effects*
Genetic studies using molecular markers have confirmed the oligogenic nature of inheritance for quantitative resistance traits. Between one and ten significant QTLs for resistance have been found for a wide range of different host/pathogen systems explaining between 9 and 90% of the genotypic variance. This is in agreement with the estimates based on quantitative genetic studies using the biometrical approach (Geiger and Heun, 1989). However, most QTL analyses have been performed with only one population, so that the number of identified QTLs is limited to the number of genes that differ between the parents of this specific cross. In addition, in most cases the amount of phenotypic variance explained by all QTLs simultaneously is smaller than the amount which is explained by the genotypic variance given by the heritability estimate (h^2). They should be identical if all genetic factors involved in the resistance have been identified. Some experiments have been conducted with a limited number of segregating genotypes (<100), a limited number of disease assessments (only one

environment) or a limited number of markers (<100) which reduces the chances of finding QTLs with small effects. Therefore, one to ten QTLs represent minimal numbers of genes involved in the expression of resistance in the total host germplasm, but it seems to be a reasonable estimate for the number of genes segregating in a cross between two breeding lines.

About one third of the studies revealed the presence of at least one QTL with a major effect, which accounted for 25 to up to 90% of the phenotypic variation. For example, Dion *et al.* (1995) found one major QTL that explained between 57 to 84% of the phenotypic variation for blackleg disease resistance in rape seed. These results strongly indicate the presence of major resistance genes which are probably modified by the interaction with the environment or other genes with minor effects. For many host/pathogen systems, both monogenic and quantitative resistance have been described, showing binomial distribution in the seedling test with a defined pathogen isolate and continuous phenotypic distribution at the adult plant stage under field conditions. QTL analyses have been performed on resistance traits showing continuous variation in segregating populations, where known or unknown major resistance genes (conferring monogenic resistance under controlled conditions) may be involved in the expression of quantitative traits. Such major genes mask the effect of genes with minor effects.

After dissection of the quantitative resistance trait into individual QTLs, the genetic effect of the different alleles can be estimated for each QTL. In order to find many genomic regions involved in the expression of quantitative resistance, crosses were usually made between highly resistant (assumed to carry many positive alleles) and highly susceptible lines (presumably without any positive alleles). Nevertheless, in most studies, QTLs with positive and negative additive effects were found in both parents, indicating that both can contribute alleles for improved resistance. For example, Chang *et al.* (1996) observed four QTLs in an F_5 population for resistance of soybean against sudden death syndrome caused by the fungal pathogen *Fusarium solani*. Three QTLs with the positive allele from the resistant parent reduced the mean disease incidence by 30% and the QTL with the positive allele from the susceptible parent by 10%. If both parents contribute to resistance, transgressive segregation is expected, i.e. some of the progeny should be more resistant or more susceptible than the parents. This holds true for the latter example. While the resistant parent showed 16% and the susceptible parent 59% disease incidence for sudden death syndrome, the F_5 progeny ranged between 5 and 95% (Hnetkovsky *et al.*, 1996).

When homozygous doubled haploid (DH), selfed progeny (F_5, F_6, F_7) or backcross (BC) populations are used for QTL mapping, only additive and epistatic effects can be estimated, whereas in the case of F_2 derived progeny ($F_{2:3}$, $F_{2:4}$) one can also estimate the dominance effects for each QTL. In contrast to monogenic resistance genes that often show complete or partial dominant gene action (i.e. the heterozygous plants are nearly as resistant as the homozygous resistant parent) usually all kinds of allelic interactions at the different QTLs were found within the same host/pathogen systems. For example, partial or complete dominance for resistance, overdominance for resistance (heterozygous class is more resistant than each of the homozygous classes), partial dominance towards susceptibility as well as an intermediate reaction (heterozygotes are as resistant as the mean of the parents) were reported at individual QTLs for resistance of maize against Northern corn leaf blight (Freymark *et al.*, 1994) and for resistance of bean against common bacterial blight (Nodari *et al.*, 1993). Dominance or overdominance was mainly observed at QTLs with major effects for resistance against different populations of downy mildew in pearl millet (Jones *et al.*, 1995a).

Averaged over all QTLs, the additive effects were predominant in most host/pathogen systems studies so far, e.g. for resistance of maize against sugar cane borer (Bohn *et al.*, 1996) and gray leaf spot (Maroof *et al.*, 1996) or resistance in French bean against common bacterial blight (Nodari *et al.*, 1993) and in pea against *Ascochyta pisi* (Dirlewanger *et al.*, 1994). Additive and dominance effects of similar magnitude were reported for resistance in maize against Northern corn leaf blight (Freymark *et al.*, 1994) and in broccoli against club root (Dion *et al.*, 1995). In contrast, Li *et al.* (1995) found greater dominance effects than additive effects for resistance of rice against sheath blight.

Epistasis is defined as non-allelic or interlocus interaction between different genes (Mather and Jinks, 1971). Epistasis occurs if different loci do not act additively together but influence each other in their phenotypic expression. Two extreme cases of epistasis are known from the interaction of monogenic resistance genes: in the case of complementary type of gene action, two genes need to be present to result in a resistance reaction whereas in the case of duplicated type action the presence of one gene is sufficient to confer resistance. There, two genes can replace each other but do not improve the resistance phenotype if they are combined. The power of detecting epistasis between genes influencing quantitative resistance in general is low and large population sizes are needed to find significant differences between the respective classes of gene combinations. Therefore, in only a few studies were digenic epistatic effects between pairs of QTLs tested. Significant interlocus interaction were found between QTLs for resistance against rice blast in rice (Wang *et al.*, 1994) and against downy mildew in pearl millet (Jones *et al.*, 1995a). Maroof *et al.* (1996) illustrated the epistatic effects between two QTLs for gray leaf spot resistance in maize. With a large population size they were able to demonstrate that QTL4 had no or only little effect on resistance when QTL1 was homozygous for the allele of the resistant parent similar to a duplicate type of interlocus interaction. While in most studies, interlocus interaction was only tested between QTL with additive effects, Lefebvre and Palloix (1996) tested all pairs of markers in order to detect genomic regions acting in epistasis for resistance of pepper against *Phytophthora capsici*. For root rot resistance they found three significant digenic interactions (six loci) that explained together 62% of the phenotypic variance. Only one of these six loci showed an additive effect. This indicates that some QTLs may only be effective in the presence of other QTLs (Lefebvre and Palloix, 1996) similar to the complementary type of interlocus interaction. One QTL without significant additive effect but significant interaction with another QTL was also reported for resistance against tan spot in wheat (Faris *et al.*, 1997). These results indicate that interlocus interaction might be very important for the inheritance of quantitative resistance traits. However, epistatic effects are difficult to employ in breeding programs since positive gene combinations of the parental lines will be dissociated in the progeny during cultivar development.

2.   Components of Quantitative Resistance

Quantitative resistance for a given host/pathogen system is often measured in terms of disease severity and results from different components of resistance such as infection frequency, latent period, rate of spore production or infectious period. Different components of resistance might be based on genes operating at different stages of infection. In most host/pathogen systems these components of resistance are correlated to a certain extent indicating that they are controlled by overlapping sets of partial resistance genes. The genetic basis of different components of quantitative resistance was analysed for resistance of maize against Northern corn leaf blight. Freymark et al. (1994) measured the average number of lesions per leaf (number), the percentage of diseased leaf tissue (severity) and the average lesion size (size) at one location in Iowa after artificial inoculation with Setosphaeria turcica race 0. The number of lesions was highly correlated with disease severity, whereas lesion size was only weakly correlated with number of lesions. Across all components of resistance they detected six different QTLs: one QTL for all three traits, two QTLs for number of lesions and disease severity, whereas three QTLs were specific for either disease severity or for lesion size. Dingerdissen et al. (1996) examined incubation period and area under the disease progress curve (AUDPC) according to Shaner and Finney (1980) of the same population at three locations in Kenya inoculated with local pathogen races. They found four QTLs for AUDPC that overlapped with the QTLs detected for severity by Freymark et al. (1994) and two QTLs for incubation period. Although the AUDPC based on 5 ratings of disease severity was moderately correlated with the incubation period, only one QTL was in common, explaining about 10% of the variance for both traits. The other QTL was specific for incubation period and explained 38% of the phenotypic variance. This indicates that lesion number and disease severity might be controlled by the same QTL whereas lesion size and incubation period are under the control of different genes.

Wang et al. (1994) mapped QTLs conferring complete resistance and partial resistance in rice against rice blast. They assessed the partial resistance in terms of lesion number, lesion size and diseased leaf area in the greenhouse and diseased leaf area in the field at two locations. Under greenhouse conditions, five genomic regions were found for complete resistance. For partial resistance, two QTLs were involved in lesion number, diseased leaf area and lesion size, seven QTLs contributed both to lesion number and diseased leaf area whereas three QTLs were specific for lesion number. Thus, a large set of genes determines the number of lesions, whereas no specific genes affecting only the lesion size or diseased leaf area were found. Eight QTLs were found for resistance in the field under natural infection pressure: four of them coincided with genes for complete resistance and three of them with QTLs for partial resistance under greenhouse conditions.

Molecular markers have also been used to examine the genetic basis of the resistance phenotype dependent on the developmental stages or particular plant organs. Steffenson et al. (1996) studied the difference between seedling and adult resistance of barley after infection with a single conidial isolate of Pyrenophora teres causing net blotch and Cochliobolus sativus causing spot blotch. Seven QTLs were found for adult plant resistance against net blotch, explaining 68% of the variance in the field, while three QTLs explained 50% of the phenotypic variance at seedling resistance. Two of these

QTLs were in common but less effective at the adult stage in the field. For spot blotch, one QTL with major effect was detected for seedling resistance tested in the growth chamber. The same QTL was found for adult plant resistance but explained only 9% of the variation, whereas a second QTL explaining 62% of the field resistance was not detected at the seedling stage. By testing different inoculation methods, Danesh *et al.* (1994) could show that the infection site has an influence on the effectiveness of the QTLs detected for bacterial wilt in tomato.

Such a detailed analysis of the genetic basis of different resistance components

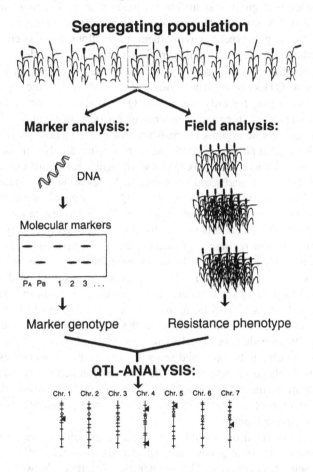

Figure 13. Basic concept of quantitative trait loci (QTL) analysis. Each genotype of a segregating population has to be tested in replicated field trials to determine the resistance phenotype. At the same time DNA is isolated from each genotype and tested with molecular markers. At each marker locus it is determined if the genotype (1, 2, 3, ...) has inherited the allele of parent A (P_A), parent B (P_B) or both in the case of heterozygosity. Based on linkage analysis of the marker data a genetic map can be constructed and markers can be aligned in linkage groups that correspond to the chromosomes (Chr.). The horizontal bars indicated the genetic location of the molecular markers. In QTL analysis the phenotypic resistance data are combined with the marker data to determine the location of the QTLs (indicated by the triangles) involved in the expression of resistance.

reveals at which stage of disease development single QTLs are involved. This also allows the formulation of a testable hypothesis about the physiological function of a particular QTLs in plant defense.

3.   *Consistency of QTLs in Different Environments and Different Hosts*

The resistance phenotype may also vary between tests performed under controlled conditions in growth chambers or in field experiments where the plants are exposed to different environmental effects. For example, Ferreira *et al.* (1995) found that none of the genomic regions responsible for resistance of *Brassica napus* against blackleg disease in the growth chamber was associated with resistance observed in the field. Significant genotype x environment interactions have been reported for several host/pathogen systems. If this variation explains a considerable amount of the phenotypic variance, environmentally specific QTLs or environmentally specific effects of common QTLs are expected. Bohn *et al.* (1996) detected significant QTL x environment interaction for only one of ten QTLs for resistance of maize against sugar cane borer tested at three environments in Mexico. In contrast, Bubeck *et al.* (1993) found nine genomic regions for resistance of maize against gray leaf spot, but only one was observed at all three environments in Virginia and North Carolina, two at two environments and the remainig six QTLs were significant in just one environment. QTLs for powdery mildew and scald resistance of barley were very inconsistent over the years (Thomas *et al.*, 1995b). Only one of six QTLs for powdery mildew resistance was detected in both years and only two of six QTLs for scald resistance were detected in two or three years. On the other hand, results of QTL analysis were very consistent in different environments for resistance of maize against Northern corn leaf blight tested in Iowa and Kenya (Freymark *et al.*, 1994; Dingerdissen *et al.*, 1996), for resistance of rice against sheath blight (Li *et al.*, 1995) and rice blast tested on the Philippines and Indonesia (Wang *et al.*, 1994), and for resistance of soybean against sudden death syndrome (SDS) tested at 5 locations in Illinois (Chang *et al.*, 1996). However, the consistency of QTLs depends not only on the host/pathogen system, but also on the level of significance that was reached in each study. QTLs with large effects or QTLs with medium effects but assessed several times on many genotypes (high LOD scores) are more likely to be detected in different environments, whereas QTLs with small effects might be missed because the LOD score was just below the threshold. Moreover, in the case of natural infection, the pathogen population could be quite variable between environments.

Since QTLs are detected on a single cross basis it remains unclear if these QTLs total the maximum number of genes involved in resistance and if they are consistent in different genetic backgrounds. The greater the difference between the parental lines for resistance, the higher the chances that all genes involved in resistance might segregate in this population and might be detected by QTL analysis. Bubeck *et al.* (1993) analyzed three different maize populations for resistance against gray leaf spot. Overall they found ten genomic regions on all maize chromosomes that were involved in resistance, but only one QTL was significant in all three populations, and eight QTLs in at least two populations. On the other hand, Maroof *et al.* (1996) found five QTLs for gray leaf spot resistance and all of them coincided with the QTLs detected by Bubeck *et al.* (1993) in at least two populations. Two different soybean populations were analysed for resistance against cyst nematode. Webb *et al.* (1995) found three QTLs on linkage group G, A and M against race 3, while Concibido *et al.* (1996) found two QTLs on group G and J against race 3a. However, the studies of QTLs for resistance in different host populations are rather limited and mostly conducted under different environments and different screening methods. Therefore, no general conclusions can be drawn at the moment.

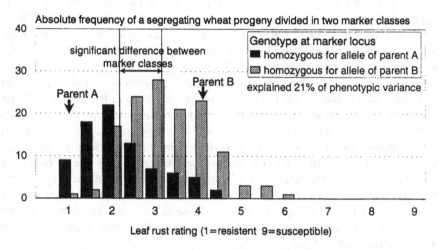

Figure 14. Phenotypic distribution of wheat progeny for leaf rust resistance divided into two classes according to their parental alleles observed for a molecular marker locus linked with a QTLs for leaf rust. The vertical lines indicate the average resistance level of each marker class. The arrows indicate the level of resistance of the parental lines.

4. Race Specificity of QTLs

With the concept of vertical and horizontal resistance, Van der Plank (1968) distinguished the quantitative resistance, assumed to be non race-specific and therefore durable, from the qualitative resistance which is per definition race-specific and confers complete resistance. However, Parlevliet and Zadoks (1977) questioned the different nature of qualitative and quantitative resistance and postulated that genes with minor effects in the host could also act in a gene-for-gene type of interaction with minor genes in the pathogen. They tested their hypothesis on a simulation model assuming the presence of five resistance genes that act either in an additive way (characteristic for minor genes) or as a combination of vertical resistance genes. In both models the additive gene action was predominant and explained most of the variance for the resistance phenotype. They concluded that race-specific genes might be involved in quantitative resistance, but will not be detected easily because additive effects will mask host-pathogen interactions at the single gene level.

Only a few studies have been undertaken to study the race specificity of QTLs for partial resistance. Jones *et al.* (1995a) studied the influence of four different *Scelerospora graminicola* pathogen populations on resistance in pearl millet under controlled green house conditions. The effects of pathogen populations and host x pathogen interactions were highly significant. They found different QTLs for the resistance against the pathogen populations: from India on linkage group (LG) 1, 6 and 7, from Nigeria on LG 1 and 4, from Niger on LG 4 and from Senegal on LG 2, 6 and 7. The authors concluded that pathotype-specific resistance is a major mechanism for downy mildew resistance, since none of the detected QTLs was effective against all pathogen populations. They speculated that QTLs with major effects might be involved in specific recognition of the pathogens, while QTLs with smaller effects could be the result of modifier genes that affect fungal growth and symptom expression. Race specific and non race-specific QTLs were reported by Concibido *et al.* (1996) for resistance of soybean against cyst nematode caused by different races of *Heterodera glycines*. Besides a QTL on chromosome G that had a major effect on resistance against all three races, they found three QTLs on chromosome D, J, and L, that were significant only for race 6, race 3a and race 1b, respectively. Leonards-Schippers *et al.* (1994) studied the quantitative resistance of an F_1 cross of heterozygous potato plants (4 alleles per locus) against race 0 and race 1 of *Phytophthora infestans* causing potato late blight disease (Fig. 15). Although both parents were susceptible to race 0, indicating the absence of major resistance genes conferring complete resistance, they found three types of QTLs: five were non race-specific, seven specific for race 0 and seven specific for race 1, including the most significant QTLs. These findings provide strong evidence that quantitative resistance can result from the combination of both race-specific and non race-specific resistance genes.

Figure 15. Symptoms of potato leaf blight caused by the pathogen *Phytophthora infestans*. There is both race-specific resistance as well as quantitative resistance against this pathogen in potato.

5.   *Coincidence of QTLs Involved in Resistance with Major Genes, Candidate*
     Genes and QTLs for Other Traits

Nelson *et al.* (1970) studied the interaction of genes of the pathogen *Setosphaeria turcica* with different numbers of genes for vertical resistance in maize and came to the conclusion that vertical and horizontal resistance are governed by the same genes. A more comprehensive theory was postulated by Robertson (1989) that qualitative mutants are extreme alleles of loci affecting quantitative traits, causing a very obvious deviation from the wild-type. Small deviations among the wild-type would result from more moderate alleles of the whole spectrum of alleles at this locus. According to this hypothesis, qualitative and quantitative traits would result from different alleles of the same loci, i.e. major and minor gene loci would be identical. Therefore, several authors have compared the position of the QTLs for quantitative resistance with the location of major resistance loci. QTLs with small effects for resistance in maize against Northern corn leaf blight were mapped close to the major resistance genes *Ht1* and *Ht2*, which were absent in both parents (Freymark *et al.*, 1994; Dingerdissen *et al.*,1996). One of the three QTLs for resistance against soybean cyst nematode was at the same position as the major gene *Rhg4* (Webb *et al.*, 1995). Likewise, Leonards-Schippers *et al.* (1994) mapped a QTL for resistance against *Phytophthora infestans* in both heterozygote potato parents to the same interval as the race-specific *R1* resistance gene. Lefebvre and Palloix (1996) detected a QTL for resistance in pepper against *Phytophthora capsici* in the vincinity of the *L* gene conferring resistance to the tobacco mosaic virus. They speculated that the same defense factor might be active against different pathogens. Maroof *et al.* (1996) found four QTLs for resistance against gray leaf spot in maize, one coincided with *hm1*, a gene conferring qualitative resistance against *Cochliobolus carbonum*, and one with *Ht2* and *Htn1* against Northern corn leaf blight. These results support the theory of Robertson (1989) that qualitative mutant phenotypes are extreme alleles at a QTLs. However, further research also at the molecular level is needed to verify if the observed coincidence of QTLs and major genes is due to different alleles of the same locus or due to close linkage of different loci.

It has been postulated that race-specific major genes that lost their effectiveness due to adaptation by the pathogen might have residual effects and contribute to quantitative resistance (Nelson, 1978). Thomas *et al.* (1995b) studied the inheritance of resistance against leaf rust in barley in a doubled haploid population between 'Blenheim' and breeding line E224/3. They found one resistance QTL from 'Blenheim' in the vicinity of the major resistance gene *Rph12*. *Rph12* is a major gene against leaf rust which is present in 'Triumph', a parent of 'Blenheim', and probably also in 'Blenheim'. However, for the assessment of leaf rust resistance they inoculated with a leaf rust race which is virulent on 'Triumph' to overcome the complete resistance caused by *Rph12*. Assuming that the detected QTL from 'Blenheim' is identical to the *Rph12* gene, this would indicate that the *Rph12* gene still has a positive effect on quantitative resistance although it was tested with a virulent race. The residual effect of so called defeated major genes was also studied by Heun (1992). He analyzed a barley population segregating for the major resistance gene *Mla12* with a powdery mildew isolate virulent on *Mla12*. In contrast to the results of Thomas *et al.* (1995b), no residual effect of the *Mla12* locus was found for the components of quantitative resistance.

Interestingly, several studies showed a coincidence of QTLs for quantitative resistance with candidate genes involved in biochemical or physiological processes related to resistance. Four of ten QTLs for resistance in maize against sugar cane borer map to genomic regions known to carry genes involved in the synthesis of cell wall components that might be important for insect resistance (Bohn et al., 1996): brittle stalk2 (bk2) cause easy leaf breakage, bm1 and bm2 cause brown midrib and lower lignin content, and peroxidase1 (bx1) is involved in dehydrodiferulic acid production. Ferreira et al. (1995) found one QTL for resistance in Brassica napus against Leptosphaeria maculans in the same interval as a locus homologous to PR2 which encodes a pathogenesis-related protein in Arabidopsis. In A. thaliana, PR2 expression is increased by inoculation with Pseudomonas solanacearum pv. tomato or by exposure to 2,6-dichloroisonicotinic acid (INA) or salicylic acid (Uknes et al., 1992). The PR2 protein is structurally similar to β-1,3-glucanase and probably associated with an increased chitinase production (Uknes et al., 1992). Chitinases are reported to have antifungal activity in vitro and could function in resistance by hydrolysing fungal cell walls (Mauch et al., 1988; Roberts and Selitrennikoff, 1988). Some minor QTLs for resistance in potato against Phytophthora infestans were associated with genetic loci for some proteins involved in the reaction of plants after pathogen infection: PRP1, phenylalanine ammonia-lyase (PAL) and 4-coumarate CoA ligase (4-CL). The latter two are enzymes involved in the phenylpropanoid pathway (Fritzemeier et al., 1987). The kinetics of their induction varies quantitatively after infection with the fungi and it is possible that these genes cause differences in the quantitative resistance response against Phytophthora infestans (Leonards-Schippers et al., 1994).

With the aid of QTL analysis it is not only possible to determine the number of genes involved in quantitative resistance response but also to determine the genetic basis of phenotypic correlations observed between the level of resistance and other agronomic traits. For instance, resistance of rice against sheath blight is often associated with plant height and heading date (Li et al., 1995). Five of the six QTLs detected for resistance against sheath blight were mapped to chromosome regions associated with increased plant height and in three cases with delayed heading, while only one QTL was independent of changes in morphological traits. Figdore et al. (1993) mapped two of three QTLs for resistance in Brassica oleracea against clubroot in the vincinity of QTLs for heading type, heading color or heading time. Since the resolution of QTL mapping is limited, it can not be distinguished whether the phenotypic correlation is due to close linkage of a resistance gene with a QTLs for these morphological traits, to pleiotropic effects of the resistance QTLs or whether the increased resistance is an indirect effect of the morphological differences. However, in each case the consequence in terms of breeding is that selection based on these QTLs will change both resistance and morphological traits simultaneously. In contrast, the QTLs independent of morphological traits might be directly involved in resistance response mechanisms and allow a selection towards increased resistance without altering other traits.

McMullen and Simcox (1995) undertook a survey to determine to what extent disease and insect resistance genes are clustered in maize. Results of resistance of maize against 15 different pathogens revealed that resistance genes (major genes and QTLs) were not randomly distributed across the ten chromosomes but clustered within each linkage group. For example, a QTL for resistance against the second generation of European corn borer, another QTL for resistance against *Fusarium* stalk rot, as well as the major genes *rp3* against common rust, *mv1* against maize mosaic virus and *wsm2* against wheat streak virus were found on chromosome 3. On chromosome 4 several QTLs were reported for resistance against Northern corn leaf blight, against European corn borer, against gray leaf spot and against *Fusarium* stalk rot as well as the major gene *rp4* against common rust (McMullen and Simcox, 1995). Since resistance genes within a cluster have different specificities and are effective against unrelated pathogens, it is unclear whether they are functionally related or not. An indication for a common functional basis for the QTL on chromosome 4 is the identification of an antimicrobial compound 2,4-dihydroxy-7-methoxy-1,4-benzoxazin 3-one (DIMBOA) which was associated with feeding resistance against European corn borer and reduced lesion expansion of Northern corn leaf blight. The accumulation of DIMBOA is regulated by the gene *bx1* which was mapped to the same interval on chromosome 4. Therefore, *bx1* could be the genetic basis for the QTL against the European corn borer and Northern corn leaf blight. However, it is unlikely that this compound is involved in the hypersensitive response mediated by *rp4* against common rust.

VII. Durable Resistance can have a Polygenic or a Monogenic Basis

The term "durable resistance" was defined by Johnson (1983) as follows: "Durable resistance is resistance that remains effective in a cultivar that is widely grown for a long period of time in an environment favourable to the disease." Although the terms "horizontal", "quantitative" and "durable resistance" are often used as synonyms, there are some examples of durable resistance genes which map to a single Mendelian locus and are possibly due to a single gene. Among those there is the very unusual *mlo* resistance against powdery mildew in barley. *mlo* resistance was induced repeatedly by mutation and several distinguishable alleles were characterized. In addition to the *mlo* alleles obtained by mutational treatment, there is at least one natural source of *mlo* in a land-race from Ethiopia (Jørgensen, 1994). This gene is an important source of resistance for spring barley breeding in Europe. *mlo* is a recessive resistance gene indicating that loss of function confers resistance. Physiologically, the *mlo* resistance is not due to a hypersensitive response but is based on the extensive formation of cell wall appositions which form a barrier against further penetration by the fungus (Skou, 1985). There are also durable monogenic resistances against nematode and viral diseases. In tomato, the *Mi* resistance gene against root-knot nematodes, *Meloidogyne incognita* and other subspecies was derived from the wild tomato *Lycopersicon peruvianum*. It has conferred resistance to tomato for many years without being overcome. In tobacco, the tobacco mosaic virus resistance gene *N* has conferred durable resistance. The *Lr34* single gene confers partial, but durable resistance to leaf rust in wheat. It is used very frequently in wheat breeding and has not lost its efficiency. The *Lr34* gene does not result in a hypersensitive reponse characteristic for race-specific genes but it decreased infection frequency and increased latency period in adult plants (Rubiales and Niks, 1995). *Lr34* is genetically completely linked with leaf tip necrosis (Singh, 1992; see Fig. 7). There is also very close genetic linkage between *Lr34* and *Yr18*, a gene conferring partial but durable adult plant resistance against stripe rust as well as a gene for tolerance against the barley yellow dwarf virus (Singh, 1993). At the moment it is not known whether these different effects are due to very close linkage or to pleiotropy. As the resistance conferred by *Lr34* is not sufficient for most agricultural environments, it is mostly used in combination with other genes to confer a more complete resistance. In conclusion, there are a few monogenic resistances that are durable but these are clearly exceptions to the rule.

Most of the durable resistances are quantitative. The wheat line "Frontana" e.g. from which the *Lr34* was first characterized has a durable resistance which is based on a small number (probably 4 to 5) of genes. Each of these genes has only minor effects but together there is an additive gene action resulting in sufficient resistance for the wheat breeder. A breeding strategy based on minor genes and horizontal resistance is followed for rust resistance in wheat in many breeding programs, among them the CIMMYT (International Maize and Wheat Improvement Center in Mexico) breeding program (Van Ginkel and Rajaram, 1993). This breeding program is one of the most important in the world, particularly for developing countries. The CIMMYT program relies on a number of genes which have additive gene actions and confer "slow rusting" against rust diseases. Good horizontal resistance is achieved by combining three or four of these additive resistance genes. This "slow rusting" is due to longer latency periods, fewer uredia, smaller uredium size and slower disease development. Thus, this type of resistance is not complete but slows down the epidemic enough to result in a good crop yield by preventing early damage to the leaves.

There are a number of possibilities to explain why certain resistances are durable. First, it could be that the products of avirulence genes are essential for the pathogen and can not be functionally lost without loss of viability. A second possibility is that the resistant cultivar carries two or more resistance genes and the growth of the pathogen would require two simultaneous mutations which is very unlikely. Finally, there are certainly resistances which are very difficult to overcome by the pathogen because they require complex physiological changes for growth of the pathogen.

VIII. Cytoplasmic Resistance

All the resistance genes described until now are inherited in a Mendelian way. They are localized on the chromosomes in the nucleus. However, there are a few exceptions to the Mendelian inheritance of resistance genes. These resistances are inherited via the nonnuclear, cytoplasmic component of the cell. They are encoded by chloroplast or mitochondrial genomes and are transmitted through the mother plant via the female gamete. There is one famous example for cytoplasmic disease resistance in maize (Pring and Lonsdale, 1989). In 1969 and 1970, particular races of the two fungal diseases southern corn leaf blight (*Cochliobolus heterostrophus* or *Helminthosporium maydis*) and yellow corn leaf blight (*Mycosphaerella zeae-maydis*) caused high losses in maize production. The losses were very high in hybrid varieties which contained the T (Texas) source of cytoplasmic male sterility which was used for an efficient production of hybrid seed. About 17% of the US maize crop was destroyed in 1970. The susceptibility to the disease was inherited maternally and was later found to be due to a mutation in the mitochondrially encoded gene T-*urf13*. In male fertile and disease resistant lines, the T-*urf13* gene product is truncated by a frame-shift mutation or deletions. However, when the gene is in a form where it encodes a 13 kD protein, this protein confers sensitivity to the toxins produced by southern corn leaf blight and yellow corn leaf blight. At that time many lines had the same cytoplasm and this genetic uniformity of the maize lines in terms of susceptibilty to the pathogen resulted in high losses. The experience of the Texas T cytoplasm and its consequences led to a better awareness of the problem of genetic uniformity and its negative impacts. It is now recognized that the use of genetically diverse material is an essential strategy to prevent such epidemics and the value of conserving genetic resources is widely accepted. In addition, the strategies for the use of these valuable resources of resistance are also much debated. The use of monogenic resistances as a single source of resistance is considered to be very problematic as it risks wasting valuable genetic resources.

Maternal inheritance occurs for traits encoded in the chloroplast or mitochondrial genomes. However, there is also the possibility for paternal effects on inheritance of pathogen resistance. In the common morning glory (*Ipomoea purpurea*), the genetic variance of resistance to anthracnose (caused by *Colletotrichum dematium*) was found to be largely determined by a paternal effect (Simms and Triplett, 1996). The molecular basis of this paternal inheritance is not known.

IX.   Genetic Resources for Resistance

Due to the increasing demands on agricultural resources through an increasing world population the putative sources of resistance have become even more valuable material for plant breeding. In conventional resistance breeding, the major source of resistant germplasm comes from the gene pool of the crop plant itself. Resistances can be found in lines from other breeding programs in the same or in different geographic areas and can be crossed into lines with the desired genetic background. Gene banks containing many different plant accessions, collected from various geographic regions, can also be a very valuable resource for resistant germplasm. There are many examples of resistances which were found in cultivars in different parts of the world. A famous example is the durable leaf rust resistance which is present in the South American wheat variety "Frontana" described above. As wheat is not an endogenous plant in South America, this resistance must have either appeared spontaneously or it was possibly imported from Europe or North America in the form of a land race. Landraces in general are very important sources of genetic variability for resistance breeding and usually consist of mixtures of various genotypes. The conservation of the genetic diversity present in this genetic material is one of the most important tasks not only for plant breeders but for society in general.

An interesting case of a resistance that can be selected from the same cultivar is found in sorghum. There, the resistance to the Milo disease (caused by *Periconia circinata*) occurs spontaneously in 1 in 8000 plants. An additional source of resistance is represented in the so called "primary gene pool". Thus, close relatives of a crop plant belonging to a different cultivated species, but which can be freely crossed with the cultivated plant, form this gene pool. The hexaploid wheat (*Triticum aestivum*) and spelt (*Triticum spelta*, an old hexaploid plant similar to wheat which is mainly grown in middle Europe) are other examples of this type of relationship. Additional examples are found among diploid progenitors of polyploid crop plants. E.g., *Triticum monococcum*, a close relative to the donor of the A-genome of wheat or *T. tauschii*, donor of the D-genome can be used as sources for resistance breeding in wheat (Fig. 16).

Figure 16. Wild relatives of crop plants as donors for resistance genes. Wild germplasm contains valuable genetic material for resistance breeding. Three wild grasses are shown. From all of them a number of rust resistance genes have been introgressed into the cultivated wheat by techniques such as embryo rescue and irradiation.

A last and increasingly important class of genetic resources for resistance breeding is found in the wild relatives of crop plants. The genus *Lycopersicon* comprises wild relatives of the cultivated tomato. Several of the wild relatives have been used as donors of important resistance genes for tomato: *L. hirsutum*, *L. peruvianum* and *L. pimpinellifolium* were the donors of fungal resistance, *L. chinese* and *L. peruvianum* for virus resistance and *L. peruvianum* also for nematode resistance. As most of the genomes of these wild plants recombine more or less freely with the genome of the cultivated tomato, the introgression of the resistance genes was not too difficult. This is also true for barley, where the wild relative *H. spontaneum* crosses very well with the cultivated barley *H. vulgare* and was a donor for powdery mildew and rust resistance genes. The diversity of resistance genes and physiological reaction types in *H. spontaneum* seems to be much broader than in cultivated barley. Besides these wild relatives which can be crossed with the cultivated species, there are also wild relatives of crop plants that can only be used as donors of resistant germplasm if cytogenetic manipulations such as irradiation and embryo rescue are used. Examples for such donor plants are relatives of wheat which belong to the wild grasses (Table 3, Fig. 16). The introgression of resistance genes from such species normally results in introgressions of large chromosomal segments as it was found in the introgression of the *Lr24* leaf rust resistance gene from *Agropyron elongatum* into wheat (Fig. 17, Schachermayr *et al.*, 1995). In this case, more than half of the long arm of chromosome 3D was found to be replaced by *Agropyron* DNA. Such large chromosomal introgressions usually carry a large number of genes and many of the alleles are inferior to the alleles in an adapted, commercially useful variety of a crop plant. Sometimes the genetic background of a line can compensate for these negative traits but the way to adapt the background for this purpose is time-consuming. The molecular isolation of the resistance gene out of the wild plant and subsequent transformation of the single gene into adapted germplasm would allow an efficient use of resistant germplasm without these side effects. A schematic representation of the advantages of such methods is shown in Fig. 18. In fact, only genetic transformation will allow the efficient use of such wild germplasm in many different breeding programs. The classical techniques are too slow and too cumbersome to use these resistance genes in adapted varieties. Therefore, a lot of the variation of resistances present in wild germplasm is currently not used due to technical limitations.

Table 3. Wild grasses as donors for rust and eyespot disease resistance in the hexaploid wheat (incomplete list, McIntosh *et al.*, 1995).

Donor	Gene	Disease
Triticum ventricosum	*Lr37, Pch, Yr17, Sr38,*	Leaf rust, eye spot (*Pseudocercosporella herpotrichoides*), yellow rust, stripe rust
Agropyron elongatum	*Lr24, Lr19, Lr29, Sr24, Sr26,*	Leaf rust, stem rust
Triticum umbellulatum	*Lr9*	Leaf rust
Triticum speltoides	*Lr35, Sr32*	Leaf rust, stem rust
Triticum comosum	*Sr34, Yr8*	Stem rust, stripe rust
Triticum timopheevii	*Sr36*	Stem rust

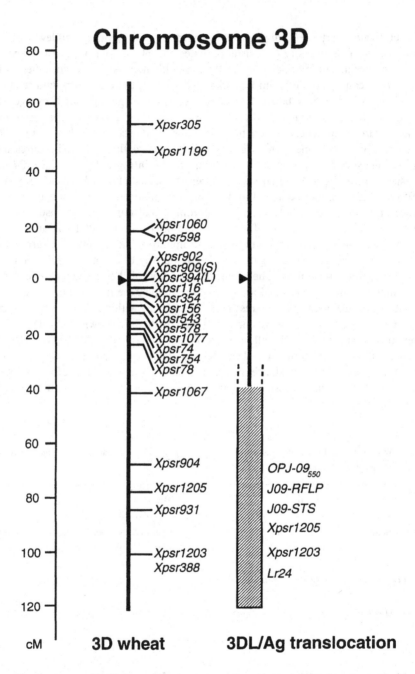

Figure 17. Chromosome 3D of wheat and the introgressed chromosome segment from *Agropyron elongatum* carrying the *Lr24* leaf rust resistance gene. At the left, the genetic distance is given in centiMorgan (cM). In the middle, the genetic map with some of the mapped markers is shown. At the right, the size of the introgressed segment from *A. elongatum* is indicated in the genetic map. The arrow shows the position of the centromer.

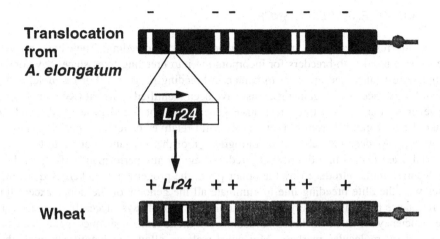

Figure 18. Comparison of the current strategy to introgress genes from wild grasses into wheat (top) and the planned application of gene technology (bottom) with the *Lr24* leaf rust resistance gene as an example. With current techniques, large chromosomal segments carrying the desired resistance gene as well as negative traits (-) for wheat breeding are introgressed into wheat. The isolation of the resistance gene would allow the removal of these negative traits. The isolated gene could then be transformed into genetically optimized wheat varieties. These varieties already carry positive alleles for agronomically important traits (+). The introduction of the *Lr24* resistance gene would improve these varieties specifically for leaf rust resistance, without compromising other properties.

X. Breeding for Resistance

Breeding for resistance is one of the most ecological and economic means of crop protection. The goal of each breeding program is to provide farmers with cultivars which combine excellent agronomic performance and high product quality with a high level of resistance against a variety of different diseases and pests. Selection has to be done for each pathogen separately, as resistance genes are not universal but mostly pathogen specific. Moreover, the level of resistance is always the result of the complex host-pathogen interactions and not just dependent on the host plant genotype. Selection for very high level of resistance of the host plant will increase the selection pressure on the pathogen. This will change the composition of the pathogen population favoring the accumulation of compatible pathogen races. As described above, it has been frequently observed that cultivars with a monogenic, complete resistance against a certain pathogen lost their resistance within some years when they were widely grown. Due to the coevolution of host and pathogen (see Chapter 3) resistant cultivars are always threatened to be overcome by the pathogen. Therefore, breeding for resistance is a continuous process. New strategies have to be developed to achieve a more durable resistance.

A.   MONOGENIC RESISTANCE

Monogenic resistance genes have been widely used in breeding. They have always been very attractive to breeders for incorporation into breeding lines since they confer complete resistance and are easy to handle in breeding programs. For example, more than 40 resistance genes against leaf rust have been identified in wheat (Kolmer, 1996). However, in many cases these resistance genes were not found in adapted breeding material but originated from old land-races, wild relatives or related species (Jones *et al.*, 1995b). As described above, the introgression of the resistance gene into breeding material often results in a dramatically reduced agronomic performance. To get rid of the negative traits introduced by the donor plant, the progeny are backcrossed several times with the elite breeding line to eliminate all the genome of the donor except for the resistance gene. This is very time consuming and not always successful (see above). The efficiency of such backcross breeding for monogenic resistance genes can be improved by molecular markers. Molecular markers allow the identification of the resistance gene in different plant material, the selection of such progeny with less linkage drag and a high proportion of the genome of the recurrent parent. Finally, the cloning of resistance genes will enlarge the number of useful resistance genes since it will allow the utilization of genes across species.

Most of the single resistance genes are race specific and can be overcome by the pathogen within a few years of widespread cultivation. In the long run it is not a promising approach to use these genes individually, because one resistance gene after another has to be incorporated into the plant to overcome the problem of pathogen adaptation to monogenic resistance. Considering that the number of useful resistance genes is limited (even with gene technology), they should be employed carefully. One strategy to delay the adaptation of the pathogen, is the pyramiding of resistance genes, i.e. several resistance genes are incorporated into a single cultivar (Brown, 1995; Pedersen and Leath, 1988). The pathogen needs to overcome each of these resistance genes simultaneously to be able to grow on the host. The probability that such a pathogen would arise by three or four independent mutations is extremely small. It is essential for the success of this strategy that each of these genes is still effective and can be identified individually.

Another strategy to prolong the effectiveness of newly incorporated resistance genes is the development of multilines or variety mixtures. Multilines are mixtures of different lines of a single variety each possessing a different resistance gene, whereas in variety mixtures each resistance gene is incorporated in a different genetic background. The idea behind this approach is to avoid an epidemic spread of a compatible race by increasing the genetic diversity of the host plants. The major disadvantage is that the development of multilines by backcrossing with one elite line takes a lot of time and effort and shows a genetic improvement for just one resistance trait. Therefore, the multilines will probably no longer be able to compete with newly released cultivars concerning other agronomic traits. In addition, the variety mixtures are often not accepted by the market if the product quality is not as homogeneous as for a single variety.

B.   QUANTITATIVE RESISTANCE

Quantitative resistance genes have the advantage that they reduce the sporulation of the pathogen but do not prevent it. Consequently, selection pressure on the pathogen population is less extreme and adaptation of the pathogen population is slowed down. Therefore, quantitative resistance is expected to be a lot more durable than complete resistance. However, breeding for an increased level of quantitative resistance is a difficult task. First of all the breeder has to define which traits should be measured to determine the level of resistance. For the farmer the reduction of yield loss caused by the respective diseases or pests is of primary importance. Therefore, breeding for a degree of tolerance, defined as reduced yield loss under severe infection pressure compared to a disease-free control, is one possible trait to select for. The greatest challenge for the breeder is to estimate the genotypic value of a breeding line. As pointed out before, the continuous variation for disease severity depends on the interactions between host genotype, pathogen population, and the environment. To eliminate confounding environmental effects, breeding lines must be tested in replicated trials in different environments (over several locations and years) to assess their genotypic level of resistance. Artificial inoculations are often employed to guarantee a homogeneous infection pressure and a good differentiation for disease severity each year. Since field testing is very laborious and time-consuming, other indirect methods were developed that allow a prediction of the resistance level under field conditions. For many pathosystems, easier screening methods have been established to assess components of partial resistance. Infection of seedlings or detached leaves can be performed under controlled conditions in order to assess differences in infection frequency, incubation and latent period. Response of immature embryos or microspores to toxins produced by the pathogen can be tested in in vitro culture. If these methods are moderately correlated with the resistance level of adult plants in the field, they can be used for preselection in early generations. However, the predictions are currently not so reliable that they can completely substitute the field testing. Until recenty, breeders did not know how many genes are involved in the expression of the resistance level in the field. To improve quantitative resistance parental lines are selected for high level of resistance combined with good agronomic performance and crossed with each other. However, if both parents carry the same resistance genes it is not possible to obtain progeny that are more resistant than either parent (so called transgressive progeny) and no breeding gain can be achieved. When different resistance genes can be combined in a cross, more resistant progeny can be selected. Depending on the importance of additive effects, these progeny can be detected in early generations. However, if dominance effects play a major role or the heritability of the resistance trait is low, successful selection can only be performed at the homozygous stage in later generations with replicated trials. With the aid of molecular markers, it is now possible to dissect the quantitative resistance into individual genes. This information will assist breeders to choose the appropriate crossing partners in order to combine the different sources of resistance in one genotype. Promising progeny can be selected on the genotypic level in early generations with molecular markers which are linked with such resistance QTLs. The increase of the resistance level is often negatively correlated with other agronomic traits such as plant height or earliness. The genetic basis of such correlations can be elucidated with molecular markers by comparing the location of QTLs for these traits. If the negative correlation is not due to a pleiotropic effect (a resistance gene has a direct effect on other traits) but due to linkage of some QTLs, breeders can concentrate on QTLs in genomic regions which do not affect other traits or try to find recombinants.

The probably most promising strategy to obtain durable resistance is the combination of both monogenic and quantitative resistance genes. Until now it has not been possible to select for improved quantitative resistance in the presence of genes conferring complete resistance. Molecular markers now allow the simultaneous selection of genotypes that carry QTLs for partial resistance as well as major genes for complete resistance (Cox, 1995). If a pathogen race can overcome the complete resistance, its distribution will be hindered by the high level of partial resistance and so an epidemic spread can be avoided. However, large number of progeny have to be screened to find the rare genotypes that carry the desired allele at each resistance locus.

XI. Outlook

The genetic analysis of disease resistance has thrown light on some of the basic principles of plant-pathogen interactions. It has had an enormous impact on plant breeding and provided the scientific basis for the selection of resistant plants. With the development of molecular markers, quantitative resistance can now be analyzed in great detail and we are starting to understand the basis of this complex trait. In the future, more and more genes involved in monogenic and polygenic resistance will be isolated. The understanding of the molecular mechanisms contributing to disease resistance will enable the plant breeder to further improve the strategies for breeding varieties that are no longer dependent on pesticides and contribute to food security, particularly in developing countries.

XII. References

Adam-Blondon AF, Sevignac M, Bannerot H and Dron M (1994) SCAR, RAPD, RFLP markers linked
    to a dominant gene (*Are*) conferring resistance to anthracnose in common bean. Theor Appl Genet
    88: 865—870
Ahn S and Tanksley SD (1993) Comparative linkage maps of the rice and maize genomes. Proc Natl
    Acad Sci USA 90: 7980—7984
Ausemus ER, Harrington JB, Reitz LP and Worzella WW (1946) A summary of genetic studies in hexaploid
    and tetraploid wheats. J Am Soc Agron 38: 1082—1099
Baker EP (1966) Isolation of complementary genes conditioning crown rust resistance in the oat variety
    Bond. Euphytica 15: 313—318
Balint-Kurti PJ, Dixon MS, Jones DA, Norcott KA and Jones JDG (1994) RFLP linkage analysis of the *Cf*-4
    and *Cf*-9 genes for resistance to *Cladosporium fulvum* in tomato. Theor Appl Genet 88: 691—700
Bennetzen J, Richter T, Hu G, SanMiguel P, Hong K, Frederick R and Hulbert S (1994) organization and
    hyperevolution of rust resistance genes in maize. In: Daniels MJ *et al.* (eds) Advances in Molecular

Genetics of Plant-Microbe Interactions 3: 261—266. Kluwer, Dordrecht

Bent AF (1996) Plant disease resistance genes: function meets structure. Plant Cell 8: 1757—1771

Bentolila S, Guiton C, Bouvet N, Sailand A, Nykaza S and Freyssinet G (1991) Identification of RFLP marker tighly linked to the *Ht1* gene in maize. Theor Appl Genet 82: 393—398

Biffen RH (1905) Mendel's laws of inheritance and wheat breeding. J Agric Sci 1: 4—48

Bohn M, Khairallah M, Gonzales-de-Leon D, Hoisington A, Utz HF, Deutsch JA, Jewell DC, Mihm JA and Melchinger AE (1996) QTL mapping in tropical maize: I. genomic regions affecting leaf feeding resistance to sugar cane borer and other traits. Crop Sci 36: 1352—1361

Brown JKM (1995) Pathogens' response to the management of disease resistance genes. Advances in Plant Pathology 11: 73—102

Bubeck DM, Goodman MM, Beavis WD and Grant D (1993) Quantitative trait loci controlling resistance to gray leaf spot in maize. Crop Sci 33: 838—847

Chang SJC, Doubler TW, Kilo V, Suttner R, Klein J, Schmidt ME, Gibson PT and Lightfoot DA (1996) Two additional loci underlying durable field resistance to soybean sudden death syndrome (SDS). Crop Sci 36: 1684—1688

Chao S, Sharp PJ, Worland AJ, Warham EJ, Koebner RMD and Gale MD (1989) RFLP-based genetic maps of wheat homoeologous group 7 chromosomes. Theor Appl Genet 78: 495—504

Concibido VC, Denny RL, Lange DA, Orf JH and Yound ND (1996) RFLP mapping and marker-assisted selection of soybean cyst nematode resistance in PI 209332. Crop Sci 36: 1643—1650

Cox TS (1995) Simultaneous selection for major and minor resistance genes. Crop Sci 35: 1337—1346

Danesh D, Aarons S, McGill GE and Young ND (1994) Genetic dissection of oligogenic resistance to bacterial wilt in tomato. Mol Plant Mic Int 7: 464—471

Devos KM, Atkinson MD, Chinoy CN, Francis HA, Harcourt RL, Koebner RMD, Liu CJ, Masojc P, Xie, DX and Gale MD (1993a) Chromosomal rearrangements in the rye genome relative to that of wheat. Theor Appl Genet 85: 673—680

Devos KM, Millian T and Gale MD (1993b) Comparative RFLP maps of homoeologous group 2 chromosomes of wheat, rye and barley. Theor Appl Genet 85: 784—792

Dingerdissen AL, Geiger HH, Lee M, Schechert A and Welz HG (1996) Interval mapping of genes for quantitative resistance of maize to *Setosphaeria turcica*, cause of northern corn leaf blight, in a tropical environment. Mol Breed 2: 143—156

Dion Y, Gugel RK, Rankow GFW, Séguin-Swartz and Landry BS (1995) RFLP mapping of resistance to the blackleg disease [causal agent, *Leptosphaeria maculans* (Desm.) Ces. et de Not.] in canola (*Brassica napus* L.). Theor Appl Genet 91: 1190—1194

Dirlewanger E, Isaac PG, Ranade S, Belajouza M, Cousin R, and de Vienne D (1994) Restriction fragment length polymorphism analysis of loci associated with disease resistance genes and developmental traits in *Pisum sativum* L. Theor Appl Genet 88: 17—27

Dunford RP, Kurata N, Laurie DA, Money TA, Minobe Y and Moore G (1995) Conservation of fine-scale DNA marker order in the genomes of rice and the Triticeae. Nucl Acids Res 23: 2724—2728

Eastwood RF, Lagudah ES and Appels R (1994) A direct search for DNA sequences tightly linked to cereal cyst nematode resistance genes in *Triticum tauschii*. Genome 37: 311—319

Ellingboe AH (1976) Genetics of host-parasite interactions. In: Heitefuss R and Williams PH (eds) Physiological Plant Pathology, pp 761—788. Springer, Berlin

Ellis JG, Lawrence GJ, Finnegan EJ and Anderson PA (1995) Contrasting complexity of two rust resistance loci in flax. Proc Natl Acad Sci USA 92: 4185—4188

Ellis JG, Lawrence GJ, Peacock WJ and Pryor AJ (1988) Approaches to cloning plant genes conferring resistance to fungal pathogens. Ann Rev Phytopathol 26: 254—263

Falconer DS (1981) Introduction to Quantitative Genetics. 2nd edition, Longman Group Limited, London

Faris JD, Anderson JA, Francl LJ and Jordahl JG (1997) RFLP mapping of resistance to chlorosis induction by *Pyrenophora tritici-repentis* in wheat. Theor Appl Genet 94: 98—103

Farrara BF, Ilott TW and Michelmore RW (1987) Genetic analysis of genes for resistance to downy mildew (*Bremia lactucae*) in species of lettuce (*Lactuca sativa* and *L. serriola*). Plant Pathol 36: 499—514

Ferreira ME, Rimmer SR, Williams PH and Osborn TC (1995) Mapping loci controlling *Brassica napus* resistance to *Leptosphaeria maculans* under different screening conditions. Phytopathology 85: 213—217

Fett WF and Sequeira L (1981) Further characterization of the physiological races of *Pseudomonas glycinea*. Can J Bot 59: 283—287

Feuillet C, Messmer M, Schachermayr G and Keller B (1995) Genetic and physical characterisation of the *Lr1* leaf rust resistance locus in wheat (*Triticum aestivum* L.). Mol Gen Genet 248: 553—562

Feuillet C, Schachermayr G and Keller B (1997) Molecular cloning of a new receptor-like kinase gene encoded at the *Lr10* disease resistance locus of wheat. Plant J 11: 45—52

Figdore SS, Ferreira ME, Slocum MK and Williams PH (1993) Association of RFLP markers with trait loci affecting clubroot resistance and morphological characters in *Brassica oleracea* L. Euphytica 69: 33—44

Flor HH (1942) Inheritance of pathogenicity in Melampsora lini. Phytopathology 32: 653—669

Flor HH (1955) Host-parasite interaction in flax rust — its genetics and other implications. Phytopathology 45: 680—685

Flor HH (1971) Current status of the gene-for-gene concept. Annu Rev Phytopathol 9: 275—296

Freialdenhoven A, Scherag B, Hollricher K, Collinge DB, Thordal-Christensen H and Schulze-Lefert P (1994) *Nar-1* and *Nar-2*, two loci required for Mla_{12}-specified race-specific resistance to powdery mildew in barley. Plant Cell 6: 983—994

Freymark PJ, Lee M, Martinson CA and Woodman WL (1994) Molecular-marker-facilitated investigation of host-plant response to *Exserohilum turcicum* in maize (*Zea mays* L.): components of resistance. Theor Appl Genet 88: 305—313

Fritzemeier K-H, Cretin C, Kombrink E, Rohwer F, Taylor J *et al.* (1987) Transient induction of phenylalanine ammonia-lyase and 4-coumarate: CoA ligase mRNAs in potato leaves infected with virulent or avirulent races of phytophthora infestans. Plant Physiol. 85: 34—41

Geiger HH and Heun M (1989) Genetics of quantitative resistance to fungal diseases. Annu Rev Phytopathol 27: 317—341

Geldermann H (1975) Investigations on inheritance of quantitative characters in animals by gene markers. I. Methods. Theor Appl Genet 46: 319—330

Giovannoni JJ, Wing RA, Ganal MW and Tanksley SD (1991) Isolation of molecular markers from specific chromosomal intervals using DNA pools from existing mapping populations. Nucl Acids Res 19: 6653—6658

Green GJ, Knott DR, Watson IA and Pugsley AT (1960) Seedling reaction to stem rust of lines of wheat with substituted genes for rust resistance. Can J Plant Sci 40: 524—538

Habgood RM (1974) The inheritance of resistance to *Rhynchosporium secalis* in some European spring barley cultivars. Annals of Applied Biology 77: 191—200

Haley CS and Knott SA (1992) A simple regression method for mapping quantitaitve trait loci in line crosses using flanking markers. Heredity 69: 315—324

Hallauer AR and Miranda Fo JB (1981) Quantitative Genetics in Maize Breeding. Iowa State University Press, Ames

Hammond-Kosack KE, Jones DE and Jones JDG (1994) Identification of two genes required in tomato for full *Cf-9*-dependent resistance to *Cladosporium fulvum*. Plant Cell 6: 361—374

Hartl L, Weiss H, Zeller FJ and Jahoor A (1993) Use of RFLP markers for identification of the alleles of the *Pm3* locus confering powdery mildew resistance in wheat (*Triticum aestivum L.*). Theor Appl Genet 86: 959—963

Hatchett J and Gallun R (1970) Genetics of the ability of the Hessian fly; *Mayetiola destructor*, to survive on wheat having different genes for resistance. Ann Entomol Soc Am 63: 1400—1407

Heun M (1992) Mapping quantitative powdery mildew resistance of barley using a restriction fragment length polymorphism map. Genome 35: 1019—1025

Hinze K, Thomson RD, Ritter E, Salamini F and Schulze-Lefert P (1991) Restriction fragment length polymorphism-mediated targeting of the *ml-o* resistance locus in barley (Hordeum vulgare). Proc Natl Acad Sci USA 88: 3691—3695

Hnetkovsky N, Chang SJC, Doubler TW, Gibson PT and Lightfoot DA (1996) Genetic mapping of loci underlying field resistance to soybean sudden death syndrome (SDS) Crop Sci 36: 393—400

Jenkins MT and Robert AL (1961) Further genetic studies of resistance to *Helminthosporium turcicum* Pass. by means of translocations. Crop Sci 1: 450—455

Jenkins MT, Robert AL and Findley WR Jr (1957) Genetic studies of resistance *to Helminthosporium turcicum* in maize by means of chromosomal translocations Agron J 49: 197—201

Johal GS and Briggs SP (1992) Reductase activity encoded by the *HM1* disease resistance gene in maize. Science 258: 985—987

Johnson R (1983) Genetic background of durable resistance. In: Lamberti F, Waler Jm, Van der Graaff NA (eds) Durable resistance in crops. Plenum Press, New York, pp 5—26

Jones DA and Jones JDG (1997) The roles of leucine-rich repeat proteins in plant defences. Adv Bot Res Adv Plant Pathol 24: 89—117

Jones ES, Liu CJ, Gale MD and Hash CT (1995a) Mapping quantitative trait loci for downy mildew resistance in pearl millet. Theor Appl Genet 91: 448—456

Jones SS, Murray TD and Allan RE (1995b) Use of alien genes for the development of disease resistance in wheat. Annu Rev Phytopathol 33: 429—443

Jørgensen JH (1994) Genetics of powdery mildew resistance in barley. Critical Review in Plant Science 13: 97—119

Kanazin V, Marek LF and Shoemaker RC (1996) Resistance gene analogs are conserved and clustered in soybean. Proc Natl Acad Sci USA 93: 11746—11750

Kesseli R, Witsenboer H, Stanghellini M, Vandermark G and Michelmore R (1993) Recessive resistance to *Plasmopara lactucae-radicis* maps by bulked segregant analysis to a cluster of dominant disease resistance genes in lettuce. Mol Plant Mic Int 6: 722—728

Kilian A, Kudrna DA, Kleinhofs A, Yano M, Kurata N, Steffenson B and Sasaki T (1995) Rice-barley synteny and its application to saturation mapping of the barley *Rpg1* region. Nucl Acids Res 23: 2729—2733

Klein-Lankhorst RM, Vermunt A, Weide T, Liharska T and Zabel P (1991) Isolation of molecular markers for tomato (*L. esculentum*) using random amplified polymorphic DNA (RAPD). Theor Appl Genet 83: 108—114

Knott DR (1989) The wheat rusts — breeding for resistance. Monographs on Theoretical and Applied Genetics. Springer, Berlin

Kolmer JA (1996) Genetics of resistance to wheat leaf rust. Annu Rev Phytopathol 34: 435—455

Kunkel BN, Bent AF, Dahlbeck D, Innes RW and Staskawicz BJ (1993) *RPS2*, an *Arabidopsis* disease resistance locus specifiying recognition of *Pseudomonas syringae* strains expressing the avirulence gene *avrRpt2*. Plant Cell 5: 865—875

Kurata N, Moore G, Nagamura Y, Foote T, Yano M, Minobe Y and Gale M (1994) Conservation of genome structure between rice and wheat. Biotechnology 12: 276—278

Lamb CJ (1994) Plant disease resistance genes in signal perception and transduction. Cell 76: 419—422

Lander ES and Botstein D (1989) Mapping Mendelian factors underlying quantitative traits using RFLP linkage maps. Genetics 121: 185—199

Lee M (1995) DNA markers and plant breeding programs. Advances in Agronomy 55: 265—344

Lefebvre V and Palloix A (1996) Both epistatic and additive effects of QTLs are involved in polygenic induced resistance to disease: a case study, the interaction pepper — *Phytophthora capsici* Leonian. Theor Appl Genet 93: 503—511

Leister D, Ballvora A, Salamini F and Gebhardt C (1996) A PCR based approach for isolating pathogen resistance genes from potato with potential for wide application in plants. Nature Genet 14: 421—429

Leonards-Schippers C, Gieffers W, Sch˘fer-Pregl R, Ritter E, Knapp SJ, Salamini F, and Gebhardt C (1994) Quantitative resistance *to Phytophthora infestans* in potato: a case study for QTL mapping in an allogamous plant species. Genetics 137: 67—77

Li Z, Pinson SRM, Marchetti MA, Stansel JW and Park WD (1995) Characterization of quantitative trait loci (QTLs) in cultivated rice contributing to field resistance to sheath blight (*Rhizoctonia solani*). Theor Appl Genet 91: 382—388

Maroof MAS, Yue YG, Xiang ZX, Stromberg EL and Rufener GK (1996) Identification of quantitative trait loci controlling resistance to gray leaf spot disease in maize. Theor Appl Genet 93: 539—546

Martin GB, Williams JGK and Tanksley SD (1991) Rapid identification of markers linked to Pseudomonas resistance gene in tomato using random primers and near isogenic lines. Proc Natl Acad Sci USA 88: 2336—2340

Martin GB, Carmen de Vicente M and Tanksley SD (1993a) High resolution linkage analysis and physical characterization of the *Pto* bacterial resistance locus in tomato. Mol Plant Mic Int 6: 26—34

Martin GB, Brommonschenkel SH, Chunwongse J, Frary A, Ganal MW, Spivey R, Wu T, Earle ED and Tanksley SD (1993b) Map-based cloning of a protein kinase gene conferring disease resistance in tomato. Science 262: 1432—1436

Mather K and Jinks JL (1971) Biometrical Genetics. 2n edition, Chapman and Hall, London

Matzke MA and Matzke AJM (1995) How and why do plants inactivate homologous (trans)genes? Plant Physiol 107: 679—685

Mauch F, Mauch-Mani B and Boller T (1988) Antifungal hydrolases in pea tissue. II. Inhibition of fungal growth by combination of chitinase and β-1,3-glucanase. Plant Physiol 88: 936—942

McIntosh RA (1988) Catalogue of gene symbols for wheat. Proceedings of the Seventh International Wheat Genetics Symposium, Cambridge, 13-19 July 1988.

McIntosh RA, Wellings CR and Park RF (1995) Wheat rust: an atlas of resistance genes. CSIRO Australia and Kluwer Academic Publishers, Dordrecht, The Netherlands

McMillin DE, Allan RE and Roberts DE (1986) Association of an isozyme locus and strawbreaker foot rot resistance derived from *Aegilops ventricosa* in wheat. Theor Appl Genet 72: 743—747

McMullen MD and Simcox KD (1995) Genomic organization of disease and insect resistance genes in maize. Mol Plant Mic Int 8: 811—815

Melake Berhan A, Hulbert SH, Butler LG and Bennetzen JL (1993) Structure and evolution of the genomes of *Sorghum bicolor* and *Zea mays*. Theor Appl Genet 86: 598—604

Michelmore R (1995) Molecular approaches to manipulation of disease resistance genes. Ann Rev Phytopath 15: 393—427

Michelmore RW, Paran I and Kesseli R (1991) Identification of markers linked to disease resistance genes by bulked segregant analysis: A rapid method to detect markers in specific genomic regions by using segregating populations. Proc Natl Acad Sci USA 88: 9828—9832

Mohan M, Nair S, Bhagwat A, Krishna TG, Yano M, Bhatia CR and Sasaki T (1997) Genome mapping, molecular markers and marker-assisted selection in crop plants. Mol Breed 3: 87—103

Mohler V and Jahoor A (1996) Allele-specific amplification of polymorphic sites for the detection of powdery mildew resistance loci in cereals. Theor Appl Genet 93: 1078—1082

Nelson RR (1978) Genetics of horizontal resistance to plant diseases. Ann Rev Phytopathol 16: 359—378

Nelson RR, MacKenzie DR and Scheifele GL (1970) Interaction of genes for pathogenicity and virulence in *Trichometasphaeria turcica* with different numbers of genes for vertical resistance in *Zea mays*. Phytopathology 60: 1250—1254

Nicholson P, Rezanoor HN and Worland AJ (1993) Chromosomal location of resistance to *Septoria nodorum* in a synthetic hexaploid wheat determined by the study of chromosomal substitution lines in "Chinese Spring' wheat. Plant Breeding 110: 177—184

Nodari RO, Tsai SM, Guzman P, Gilbertson RL and Gepts P (1993) Towards an integrated linkage map of common bean. III. Mapping genetic factors controlling host-bacteria interactions. Genetics 134: 341—350

Paran I and Michelmore R (1993) Development of reliable PCR-based markers linked to downy mildew resistance genes in lettuce. Theor Appl Genet 85: 985—993

Paran I, Kesseli R and Michelmore R (1991) Identification of restriction fragment length polymorphism and random amplified polymorphic DNA markers linked to downy mildew resistance genes in lettuce, using near-isogenic lines. Genome 34: 1021—1027

Parlevliet JE (1979) Components of partial resistance that reduce the rate of epidemic development. Ann Rev Phytopathol 17: 203—222

Parlevliet JE and Zadoks JC (1977) The integrated concept of disease resistance; a new view including horizontal and vertical resistance in plants. Euphytica 26: 5—21

Paterson AH, Tanksley SD and Sorrells ME (1991) DNA markers in plant improvement. Adv in Agr 46: 39—90

Patterson F, Foster J, Ohm H, Hatchett J and Taylor P (1992) Proposed system of nomenclature for biotypes of Hessian fly (Diptera: Cecidomyiidae) in North America. J Econ Entomol 85: 307—311

Pedersen WL and Leath S (1988) Pyramiding major genes for resistance to maintain residual effects. Ann Rev Plant Pathol 26: 369—378

Penner G, Chong J, Levesque M, Molnar S and Fedak G (1993) Identification of a RAPD marker linked to oat stem rust gene *Pg3*. Theor Appl Genet 85: 702—705

Powell W, Morgante M, Andre C, Hanafey M, Vogel J, Tingey S and Rafalski A (1996) The comparison of RFLP, RAPD, AFLP and SSR (microsatellite) markers for germplasm analysis. Mol Breed 2: 225—238

Pring DR and Lonsdale DM (1989) Cytoplasmic male sterility and maternal inheritance of disease susceptibility in maize. Annu Rev Phytopathol 27: 483—502

Roberts WK and Selitrennikoff CP (1988) Plant and bacterial chitinases differ in antifungal activity. J Gen Microbiol 134: 169—176

Robertson DS (1989) Understanding the relationship between qualitative and quantitative genetics. In: Helentjaris T and Burr B (eds): Development of Application of Molecular Markers to Problems in Plant Genetics, Cold Spring Harbor Laboratory, Cold Spring Harbor, NY, pp. 81—87

Robinson RA (1969) Disease resistance terminology. Rev Appl Mycol 48: 593—606

Röder MS, Plaschke J, König SU, Börner A, Sorrells ME, Tanksley SD and Ganal MW (1995) Abundance, variability and chromosomal location of microsatellites in wheat. Mol Gen Genet 246: 327—333

Ronald PC, Albano B, Tabien R, Abenes L, Wu KS, McCouch S and Tanksley SD (1992) Genetic and physical analysis of the rice blight disease resistance locus *Xa21*. Mol Gen Genet 236: 113—120

Rubiales D and Niks RE (1995) Characterization of Lr34, a major gene conferring nonhypersensitive resistance to wheat leaf rust. Plant Dis 79: 1208—1212

Saghai-Maroof MA, Yang GP, Biyashev RM, Maughan PJ and Zhang Q (1996) Analysis of the barley and rice genomes by comparative RFLP linkage mapping. Theor Appl Genet 92: 541—551

Sax K (1923) The association of size differences with seed-coat pattern and pigmentation in Phaseolus vulgaris. Genetics 8: 552—560

Saxena KMS and Hooker AL (1968) On the structure of a gene for disease resistance in maize. Proc Natl Acad Sci USA 61: 1300—1305

Schachermayr G, Siedler H, Gale, MD, Winzeler H, Winzeler M and Keller B (1994) Identification and localization of molecular markers linked to the *Lr9* leaf rust resistance gene of wheat. Theor Appl Genet 88: 110—115

Schachermayr GM, Messmer MM, Feuillet C, Winzeler H, Winzeler M and Keller B (1995) Identification of molecular markers linked to the *Agropyron elongatum*-derived leaf rust resistance gene *Lr24* in wheat. Theor Appl Genet 90: 982—990

Schachermayr G, Feuillet C and Keller B (1997) Molecular markers for the detection of the wheat leaf rust resistance gene *Lr10* in diverse genetic backgrounds. Mol Breeding 3: 65—74

Schüller C, Backes G, Fischbeck G and Jahoor A (1992) RFLP markers to identify the alleles on the *Mla* locus confering powdery mildew resistance in barley. Theor Appl Genet 84: 330—338

Schumann GL (1991) Plant diseases: their biology and social impact. The American Phytopathological Society. St. Paul, Minnesota

Shaner G and Finney RE (1980) New sources of slow leaf rusting resistance in wheat. Phythopathology 70: 1183—1186

Simms EL and Triplett JK (1996) Paternal effects in inheritance of a pathogen resistance trait in *Ipomoea purpurea*. Evolution 50: 2178—2186

Singh RP (1992) Association between gene *Lr34* for leaf rust resistance and leaf tip necrosis in wheat. Crop Sci 32: 874—878

Singh RP (1993) Genetic association of gene *Bdv1* for tolerance to barley yellow dwarf virus with genes *Lr34* and *Yr18* for adult plant resistance to rusts in bread wheat. Plant Dis 77: 1103—1106

Singh RP and McIntosh (1984) Complementary genes for reaction to *Puccinia recondita tritici*. II Cytogenetic studies. Can J Genet Cytol 26: 736—742

Singh RP, Ma H and Autrique E (1996) Suppressors for leaf rust and stripe rust resistance in interspecific crosses. In: Kema GHJ, Niks RE and Daamen RA (eds) Proceeding of the 9th European and Mediterranean Cereal Rusts & Powdery mildews Conference. Lunteren, The Netherlands.

Skou JP (1985) On the enhanced callose deposition in barley with *ml-o* powdery mildew resistance genes. Phytopathol Z 112: 207—216

Soller M and Beckmann JS (1990) Marker-based mapping of quantitative trait loci using replicated progenies. Theor Appl Genet 80: 205—208

Song, WY, Wang GL, Chen LL, Kim HS, Pi LY, Holsten T, Gardner J, Wang B, Zhai WX, Zhu LH, Fauquet C and Ronald PC (1995) A receptor kinase-like protein encoded by the rice disease resistance gene, *Xa-21*. Science 270: 1804—1806

Sprague GF (1983) Heterosis in maize: theory and practice. In: Frankel R (ed) Heterosis, Monographs on Theor Appl Genet, Vol 6, pp 46—70. Springer Verlag, Berlin

Staskawicz BJ, Ausubel FM, Baker BJ, Ellis JG and Jones JDG (1995) Molecular genetics of plant resistance. Science 268: 661—667

Steffenson BJ, Hayes PM and Kleinhofs A (1996) Genetics of seedling and adult plant resistance to net blotch (*Pyrenophora teres f. teres*) and spot blotch (*Cochliobolus sativus*) in barley. Theor Appl Genet 92: 552—558

Sudupak MA, Bennetzen JL and Hulbert SH (1992) Unequal exchange and meiotic instability of *Rp1* region disease resistance genes in maize. Genetics 133: 119—125

Tanksley SD, Ganal MW, Prince JP, de Vincente MC, Bonierbale MW, Broun P, Fulton TM, Giovannoni JJ, Grandillo S, Martin GB, Messeguer R, Miller JC, Miller L, Paterson AH, Pineda O, Röder MS, Wing RA, Wu W and Young ND (1992) High density molecular linkage maps of the tomato and potato genomes. Genetics 132: 1141—1160

Tanksley SD and Nelson JC (1996) Advanced backcross QTL analysis: a method for the simultaneous discovery and transfer of valuable QTLs from unadapted germplasm into elite breeding lines. Theor Appl Genet 92: 191—203

Thoday JM (1961) Location of polygenes. Nature 191: 368—370

Thomas CM, Vos P, Zabeau M, Jones DA, Norcott KA, Chadwick BP and Jones JDG (1995a) Identification of amplified restricion fragment polymorphism (AFLP) markers tighly linked to the tomato *Cf-9* gene for resistance to *Cladosporium fulvum*. Plant J 8: 785—794

Thomas WTB, Powell W, Waugh R, Chalmers KJ, Barua UM, Jack P, Lea V, Forster BP, Swanston JS, Ellis RP, Hanson PR and Lance RCM (1995b) Detection of quantitative trait loci for agronomic, yield, grain and disease characters in spring barley (*Hordeum vulgare* L.). Theor Appl Genet 91: 1037—1047

Uknes S, Mauch-Mani B, Moyer M, Potter S, Williams S, Dincher S, Chandler D, Slusarenko A, Ward E and Ryals J (1992) Acquired resistance in *Arabidopsis*. Pant Cell 4: 645—656

Ullstrup AJ (1977) Diseases of corn. In: Sprague GF and Dudely JW (eds) Corn and Corn Improvement, 2nd ed., pp 391—500, Am Soc Agron, Madison, Wisconsin

Uphoff H and Wricke G (1992) Random amplified polymorphic DNA (RAPD) markers for sugar beet (*Beta vulgaris* L.): mapping the genes for nematode resistance and hypocotyl colour. Plant Breeding 109: 168—171

Van der Plank JE (1963) Plant diseases: epidemics and control. Academic Press, New York

Van der Plank JE (1968) Disease resistance in plants. Academic Press, New York

Van der Plank JE (1978) Genetic and molecular basis of plant pathogenesis. Springer Verlag, Berlin

Van Deynze AE, Nelson JC, O'Donoughue S, Ahn SN, Siripoonwiwat W, Harrington SE, Yglesias ES, Braga DP, McCouch SR and Sorrells ME (1995a) Comparative mapping in grasses. Oat relationships. Mol Gen Genet 249: 349—356

Van Deynze AE, Nelson JC, Yglesias ES, Harrington SE, Braga DP, McCouch SR and Sorrells ME (1995b) Comparative mapping in grasses. Wheat relationships. Mol Gen Genet 248: 744—754

Van Ginkel M and Rajaram S (1993) Breeding for durable resistance to diseases in wheat: an international perspective. In: Jacobs T and Parlevliet JE (eds) Durability of disease resistance, pp. 259—272. Kluwer Academic Publishers, The Netherlands

van Ooijen JW (1992) Accurracy of mapping quantitative trait loci in autogamous species. Theor Appl Genet 84: 803—811

Vos P, Hogers R, Bleeker M, Reijans M, van de Lee T, Hornes M, Frijters A, Pot J, Peleman J, Kuiper M and Zabeau M (1995) AFLP: A new technique for DNA fingerprinting. Nucl Acids Res 23: 4407—4414

Wang GL, Mackill DJ, Bonman JM, McCouch SR, Champoux MC and Nelson RJ (1994) RFLP mapping of genes conferring complete and partial resistance to blast in a durably resistant rice cultivar. Genetics

136: 1421—1434

Wastie RL (1991) Breeding for resistance. Advances in Plant Pathology 7: 193—224

Webb DM, Baltazar BM, Rao-Arelli AP, Schupp J, Clayton K, Keim P and Beavis WD (1995) Genetic mapping of soybean cyst nematode race-3 resistance loci in the soybean PI 437.654. Theor Appl Genet 91: 574—581

Weeden NF and Provvidenti R (1987) A marker locus, *Adh-1*, for resistance to pea enation mosaic virus. Pisum Newl 19: 82—83

Weeden NF, Provvidenti R and Marx GA (1984) An isozyme marker for resistance to bean yellow mosaic virus in *Pisum sativum*. J Hered 75: 411—412

Wilcoxon RD (1981) Genetics of slow rusting in cereals. Phythopatholoy 58: 605—608

Williams JGK, Kubelik AR, Livak KJ, Rafalski JA and Tinguey SV (1990) DNA polymorphisms amplified by arbitrary primers are useful as genetic markers. Nucl Acids Res 18: 6531—6535

Williams KJ, Fisher JM and Langride P (1996a) Development of a PCR-based allele-specific assay for an RFLP probe linked to resistance to cereal cyst nematode in wheat. Genome 39: 798—801

Williams CE, Wang B, Holsten TE, Scambray J, de Assis Goes da Silva F and Ronald P (1996b) Markers for selection of the rice *Xa21* disease resistance gene. Theor Appl Genet 93: 1119—1122

Winter P and Kahl G (1995) Molecular marker technologies for plant improvement. World J of Microbiol and Biotech 11: 438—448

Winzeler M, Winzeler H and Keller B (1995) Endopeptidase polymorphism and linkage of the *Ep-D1c* null allele with the *Lr19* leaf rust resistance gene in hexaploid wheat. Plant Breeding 114: 24—28

Wise RP and Ellingboe AH (1985) Fine structure and instability of the *Ml-a* locus in barley. Genetics 111: 113—130

Yoshimura S, Yoshimura A, Saito A, Khishimoto N, Kawase M, Yano M, Yano M, Nakagahra M, Ogawa T and Iwata N (1992) RFLP analysis of introgressed chromosomal segments in three near-isogenic lines of rice for bacterial blight resistance genes, *Xa-1, Xa-3 and Xa-4*. Jpn J Genet 67: 29—37

Young ND (1996) QTL mapping and quantitative disease resistance in plants. Annu Rev Phytopathol 34: 479—501

Young ND and Tanksley SD (1989) RFLP analysis of the size of chromosomal segments retained around the *Tm2* locus of tomato during backcross breeding. Theor Appl Genet 77: 353—359

Zabeau M and Vos P (1993) Selective restriction fragment amplification: A general method for DNA fingerprinting. Eur Pat App 92402629.7 (Publ. Number 0 534858 A1)

Zhang G, Angeles ER, Abenes MLP, Khush GS and Huang N (1996) RAPD and RFLP mapping of the bacterial blight resistance gene *Xa-13* in rice. Theor Appl Genet 93: 65—70

Chapter 3

RESISTANCE IN POPULATIONS

J. FRANTZEN

Department of Biology, University of Fribourg,
3, rue Albert-Gockel, 1700 Fribourg, Switzerland
(jozef.frantzen@unifr.ch)

Summary

Knowledge of resistance at the molecular, cellular and organismal level is a basis for understanding resistance at the population level, but does not completely explain resistance at the population level. Other processes or mechanisms operate at the population level increasing or decreasing the significance of resistance. This chapter deals with the various mechanisms operating at the population level influencing host - pathogen interactions. The role of the mechanism of resistance is clarified considering the set of mechanisms determining host - pathogen interactions at the population level.

Populations and their attributes are described in general terms to allow development of an understanding of what is meant by the phenomenon "population". Subsequently, escape, avoidance, and the paired effects of predisposition and immunisation are described as mechanisms operating before resistance comes into play. Compensation and tolerance are described as mechanisms operating after resistance and relevant to the phenomena of selection and fitness, which are treated subsequently.

Some thoughts about management of resistance in crops are developed by considering all the mechanisms involved in host - pathogen interactions at the population level, and by contrasting wild and crop pathosystems.

I.   Introduction

Knowledge of resistance mechanisms at the molecular, cellular and organismal level is a basis for understanding resistance at the population level, but this knowledge of resistance at the lower levels does not explain all aspects of resistance at the population level. Other processes or mechanisms operate at the population level to increase or decrease the significance of resistance. This chapter deals with various mechanisms operating at the population level that influence host - pathogen interactions. The role of the mechanism of resistance will be clarified by considering the set of mechanisms determining host - pathogen interactions at the population level.

The phenomenon "population" is central to this chapter, but what is a population? The literature dealing with populations is vast, but in the majority of papers the term is

A. Slusarenko, R.S.S. Fraser, and L.C. van Loon (eds), Mechanisms of Resistance to Plant Diseases, 161-187.

used without definition. Defining the phenomenon is troublesome. Zadoks and Schein (1979) defined a population as "any set of elements that have at least one attribute in common" - a rather vague definition. To define a biological population they also stipulated that all elements should belong to the same species and that the population should have certain limits in space and time. The definition remains vague and they concluded that "research workers must identify the particular population they are studying by stating an operational definition in which the relevant attributes are specified".

To identify a particular population and to specify relevant attributes requires a feeling for the term population and knowledge about the attributes that are commonly used to describe and quantify a population. In section II, populations and their attributes are described in general terms to get such a feeling. Having established this basis, host - pathogen interactions at the population level are then described in the subsequent sections. Escape and avoidance (section III. A), and predisposition and immunisation (section III. B) are described as mechanisms operating before resistance (section III. C). Compensation and tolerance (section III. D) are described as mechanisms operating after resistance and relevant to the phenomena of selection and fitness (section IV). In the final section (V), some thoughts are presented about management of resistance in crops, which are based on the preceding sections.

This chapter aims to provide a general overview of the essential elements of the whole complex story of resistance at the population level. The objective is to transmit understanding of the essential elements of resistance at the population level rather than an exhaustive review.

II.   Populations of Host and Pathogen

Consider a group of spores, or seeds, placed regularly in a Petri dish to germinate. We determine for each spore whether it germinates and the time at which it germinates. If we plot the fraction of spores germinated against time we will see that not all spores germinate at the same time (Fig. 1). A few spores germinate rather quickly and a few spores rather slowly. For one spore we have only one germination time, for a group of spores we have a range of germination times. This range of germination times may be summarised as a mean and standard deviation, or any other relevant statistical parameter.

> A population has one, or more, common attributes and has variation with respect to other attributes.

The population of spores was clearly defined in the above example: it consisted of all the spores inside the Petri dish.

Consider now another population that is limited in space: a wheat field surrounded by roads that separate it from fields with potatoes. Assuming that the farmer did his work well, the distribution of the wheat plants is regular with equal distances between plants within the rows, and between the rows: and the population of wheat plants has an uniform

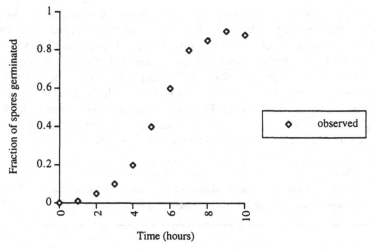

Figure 1. Fraction of spores germinated versus time after placing spores on water agar. The curve is sigmoidal and quite common for germination of spores. Fictional data are used.

distribution. A weed occurring in the field, e.g. *Galium aparine* (catchweed), will show a less regular distribution varying from random to clumped or aggregated.

The distribution of a pathogen, e.g. *Puccinia recondita* (brown leaf rust) may depend on time of observation. The rust fungus may infect a small group of plants at the start of an epidemic, i.e. a clumped pattern, but may infect nearly all plants at the end of an epidemic, achieving a nearly uniform pattern. Estimation of the distribution of the rust becomes more troublesome if we include also the spores. These are present on the soil surface, on the plants and in the air. We have therefore to include a third dimension to estimate adequately the distribution of the rust population. The question is then to which height should we trap spores? Or, where is the border of this population. We will return to this question later.

> A population has a certain distribution of the members belonging to it and this distribution is in the range of uniform to aggregated.

Consider again the wheat field. If we count the number of wheat plants and divide it by the area of the field we find the density of the wheat population. In fact, we don't have to count all plants because the distribution of the plants is uniform. Counting plants on a small sample area only can give an accurate measure of the density of the population. The story is different for a weed. The distribution of a weed may be patchy with a rather high density inside the patches and a low density outside the patches. To estimate the population density of the weed we would have to count the number of weeds of a large sample area, or even of the whole field. Estimation of the density of the rust population is even more complicated and would require both counting diseased plants on a certain area and sampling a certain air volume to count spores.

The example of the rust points to another problem in the estimation of population density. The members of a population are never 100% similar. In the case of the rust we counted diseased plants, which reflect rust mycelium, and spores. Even if we translate diseased plants to a number of spores infecting each plant, the fact remains that the pathogen is present in two different forms and stating a density for the rust population does mask this difference. The differences between individuals are also present in the wheat population, but are more subtle. There will be small genetic differences between plants, and each plant has its own micro-environment shaping the phenotype. And so, we are back at the variation within a population mentioned above.

> Density is a crude estimate masking the type of distribution of population members and masking differences between members of the population.

Estimation of the population density may sometimes be used for investigating processes in populations with a rather low amount of variation, like a crop variety, and an uniform spatial distribution. If the spatial distribution is heterogeneous, the density of a population should be indicated together with information about the spatial distribution. If the population is heterogeneous with respect to genotypes or phenotypes, the density of a population should be indicated together with information about the frequency distribution of genotypes or phenotypes.

Populations have their dynamics. The dynamics is predictable for the wheat population and depend completely on man. The dynamics of a weed is less predictable. Within the limits set by agriculture, density and distribution of the weed population change as a result of seed dispersal. Dispersal of spores changes the density and distribution of the rust population. Dispersal is referred to here as the movement of propagules within a population and as the movement causing extension of the population (Fig. 2). The latter may result in the foundation of a new population if the dispersal distance is relatively large. In contrast, migration is the movement of propagules between existing populations.

Dispersal and migration do not only cause changes in density and distribution. Independent of the type of propagule, be it spores, pollen, seeds or plant parts, genes are also transferred, resulting in changes in the genetic constitution of a population. This gene flow may change both the frequencies of genes in a population and the spatial distribution of genes within a population.

> Dispersal and migration cause changes in the density, the spatial distribution of the members, and the genetic constitution of a population.

Consider two adjacent wheat fields separated by a road and belonging to two different farmers. The direct influence of the wheat in one field on the wheat in the other field is negligible and it is useful to consider them as separate populations.

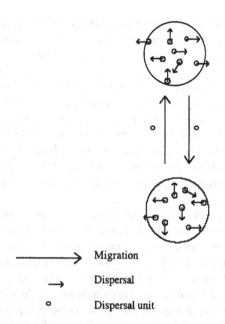

Migration

Dispersal

Dispersal unit

Figure 2. Dispersal and migration of dispersal units within and between populations. Dispersal is within and at the border of a population, migration is between populations.

The situation is less clear for a weed or rust occurring in the two fields. Most weed seeds may not be able to pass the road, but a few may do. Gene flow may occur at a relatively low rate. The rust, however, may easily pass the road and the rate of gene flow may be rather high. The distinction between the populations in the two fields then becomes vague. This again raises the question about the border of a population. But we now have an additional tool to determine the border of a population, the rate of gene flow. Two populations of a species may be considered distinct if the gene flow between them is absent, low, or infrequent; although this leaves the terms "low" and "infrequent" to be defined. We have, however, arrived at a measurable variable that can be used to define the border of a population. The availability of DNA markers facilitates the estimation of gene flow and may give research on gene flow a more pronounced place in studies on plant diseases (McDermott and McDonald, 1993).

We will now widen our view of populations. The example of brown rust has already suggested that the scale of a population may be larger than a field. For example, Wolfe *et al.* (1992) studied *Erysiphe graminis* f.sp. *hordei* (barley mildew) in Europe and considered it as one large population, made up of subpopulations. Subpopulations were not isolated and gene flow between them was possible. Despite this, subpopulations could be distinguished on the basis of differing virulence of isolates taken from the different subpopulations, and also by DNA markers and sensitivity to fungicides. This suggested that local factors acting on the subpopulations were more important for the genetic constitution of a subpopulation than genes transferred between the subpopulations. In similar work the genetic variation of three Danish local populations of *E. graminis* f.sp. *hordei* was studied (Damgaard and Giese, 1996). There was no indication of genetic differentiation amongst the populations. In fact, the overall Danish population is likely to be part of the European population and may be considered as a subpopulation of it.

The example of *E. graminis* f.sp. *hordei* makes clear that (1) the border of a population is a priori determined more by the working range of the researcher than by biological criteria, (2) gene flow may occur over rather large distances increasing the population size to one covering a very large area, and (3) local factors may still influence parts of the population relatively strongly. There is consequently a tension between increasing the size of a population to take gene flow into account, and restricting size to account for local factors. Moreover, a researcher may prefer to study a smaller population for technical and indeed financial reasons.

Gene flow is not always as extensive as in *E. graminis* f.sp. *hordei*. Most plant species and pathogens such as soil-borne fungi have limited gene flow and it is easier to distinguish populations. But here another problem is encountered: small populations have a relatively high probability of extinction. This does not inevitably mean extinction of the species as new populations may be founded. One spore, or one seed, may be enough to found a new population at a suitable site. This single propagule will largely shape the genetic constitution of the population in later generations, if the population remains isolated. The strong dependency of the genetic constitution of a population on the original individual(s) is called founder-effect (Falconer, 1989).

If we look at the parent population from which the colonisers originate, many genotypes may be present and of these, many may disappear at the time of extinction of the population, others may found a new population, or enter existing populations. There is therefore a dynamic situation of extinction and appearance of populations with an infrequent gene flow. Against this pattern of instability at the population level, the species does persist at a higher level, the metapopulation (Olivieri *et al.*, 1990). The concept of metapopulation takes into account local factors acting on individual populations and also gene flow between populations.

The concept of metapopulation was developed for endangered plant and animal species. Plant pathologists do, however, recognise immediately the analogy with various pathogens. A crop disease may be adequately controlled in a region by a resistant cultivar, leading to local extinction of the pathogen population. However, one virulent spore arriving from another region may build up a new population rendering the formerly resistant cultivar futile. The concept of metapopulation is a formalization of what had already been recognised for a long time by plant pathologists. Fry *et al.* (1992) successfully applied the concept of metapopulation to *Phytophthora infestans* causing potato late blight.

The rate of gene flow is a major criterion to distinguish populations.
The importance of gene flow is expressed in the concept of metapopulation.

Recent developments in modelling using the metapopulation concept (Hanski, 1994), and in analysis of genetic variation using DNA markers (McDermott and McDonald,

1993) offer new opportunities to study populations and their interconnections on a large scale. The relative importance of studies at a smaller, local, scale may decrease, but understanding of processes at a small scale will still be required to underpin understanding of processes at a higher level.

The phenomenon of population will remain central in the following sections where host - pathogen interactions are considered at the population level, but there is an additional problem in considering host - pathogen interactions: the scale of the host population often does not fit the scale of the pathogen population. This drawback must be tackled in studies of host - pathogen interactions.

III.   Host - Pathogen Interactions

In the following section we will compare different types of pathosystem to illustrate the whole set of mechanisms involved in host - pathogen interactions at the population level. A pathosystem is defined as any sub-system of the ecosystem that involves parasitism (Robinson, 1976). He distinguished between wild pathosystems, for pathogens with a wild plant as host, and crop pathosystems for pathogens with a crop as host. Later, a weed pathosystem was defined as any sub-system of the ecosystem that involves parasitism causing diseases of weeds (Frantzen, 1994a). The term "interactions" has to be treated with care because it also encompasses mechanisms that prevent an interaction sensu stricto between host and pathogen.

A.   ESCAPE AND AVOIDANCE

A plant will become diseased after contact with a pathogen in an infectious state. This may appear a very obvious statement. However, the occurrence of the contact between a plant in a susceptible state and a pathogen in an infectious state is by no means straightforward. The pathogen has first of all to arrive at the plant population. A study of the wild pathosystem *Viscaria vulgaris - Ustilago violacea* may be used to illustrate this point. The incidence of the smut fungus was determined in *V. vulgaris* patches (Jennersten *et al.*, 1983). The patches may be considered as populations. The smaller the patches, the lower the incidence of the smut fungus found. No smut was found in patches with less than 35 host plants. In this case, the authors suggested that insects transferring the smut did not visit the small patches. It may be stated as a general principle that small plant populations have a lower probability of being "detected" by pathogens than larger ones. Because small populations rather than larger populations are the rule for wild plants, escape of disease may be frequent in wild pathosystems. In contrast, crops are often planted over whole regions enabling a relatively easy "detection" by pathogens.

> The probability of dispersal of a pathogen to a host population decreases with decreasing size of the host population.

Disease escape is also possible even though the pathogen is present in a host population. A good example is provided by the attack of *Ammophila arenaria* (marram grass) by pathogenic soil organisms (Van der Putten and Troelstra, 1990). This clonal plant grows on the beach and dunes in The Netherlands. The plants grew well at sites where fresh, windblown sand was deposited continuously from the sea side because soil pathogens were not able to develop in the fresh sand. In contrast, plants growing at neighbouring sites without continuous sand deposition degenerated as soil pathogens developed well at these sites. Thus escape of plants from disease depended on factors governing sand deposition.

Disease escape is not uncommon in wild and weed pathosystems. Plants may use their seeds to escape from disease (Augspurger, 1983; Parker, 1987). Seeds are dispersed and offspring may establish at some distance from the mother plant. If the mother plant is infected by a pathogen, the pathogen may not be able to bridge the distance between mother plant and offspring and so the offspring may escape from disease. Plants may also use their ability for clonal growth to escape from disease (Wennström and Ericson, 1992; Frantzen, 1994b). New ramets may be produced at some distance from the mother ramet. If the mother ramet is infected by a pathogen, the pathogen may be not able to bridge the distance between mother ramet and daughter ramets and, so, the new ramets may escape from disease.

Disease escape also exists in crop pathosystems. Epidemics in crops may start from a relatively small amount of inoculum surviving within the field from a previous cropping period, or coming from outside; these are called focal epidemics (Zadoks and Schein, 1979). Plants in the vicinity of the inoculum are most likely to be infected whereas plants further away may escape from disease. The epidemic will progress and the rate of progress will be governed by, among other factors, environmental conditions and dispersal capacity of the pathogen. Thus the epidemic rate and the position of a plant relative to the inoculum source determine the probability and time of infection of that plant.

Disease escape has particular importance in multiline cultivars and variety mixtures. The rationale behind the use of multilines and variety mixtures is to create a heterogeneous crop with respect to disease resistance characters, while at the same time maintaining homogeneity with respect to characters determining quality and quantity of the yield (Wolfe, 1985). Consider the case where inoculum of one race of a pathogen enters the population and infects a susceptible plant (Fig. 3). Spores will be produced on this plant, and then encounter two subsequent problems. Firstly the distance to the next susceptible plant is relatively large compared to that in a homogeneous susceptible crop. Secondly, the spores may be caught by the surrounding, resistant plants: this is called the "barrier effect" (Wolfe, 1985). The net result is that susceptible plants may escape from disease. Such disease escape is independent of the race arriving or present in the crop, unless that race is pathogenic to all plant genotypes. We will discuss the likelihood of such a super race later.

> Disease escape is the rule rather than the exception in wild, weed and crop pathosystems. The concept of multilines and variety mixtures is based on it.

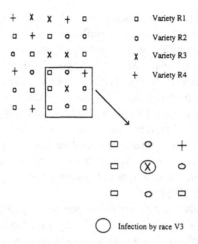

Figure 3. Concept of multilines / variety mixture illustrated by a plant mixture of which each plant has one resistance gene (R1, R2, R3, R4) and the crop is attacked by one virulent pathogen race (V3).

Until now, we have considered escape as a chance event. Whether infection occurs or not depends on environmental conditions and the fortuitous position of a plant. A form of escape may also depend on the genotype of the plant, in which case it is known as avoidance. Examples of avoidance were given by Agrios (1980), although he did not use the term. Later, Dinoor and Eshed (1984) noted that "under the selection pressure imposed by pathogens, early- or late-maturing plants may have an advantage in that they avoid the impact of disease". The terms escape and avoidance were distinguished explicitly by Burdon (1987a). In his opinion escape is the result of a fortuitous set of circumstances whereas disease avoidance has a genetic base. Because of this genetic base, he classified avoidance as a (passive) resistance mechanism.

Avoidance may be based on physical protection of susceptible tissue against a pathogen, or on differences between genotypes in rate of development. Classical examples of the former are varieties of barley and wheat in which the flowers remain closed, reducing the probability of contact between the stigmas and the pathogen *Ustilago nuda*, a smut disease. Open flowering varieties are more prone to disease. There is some doubt whether this type of avoidance can justifiably be considered as separate from mechanisms of resistance in the strict sense (see section III. C). The physical protection could be considered as falling within constitutive resistance. In contrast, the second type of avoidance, based on differences between genotypes in growth rate is more distinct from resistance sensu stricto. We will turn first to weed pathosystems to illustrate this distinction.

The perennial *Silene alba* is frequently infected by the anther-smut *Ustilago violacea* (more recently called *Microbotryum violaceum*) and teliospores are florally transmitted by insects. Biere and Antonovics (1996) observed that plant families differed significantly in time of onset of flowering and plants flowering late had a lower probability of becoming infected, i.e. avoided disease. The disease avoidance mechanism therefore had two components, (1) differences between plant stages in susceptibility to the pathogen (only susceptible in the flowering stage), and (2) differences between plant families (genotypes) in growth rate. A similar phenomenon was observed for the annual plant *Senecio vulgaris* and the rust fungus *Puccinia lagenophorae* (J. Frantzen and H. Müller-Schärer, unpublished). Plants at a young stage were less susceptible than plants at an older stage, probably because spores adhered less to the leaves of young plants. The plant families used in the experiments differed in rate of development and, therefore, the number of plants present in a susceptible state at the time of inoculation differed between families. Two plant families avoided disease compared to the third family. Again, this system showed the two components of the avoidance mechanism: the differences between plant stages in susceptibility, and the differences in rate of development.

The rate of development also may govern avoidance in crop pathosystems (Agrios, 1980). For example, timing of blossoming and fruit set may vary between apple varieties and some of the varieties may blossom at a time when a pathogen is not infectious. The use of the mechanism of avoidance in agriculture may, however, be constrained by factors not related to crop protection, e.g. feasibility of harvest, and yield quality.

> Avoidance is a disease defence mechanism that can be distinguished from resistance. It may be quite common. Its use in agriculture may be constrained by factors not related to crop protection.

B.   PREDISPOSITION AND IMMUNISATION

Plants are exposed to environmental conditions, both biotic and abiotic. These conditions may predispose them to disease making them susceptible to a pathogen before the contact with the pathogen (Schoeneweiss, 1975). The reverse may also occur making plants immune to a pathogen (Dean and Kuç, 1987). The mechanisms of predisposition and immunisation are non-specific with respect to genotypes of host and pathogen. Other terms in use for immunisation are induced resistance and systemic acquired resistance (SAR). I prefer the term immunisation to avoid confusion with the term resistance sensu stricto treated in the next section. (See, however, chapter 10 for an extended discussion of the terminology.)

The mechanisms of immunisation at the individual plant level are the subject of extensive research. Less is know about the role of immunisation in populations. The role of immunisation was studied for the weed pathosystem *Rumex crispus - Uromyces rumicis* (Hatcher *et al.*, 1994). A beetle infests *Rumex* plants in spring and early summer before epidemics of *U. rumicis* start. This feeding results in immunisation of the plants to subsequent infection by the rust fungus. The beetle has a negative impact on the plants by feeding on them, but also prevents plant disease. The immunisation is, however, not complete and the rust may develop in relatively small amounts on plants.

The mechanism of immunisation was not investigated in detail, but probably involved development of the normal cascade of resistance responses following attack.

Immunisation may play a role in the concept of multilines or variety mixtures. The spread of *Erysiphe graminis* f. sp. *hordei* Marchal was quantified for a barley population consisting of four varieties (Chin and Wolfe, 1984). The results suggested that spores of a non-virulent pathogen biotype landing on a resistant variety turned on defence mechanisms making the plants immune to the virulent pathogen biotypes. The protection was not 100% as some reproduction of the virulent biotype occurred. The epidemic was slowed down, but not stopped. In general, incomplete protection of the plants seems to be a characteristic of the mechanism of immunisation (Dean and Kuç, 1987).

Predisposition and immunisation are mechanisms likely to operate in populations, but it is difficult to estimate their significance in the development of epidemics in populations. A set of environmental conditions may be present in a population triggering both predisposition and immunisation. Which one will win? In the study by Chin and Wolfe (1984) is was likely that immunisation won. However, environmental conditions, biotic and abiotic, are dynamic and it might be difficult to predict the effects of predisposition and immunisation, respectively, on disease epidemics. Moreover, environmental conditions are not uniform in plant populations, especially in wild plant populations, and this makes predictions even more difficult.

> The effects of the mechanisms of predisposition and immunisation on disease epidemics are hard to predict because of the stochastic nature of the environmental factors, biotic and abiotic, governing these mechanisms.

C.   RESISTANCE

We now arrive at the moment when a pathogen in an infectious state contacts a plant in a susceptible stage. It is assumed that the plant is predisposed to disease by the environmental conditions and the environmental conditions at the point of contact are suitable for the process of infection. In short, all is fine for development of disease. Plant genotypes may, however, react in different ways to the pathogen. Some may inhibit its growth completely, others may inhibit growth incompletely, and some may not inhibit development of the pathogen at all, reflecting complete resistance, incomplete resistance and susceptibility, respectively. This subdivision of resistance is relevant to epidemics in plant populations. An epidemic does not progress in a population with only resistant plants, it is delayed in a population with plants having incomplete resistance, and progresses without any restriction in a population with only susceptible plants.

The genotypes of the pathogen may differ in their pathogenicity to plants. A genotype may develop on a plant without restriction, development may be limited or the genotype may not develop at all. This reflects virulence, the degree of aggressiveness, and avirulence, respectively.

If a host encompasses genotypes with reaction types in a continuous range of susceptibility to incomplete resistance the effect is called quantitative resistance. The counterpart of quantitative resistance is aggressiveness of the pathogen. If the host reactions are either (close to) completely resistant or susceptible the effect is referred to as qualitative resistance. The pathogen is virulent or avirulent. As pointed out by Nelson (1978), this division into qualitative and quantitative resistance is relevant to understanding the development of epidemics. Qualitative resistance inhibits the infection process and prevents production of inoculum for progress of the epidemic. Quantitative resistance does not inhibit the infection process completely and allows production of inoculum. The production may, however, be delayed (longer latent period) or may be reduced. The consequences are a delay of the epidemic and reduction of disease levels in a population.

There may be specific interactions between genotypes of the host and genotypes of the pathogen. Using agricultural terms, we are talking about race-specific resistance and cultivar-specific pathogenicity, respectively. Specific interactions between genotypes of host and pathogen may be determined in both qualitative and quantitative resistance.

Qualitative resistance inhibits development of an epidemic. Quantitative resistance reduces the rate of development of an epidemic. The counterpart of qualitative resistance is virulence of the pathogen. The counterpart of quantitative resistance is aggressiveness of the pathogen. Specific interactions between genotypes of the host and genotypes of the pathogen are possible.

The reader should be aware of the confusion in the literature concerning the term resistance and the types of resistance. The terminology adopted here is basically that of Parlevliet (1989) because it is suitable for describing processes at the population level, in which resistance is involved. Additionally, the term qualitative is used here as a logical counterpart of quantitative resistance. The definitions of virulence and avirulence deviate from the terminology of Parlevliet (1989), but seem logical and fit the common use of these terms (Ennos, 1992).

After this consideration of terminology, we now examine disease development in wild pathosystems. The perennial plant *Plantago lanceolata* is infected by the necrotrophic fungus *Phomopsis subordinaria* (teleomorph *Diaporthe adunca*). An insect vector is involved in the transmission of the fungus (de Nooij, 1988). Transmission is also possible by splash dispersal (Linders *et al.*, 1996). Variation of resistance in host populations was investigated (de Nooij and van Damme, 1988a). The study is summarised in the box below.

Resistance of *Plantago lanceolata* to *Phomopsis subordinaria*

- Plants sampled in 3 populations
- Sampling along a transect at distances of 5 m
- Clones inoculated with *P. subordinaria*
- Lesion size was between 193 mm and 246 mm
- Significant differences between populations
- No significant differences within populations
- Survival decreased with increasing lesion size

Differences in resistance to *P. subordinaria* were detected between populations and not within populations. The differences in resistance were relevant to survival of plants. More resistant plants had a higher probability of survival.

Variation in pathogenicity was also investigated using the same plant populations (de Nooij and van Damme, 1988b). Differences in pathogenicity were detected between and within populations (see box below). The differences within populations were detected on a rather small scale. The differences determined were relevant to survival of the host plant.

Pathogenicity of *P. subordinaria*

- Isolates collected in 3 populations
- Isolates collected from 12 plants in a population
- Three clones inoculated with isolates
- Lesion size was between 176 mm and 314 mm
- Significant differences between populations
- Significant differences within populations
- Differences on a scale of 1.5 m

The work of de Nooij and van Damme (1988b) is a notable exception (Ennos, 1992) among studies on wild and weed pathosystems for measuring the distribution of pathogenicity in the pathogen population on such a small scale. The studies of the wild pathosystem *P. lanceolata - P. subordinaria* illustrate some phenomena of interest.

- The presence of quantitative resistance of the host and the corresponding aggressiveness of the pathogen.
- The relevance of quantitative resistance to plant survival.
- Variation of quantitative resistance and aggressiveness on a rather small spatial scale.
- Specific interactions between host plants and isolates were present suggesting a gene-for-gene interaction.
- The variation in aggressiveness of the pathogen seemed to be larger than that of quantitative resistance of the host. This suggested that the variation in aggressiveness was not only the result of a reaction to the variation in quantitative resistance, but

also the result of isolation phenomena in the pathogen population. Isolation might be explained by the limitations of the transmission of the fungus by way of an insect vector and by way of splash dispersal.

The genetic basis of quantitative resistance and aggressiveness was not unravelled for the pathosystem *P. lanceolata - P. subordinaria*. This is a general drawback of studies on wild and weed pathosystems. Experimental crosses are a prerequisite for a classical genetic analysis, but are often not feasible. One of the few exceptions is work by Burdon (1987b). Plants of the perennial *Glycine canescens* were sampled in two populations and tested for resistance to nine races of the rust fungus *Phakopsora pachyrhizi*. Both quantitative and qualitative resistance were determined and resistance was race-specific. Subsequent crosses between plant lines indicated the presence of one, two or three resistance genes (factors) in the plant lines. The identity of the resistance genes in the plant lines remained unknown. Application of the principles of the gene-for-gene theory suggested the presence of at least 10-12 resistance genes in each of the *G. canescens* populations.

The major point of interest in the study by Burdon (1987b) was that race-specific resistance may be present in wild pathosystems and the gene-for-gene theory, more recently called allele-for-allele hypothesis (e.g. Newton and Andrivon, 1995), may apply. Race-specific resistance and the suitability of the gene-for-gene theory was also demonstrated for the weed pathosystem *Senecio vulgaris - Erysiphe fischeri* (Harry and Clarke, 1986, 1987). Neither Harry and Clarke (1986, 1987) nor Burdon (1987b) distinguished qualitative and quantitative resistance in their analysis, although the data presented indicated both types of resistance.

> Both qualitative and quantitative resistance are present in wild and weed
> pathosystems. The variation of resistance in the host populations and the
> variation of pathogenicity in the pathogen populations may be relatively
> large. The underlying genetics are rather poorly understood.

The variation of resistance and pathogenicity in host and pathogen populations, respectively, may be dynamic. This was demonstrated for the pathogenicity of the rust fungus *Melampsora lini* infecting plants of *Linum marginale* (Burdon and Jarosz, 1992). Nine populations of the rust were sampled during 2, 3, or 4 years. The date of sampling and the number of dates of sampling differed between years and populations. The data were, however, analysed on a year to year base. The isolates sampled were tested on a differential set of host lines and classified to five categories. The distribution of isolates over these five categories changed significantly from year to year for each rust population indicating relatively strong dynamics. The authors explained the temporal changes by metapopulation dynamics (genetic drift, extinction / recolonisation events) rather than by considering the changes as a response to the variation of resistance in the host populations.

We turn now to crop pathosystems and the picture becomes simpler with respect to variation of resistance in the host population. The variation is nearly absent, or at least low compared to wild and weed pathosystems. In so far as it is present, variation is nearly completely managed by man. The reason for this low variation is well understood and of long standing (see e.g. Zadoks and Schein, 1979). Crops should be high yielding and as uniform as possible to ease the harvest. The selection force exerted by man during centuries of plant breeding has resulted in rather uniform crops with few resistances to diseases. The frequency of resultant epidemics, like those of yellow rust, then stimulated breeding for disease resistance (Johnson, 1992). As pointed out by Johnson, breeding for resistance to a number of diseases affecting crops is a difficult task given the constraints of agronomic demands (e.g. yield quality). Moreover, resistance should be durable. Durability is, however, the exception rather than the rule in crops. This may be explained by the "flexibility" of the pathogens as outlined next.

The dynamics of the population of *Phytophthora infestans* was studied in The Netherlands (Drenth *et al.*, 1994), using isolates collected before and after 1980 (see Box).

Dutch population of *P. infestans*
- Isolates collected before 1980 ("old")
- Isolates collected after 1980 ("new")
- Differential set to test for pathogenicity
- DNA fingerprinting using the probe RG-57
- 148 "old" isolates belonged to 8 races
- "old" isolates were same fingerprint genotype
- 253 "new" isolates belonged to 73 races
- "new" isolates were 134 distinct genotypes

Two explanations for the change of genetic variation around 1980 were presented by the authors: a new population was introduced into The Netherlands from Mexico consisting of a large number of heterogeneous isolates, and the introduction of the A2 mating type, also from Mexico, into the former A1 mating type Dutch population allowed sexual reproduction and an increase in genetic diversity. In common to both explanations is that gene flow from Mexico to The Netherlands had a tremendous effect on the genetic constitution of the Dutch population, which might have severe consequences for the protection of potatoes against late blight. As pointed out by Fry *et al.* (1992) the devastating potato late blight outbreak in Europe in 1845 might have been caused by a previous gene flow of *P. infestans* from Mexico to Europe and, following Andrivon (1996), that gene flow may have passed from Mexico via South America to Europe.

The relevant point in the potato late blight story is that a reservoir exists from which races disperse to vacant sites, which are crops not yet occupied by the pathogen due to resistance or any other adequate control measure. It is a matter of trial and error for the pathogen to "send out" the right race to break the control measure. Looking at the size of the metapopulation of *P. infestans*, or the European population of *E. graminis* f.sp. *hordei,* it is clear that one or more reservoirs exist. These reservoirs may exist as (sub)populations on crop populations not protected against the pathogen, as (sub) populations on populations of alternate hosts, or as survival units outside a host. Migration of races is assured by various vectors. Even pathogens like soil-borne fungi and viruses may be transported over large distances by man acting as a vector.

> The sizes of (meta)populations of pathogens are large compared to crop populations and provide enough reservoirs to develop new races to confront crop populations frequently with races never encountered before.

We may assume that a crop population is confronted frequently with races of a pathogen not encountered before. This triggers the question how often? This question will be returned to in a later section (IV. Selection and Fitness).

If crops are confronted frequently with new races of a pathogen, a reaction is required to minimise crop losses due to disease. Until now, the reaction of man has been primarily in breeding crops for resistance based on the gene-for-gene hypothesis, that is in deploying race-specific resistance. To summarise a long story, a plant variety has a factor enabling recognition of a corresponding factor in a pathogen race. These factors are called commonly R gene and Avr gene, respectively. More details are given elsewhere in this book (chapter 2). The quintessence is to have the right R-genes in the crop plants to recognise the pathogenic races arriving at the population. And, as outlined above, the race arriving at the crop may change with time. Some varieties may resist for quite a long time and are called durably resistant (Johnson, 1992), other varieties do not resist for long.

Three strategies may be followed to increase the chance of durability of resistance. The first strategy is "pyramiding" of R-genes in one variety, reducing the R-gene homogeneity to which the pathogen population is exposed, and integrated control (Crute and Pink, 1996). The strategy of "pyramiding" has, in general, not resulted in more durability. This might be a result of the constraints set by using traditional breeding methods. Consequently, R genes were used for which the matching virulence-genes were already present in the pathogen (meta)population. The second strategy may be based on rotating cultivars with different R-genes over seasons, planting cultivars with different R-gene complements in adjacent fields, or using the concept of variety mixtures / multilines, as described in previous sections (III. A and B). An example of the third strategy is the integration of race-specific resistance in chemical control. This strategy was, for example, forced on lettuce growers as races of *Bremia lactucae* in the UK became insensitive to fungicides (Crute and Pink, 1996).

A whole cascade of defence reactions may be turned on after recognition of the pathogen race by (the product of) one R-gene as outlined above. These genes are often called major genes (but see comments by Nelson, 1978). The type of resistance triggered is often complete, qualitative resistance, but may also be incomplete, quantitative resistance. In the case of quantitative resistance, several genes may be involved in the start of the defence reaction of the host: each such minor gene contributes a part to the reaction. If minor genes are involved, the type of resistance is called partial

(Parlevliet, 1989). As pointed out by Parlevliet, partial resistance might be more durable than resistance based on major genes: he referred to the durability of partial resistance of corn to *Puccinia sorghi* and of barley to *Puccinia hordei* as examples. He also pointed to the problems involved in breeding for partial resistance, which are primarily based on the difficulties of separating the effects of major genes and minor genes as both often go hand in hand. Also, the protection achieved by partial resistance may depend on the severity of an epidemic, which in turn may depend on environmental conditions. This indicates a point relevant to employing partial resistance and the multiline / variety mixture concept to protect crops from diseases. Epidemics will still occur although at a relatively low level. This might be acceptable from the point of view of durability, but what about the damage caused to the crops by the low level of epidemics? This question will be central to the next section.

D.   COMPENSATION AND TOLERANCE

This section deals with the mechanisms of compensation and tolerance. Compensation may operate at the individual plant level and at the population level. Compensation at the individual plant level might be the basis for the mechanism of tolerance. First, compensation at the individual plant level will be described for the weed pathosystem *Senecio vulgaris - Puccinia lagenophorae*.

The effects of the rust *P. lagenophorae* on plants of the annual weed *S. vulgaris* were quantified by Paul and Ayres (1984). The sixth leaf (third leaf pair on the stem) of plants was inoculated with the fungus, or the leaf was not inoculated and plants served as control. After establishment of the rust, the net photosynthesis was measured in the eighth leaf (fourth leaf pair on the stem). Net photosynthesis of leaf eight on plants with an infected sixth leaf was significantly higher than the net photosynthesis of leaf eight on the control plants, compensating partially for the losses in net photosynthesis of the infected sixth leaf. Compensation of losses by increased photosynthesis rate was also demonstrated for crops such as broad beans (Murray and Walters, 1992, and references therein).

If the capacity for compensation differs between two plant genotypes, compensation is an underlying mechanism of tolerance and tolerance is defined as "that capacity of a cultivar resulting in less yield or quality loss relative to disease severity or pathogen development when compared with other cultivars" (Schafer, 1971). Later on, Clarke (1986) argued that demonstration of tolerance is impossible because disease severity can not be measured precisely. A theoretical exercise may, however, make it clear that tolerance is measurable.

Consider the case of two crop varieties. The difference between the two varieties in crop yield is determined if disease is absent (δy). Subsequently, we estimate the disease severity - crop yield relation for each cultivar. The absolute difference between the curves of the varieties is constant, i.e. independent of the disease severity, and equals:

$$\delta y_{tot} = \delta y + \delta y_{tol}$$

where δy_{tol} is the crop yield difference caused by the difference in tolerance between the cultivars. Disease severity - crop yield curves are used to give a more precise estimate of δy_{tot} than obtainable by comparing the cultivars only for one level of disease severity. Moreover, it might be difficult to obtain a similar disease severity on the two varieties. A basic assumption is that disease severity is measured appropriately. If not, δy_{tol} also may include a resistance effect. To control, for this bias, the yield curves of the cultivars should be compared using more than one pathogen strain. By definition, tolerance is independent of the pathogen strain used and we should find the same δy_{tol} for each pathogenic strain.

So far as is known, tolerance has not yet been demonstrated unequivocally in practice, but the mechanism has to be mentioned as a possibility. If it does exist, breeding for tolerance might produce crop varieties, which would be durable as no pathogen race can break tolerance. Of course, the question is how tolerant a crop plant can be to disease?

> Losses due to infection may be compensated by uninfected parts of a plant. If genotypes exist with a different capacity of compensation, breeding for tolerance might be possible. If it exists, tolerance would be race non-specific.

We return now to the question posed at the end of the previous section, are low levels of epidemics acceptable for the farmer? The answer is no if we consider yield quality. Each apple infected with a small amount of *Venturia inaequalis* is unmarketable because most consumers would not accept blemished fruit, even if the disease symptoms are superficial and do not influence the taste. The story becomes different if we consider yield quantity.

The yield, or biomass, depends firstly on the availability of resources (light, energy, nutrients, water) at a specific site. The availability of resources differs between regions, fields, and (micro-)sites within a field. The availability of resources also has a heterogeneous distribution in time. One year has more sunny days than another, fertilisation is concentrated at some times within a growing season, rainfall varies, etc. Secondly, stresses are imposed on the crop and may reduce the biomass. Stresses are abiotic like extreme rainfall, or dry and hot weather, and biotic like pests and diseases. The stresses are distributed heterogeneously in both space and time. The effect of each of the stresses on the host depends on the tolerance of the crop to these stresses. The influence of one stress on a crop will, however, also depend on the presence of other stresses, and the availability of resources. It should by now be clear how difficult it will be to predict what level of disease can be accepted by a farmer.

A third factor, plant competition, has also to be considered. Each plant has to compete with its neighbours for resources and this competition will be stronger the more similar a plant's resource requirements are to those of its neighbours. Resource requirements are more similar the more similar the genetic constitution of plants. Thus the resource requirements of plants belonging to the same species are more similar than those plants belonging to different species, or those of plants belonging to the same variety are more similar than plants belonging to different varieties.

Competition is the motor behind the mechanism of compensation at the population level. Resources not used by a diseased plant are available for neighbours and these may grow better, i.e. produce more biomass. The net result may be that the total biomass of the population (crop) is similar for a diseased and a non-diseased population (Zadoks & Schein, 1979). In fact, the biomass of the diseased population might even be higher if disease changes the resource requirements of a plant by mimicking a change from intravariety competition into intervariety competition. Stresses other than disease may cause similar compensation effects.

Of the three main factors explaining crop yield: resource availability, competition and stresses, disease is just one part of the whole story. In general, crop yield may be more stable in variety mixtures than in monocultures (Schutz and Brim, 1971). In their study, the yield obtained from variety mixtures of soybean varied less over years and planting sites than the yield of monocultures of the individual varieties. The authors attributed this stability largely to compensation effects.

> Local biotic and abiotic factors determine crop yield. Depending on these factors, crop losses due to disease may be more or less compensated at the population level.

IV. Selection and Fitness

The process of (natural) selection may operate if, (1) a population has variation among individuals in some attribute, (2) there is a consistent relationship between that attribute and survivorship or fecundity, i.e. fitness differences, and (3) there is a consistent relationship, for that attribute, between parents and their offspring, which is at least partially independent of common environmental effects, i.e. inheritance (Endler, 1986). If the process of selection occurs, i.e. the conditions (1) to (3) are present, the attribute frequency of all offspring in the population will be predictably different from that of all parents.

Demonstration of variation among individuals of a population in some attribute is relatively easily. Variation relevant to host - pathogen interactions involves avoidance, resistance and tolerance on the host side. Variation in pathogenicity is the counterpart of resistance on the pathogen side.

The most fundamental definition of fitness is the expected time to extinction of a unit of interest, which may be a gene, a genotype, a phenotype, or a complete species (Cooper, 1984). This is a logical definition with respect to evolution, but intractable to research. Therefore, time scales of one or a few generations are commonly used to study fitness (Antonovics and Alexander, 1989). A time scale of one generation is used in the "predicted fitness" method, where the contribution of an individual to the next generation is estimated by measuring traits (number of spores, seed size, etc.) that may influence that contribution. A longer time scale is used in the "realised fitness" method, where the observed change in frequency of a genotype, or phenotype, across generations is related to differences in fitness.

The inheritance of an attribute is demonstrated classically by crosses. Crosses are, however, impossible if the target organism lacks a stage of recombination, or the recombination is not open to manipulation by the researcher. If so, inheritance of an attribute may be ascertain by studying several generations of the organism under controlled conditions, or by manipulating the genome using mutagenic factors. The availability of methods like RFLPs offer new possibilities to study inheritance in asexual reproducing organisms by excluding the factor "environmental conditions" in inheritance studies. Considering host - pathogen interactions, the attributes avoidance, resistance, and pathogenicity are inherited.

Individuals in a population differ in many attributes and each attribute contributes to the fitness of the individual. It is likely that an attribute contributing more to the fitness of an individual is more important with respect to the process of selection. For example, pathogenicity may contribute more to the fitness of an obligate biotrophic fungus than to the fitness of a facultative biotrophic fungus. A virulent race of a biotrophic fungus has a higher fitness than an avirulent one because it can reproduce and the avirulent one cannot. In contrast, a virulent race of a facultative fungus may be outcompeted by an avirulent race because the latter might reproduce outside a host at a higher rate than the virulent race on the host.

Looking at host - pathogen interactions the fitness story is more complicated than just resistance and pathogenicity. Survival and reproduction of both host and pathogen are influenced by many biotic and abiotic factors. The complexity of the fitness story may be illustrated by a rather simple example of two races of *Puccinia graminis* f.sp. *tritici* with a different degree of aggressiveness and a different response to temperature (Katsuya and Green, 1967).

Fitness of two *P. graminis* races
- Plants of wheat cultivar Little Club
- Inoculation with races 15B-1 and 56
- Mixture of 50: 50 at first inoculation
- Incubation at 15 or 25°C
- Collection of spores and re-inoculation
- Same procedure for 10 rust generations
- Race 56 outcompeted 15B-1 at 25°C
- Race 15B-1 outcompeted 56 at 15°C

The frequency of race 56 increased from generation to generation at 25°C outcompeting race 15B-1 (see Box). In contrast, the frequency of race 15B-1 increased at 15°C outcompeting race 56. Thus the response to temperature was more important in this study with respect to fitness than aggressiveness. Temperature and not resistance of the host was the major selection factor.

The previous paragraph added a fourth factor to the process of selection, a selection factor. Fitter individuals in a population have to be selected out by a specific factor. In the example, temperature was the factor selecting out the fittest race, but temperature also determined fitness and the argument runs into a circle. An individual is only fitter if the selection factor is present and favourable to it. In the example, temperature was the only variable in a controlled environment. In reality, the overall fitness of an individual in a population is determined by an array of attributes and an array of selection factors interacting with it. Therefore, fitness has always to be defined in the context of the environmental conditions.

> Selection operates if a population has variation in an attribute, the attribute
> is heritable, and a selection factor transforms the variation in the attribute
> into a variation in fitness.

A pathogen may be a selection factor by selecting out resistant genotypes of the host. The reverse may also occur: the host selects out virulent, or aggressive, genotypes of the pathogen. If this reciprocal selection occurs, the basis for coevolution is present. In this context, it is worthwhile to consider firstly to what extent a pathogen is a selection factor for the host and to what extent the host is a selection factor for the pathogen. We will return to some examples presented in the previous section (III. C. Resistance) to estimate the extent of host and pathogen as selection factors.

Populations of *Plantago lanceolata* had variation among plants in resistance to the fungus *Phomopsis subordinaria*. Plants more resistant to the fungus had a higher probability of survival and may be called fitter. Inheritance of resistance was not studied, but is likely. So, all conditions seemed to be present for the process of selection. The disease incidence in the study populations was, however, contradictionary to the results of the pathogenicity tests indicating that environmental factors were more important for the development of the disease than pathogenicity of the fungus and resistance of the plant (de Nooij and van Damme, 1988b). In this case, a weevil functioning as vector seemed to be an important environmental factor. In general, a pathogen is a selection factor of plants if the conditions are suitable for development of an epidemic and the various genotypes of the host have a similar probability to come into contact with the pathogen. The same conclusion holds for a host to be a selection factor for a pathogen.

The wild pathosystem *Linum marginale - Melampsora lini* provided information about the temporal and spatial scale of selection processes in host - pathogen interactions (Burdon and Jarosz, 1992). The genetic constitution of a pathogen population present on a host population was dynamic. This pointed to a relatively high rate of gene flow between various pathogen populations, whereas this flow was far more limited for the host. Considering the various pathogen populations as a metapopulation, the size of the pathogen population was far larger than the size of the host population. This is a phenomenon also present in crop pathosystems as illustrated by potato late blight on potatoes (Fry *et al.*, 1992).

We will now take a closer look at the consequences of the metapopulation concept for selection processes, focusing on crop pathosystems. The story is, however, similar for wild and weed pathosystems. The only, though not unimportant, difference is that the variation of the crop with respect to resistance is determined by man, whereas wild and weed pathosystems have their own dynamics with respect to variation in resistance.

Consider a crop monogenic with respect to resistance. A pathogen will not establish until a race arrives with the corresponding virulence gene. This race may come from any part of the pathogen metapopulation. This may be from another population of the crop, a wild population of the crop, a population of another host, or a reservoir outside any host. The probability of dispersal of the virulence gene to the host population depends on the presence of the gene somewhere in the metapopulation, and on the dispersal rate of the pathogen. After arrival at the host population, conditions have to be suitable for infection. If so, the pathogen starts to produce dispersal units with the virulence gene, which may leave the crop and spread to other parts of the metapopulation. A dispersal unit with the virulence gene has a certain probability to leave the crop and to find a susceptible host plant elsewhere.

The history of crop diseases makes it clear that the probability of dispersal of a virulence gene to a monogenic resistant crop is high enough frequently to break resistance. Pathogens may therefore be selection factors for host populations. The inverse, hosts as selection forces for pathogens, is less clear. What is the probability of a virulence gene returning from a crop to the relevant sites in the metapopulation? Does a virulence gene of *Phytophthora infestans* return from Europe to the site of origin in Mexico in such an amount to replace other genotypes in the pathogen population there? Or can the European potato population be a selection factor for the *P. infestans* population in Mexico? It seems unlikely.

Matters are different if we consider the case of a crop with more than one resistance gene present due to the use of a cultivar mixture, or multilines. Various virulence genes may arrive at the crop and establish. Establishment of a pathogen is more likely in the case of multilines than in the case of a monogenic resistant crop. After establishment of the pathogen in the multigenic resistance crop, a selection process might start. In theory, a pathogen race breaking all types of resistance inside the crop would maximise reproduction and may outcompete other races. Such a "super race" might arise by mutation or recombination of the pathogen. The question is then whether a super race may indeed arise. This question has triggered various studies. One will be considered in detail.

Selection of virulence of wheat stem rust *P. recondita* f.sp. *tritici* against five resistance genes was studied in the greenhouse (Kolmer 1995). Pathogen phenotypes virulent to three resistance genes became dominant in three populations whereas phenotypes virulent to all five resistance genes remained at low levels, i.e. less that 2% of the isolates sampled, throughout the generations (see Box). The results of this study suggested that selection may occur in the direction of more complex pathogen races, with respect to virulence, but that there is an upper limit to the accumulation of virulences inside one race.

Selection of virulence of *P. recondita*
- Study in the greenhouse
- Three populations of wheat
- One with five isogenic lines, and each line with one specific resistance gene
- One with the same five isogenic lines, plus a susceptible cultivar
- One with the same five isogenic lines, plus the susceptible cultivar in a higher frequency
- Inoculation with mixture of virulence phenotypes
- Collection of spores and re-inoculation
- Same procedure for 12 rust generations
- Phenotypes virulent to three resistance genes became dominant
- The frequency of a "super race" remained low

A crop with more than one resistance gene present due to a cultivar mixture, or multilines, seems not to be a risk in terms of selecting out a "super race". However, various pathogen races may establish and in the end reach a situation similar to that in wild pathosystems, i.e. variation of resistance within the host population and variation of pathogenicity within the pathogen population. As in the case of a monogenic resistant crop, the crop determines the local pathogen population on it, but it seems unlikely that the crop is a selection factor for the metapopulation of the pathogen for the same reasons as described for the monogenic resistant crop.

A pathogen may be a selection factor for a host population if there is contact between host and pathogen and the conditions are favourable for infection. A host only may be a selection factor for a pathogen population at a local scale.

We return now to the topic of coevolution, defined as reciprocal evolution in interacting species. The hypothesis of gene-for-gene coevolution assumes that each gene for resistance in a plant population is matched over evolutionary time by a corresponding specific gene for pathogenicity in a pathogen population through reciprocal selection of the plants and pathogens (Thompson, 1990). Two predictions can be derived from this hypothesis. One, genes for resistance and pathogenicity require the presence of one another within a population to be maintained by natural selection. Two, polymorphisms will develop for resistance within host populations and for pathogenicity within parasite populations. The hypothesis applies to race-specific resistance and cultivar-specific pathogenicity. No prediction is made about speciation on the host or pathogen side.

The gene-for-gene hypothesis has been applied to various pathosystems, as described before (section III. C.), supporting the idea of coevolution of host and pathogen. Strictly speaking, coevolution is not possible in crop pathosystems because the dynamics of the genetic constitution of a crop are completely determined by man. Neglecting this objection, there is nevertheless a need for care in considering coevolution in pathosystems. Firstly, the size of pathogen (meta)populations always seems to be larger than the size of the host populations, making it unlikely that the host is a selection factor for the whole pathogen population. Secondly, resistance and pathogenicity are just one of the sets of characteristics determining the fitness of host and pathogen, respectively. Thirdly, resistance and pathogenicity are only two of many factors determining host - pathogen interactions at the population level as outlined in the preceding sections.

V.   Management of Resistance

Various mechanisms at the population level are involved in host - pathogen interactions. Escape and avoidance are likely to have a more significant role in wild and weed pathosystems than in crop pathosystems. The role of immunisation and predisposition at the population level is less clear, but might be more pronounced in wild pathosystems due to the relatively large spatial and temporal heterogeneity of biotic and abiotic factors inducing these mechanisms. Compensation and tolerance might also have a more pronounced role in wild and weed pathosystem than in crop pathosystems, although these mechanisms are relatively poorly studied. Comparing wild, weed and crop pathosystems, the mechanism of resistance has a different magnitude of importance in crop pathosystems from that in the wild counterparts. Resistance has become of utmost importance in crops because the other mechanisms that may reduce the impact of pathogens on plants are nearly absent in crop pathosystems. Reinforcement of these mechanisms in crop pathosystems may reduce the reliance on resistance to protect crops against disease. Is reinforcement, however, feasible?

Reinforcement of the escape mechanism would require reduction of the sizes of crop populations in concert with an increase in the number of crops planted in a region. The host populations should become "undetectable" to the pathogens. At a smaller scale, the escape mechanism is reinforced by the use of multilines and cultivar mixtures. Reinforcing the mechanism of avoidance also requires the use of multilines or cultivar mixtures.

Immunisation of the crop seems attractive because it may be race non-specific. But deployment of immunisation may have two major constraints. All the known examples of immunisation are quantitative, so a crop will be incompletely protected against pathogens. For some crops, this may be acceptable, but for others it would not. Immunisation is inducible and can turn on and off under natural conditions suggesting that a plant cannot remain in an immune state. It remains to be seen whether the physiology of a plant might allow a permanent immune state to be induced.

A similar question may be posed about tolerance. How tolerant can a plant be to disease? Without an answer to this question, the potential value of tolerance in crop protection remains unclear. In contrast, compensation at the population level can be employed using cultivar mixtures / multilines, and may be useful at relatively low disease levels in a crop, but not at relatively high levels. A disadvantage of deploying the concepts of tolerance and compensation is that diseased plants are present in the crop functioning as inoculum sources for further spread of disease.

At the end, the control of disease comes back to the mechanism of resistance. If all other mechanisms that may reduce the impact of a pathogen are of minimal effect, a pathogen has "only" to break resistance to cause a relatively high crop loss. If monogenic resistant crops are grown, knowledge of the (meta)population dynamics of a pathogen is basic to predicting the pathotypes arriving at a crop and thus to choosing the right variety to resist the pathogen. Using cultivar mixtures, or multilines implies accepting disease in the crop, but yield loss may be less dramatic than that resulting from the breakdown of monogenic resistance. Furthermore, compensation effects may minimise crop losses in the case of variety mixtures or multilines. Employment of quantitative resistance also implies accepting a certain level of disease, but the risk of a dramatic breakdown as in monogenic resistance seems to be low. Finally, pyramiding may be promising if the number of resistance genes pyramided in a crop exceeds the maximum number of virulence genes that can be accumulated in one pathogen race.

Decisions in crop protection have to be taken in concrete situations. This chapter provides a general overview of the whole complex story. The aim in writing it was to transmit a broad understanding of mechanisms at the population level to the reader, rather than to compile an exhaustive review.

Acknowledgements

I wish to thank J.C. Zadoks for comments on a previous version of the manuscript and U. Brändle for encouraging discussions. Many thanks to Ron Fraser for his editorial work.

References

Agrios GN (1980) Escape from disease. In: Horsfall JG and Cowling EB (eds) Plant Disease: an Advanced Treatise, Vol 5: How Plants Defend Themselves, pp 17-37. Academic Press, New York / London

Andrivon D (1996) The origin of *Phytophthora infestans* populations present in Europe in the 1840s: a critical review of historical and scientific evidence. Plant Pathol 45: 1027—1035

Antonovics J and Miller-Alexander H (1989) The concept of fitness in plant-fungal pathogen systems. In: Leonard KJ and Fry WE (eds) Plant Disease Epidemiology, Vol 2: Genetics, Resistance and Management, pp 185—214. McGraw-Hill Publishing Company, New York

Augspurger CK (1983) Seed dispersal of the tropical tree, *Platypodium elegans*, and the escape of its seedlings from fungal pathogens. J Ecol 71: 759—771

Biere A and Antonovics J (1996) Sex-specific costs of resistance to the fungal pathogen *Ustilago violacea* (*Microbotryum violaceum*) in *Silene alba*. Evolution 50: 1098—1110

Burdon JJ (1987a) Diseases and Plant Population Biology. Cambridge University Press, Cambridge

Burdon JJ (1987b) Phenotypic and genetic patterns of resistance to the pathogen *Phakopsora pachyrhizi* in populations of *Glycine canescens*. Oecologia (Berlin) 73: 257—267

Burdon JJ and Jarosz AM (1992) Temporal variation in the racial structure of flax rust (*Melampsora lini*) populations growing on natural stands of wild flax (*Linum marginale*): local versus metapopulation dynamics. Plant Pathol 41: 165—179

Chin KM and Wolfe MS (1984) The spread of *Erysiphe graminis* f. sp. *hordei* in mixtures of barley varieties. Plant Pathol 33: 89—100

Clarke DD (1986) Tolerance of parasites and disease in plants and its significance in host-parasite interactions. Adv Plant Pathol 5: 161—197

Cooper WS (1984) Expected time to extinction and the concept of fundamental fitness. J Theor Biol 107: 603—629

Crute IR and Pink DAC (1996) Genetics and utilization of pathogen resistance in plants. Plant Cell 8: 1747—1755

Damgaard C. and Giese H (1996) Genetic variation in Danish populations of *Erysiphe graminis* f.sp. *hordei*: estimation of gene diversity and effective population size using RFLP data. Plant Pathol 45: 691—696

Dean RA and Kuç JA (1987) Immunisation against disease: the plant fights back. In: Pegg GF and Ayres PG (eds) Fungal Infection of Plants, pp 383—410. Cambridge University Press, Cambridge

Dinoor A and Eshed N (1984) The role and importance of pathogens in natural plant communities. Annu Rev Phytopathol 22: 443—466

Drenth A, Tas ICQ and Govers F (1994) DNA fingerprinting uncovers a new sexually reproducing population of *Phytophthora infestans* in the Netherlands. Eur J Plant Pathol 100: 97—107

de Nooij MP (1988) The role of weevils in the infection process of the fungus *Phomopsis subordinaria* in *Plantago lanceolata*. Oikos 52: 51—58

de Nooij MP and van Damme JMM (1988a) Variation in host susceptibility among and within populations of *Plantago lanceolata* L. infected by the fungus *Phomopsis subordinaria* (Desm.) Trav. Oecologia (Berlin) 75: 535—538

de Nooij MP and van Damme JMM (1988b) Variation in pathogenicity among and within populations of the fungus *Phomopsis subordinaria* infecting *Plantago lanceolata*. Evolution 42: 1166—1171

Endler JA (1986) Natural Selection in the Wild. Princeton University Press, Princeton.

Ennos R (1992) Ecological genetics of parasitism. In: Berry RJ, Crawford TJ and Hewitt GM (eds) Genes in Ecology, pp 255—279. Blackwell Scientific Publications, Oxford

Falconer DS (1989) Introduction to Quantitative Genetics. Longman Scientific & Technical, Harlow

Frantzen J (1994a) Studies on the weed pathosystem *Cirsium arvense - Puccinia punctiformis*. Thesis Agricultural University Wageningen, Wageningen. ISBN 90-5485-211-9

Frantzen J (1994b) The role of clonal growth in the pathosystem *Cirsium arvense - Puccinia punctiformis*. Can J Bot 72: 832—836

Fry WE, Goodwin SB, Matuszak JM, Spielman LJ and Milgroom MG (1992) Population genetics and intercontinental migrations of *Phytophthora infestans*. Annu Rev Phytopathol 30: 107—129

Hanski I (1994) A practical model of metapopulation dynamics. J Anim Ecol 63: 151—162

Harry IB and Clarke DD (1986) Race-specific resistance in groundsel (*Senecio vulgaris*) to the powdery mildew *Erysiphe fischeri*. New Phytol 103: 167—175

Harry IB and Clarke DD (1987) The genetics of race-specific resistance in groundsel (*Senecio vulgaris* L.) to the powdery mildew fungus, *Erysiphe fischeri* Blumer. New Phytol 107: 715—723

Hatcher PE, Paul ND, Ayres PG and Whittaker JB (1994) Interactions between *Rumex* spp., herbivores and a rust fungus: *Gastrophysa viridula* grazing reduces subsequent infection by *Uromyces rumicis*.

Func Ecol 8: 265—272

Jennersten O, Nilsson SG and Wästljung U (1983) Local plant populations as ecological islands: the infection of *Viscaria vulgaris* by the fungus *Ustilago violacea*. Oikos 41: 391—395

Johnson R (1992) Reflections of a plant pathologist on breeding for disease resistance, with emphasis on yellow rust and eyespot of wheat. Plant Pathol 41: 239—254

Katsuya K and Green GJ (1967) Reproductive potentials of races 15B and 56 of wheat stem rust. Can J Bot 45: 1077—1091

Kolmer JA (1995) Selection of *Puccinia recondita* f.sp. *tritici* virulence phenotypes in three multilines of Thatcher wheat lines near isogenic for leaf rust resistance genes. Can J Bot 73: 1081—1088

Linders EGA, van Damme JMM and Zadoks JC (1996) Transmission and overseasoning of *Diaporthe adunca* on *Plantago lanceolata*. Plant Pathol 45: 59—69

McDermott JM and McDonald BA (1993) Gene flow in plant pathosystems. Annu Rev Phytopathol 31: 353—373

Murray DC and Walters DR (1992) Increased photosynthesis and resistance to rust infection in upper, uninfected leaves of rusted broad bean (*Vicia faba* L.). New Phytol 120: 235—242

Nelson RR (1978) Genetics of horizontal resistance to plant diseases. Annu Rev Phytopathol 16: 359—378

Newton AC and Andrivon D (1995) Assumptions and implications of current gene-for-gene hypotheses. Plant Pathol 44: 607—618

Olivieri I, Couvet D and Gouyon P-H (1990) The genetics of transient populations: research at the metapopulation level. TREE 5: 207 —210

Parker MA (1987) Pathogen impact on sexual vs. asexual reproductive success in *Arisaema triphyllum*. Am J Bot 74: 1758—1763

Parlevliet JE (1989) Identification and evaluation of quantitative resistance. In: Leonard KJ and Fry WE (eds) Plant Disease Epidemiology, Vol 2: Genetics, Resistance and Management, pp 215—248. McGraw-Hill Publishing Company, New York

Paul ND and Ayres PG (1984) Effects of rust and post-infection drought on photosynthesis, growth and water relations in groundsel. Plant Pathol 33: 561—569

Robinson (1976) Plant Pathosystems. Springer Verlag, Berlin

Schutz WM and Brim CA (1971) Inter-genotypic competition in soybeans. III. An evaluation of stability in multiline mixtures. Crop Sci 11: 684—689

Schafer JF (1971) Tolerance to plant disease. Annu Rev Phytopathol 9: 235—252

Schoeneweiss DF (1975) Predisposition, stress, and plant disease. Annu Rev Phytopathol 13: 193—211

Thompson JN (1990) Coevolution and the evolutionary genetics of interactions among plants and insects and pathogens. In: Burdon JJ and Leather SR (eds) Pests, Pathogens and Plant Communities, pp 249—271. Blackwell Scientific Publications, Oxford

Van der Putten WH and Troelstra SR (1990) Harmful soil organisms in coastal foredunes involved in degeneration of *Ammophila arenaria* and *Calammophila baltica*. Can J Bot 68: 1560—1568

Wennström A and Ericson L (1992) Environmental heterogeneity and disease transmission within clones of *Lactuca siberica*. J Ecol 80: 71—77

Wolfe MS (1985) The current status and prospects of multiline cultivars and variety mixtures for disease resistance. Annu Rev Phytopathol 23: 251—273

Wolfe MS, Brändle U, Koller B, Limpert E, McDermott JM, Müller K and Schaffner D (1992) Barley mildew in Europe: population biology and host resistance. Euphytica 63: 125—139

Zadoks JC and Schein RD (1979) Epidemiology and Plant Disease Management. Oxford University Press, New York / Oxford

RESISTANCE GENES AND THE PERCEPTION AND TRANSDUCTION OF ELICITOR SIGNALS IN HOST-PATHOGEN INTERACTIONS

THOMAS BOLLER[1] & NOEL T. KEEN[2]

[1] *Botanisches Institut der Universität, Hebelstrasse 1*
  CH-4056 Basel, Switzerland (boller@ubaclu.unibas.ch)
[2] *Department of Plant Pathology*
  University of California, Riverside, CA 92521, U.S.A.

Summary

Plants lack immune systems of the types known in animals, but nevertheless are resistant to most potential pathogens. Like in animals, resistance is based on an active response of the plant to pathogen attack. Activated defense responses most often culminate in the so-called hypersensitive response in which cells exposed to the pathogen undergo rapid cell death and prevent further invasion. Also similar to animals, this reaction depends primarily on recognition of the invading pathogen. Disease resistance genes play a pivotal role in the recognition process. Several resistance genes have been cloned, and current evidence suggests that their products physically interact with the products of microbial avirulence genes, named specific elicitors. In addition to these highly specific recognition phenomena, based on matching genes in plant and pathogen, plants also have exquisitely sensitive perception systems for so-called general elicitors, i.e. substances characteristic of whole groups of micro-organisms, such as microbial glycopeptides, cell wall fragments, and sterols. The substances recognized occur not only in pathogens, but also in saprophytes and even in symbiotic microorganisms. Chemoperception of these substances may trigger only some reactions associated with defense responses, thus providing an early warning for the presence of a foreign organism, or contribute substantially to reactions associated with the hypersensitive response, depending on plant species and developmental stage. Transduction of microbial signals in plants has been extensively studied after treatment with general elicitors. It remains an open question, however, how the signals generated by the interaction between avirulence gene products and resistance gene products are related to those generated by the perception of general elicitors.

A. Slusarenko, R.S.S. Fraser, and L.C. van Loon (eds), Mechanisms of Resistance to Plant Diseases, 189-229.
© 1999 *Kluwer Academic Publishers. Printed in the Netherlands.*

Abbreviations

CD	circular dichroism
cGMP	cyclic guanosine monophosphate
DPI	diphenylene iodonium
FSH	follicle-stimulating hormone
FTIR	Fourier-transform infrared spectroscopy
GUS	-glucuronidase
HR	hypersensitive response
JA	jasmonic acid
LH/CG	luteinizing hormone/chorionic gonadotropin
LRR	leucine-rich repeat
MHC	major histocompatibility complex
NO	nitric oxide
NOS	nitric oxide synthase
PAL	phenylalanine ammonia-lyase
PCR	polymerase chain reaction
PRs	pathogenesis-related proteins
SA	salicylic acid
SAR	systemic acquired resistance.

I.   Introduction: recognition and active defense

When passive plant defense mechanisms such as preformed structural barriers and inhibitory compounds fail to arrest an invading pathogen, the last resort is an active defense mechanism, classically known as the hypersensitive response or HR (see also chapter 6). The HR is considered an active defense mechanism because it is only invoked in response to intimate contact of the pathogen with plant cells. Its most notable feature — the death of variable numbers of plant cells at and surrounding the pathogen infection site — remains poorly understood with respect to its role, if any, in the resistance process. Two general types of HR are recognized: general (or non-host) resistance, which involves activation of the HR by microorganisms or viruses that are pathogenic on other plant species; and specific (or host) resistance, in which some but not all isolates of a single pathogen species elicit the HR only in particular plant cultivars. Resistant plant cultivars harbor single disease resistance genes and races or strains of the pathogen that activate them express complementary or matching avirulence genes. Since genetic variation occurs in both the pathogen and the host, these are called gene-for-gene interactions (Flor, 1971).

As we will discuss later, plant defense reactions associated with the HR are initiated by pathogen-produced signal molecules, called elicitors. Two types of elicitors are recognized: general (or non-specific) elicitors, which do not significantly differ in their effect on different cultivars within a plant species, and may therefore be involved in general resistance; and specific elicitors, which are special to the pathogenic race or strain and function only in plant cultivars carrying a matching disease resistance gene, thereby accounting for specific resistance (Boller, 1995; Hahn, 1996). General elicitors include substances typically associated with basic microbial metabolism, such as cell wall glucans, chitin oligomers, fatty acids and sterols and glycopeptides. Chemoperception of such general elicitors appears to be based on specific high-affinity receptors, which presumably initiate an intracellular signal cascade that eventually results in the coordinate transcription of a large number of defense response genes. These signaling cascades are not completely understood, and increasing evidence suggests that unique cascades may operate with different elicitors, even in the same plant (Penninckx *et al.*, 1996).

Unlike general elicitors, the formation of specific elicitors requires the function of pathogen avirulence genes. Specific elicitors in general also have more unique structures (proteins, peptides and the syringolides to be discussed later). The isolation of avirulence gene-specific elicitors which exhibit the same specificity as the infecting pathogen also supports the elicitor-receptor model (Keen, 1990; De Wit, 1992). This model states that plants possess receptors for pathogen-produced elicitors which are controlled by the cognate disease resistance genes.

The products of defense response genes are responsible for the physiological processes that collectively cause disease resistance. These include the production of molecules which are directly toxic to, or physically encase the pathogen to retard further colonization of the plant. There are synergistic interactions between the various defense responses, as exemplified by the effects of glucanases and chitinases on the growth of various fungal pathogens (Mauch *et al.*, 1988; Sela-Buurlage *et al.*, 1993; Zhu *et al.*, 1994). Nonetheless, certain components of the array of defense responses appear targeted to particular types of pathogens. For instance, phytoalexins generally inhibit fungal pathogens and, to a degree, bacterial pathogens and nematodes, but are unlikely to have direct effects on viral replication or movement. Similarly, glucanases and chitinases inhibit fungi, and some chitinases also inhibit bacteria, but these enzymes would not be expected to affect other pathogens. However, reduction of β-glucanase activity in antisense tobacco plants led to unexpected reductions in the sizes of the lesions produced by tobacco mosaic virus (TMV) (Beffa *et al.*, 1996). This implies that β-glucans, such as callose, may be involved in the prevention of viral spread during the HR, and that the induction of β-glucanase by the plant, which may be directed against fungi, is actually reducing the efficiency of this antiviral defense. This illustrates that much remains to be learned about the deployment of HR defense responses. While these reactions are nominally localized to the pathogen infection site, cooperative effects and the role of healthy, neighboring cells are also clearly important (Graham and Graham, 1991).

Hypersensitive cell death has recently been shown to be a kind of programmed cell death (Greenberg *et al.*, 1994; Dangl *et al.*, 1996), extensively studied in animals and known as apoptosis (Gerchenson and Totello, 1992). However, evidence from several plant-pathogen systems argues that hypersensitive plant cell death *per se* may have little role in resistance (Jakobek and Lindgren, 1993; Hammond-Kosack *et al.*, 1996; Mittler *et al.*, 1996). Although often argued to be of importance in resistance to obligate biotrophs, which require living host cells, cell death in the HR may be a serendipitous consequence of the concerted stimulation of the cells around the infection site, which all react to pathogen attack and produce signals to activate defense responses (Gross et al., 1993).

As discussed in more detail later, induction of the HR appears to involve the production of active oxygen species (for review, see Baker and Orlandi, 1995) and these may also be directly toxic to microbial pathogens (e.g. Peng and Kuc, 1992). Certain active oxygen species, most notably H_2O_2 (Brisson et al., 1994), may contribute to structural reinforcement of the plant cell wall by cross-linking wall components. There is, however, no good evidence suggesting the importance of these mechanisms in inhibiting the pathogen during the HR, because they are generally produced early in the interactions while observed cessation of pathogen development occurs much later. It seems more likely that the importance of active oxygen generation in the HR is related to signal generation (Dangl et al., 1996).

Another consequence of the HR related to signaling is the initiation of <u>systemic acquired resistance</u> (SAR) (Neuenschwander et al., 1996; Ryals et al., 1996). SAR is the phenomenon that plants become more resistant to subsequent inoculation with the same or different pathogen species, a condition which generally remains effective for weeks. Unfortunately, we know neither how the HR initiates SAR, nor what the systemic signaling mechanism is (see also chapter 10).

II.   General elicitors and their perception

An essential condition for the activation of defense responses is <u>recognition</u>, i.e., perception of the presence of a potential pathogen. Although such recognition can be based on physical contact, e.g. when a pathogen exerts pressure during penetration, it is likely that the primary mode of recognition is through chemical signals produced by invading microbes. All microbial signals that are perceived by plant cells and induce defensive responses are considered elicitors (Keen and Bruegger, 1977), and the concept of elicitor perception has been studied extensively at the biochemical and molecular level (for recent reviews see Boller, 1995; Hahn, 1996). Often, plant cell cultures have been used as models, as they offer the advantage of reacting rapidly and uniformly to elicitors (Boller and Felix, 1996). However, in such model systems it is difficult to link the observed reactions to plant-pathogen interactions, and students of elicitor perception should be aware that many of the general elicitors studied in cell culture systems are produced not only by pathogens, but also by harmless saprophytes and even by symbionts. It has been found that general elicitors comprise very different types of compounds which appear to be non-specifically recognized by plants and can elicit from only some of the responses associated with active defense up to cell death similar to the HR. Their actual contributions to the initiation of defense reactions may also differ depending on plant species and the rate at which defense develops upon pathogen attack, as well as on the type of assay used.

The biochemistry of elicitor perception has been studied extensively, starting from the hypothesis that the selective and highly sensitive perception of elicitors is based on recognition by specific receptors, and that these receptors are located at the outer surface of the plasma membrane, at least for the hydrophilic elicitors. The putative receptors are thus identified in plant cell membranes as <u>high-affinity binding sites</u>, using radioactively labeled elicitors for binding studies. It is important to verify that the selectivity of the binding site for various structural analogues of the elicitor corresponds to the biological activity of the analogues. Subsequently, the binding site can be solubilized and purified, although this procedure is rendered difficult by the fact that such binding sites have a low abundance and often represent less than 0.001 % of the membrane protein.

Using these approaches, several studies have identified high-affinity binding sites with the expected selectivity and specificity for some of the elicitors. It is probable that these binding sites represent the receptors although functional proof for this is still lacking. In the following sections, we discuss the most important general elicitors derived from fungi and summarize current information on the putative receptors for general elicitors.

A. CELL WALL POLYSACCHARIDES

1. *Glucans*
A classic elicitor is the branched β-1,3,β-1,6-heptaglucoside (Fig. 1a), isolated as the smallest elicitor-active compound from cell walls of *Phytophthora megasperma* f.sp. *glycinea*, a fungal pathogen of soybean (Darvill and Albersheim, 1984). As demonstrated in the early studies with oligosaccharides derived from *Phytophthora* cell walls, many seemingly similar oligomers of glucose, with slightly different linkage patterns, are inactive as elicitors, attesting to the high selectivity and specificity of the perception system (Darvill and Albersheim, 1984; Cheong *et al.*, 1991; Ebel and Cosio, 1994; Hahn, 1996). Based on the initial finding of a high-affinity binding site for *Phytophthora* glucans in soybean plasma membranes (Schmidt and Ebel, 1987), this putative receptor was investigated by several groups. Extensive studies with structural analogues showed a close correlation between elicitor activity *in vivo* and the capacity to compete with elicitor binding *in vitro* (Cheong *et al.*, 1991), indicating that the binding site functions indeed as a receptor. Also, the presence of related binding sites in membranes of various legumes and their affinities correlated with the responsiveness of the legumes to the glucan elicitors (Cosio *et al.*, 1996). The binding site has been solubilized (Cosio *et al.*, 1990; Cheong *et al.*, 1993) and partially purified (Frey *et al.*, 1993), and a cDNA encoding the binding site has recently been cloned (Umemoto *et al.*, 1997), allowing proof of its receptor function in the near future.

While soybean and other related legumes are sensitive to the glucan elicitors, many other plants are insensitive, indicating a diversity of recognition systems in different plants (Ebel and Cosio, 1994; Hahn, 1996). It is worth noting that glucans occur in fungal cell walls in insoluble form, and that elicitor-active fragments must first be liberated before recognition can occur. It is likely that β-1,3-glucanases, in addition to their direct role as antifungal enzymes, also function in the release of elicitors (see Hahn, 1996 for a recent review).

2. *Chitin fragments*
Chitin, a β-1,4-linked linear polymer of *N*-acetylglucosamine, is a major constituent of the cell walls of most higher fungi, and it also occurs in arthropods and many other invertebrates but is not present in plants. Plant cells have a highly sensitive perception

system for chitin fragments (Fig. 1a), as studied in some detail in tomato (Felix *et al.*, 1993). Chitin oligomers with four or more *N*-acetylglucosamine units are recognized by the plant cells at threshold concentrations of about 1 pM; the trisaccharide is about 1000-fold less active, and the dimer and monomer, which potentially could be present in plants, are essentially inactive (Felix *et al.*, 1993).

Tomato cells and membranes possess a high affinity binding site for chitin oligomers with four or more *N*-acetylglucosamine units (K_D of ~ 1-3 x 10^{-9} M for intact tomato cells) but a ~ 300-fold and ~ 300,000-fold lower affinity for the trimer and dimer, respectively (Baureithel *et al.*, 1994). The relative affinities of these compounds for the binding site are similar to their relative biological activities, and distinguish the binding site from chitin-binding lectins (Baureithel *et al.*, 1994). A high-affinity binding site for chitin oligomers has also been found in rice cells (Shibuya *et al.*, 1993). Compared to the chitin-binding site of tomato, it has a lower affinity for the smaller chitin fragments and reaches its highest affinity only with oligomers of eight *N*-acetylglucosamine units (Shibuya *et al.*, 1996). Chitin perception appears to be a typical non-self recognition system (Boller, 1995). As in the case of glucans, chitin itself is highly insoluble, and for recognition to take place, it is necessary for the plant to secrete chitinases that release chitin fragments from invading fungi (Felix *et al.*, 1993). Thus, as in the case of glucanases, chitinases may function in the release of elicitors in addition to their direct antifungal activity (Boller, 1995).

In some groups of fungi, chitin occurs in association with chitosan, a deacetylated form of chitin. Also chitosan oligosaccharides (Fig. 1a) can act as elicitors, but in contrast to chitin, at least seven residues are required for biological activity (Côté and Hahn, 1994).

B.   SECRETED PROTEINS AND GLYCOPEPTIDES

1.   *Pectolytic enzymes*
Pectolytic enzymes had early been found to induce defense responses, and it became clear that the elicitor-active principle was not in the pectolytic enzymes themselves, but in fragments that they released from the plant cell wall (see West, 1981 for an early review). Oligogalacturonides (Fig. 1b) with a degree of polymerization between 10 and 20 residues have been found to be biologically active. Because they are plant-derived, they are called endogenous elicitors (Hahn *et al.*, 1981). Recognition of such elicitors is an example of an indirect mode for perception of microbial attack. One could say that the primary recognition event is actually the interaction of polygalacturonase or pectate lyase with the pectic component of plant cell walls, and that the degradation products generated, the endogenous elicitors, represent already second messengers. Interestingly, many plants produce inhibitors of the pectic enzymes of microbes, and these inhibitors not only reduce the enzymatic activity but also cause a shift in the chain length of the fragments generated, so that the proportion of highly active endogenous elicitor molecules is enhanced (Cervone and Albersheim, 1989).

2.   *Xylanase*
Xylanase from the saprophyte *Trichoderma viride* acts as a potent elicitor in tobacco leaves (Sharon *et al.*, 1993), as well as in tobacco and tomato cells (reviewed in

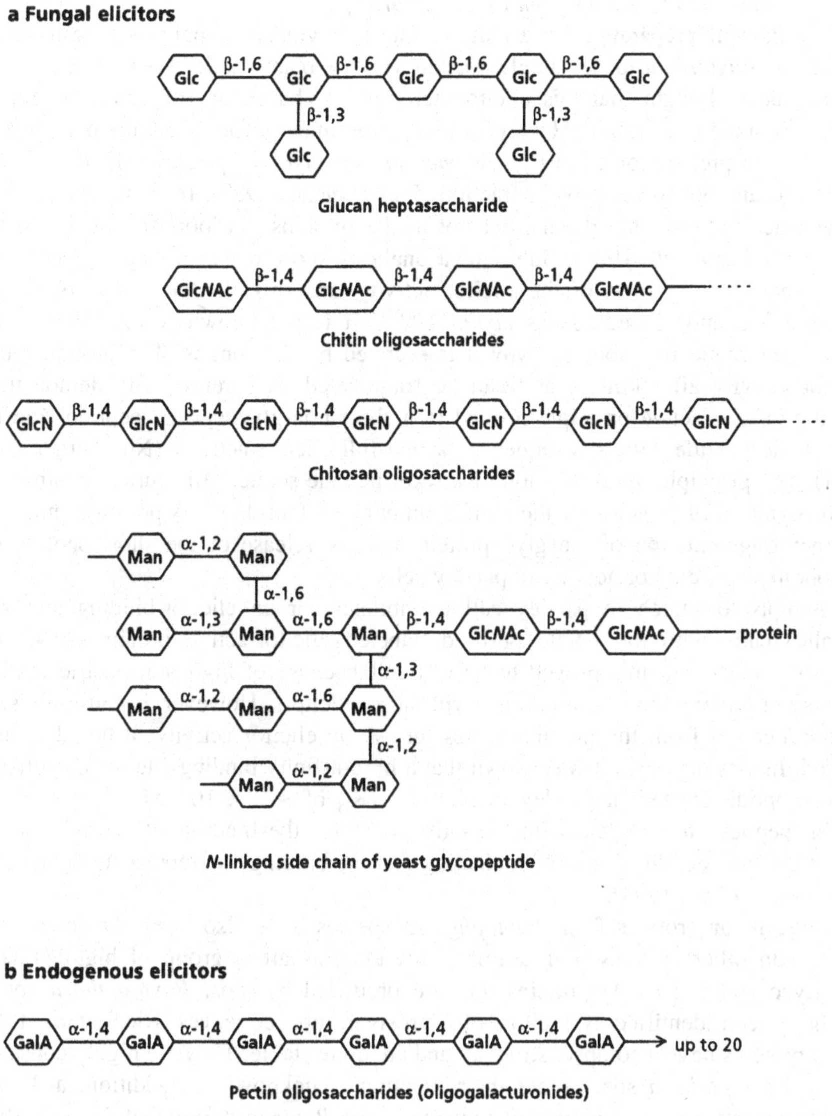

a Fungal elicitors

Glucan heptasaccharide

Chitin oligosaccharides

Chitosan oligosaccharides

N-linked side chain of yeast glycopeptide

b Endogenous elicitors

Pectin oligosaccharides (oligogalacturonides)

Figure 1. Schematic structures of carbohydrate elicitors of defense reactions derived from (a) fungal components and (b) plant cell walls. The types of linkage between individual residues are indicated. Glc, glucose; GlcNac, N-acetylglucosamine; GlcN, glucosamine; Man, mannose; GalA, galacturonic acid.

Boller and Felix, 1996). Although it was initially thought that xylanase acts through the liberation of xylan fragments acting as endogenous elicitors, current evidence indicates that this is not the case (Sharon *et al.*, 1993). Hence, plants are able to detect the presence of microbial enzymes either indirectly, through the products formed, as in the case of pectic enzymes, or by direct recognition of the protein, as in the case of xylanase.

3.   *Extracellular proteins from Phytophthora species*

Crude cell wall preparations and partially purified glucan preparations from several species of *Phytophthora* act as elicitors in various plant model systems, and it had initially been thought that this elicitor activity was based on the cell wall glucans contained in the preparation. However, it was subsequently found that upon application of an elicitor preparation from *P. megasperma* f.sp. *glycinea* parsley cells do not react to the glucans but to the protein fraction (Parker *et al.*, 1988). Interestingly, soybean tissue reacts only to the glucans but not to the proteins, the opposite of the parsley pattern (Parker *et al.*, 1988). Subsequent analyses showed that a single glycoprotein was responsible for most of the elicitor activity, and this elicitor-active protein was later purified and cloned (Sacks *et al.*, 1995). It is not known whether the protein has an enzymatic function and why it is secreted by the fungus. The protein retains elicitor activity after boiling and can be fragmented. A careful study demonstrated that the elicitor-active principle resided in a short non-glycosylated peptide, and that a synthetic peptide with this sequence carried full elicitor activity (Nürnberger *et al.*, 1994). The principle of recognition of a short peptide sequence is curiously similar to the recognition of proteins by the animal immune system, but it is presently unknown whether fragmentation of the glycoprotein and the release of the short peptide is a precondition for elicitor activity in parsley cells.

Attempts to search in parsley cell membranes for an elicitor-binding site were initially made with the whole purified, labeled elicitor-active protein of 42 kDa molecular mass, but this proved to be difficult because of high nonspecific binding. This is not uncommon when working with large elicitors. However, because the small peptide derived from the protein carries the whole elicitor activity, it could be used instead. In elegant work, it was shown that a high-affinity binding site for this elicitor-active peptide existed in parsley membranes (K_D of ~ 2 x 10^{-9} M). The ability of similar peptides to compete with the radioligand for the binding site correlated with their biological activity, strongly indicating that the binding site represents the receptor (Nürnberger *et al.*, 1994).

Extracellular proteins from *Phytophthora* species have also been characterized as elicitors in other systems. For example, the elicitins are a group of highly related, non-glycosylated 10 kDa proteins that are produced by most *Phytophthora* species and have been identified as elicitors of necrosis in tobacco leaves (Ricci *et al.*, 1989). These proteins have a compact structure and are quite stable. They are highly conserved among *Phytophthora* species but their function is unknown. In addition, a 32 kDa elicitin-related glycoprotein has been purified from *P. megasperma* that likewise elicits necrosis in tobacco leaves (Baillieul *et al.*, 1995). Both proteins are active at nanomolar levels in appropriate bioassays, documenting the very high sensitivity of tobacco plants to these particular stimuli. Interestingly, tomato cells recognize neither the glycoprotein with elicitor activity in parsley, nor elicitins, but are sensitive to another glycoprotein in a crude *P. megasperma* cell wall preparation (Boller and Felix, 1996).

Thus, different plant species have different capabilities to recognize extracellular proteins of *Phytophthora*, each species apparently recognizing one specific protein with exceedingly high sensitivity. It is characteristic that isolates of *Phytophthora* that are able to colonize tobacco as pathogens do not express the elicitins recognized by tobacco. This implies that specific recognition of such proteins is an important factor in the resistance of tobacco against other *Phytophthora* species (Bonnet *et al.*, 1994).

4.   *Glycopeptides*

Also acting as general elicitors are glycopeptides derived from extracellular proteins of yeast which were found to be recognized at nanomolar concentrations by tomato cells (Basse and Boller, 1992; Basse *et al.*, 1992). In this case, the structure of the *N*-linked glycan side chains is decisive for activity: glycopeptides with 8 mannosyl residues have little elicitor activity but those with 9-11 residues are potent elicitors (Fig. 1a). To study rare high-affinity binding sites such as elicitor receptors, it is necessary to obtain ligands with very high specific activity, such that binding can be studied in the presence of 10^{-10} M concentrations of the ligand or less. To achieve this, chemically defined glycopeptides were labeled with ^{35}S, yielding radioligands with specific activities of 1000 Ci/mmol. Using this material, it was possible to demonstrate a high-affinity binding site for the yeast glycopeptides, both on intact tomato cells and in tomato membranes (Basse *et al.*, 1993). Analysis of the binding data yielded a K_D of about 3 x 10^{-9} M, a value which corresponds to the concentration required for half-maximal elicitor activity (Basse *et al.*, 1993). Glycopeptides with *N*-linked side chains of 9-11 mannosyl residues have a high affinity for the binding site while those with side chains of 8 mannosyl residues have low affinity, as would be expected for a binding site representing the biological receptor (Basse *et al.*, 1993). This binding site has been solubilized and partially purified (Fath and Boller, 1996), using various chromatographic techniques. The binding site appeared to be stable after solubilization and maintained its binding characteristics, but further progress was rendered difficult because after binding to an affinity column containing the elicitor as a ligand, the binding protein could not be released in a functional form.

Interestingly, the structure with 8 mannosyl residues, which are present in the inactive glycopeptides, forms a core which is present in all eukaryotes, including plants. However, all the larger chains, which are present in the active glycopeptides, have their nineth mannosyl residue attached to this core at a position unique for fungi, indicating that this perception system is specific for non-self molecules (Basse *et al.*, 1992). One particularly intriguing aspect of this recognition system is the fact that the oligosaccharides present in the active elicitors act as competitive antagonists of elicitation when they are liberated from the glycopeptide. Thus, the free *N*-linked glycans, after cleavage from the peptide, counteract the activity of the corresponding elicitors and functionally act as suppressors (Basse *et al.*, 1992; Basse and Boller, 1992). Suppressors may be important in modulating the action of elicitors, and plant pathogens may accordingly need suppressors of this or other kinds to avoid recognition by the plant's general surveillance system (see Boller, 1995, for discussion).

C.   MEMBRANE CONSTITUENTS

1.   *Arachidonic acid*

Arachidonic acid, a fatty acid not normally present in plants but a major constituent of membrane lipids of *Phytophthora infestans*, acts as an elicitor in potato (Bostock et al., 1981). It is possible that the active compound recognized by the plant cells is an oxidized form of arachidonic acid generated by lipoxygenase (Bostock et al., 1992). Thus, as in the case of the oligosaccharide elicitors mentioned above, processing and metabolism of fungal membrane components is required to release the elicitor-active principle and allow its recognition by the plant.

2.   Ergosterol

Tomato cells have an extremely sensitive perception system for ergosterol, the main sterol in most higher fungi (Granado et al., 1995). This chemosensory system recognizes ergosterol at threshold levels below 1 pM, whereas plant sterols are completely inactive even at micromolar concentrations, exemplifying once again a characteristic system for non-self recognition. The ergosterol released from as few as 100 fungal spores suspended in water is sufficient to stimulate about 100,000 plant cells. This suggests a scenario in which a plant cell might recognize a single spore landing on its surface through ergosterol leakage during germination (Granado et al., 1995).

III.   Avirulence genes and specific elicitors

Specific elicitors act in a fungal race- or bacterial strain-specific manner in triggering resistance according to the gene-for-gene relationship (see chapter 2), and are taken to be the products of avirulence genes in the pathogen. More than 40 pathogen avirulence genes have been cloned and characterized (Leach and White, 1996) and the specific elicitors they direct have been identified in several cases (Table 1). Somewhat surprisingly, avirulence gene sequences generally do not resemble known genes in the databases. In the case of bacterial avirulence genes, the deduced protein products also usually lack identifiable signal sequences directing secretion, suggesting that they remain inside the cells of the pathogen or are secreted by mechanisms other than the general secretory pathway (see section B). Certain fungal primary avirulence gene products, however, are processed and extracellularly secreted, where they function as elicitors (De Wit, 1992). In addition, viral avirulence genes have been identified as pleiotropic functions of known viral genes, such as those encoding capsid protein (Table 1; see also chapters 1 and 9). This section will focus mainly on bacterial avirulence genes and elicitors because Cladosporium and TMV elicitors have been discussed in detail in chapter 1.

The physiological role of avirulence genes in pathogens is not well understood. However, it is likely that they do have important functions because of their evolutionary conservation in pathogen populations. Indeed, several avirulence genes in bacterial plant pathogens of the Xanthomonas campestris (e.g. Kearney and Staskawicz, 1990; Yang et al., 1996) and Pseudomonas syringae groups (e.g. Lorang et al., 1994; Ritter and Dangl, 1995) are required for full virulence on host cultivars that lack the cognate resistance gene. Thus, avrA and avrE in P. syringae pv. tomato strain PT23 are required for full virulence and bacterial multiplication in tomato leaves of susceptible cultivars. In contrast, mutation experiments have failed to disclose a virulence role for avrD or avrPto in strain PT23 or for the avrE homologue of P. syringae pv. tomato strain

DC3000 (Lorang *et al.*, 1994). It is possible that batteries of pathogen avirulence genes may exhibit cross-complementary virulence functions, so that the loss of any one virulence function by mutation might not be detrimental to the activity of the pathogen on plants. This situation would be appealing from the pathogen point of view in order to minimize virulence losses when particular avirulence genes were mutated to escape plant resistance gene surveillance. Similar to the role of bacterial avirulence genes in virulence, the fungal avirulence gene *Nip1* from *Rynchosporium secalis* is also required for full virulence (Knogge, 1996) as are all known viral avirulence genes. These findings immediately rationalize the observed conservation of avirulence genes in pathogen populations. With the exception of the viral genes, however, we do not yet understand the nature of the virulence functions of avirulence genes.

Table 1.   Selected avirulence genes and the specific elicitors they direct

Avirulence gene	Cloned from	Elicitor	Reference
Bacteria			
avrA, B, C, Rpm, Rpt2,			
Pph3 and *Pto*	*Pseudomonas syringae*	Protein products?	Pirhonen *et al.*, 1996
avrBs2	*Xanthomonas campestris*	Enzymatic product?	Swords *et al.*, 1996
avrBs3 family	*Xanthomomas campestris*	Protein products?	Yang and Gabriel, 1995
avrD family	*Pseudomonas syringae*	Syringolides	Midland *et al.*, 1993
avrRps4	*Pseudomonas syringae*	Protein product?	Hinsch and Staskawicz, 1996
hrmA	*Pseudomonas syringae*	Protein product?	Alfano *et al.*, 1997
PopA1	*Pseudomonas solanacearum*	Protein product?	Arlat *et al.*, 1994
Fungi			
AVR2-YAMO	*Magnaporthe grisea*	Protease?	Howard and Valent, 1996
avr4	*Cladosporium fulvum*	Peptide elicitor	Joosten *et al.*, 1994
avr9	*Cladosporium fulvum*	Peptide elicitor	Van den Ackerveken *et al.*, 1993
nip1	*Rynchosporium secalis*	Peptide elicitor	Knogge, 1996
Pwl	*Magnaporthe grisea*	Unknown	Kang *et al.*, 1995
Pwl2	*Magnaporthe grisea*	Unknown	Sweigert *et al.* 1995
-	*Phytophthora* sp.	Elicitins	Kamoun *et al.*, 1993
-	*Uromyces vignae*	Peptide elicitor	DSilva and Heath, 1997
Viruses			
Capsid protein	Potato virus X	Protein	Bendahmane *et al.*, 1995
Capsid protein	TMV	Protein	Taraporewala and Culver, 1996
Movement protein	TMV	Protein	Weber *et al.*, 1993
Replicase	TMV	Protein	Padgett and Beachy, 1993

The only avirulence gene known to direct production of a low-molecular-weight, non-proteinaceous specific elicitor is *P. syringae avrD*. However, Swords *et al.* (1996) noted interesting similarities between the cloned *X. campestris avrBs2* gene and agrocinopine synthase. This may suggest that *avrBs2* exhibits an enzymatic activity required for synthesis of a bacterial specific elicitor, but such a molecule has not yet been isolated. Several low-molecular-weight fungal metabolites also have properties of specific elicitors, although they are commonly referred to as host-selective toxins. For example, peritoxin (Dunkle and Macko, 1995), produced by *Periconia circinata*, is responsible for pathogenicity on certain cultivars of sorghum. Sensitivity to peritoxin and disease susceptibility are conferred by the dominant gene, *Pc*. The responses of *Pc*-containing sorghum plants to peritoxin are remarkably similar to disease defense responses in other plant species, including the requirement for protein synthesis, phosphorylation/ dephosphorylation reactions and accumulation of small pathogenesis-related proteins (PRs). Therefore, it is conceivable that peritoxin is in fact an elicitor of defense responses in *Pc*-containing sorghum plants but that the fungus can tolerate those.

A.   *AVRD* AND THE SYRINGOLIDE SPECIFIC ELICITORS

The *avrD* gene was cloned from *P. syringae* pv. *tomato* and functions in *P. syringae* pv. *glycinea* to initiate the HR in soybean cultivars carrying the *Rpg4* disease resistance gene (Kobayashi *et al.*, 1990a; Keen and Buzzell, 1991). Curiously, all of eight investigated *P. syringae* pv. *glycinea* isolates, representing six different races, also contained *avrD* homologues, but these were uniformly non-functional as avirulence genes on soybean (Kobayashi *et al.*, 1990b; Yucel and Keen, 1994; Keith *et al.*, 1997). All of these *avrD* alleles except one contained one or more missense mutations affecting a relatively small number of amino acids; the other one contained a nonsense mutation that truncated the carboxyl terminus of the protein by 17 amino acids. Thus, all of the *P. syringae* pv. *glycinea avrD* alleles contain complete or nearly complete open reading frames. This conservation suggests that *avrD* has a selected, but currently unknown function in *P. syringae* pv. *glycinea* (Keith *et al.*, 1997).

The *avrD* gene is plasmid-borne in most but not all *P. syringae* strains (Murillo *et al.*, 1994; Keith *et al.*, 1997) and encodes a 34 kDa protein of 311 amino acids. The promoter of all studied *avrD* alleles contains a σ^{54} promoter consensus sequence and also shares a conserved upstream sequence with other *P. syringae* avirulence genes (Kobayashi *et al.*, 1990a; Shen and Keen, 1993). When *avrD* is expressed in *E. coli*, it enables the cells to elicit a HR on *Rpg4* soybean cultivars. Functional *avrD* alleles also result in the release of elicitor activity in *E. coli* culture fluids (Keen *et al.*, 1990) and this observation permitted the isolation and elucidation of the structure of syringolides as the biologically active compounds (Midland *et al.*, 1993). Naturally-occurring syringolides differ in the length of the alkyl chain but are otherwise identical (Fig. 2) and have indistinguishable elicitor activity (Midland *et al.*, 1993; Yucel *et al.*, 1994b). Their proposed structures have been recently confirmed and the absolute stereochemistry established by several independent syntheses (Henschke and Rickards, 1996 and references therein).

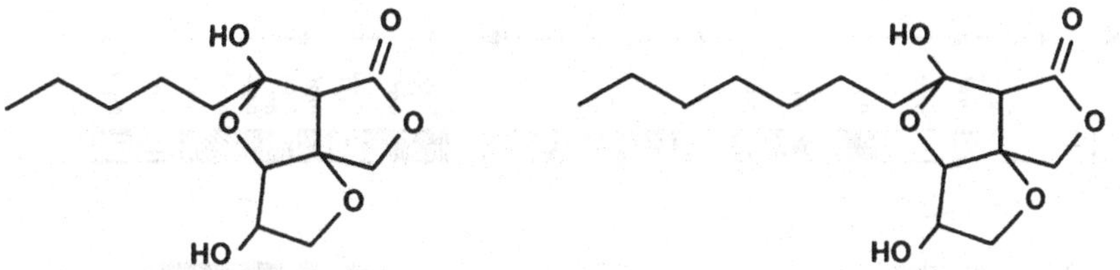

Syringolide 1   **Syringolide 2**

Figure 2. Structures of syringolide products specified by *avrD* alleles in *Pseudomonas syringae*.

The syringolides are specific elicitors directing the HR only in soybean cultivars carrying the dominant disease resistance gene *Rpg4* (Keen and Buzzell, 1991). Two classes of functional *avrD* alleles occur in *P. syringae* (Yucel *et al.*, 1994a; C. Boyd and N. Keen, unpublished data). These classes can be distinguished by particular amino acid substitutions in the AvrD proteins, but most notably they direct different syringolide products: class I alleles direct syringolides with C8 or C10 alkyl chains, whereas class II alleles direct syringolide products with C6 or C8 alkyl chains (Yucel *et al.*, 1994b). Several amino acids have been identified as essential for the functioning of *avrD* (Yucel and Keen, 1994). It is these residues that have often been naturally mutated in *P.s.* pv. *glycinea* isolates which do not produce syringolides and are devoid of the avirulence phenotype (Keith *et al.*, 1997). In this way, these isolates evade recognition by *Rpg4*-containing hosts.

B.   *PSEUDOMONAS SYRINGAE* AVIRULENCE GENES AND THEIR RELATIONSHIP TO THE *HRP* GENES

Lindgren *et al.* (1986) discovered a large (ca. 22 kb) chromosomal gene cluster in *P. syringae* pv. *phaseolicola* that was required for pathogenesis on bean plants and also directed induction of the HR on non-host plants such as tobacco. This cluster was called the *hrp* (hypersensitive response and pathogenicity) gene cluster (for review, see Alfano and Collmer, 1996). Although *hrp* genes are assumed to function only in pathogens, homologues have been detected in saprophytic bacteria (e.g. Mulya *et al.*, 1996). *hrp* gene clusters in *P. syringae* pathovars and other plant pathogenic bacteria contain several transcriptional units, each with one or more cistrons (Fig. 3). Many *hrp* genes encode components of a type III secretion pathway for extracellular proteins similar to that employed by vertebrate pathogens belonging to the genera *Yersinia*, *Salmonella* and *Shigella* to transfer toxic proteins into host cells. The type III protein secretion system contains components similar to those involved in flagellar biosynthesis, and it is assumed that secreted proteins are transferred across the inner and outer membranes by direct injection into the host cell through some kind of pilus structure (Salmond, 1994). By analogy, it is taken that the *hrp* gene cluster codes for the type III protein secretion machinery to deliver pathogenicity factors into host plant cells. Indeed, DspA, an essential pathogenicity factor of *Erwinia amylovora*, has been shown to be secreted via the Hrp secretion pathway (Gaudriault *et al.*, 1997).

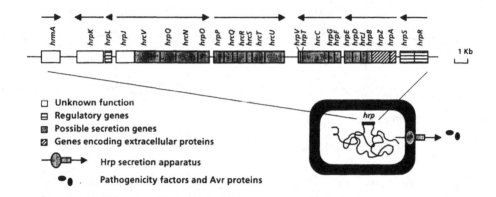

Figure 3. The organisation of the *hrp* gene cluster of *Pseudomonas syringae*. The cluster consists of seven operons encoding proteins required for type III protein secretion as well as putative pathogenicity factors. The Hrp secretion apparatus functions in the secretion of pathogenicity factors and avirulence proteins.

Bacterial plant pathogens possess genes flanking the formal *hrp* gene clusters that are also *hrp* regulated and may be involved in bacterial virulence. This is similar to the occurrence of "pathogenicity islands", frequently observed in vertebrate pathogens (Groisman and Ochman, 1996), wherein large clusters of genes important for virulence are clustered at particular locations in the genome. For example, the *hrmA* gene of *P. syringae* pv. *syringae* is located at the extreme left of the *hrp* gene cluster (Fig. 3) and appears to have an avirulence gene function (Alfano *et al.*, 1997). Similarly, it was shown that *avrPphE* occurred outside the left margin of the *hrp* gene cluster of *P. syringae* pv. *phaseolicola* (Mansfield *et al.*, 1994), and the *popA1* gene, specifying a hypersensitive-like response in specific Petunia genotypes, is similarly situated in *Pseudomonas solanacearum* (Arlat *et al.*, 1994). *hrp*-related genes have also been observed at the right side of the formal *hrp* gene cluster. For example, Lorang and Keen (1995) observed that two large transcriptional units, called collectively *avrE*, which occurred to the right of *hrpRS* in *P. syringae* pv. *tomato*, were required for an avirulence phenotype in soybean. At least two further transcriptional units were also identified in the vicinity of *P. syringae* pv. *tomato avrE* that were *hrp*-regulated, and one of them was recently shown to encode a 60 kDa protein that is secreted by the *hrp* system (S-Y He, personal communication). Furthermore, A. Collmer and collaborators (personal communication) have made the interesting finding that the *hrp* gene cluster of *Erwinia chrysanthemi* contains two hemolysin genes at one border and a phospholipase C gene at the other border. The phospholipase gene had previously been cloned (Keen *et al.*, 1992). When overexpressed in *E. coli* cells, the enzyme was secreted efficiently into the periplasm, but was not processed at the amino terminus, consistent with the absence of a conventional leader peptide sequence. These observations suggest that the phospholipase is secreted by the *E. chrysanthemi hrp* system during pathogenesis.

The *P. syringae hrp* operons and at least 12 different avirulence genes have a conserved consensus sequence in their promoters (Kobayashi *et al.*, 1990a; Innes *et al.*, 1993; Salmeron and Staskawicz, 1993; Shen and Keen, 1993; Xiao *et al.*, 1994; Xiao and Hutcheson, 1994), called the "*avr/hrp* box" (GGAACC-N15/16-CCAC). The co-regulation of *hrp* and *avr* genes suggests that they may also be functionally related and fostered the idea that the function of certain avirulence genes may require the *hrp* gene cluster. Indeed, Pirhonen *et al.* (1996) recently showed that *E. coli* MC4100 or *Pseudomonas fluorescens* cells carrying a cloned *P. syringae hrp* gene cluster in addition to any one of seven different cloned *P. syringae avr* genes (*avrA, avrB, avrC, avrPph3, avrRpt2, avr Rpm1* or *avrPto*) elicited the HR only in soybean, tomato or Arabidopsis cultivars carrying the complementary disease resistance genes. *E. coli* or *P. fluorescens* cells expressing only the various cloned avirulence genes elicited no detectable reaction on any plant cultivar tested, demonstrating the necessity of the *hrp* gene cluster for HR competence by these bacteria, which otherwise behave as saprophytes in plants. As expected, *E. coli* or *P. fluorescens* cells expressing *avrD* produced the expected syringolides independently of the *hrp* gene cluster. These results clearly indicated that the Hrp type III secretion pathway might be instrumental in delivering certain *P. syringae* avirulence gene proteins directly to or into plant cells. Support for this idea was obtained in several laboratories (Gopalan *et al.*, 1996; Yuan and He, 1996; Roine *et al.*, 1997). It was demonstrated that, indeed, the *hrp* system in *P. syringae* encodes a specialized pilus system, that may deliver certain *hrp*-secreted proteins to plant cells, where they appear to function as elicitors. Scofield *et al.* (1996) and Tang *et al.* (1996) have also shown that the protein product of *avrPto* interacts with the cognate product of the resistance gene, *Pto*, in the yeast two-hybrid system, therefore suggesting that AvrPto is transferred into the plant cell. Several different bacterial avirulence proteins could accordingly be delivered directly to the plant cell where they may function as virulence factors, or, in the case of resistant plants, are recognized as elicitors of the HR.

Leister *et al.* (1996) have provided additional support for the idea that certain *P. syringae* avirulence gene proteins function as elicitors inside plant cells. These workers devised a clever assay in which avirulence genes cloned into plant expression plasmids were mixed with GUS expression plasmids and these DNAs were transformed into leaf cells using particle bombardment. When *avrRpm1* or *avrRpt2* plus GUS were transformed into Arabidopsis leaves carrying the cognate disease resistance genes, *Rpm1* or *Rps2*, respectively, few or no blue GUS spots were subsequently seen. This was interpreted to result from hypersensitive cell death of transformed cells such that they could not express the GUS gene. Supporting this interpretation, transformation of GUS plus *avrRpm1* or *avrRps2* into Arabidopsis lines lacking the cognate resistance genes resulted in a normal number of GUS spots, similar to the case when no avirulence gene DNA was added. Accordingly, these results support the idea that the avirulence gene proteins function inside the plant cell as specific elicitors.

C.   *XANTHOMONAS AVRBS3* AND ITS PROTEIN ELICITOR

The avirulence gene *avrBs3* was cloned from *X. campestris* pv. *vesicatoria*, the causal agent of bacterial spot disease of pepper and tomato (Bonas *et al.*, 1989). The *avrBs3* gene is complementary to the resistance gene *Bs3* in pepper and contains a large open reading frame encoding a protein of 122 kDa. The central portion of this protein contains 17.5 repeats of an almost identical stretch of 34 amino acids. Deletion of certain of these repeats made radical changes in the specificity of the avirulence phenotype. The most striking was a deletion of four repeat units from the AvrBs3 protein, which made that it was no longer recognized by the pepper *Bs3* genotype, but instead elicited the HR in the susceptible *bs3* genotype (Herbers *et al.*, 1992)! Apparently, the truncated AvrBs3 protein had unmasked an as yet undiscovered resistance gene.

The *avrBs3* gene is the prototype of a large family of related avirulence genes in various *X. campestris* pathovars, several of which are important for bacterial virulence (e.g. Yang *et al.*, 1996). These results raised the possibility that *avrBs3* family member proteins might be targeted to the plant cell in the same way as the avirulence proteins of *P. syringae*. The first experimental evidence supporting this notion was supplied by Yang *et al.* (1995), who observed the occurrence of three nuclear localization signals near the carboxyl terminus of an *avrBs3* family member. These signals were functional: they directed the protein to the nucleus as determined by fusion to the GUS reporter gene. Furthermore, mutation of the nuclear localization signals destroyed nuclear localization. Van den Ackerveken *et al.* (1996) confirmed these results and showed that transient expression of *avrBs3* in pepper leaves, using an *Agrobacterium* vector, resulted in the HR only in leaves of pepper plants containing the cognate *Bs3* resistance gene. These observations support the idea that *avrBs3* family member proteins are delivered into plant cells, presumably by the bacterial *hrp* gene system, and targeted to the nucleus where they may function as virulence factors, or as avirulence determinants if the plant carries the matching disease resistance gene.

IV.   Plant disease resistance gene products and their functions

Disease resistance genes provide surveillance for pathogens which might otherwise seriously reduce plant populations. The apparently universal distribution of disease resistance genes targeting a large array of potential pathogens *a priori* establishes that resistance has been a high evolutionary priority in plants. We lack the space for an exhaustive discussion of the characteristics of plant disease resistance genes, but refer readers to an excellent review by Bent (1996).

The first cloning of a resistance gene was achieved by transposon tagging of *Hm1* from maize, which confers resistance to race 1 strains of the fungus *Cochliobolus carbonum* (Johal and Briggs, 1992). The *Hm1* gene encodes an enzyme that inactivates the host-specific HC-toxin produced by the fungus. However, this relationship is very different from the interaction between fungal *Avr* and plant *R* genes that confer gene-for-gene complementarity. When finally isolated, such *R* genes had strikingly different structures. A common feature of these latter disease resistance genes is that they are frequently clustered and undergo recombination events, sometimes at high frequencies. For example, Richter *et al.* (1995) studied selected recombinants of corn plants carrying phenotypically different disease resistance genes in the *Rp1* cluster, conferring race-specific resistance to *Puccinia sorghi*. It has been known for some time that the *Rp1* locus is meiotically unstable, due at least in part to unequal crossing over and gene conversion. Richter *et al.* (1995) identified recombinants with altered resistance to various members of a collection of corn rust races. Most significantly, they identified four recombinants that exhibited resistance specificities that were not present in either parent. The generation of such new recognitional characteristics strongly argues that recombination among genes of the *Rp1* family is the mechanism by which new recognitional specificities arise. These are important findings, because they provide support for the idea that recombination or gene conversion between tandem clusters of resistance genes may be a general feature in the generation of novel specificities that complement new pathogen avirulence genes. This mechanism is also reminiscent of somatic recombination leading to new antibody specificities (Tonegawa, 1983), the major difference being that in plants recombination occurs meiotically.

There is a great deal of circumstantial evidence indicating that plant surveillance of particular pathogens differs in stringency. Stringent surveillance means that a plant species or population carries a large number of different resistance gene specificities targeted to all or most biotypes of a single pathogen species. Individual plants will probably carry only a subset of these genes, but the composite plant population will have evolved surveillance of a sufficient number of avirulence genes to effectively target all extant genotypes of the pathogen. Non-stringent surveillance, on the other hand, implies the other extreme, in which a plant population possesses few or no resistance genes to a certain pathogen species. Stringent surveillance probably means that the plant and the pathogen have co-existed for a long period of time, while non-stringent implies that the pathogen and the plant have only recently come into contact, or that the pathogen has recently acquired new genes which markedly increase pathogenicity. There are abundant examples of these situations. Root rotting fungi, for example, frequently parasitize hosts that have few or no HR resistance genes. In many cases, this situation stems from recently contrived environmental conditions stimulating disease as a result of modern agricultural practices e.g. monoculture, high nitrogen fertilization, no-tillage etc. An acquired ability to produce a toxin similarly provides new-found pathogenicity and plants find themselves without resistance genes targeted to such an unexpected pathogen. At the other extreme, the hosts of many rust fungi exemplify stringent plant surveillance, with most studied plant species containing large numbers of resistance genes against a single rust species, sometimes more than 60 (see also Chapter 2).

The cell death that is the hallmark of the HR may be considered the result of effective surveillance. This phenotype is reflected in lesion mimic mutants, in which surveillance appears excessively stringent such that spontaneous necrosis occurs or the intensity of the HR in response to pathogen infection is abnormally high. While such accelerated cell death may involve several different genes, Dangl *et al.* (1996) refer to them as 'rogue' resistance genes. Some of these genes are indeed turning out to be modified disease resistance genes but at least two lesion mimics involve quite different genes. The recessive *mlo* gene in barley, which confers resistance to all known races of the powdery mildew fungus *Erysiphe graminis*, determines a defensive response resulting in thickening and apposition of cell wall material aty the location of attempted penetration by fungal hyphae. Although cell wall reinforcement is not normally accompanied by cell death, lesions can appear on *mlo*-containing plants at low temperatures in the absence of the pathogen. Mutations at the *mlo* locus can confer a lesion mimic phenotype with spontaneous leaf cell death and increased resistance to a range of plant pathogens. Buschges *et al.* (1997) recently cloned and sequenced a barley *mlo* gene and found that its protein product contained at least six putative membrane spanning domains but showed no homology to known disease resistance genes. Thus, *mlo* does not seem to be a resistance gene but instead may modulate intensity of the HR in some unknown way. Similarly, Gray *et al.* (1997) have shown that the *Lls1* lesion mimic gene in corn has significant homology to enzymes suspected to degrade aromatic compounds. They accordingly speculated that *Lls1* function may involve detoxification of phenolic compounds which might otherwise cause cell death.

Some lesion mimic mutants do appear to involve established disease resistance genes. Hu *et al.* (1996) identified examples of lesion mimic mutants at the *Rp1* locus in corn, discussed earlier in this section. Recombination events occurring between the *Rp1* resistance genes resulted in recombinants in which the stringency of surveillance was increased to the extent that resistance was expressed against all rust biotypes, or even diffuse necrosis developed spontaneously. The lesson of these considerations on stringency is that during natural plant evolution greater infection pressure increases the selection of recombinational events, leading to new resistance genes which increase the stringency for pathogen detection (cf. McDowell *et al.*, 1998). On the other hand, if disease pressure decreases due to environmental or other changes, it is likely that recombinant forms of resistance genes are generated which have lost the ability to recognize the pathogen — that is, resistance becomes less stringent. As large scale DNA sequencing of resistance gene clusters proceeds over the next few years, and recombinant resistance genes are constructed *in vitro*, we should develop an understanding of the molecular factors governing surveillance stringency.

A.   CLONED DISEASE RESISTANCE GENES

Several disease resistance genes have already been cloned and characterized from various plant species (for reviews see Martin, 1995; Michelmore, 1995; Staskawicz *et al.*, 1995; Bent, 1996; Hammond-Kosack and Jones, 1997). Although it was speculated that these genes might resemble those of vertebrate immune systems, they in fact most closely resemble components of generic eucaryote signaling systems. Plant disease resistance genes encode proteins that fall into three general classes with common structural motifs (Table 2; Fig. 4): (i) proteins consisting exclusively of a protein kinase domain, the only known example being the tomato *Pto* gene product (Martin *et al.*, 1993); (ii) proteins with leucine-rich repeat (LRR) domains, also containing leucine zipper or Toll/interleukin receptor-like and nucleotide binding site domains and/or a membrane-spanning domain, such as the Arabidopsis Rpm1 and *Rps2* genes (Staskawicz *et al.*, 1995) and the tobacco *N* gene (Whitham *et al.*, 1994; Dinesh-Kumar *et al.*, 1995); (iii) a hybrid with leucine zipper, LRR and protein kinase domains in the same protein, exemplified by the rice *Xa-21* gene product (Song *et al.*, 1995). Recent reviews skillfully discuss resistance gene proteins and their similarity with other proteins of known function (Dangl, 1995; Bent, 1996; Hammond-Kosack and Jones, 1997).

Table 2.   Cloned plant disease resistance genes that modulate the HR

Gene	Plant species	Targeted against	Type	Reference
Cf-2	Tomato	Cladosporium fulvum	LRR II	Dixon et al. 1996
Cf-4	Tomato	Cladosporium fulvum	LRR II	Hammond-Kosack and Jones, 1997
Cf-5	Tomato	Cladosporium fulvum	LRR II	Hammond-Kosack and Jones, 1997
Cf-9	Tomato	Cladosporium fulvum	LRR II	Jones et al., 1994
HS1[pro-1]	Sugar beet	Beet cyst nematode	LRR II	Cai et al., 1997
I2C	Tomato	Fusarium oxysporum	LRR I	Ori et al., 1997
L6	Flax	Melampsora lini	LRR I	Lawrence et al., 1995
M	Flax	Melampsora lini	LRR I	Anderson et al., 1997
Mi	Tomato	Root knot nematode	LRR I	I. Kaloshian & V. Williamson, pers. comm.
N	Tobacco	Tobacco mosaic virus	LRR I	Whitham et al., 1994
Prf	Tomato	Pseudomonas syringae pv. tomato	LRR I	Salmeron et al., 1996
Pto	Tomato	Pseudomonas syringae pv. tomato	Kinase	Martin et al. 1993
Rp1-D	maize	Puccinia sorghi	LRR I	T. Pryor and S. Hulbert, pers. comm.
Rpm1	Arabidopsis	Pseudomonas syringae pv. maculicola	LRR I	Grant et al., 1995
Rpp5	Arabidopsis	Peronospora parasitica	LRR I	Parker et al., 1997
Rpp8	Arabidopsis	Peronospora parasitica	LRR I	McDowell et al., 1998
Rps2	Arabidopsis	Pseudomonas syringae pv. tomato	LRR I	Bent et al., 1994; Mindrinos et al., 1994
Rps5	Arabidopsis	Pseudomonas syringae pv. tomato	LRR I	R. Innes, pers. comm.
Xa1	Rice	Xanthomonas campestris pv. oryzae	LRR I	Yoshimura et al., 1998
Xa21	Rice	Xanthomonas campestris pv. oryzae	LRR II/kinase	Song et al., 1995

Type I LRR proteins are defined as putative intracellular proteins possessing nucleotide binding sites while type II LRR proteins are postulated to be plasma membrane associated and lack nucleotide binding sites.

To date, disease resistance genes have been cloned by laborious map-based approaches or by transposon tagging, neither of which are applicable to most economically important plant species, such as polyploidous wheat or potato. This situation is changing rapidly, however, since PCR approaches utilizing conserved sequences from cloned resistance genes have permitted the rapid cloning of genes from soybean and potato that resemble known disease resistance genes (for review, see Michelmore, 1996). Thus, two different groups utilized PCR amplification to identify a large number of putative resistance genes from soybean (Kanazin, 1996; Yu *et al.*, 1996). Many of these amplified genes turned out to be clustered, a cardinal characteristic of disease resistance genes, as discussed earlier. Most of the recovered sequences mapped to regions of the soybean genome already known to harbor resistance genes by genetic studies. Finally, Yu *et al.* (1996) found that at least one of their PCR sequences was contiguous with an LRR domain typical of known resistance genes. Similarly, Feuillet *et al.* (1997) utilized a serine/threonine protein kinase clone to detect a polymorphism which mapped at the *Lr10* leaf rust resistance locus in wheat. While it is uncertain whether the kinase itself constitutes *Lr10*, it provides an avenue to investigate the structure of the *Lr10* locus. These kinds of developments portend the isolation of large numbers of resistance genes from important crop plants in the future.

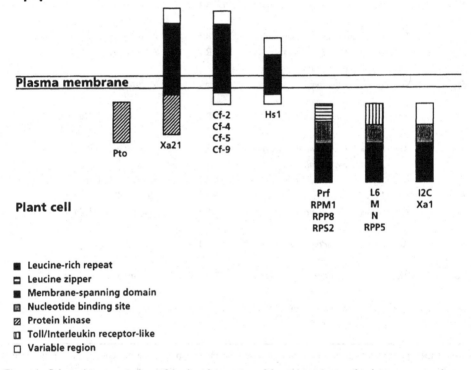

Figure 4. Schematic representations of the domain structure of the various classes of resistance gene products and their transmembranous or cytoplasmic localization. See Table 2 for details of the genes listed.

B.   HOW RESISTANCE GENES FUNCTION

Although disease resistance genes have been cloned for half a decade, surprisingly little is known about how they function. One of the better understood resistance genes is *Pto* in tomato. The *Pto* gene encodes a serine/threonine protein kinase and is the only known resistance gene that does not contain LRR (Fig. 4). Although *Pto* resistance is effective against isolates of *P. syringae* pv. *tomato* carrying the *avrPto* avirulence gene, Salmeron *et al.* (1994) showed that *Pto* function requires the presence of a second gene, called *Prf*. When this gene was cloned and sequenced (Salmeron *et al.*, 1996), it turned out to be a member of the LRR class of resistance gene proteins. *Prf* is also required for the function of *Fen*, a protein kinase homologue of *Pto* that confers sensitivity in tomato to the insecticide fenthion (Martin *et al.*, 1994; Rommens *et al.*, 1995a). Current models (cf. Fig. 5) assume that both Pto and Fen can physically interact with Prf, initiating reactions leading to hypersensitive necrosis. Using the yeast two-hybrid system, Martin and colleagues showed that Pto but not Fen interacted with another protein kinase, called *Pti1* (Zhou *et al.*, 1995). Further proteins interacting with Pto, termed Pti4, Pti5 and Pti6 have been found to strongly resemble ethylene-responsive element-binding proteins (EREBPs) of tobacco. Pti5 and Pti6 have been shown to bind promoter sequences of defense-related PR genes and, thus, appear to function as transcription factors. One can assume, therefore, that Pto and Pti1 are part of a stereotypical kinase signaling cascade that eventually activates defense response gene promoters (Fig. 5).

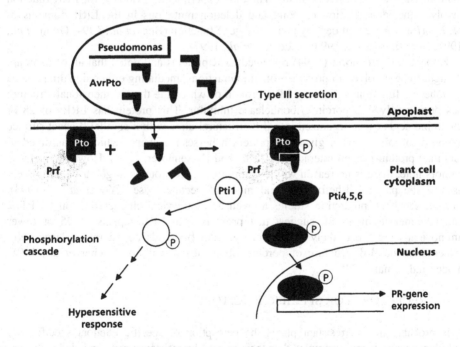

Figure 5. A current model for the signaling pathways initiated upon binding of the bacterial avirulence gene product AvrPto to the plant disease resistance gene product Pto. Pto interacts with Prf, Pti1 and a complex containing the transcription factors Pti4, Pti5 and Pti6. Upon binding AvrPto, Pto autophosphorylates and then phosphorylates Pti1 and at least one of the transcription factors (indicated by P). Thereupon, Pti1 is involved in the the signal transduction cascade leading to the hypersensitive response, whereas Pti 4, 5 and 6 interact with promoters of genes encoding pathogenesis-related proteins to initiate PR-gene expression.

Using the yeast two-hybrid system, Scofield *et al.* (1996) and Tang *et al.* (1996) presented evidence that the *Pto* resistance gene protein and the protein product of *avrPto* also directly interact. While not yet verified biochemically, the results support the indications discussed earlier that an avirulence gene protein can function as an elicitor *per se* by physical interaction with its cognate resistance gene protein. Jia *et al.* (1997) cloned a nonfunctional *pto* allele from tomato and showed that it was 87% identical to *Pto*. Although this protein exhibited protein kinase activity, it did not interact with Pti1 in the yeast two-hybrid system. The authors therefore speculated that *pto* may be nonfunctional due to defects in complexing with Pti1 or possibly other required proteins, such as Prf or even AvrPto.

Except for *Pto*, all known plant disease resistance genes are of the LRR type (Fig. 4). It is postulated that the LRR domains are involved in elicitor recognition and activate protein kinases or phosphorylation reactions to initiate a signal cascade eventually resulting in the activation of defense response genes (Dangl, 1995; Bent, 1996). Deletion and domain-swapping experiments proved that extracellular LRR domains of the vertebrate LH/CG and FSH receptors do function as specific binding sites for peptide hormones (Braun *et al.*, 1991), but no biochemical proof is yet available for plant resistance gene LRR domains. The only experimental evidence for LRR function involves the identification of point and deletion mutations in the LRR domains of *M*, *Rpm1* and *Rps2* that destroy resistance gene function (Bent *et al.*, 1994; Grant *et al.*, 1995; Mindrinos *et al.*, 1994; Anderson *et al.*, 1997).

Kobe and Deisenhofer (1994) reviewed LRR proteins and noted that all of them are thought to be involved in protein-protein interactions, mediating either signaling events or adhesive functions. The only LRR protein of which the three-dimensional structure has been resolved is porcine ribonuclease inhibitor. This protein has LRRs of 28 or 29 amino acids each, organized into β-β structural units with a parallel β-sheet surface exposed to solvent. This gives the protein a horseshoe shape, which is assumed to facilitate protein-protein interactions (Kobe and Deisenhofer, 1994). However, as these authors note, there is no reason why certain other LRR proteins might not alternatively assume the parallel β-helix structural motif of pectate lyases (Yoder *et al.*, 1993). Indeed, sequence prediction studies as well as biophysical characterization by FTIR and CD measurements all suggest that proteins with LRR repeats of 25 or fewer amino acids are most likely folded into parallel β-strands coiled into a large helix rather than the β-β structure of porcine ribonuclease inhibitor (Sieber *et al.*, 1995; Yoder and Jurnak, 1995).

C.   RECEPTORS FOR SPECIFIC ELICITORS

It is probable that in resistant plants the perception of specific elicitors specified by avirulence genes is more complex than their binding by resistance gene-encoded receptors. Dangl (1992) noted similarities between active defense in higher plants and immune systems in higher vertebrates, particularly the major histocompatibility complex (MHC). One especially interesting feature of the MHC is antigen presentation to T cells. This is a complex process in which antigens are bound to class I MHC molecules and displayed to antigen-specific T-cells. While it is premature to speculate about analogies with elicitor recognition by plants expressing particular disease resistance genes, it is nonetheless likely that elicitor display is important in plant defense. Indeed, in two cases, ^{125}I-labeled specific elicitors were shown to bind with similar affinity to cell fractions from both resistant and susceptible genotypes. In the case of the *avr9* peptide elicitor from *Cladosporium fulvum*, Kooman-Gersmann *et al.* (1996) observed saturable, ligand-displaceable binding to plasma membrane preparations from either *Cf9* or *cf9* tomato genotypes. Likewise, labeled syringolides, produced by bacteria expressing avirulence gene *avrD*, bound ligand-specifically to a site in the soluble fraction of soybean leaves; again, however, no significant differences were observed in binding to extracts from soybean cultivars containing or lacking the cognate disease resistance gene, *Rpg4* (Ji *et al.*, 1997). These results suggest that the *Cf9* and *Rpg4* resistance gene products are not directly involved with elicitor binding but instead may be components of recognitional complexes or signal transduction pathways leading to defense response gene activation. In this regard, the *Cf9* and *Rpg4* cases are also reminiscent of the relationship of the *Pto* and *Prf* genes in tomato, discussed earlier.

D. TRANSFER OF CLONED DISEASE RESISTANCE GENES TO OTHER PLANT SPECIES

One of the major compulsions to clone plant disease resistance genes is to transfer them into other plant species which lack adequate resistance to a pathogen. Early indications suggest that this strategy may succeed, since Rommens *et al.* (1995b) and Thilmony *et al.* (1995) showed that transgenic tobacco plants carrying the *Pto* resistance gene from tomato exhibited resistance to *P. syringae* pv. *tabaci* expressing *avrPto*. Since *P. syringae* pv. *tabaci* does not appear to harbor *avrPto*, the practical impact of these findings is limited, but they do indicate that other required signal transduction machinery required for *Pto* function is present in tobacco. More appealing from a practical point of view is the work of Whitham *et al.* (1996) in which tomato plants were transformed with the tobacco *N* gene for resistance to TMV. This gene functioned in the transgenic tomato plants to confer virus resistance. Because TMV is a natural pathogen on tomato, the results may have practical impact. They also encourage similar exercises with other resistance genes, particularly with plants where good sources of resistance are not readily available. For example, *Phytophthora* sp. and *Fusarium* sp. are important root and stem rot pathogens on a wide range of plant species, many of which have few if any known disease resistance genes. However, genes conferring resistance against these same pathogens may be cloned from other plant species (e.g. the tomato *I2C* resistance gene against *F. oxysporum*, Table 2). It would be appealing to clone several such resistance genes and transfer them to susceptible plant species in order to attempt introducing resistance. As described earlier, several pathogen avirulence genes function in virulence of the pathogen on hosts lacking the corresponding resistance genes. It will accordingly be appealing to utilize disease resistance genes in transgenic plants that complement pathogen avirulence genes with an important role in virulence. Such resistance genes have been known for decades as "strong" genes, since to "overcome" them by losing the matching avirulence function, the pathogen also suffers a large decrease in virulence or fitness (Van der Plank, 1968).

V.   Signal transduction and early responses

Key questions in the analysis of the plant defense response deal with signal transduction. What are the signal transduction pathways following the primary interaction between an elicitor and a receptor or between a resistance gene product and an avirulence gene product? Do the two forms of recognition, by general elicitors and by specific elicitors, utilize similar or different transduction pathways? It has been argued that the signal transduction pathways must be short or converge quickly, because of the high number of resistance genes known. Recent studies indicate that transduction of elicitor signals in plants proceeds along pathways similar to animal and microbial chemoperception systems. However, there are considerable deficiencies in our understanding of signal transduction pathways in plants.

A.   CHANGES IN MEMBRANE POTENTIAL AND INTRACELLULAR ION CONCENTRATIONS

One of the earliest effects often observed in response to elicitors is a rapid depolarization of the plasma membrane (Mathieu *et al.*, 1991). In cultured plant cells, this is accompanied by rapid changes in ion concentrations on the outside of the cells, indicative of rapid ion fluxes, such as an alkalinization of the growth medium (Felix *et al.*, 1993; Granado *et al.*, 1995) and a corresponding efflux of K^+ (Mathieu *et al.*, 1991) (Fig. 6). Calcium also takes part in these ion fluxes, probably through voltage dependent calcium channels (Thuleau *et al.*, 1994). An elegant demonstration of the elicitor-mediated increase in cytoplasmic calcium levels has been obtained with transgenic plants expressing aequorin, a protein that emits light in a calcium-dependent fashion; plants treated with elicitors show increased light emission (Knight *et al.*, 1991).

Are such increases in cytoplasmic calcium necessary and/or sufficient for the transduction of the elicitor signal? We do not know yet with any certainty, primarily because the duration of the calcium fluxes is much shorter than the typical active defense responses, which are usually measured after hours or even days. However, there is indirect evidence from work with calcium chelators, calcium channel blockers and ionophores indicating that calcium is required for activation of defense genes and phytoalexin accumulation (reviewed in Ebel and Cosio, 1994; Boller, 1995).

B.   PROTEIN PHOSPHORYLATION

Protein phosphorylation is commonly involved in animal and microbial signal transduction pathways, and the finding that some resistance genes encode homologues of protein kinases strongly indicates that the same is true for pathways leading to active defense. Studies with drugs also provide evidence for an involvement of protein kinases. For example, K-252a and staurosporine, two related inhibitors of protein kinases, rapidly block elicitor-induced responses, but they do not affect basal metabolism and protein synthesis (Grosskopf *et al.*, 1990; Felix *et al.*, 1991). Derivatives of staurosporine are available which differ in their potential to inhibit protein kinases *in vitro*. These derivatives similarly differ in their potential to inhibit elicitor responses *in vivo* (Felix *et al.*, 1991).

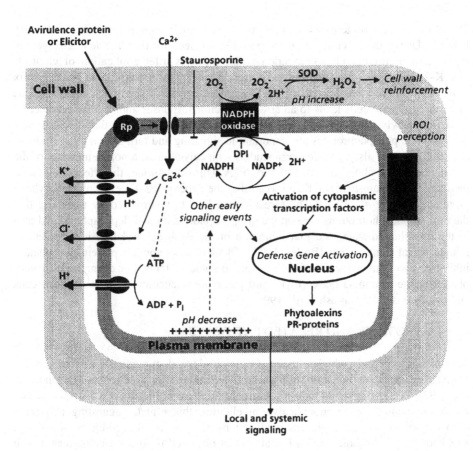

Figure 6. Early signal transduction events in incompatible host-pathogen interactions. Binding of an avirulence protein or elicitor to a resistance gene product or receptor (Rp) results in Ca^{2+} influx, Cl^- efflux, K^+/H^+ exchange and acidification of the cytosol/alkalinization of the apoplast. Activation of a plasma membrane NADPH oxidase generates active oxygen species which, through the action of superoxide dismutase (SOD), give rise to H_2O_2, that is used by peroxidases to produce lignin-like compounds to reinforce cell walls. Increased intracellular Ca^{2+}, perception of reactive oxygen intermediates (ROI) and decreased cytosolic pH, together with other signaling compounds generated, lead to defense gene activation. As a result phytoalexins are produced, PRs start to accumulate and, upon hypersensitive cell death, further local and systemic signals move to neighbouring cells and to more distant plant parts. At several stages phosphorylation events are required which can be blocked by application of staurosporine.: established stimulation; proposed stimulation;: inhibition.

While these studies indicate that the activity of certain protein kinases is required for the elicitor response, there is also direct evidence that the pattern of protein phosphorylation changes rapidly after elicitor addition (Dietrich *et al.*, 1990; Felix *et al.*, 1991). A useful technique to reveal these changes is in short pulse-labeling with radioactive phosphate (Felix *et al.*, 1991). Plant cells take up the radioactive phosphate very rapidly, leading to rapid and strong labeling of the exchangeable phosphate groups in ATP.

Therefore, phosphoproteins with a rapid turnover of their phosphate group are strongly labeled. Using this technique, it is possible to see drastic, transient changes in the pattern of protein phosphorylation within minutes after application of elicitors, and K-252a blocks the appearance of newly labeled phosphoprotein bands (Felix *et al.*, 1991, 1993). Application of K-252a in mid-course of the alkalization response of cultured tomato cells leads to an arrest of elicitor responses and to disappearance of label from newly phosphorylated proteins within minutes, indicating that the phosphate groups in the phosphoproteins are important in signaling and turn over rapidly (Felix *et al.*, 1991). This is also apparent from studies with calyculin A, a potent inhibitor of the two major protein phosphatases, PP1 and PP2A. This drug mimics the effect of elicitors, inducing rapid changes in ion fluxes as well as the appearance of newly phosphorylated proteins (Felix *et al.*, 1994). These results can best be interpreted by assuming that the critical phosphoproteins are continually phosphorylated and dephosphorylated also in the non-elicited state, and that inhibition of the dephosphorylation is sufficient to initiate signal transduction (Boller, 1995). Okadaic acid, another protein phosphatase inhibitor, also mimics the effects of elicitors in some systems. For example, it induces phenylalanine ammonia-lyase (PAL) and phytoalexin accumulation in soybean cells and cotyledons (MacKintosh *et al.*, 1994).

C.   G PROTEINS AND RELATED PATHWAYS

Mastoparan, a peptide which activates G protein-dependent signaling in animal cells, induces an oxidative burst in soybean cells a similar way as elicitors (Legendre *et al.*, 1992). Pectic fragments induce a transient increase in inositol 1,4,5-trisphosphate, and mastoparan is even more active in eliciting this effect, suggesting activation of phospholipase C in a process mediated by G proteins (Legendre *et al.*, 1993). Breakdown of phosphoinositides has also been observed in tobacco cells treated with pathogenic bacteria (Atkinson *et al.*, 1993) and in elicitor-treated pea tissues and membranes (Toyoda *et al.*, 1993).

D.   PRODUCTION OF ACTIVE OXYGEN SPECIES, NITRIC OXIDE, AND LIPID OXIDATION

Rapid production of active oxygen species such as the superoxide radical and hydrogen peroxide, the so-called oxidative burst, is a hallmark of the activation of lymphocytes and is also typical of the HR in plants. It is often one of the earliest reactions of plant cells to elicitors (Apostol *et al.*, 1989; Bradley *et al.*, 1992; Mehdy, 1994; Baker and Orlandi, 1995). As a result of prior plasma membrane depolarization an NADPH oxidase is activated (Pugin *et al.*, 1997; Keller *et al.*, 1998), which is held responsible for the transient production of the active oxygen species (Fig. 6). In plant cells, like in the animal immune system, the active oxygen species may be toxic for pathogens directly (Mehdy, 1994), but they may also lead to the cross-linking of cell wall components, rendering the walls more resistant to attack by fungal enzymes (Bradley *et al.*, 1992; Brisson *et al.*, 1994). Thus, the oxidative burst appears to be part of the active defense against pathogens. In addition, however, the active oxygen species may act themselves as second messengers (see chapter 6, The hypersensitive response, for details). They may even be necessary for the initiation of host cell death (Levine *et al.*, 1994).

The generation of active oxygen species can be inhibited by the drug diphenylene iodonium (DPI), an inhibitor of the superoxide anion-forming NADPH oxidase in lymphocytes. In elicitor-treated soybean cells, DPI blocks both the generation of active oxygen species and the production of phytoalexins in response to elicitors (Tenhaken *et al.*, 1995). However, when soybean or cotton cells were treated with a glycoprotein elicitor from *Verticillium dahliae*, the carbohydrate portion of the elicitor appeared to be responsible for the induction of the oxidative burst, whereas the protein moiety acted as the specific extracellular signal for phytoalexin induction (Davis *et al.*, 1993). Thus, the oxidative burst and phytoalexin formation can be induced independently of each other. this conclusion is supported by results obtained from parsley cells, in which DPI blocks the generation of active oxygen species but does not interfere with the induction of phytoalexins (Hahlbrock *et al.*, 1995). Similarly, induction of phytoalexins did not seem to depend on active oxygen species in white clover suspension cultures treated with incompatible *Pseudomonas corrugata* or the abiotic elicitor mercuric chloride (Devlin and Gustine, 1992). However, in intact plants the oxidative burst may induce other defense responses or stimulate defense-related reactions. Thus, active oxygen species may or may not be involved in signal transduction, depending on the plant species.

Recent evidence suggests that nitric oxide (NO) potentiates the action of active oxygen species and also functions independently of such intermediates to induce defense genes. Inoculation of soybean cell suspensions with avirulent *P. syringae* pv. *glycinea* stimulated NO production as well as induction of PAL. The latter was blocked by the nitric oxide synthase (NOS) inhibitors N^{β}-nitro-L-arginine (L-NNA) and *S,S'*-1,3-phenylene-bis(1,2-ethanediyl)-bis-isothiourea (PBITU). Both inhibitors also blocked the HR of Arabidopsis to avirulent *P. syringae* pv. *maculicola* and promoted the spreading chlorosis characteristic of infection with virulent bacteria (Delledonne *et al.*, 1998). Also in tobacco reacting hypersensitively to TMV NOS activity is increased and injection with NOS or NO donors induced defense gene activation, possibly mediated by cGMP (Durner *et al.*, 1998). Because in the vertebrate immune system active oxygen species and NO often function together, for instance in macrophage killing of bacteria or tumor cells, a role of NO in plant defense signalling strengthens even more the similarities between plant and animal defenses.

A lipid oxidation product in plants that has received much attention recently, namely *cis*-jasmonic acid (JA), a product of the oxidative breakdown of linolenic acid (see Blechert *et al.*, 1995 for a review). Rapid synthesis of JA occurs in response to elicitors (Gundlach *et al.*, 1992; Müller *et al.*, 1993; Nojiri *et al.*, 1996), and application of JA can induce phytoalexin accumulation in suspension-cultured cells of several plant species (Gundlach *et al.*, 1992; Nojiri *et al.*, 1996), suggesting that JA acts as a second messenger to induce phytoalexins in this model system. However, in intact plants JA appears to be involved in other signal transduction pathways such as the response to mechanical wounding (Farmer *et al.*, 1994) or to UV radiation (Conconi *et al.*, 1996), responses which typically do not involve phytoalexin production. Thus, JA is probably not a second messenger tied uniquely to defense responses to pathogens. Nevertheless,

it is interesting that JA may act synergistically with other signaling compounds such
as ethylene in the induction of defense responses, as exemplified by the induction of
genes for osmotin, a PR-protein in potato (Xu *et al.*, 1994).

E.   SALICYLIC ACID AND SYSTEMIC SIGNALS

Another compound implicated in plant defense responses is salicylic acid (SA)
(see Durner *et al.*, 1997 for a review). In an elegant series of experiments, Gaffney
et al. (1993) have used transgenic tobacco plants expressing the *NahG* gene, which
encodes the enzyme salicylate hydroxylase from *Pseudomonas putida*, to convert SA
into catechol. *NahG*-containing tobacco and Arabidopsis plants fail to accumulate
salicylic acid and are strongly inhibited in their capacity to produce pathogenesis-
related proteins (PRs) in response to a pathogen infection. Thus, it has become clear
that SA potentiates defense responses upon primary infection and is instrumental in
the activation of genes encoding PRs (Delaney *et al.*, 1994). PAL catalyzes the first
step in SA biosynthesis. Arabidopsis plants become susceptible to avirulent fungal
pathogens when PAL activity is specifically inhibited, but resistance can be restored
by application of SA (Mauch-Mani and Slusarenko, 1996), indicating that synthesis
and accumulation of SA are required for the expression of resistance.

As discussed earlier, the whole plant may show systemic acquired resistance (SAR)
after the hypersensitive response has been elicited locally by elicitors or infections
causing necrosis (Neuenschwander *et al.*, 1996; Ryals *et al.*, 1996). SAR is accompanied
by the systemic accumulation of PRs (see chapter 8), and evidence has been provided
that an increase in SA levels in the tissue as well as in the phloem precedes induction
of SAR and PRs (Malamy *et al.*, 1990; Métraux *et al.*, 1990; Rasmussen *et al.*, 1991).
In *NahG*-containing plants no SAR develops upon local stimulation, implicating SA
as an important second messenger in this response (Gaffney *et al.*, 1993). However,
the systemic signal moves out of a leaf before salicylic acid accumulates (Rasmussen
et al., 1991), and grafting of wild-type plants onto plants expressing salicylate
hydroxylase showed that systemic signaling is not affected by the absence of SA
accumulation in leaves showing local hypersensitive responses (Vernooij *et al.*, 1994).
Thus, contrary to the initial expectations, salicylic acid is not the transmitted signal,
or at least not the only one, and at least one systemic signal for SAR still remains to
be discovered (see also Chapter 10).

Can elicitors themselves move systemically? Xylanase (Sharon *et al.*, 1992) and
elicitins (Devergne *et al.*, 1992) are translocated systemically under certain conditions,
apparently by a hydraulic mechanism in the xylem (Sharon *et al.*, 1992). In general,
however, systemic signaling appears to be based primarily on mobile plant factors
formed endogenously rather than on mobile microbial elicitors (Ryals *et al.*, 1996).

Conclusions

One of the particularly intriguing aspects of the active defense response of plants is
the possible interplay between recognition systems based on generally occurring microbial

molecules, exemplified by the general elicitors, and those based on highly specific molecules occurring only in a single race of a pathogen, exemplified by the specific elicitors. Due to the different approaches used to study these two types of recognition systems, information about them is not easy to match: the chemoperception systems for the general elicitors have mainly been studied with respect to biochemistry but we know little about the genes involved; in contrast, the resistance genes (and the matching avirulence genes) have mainly been investigated genetically and more recently with the tools of molecular biology, so that we know the genes but not the biochemical functions linked to them. Therefore, we do not know for the moment whether elicitor receptors and resistance gene products are sensors of the same surveillance system, comparable to different antibodies in the immune system with different specificities, or whether they belong to two separate, different surveillance systems — an early warning system based on highly sensitive perception of general elicitors, and a more specialized recognition system dedicated to the recognition of the most well-adapted pathogens that avoid or suppress the early warning system.

The paradigm of resistance genes indicates that active defense is based on a single, all-important recognition phenomenon, namely the interaction between the resistance gene product and the avirulence gene product: the presence or absence of a single resistance gene appears to make all the difference between full resistance and full susceptibility in many pathosystems (Bent, 1996). However, these seemingly simple one-dimensional black-and-white systems are probably just the tip of an iceberg of chemical signaling. A given plant has an array of highly sensitive chemoperception systems both for molecules unique to a microbial species or genus and for molecules common and typical for a whole class of microbes. It would be surprising if the plant did not make use of this potential to combine and integrate microbial signals, allowing it to respond to an approaching microbe with a differentiated reaction on a gray scale between full acceptance and outright defense (cited from Boller, 1995). The recently characterized mutants of Arabidopsis that spontaneously form necrotic lesions illustrate the point: these mutants have the appearance of hypersensitively reacting plants even in the absence of a pathogen, and it has been shown that they react locally with a hypersensitive defense response against micro-organisms that do not elicit a visible response in the wild type. This implies that the plant has the potential to perceive these micro-organisms but does not normally react with the hypersensitive response (Dietrich *et al.*, 1994; Greenberg *et al.*, 1994). Another surprising result came from a cytological study of the infection process of *Phytophthora infestans* on potato leaves carrying or lacking a resistance gene against the particular isolate of the pathogen (Freytag *et al.*, 1994). At the macroscopic level, the interaction is totally different: the leaves appear fully resistant when they carry the resistance genes but are fully susceptible when they lack it. At the microscopic scale, however, there were no statistically significant differences in the cytology after infection, despite a highly refined evaluation scheme. In both the susceptible and resistant tissue, most of the invading hyphae were apparently recognized, induced rapid defense responses and were effectively stopped from further invasion (Freytag *et al.*, 1994).

In the context of this chapter, perception of microbial elicitors has been considered only from the point of view of plant-pathogen interactions. It should be noted, however, that plant-microbe interactions lead to disease only relatively rarely. It is true that plants are continuously exposed to potential pathogens, as the textbooks say. However, they are equally exposed to harmless saprophytes, and furthermore they continuously interact with mutualistic symbionts such as mycorrhizal fungi, beneficial endophytes, or growth-promoting bacteria. The recognition systems that we have described may also have significance for these interactions. For example, recognition of the symbiotic nitrogen-fixing rhizobia by legume roots is based on a signal molecule produced by the bacteria, the Nod factor, which is similar in structure to the chitin elicitors discussed above (see Boller, 1995). In fact, Nod factors are recognized in the same way as chitin fragments in cells of the non-host plant, tomato (Staehelin et al., 1994), illustrating the possibility that one and the same "elicitor" may be recognized as a signal for mutualistic symbiosis in one plant species and as a signal for defense in another. This is one of the important challenges for breeders and genetic engineers interested in enhancing plant resistance against pathogens: enhanced resistance against pathogens should not go along with a greater sensitivity to harmless saprophytes or with a reduced potential for symbiotic interactions.

The recognition of pathogens, which we described as a typical non-self recognition phenomenon, also has parallels to the recognition of endogenous plant signals. An example is the pectic fragments released by microbial pectolytic enzymes which act as endogenous elicitors. Similar fragments may also be released by the plant's own enzymes and act as hormone-like endogenous signals, as postulated in the "oligosaccharin" concept (Darvill et al., 1992; Côté and Hahn, 1994). Indeed, fruit ripening in tomato can be induced by pectic fragments resembling the endogenous elicitors (Melotto et al., 1994). Similarly, the elicitor-active peptides derived from *Phytophthora megasperma* (Nürnberger et al., 1994) can be likened to systemin, a peptide hormone involved in the systemic induction of proteinase inhibitors in response to wounding (Pearce et al., 1991). Furthermore, some of the cloned resistance genes resemble the family of the so-called receptor kinases (see Bent, 1996) and have their closest homologues among the receptor kinases involved in sporophytic self-incompatibility, a self-recognition process preventing self-fertilization and thereby promoting outbreeding (Stein et al., 1991). Thus, the recognition systems for self and non-self may have a similar basis, and it is an intriguing question how they relate to each other in evolutionary terms. Which was first, self-recognition or non-self-recognition? Also, the similarities between non-self recognition and self recognition present another challenge to breeders and genetic engineers aiming at increasing plant disease resistance: modulation of the non-self recognition systems for pathogens may also interfere with self-recognition processes important in growth and development.

References

Alfano JR, Kim H-S, Delaney TP and Collmer A (1997) Evidence that the *Pseudomonas syringae* pv. *syringae hrp*-linked *hrmA* gene encodes an Avr-like protein that acts in an *hrp*-dependent manner within tobacco cells. Mol Plant-Microbe Interact 10: 580—588

Alfano JR and Collmer A (1996) Bacterial pathogens in plants: life up against the wall. Plant Cell 8: 1683—1698

Anderson PA, Lawrence GJ, Morrish BC, Ayliffe MA, Finnegan EJ and Ellis JG (1997) Inactivation of the flax rust resistance gene *M* associated with loss of a repeated unit within the leucine-rich repeat coding region. Plant Cell 9: 641—651

Apostol I, Heinstein PF and Low PS (1989) Rapid stimulation of an oxidative burst during elicitation of cultured plant cells. Role in defense and signal transduction. Plant Physiol 90: 109—116

Arlat M, Van Gijsegem F, Huet J-C, Pernollet J-C and Boucher CA (1994) PopA1, a protein which induces a hypersensitivity-like response on specific *Petunia* genotypes, is secreted via the Hrp pathway of *Pseudomonas solanacearum*. EMBO J 13: 543—553

Atkinson M, Bina J and Sequeira L (1993) Phosphoinositide breakdown during the K$^+$/H$^+$ exchange response of tobacco to *Pseudomonas syringae* pv. *syringae*. Mol Plant-Microbe Interact 6: 253—260

Baillieul F, Genetet I, Kopp M, Saindrenan P, Fritig B and Kauffmann S (1995) A new elicitor of the hypersensitive response in tobacco: a fungal glycoprotein elicits cell death, expression of defence genes, production of salicylic acid, and induction of systemic acquired resistance. Plant J 8: 551—560

Baker CJ and Orlandi EW (1995) Active oxygen in plant pathogenesis. Annu Rev Phytopathol 33: 299—321

Basse CW and Boller T (1992) Glycopeptide elicitors of stress responses in tomato cells. *N*-linked glycans are essential for activity but act as suppressors of the same activity when released from the glycopeptides. Plant Physiol 98: 1239—1247

Basse CW, Bock K and Boller T (1992) Elicitors and suppressors of the defense response in tomato cells. Purification and characterization of glycopeptide elicitors and glycan suppressors generated by enzymatic cleavage of yeast invertase. J Biol Chem 267: 10258—10265

Basse CW, Fath A and Boller T (1993) High affinity binding of a glycopeptide elicitor to tomato cells and microsomal membranes and displacement by specific glycan suppressors. J Biol Chem 268: 14724—14731

Baureithel K, Felix G and Boller T (1994) Specific, high affinity binding of chitin fragments to tomato cells and membranes. Competitive inhibition of binding by derivatives of chitooligosaccharides and a Nod factor of *Rhizobium*. J Biol Chem 269: 17931—17938

Beffa RS, Hofer R-M, Thomas M and Meins F, Jr. (1996) Decreased susceptibility to viral disease of β-1,3-glucanase-deficient plants generated by antisense transformation. Plant Cell 8: 1001—1011

Bendahmane A, Kohm BA, Dedi C and Baulcombe DC (1995) The coat protein of potato virus X is a strain-specific elicitor of *Rx1*-mediated virus resistance in potato. Plant J 8: 933—941

Bent AF (1996) Plant disease resistance genes: function meets structure. Plant Cell 8: 1757—1771

Bent AF, Kunkel BN, Dahlbeck D, Brown KL, Schmidt R, Giraudat J, Leung J and Staskawicz BJ (1994) *RPS2* of *Arabidopsis thaliana*: a leucine-rich repeat class of plant disease resistance genes. Science 265: 1856—1860

Blechert S, Brodschelm W, Holder S, Kammerer L, Kutchan TM, Mueller MJ, Xia ZQ and Zenk MH (1995) The octadecanoic pathway: signal molecules for the regulation of secondary pathways. Proc Natl Acad Sci USA 92: 4099—4105

Boller T (1995) Chemoperception of microbial signals in plant cells. Annu Rev Plant Physiol Plant Mol Biol 46: 189—214

Boller T and Felix G (1996) Olfaction in plants: specific perception of common microbial molecules. In: Stacey G, Mullin B and Gresshoff PM (eds) Biology of plant-microbe interactions, pp 1—8. International Society of Molecular Plant-Microbe Interactions, Knoxville

Bonas U, Stall RE and Staskawicz B (1989) Molecular and structural characterization of the avirulence gene *avrBs3* from *Xanthomonas campestris* pv. *vesicatoria*. Mol Gen Genet 218: 127—136

Bonnet P, Lacourt I, Venard P and Ricci P (1994) Diversity in pathogenicity to tobacco and in elicitin production among isolates of *Phytophthora parasitica*. J Phytopathol 141: 25—37

Bostock RM, Kuc JA and Laine RA (1981) Eicosapentaenoic and arachidonic acids from *Phytophthora infestans* elicit fungitoxic sesquiterpenes in the potato. Science 212: 67—69

Bostock RM, Yamamoto H, Choi D, Ricker KE and Ward BL (1992) Rapid stimulation of 5-lipoxygenase activity in potato by the fungal elicitor arachidonic acid. Plant Physiol 100: 1448—1456

Bradley DJ, Kjellbom P and Lamb CJ (1992) Elicitor-induced and wound-induced oxidative cross-linking of a proline-rich plant cell wall protein - a novel, rapid defense response. Cell 70: 21—30

Braun T, Schofield PR and Sprengel R (1991) Amino-terminal leucine-rich repeats in gonadotropin receptors determine hormone specificity. EMBO J 10: 1885—1890

Brisson LF, Tenhaken R and Lamb C (1994) Function of oxidative cross-linking of cell wall structural proteins in plant disease resistance. Plant Cell 6: 1703—1712

Buschges R, Hollricher K, Panstruga R, Simons G, Wolter M, Frijters A, Van Daelen R, Van der Lee T, Diergaarde P, Groenendijk J, Topsch S, Vos P, Salamini F and Schulze-Lefert P (1997) The barley *Mlo* gene: a novel control element of plant pathogen resistance. Cell 88: 695—705

Cai D, Kleine M, Kifle S, Harloff H-J, Sandal NN, Marcker KA, Klein-Lankhorst RM, Salentijn EMJ, Lange W, Stiekema WJ, Wyss U, Grundler FMW and Jung KC (1997) Positional cloning of a gene for nematode resistance in sugar beet. Science 275: 832—834.

Cervone F and Albersheim P (1989) Host-pathogen interactions. XXXIII. A plant protein converts a fungal pathogenesis factor into an elicitor of plant defense responses. Plant Physiol 90: 542—548

Cheong J-J, Alba R, Côté F, Enkerli J and Hahn MG (1993) Solubilization of functional plasma membrane-localized hepta-β-glucoside elicitor-binding proteins from soybean. Plant Physiol 103: 1173—1182

Cheong J-J, Birberg W, Fügedi P, Pilotti A, Garegg PJ, Hong N, Ogawa T and Hahn MG (1991) Structure-activity relationships of oligo-β-glucoside elicitors of phytoalexin accumulation in soybean. Plant Cell 3: 127—136

Conconi A, Smerdon MJ, Howe GA and Ryan CA (1996) The octadecanoid signalling pathway in plants mediates a response to ultraviolet radiation. Nature 383: 826—829

Cosio EG, Frey T and Ebel J (1990) Solubilization of soybean membrane binding sites for fungal β-glucans that elicit phytoalexin accumulation. FEBS Letters 264: 235—238

Cosio EG, Feger M, Miller CJ, Antelo L and Ebel J (1996) High-affinity binding of fungal beta-glucan elicitors to cell membranes of species of the plant family Fabaceae. Planta 200: 92—99

Côté F and Hahn MG (1994) Oligosaccharins: structures and signal transduction. Plant Mol Biol 26: 1379—1411

Dsilva I and Heath MC (1997) Purification and characterization of two novel hypersensitive response-inducing specific elicitors produced by the cowpea rust fungus. J Biol Chem 272: 3924—3927

Dangl JL (1992) The major histocompatibility complex a la carte: Are there analogies to plant disease resistance genes on the menu? Plant J 2: 3—11

Dangl JL (1995) Pièce de résistance: novel classes of plant disease resistance genes. Cell 80: 363—366

Dangl JL, Dietrich RA and Richberg MH (1996) Death don't have no mercy: cell death programs in plant-microbe interactions. Plant Cell 8: 1793—1807

Darvill AG and Albersheim P (1984) Phytoalexins and their elicitors — a defense against microbial infection of plants. Annu Rev Plant Physiol 35: 243—275

Darvill A, Augur C, Bergmann C, Carlson RW, Cheong J-J, Eberhard S, Hahn MG, Lo VM, Marfa V, Meyer B, Mohnen D, O'Neill MA, Spiro MD, Van Halbeek H, York WS and Albersheim P (1992) Oligosaccharins — oligosaccharides that regulate growth, development and defence responses in plants. Glycobiology 2: 181—198

Davis D, Merida J, Legendre L, Low PS and Heinstein P (1993) Independent elicitation of the oxidative burst and phytoalexin formation in cultured plant cells. Phytochemistry 32: 607-611

De Wit PJGM (1992) Molecular characterization of gene-for-gene systems in plant-fungus interactions and the application of avirulence genes in control of plant pathogens. Annu Rev Phytopathol 30: 391—418

Delaney T, Uknes S, Vernooij B, Friedrich L, Weymann K, Negrotto D, Gaffney T, Gut-Rella M, Kessmann H, Ward E and Ryals J (1994) A central role of salicylic acid in plant disease resistance. Science 266: 1247—1250

Delledonne M, Xia Y, Dixon RA and Lamb C (1998) Nitric oxide functions as a signal in plant disease resistance. Nature 394: 585-588

Devergne J-C, Bonnet P, Panabières F, Blein J-P and Ricci P (1992) Migration of the fungal protein cryptogein within tobacco plants. Plant Physiol 99: 843—847

Devlin WS and Gustine DL (1992) Involvement of the oxidative burst in phytoalexin accumulation and the hypersensitive reaction. Plant Physiol 100: 1189-1195

Dietrich A, Mayer JE and Hahlbrock K (1990) Fungal elicitor triggers rapid, transient and specific protein phosphorylation in parsley cell suspension cultures. J Biol Chem 265: 6360—6368

Dietrich RA, Delaney TP, Uknes SJ, Ward ER, Ryals JA and Dangl JL (1994) Arabidopsis mutants simulating disease resistance response. Cell 77: 565—577

Dinesh-Kumar SP, Whitham S, Choi D, Hehl R, Corr C and Baker B (1995) Transposon tagging of tobacco mosaic virus resistance gene *N*: its possible role in the TMV-N-mediated signal transduction pathway. Proc Natl Acad Sci USA 92: 4175—4180

Dixon MS, Jones DA, Keddie JS, Thomas CM and Jones JDG (1996) The tomato *Cf-2* disease resistance locus comprises two functional genes encoding leucine-rich repeat proteins. Cell 84: 451—459

Dunkle LD and Macko V (1995) Peritoxins and their effects on *Sorghum*. Can J Bot 73: S444—S452

Durner J, Shah J and Klessig DF (1997) Salicylic acid and disease resistance in plants. Trends Plant Sci 2: 266—274

Durner J, Wendehenne D, Klessig DF (1998) Defense gene induction in tobacco by nitric oxide, cyclic GMP, and cyclic ADP-ribose. Proc Natl Acad Sci USA 95: 10328—10333

Ebel J and Cosio EG (1994) Elicitors of plant defense responses. Int Rev Cytol 148: 1—36

Farmer EE, Caldelari D, Pearce G, Walker-Simmons MK and Ryan CA (1994) Diethyldithiocarbamic acid inhibits the octadecanoid signaling pathway for the wound induction of proteinase inhibitors in tomato leaves. Plant Physiol 106: 337—342

Fath A and Boller T (1996) Solubilization, partial purification, and characterization of a binding site for a glycopeptide elicitor from microsomal membranes of tomato cells. Plant Physiol 112: 1659—1668

Felix G, Grosskopf DG, Regenass M and Boller T (1991) Rapid changes of protein phosphorylation are involved in transduction of the elicitor signal in plant cells. Proc Natl Acad Sci USA 88: 8831—8834

Felix G, Regenass M and Boller T (1993) Specific perception of subnanomolar concentrations of chitin fragments by tomato cells: induction of extracellular alkalinization, changes in protein phosphorylation, and establishment of a refractory state. Plant J 4: 307—316

Felix G, Regenass M, Spanu P and Boller T (1994) The protein phosphatase inhibitor calyculin A mimics elicitor action in plant cells and induces rapid hyperphosphorylation of specific proteins as revealed by pulse labeling with [^{33}P]phosphate. Proc Natl Acad Sci USA 91: 952—956

Feuillet C, Schachermayr G and Keller B (1997) Molecular cloning of a new receptor-like kinase gene encoded at the *Lr10* disease resistance locus of wheat. Plant J 11: 45—52

Flor HH (1971) Current status of the gene-for-gene concept. Annu Rev Phytopathol 9: 275—296

Frey T, Cosio EG and Ebel J (1993) Affinity purification and characterization of a binding protein for a hepta-β-glucoside phytoalexin elicitor in soybean. Phytochemistry 32: 543—549

Freytag S, Arabatzis N, Hahlbrock K and Schmelzer E (1994) Reversible cytoplasmic rearrangements precede wall apposition, hypersensitive cell death and defense-related gene activation in potato/*Phytophthora infestans* interactions. Planta 194: 123—135

Gaffney T, Friedrich L, Vernooij B, Negrotto D, Nye G, Uknes S, Ward E, Kessmann H and Ryals J (1993) Requirement of salicylic acid for the induction of systemic acquired resistance. Science 261: 754—756

Gaudriault S, Malkandrin L, Paulin J-P and Barry M-A (1997) DspA, an essential pathogenicity factor of *Erwinia amylovora* showing homology with AvrE of *Pseudomonas syringae*, is secreted via the Hrp secretion pathway in a DspB-dependent way. Mol Microbiol 26: 1057—1069

Gerchenson LE and Totello RJ (1992) Apoptosis: a different type of cell death. FASEB J 6: 2450—2455

Gopalan S, Bauer DW, Alfano JR, Loniello AO, He SY and Collmer A (1996) Expression of the *Pseudomonas syringae* avirulence protein AvrB in plant cells alleviates its dependence on the hypersensitive response and pathogenicity (Hrp) secretion system in eliciting genotype-specific hypersensitive cell death. Plant Cell 8: 1095—1105

Graham TL and Graham MY (1991) Cellular coordination of molecular responses in plant defense. Mol Plant-Microbe Interact 4: 415—422

Granado J, Felix G and Boller T (1995) Perception of fungal sterols in plants. Subnanomolar concentrations of ergosterol elicit extracellular alkalinization in tomato cells. Plant Physiol 107: 485—490

Grant MR, Godiard L, Straube E, Ashfield T, Innes R and Dangl JL (1995) Structure of the *Arabidopsis Rpm1* gene enabling dual specificity disease resistance. Science 269: 843—846

Gray J, Close PS, Briggs SP and Johal GS (1997) A novel suppressor of cell death in plant encoded by the *Lls1* gene of maize. Cell 89: 25—31

Greenberg JT, Guo A, Klessig DF and Ausubel FM (1994) Programmed cell death in plants: a pathogen-triggered response activated coordinately with multiple defense functions. Cell 77: 551—563

Groisman EA and Ochman H (1996) Pathogenicity islands: bacterial evolution in quantum leaps. Cell 87: 791—794

Gross P, Julius C, Schmelzer E and Hahlbrock K (1993) Translocation of cytoplasm and nucleus to fungal penetration sites is associated with depolymerization of microtubules and defense gene activation in infected, cultured parsley cells. EMBO J 12: 1735—1744

Grosskopf DG, Felix G and Boller T (1990) K-252a inhibits the response of tomato cells to fungal elicitors *in vivo* and their microsomal protein kinase *in vitro*. FEBS Letters 275: 177—180

Gundlach H, Müller MJ, Kutchan TM and Zenk MH (1992) Jasmonic acid is a signal transducer in elicitor-induced plant cell cultures. Proc Natl Acad Sci USA 89: 2389—2393

Hahlbrock K, Scheel D, Logemann E, Nürnberger T, Parniske M, Reinold S, Sacks WR and Schmelzer E (1995) Oligopeptide elicitor-mediated defense gene activation in cultured parsley cells. Proc Natl Acad Sci USA 92: 4150—4157

Hahn MG (1996) Microbial elicitors and their receptors in plants. Annu Rev Phytopathol 34: 387—412

Hahn MG, Darvill A and Albersheim P (1981) Host-pathogen interactions. XIX. The endogenous elicitor, a fragment of a plant cell wall polysaccharide that elicits phytoalexin accumulation in soybeans. Plant Physiol 68: 1161—1169

Hammond-Kosack KE and Jones JDG (1997) Plant disease resistance genes. Annu Rev Plant Physiol Plant Mol Biol 48: 573—607

Hammond-Kosack KE, Silverman P, Raskin I and Jones JDG (1996) Race-specific elicitors of *Cladosporium fulvum* induce changes in cell morphology, and ethylene and salicylic acid synthesis, in tomato cells carrying the corresponding *Cf*-disease resistance gene. Plant Physiol 110: 1381—1394

Henschke JP and Rickards RW (1996) Biomimetic synthesis of the microbial elicitor syringolide 2. Tetrahedron Lett 37: 3557—3560

Herbers K, Conrads-Strauch J and Bonas U (1992) Race-specificity of plant resistance to bacterial spot disease determined by repetitive motifs in a bacterial avirulence protein. Nature 356: 172—174

Hinsch M and Staskawicz B (1996) Identification of a new *Arabidopsis* disease resistance locus, *Rps4*, and cloning of the corresponding avirulence gene, *avrRps4*, from *Pseudomonas syringae* pv. *pisi*. Mol Plant-Microbe Interact 9: 55—61

Howard RJ and Valent B (1996) Breaking and entering—host penetration by the fungal rice blast pathogen *Magnaporthe grisea*. Ann Rev Microbiol 50: 491—512

Hu G, Richter TE, Hulbert SH and Pryor T (1996) Disease lesion mimicry caused by mutations in the rust resistance gene *rp1*. Plant Cell 8: 1367—1376

Innes RW, Bent AF, Kunkel BN, Bisgrove SR and Staskawicz BJ (1993) Molecular analysis of avirulence gene *avrRpt2* and identification of a putative regulatory sequence common to all known *Pseudomonas syringae* avirulence genes. J Bacteriol 175: 4859—4869

Jakobek JL and Lindgren PB (1993) Generalized induction of defense responses in bean is not correlated with the induction of the hypersensitive reaction. Plant Cell 5: 49—56

Ji C, Okinaka Y, Takeuchi Y, Tsurushima T, Buzzall RI, Sims JJ, Midland SL, Slaymaker D, Yoshikawa M, Yamaoka N and Keen NT (1997) Specific binding of the syringolide elicitors to a soluble fraction from soybean leaves. Plant Cell 9: 1425—1433

Jia Y, Loh Y-T, Zhou J and Martin GB (1997) Alleles of *Pto* and *Fen* occur in bacterial speck-susceptible and fenthion-insensitive tomato cultivars and encode active protein kinases. Plant Cell 9: 61—73

Johal GS and Briggs SP (1992) Reductase activity encoded by the *HM1* disease resistance gene in maize. Science 258: 985—987

Jones DA, Thommas CM, Hammond-Kosack KE, Balint-Kurti PJ and Jones JDG (1994) Isolation of the tomato *Cf-9* gene for resistance to *Cladosporium fulvum* by transposon tagging. Science 266: 789—793

Joosten, MHAJ, Cozjinsen TJ and De Wit PJGM (1994) Host resistance to a fungal tomato pathogen lost by a single base-pair change in an avirulence gene. Nature 367: 384—386

Kamoun S, Young M, Glascock CB and Tyler BM (1993) Extracellular protein elicitors from *Phytophthora*: host-specificity and induction of resistance to bacterial and fungal pathogens. Mol Plant-Microbe Interact 6: 15—25

Kanazin V, Marek LF and Shoemaker RC (1996) Resistance gene analogs are conserved and clustered in soybean. Proc Nat Acad Sci USA 93: 11746—11750

Kang S, Sweigard JA and Valent B (1995) The *Pwl* host specificity gene family in the blast fungus *Magnaporthe grisea*. Mol Plant-Microbe Interact 8: 939—948

Kearney B and Staskawicz BJ (1990) Widespread distribution and fitness contribution of *Xanthomonas campestris* avirulence gene *avrBs2*. Nature 346: 385—386

Keen NT (1990) Gene-for-gene complementarity in plant-pathogen interactions. Annu Rev Genet 24: 447—463

Keen NT, Tamaki S, Kobayashi D, Gerhold D, Stayton M, Shen H, Gold S, Lorang J, Thordal-Christensen H, Dahlbeck D and Staskawicz BJ (1990) Bacteria expressing avirulence gene D produce a specific elicitor of the soybean hypersensitive reaction. Mol Plant-Microbe Interact 3: 112—121

Keen NT and Bruegger B (1977) Phytoalexins and chemicals that elicit their production in plants. In: Hedin PA (ed) Host Plant Resistance to Pests, pp 1—26. American Chemical Society, Washington DC

Keen NT and Buzzell RI (1991) New disease resistance genes in soybean against *Pseudomonas syringae* pv.

glycinea: evidence that one of them interacts with a bacterial elicitor. Theor Appl Genet 81: 133—138

Keen NT, Ridgway D and Boyd C (1992) Cloning and characterization of a phospholipase gene from *Erwinia chrysanthemi* EC16. Mol Microbiol 6: 179—187

Keith LM, Boyd C, Keen NT and Partridge JE (1997) Comparison of *avrD* alleles from *Pseudomonas syringae* pv. *glycinea*. Mol Plant-Microbe Interact 10: 416—422

Keller T. Damude HG, Werner D, Doerner P, Dixon RA and Lamb C (1998) A plant homolog of the neutrophil NADPH oxidase gp91^{phox} subunit gene encodes a plasma membrane protein with calcium binding motifs. Plant Cell 10: 255-266

Knight MR, Campbell AK and Trewavas AJ (1991) Transgenic plant aequorin reports the effects of touch and cold-shock and elicitors on cytoplasmic calcium. Nature 352: 524—526

Knogge W (1996) Fungal infection of plants. Plant Cell 8: 1711—1722

Kobayashi DY, Tamaki S and Keen NT (1990a) Molecular characterization of avirulence gene D from *Pseudomonas syringae* pv. *tomato*. Mol Plant-Microbe Interact 3: 94—102

Kobayashi DY, Tamaki S, Trollinger DJ, Gold S and Keen NT (1990b) A gene from *Pseudomonas syringae* pv. *glycinea* with homology to avirulence gene D from *P.s.* pv. *tomato* but devoid of the avirulence phenotype. Mol Plant-Microbe Interact 3: 103—111

Kobe B and Deisenhofer J (1994) The leucine-rich repeat: a versatile binding motif. Trends Biochem Sci 19: 415-421

Kooman-Gersmann M, Honée G, Bonnema G and De Wit PJGM (1996) A high-affinity binding site for the AVR9 peptide elicitor of *Cladosporium fulvum* is present on the plasma membranes of tomato and other solanaceous plants. Plant Cell 8: 929—938

Lawrence GJ, Finnegan EJ, Ayliffe MA and Ellis JG (1995) The *L6* gene for flax rust resistance is related to the *Arabidopsis* bacterial resistance gene *Rpg2* and the tobacco viral resistance gene *N*. Plant Cell 7: 1195—1206

Leach JE and White FF (1996) Bacterial avirulence genes. Annu Rev Phytopathol 34: 153—179

Legendre L, Heinstein PF and Low PS (1992) Evidence for participation of GTP-binding proteins in elicitation of the rapid oxidative burst in cultured soybean cells. J Biol Chem 267: 20140—20147

Legendre L, Yueh YG, Crain R, Haddock N, Heinstein PF and Low PS (1993) Phospholipase-C activation during elicitation of the oxidative burst in cultured plant cells. J Biol Chem 268: 24559—24563

Leister RT, Ausubel FM and Katagiri F (1996) Molecular recognition of pathogen attack occurs inside of plant cells in plant disease resistance specified by the *Arabidopsis* genes *RPS2* and *RPM1*. Proc Natl Acad Sci USA 93: 15497—15502

Levine A, Tenhaken R, Dixon RA and Lamb C (1994) H_2O_2 from the oxidative burst orchestrates the plant hypersensitive response. Cell 79: 583—593

Lindgren PB, Peet RC and Panopoulos NJ (1986) Gene cluster of *Pseudomonas syringae* pv. *phaseolicola* controls pathogenicity on bean plants and hypersensitivity on nonhost plants. J Bacteriol 168: 512—522

Lorang JM and Keen NT (1995) Characterization of *avrE* from *Pseudomonas syringae* pv. *tomato*: a *hrp*-linked avirulence locus consisting of at least two transcriptional units. Mol Plant-Microbe Interact 8: 49—57

Lorang JM, Shen H, Kobayashi D, Cooksey D and Keen NT (1994) *avrA* and *avrE* in *Pseudomonas syringae* pv. *tomato* PT23 play a role in virulence on tomato plants. Mol Plant-Microbe Interact 7: 508—515

MacKintosh C, Lyon GD and MacKintosh RW (1994) Protein phosphatase inhibitors activate anti-fungal defence responses of soybean cotyledons and cell cultures. Plant J 5: 137—147

Malamy J, Carr JP, Klessig DF and Raskin I (1990) Salicylic acid: a likely endogenous signal in the resistance response of tobacco to viral infection. Science 250: 1002—1004

Mansfield J, Jenner C, Hockenhull R, Bennett MA and Stewart R (1994) Characterization of *avrPphE*, a gene for cultivar-specific avirulence from *Pseudomonas syringae* pv *phaseolicola* which is physically linked to *hrpY*, a new *hrp* gene identified in the halo-blight bacterium. Mol Plant-Microbe Interact 7: 726—739

Martin GB (1996) Molecular cloning of plant disease resistance genes. In: Stacey G and Keen NT (eds) Plant-Microbe Interactions Vol 1, pp 1—32. Chapman and Hall, New York

Martin GB, Brommonschenkel SH, Chunwongse J, Frary A, Ganal MW, Spivey R, Wu T, Earle ED and Tanksley SD (1993) Map-based cloning of a protein kinase gene conferring disease resistance in tomato. Science 262: 1432—1436

Martin GB, Frary A, Wu TY, Brommonschenkel S, Chunwongse J, Earle ED and Tanksley SD (1994) A member of the tomato *Pto* gene family confers sensitivity to fenthion resulting in rapid cell death. Plant Cell 6: 1543—1552

Mathieu Y, Kurkdjian A, Xia H, Guern J, Koller A, Spiro MD, O'Neill M, Albersheim P and Darvill A (1991) Membrane responses induced by oligogalacturonides in suspension-cultured tobacco cells. Plant J 1: 333—343

Mauch F, Mauch-Mani B and Boller T (1988) Antifungal hydrolases in pea tissue. 2. Inhibition of fungal growth by combinations of chitinase and β-1,3-glucanase. Plant Physiol 88: 936—942

Mauch-Mani B and Slusarenko AJ (1996) Production of salicylic acid precursors is a major function of phenylalanine ammonia-lyase in the resistance of Arabidopsis to *Peronospora parasitica*. Plant Cell 8: 203—212

McDowell JM, Dhandaydham M, Long TA, Aarts MGM, Goff S, Holub EB and Dangl JL (1998) Intragenic recombination and diversifying selection contribute to the evolution of downy mildew resistance at the *RPP8* locus of Arabidopsis. Plant Cell 10: 1861—1874

Mehdy MC (1994) Active oxygen species in plant defense against pathogens. Plant Physiol 105: 467—472

Melotto E, Greve LC and Labavitch JM (1994) Cell wall metabolism in ripening fruit. VII. Biologically active pectin oligomers in ripening tomato (*Lycopersicon esculentum* Mill) fruits. Plant Physiol 106: 575—581

Métraux J-P, Signer H, Ryals J, Ward E, Wyss-Benz M, Gaudin J, Raschdorf K, Schmid E, Blum W and Inverardi B (1990) Increase in salicylic acid at the onset of systemic acquired resistance in cucumber. Science 250: 1004—1006

Michelmore R (1995) Molecular approaches to manipulation of disease resistance genes. Annu Rev Phytopathol 33: 393—427

Michelmore R (1996) Flood warning — resistance genes unleashed. Nature Genetics 14: 376—378

Midland SL, Keen NT, Sims JJ, Midland MM, Stayton MM, Burton V, Smith MJ, Mazzola EP, Graham KJ and Clardy J (1993) The structures of syringolide-1 and syringolide-2, novel C-glycosidic elicitors from *Pseudomonas syringae* pv. *tomato*. J Org Chem 58: 2940—2945

Mindrinos M, Katagiri F, Yu G-L and Ausubel FM (1994) The *A. thaliana* resistance gene *RPS2* encodes a protein containing a nucleotide-binding site and leucine-rich repeats. Cell 78: 1089—1099

Mittler R, Shulaev V, Seskar M and Lam E (1996) Inhibition of programmed cell death in tobacco plants during a pathogen-induced hypersensitive response at low oxygen pressure. Plant Cell 8: 1991—2001

Müller MJ, Brodschelm W, Spannagl E and Zenk MH (1993) Signaling in the elicitation process is mediated through the octadecanoid pathway leading to jasmonic acid. Proc Natl Acad Sci USA 90: 7490—7494

Mulya K, Takikawa YK and Tsuyumu S (1996) The presence of regions homologous to *hrp* cluster in *Pseudomonas fluorescens* PF632R. Ann Phytopathol Soc Japan 62: 353—359

Murillo J, Shen H, Gerhold D, Sharma A, Cooksey DA and Keen NT (1994) Characterization of pPT23B, the plasmid involved in syringolide production by *Pseudomonas syringae* pv. *tomato* PT23 Plasmid

31: 275—287

Neuenschwander U, Lawton K and Ryals J (1996) Systemic acquired resistance. In: Stacey G and Keen NT (eds) Plant-Microbe Interactions Vol 1, pp 81—106. Chapman and Hall, New York

Nojiri H, Sugimori M, Yamane H, Nishimura Y, Yamada A, Shibuya N, Kodama O, Murofushi N and Omori T (1996) Involvement of jasmonic acid in elicitor-induced phytoalexin production in suspension-cultured rice cells. Plant Physiol 110: 387—392

Nürnberger T, Jabs T, Nennstiel D, Sacks WR, Hahlbrock K and Scheel D (1994) High affinity binding of a fungal oligopeptide elicitor to parsley plasma membranes triggers multiple defense responses. Cell 78: 449—460

Ori N, Eshed Y, Presting G, Aviv D, Tanksley S, Zamir D and Fluhr R (1997) The *I2C* family from the wilt disease resistance locus *I2* belongs to the nucleotide binding, leucine-rich repeat superfamily of plant resistance genes. Plant Cell 9: 521—532

Padgett HS and Beachy RN (1993) Analysis of a tobacco mosaic virus strain capable of overcoming *N* gene-mediated resistance. Plant Cell 5: 577—586.

Parker JE, Hahlbrock K and Scheel D (1988) Different cell-wall components from *Phytophthora megasperma* f. sp. *glycinea* elicit phytoalexin production in soybean and parsley. Planta 176: 75—82

Parker JE, Coleman MJ, Szabo V, Frost LN, Schmidt R, Van der Biezen EA, Moores T, Dean C, Daniels MJ and Jones JDG (1997) The Arabidopsis downey mildew resistance gene *RPP5* shares similarity to the Toll and interleukin-1 receptors with *N* and *L6*. Plant Cell 9: 879—894

Pearce G, Strydom D, Johnson S and Ryan CA (1991) A polypeptide from tomato leaves induces wound-inducible proteinase inhibitor proteins. Science 253: 895—898

Peng M and Kuc J (1992) Peroxidase-generated hydrogen peroxide as a source of antifungal activity *in vitro* and on tobacco leaf disks. Phytopathology 82: 696—699

Penninckx IAMA, Eggermont K, Terras FRG, Thomma BPJ, De Samblanx GW, Buchala A, Métraux J-P, Manners JM and Broekaert WF (1996) Pathogen-induced systemic activation of a plant defensin gene in *Arabidopsis* follows a salicylic acid-independent pathway. Plant Cell 8: 2309—2323

Pirhonen MU, Lidell MC, Rowley DL, Lee SW, Jin SM, Liang YQ, Silverstone S, Keen NT and Hutcheson SW (1996) Phenotypic expression of *Pseudomonas syringae avr* genes in *E. coli* is linked to the activities of the *hrp*-encoded secretion system. Mol Plant-Microbe Interact 9: 252—260

Pugin A, Franchisse J-M, Tavernier E, Bligny R, Gout E, Douce R and Guern J (1997) Early events induced by the elicitor cryptogein in tobacco cells: involvement of a plasma membrane NADPH oxidase and activation of glycolysis and the pentose phosphate pathway. Plant Cell 9: 2077-2091

Rasmussen JB, Hammerschmidt R and Zook MN (1991) Systemic induction of salicylic acid accumulation in cucumber after inoculation with *Pseudomonas syringae* pv. *syringae*. Plant Physiol 97: 1342—1347

Ricci P, Bonnet P, Huet J-C, Sallantin M, Beauvais-Cante F, Bruneteau M, Billard V, Michel G and Pernollet J-C (1989) Structure and activity of proteins from pathogenic fungi *Phytophthora* eliciting necrosis and acquired resistance in tobacco. Eur J Biochem 183: 555—563

Richter TE, Pryor TJ, Bennetzen JL and Hulbert SH (1995) New rust resistance specificities associated with recombination in the *Rp1* complex in maize. Genetics 141: 373—381

Ritter C and Dangl JL (1995) The *avrRpm1* gene of *Pseudomonas syringae* pv. *maculicola* is required for virulence on *Arabidopsis*. Mol Plant-Microbe Interact 8: 444—453

Roine E, Wei W, Yuan J, Nurmiaho-Lassila L, Kalkkinen N, Romantschuk M and He SY (1997) Hrp pilus: a novel *hrp*-dependent bacterial surface appendage produced by *Pseudomonas syringae* DC3000. Proc Nat Acad Sci USA 94: 3459—3464

Rommens CMT, Salmeron JM, Baulcombe DC and Staskawicz BJ (1995a) Use of a gene expression system based on potato virus X to rapidly identify and characterize a tomato *pto* homolog that controls fenthion sensitivity. Plant Cell 7: 249—257

Rommens CMT, Salmeron JM, Oldroyd GED and Staskawicz BJ (1995b) Intergeneric transfer and functional expression of the tomato disease resistance gene *Pto*. Plant Cell 7: 1537—1544

Ryals JA, Neuenschwander UH, Willits MG, Molina A, Steiner H-Y and Hunt MD (1996) Systemic acquired resistance. Plant Cell 8: 1809—1819

Sacks WR, Nürnberger T, Hahlbrock K and Scheel D (1995) Molecular characterization of nucleotide sequences encoding the extracellular glycoprotein elicitor from *Phytophthora megasperma*. Mol Gen Genet 246: 45—55

Salmeron JM and Staskawicz BJ (1993) Molecular characterization and *hrp*-dependence of the avirulence gene *avrPto* from *Pseudomonas syringae* pv. *tomato*. Mol Gen Genet 239: 6—16

Salmeron JM, Barker SJ, Carland FM, Mehta AY and Staskawicz BJ (1994) Tomato mutants altered in bacterial disease resistance provide evidence for a new locus controlling pathogen recognition. Plant Cell 6: 511—520

Salmeron JM, Oldroyd GED, Rommens CMT, Scofield SR, Kim HS, Lavelle DT, Dahlbeck D and Staskawicz BJ (1996) Tomato *Prf* is a member of the leucine-rich repeat class of plant disease resistance genes and lies embedded within the *Pto* kinase gene cluster. Cell 86: 123—133

Salmond GPC (1994) Secretion of extracellular virulence factors by plant pathogenic bacteria. Annu Rev Phytopathol 32: 181—200

Schmidt WE and Ebel J (1987) Specific binding of a fungal glucan phytoalexin elicitor to membrane fractions from soybean *Glycine max*. Proc Natl Acad Sci USA 84: 4117—4121

Scofield SR, Tobias CM, Rathjen JP, Chang JH, Lavelle DT, Michelmore RW and Staskawicz BJ (1996) Molecular basis of gene-for-gene specificity in bacterial speck disease of tomato. Science 274: 2063—2065

Sela-Buurlage MB, Ponstein AS, Bres-Vloemans SA, Melchers LS, Van den Elzen PJM and Cornelissen BJC (1993) Only specific tobacco (*Nicotiana tabacum*) chitinases and β-1,3-glucanases exhibit antifungal activity. Plant Physiol 101: 857—863

Sharon A, Bailey BA, McMurtry JP, Taylor R and Anderson JD (1992) Characteristics of ethylene biosynthesis-inducing xylanase movement in tobacco leaves. Plant Physiol 100: 2059—2065

Sharon A, Fuchs Y and Anderson JD (1993) The elicitation of ethylene biosynthesis by a *Trichoderma* xylanase is not related to the cell wall degradation activity of the enzyme. Plant Physiol 102: 1325—1329

Shen H and Keen NT (1993) Characterization of the promoter of avirulence gene D from *Pseudomonas syringae* pv. *tomato*. J Bacteriol 175: 5916—5924

Shibuya N, Kaku H, Kuchitsu K and Maliarik MJ (1993) Identification of a novel high-affinity binding site for *N*-acetylchitooligosaccharide elicitor in the membrane fraction from suspension-cultured rice cells. FEBS Letters 329: 75—78

Shibuya N, Ebisu N, Kamada Y, Kaku H, Cohn J and Ito Y (1996) Localization and binding characteristics of a high-affinity binding site for *N*-acetylchitooligosaccharide elicitor in the plasma membrane from suspension-cultured rice cells suggest a role as a receptor for the elicitor signal at the cell surface. Plant & Cell Physiol 37: 894—898

Sieber V, Jurnak F and Moe GR (1995) Circular dichroism of the parallel β helical proteins pectate lyase C and E. Proteins 23: 32—37

Song WY, Wang GL, Chen LL, Kim HS, Pi LY, Holsten T, Gardner J, Wang B, Zhai WX, Zhu LH, Fauquet C and Ronald P (1995) A receptor kinase-like protein encoded by the rice disease resistance gene, *Xa21*. Science 270: 1804—1806

Staehelin C, Granado J, Müller J, Wiemken A, Mellor RB, Felix G, Regenass M, Broughton WJ and Boller
    T (1994) Perception of *Rhizobium* nodulation factors by tomato cells and inactivation by root chitinases.
    Proc Natl Acad Sci USA 91: 2196—2200

Staskawicz BJ, Ausubel FM, Baker BJ, Ellis JG and Jones JDG (1995) Molecular genetics of plant disease
    resistance. Science 268: 661—667

Stein JC, Howlett B, Boyes DC, Nasrallah ME and Nasrallah JB (1991) Molecular cloning of a putative
    receptor protein kinase gene encoded at the self-incompatibility locus of *Brassica oleracea*. Proc Natl
    Acad Sci USA 88: 8816—8820

Sweigert JA, Carroll AM, Kang S, Farrall L, Chumley FG and Valent B (1995) Identification, cloning,
    and characterization of *PWL2*, a gene for host species specificity in the rice blast fungus. Plant
    Cell 7: 1221—1233

Swords KMM, Dahlbeck D, Kearney B, Roy M and Staskawicz BJ (1996) Spontaneous and induced
    mutations in a single open reading frame alter both virulence and avirulence in *Xanthomonas campestris*
    pv. *vesicatoria avrBs2*. J Bacteriol 178: 4661—4669

Tang X, Frederick RD, Zhou J, Halterman DA, Jia Y and Martin GB (1996) Initiation of plant disease
    resistance by physical interaction of AvrPto and Pto kinase. Science 274: 2060—2063

Taraporewala ZF and Culver JN (1996) Identification of an elicitor active site within the three-dimensional
    structure of the tobacco mosaic tobamovirus coat protein. Plant Cell: 169—178

Tenhaken R, Levine A, Brisson LF, Dixon RA and Lamb CJ (1995) Function of the oxidative burst in
    hypersensitive disease resistance. Proc Natl Acad Sci USA 92: 4158—4163

Thilmony RL, Chen ZT, Bressan RA and Martin GB (1995) Expression of the tomato *Pto* gene in tobacco
    enhances resistance to *Pseudomonas syringae* pv. *tabaci* expressing *avrPto*. Plant Cell 7: 1529—1536

Thuleau P, Ward JM, Ranjeva R and Schroeder JI (1994) Voltage-dependent calcium-permeable channels in
    the plasma membrane of a higher plant cell. EMBO J 13: 2970—2975

Tonegawa S (1983) Somatic generation of antibody diversity. Nature 302: 575—581

Toyoda K, Shiraishi T, Yamada T, Ichinose Y and Oku H (1993) Rapid changes in polyphosphoinositide
    metabolism in pea in response to fungal signals. Plant & Cell Physiol 34: 729—735

Umemoto N, Kakitani M, Isamatsu, A, Yoshikawa M, Yamaoka N and Ishida I (1997) The structure and
    function of a soybean β-glucan-elicitor-binding protein. Proc Nat Acad Sci USA 94: 1029-1034

Van den Ackerveken GFJM, Vossen P and De Wit PJGM (1993) The Avr9 race-specific elicitor of
    Cladosporium fulvum is processed by endogenous and plant proteases. Plant Physiol. 103: 91—96

Van den Ackerveken GF, Marois E and Bonas U (1996) Recognition of the bacterial avirulence protein
    AvrBs3 occurs inside the host plant cell. Cell 87: 1307—1316

Van der Plank JE (1968) Disease Resistance in Plants. Academic Press, New York

Vernooij B, Friedrich L, Morse A, Resit R, Kolditz-Jawhar R, Ward E, Uknes S, Kessmann H and Ryals J
    (1994) Salicylic acid is not the translocated signal responsible for inducing systemic acquired resistance
    but is required in signal transduction. Plant Cell 6: 959—965

Weber H, Schultze S, Pfitzner AJ (1993) Two amino substitutions in the tomato mosaic virus 30-kilodalton
    movement protein confer the ability to overcome the *Tm2²* resistance gene in the tomato. J Gen
    Virol 67: 6432—6438

West CA (1981) Fungal elicitors of the phytoalexin response in higher plants. Naturwissenschaften 68:
    447—457

Whitham S, Dinesh-Kumar SP, Choi D, Hehl R, Corr C and Baker B (1994) The product of the tobacco
    mosaic virus resistance gene *N*: similarity to toll and the interleukin-1 receptor. Cell 78: 1101—1115

Whitham S, McCormick S and Baker B (1996) The *N* gene of tobacco confers resistance to tobacco mosaic virus in transgenic tomato. Proc Natl Acad Sci USA 93: 8776—8781

Xiao YH and Hutcheson SW (1994) A single promoter sequence recognized by a newly identified alternate sigma factor directs expression of pathogenicity and host range determinants in *Pseudomonas syringae*. J Bacteriol 176: 3089—3091

Xiao YH, Heu S, Yi J, Lu Y and Hutcheson SW (1994) Identification of a putative alternate sigma factor and characterization of a multicomponent regulation cascade controlling the expression of *Pseudomonas syringae* pv. *syringae* Pss61 *hrp* and *hrmA* genes. J Bacteriol 176: 1025—1036

Xu Y, Chang PFL, Liu D, Narasimhan ML, Raghothama KG, Hasegawa PM and Bressan RA (1994) Plant defense genes are synergistically induced by ethylene and methyl jasmonate. Plant Cell 6: 1077—1085

Yang YN and Gabriel DW (1995) *Xanthomonas* avirulence/pathogenicity gene family encodes functional plant nuclear targeting signals. Mol Plant-Microbe Interact 8: 627—631

Yang YN, Yuan QP and Gabriel DW (1996) Watersoaking function(s) of XcmH1005 are redundantly encoded by members of the *Xanthomonas avr/pth* gene family. Mol Plant-Microbe Interact 9: 105—113

Yoder MD and Jurnak F (1995) The parallel β helix and other coiled folds. FASEB Journal 9: 335—342

Yoder MD, Keen NT and Jurnak F (1993) New domain motif: the structure of pectate lyase C, a secreted plant virulence factor. Science 260: 1503—1507

Yoshimura S, Yamanouchi U, Katayose Y, Toki S, Wang Z-X, Kono I, Kurata N, Yano M, Iwata N and Sasaki T (1998) Expression of *Xa 1*, a bacterial blight-resistance gene in rice, is induced by bacterial inoculation. Proc Natl Acad Sci USA 95: 1663-1668

Yu YG, Buss GR and Maroof MAS (1996) Isolation of a superfamily of candidate disease-resistance genes in soybean based on a conserved nucleotide-binding site. Proc Natl Acad Sci USA 93: 11751—11756

Yuan J and He SY (1996) The *Pseudomonas syringae* hrp regulation and secretion system controls the production and secretion of multiple extracellular proteins. J Bacteriol 178: 6399—6402

Yucel I and Keen NT (1994) Amino acid residues required for the activity of *avrD* alleles. Mol Plant-Microbe Interact 7: 140—147

Yucel I, Boyd C, Debnam Q and Keen NT (1994a) Two different classes of *avrD* alleles occur in pathovars of *Pseudomonas syringae*. Mol Plant-Microbe Interact 7: 131—139

Yucel I, Midland SL, Sims JJ and Keen NT (1994b) Class I and class II *avrD* alleles direct the production of different products in gram-negative bacteria. Mol Plant-Microbe Interact 7: 148—150

Zhou JM, Loh Y-T, Bressan RA and Martin GB (1995) The tomato gene *Pti1* encodes a serine/threonine kinase that is phosphorylated by Pto and is involved in the hypersensitive response. Cell 83: 925—935

Zhu Q, Maher EA, Masoud S, Dixon RA and Lamb CJ (1994) Enhanced protection against fungal attack by constitutive co-expression of chitinase and glucanase genes in transgenic tobacco. Bio/Technology 12: 807—812

CHAPTER 5

STRUCTURAL ASPECTS OF DEFENSE

BRUNO MOERSCHBACHER[1] & KURT MENDGEN[2]

[1] *Institut für Biochemie und Biotechnologie der Pflanzen*
Westfälische Wilhelms-Universität Münster
Hindenburgplatz 55, D-48143 Münster, Germany
(moersch@unimuenster.de)
[2] *Lehrstuhl Phytopathologie, Universität Konstanz, Fakultät für Biologie*
Universitätsstr. 10, D-78434 Konstanz, Germany

Summary

Plants can defend themselves very efficiently against phytopathogens. This resistance can be based on preformed resistance factors or it may be the result of infection-induced resistance reactions. Preformed and induced resistance mechanisms can be structural or chemical in nature. Clearly, the resistance of many plants to attack by potentially pathogenic micro-organisms is due to preformed structural properties such as an inappropriate surface hydrophobicity or topography which fails to supply the signals required for microbial ingress in addition to cell walls resilient to physical and chemical attack. Should a micro-organism overcome these preformed barriers, plants will almost invariably fall back on their second line of defense, i.e., active resistance reactions will be induced. Again, these may have structural aspects, such as local cell wall thickening and reinforcement, encapsulation of the penetrating pathogen in dead and often lignified or suberized cells, and even the formation of new meristems forming new layers of cells around the site of attempted microbial ingress. In this chapter, we will learn how plant cells build a strong cell wall as a protection against phytopathogens, and we will get to know the strategies that micro-organisms have adopted to circumvent or breach these cell walls. Finally, we shall explore the many ways plants have evolved to counteract microbial attacks - including preformed structural resistance factors and infection-induced structural resistance reactions. We will close the chapter with a discussion of the ways open to plant pathologists to unravel cause/consequence-relationships between plant defense mechanisms and disease.

Abbreviations

AGP	arabinogalactan protein
CAD	cinnamyl alcohol dehydrogenase
CCR	cinnamoyl-CoA reductase

A. Slusarenko, R.S.S. Fraser, and L.C. van Loon (eds), *Mechanisms of Resistance to Plant Diseases*, 231-277.
© 2000 *Kluwer Academic Publishers. Printed in the Netherlands.*

CWDE	cell wall degrading enzyme
GAX	glucuronoarabinoxylan
GRP	glycine-rich protein
HRGP	hydroxyproline-rich glycoprotein
PAL	phenylalanine ammonia-lyase
PGIP	polygalacturonase inhibiting protein
PRP	proline-rich protein
THRGP	threonine hydroxyproline-rich glycoprotein

I.   Introduction

Plants are completely resistant to the vast majority of the myriad of potentially pathogenic micro-organisms. Only a handful of microbes are able to overcome the multiple resistance barriers - preformed or induced - that protect a given plant species from pathogen attack. This so-called 'non-host resistance' of plants is often due to preformed resistance factors, which can be structural barriers or preformed chemical resistance factors. Only those microbes capable of breaching this line of non-specific resistance can colonize the plant successfully, i.e., the plant becomes a host species for this pathogen and so-called 'basic compatibility' is achieved. The ensuing selection pressure can lead to coevolution between host and pathogen and to the occurrence of race/cultivar-specific resistance phenomena: some cultivars of the host plant species develop resistance to some races or strains of the pathogen. This race/cultivar-specific resistance is always based on infection-induced active resistance mechanisms which may be structural or chemical in nature.

Clearly then, there are two different structural aspects of defense in plants against microbial pathogens: structural barriers as **preformed resistance factors** and structural barriers as **induced resistance mechanisms**. The former usually act at the level of host recognition or primary penetration of the pathogen, while the latter usually interfere either with primary penetration or with the establishment of a functional host/pathogen-relationship and the spread of the pathogen in the host tissue. A look at the recognition and penetration strategies of fungal and bacterial plant pathogens pinpoints the crucial steps where structural resistance factors or mechanisms may interfere with pathogen ingress and growth.

After having passively reached, or actively found, a suitable host plant, the pathogen has to recognize it, possibly orient itself on the surface, and initiate the penetration process in order to be able to use the plant as a source of nutrients. These processes are all vitally necessary components of infection, and failure of any of these invariably leads to resistance of the plant to the micro-organism.

Except for the rather few cases where the pathogen is motile and, thus, capable of actively searching out a suitable host plant (e.g., by the directional movement of oomycete zoospores, attracted by leaking root exudates, towards plant roots), or the inoculation of host plants by mobile vectors (e.g., viral inoculation via insects feeding on plants), contact with a host plant is accidental. Attachment of a fungal spore to a potential host plant and ensuing germination are usually thought to be rather non-specific processes. Little is known at present concerning the specificity of surface attachment and whether chemical or structural surface properties - such as hydrophobicity - influence the process. Spore germination is usually triggered by the presence of moisture in the form of free water, or at least high humidity. In some cases, germination seems to be stimulated chemically by the presence of nutrients or even more specific host plant metabolites. Recently, surface hydrophobicity has been shown to be a possible structural trigger of spore germination (Shaw *et al.*, 1998).

In order to penetrate into the host tissue, the emerging germ tube has to breach the host plant's surface (Fig. 1). Penetration can be *via* natural openings such as stomata or lenticels or *via* self-inflicted wounds (e.g., during the emergence of lateral roots) or after traumatic damage (e.g., by agricultural practices or animal feeding). Alternatively, some pathogens penetrate directly through the outer periclinal epidermal cell wall into an epidermal cell or *via* the middle lamella between anticlinal epidermal cell walls into the underlying intercellular spaces between the mesophyll cells. In most, and possibly all, of these cases, penetration does not occur at random but only after recognition of an appropriate site such as a stoma or an underlying anticlinal cell wall. Recognition most likely occurs at the tip of the growing germ tube, and the triggering signals may be chemical or physical in nature. Looking for possible structural defense factors of plants against microbial infection, it can be concluded that an inappropriate surface hydrophobicity or topography leading to failure of attachment, germination, or correct recognition may present effective structural defense factors.

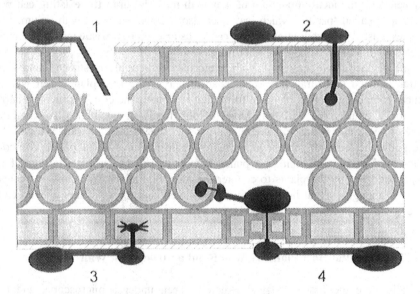

Figure 1. Penetration strategies of microbial plant pathogens
In contrast to bacterial pathogens, which, due to a lack of cutinases, can only penetrate into host plant tissues via natural openings, fungal pathogens have evolved a number of different ways to gain access to the nutrient pool of their host plants. Fungal penetration may be via wounds (1), via the cuticle and the middle lamella of an anticlinal epidermal cell wall (2), directly via the cuticle and the outer periclinal epidermal cell wall (3), or via stomata (4).

If the pathogen penetrates via the epidermal cell walls, then first the cuticle and later the middle lamella or the entire cell wall must be breached. If the initial penetration allows the pathogen access to the intercellular spaces of the host tissue, a secondary penetration through the cell walls of mesophyll cells, or the dissolution of the host cell walls, is often required in order for the pathogen to reach the intracellular nutrients of its host plant. Both cuticular and cell wall penetration may be either by chemical means (i.e., by the action of hydrolytic enzymes secreted by the pathogen) or by physical means (i.e., by the exertion of a highly localized pressure). Both the chemical and the physical structure of the cell wall, thus determine its resistance to microbial penetration, and **resilient cell walls** are therefore the major line of structural defense of plants against microbial pathogens.

If the host plant's cell wall in its native, pre-infection state is not completely resistant to penetration, the host cell may attempt to reinforce it during the actual process of penetration. In contrast to the above described preformed or passive structural barriers to resistance, this is an active process induced by the infection of the plant by the pathogen. The process does not necessarily involve the production of new cell wall material. Instead, re-arrangement or **cross-linking of pre-existing cell wall components** may render them more resistant to enzymatic hydrolysis or physical pressure. Usually, though, the **incorporation of new components into the existing wall** is concomitantly induced, and these are typically components that are easily cross-linked or polymerized, such as phenolics, offering additional strength to the wall. In many cases, the cell wall is further strengthened by the **local apposition of new wall material onto the existing cell wall**, forming a so-called 'papilla' which may later also become encrusted with polymerizing molecules. Lastly, the **production of new cell walls or even new tissue** may be attempted to isolate the infecting pathogenic micro-organism.

In all of the above cases of both preformed structural resistance factors and infection-induced structural defense mechanisms, it is the host plant's cell wall, its chemical composition or its physical form, that represents the crucial structural barrier to microbial attack. We will thus have to look firstly at the highly sophisticated, complex three-dimensional 'extracellular organelle' that is the cell wall. We will then get to know the tools phytopathogenic micro-organisms have developed to overcome the structural barriers of their host plants before we proceed to discuss the defense strategies adopted by the plants to counter the microbial attack. We close the chapter with a few thoughts about how to prove a presumed causal role of any one structural feature of a plant in resistance against a given pathogen.

II.   The Potential Host Plant: How to Build a Strong Cell Wall

Cell walls were the first biological structures seen under a microscope, and they have continued to be the focus of detailed microscopic work. At the beginning of this century, plant cell walls were described as being composed of a middle lamella, primary and, sometimes, secondary and even tertiary cell wall layers in many forms. This century has seen an explosion of knowledge on the biochemistry of the diverse cell wall components that interact to elaborate a plant cell wall. Plant cells are surrounded and shaped by a cell wall which physically and chemically protects individual cells and separates and connects neighboring cells. In plant cell walls, **cellulose** fibers take up the tensile forces while **hemicelluloses** and **pectins** make up the pressure-bearing matrix. Apart from these components, which appear to be present in every plant cell wall, additional molecules have long known to be incorporated into, or layered onto, some cell walls in order to provide specialized functions: **lignin** to give strength to dead, turgor-less cells, e.g., in the xylem; **suberin** to impregnate cell walls and to make them impermeable to water, e.g., in cork cells and in the Casparian band of the endodermis; **cutin** and the epicuticular **waxes** of the cuticle to protect the above ground parts of land plants from desiccation. In addition, it is becoming increasingly apparent that cell walls also contain many different proteins and **glycoproteins** which fulfill both structural and enzymatic functions. All of these components have been investigated in great detail for a long time, but they still continue to yield new and surprising features. Based on our knowledge of the biochemistry of cell wall components, forged mainly from studies on plant cells growing in liquid suspension culture, several models have been proposed for their interactions to form a primary cell wall (Keegstra *et al.*, 1973; Lamport, 1986; Carpita and Gibeaut, 1993; Iiyama *et al.*, 1994). The advent of new and sophisticated microscopic methods in the past few years has kindled a new interest to look at the extremely more complex cell walls of intact plant tissues (Hoson, 1991, McCann *et al.*, 1995). In spite of this long standing interest, our current models of plant cell wall architecture remain approximate sketches at best (Fig. 2).

A.   PRIMARY CELL WALLS OF SUSPENSION-CULTURED PLANT CELLS

The first models of plant cell wall architecture resulted from data obtained by investigating the primary walls of *Acer* cells rapidly growing in liquid culture. These can be regarded as a prototype for the primary cell wall of dicot plants. In contrast, the primary walls of some of the monocot plants, most notably the grasses, are made up of somewhat different components and, consequently, their architecture differs to some extent from that of dicots. Since cereals are monocots and play a major role in human nutrition, this necessitates an equally detailed interest in monocot cell walls.

1.   *Dicot Plants*
Plant cell walls are made up of a fibrillar component - cellulose microfibrils (Fig. 3) - embedded in a complex matrix. Cellulose accounts for about one quarter of the dry weight of primary cell walls. Each cellulose microfibril, with a diameter of about 10 nm, is thought to be completely wrapped in an envelope of xyloglucan chains (Fig. 3) which mediate the surface interaction between the fibrillar component of the cell wall and the surrounding matrix. Neighboring microfibrils are about 20 to 40 nm apart, and individual xyloglucan molecules of about 200 nm length are thought to span the interfibrillar space, their ends hydrogen-bonding to different cellulose microfibrils. The cellulose and xyloglucan molecules thus form a three dimensional network in the cell wall, accounting for about one half of its dry weight (Fig. 2).

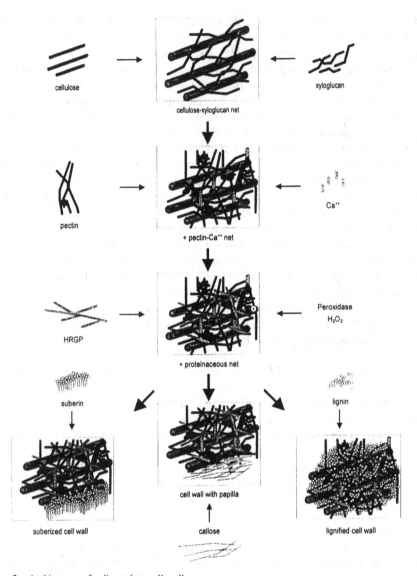

Figure 2. Architecture of a dicot plant cell wall

The typical primary cell wall of a dicot plant cell is thought to be composed of three independent networks. Cellulose microfibrils are covered by xyloglucans - individual xyloglucan molecules can hydrogen-bond to different microfibrils, thus cross-linking the fibrils to a cellulose-xyloglucan network. Pectic polymers built from linear (smooth) homogalacturonan and branched (hairy) rhamnogalacturonan domains are oriented perpendicular to the cellulose fibers - Ca^{++}-mediated egg box structures crosslink pectic molecules to the second, independent network in the cell wall. When cell growth has ceased, the final shape of the cell is fixed by the knitting of a third, proteinaceous network from hydroxyproline and proline rich glycoproteins through peroxidative formation of intermolecular isodityrosine bridges. In some cells fulfilling specialized functions, additional cell wall layers are laid down onto the inner surface of the primary cell wall, and this secondary wall can be modified, e.g., by the accrustation of suberin or the incrustation of lignin. Upon wounding, breaches in the cell wall may be sealed by the rapid apposition of callose.

Cellulose

- Cellulose is a linear chain of glucose (ß-D-glucopyranose) units connected in 1,4-glycosidic linkages.
- Around 4000-8000 cellobiose units make up a cellulose molecule of about 5-10 μm length.
- Around 30-100 cellulose molecules tightly self-assemble *via* intermolecular hydrogen bonds to a linear cellulose microfibril of about 10 nm diameter. As the cellulose molecules in a fibril are staggered, the resulting microfibril is much longer than the individual molecules.
- The tensile strength of the microfibril is comparable to a steel wire of identical diameter.

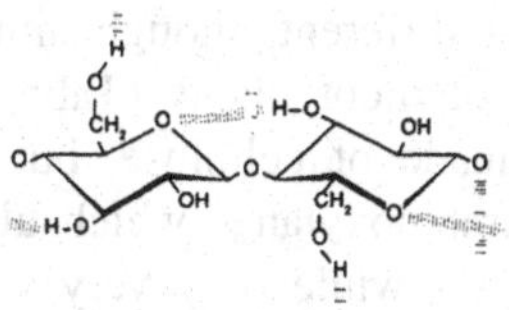

Hemicelluloses

- Hemicelluloses are the non-pectic components of the cell wall matrix.

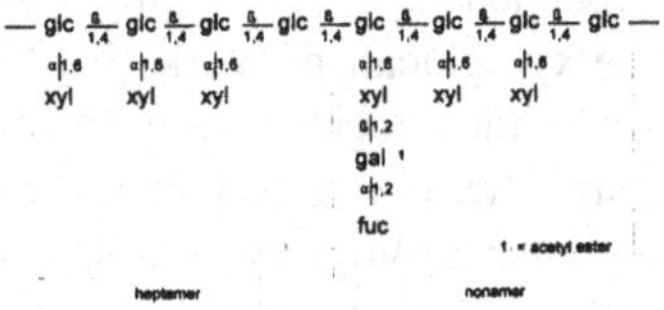

- Typical of dicot cell walls is *xyloglucan* which consists of a cellulosic backbone with three out of four glucose residues bearing an α-xylose residue which may be further substituted by a usually acetylated ß-galactose which may in turn carry an α-fucose. These different side chains build regular heptasaccharide and nonasaccharide repeat units.
- The trisaccharide side chain of the nonasaccharide repeat folds back on the glucan backbone stabilizing a planar configuration of the molecule where all side chains are on one side of the backbone so that the opposite side can hydrogen-bond to the surface of a cellulose microfibril.
- Some stretches of the xyloglucan molecules are thought to completely wrap the cellulose micrifibrils thus ensuring the individual identity of separate fibrils while other stretches may span the interfibrillar spaces thus crosslinking the microfibrils to a cellulose/xyloglucan-network.

Pectins

- Pectins, in concert with hemicelluloses, form the matrix of the cell wall.
- Pectins are defined as cell wall components containing galacturonic acid residues.
  The carboxyl groups convey a very hydrophilic character enabling pectins to attract large amounts of water which ensures the "apoplastic free space" for diffusion, the gelling properties and the pressure-bearing characteristics of the matrix.
- There are two basic types of pectic polysaccharides built on different backbones:
  Homogalacturonan consists of a linear unbranched chain of α-1,4-glycosidically linked galacturonic acid residues; 200 residues are thought to form a chain of around 100 nm length.
  Stretches of homogalacturonan molecules can dimerize either *via* divalent calcium cations crosslinking the carboxyl groups or *via* hydrophobic interactions dimerizing two highly methyl-esterified stretches of homogalacturonan. Partial dimerization leads to the formation of the pectic network.
  Rhamnogalacturonan I has a backbone of unknown length made of strictly alternating galacturonic acid and rhamnose residues and might be flanked on both sides by short consecutive stretches of highly methyl-esterified galacturonic acids.
  About every other rhamnose residue acts as a branching point by bearing an arabinan, galactan, or arabinogalactan side chain. Due to its strongly branched nature, rhamnogalacturonan I is described as the "hairy region" of pectin, as opposed to the "smooth region" of homogalacturonan.
- Most likely, homogalacturonan and rhamnogalacturonan I form different domains of the same pectic molecule.

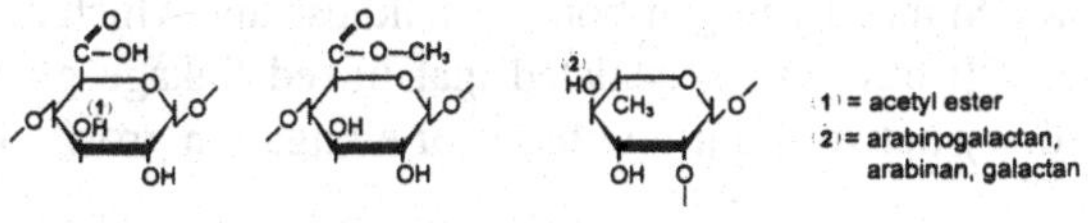

Figure 3. Constitutive components of dicot plant cell walls

All primary and secondary cell walls of dicot plants are built from a set of major polysaccharide components: cellulose, xyloglucan, and pectin.

This cellulose/xyloglucan-network is embedded in the pectic matrix which makes up about one third of the dry weight of the cell wall. The pectins (Fig. 3) probably form a second, independent network in the cell wall which may not be covalently linked to the cellulose/xyloglucan network (Fig. 2). The intermolecular cross-links knitting pectic polysaccharides together are most likely Ca^{++}-egg box structures between non-esterified stretches of homogalacturonan, hydrophobic interactions between methyl-esters of homogalacturonan stretches, and, possibly, boron diesters between two rhamnogalacturonan II-stretches.

The structural glycoproteins (Fig. 4) - roughly 10 to 20% of the dry weight of the wall - may form a third independent network in the primary plant cell wall of dicot plants (Fig. 2). The major component appears to be the hydroxyproline-rich glycoprotein (HRGP) extensin, but proline-rich proteins may be involved as well. The only cross-links proposed so far to hold the protein network together are intermolecular isodityrosine bridges, but their existence still awaits experimental support.

2.   Grasses

Primary cell walls of grasses are made up of somewhat different, though similar components compared to the typical primary cell walls of dicot plants (Table 1). The fibrillar component of both types of cell walls is made of cellulose, but the major hemicellulose of the grass cell wall is glucuronoarabinoxylan - which alone can account for about half of the dry weight of the walls - while only very small amounts of xyloglucan are present. In grasses, glucuronoarabinoxylan chains are believed to coat individual cellulose microfibrils and to span the interfibrillar spaces thus forming a cellulose/glucuronoarabinoxylan-network functionally equivalent to the cellulose/xyloglucan-network of the dicot cell wall.

The grass cell wall is rather poor in pectin - around 10% compared to about 30% in dicot walls - and the negatively charged glucuronoarabinoxylans are thought to partly substitute for the missing pectins. It can be expected that the same types of interpectic cross-links exist in the grass cell wall as discussed above for the dicot wall. Many of the arabinose residues of the glucuronoarabinoxylan carry ferulic acid or p-coumaric acid esters, and these may dimerize oxidatively to form diphenolic acid cross-links possibly substituting for missing interpectic links. Thus, a pectin and/or a glucuronoarabinoxylan/phenolic acid-network may exist in grass cell walls.

Grass cell walls are relatively poor in structural glycoproteins, which account for only about 2 to 10% of the cell wall dry weight. The relative importance of a presumed proteinaceous network in the grass cell wall is not known at present. However, grass cell walls may contain an additional polysaccharide component - a mixed linkage β-1,3-β-1,4-glucan - which may hydrogen-bond to cellulose and which may be covalently cross-linked to proteins. It has been speculated that mixed linkage glucan chains form an important intermediary network during phases of elongation growth of the cell.

Proteins, Glycoproteins, and Proteoglycans

- HRGPs, PRPs, GRPs, and AGPs are structural proteins.
- *Hydroxyproline-rich glycoproteins* (HRGPs, extensins) are characterized by post-translationally hydroxylated proline (hydroxyproline, hyp), which makes up about 40 % of the amino acid residues and often occurs in the repeated sequence ser-hyp-hyp-hyp-hyp.

  Some serins are substituted with a galactose residue while most hydroxyprolines carry side chains containing one to four arabinoses stabilizing the protein's helical secondary structure and giving it a stiff rod-like conformation.

— ser —	hyp —	hyp —	hyp —	hyp —
α|1.	β|1.	β|1.	β|1.	β|1.
gal	ara	ara	ara	ara
	β|1,2	β|1,2	β|1,2	
	ara	ara	ara	
	β|1,2	β|1,2		
	ara	ara		
	α|1,3			
	ara			

HRGPs might form a defined scaffold which determines the spacing of the cellulose microfibrils; intermolecular diphenylether isodityrosine bridges may cross-linkHRGPs creating a three dimensional proteinaceous network.

- *Proline-rich proteins* (PRPs) have been implicated to be involved in the termination of cell growth when they may form a heteroprotein polymer with HRGPs.
- The synthesis of *glycine-rich proteins* (GRPs) appears to be correlated with the process of lignification of cell walls.

  GRPs most probably adopt a ß-pleated sheet secondary structure with one side orientated toward the plasma membrane and the other side - which may contain aromatic amino acids serving as anchors for the phenolic lignin polymer - facing the cell wall.
- Some *arabinogalactan proteins* (AGPs) seem to be embedded in the plasma membrane, others are located in the middle lamella. They are thought to play important roles in the signal exchange between the cytoplasm and the extraplasmatic wall of a cell, or between adjacent cells, respectively.

  The protein part of the AGPs accounts for only around 2 % of their molecular weight; the rest is made up of highly branched arabinogalactans. AGPs, thus, are proteoglycans rather than glycoproteins.

Lignin

- *Lignin* is a polymer of phenylpropenols consisting of three slightly different monomeric units which are connected by a variety of covalent linkages thought to be produced in a random polymerization process.

  Lignin, as a random polymer, is extremely difficult to degrade enzymatically.
- Lignin of gymnosperm plants is made up of coniferyl (= ferulyl) alcohol, dicot angiosperm lignin contains in addition sinapyl alcohol, while monocot plants incorporate also p-coumaryl alcohol and phenolic acids.

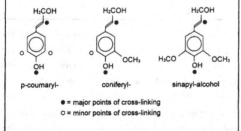

p-coumaryl-   coniferyl-   sinapyl-alcohol

● = major points of cross-linking
○ = minor points of cross-linking

Suberin, Cutin, and Waxes

- *Suberin*, a lignin-like phenolic acid polymer, and *cutin*, a polyester and polyether of hydroxy-, epoxy-, and oxo-acids and -alcohols, are laid down in layers alternating with layers of waxes.

  Both are extremely difficult to degrade enzymatically.
- *Waxes*, complex mixtures of esters of long chain fatty acids and alcohols with other components such as long chain hydrocarbons and fatty aldehydes, form water-insoluble barriers.

Callose

- Callose is a linear chain of ß-1,3-linked glucose residues which can be produced extremely fast to reinforce an existing cell wall.
- Of equal importance may be the ease with which callose can be degraded by plant enzymes should the need arise.

Figure 4. Induced components of dicot plant cell walls
Both primary and secondary walls of individual plant cells contain components that are specific to a certain cell type or to a certain developmental stage of a cell, and that help the cell to fulfill its specific function: e.g., proteins, glycoproteins, and proteoglycans; lignin, suberin, cutin, and waxes; and callose.

Table 1.   Typical primary cell walls of grasses as compared to typical primary cell walls of dicot plants.

	grass cell walls	dicot cell walls
microfibrils		
cellulose	~ 20%	~ 20%
matrix		
major hemicellulose	glucuronoarabinoxylan ~ 50%	xyloglucan ~ 20%
	The backbone of glucuronoarabinoxylan (β-1,4-xylose) and of xyloglucan (β-1,4-glucose) closely resembles or is identical to cellulose, respectively. Both can, therefore, hydrogen-bind to the surface of cellulose microfibrils.	
pectins	homo- and rhamnogalacturonan ~ 10%	homo- and rhamnogalacturonan ~ 30%
	The rather small amount of pectins present in monocot plants may lead to either a higher or a lower importance of the individual pectic cross-links in the grass wall compared to the dicot wall. A decision will have to await further experimentation.	
structural proteins	HRGP, PRP, GRP, and AGP ~ 2-10%	THRGP, PRP, GRP, and AGP ~ 10-20%
low molecular weight phenolics	ferulic and p-coumaric acid esterified and etherified to GAX and lignin ~ 2-10%	ferulic and p-coumaric acid esterified to pectins ~ 0-5%
lignin	sinapyl, coniferyl, and p-coumaryl monolignols; ferulate and p-coumarate esters and ethers	sinapyl and coniferyl monolignols

Primary cell walls of grasses - such as our most important crop plants, the cereals, e.g., wheat and rice - are built of somewhat different components compared to the typical primary cell wall of a dicot plant. While cellulose is the fibrous material in both cases, the major hemicellulose of grass cell walls is glucuronoarabinoxylan, while only very little xyloglucan is present. The negatively charged glucuronoarabinoxylan also at least partially takes over the role that pectins play in dicot plants. While the same pectic molecules appear to be present in both types of cell walls, they are clearly less abundant in grasses. Also, grass cell walls appear to be rather poor in glycoproteins. The fixation of shape once cell growth has ceased is most likely brought about by a peroxidative formation of diferulate crosslinking glucuronoarabinoxylan chains to which the monomeric ferulic acids are esterified. Lastly, the lignin found in grass cell walls is more complex in

3.   *Complex Cell Walls of Plant Tissues*

In contrast to the artificial situation of plant cells growing in liquid suspension where all cells are permanently in the same state of differentiation, namely rapid spherical growth, cells of many different types, and in many different stages of differentiation, interact with each other to form a highly complex tissue in the whole plant. Cultured plant cells are surrounded only by a primary cell wall. In a plant tissue, all cells still in the process of growing and even many fully differentiated cells, such as leaf mesophyll cells, also contain only a primary and no secondary cell wall. However, all the different cells that make up a tissue are 'glued' together by the middle lamella, a structure which may not be present in cultured cells, at least if they grow as a very fine suspension of individual cells. Furthermore, it appears that the primary wall of a cell in a tissue is much less uniform than that of a suspension cultured cell.

The mature primary cell wall of about 100 nm thickness consists of only about three layers of cellulose microfibrils. Each layer, thus, resides in a different micro-environment, the outer layer facing the pectic middle lamella, the middle layer surrounded by the other two, and the inner layer facing the plasma membrane or, later, the secondary cell wall layers. Virtually nothing is known on the interaction of the primary cell wall with these neighboring layers. Immunocytological electron microscopic studies using antibodies to different cell wall components, however, revealed striking differences in the composition of the different parts of the primary cell wall. Antibodies cross-reacting with non-methyl-esterified epitopes of pectin stained the middle lamella, especially in cell corners, but also lined the inner surface of the primary cell walls along the plasma membrane. Antibodies cross-reacting with highly methyl-esterified pectic components, in contrast, evenly showed up across the whole width of the primary cell wall. Cell wall angles where three or more different cells meet are distinct from cell wall stretches that contact a single neighboring cell and these again differ from areas that face the intercellular space. These and related approaches will certainly shed new light on our understanding of the plant cell wall, but they can be expected to pose more questions initially than will be answered - e.g., how is this enormous complexity outside the cytoplasm regulated?

The wall of an individual cell is a highly differentiated organelle with domains differing in structure and - most likely - also in function. Cells are formed in meristematic zones where cell division occurs, they undergo elongation, differentiation, and eventually may senesce. Clearly, the primary wall of a cell constantly changes during these processes, but even today little is known of what these changes entail.

Moreover, many cells that fulfill special functions within a plant tissue produce unusually, and often unevenly, thick primary cell walls or develop highly sophisticated secondary cell wall layers upon differentiation after elongation growth has stopped - e.g., thickening and cuticle excretion of the outer epidermal cell walls, and localized thickenings and lignification in tracheary elements.

An example of unevely thickened primary cell walls is given by the outer walls of epidermal cells. In addition, cutin monomers and cuticular and epicuticular waxes (Fig. 4) are produced in epidermal cells and transported to the outer surface of the outer periclinal cell wall to reach the surface where they are laid down in layers and where the cutin is polymerized *in situ*. Besides protecting from transpiration loss of water, the cuticle is also of paramount importance for warding off potentially phytopathogenic micro-organisms (Kerstiens, 1996).

One of the most prominent examples of secondary cell walls is certainly the intricate elaboration of cell wall thickenings in tracheary elements of the xylem where additional cellulose fibers and embedding matrix polymers are laid down only on certain parts of the cell walls. Eventually, lignin is incorporated into the walls of these cells (Fig. 4). The polymerization process of the monolignols taking place *in muro* leads to a growing lignin polymer infiltrating the primary cell wall, displacing the water as it does so. Eventually, when the polymerization process is complete, the lignin polymer serves to cement in place all the other components of the cell wall. The wall is then no longer permeable to water, and the mature tracheid dies. The ensuing loss of turgor pressure would result in a shrinking of the dead cell and most likely a collapse when the negative pressure (tension) of the transpiration stream is taken into account, were it not for the lignin impregnating and stiffening the cell wall of the tracheid.

III.   The Pathogen: How to Breach a Plant Cell Wall

Almost any potential pathogen of above ground plant organs will first meet the waxes and the cutin of the cuticle as an obstacle in front of the nutrient pool within the plant cells (Fig. 1). If this barrier is crossed, most pathogens will then be confronted with the middle lamella of the plant cell walls, and many pathogens appear to prefer the middle lamella portion of their host cell walls for further ingress; penetrating into the leaf at the anticlinal junction between two adjacent cells. Most pathogens will eventually have to deal with the primary cell wall enveloping every host cell. Clearly, there are two ways of breaching through a cell wall. The more 'biological' way appears to be enzymatic hydrolysis of some or all of the cell wall components. But mere physical pressure is another way apparently adopted by some fungal pathogens.

A.   PENETRATION WITH THE AID OF CELL WALL HYDROLYZING ENZYMES

Plant pathogenic micro-organisms are typically able to produce a wide array of plant cell wall degrading enzymes (Walton, 1994). This is certainly true for necrotrophic pathogens such as plant pathogenic bacteria, e.g., *Erwinia carotovora,* and many fungi, e.g., *Botrytis cinerea.* It is less true for biotrophic pathogens such as the obligately biotrophic rust and mildew fungi. But even these have to degrade the walls of their host cells very locally in order to come in intimate contact with the protoplast within.

Cell wall degrading enzymes are usually not produced constitutively in great amounts but they are induced by the appropriate substrates or small amounts of the breakdown products. As the pathogen finds its way into the plant tissue, different components of the plant cell wall are met consecutively, and the degrading enzymes are thus induced sequentially. We will organize our description of the relevant cell wall hydrolyzing enzymes according to this natural sequence.

1. *Cutinase*

Pathogens of above ground plant parts will first encounter the epicuticular wax layer covering the surface of land plants. Little is known on interactions between a potential pathogen and these waxes although an influence of a pathogen landing or growing on the outermost layer of a plant surface is clearly visible as an imprint when the pathogen is removed. Whether or not an enzymatic interaction or rather a physical process is causing the imprint, is mostly unknown. As the components of the epicuticular wax layer are rather low molecular weight compounds and are not polymerized, the wax layer appears to be easily breached, and no enzyme is thought to be necessary or known to be produced by plant pathogenic fungi.

Unlike the wax layer, the underlying cutin layers are complex polyesters and polyethers of different monomers. In order for a pathogen to penetrate this layer, either enzymatic hydrolysis or a considerable physical pressure can be expected to be necessary. Cutinases are esterases with a greater or lesser degree of specificity able to saponify the ester linkages of cutin. No enzyme is known that would be capable of cleaving the ether linkages of cutin but it would appear as if hydrolysis of the ester linkages suffices to weaken the cutin to a point where microbial ingress is possible. Cutinases are produced by many phytopathogenic fungi but they appear to be absent in bacteria. This may explain why bacteria always require pre-existing openings such as wounds or stomata to enter plant tissues.

The induction process of fungal cutinases has been studied in some detail (Fig. 5) (Podila *et al.*, 1988). Some fungi appear to produce low amounts of cutinase constitutively. When a fungal spore lands on a plant surface, the enzyme starts to degrade the cutin, producing cutin monomers. These in turn are taken up by the fungus and act as inducing signals to activate transcription of the cutinase gene. More enzyme is produced, and this autocatalytic cycle leads to rapid degradation of the plant cutin. Most likely this induction strategy is typical of many other cell wall hydrolyzing enzymes of microbial plant pathogens.

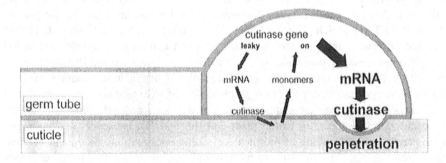

Figure 5. Induction process of fungal cutinases

Cutinases, similar to other cell wall degrading enzymes, are often induced by low molecular weight products of their own enzymatic action. The cutinase gene of some fungi is leaky so that low amounts of cutinase are produced constitutively. These begin to degrade the cuticle of a potential host plant as soon as the fungal spore has landed on a plant surface. The cutin monomers produced are taken up by the fungus and strongly induce the cutinase gene. More cutinase is produced so that the fungus can eventually degrade and penetrate the plant cuticle (Podila *et al.*, 1988).

It is important to note that some pathogens never penetrate the cuticle. The large group of root pathogens is not likely to ever encounter a cuticle proper, and some leaf or shoot pathogens avoid the cuticle by penetrating through wounds or through natural openings - as mentioned above for bacteria. Such a penetration mechanism, however, requires the successful location and recognition of stomata, a process that we will investigate in more detail later in this chapter.

2.   Pectic Enzymes

Whether a pathogen has penetrated the cuticle to gain access to the plant tissue or whether it has entered the intercellular space of the host tissue via one of the alternative routes, it will invariably then be confronted with pectic components of the host cell wall. Some pathogenic fungi, e.g., *Claviceps purpurea*, initially penetrate only the cuticular membrane and then grow just below the cuticle, within the outermost, pectin-rich layer of the outer periclinal epidermal cell wall (Fig. 6). Other fungi penetrate deeper into the host tissue within the middle lamella of an anticlinal epidermal cell wall. The middle lamella, which is also the outermost layer of all 'internal' plant cells, is composed chiefly of pectins, at least in the case of dicot plants, and consequently, pectin degrading enzymes are produced early during the penetration process by many plant pathogenic organisms (Fig. 6). The complexity of the pectic polysaccharide fraction in cell walls is mirrored by the broad spectrum of the enzymes pathogens produce to degrade it (Fig. 7), although to date, only enzymes capable of depolymerizing homogalacturonan have been identified from pathogenic organisms (Alghisi and Favaron, 1995).

Any one plant pathogenic micro-organism usually produces several of these possible enzymes. Moreover, the enzymes are usually produced as families of isoenzymes so that the pectic polysaccharide component of a plant cell wall is attacked by a wide array of different enzymes. The reason for this immense complexity is not yet fully understood. It might be that we understand only too little of the complexity of the substrate. Owing to their exposed position in the plant cell wall, many specialized features may have evolved that help to protect the pectic molecules from enzymatic degradation, necessitating equally specialized enzyme activities. Moreover, it has become apparent that the pectins of the cell wall are a rich source of biologically active signal molecules. It should not be too surprising that surface molecules of plant cells act as signals recognized by neighboring cells in order to build a functionally intact tissue. But apparently, plant cells have also learned to interpret some of the degradation products of their pectins as a sign of alarm. Medium length oligomers of galacturonic acid, as produced by the action of endo-pectate hydrolase (also termed endopolygalacturonase), act as endogenous elicitors signaling to the plant cell the potential presence of a pathogen and triggering induced resistance mechanisms. Some micro-organisms appear to have counteracted this signaling mechanism by developing extremely active pectic enzymes which rapidly degrade pectic polymers down to very small oligomers, galacturonic acid dimer or trimer, which do not act as endogenous elicitors. Then again, plants contain polygalacturonase inhibiting proteins (PGIP) in their cell walls which slow down the activity of the enzymes so that medium length oligomers might be present long enough to be recognized by the plant cell. Another way around the problem of recognition might be for the pathogen to produce exo-cleaving enzymes, as these would only produce

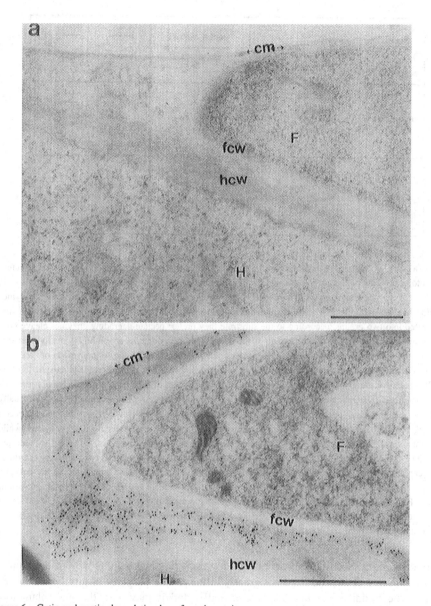

Figure 6. Cutin and pectin degradation by a fungal parasite

Many phytopathogenic fungi are able to enzymatically degrade cutin and they can thus penetrate intact cuticles of potential host plants. Some fungi immediately continue to breach the underlying pectin rich region of the outer periclinal epidermal cell wall or they digest their way down the pectin rich-middle lamella of an anticlinal epidermal cell wall. Some fungi, however, after having penetrated the plant cuticle, exhibit a phase of subcuticular growth, as seen here with *Claviceps purpurea*, the causal agent of ergot, on rye: the fungal hypha (F) grows just below the cuticular membrane (cm) of the host cell (H) (a). A monoclonal antibody reacts with non-methylesterified regions of pectin in the host cell wall (hcw) only just in advance of the fungal cell wall (fcw), presumably due to the action of a fungal pectin methyl-esterase (b) (courtesy of K. Tenberge, Münster, Germany).

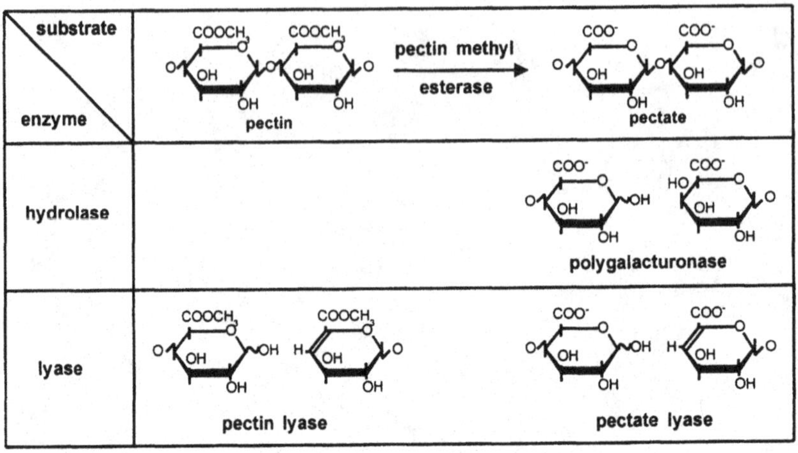

Figure 7. Pectin and pectate degrading enzymes

Pectin is a complex mixture of linear homogalacturonans and branched rhamnogalacturonans. The structure of both molecules may be further complicated by the modification with methyl and acetyl esters. Accordingly, microbial enzyme systems aimed at degrading plant pectins are fairly complex. Today, only enzymes degrading homogalacturonans have been studied with respect to plant-microbe interactions. These can first be separated into two groups, according to the substrate they prefer: pectate-degrading enzymes cleave non-methylesterified regions of homogalacturonan, while pectin-degrading enzymes act on highly methylesterified regions of the same molecules. Secondly, the enzymes can be grouped according to their mode of cleavage: hydrolases perform a typical hydrolytic cleavage where a molecule of water is added to the glycosidic linkage, while lyases cleave the glycosidic bond without the addition of water thus forming a double bond at the new non-reducing end. Theoretically, this makes already for four different enzymes: pectate hydrolase (usually named polygalacturonase), pectin hydrolase, pectate lyase, and pectin lyase. However, the second one, pectin hydrolase, has not been found so far. Polygalacturonase and pectin lyase are typical fungal enzymes, while bacteria are more likely to produce pectate lyases. However, this is not the end of the story. In addition, the enzymes can either cleave internal glycosidic bonds in which case they are named endo-cleaving enzymes, or they can act on the ultimate or pen-ultimate glycosidic bond near the non-reducing end of the polymer which makes them exo-cleaving enzymes. A pectin lyase can, thus, be an endo-pectin lyase or an exo-pectin lyase, etc. Pathogens with only pectate-degrading enzymes but devoid of pectin-cleaving activities would have problems to degrade highly methylesterified pectins. However, such organisms usually also produce a pectin methyl-esterase converting pectin to pectate which can then easily be degraded.

dimers or monomers of galacturonic acid. However, this is a tedious way of degrading a polymer (a bite at the end does not greatly influence the property of a polymer, while a bite in the middle cuts the molecule in half!) and it may give the host plant too much time to organize its defense. We seem to be witnesses to an ongoing, interactive process, where depending on the host/pathogen-interaction under investigation, either the pathogen or the host plant comes out as the winner.

Similar to cutinase, pectic enzymes are not usually produced constitutively in large amounts. Again, small amounts may be present to start the degradation of the substrate, and the degradation products - usually galacturonic acid oligomers which may or may not bear a double bond at their non-reducing end - are taken up by the micro-organism. Subsequent cytoplasmic degradation products then act as signals inducing transcription of the appropriate genes for the pectic enzymes (Hugouvieux-Cotte-Pattat *et al.*, 1996). This autocatalytic, 'substrate induction' cycle guarantees rapid production of enough enzyme to degrade the pectic component of the middle lamella and of the primary cell wall of the host plant.

3. *Hemicellulases*

Pectic enzymes are by far the best studied cell wall degrading enzymes of phytopathogenic organisms but degradation of pectin appears not to be enough to allow cell wall penetration or to lead to bursting of the host cells. This is not surprising regarding the architecture of the plant primary cell wall which appears to consist of three independent networks (Fig. 2). When the pectic network is degraded and, thus, the matrix is partly removed from the cell wall, the load-bearing cellulose/xyloglucan-network is still in place. This network can rather easily be weakened significantly by cleavage of only a few xyloglucan bonds. If a (xylo)glucanase snips the interfibrillar domains of xyloglucan, this might weaken the network to the point where the cell wall can no longer withstand the internal turgor pressure. The bursting of the host cell not only releases its cytoplasmic content to the pathogen, but also renders any attempt at an induced, active resistance mechanism of this cell impossible (this, of course, would not be a good strategy for a biotrophic pathogen!). It may well be that one of the main reasons for the primary degradation of pectins is to expose the xyloglucan chains to enzymatic attack.

Xyloglucanases are typically endo-acting β-1,4-glucanases which can degrade the xyloglucan backbone only next to a glucose residue that is not substituted by a xylose. Owing to the rather regular organization of xyloglucan, this mode of cleavage gives rise to predominantly heptamers and nonamers of xyloglucan. These are presumed to be biologically active oligomers involved in developmental processes of the plant, but their participation in host/pathogen-interactions, either as pathogenicity factors or as endogenous elicitors of induced plant defense responses, has not been described.

In the case of grasses as potential hosts, the situation is a little different. Firstly, there are less pectins and, consequently, pathogens of grasses usually do not produce significant amounts of pectic enzymes early in the infection process. Also, xyloglucanases would probably be of limited value. It has been shown in some cases that grass pathogens produce a set of enzymes capable of degrading glucuronoarabinoxylan. This involves an acetyl esterase, an α-arabinosidase, and an endo-β-1,4-xylanase. However, little is known on the action of these enzymes on native glucuronoarabinoxylan and on the products formed.

4. *Cellulase*

Degradation of the cellulose microfibrils appears not to be a prerequisite for cell wall penetration, and we have seen why this is so. Moreover, the crystalline, water-insoluble nature of cellulose makes for a very difficult substrate. Nevertheless, many pathogens have evolved an enzymatic machinery for cellulose breakdown which is probably less involved in breaching structural barriers of plants but rather opens up the glucose reservoir of cellulose as a nutrient source for the pathogen. In the context of this chapter, it might therefore suffice to summarize that the cellulase complex usually consists of at least three different enzymatic activities, an endo-β-1,4-glucanase, a cellobiohydrolase, and a β-glucosidase. Acting in concert, these three enzymes are able to bind to crystalline cellulose and to degrade it to cellobiose units and, eventually, to glucose residues.

5.   *Proteases*

At this point, the careful reader should expect a section on microbial proteases able to degrade the third, proteinaceous network of the primary plant cell wall. However, virtually nothing is known about such proteases. Many phytopathogens are, of course, known producers of proteases, but these are usually enzymes able to degrade soluble proteins, and they can be envisaged as being involved in protein breakdown to produce amino acids as nutrients. The highly insoluble, structural cell wall glycoproteins, in contrast, are highly resistant to proteolytic attack. Even the soluble precursors of extensin are resistant to proteases unless the arabinose side chains are first removed. One might expect a specialized family of proteases adapted to the degradation of the insoluble cell wall proteins - similar to collagenolytic proteases - but these have received very little attention so far. One of the few reports comes from an obligately biotrophic rust fungus (Rauscher *et al.*, 1995), often falsely considered an unlikely candidate for cell wall degrading enzyme activities.

6.   *Ligno-Peroxidases*

Very few organisms are capable of degrading lignified cell walls. For most pathogens, it is rather easy to avoid the few lignified cells in a host tissue - unless the host plant counters the pathogenic attack by induced lignification! The situation is clearly different for pathogens living in heavily lignified substrates such as wood. To date, the only known pathogenic micro-organisms capable of lignin degradation are the white rot fungi (Kurek, 1992).

Lignin, as a random polymer consisting of up to three different monomers connected via at least ten different covalent linkages (Ralph and Helm, 1993) and on top of that being completely water-insoluble, makes a very difficult substrate. On the one hand, the substrate is so varied that it is virtually impossible to produce an enzyme with an active center fitting to a significant portion of the molecule so that more than just a very few linkages can be broken. In addition, the C-O-C ether linkages and, especially, the direct C-C bonds are extremely stable and difficult to break. The strategy for lignin breakdown is, therefore, an unusual one: a peroxidase called ligno-peroxidase appears to oxidize a small, readily diffusable substrate, possibly veratryl alcohol (Piontek *et al.*, 1993). The oxidized product, a veratryl alcohol radical cation, apparently diffuses into the lignin polymer and is able to oxidize and cleave the phenolic ring. This will eventually give rise to a vast array of different lignin fragments the further metabolization of which is unknown.

B.   PENETRATION BY PHYSICAL PRESSURE

While mechanical pressure is routinely cited as a possibility to breach the structural defense barriers of plants, there is little experimental evidence for this kind of host cell wall penetration. One notable and long known example is the penetration behavior of zoospores of the obligately biotrophic fungal pathogen *Plasmodiophora* into root hairs of their cruciferous host plants where they cause the clubroot disease (Williams *et al.*, 1973). Inside the zoospore encysted on the surface of a root hair, a bullet-like 'Stachel' is visible which is virtually shot into the root hair cell, most likely by turgor pressure developing in the enlarging vacuole - the actual penetration process taking only seconds. The cytoplasmic content of the spore is then injected through this fine puncture into the host cell.

Another more recently investigated example of epidermal cell wall penetration by the action of pressure is the penetration of the rice blast fungus, *Magnaporthe grisea*, a necrotrophic pathogen of rice leaves (Howard and Valent, 1996). When the fungal spore has germinated on the surface of a rice leaf, a short germ tube develops which tightly adheres to the rice cuticle. An appressorium is formed at the tip of the germ tube which is separated from the rest of the germ tube by a septum. The cell wall of the appressorium becomes heavily melanized, a process somewhat reminiscent of lignification of plant cell walls. Incorporation of melanin renders the appressorial cell wall completely impermeable to larger solutes while its permeability for water is retained. The subsequent intracellular production of large amounts of glycerol leads to the production of an enormous turgor pressure within the appressorium which has been measured to almost 80 bar! This pressure must then apparently be released at a preformed breaking point facing the outer periclinal epidermal cell wall. The mere pressure is thought to be responsible for cuticular and cell wall penetration. The question of how this pressure release is regulated, i.e., how the fungus can penetrate the tough cell wall without also penetrating the directly underlying soft plasma membrane, has not even been addressed today. That pressure alone suffices for surface penetration can be shown when the fungus is allowed to germinate on a plastic surface of suitable hydrophobicity. Provided the appressorium can attach firmly to the surface, the fungus will penetrate any artificial substrate that is not too tough. Conversely, albino mutants of the fungus not able to produce melanin any more are incapable of cuticular penetration even though their enzymatic machinery is unchanged. Of course, while these observations allow the conclusion that physical pressure is involved in cuticular penetration, it may still be that *in planta*, a combination of pressure and enzymatic hydrolysis is active.

IV. Resistant Host Plants: How to Prevent Breaching of the Cell Wall

We have seen the intricate ways in which plants build their cell walls, and the equally sophisticated means by which phytopathogens have learned to breach these structural barriers. However, as we have stated early on, most plants are resistant to the attack by most potentially pathogenic organisms. In these cases, the tools of the pathogen are obviously inadequate to deal with the barriers put up by the plant. In the following sections, we will highlight some aspects of the possible interactions between host plants and microbial pathogens. We will first present the most crucial preformed resistance factors which plants may possess in order to ward off potential pathogens passively, and we will then continue with a description of induced resistance reactions which may be triggered in an infected plant in order to actively prevent further ingress of the pathogen.

A.   PREFORMED RESISTANCE FACTORS

The successful infection of a plant tissue by a pathogenic micro-organism is a complex process involving a number of critical steps. Each one of a number of requirements has to be met in order for the pathogen to be able to recognize the host, attach to it, move and orient itself on its surface, and penetrate into its tissues. Both chemical and physical - i.e., structural - signals appear to be involved in these processes (Hardham, 1992; Read et al., 1992; Romantschuk, 1992; Gow, 1993; Schäfer, 1994; Mendgen et al., 1996). Clearly, a plant not presenting required signals 'appropriately' may consequently be resistant to the attack by a pathogen relying on that signal.

1.   Surface Hydrophobicity

Attachment and germination are *conditiones sine quibus non* for a successful infection of a plant by many phytopathogens. Already these very early events preceding host tissue penetration and colonization can fail due to structural features of the potential host plant. Perhaps the simplest of such features is the surface hydrophobicity. Generally, the contact angle of a water droplet, which is a measure of hydrophobicity, is extremely large on above ground plant surfaces (Holloway, 1969). Spores or germ tubes that can not adhere to such a hydrophobic surface cannot initiate a successful infection and the fungus will usually not be able to parasitize the plant. Pathogens, therefore, have adopted means to adjust their own surface hydrophobicity to that of the host plant, e.g., by the secretion of hydrophobins (Talbot et al., 1993; Wessels, 1996), or vice versa, e.g., by enzymatic erosion of the cuticle.

Phytopathogenic bacteria which are flagellate, enter host plants via stomata or wounds and require a film of water on the leaf surface in order to be able to move. The hydrophobicity of the cuticle, however, favors the formation of little droplets of water rather than of a continuous film, greatly reducing the chances of bacterial ingress into host leaves.

Little experimental evidence is available to judge the importance of attachment as a presumably critical surface interaction. It has been shown that some fungal pathogens require a certain surface hydrophobicity or hydrophilicity in order to attach, and attachment has sometimes been proven to be essential for subsequent penetration. This appears evident in cases of appressorial penetration, especially when physical pressure is involved in breaching the epidermal cell wall. Here, the fungal structure has to tightly adhere to the cell wall lest the pressure exerted should only lift the appressorium off the plant surface. Attachment is most likely reached by the excretion of a sticky matrix (Nicholson and Epstein, 1991) and may be aided by the presence of cutinases and other esterases (Kunoh et al., 1990; Deising et al., 1992).

Similarly, it has been shown that oriented growth of a fungal germ tube on the host surface requires tight adhesion of the germ tube to the surface (Braun and Howard, 1994). Unlike many other fungal hyphae, a germ tube typically grows tightly appressed to a surface, exhibiting a somewhat flattened profile in cross section and a 'nose down' profile in longitudinal section. An appropriate surface hydrophobicity appears to be a prerequisite for this type of growth behavior which in turn can be required to allow the fungal germling an orientation on the host surface guided by topographical features.

2. *Inappropriate Surface Topography*

While surface hydrophobicity seems critical in the attachment of spores, subsequent growth of the emanating germ tube appears to be influenced both by surface hydrophobicity and topography. A fascinating example of oriented germ tube growth and recognition of a stoma on the surface of an adequate host plant is given by the germ tubes of most rust fungi on their way to a stoma needed for penetration into the host plant tissue (Read *et al.*, 1992; Mendgen *et al.*, 1996).

a. Oriented Germ Tube Growth. We have already mentioned above the necessity for the fungal germ tube to quickly reach a host stoma before the resources of the spore are depleted. Chemo-attraction to stomata - based on pH-gradients, changes in epicuticular wax composition, or volatiles emanating from the stomata - has been proposed as a possible mechanism for oriented germ tube growth, but experimental evidence is largely missing. In contrast, convincing evidence has now accumulated for a surface orientation of the fungal germ tube based on topographical features of the outer epidermal cell walls. Using artificial ridges on plastic surfaces, it was clearly shown that the germ tube tip can sense a depression in the surface - such as naturally brought about by the underlying anticlinal cell walls of the epidermis - and orient its further growth perpendicular to this depression.

On a typical grass leaf, it is readily understandable that this simple behavior will guide the germ tube at a relatively short path to a stoma: the epidermal cells of a grass leaf are long and narrow and organized in longitudinal rows, stomata are aligned in some of these rows, and stomata in one row are staggered compared to those in the next stoma-containing row. Thus, a fungal germ tube is highly likely to meet a depression brought about by a longitudinal anticlinal cell wall and this will lead to a transverse growth of the germ tube perpendicular to the long axis of the leaf. Chances are high then that the germ tube tip will meet a stoma at the first stoma-containing row it crosses. However, should it grow in between two stomata, it is rather likely to meet one in the next stoma-containing row.

The situation is less clear on a typical leaf of a dicot plant where the pattern of epidermal cells is less evident. However, computer simulations have shown that a perpendicular re-orientation of germ tube growth at each epidermal cell border encountered will usually lead to a 'homing in' on a stoma.

Rather simple orientation strategies, thus, lead growing germ tubes of different rust fungi to stomata - a necessary prerequisite for successful penetration of the pathogen into its host tissue. It would appear reasonable to assume that a changed pattern of epidermal cells might misguide the fungus, thus rendering the plant resistant. The fungus, not finding a stoma to penetrate, would rapidly exhaust the nutrient reservoir of the spore and starve. Evidence for such a structural resistance factor has been reported for several non-host plants inoculated with different rust fungi (Heath, 1977).

b. Recognition of a Stoma.   Simply reaching a stoma, though, is not good enough for successful penetration. Upon contact with a stoma, the germ tube tip has to recognize that it has reached its destination, growth has to be stopped, and an appressorium as the first of a series of infection structures has to be differentiated (Read *et al.*, 1992; Mendgen *et al.*, 1996). Again, chemical signals have been implicated in this recognition process but experimental evidence is stronger for structural signals provided by the topography of the stoma.

It had long been known that germ tubes of some rust fungi, when growing on artificial substrates, may be induced to form appressoria by scratches in the surfaces. In an elegant series of experiments using polystyrene replicas of micro-manipulated silicon wafer templates, it was now possible to precisely identify the topographical signal that induces appressorium differentiation in the bean rust fungus (Hoch *et al.*, 1987): a single step up or down of 0.5 μm height is a necessary and sufficient signal. This is exactly the height of the stomatal lip surrounding bean stomata.

It is perhaps surprising that such a simple signal is a 'safe' enough trigger for such a crucial differentiation step of the fungus. Both failure to recognize a stoma reached and erroneous 'recognition' of a stoma not present, could be fatal events leading to fungal death. On the other hand, it would appear rather simple for a plant to slightly change the dimension of its stomatal lip and thus evade fungal recognition and become resistant. However, the observations known so far appear to indicate that recognition of a precisely dimensioned stomatal lip is a general mechanism rust fungi use to penetrate dicot plants.

Again, the situation is different on grass leaves. Rust fungi which infect grasses do not react to scratches on artificial surfaces, and variations of the dimensions of the ridges or grooves on the polystyrene surfaces did not immediately furnish a developmental signal for these fungi (Allen *et al.*, 1991). However, after overgrowing a series of narrowly spaced ridges, grass-infecting rust fungi may differentiate appressoria (Read *et al.*, 1997). Comparing this topography with the surface of a grass leaf, it was concluded that the series of narrowly spaced anticlinal walls of the cells forming a gramineaceous stomatal complex may be the signal recognized by the rust fungi of grasses. However, *in vitro*, only about half of the germ tubes developed an appressorium, and this only after growing over a dozen or so narrowly spaced ridges - far more than the number of narrowly spaced anticlinal walls in a grass stomatal complex. It has long been known that the wheat stem rust fungus can easily be chemically induced to form appressoria - a component of the epicuticular wax and a volatile substance emanating from the leaves appeared to act synergistically (Grambow and Riedel, 1977). These chemical signals may work in conjunction with the topographical signal. In fact, the combination of physical and chemical triggers led to the formation of an appressorium at the tip of almost every fungal germ tube (Moerschbacher and Read, unpublished observation).

Thus, it would appear that rust fungi which infect grasses use a combination of signals to identify with reasonable certainty that a stoma has been reached. Maybe the developmental program for building the topography of the gramineaceous stomatal complex is less fixed than in dicots and reliance on topographical features alone for stomatal recognition, therefore, is not certain enough. Interestingly, the resistance of a certain barley species, *Hordeum chilense*, to cereal leaf rusts was shown to be based on failure of the germ tubes to recognize stomata (Rubiales and Niks, 1992). The epicuticular waxes of this barley variety form finger-like protrusions reaching over the stomata. Most likely, the fungus is thus no longer able to feel the real surface of the stoma, it fails to recognize having reached its destination, and keeps on growing - a fascinating case of structural resistance of a host plant to a fungal pathogen.

We have concentrated the description of surface topography as an aspect of structural resistance of plants to pathogens on the rust fungi not because the phenomenon is restricted to this group of pathogens but because these are by far the most intensively studied. However, surface topography can be expected to play a critical role during the process of infection by many pathogens, certainly by all those which prefer a specific site of their host's surface for penetration. A rather different example of stomatal morphology acting as a determinant of resistance or susceptibility is given by citrus fruits where the conformation of the cuticle around the stoma is thought to either prevent or allow the passage of water droplets containing bacterial pathogens. This is a wide and promising field for future research.

3. *Resilient Cell Walls*

The above sections have dealt with surface attachment and recognition leading to a correct positioning of the microbial infection structures for the subsequent attempt at penetrating into the host tissue. This site of attempted penetration may be an outer periclinal wall of an epidermal cell, an anticlinal wall between two epidermal cells, or a stoma. Even in the latter case, the pathogen has only reached the intercellular spaces of a host tissue, and the formidable barrier of the plant cell wall in all its complexity is still to be overcome. We have described the cell wall in detail above, and we have also elaborated on the machinery evolved by the pathogens to breach, or partially circumvent and then breach, this barrier. In this section, we shall summarize rather briefly the counter-mechanisms plant cells have developed to specifically make their cell walls resistant to pathogenic attack.

a. Cutin and Suberin. The first line of defense and possibly the single most important is the excretion and polymerization of the cuticle at the outside of the outer periclinal plant cell walls. The waxes incorporated into the cuticle proper, and the epicuticular waxes on the very surface, hinder the diffusion of cell wall degrading enzymes into the wall (and prevent the diffusion of nutrients from the leaf cells to the micro-organisms on the surface). The underlying cutin polymer, as a polyester and polyether, is obviously difficult to degrade enzymatically.

Clearly, an intact cuticle is a paramount obstacle that can only be breached by a very limited number of micro-organisms. In plant organs exhibiting secondary growth, the epidermis is replaced by layers of cork cells, and these are impregnated with suberin. Similar to cutin, suberin is a random polymeric material (chemically related to lignin and cutin) (Schmutz *et al.*, 1993; Bernards *et al.*, 1995), and it is equally laid down in alternate layers with waxes. Consequently, it can be assumed that suberin fulfills similar functions as cutin in the warding off of phytopathogenic micro-organism, albeit this has been much less studied. Nothing is known about possible 'suberinases' but those can be expected to be produced by some fungi specialized on the infection of plant tissues protected by cork cells.

b. Substitution and Cross-Links.   Once a pathogen has breached the cuticle or cork layers, the primary cell walls of the host cells are exposed to chemical and, possibly, physical attack. The best way to protect the wall from enzymatic degradation is to chemically change the substrates for the microbial enzymes. The best way to prevent the wall from being pierced by pressure is to increase its physical strength by strategically introducing inter-polymer cross-links.

Chemical modification is a way to protect pectins and hemicelluloses from enzymatic degradation. We have seen that pectic molecules may partially be methyl-esterified or they may bear acetyl groups. The degree of substitution by and the distribution of methyl-esters on homogalacturonan chains has been shown to determine their degradability by pectic enzymes (Forrest and Lyon, 1990; Mort and Chen, 1996). However, due mainly to a lack of adequate analytical techniques to compare the substitution patterns of pectins from different plants, it is not yet clear to what extent such a substitution really is involved in successfully protecting pectins from microbial degradation in disease. Addition of phenolic acid esters or ethers to pectins and hemicelluloses is also thought to protect the substituted molecules from enzymatic depolymerization, but experimental evidence is largely lacking (except maybe for the reduced biodegradability of phenolic acid-rich cell walls by rumen micro-organisms) (Ralph and Helm, 1993; Wallace and Fry, 1994). Also, little is known about the substitution of hemicelluloses and its influence on enzymatic degradation, e.g., acetyl groups on some xylose residues in the backbone of glucuronoarabinoxylans and on some galactose residues in the di- or trisaccharide side chains of xyloglucan.

Interpolymeric cross-links to physically strengthen the plant cell wall are an integral concept in the architecture of the primary cell wall to create the three independent networks. Cross-links may be defined as being linkages that are introduced *in muro* and not during synthesis of the polymers (Iiyama *et al.*, 1994). They can occur between identical molecules or between different types of molecules and they may or may not be mediated by low molecular weight cross-linkers. Examples are the crystallization of cellulose molecules to form insoluble microfibrils, the Ca^{++}-mediated cross-links between homogalacturonans, the postulated isodityrosine bridges between two hydroxyproline-rich glycoproteins and the phenolic acid bridges between glucuronoarabinoxylans. The only indisputable examples for mixed polymer cross-links are the hydrogen-bonding of xyloglucans or arabinoxylans to cellulose microfibrils. There can be no doubt that all of these cross-links are absolutely required to build a coherent cell wall structure, but the relative contribution of one type of cross-link versus another are not known. Consequently, their individual role in strengthening the cell wall against the physical impact of microbial penetration attempts, though certainly crucial, remains undetermined.

c. Lignin.   Lignification clearly is a superb way of making a cell wall almost completely resistant to both chemical and physical attack. Moreover, lignin replaces the water in the cell wall thus making it perfectly impermeable to water and thus leakage of nutrients from the plant cells to a potential pathogen and, in the reverse direction, to cell wall degrading enzymes or toxins from the pathogen to the plant cells. Of course, there must be a serious drawback: cells with lignified cell walls will die as they are shut off from water and nutrients supplied from the rest of the plant. Thus, whereas preformed lignification is a means of stabilizing certain cell types such as xylem elements or sclerenchyma fibers in performing a duty to the whole plant physiology and morphology, wound-induced lignification, in which a ring of cells with lignified cell walls encircles a wound, is an efficient protection of the surrounding tissue from secondary infection by opportunistic micro-organisms (see below).

d. Silicon. A further way of impregnating cell walls and making them resistant - probably to chemical degradation and possibly to physical offense - is given by the deposition of silicon in the plant cell walls. However, in spite of the near ubiquity of silicon, little is known about the chemical composition or the site of deposition of silicon in plants. Clearly, the leaves of some plants, notably grasses including some cereals, can be very rich in silicon making them unpalatable for grazing animals, but convincing evidence for protection against pathogenic infection is rare (Zeyen *et al.*, 1993), most likely due to a lack of research into this question.

B.   INDUCED RESISTANCE REACTIONS

When a micro-organism has breached all the preformed barriers mentioned above, which a plant has put up as a passive, structural means of protection, and threatens to colonize the host tissue, plant cells usually exhibit typical morphological reactions as a second line of active structural defense. All of these infection-induced, active resistance mechanisms involve the formation of new cell walls or cell wall materials or a change in composition or interaction of existing cell walls or cell wall materials (Fig. 8). As early as 1926, Young described wall appositions *(callosities)* during penetration of leaf cells by numerous fungi. Similarly, root cells were shown to exhibit tuber-like wall appositions around the hyphae of pathogenic and mycorrhizal fungi (Burgeff, 1932). During the past few years, these wall appositions have been studied with renewed interest focusing on their role as a possible means of resistance. Their complex composition has been analyzed and was further characterized with the advent of new probes. It was shown that the normal plant wall, as well as the callose-appositions, could be modified by the infection-induced addition and cross-linking of phenolics, suberin, lignin, proteins, calcium, and silicon. In addition, walls may contain inhibitors of enzymes or toxins that deter a possible parasite. Numerous reviews have covered cell wall alterations such as modifications in thickness (Rice, 1945; Aist, 1976; 1983) or changes in composition after infection (Akai and Fukutomi, 1980; Vance *et al.*, 1980; Ride, 1983; Smart, 1991, Heitefuss, 1997).

1.   *Intracellular Events Before and During Penetration*
Even before actual penetration of the cell wall has been attempted, the underlying host cells can very often be shown to be reacting to the presence of the pathogen.

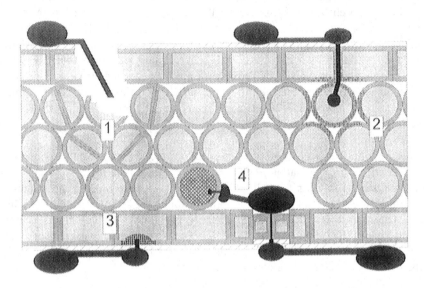

Figure 8. Induced structural resistance mechanisms of a challenged host plant
When the preformed structural and biochemical resistance factors that protect a given plant from the attack by most potentially phytopathogenic microorganisms have failed, a second line of defense is activated, including induced structural and biochemical resistance reactions. In some cases, a newly developed meristem may produce a new layer of cells which may eventually become suberized or lignified in order to localize an ingrowing pathogen (1). In other cases, the cell walls of attacked plant cells may be fortified by the incorporation of new cell wall components (2), often accompanied by the apposition of new cell wall material in the form of a papilla (3). In yet other cases, the cytoplasm of a penetrated host cell may undergo rapid structural changes which may even lead to host cell death - the hypersensitive resistance reaction HR - e. g. by the intracellular polymerization of monolignols (4). The occurrence of none of these different types of induced structural resistance reactions (1-4) is specifically linked to any one of the penetration strategies shown in the figure.

a. Permeability Changes of the Plasmalemma and Gene Induction.   Plant cells can respond rapidly to the presence of a potential pathogen. Early reactions include the depolarization of the plasma membrane (Kauss, 1990), ion fluxes and changes in the internal Ca^{++}-concentrations (Scheel, 1998) upon elicitor application, and the stimulation of an oxidative burst (Brisson *et al.*, 1994). It is not yet fully understood whether these early responses are causally related to rapid changes in gene expression that have been measured within minutes of contact with pathogens or pathogen derived signals (Lamb, 1994; Dangl *et al.*, 1996). Clearly though, induced resistance responses very often require the transcriptional activation of defense genes.

b. The Cytoplasmic Aggregate.   Light microscopy of living cells revealed cytoplasmic changes within minutes after the first contact between host and parasite. Most impressive are changes in the cytoskeleton during these early interactions (Gross *et al.*, 1993; Kobayashi *et al.*, 1994). Other responses include acceleration of cytoplasmic streaming, increase in cytoplasmic strands, and migration of the nucleus towards the area of fungal penetration. As a result, a cytoplasmic aggregate assembles near the infection site (Bushnell and Berguist, 1975). In the interaction between the powdery mildew pathogen (*Erysiphe graminis*) and barley (*Hordeum vulgare*), the number of cytoplasmic strands increased dramatically by the time appressoria form on the epidermis, i.e., about four to five hours prior to actual host penetration by the fungus (Aist and Bushnell, 1991).

c. Endo- and Exocytosis at the Penetration Site. Conventional electron-microscopic studies exhibited an accumulation of vesicles in the cytoplasmic aggregate during infection of barley by *E. graminis* f. sp. *hordei* (Zeyen and Bushnell, 1979). After ultrarapid freezing, the endo- and exocytotic nature of such vesicles could be resolved. At the penetration site of the cowpea rust fungus (*Uromyces vignae*) into epidermal cells of broad bean (*Vicia faba*), numerous coated pits and coated vesicles were observed on the plant plasma membrane suggesting greatly enhanced endocytosis and secretion of material at the penetration site (Xu and Mendgen, 1994). Most coated vesicles were filled with β-1,3-glucans (Fig. 9). These glucans might have originated from the wall of the penetrating fungus due to the activity of host plant glucanases. β-1,3-Glucans from fungal walls, which can be elicitors of plant defense responses, are recognized by soybean cells through receptor-like binding sites associated with the plasma membrane. They bind elicitor molecules in a saturable and reversible manner (Hahn, 1996; Umemoto *et al.*, 1997). Endocytosis of fungal elicitors might serve as a system to clear fungal cell wall components from the plant surface after signal transduction and elicitation of a second messenger system inducing phytoalexin synthesis or other plant defence reactions. In addition, endocytosis could be involved in recycling the elicitor receptors.

Coated vesicles have also been observed around haustoria of *Uromyces vignae* in

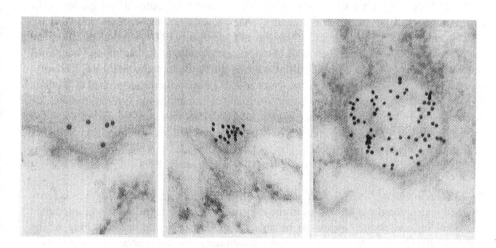

Figure 9. Endocytosis of β-1,3-glucans by coated vesicles on the plasmalemma of epidermal cells from *Vicia faba* at the penetration site of *Uromyces vignae*
Different stages of endocytosis (a, b) and multivesicular body (c). x 125 000, adapted from Xu and Mendgen (1994).

the host *Vigna sinensis* (Stark-Urnau and Mendgen, 1995) and around haustoria of
the smut fungus, *Ustacystis waldsteiniae*, in the host plant *Waldsteinia geoides* (Bauer
et al., 1995). These results show that endo- and exocytotic events may also occur at
later stages of fungal development within host cells, especially around symbiontic
structures such as haustoria.

During infection of legumes by *Rhizobia*, the plasma membrane is invaginated and
the bacteria produce an infection thread which grows through the root cortical cells.
The plant membrane along this thread is studded with coated vesicles, and smooth
vesicles are situated close by (Robertson, 1982). Obviously, the early contact zone
between plants and invading micro-organisms in this symbiontic interaction is also
characterized by endocytotic and exocytotic events which culminate in characteristic
extensions of the plasma membrane, in the modification of the existing wall and
the subsequent production of large wall appositions. Therefore, the infection site of
bacterial and fungal pathogens is an interesting area to study vesicle-mediated transport
of molecules, a field of research that has attracted much interest in the last years
(Low and Chandra, 1994; Hawes *et al.*, 1995).

2.   *Modification of Plant Cell Wall Composition after Infection*
A typical sign of plant cell wall penetration or even attempted penetration which is easily
visible with a light microscope is the halo of stainable material around the penetration
site. UV-light induces fluorescence of the halo and this is most likely due to the deposition
of phenolic material, e.g., esterified phenolic acids or lignin. Quantitative interference
microscopy on onion leaves attacked either by *Botrytis cinerea* or *Colletotrichum circinans*
suggested that if the interaction was susceptible, mass was lost or, if resistant, mass was
increased within the halo area (Russo and Pappelis, 1981). Cytochemical studies with
antibodies raised against cell wall components and against hydrolytic enzymes produced
by the pathogen suggest a very complex picture: At first, wall material is degraded at the
penetration site of the fungus and around hyphae growing within host cell walls (Fig. 6)
(Müller *et al.*, 1997; Podila *et al.*, 1995; Xu and Mendgen, 1997). As a result, the plant
starts modifying pre-existing cell wall components and also may incorporate additional
cell wall components into the cell wall halo at the site of attempted penetration. In barley,
the area in the host wall halo was shown to accumulate lignin, silicon and phenolics (Kunoh,
1990; Zeyen, 1991). Also in many other host-parasite combinations, hydroxyproline-rich
glycoproteins and thionins accumulated after infection (Bohlmann, 1994; Heitefuss, 1997;
Showalter, 1993). Pre-existing and/or newly incorporated cell wall proteins and phenolics
can be cross-linked by peroxidases. The hydrogen peroxide required for these reactions
may be provided by an oxidative burst (Fig. 10) (Brisson *et al.*, 1994).

3.   *Apposition of New Wall Material onto the Existing Cell Wall*
The above described modifications of the existing host cell wall at the site of attempted
microbial ingress do not appear to be able to prevent the eventual breaching of the
cell wall. Rather, they may slow down the penetrating micro-organism enough to give
the host cell time to produce new cell wall material to locally thicken and reinforce
the wall. Such wall appositions or papillae have been observed in leaves, stems and
roots of plants after infection by many pathogens. A recent comprehensive review

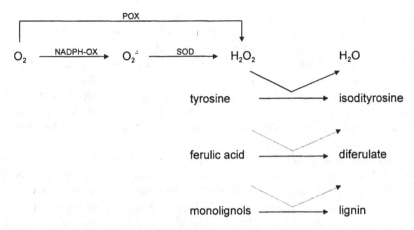

Figure 10. H₂O₂-production in the oxidative burst

Plant cells, similar to some animal cells such as human phagocytes, are able to rapidly produce active oxygen species by partially reducing molecular oxygen. A plasmamembrane-bound NADPH-oxidase (NADPH-OX) may transfer a single electron to O_2 producing the superoxide radical anion, O_2^-, which then spontaneously, or catalyzed by superoxide dismutase (SOD), converts to hydrogen peroxide, H_2O_2. Alternatively, H_2O_2 may be produced directly from O_2 by the oxidase function of a peroxidase (POX) transferring two electrons and two protons to O_2. The resulting hydrogen peroxide can act as a co-substrate in a number of peroxidase-catalyzed coupling reactions. If two tyrosine residues from different proteins are thus dimerized to isodityrosine, the two protein molecules become cross-linked in the process. Similarly, the dimerization of two ferulic acids to diferulate may cross-link two pectin or arabinoxylan molecules to which they are attached. Lastly, the polymerization of monolignols to the lignin polymer is dependent on a peroxidative radicalization of the precursors. All of these cross-links lead to the stabilization of the plant cell wall rendering it more resistant to enzymatic and physical attack.

on wall appositions by Heitefuss (1997) concludes that papillae or wall appositions may be a major factor preventing pathogen ingress into the host cell, but the speed of host reaction relative to fungal growth and the chemical composition of cell wall material seem to regulate the outcome of the interaction. Only very few examples can be cited here.

a. Leaf Pathogens.   At the penetration site of *E. graminis* f. sp. *hordei,* distinct papillae or wall appositions arise in localized areas between the plasmalemma and the cell wall of epidermal cells of barley *(Hordeum vulgare).* This system has attracted considerable interest because this pathogen can infect epidermal strips from coleoptiles. Thus, the host/parasite-interaction is extremely suitable for microscopical and cytochemical studies. With this technique, it has been shown that papillae may vary considerably in their speed of formation and in composition. One major component is β-1,3-glucan. In addition, silicon, phenolic compounds, calcium, lignin and even the secretion of hydrolytic enzymes into this structure has been observed (Aist and Bushnell, 1991). The number of components of this wall apposition increased with the advent of new or better probes. Very often, an increase of one of these components has been correlated with increased resistance to penetration by the pathogen.

Carver *et al.* (1992) were able to induce 'super-susceptibility' in compatible barley/
powdery mildew-interactions after the application of α-aminooxy-β-phenylpropionic
acid, an inhibitor of phenylalanine ammonia-lyase (PAL). Therefore, phenolic compounds appear to play a major role in penetration efficiency. Russo and Bushnell (1989)
compared papillae induced by *E. graminis* and wall appositions induced by wounding.
Both structures were positive for carbohydrates, callose, and protein, and negative
for lignin and suberin. Only wall appositions induced by wounding were positive for
cellulose and pectin whereas papillae induced by the fungus were positive for phenolics
and a basic staining material.

An exceptional case of inherited resistance is observed with the recessive alleles
(*mlo*) of the *Mlo*-locus. Each resistance allele of the locus acts in a non-race-specific
manner and confers resistance to almost all isolates of *E. graminis* f. sp. *hordei*
(Jørgensen, 1987). The development of the fungus is arrested in the papilla. This wall
apposition is very similar in morphology and composition in the susceptible and
the resistant plant. Experiments with *mlo*-resistant plants have shown that micromolar
concentrations of both 2-deoxy-D-glucose, an effective inhibitor of callose deposition
in plants, and chlorotetracyclins, a Ca^{++} chelator, drastically decrease the size of the
wall apposition in response to fungal attack. At the same time, penetration rates of
the fungus increase with these smaller wall appositions. It is obvious that also in this
case, the timing of papilla formation and its final size and composition are crucial for
the outcome of the interaction (Gold *et al.*, 1986; Smart, 1991; Zeyen and Ahlstrand,
1993; von Röpenack et al, 1998).

In rust fungus infections, wall thickenings and papillae have been observed very
often adjacent to the penetration hypha of the haustorial mother cell in incompatible
or non-host combinations (Taylor and Mims, 1991; Harder and Chong, 1991).
They can be quite variable in appearance, ranging from electron-translucent structures,
with callose as a major constituent, to electron opaque, granular deposits with silicon as
a major constituent. However, studies with silicon-depleted bean plants indicated that
reduced silicon deposition increased the frequency of haustorium formation in mature
leaves. Treatment of bean leaves with inhibitors that influence the deposition of callose,
silicon, or autofluorescent compounds very often resulted in an increase of haustorium
development in bean (*Phaseolus vulgaris*) (Heath and Stumpf, 1986).

One of the most striking features of susceptible plants infected by rust fungi is the
superficial lack of response to infection (Heath, 1997). Double inoculation experiments
with non-pathogenic fungi indicated that pathogenic rust fungi, or their haustoria, can
suppress wall depositions (Fernandez and Heath, 1991). Haustoria of *Physopella zeae*
prevented wall matrix formation around intercellular hyphae also formed by this fungus.
In contrast, death of the dikaryotic haustorium of *U. appendiculatus* (Heath, 1988)
or aging of monokaryotic haustoria of *U. vignae* (Stark-Urnau and Mendgen, 1995)
(Fig. 11) resulted in rapid encasement of the intercellular fungus. These results suggest
that a continued metabolism of the biotrophic pathogen is needed to prevent deposition
of plant cell wall components, including callose, around fungal structures.

Bacterial pathogens may also induce morphologically very similar wall appositions.
The *hrp*-genes of *Xanthomonads* are responsible for basic pathogenicity and the ability of
the bacterium to induce the hypersensitive reaction of host tissue. Strains of *Xanthomonas*

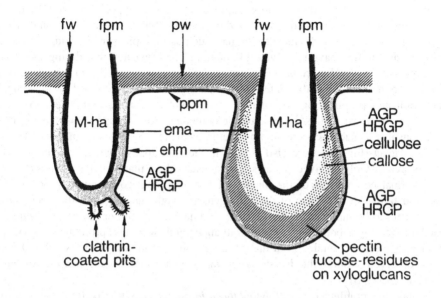

Figure 11. Deposition of plant cell wall material around monokaryotic haustorium of *Uromyces vignae* in *Vigna sinensis*

Young haustorium, exhibiting profiles of coated pits (a) and old, inactive haustorium (b) ensheathed by wall layers composed of different components. fw - fungal wall, fpm - fungal plasma membrane, M-ha - monokaryotic haustorium, ehm - extrahaustorial membrane, ema - extrahaustorial matrix, HRGP - hydroxyproline-rich glycoprotein, AGP - arabinogalactan protein.

campestris pv. *vesicatoria* with mutations in *hrp*-loci fail to multiply in the plant tissue and induce localized reactions of pepper mesophyll cells. Large wall appositions containing phenolics, callose, and hydroxyproline-rich glycoproteins are found next to the bacteria encapsulated within an amorphous matrix. The rapid and localized responses to such mutants can be suppressed by the pathogenic strains of the same bacterium. Suppression of the wall reaction requires the clustered *hrp*-loci in *X. c.* pv. *vesicatoria*. The mechanism by which such wall appositions influence bacterial multiplication is unknown (Boher *et al.*, 1996).

b. Root Colonizing Micro-Organisms. Transmission electron-microscopy has helped considerably to study the development of wall thickenings and the extent of wall appositions in epidermal and cortical cells of the root tip. Bishop and Cooper (1983) compared tomato root invasion by *Fusarium oxysporum* f. sp. *lycopersici* or *Verticillium albo-atrum* and pea root invasion by *Fusarium oxysporum* f. sp. *pisi*. Hyphae of the pathogens adhered to the host root cell walls by means of a matrix typical in morphology for each pathogen. Colonization of the root tissue occurred by intercellular hyphae. Host cells reacted with the development of wall appositions within epidermal and cortical cells. Penetration of host walls was by constricted hyphae and localized wall degradation. Continued deposition of material occurred around penetration hyphae which resulted in elongated penetration papillae. These wall appositions were lignified and resistant to enzymatic and chemical degradation. Wall appositions did not seem to prevent infection. New fixation techniques such as high-pressure freezing followed by freeze substitution have improved this picture. They revealed that during infection of cotton root tips by *F. oxysporum* f. sp. *vasinfectum,* host cytoplasm including cell organelles remained intact after infection. Cell wall synthesis proceeded during fungal penetration and proliferation within host cells. Hyphae became completely encased by wall appositions. Callose was a major component of the appositions (Fig. 12). Other constituents such as pectic homogalacturonan (Fig. 13), xyloglucan (Fig. 14) and arabinogalactan (Fig. 15) were also present in the wall appositions and were secreted into this area by Golgi-derived vesicles. These results show that wall appositions consist of the normal components of the plant wall and additional components, produced in stress situations. Most hyphae of the pathogen, although encased by wall appositions, did not seem to be inhibited in their growth and developed further. During subsequent growth of the fungus between the cells, wall appositions were less pronounced. Hyphae within cells were surrounded by a matrix, a structure reminding of the extrahaustorial matrix observed around the haustoria of many rust fungi (Rodriguez-Gálvez and Mendgen, 1995a, b).

In potato tubers infected with *Phytophthora infestans*, cessation of fungal growth in incompatible reactions is very often accompanied by wall appositions at the penetration site and by the encasement of fungal haustoria with callose-containing material. A positive correlation between resistance and relative abundance of wall appositions was found (Hächler and Hohl, 1984). Also *P. cinnamomi* induced the development of such wall appositions in root tissue of resistant *Zea mays* (Hinch *et al.*, 1985). In *P.sojae*, the speed of callose deposition was always higher in the resistance response than in susceptible roots (Enkerli *et al.*, 1997). However, the possibility that callose deposition is a secondary effect after the halt of fungal growth could not be excluded.

Also the induction of defense reactions by other microorganisms or molecules secreted by them seem to enhance plant resistance against a root pathogen. Tomato plants treated with elicitors such as fungal β-1,3-glucans or chitosan showed increased resistance to *F. oxysporum* f. sp. *radicis-lycopersici* (Benhamou *et al.*, 1996). Root tissue from such plants exhibited restriction of fungal growth and the formation of numerous wall appositions at sites of attempted penetration. These papillae varied greatly in their appearance. Cytochemical tests detected callose and pectin throughout the wall apposition whereas cellulose was concentrated over the external portion.

Even arbuscular mycorrhizal fungi can induce wall appositions in symbiontically defective plants. Mutants isolated after chemical mutagenesis of pea and *Faba* bean exhibited impaired symbiotic abilities vis-à-vis both arbuscular mycorrhizal fungi and *Rhizobium*. Such mutants elicited normal appressorium formation at the root surface, but further development of the mycorrhizal fungus was arrested at this stage. These aborted infections are characterized by an abnormally thick re-enforcement of epidermal and hypodermal cell walls in contact with the symbiont. Within these secondary structures, phenolic compounds, callose, and PR1 proteins accumulate (Gollote *et al.*, 1993). In a tomato root system partly colonised by an arbuscular mycorrhizal fungus, wall appositions were produced also in non-mycorrhizal parts of

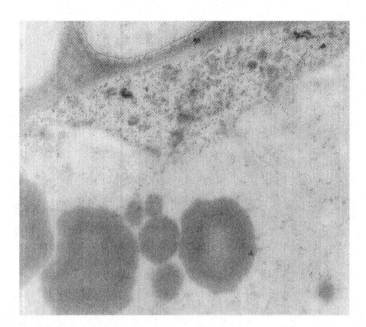

Figure 12.   Callose distribution within wall appositions in cells of cotton root tips after contact with hypha of *Fusarium oxysporum* f. sp. *vasinfectum*
detected with a polyclonal serum against β-1,3-glucan, labelled with colloidal gold. x 30 000, (courtesy of E. Rodriguez-Galvez, Piura, Peru).

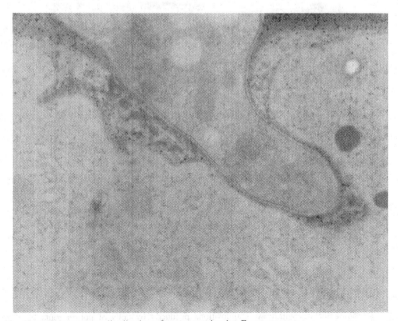

Figure 13.   Homogalacturonan distribution after penetration by *F. oxysporum*
detected with the monoclonal antibody JIM 5, labelled with colloidal gold. x 20 000, from Rodriguez and Mendgen (1995b).

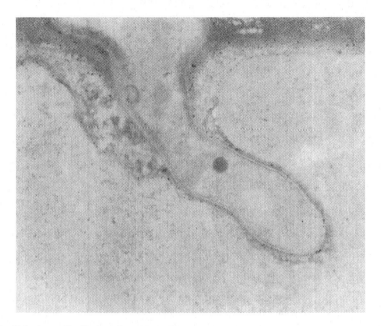

Figure 14.   Xyloglucan distribution after penetration by *F. oxysporum*
detected with the monoclonal antibody CCRC-M1, labelled with colloidal gold. x 17 000, from Rodriguez
and Mendgen (1995b).

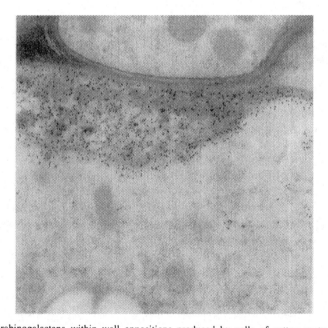

Figure 15.   Arabinogalactans within wall appositions produced by cells of cotton roots after infection
with *F. oxysporum* f. sp. *vasinfectum*
detected with the monoclonal antibody CCRC-M8, labelled with colloidal gold. x 35 000 (courtesy of E.
Rodriguez-Galvez, Piura, Peru).

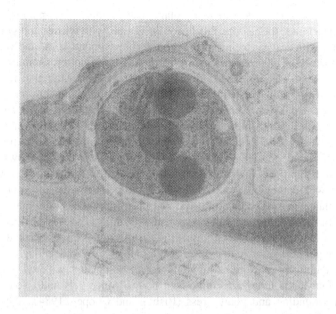

Figure 16. Rhizodermal cells of cotton (elongation zone) infected with *F. oxysporum* f. sp. *vasinfectum* Plant cell and fungal hypha appear intact. x 45 000, from Rodriguez-Galvez and Mendgen (1995a).

the root after infection with the pathogen *P. parasitica*. PR-1a protein detected within these wall appositions and in the hyphae of the pathogen correlated with induced resistance (Cordier *et al.*, 1998).

These results show that, from a morphological point of view, wall appositions in response to fungal or bacterial infections have a striking resemblance to one another. However, their composition and especially their speed of development varies considerably. Therefore, the event that initiates the production of a papilla and the process that regulates the composition of the wall apposition need to be elucidated before we can prove that wall appositions are primarily involved in plant resistance. Molecular methods are needed to solve this question. To date, it is only obvious that papillae can inhibit invading micro-organisms if speed of production and composition of the papilla adequately match the penetration strategy and machinery of the pathogen. On the other hand, the infection thread initiated by *Rhizobia* has some striking resemblance to wall appositions, but obviously supports bacterial invasion. Modification in its composition can inhibit further development of the symbiotic interaction.

4. Production of New Cells or New Tissue

The induced structural resistance mechanisms described above were highly localized and involved a single host cell only. In some cases, however, the reaction may encompass several cells, and additional cells may even be produced *via* induced mitosis in order to isolate the pathogen.

a. Tyloses. The vascular tissue of many plants can produce a morphologically very specific type of defense, the tyloses. Occlusion of vessels with tyloses and gels appears to be associated with resistance in banana, cotton, hop, tomato, elm, and many other plants to wilt diseases. The process of tylose formation is very often similar in different plants (Beckman and Talboys, 1981; Beckman *et al.*, 1982). After the development of protective layers within parenchyma cells, the pit membranes between the xylem parenchyma cells and xylem vessels are disrupted and the living cytoplasm of the parenchymatous cell expands into the vessel lumen, produces a new wall, and may fill the lumen of the vessel. The walls of tyloses apparently contain callose, pectin, phenols, and suberin (Rioux *et al.*, 1998; Robb *et al.*, 1979; Pearce and Holloway, 1984). These compounds may be involved in the protection of tylose walls against the penetration by wilt pathogens. Also vacuoles of both xylem parenchyma cells and tyloses accumulate electron opaque material and these substances, too, may contribute to the protection against further fungal growth. In addition to vessel occlusion, gel formation adds to the blocking of vessels. There is only circumstantial evidence to suggest that gel formation and tylose development are rapid enough to localize the pathogen. Pathogen propagules are only rarely associated with tyloses and gel - a surprising fact if the latter are entrapment points. Since gels and tyloses are also observed after wounding and plant stress (Bishop and Cooper, 1984), they may not represent a specific reaction against fungal pathogens.

b. Barrier Zone Formation after Infection with Ophiostoma Ulmi. The Dutch elm disease pathogen *Ophiostoma ulmi* causes serious wilting in elms, *Ulmus americana* and *U. europaeus*. A comparison of host and non-host trees inoculated with aggressive and non-aggressive isolates of the pathogen has shown that barrier zone formation is an important factor in defense of the disease (Rioux, 1996). A continuous barrier prevented further spread of the fungus completely (Rioux and Oullette, 1991). Barrier zone formation is a multi-component system. It consists of cells in phloem and xylem, mainly fiber cells, with lignified and suberized walls and vessel elements occluded by suberized tyloses. Thus, vessels in the xylem invaded by the pathogen may be completely surrounded by these barrier zone cells.

V.   Towards a Causal Role of Structural Barriers in Plant Resistance

How can we tell whether a given structural feature of a plant, whether preformed or pathogen-induced, is causally involved in making this plant resistant to some potential pathogen? Though difficult to answer, this is a question of paramount importance to our understanding of plant resistance or susceptibility and, vice versa, of pathogenicity and virulence of a micro-organism. Different strategies have to be employed in trying to solve these problems as they relate to preformed resistance factors on the one hand and to infection-induced resistance reactions on the other hand.

A. PREFORMED RESISTANCE FACTORS

The simplest way of finding out whether a given preformed structure is essential in protecting a plant from a pathogen would be to compare the susceptibility of different plants which do or do not possess the structure under investigation. One fundamental problem with this approach is the fact that most structural features of a plant serve multiple purposes, probably none can be expected to be involved in the defense against pathogens alone, e.g., the cuticle protects not only from pathogens, but also from desiccation. Thus, plants lacking such a feature altogether will be difficult to obtain, and studies using such plants would almost certainly be prone to many artifacts. Another major problem with this approach is the genetic stability of most structural traits which are usually the indirect products of many genes, greatly compounding molecular genetic approaches. Nevertheless some such studies have been undertaken with some success.

For the above reasons, it is often easier to approach the problem from the other side, i.e., to compare pathogens that have or do not have the ability to overcome the structural barrier under debate. This competence of the pathogen (e.g., the production of an enzyme degrading a certain cell wall component) is more likely to be controlled by a single gene or a manageable number of genes and, consequently, it may eventually become possible to compare wild type and mutant strains of a pathogen differing only in this or these gene/s.

1. *Plants with Enhanced or Suppressed Levels of a Presumed Resistance Factor*
In order to evaluate a structural feature's involvement in plant resistance to pathogenic attack, plants overexpressing or underexpressing a certain gene or a small set of genes thought to be involved in the building of a structural plant feature have been produced. Overexpression can be achieved under the influence of a suitable promoter, while underexpression may be reached by anti-sense or co-suppression technology. Two examples of such approaches shall be cited here.

Tomato plants were transformed with the gene for tobacco anionic peroxidase thought to be involved in lignin production in tobacco (Lagrimini *et al.*, 1993). The transgenic tomato plants exhibited higher levels of lignin in healthy plants and increased deposition of soluble phenolics and lignin upon wounding when compared to wild type tomato plants. In spite of these increases, resistance to different pathogens *(Fusarium oxysporum* f. sp. *radicis-lycopersici, F. oxysporum* f. sp. *lycopersici, Verticillium dahliae,* and TMV) was not increased in the transgenic plants. Disappointing as this result is, it certainly does not allow the general conclusion that lignin is not involved in resistance of tomato to microbial pathogens.

Using co-suppression as a strategy, tobacco plants with decreased levels of PAL and, consequently, phenylpropanoid products, were produced (Maher *et al.*, 1994). These transgenic plants showed increased susceptibility to a fungal pathogen, *Cercospora nicotianae,* indicating that PAL products are involved in the resistance of tobacco to this fungus. However, the major phenylpropanoid in these plants was soluble chlorogenic acid so that no conclusion can be drawn concerning an involvement of structural features in resistance.

2.   *Pathogens with Enhanced or Lowered Activities of Cell Wall Degrading Enzymes*
The same question can be addressed from the opposite site, i.e., that of the pathogen.
If deletion of a given cell wall degrading enzyme (CWDE, e.g., endopolygalacturonase)
in a micro-organism results in loss of its pathogenicity (i.e., if the pathogen becomes
a saprophyte and is no longer able to infect and cause disease) or loss of its virulence
(i.e., if the pathogen becomes less virulent and causes less severe symptoms) then
the cell wall component degraded by that CWDE (e.g., pectate) may tentatively
be considered a structural feature involved in pathogen resistance of this plant.
Classical studies in this field have recently been complemented by a molecular genetic
approach, where the genes for CWDEs were inactivated by site-directed mutagenesis.
However, some problems remain with this reasoning, as we will see below when we
will discuss some recent examples of this approach.

a. Cutinase.   Initial biochemical and enzymatic studies indicated that cutinase was
a key factor in the infection of pea plants by *Fusarium solani* f. sp. *pisi*. But when
cutinase-deficient mutants of the fungus were produced, the results obtained in infection
experiments were contradictory. Some authors found that the cutinase-deficient mutant
was equally virulent as the wild type and concluded that this particular cutinase is
not crucial for penetration of the outer periclinal epidermal cell walls (Stahl and
Schäfer, 1992). Other authors, using the same pathogen and host plant, found decreased
virulence of the mutant and concluded that cutinase is a virulence factor causally
involved in penetration (Rogers *et al.*, 1994). The first result would have no implication
for our basic question - is cutin critical for resistance? - as the fungus may possess
additional cutinase genes, and as we know that fungal pathogens may be able to
breach the cuticle by sheer physical force. The second result, however, would imply
that cutin and, thus, the cuticle, is a factor involved in resistance of pea plants to
potential pathogens.
    The reverse experiment has also been performed: transformation of an obligate
wound pathogen (*Mycosphaerella*) with a cutinase gene from a different pathogen
(*Fusarium*) allowed the former to penetrate the intact cuticle of its host plant (papaya)
and to cause disease (Dickman *et al.*, 1989). Apparently, cutin is a decisive resistance
factor effective against the wild type of this wound pathogen.
Clearly, the question is still open as to how important fungal cutinases and, thus,
the plant cuticle really are in deciding on pathogenicity and resistance.

b. Pectic Enzymes.   The same approach as detailed above has been used to investigate
whether pectin degrading enzymes of a number of bacterial and fungal plant pathogens
are causally involved in infection or pathogenesis. Pectin is a rather complex cell
wall component, and we have seen that there is a great variety of enzymes capable
of degrading just one component of pectin, namely homogalacturonan. Consequently,
the molecular genetic work needed to produce null-mutants no longer able to degrade
pectin is rather complex.
    The phytopathogenic bacterium most intensely studied in this respect, *Erwinia
chrysanthemi,* is known to produce a number of different pectic enzymes when grown in
liquid culture on pectin as a carbon source (Barras *et al.*, 1994; Hugouvieux-Cotte-Pattat

et al., 1996). Single and multiple mutants deficient in one, several, or all of these enzyme activities exhibited decreased virulence, but all of these mutants were still pathogenic to some extent, possibly due to a 'spare' set of pectic enzymes induced *in planta*.

These results allow the tentative conclusion that pectins are likely to be involved in resistance of plants to pathogens, as a deletion of some pectic enzymes leads to decreased virulence. The main problem with this reasoning is the possibility that pectin degradation may be essential for the pathogen's nutrition rather than for penetration of the structural defense barrier of the host plant.

B.   INDUCED RESISTANCE REACTIONS

Somewhat different strategies have to be employed if one is to find out whether a given infection-induced mechanism of structural defense is causally involved in making a plant resistant to the attack by a phytopathogen. Again, the simplest approach would be to find or make a plant that is no longer able to perform the resistance mechanism under investigation, and to compare the susceptibility of this plant with that of the wild type. However, such a mutant could principally be impaired either in the triggering or in the execution of the resistance reaction, and both cases have their drawbacks. A mutant impaired in the triggering of an induced structural resistance mechanism is almost certainly also impaired in the triggering of other resistance mechanisms, as it has become apparent that defense responses are usually complex syndromes consisting of a number of different resistance mechanisms, the triggering of which are certainly not independent (Moerschbacher and Reisener, 1997). A mutant impaired in the execution of an induced structural resistance mechanism is most likely also impaired in some fundamental aspect of plant growth or development, as the resistance mechanisms make use of biosynthesis machinery and cell wall components that are also involved in the normal building of a cell wall in a healthy plant.

1.   *Inhibitor Studies*
The approach taken therefore involves the interference with the execution of the induced resistance mechanism at the time and at the site of pathogen attack. This has not yet been achieved using a molecular genetic approach, instead, such studies have relied on the use of specific inhibitors of the process under investigation. Usually, the compound used will inhibit an enzyme involved in the biosynthesis or deposition of a cell wall component. Ideally, these can be applied only just at the time when the resistance reaction is being executed and only just at the site of attempted pathogenic attack - reducing the unwanted side effects of the inhibition on general processes of growth and development of the plant. We are still far from this 'ideal' situation described above, but novel methods of microscopy allowing *in vivo* observations and micro-applications of inhibitors should make this scenario more likely in the near future.

Another important problem to be realized is the specificity of the inhibitors involved - and the specificity of the inhibited reaction. Ideally, the inhibitor used should only inhibit one reaction, and ideally, this reaction should only be involved in the formation of one product, i.e., the structural barrier investigated. Both criteria are difficult to meet, as we will see in the examples described below.

a. Inhibitors of Lignin Biosynthesis. The inhibitor approach has repeatedly been used to investigate the relative importance of induced lignification in a given resistance reaction of a plant against a pathogenic micro-organism. We will summarize these attempts as an example to discuss the limitations of this approach.

The first experiments aimed at inhibiting induced lignification used competitive inhibitors of phenylalanine ammonia-lyase (PAL). These inhibitors are not completely specific for PAL (some also inhibit ethylene biosynthesis) and PAL (being at the beginning of the general phenylpropanoid pathway) is not specific for lignin biosynthesis. Depending on the host/pathogen-combination investigated, resistance was either partly broken by the PAL-inhibitors or the inhibitors had no influence on the outcome of the interaction. With the restrictions indicated, these results may be interpreted as indicating that direct or, more likely, indirect PAL-products are causally involved in the former cases of induced resistance reactions. They can not, though, be taken as a reliable indication that the deposition of lignin is involved in these induced resistance reactions.

Later studies used highly specific suicide inhibitors of cinnamoyl-CoA reductase (CCR) and cinnamyl alcohol dehydrogenase (CAD). Both of these enzymes are involved only in the specific branch pathway of monolignol biosynthesis (Whetten and Sederoff, 1995). The results were similar as seen with the PAL-inhibitors: some resistance reactions were partly negated by the inhibitors so that the fungus was able to colonize rather larger areas of plant tissue, while in other host/pathogen-combinations, the CAD-inhibitors did not influence the resistance reaction. The conclusion now would be that lignin biosynthesis is causally involved in the former cases of induced resistance reactions.

Interestingly, the effect of PAL- and CCR/CAD-inhibitors have been compared directly in a few host/pathogen-combinations, and the different results obtained may shed light on the relative importance of induced lignification versus other PAL-dependent resistance reactions in these host/pathogen-interactions.

In wheat plants highly resistant to the wheat stem rust fungus, inhibition of PAL and CAD both partly broke hypersensitive resistance, i.e., penetrated host cells did not lignify and collapse, and the fungus was able to develop a larger colony within the host tissue (Moerschbacher *et al.*, 1990). The effects of the CAD-inhibitors were more marked than those of the PAL-inhibitors presumably hinting at induced cellular lignification as a major factor in the hypersensitive resistance reaction, where monolignol polymerization in the cytoplasm is thought to be the method of active programmed cell death (Moerschbacher and Reisener, 1997).

The same inhibitors were employed in a series of studies on mildew resistance of barley as conditioned either by the *Mla*-gene or by the *mlo*-gene (Zeyen *et al.*, 1995). In case of the *mlo*-gene, mildew development is stopped by the rapid production of papillae at the sites of attempted epidermal penetration. In contrast, in the case of the *Mla*-gene, papillae are produced but appear not to be responsible for resistance. Instead, the fungus can usually penetrate the papilla, but the penetrated epidermal cell then undergoes hypersensitive cell death, stopping further fungal development. Again, PAL- and CAD-inhibitors had basically the same effect, namely, partly negating the *Mla*-dependent hypersensitive resistance reaction but not influencing the *mlo*-dependent papilla-involving resistance reaction. Apparently, the hypersensitive reaction requires the production of lignin while the papillae are resistant to fungal penetration even without the incrustation with lignin and other phenolics.

Recently, the PAL- and CAD-inhibitors have been used with yet another host/ pathogen-system, namely *Arabidopsis* plants inoculated with the powdery mildew fungus (Mauch-Mani and Slusarenko, 1996). In this system, the fungus can make limited growth in the host leaf before being stopped by a hypersensitive type of resistance reaction also correlated with the deposition of lignin. In this system, too, did both inhibitors have similar effects, both reducing the efficiency of the resistance reaction allowing for increased fungal growth, but the effect of the PAL-inhibitors clearly outdid that of the CAD-inhibitors, indicating that lignin biosynthesis is one, but not the major factor in resistance here. The second indirect PAL-product probably crucially involved in resistance in this system has tentatively been identified as salicylic acid.

We should finish this description of inhibitor studies with another word of caution. Even in the cases where a major causal involvement of monolignol biosynthesis has been shown, the results do not allow the conclusion that the polymerization of the phenylpropenols to the lignin polymer itself, i.e., the production of a lignified, structural barrier, is causally involved in the induced resistance reaction. Alternatively, the monolignols produced may be fungitoxic - and the subsequent polymerization of the monolignols to the lignin polymer might merely be a detoxification reaction by the plant. In that case, the defense reaction induced would be a chemical rather than a structural resistance mechanism - a distinction relevant only for this chapter, but not for plant or fungus.

Acknowledgments

We would like to very cordially thank Susanne Horn for her enthusiastic and expert help with this manuscript.

References

Aist JR (1976) Papillae and related wound plugs of plant cells. Annu Rev Phytopathol 14: 145—163

Aist JR (1983) Structural responses as resistance mechanisms. In: Bailey JA and Deveral BJ (eds.) The dynamics of host defense, pp 33—70. Academic Press, New York

Aist JR and Bushnell WR (1991) Invasion of plants by powdery mildew fungi, and cellular mechanisms of resistance. In: Cole GT and Hoch HC (eds) The fungal spore and disease initiation in plants and animals, pp 321—345. Plenum, New York

Akai S and Fukutomi M (1980) Preformed internal physical defenses. In: Horsfall JG and Cowling EB (eds) Plant disease: An advanced treatise, pp 139—159. Academic Press, New York

Alghisi P and Favaron F (1995) Pectin-degrading enzymes and plant-parasite interactions. Eur J Plant Pathol 101: 365—375

Allen EA, Hazen BE, Hoch HC, Kwon Y, Leinhos GME, Staples RC, Stumpf MA and Terhune BT (1991) Appressorium formation in response to topographical signals by 27 rust species. Phytopathology

81: 323—331

Barras F, van Gijsegem F and Chatterjee AK (1994) Extracellular enzymes and pathogenesis of soft-rot *Erwinia*. Annu Rev Phytopathol 32: 201—234

Bauer R, Mendgen K and Oberwinkler F (1995) Cellular interaction of the smut fungus *Ustacystis waldsteiniae*. Can J Bot 73: 867—883

Beckman CH and Talboys PW (1981) Anatomy of resistance. In: Bell AA and Beckman CH (eds) Fungal wilt diseases of plants, pp 487—521. Academic Press, New York

Beckman CH, Mueller WC, Tessier BJ and Harrison NA (1982) Recognition and callose deposition in response to vascular infection in *fusarium*-wilt resistant or susceptible tomato plants. Physiol Plant Pathol 20: 1—10

Benhamou N, Bélanger RR and Paulitz TC (1996) Induction of differential host responses by *Pseudomonas fluorescens* in Ri T-DNA-Transformed Pea roots. Phytopathology 86: 1174—1185

Bernards MA, Lopez ML, Zajicek J and Lewis NG (1995) Hydroxycinnamic acid-derived polymers constitute the polyaromatic domain of suberin. J Biol Chem 270: 7382—7386

Bishop CD and Cooper RM (1983) An ultrastructural study of vascular colonization in three vascular wilt diseases. I. Colonization of susceptible cultivars. Physiol Plant Pathol 23: 323—343

Bishop CD and Cooper RM (1984) Ultrastructure of vascular colonization by fungal wilt pathogens. II. Invasion of resistant cultivars. Physiol Plant Pathol 24: 277—289

Boher B, Brown I, Nicole M, Kpemoua K, Verdier V, Bonas U, Daniel JF, Geiger JP and Mansfield J (1996) Histology and cytochemistry of interactions between plants and xanthomonads. Histology, Ultrastructure and Molecular Cytology of Plant-Microorganism Interactions, pp 193—210. Kluwer Academic Publishers, Dordrecht

Bohlmann H (1994) The role of thionins in plant protection. Critical Reviews in Plant Sciences 13: 1—6

Braun EJ and Howard RJ (1994) Adhesion of fungal spores and germlings to host plant surfaces. Protoplasma 181: 202—212

Brisson LF, Tenhaken R and Lamb C (1994) Function of oxidative cross-linking of cell wall structural proteins in plant disease resistance. Plant Cell 6: 1703—1712

Burgeff H (1932) Saprophytismus und Symbiose, pp 346. Gustav Fischer, Jena

Bushnell WR and Berguist SE (1975) Aggregation of host cytoplasm and the formation of papillae and haustoria in powdery mildew of barley. Phytopathology 65: 310—318

Bushnell WR, Dueck J and Rowell JB (1967) Living haustoria and hyphae of *Erysiphe graminis* f sp *hordei* with intact and partly dissected host cells of *Hordeum vulgare*. Can J Bot 45: 1719—1732

Carpita NC and Gibeaut DM (1993) Structural models of primary cell walls in flowering plants: consistency of molecular structure with the physical properties of the walls during growth. Plant J 3: 1—30

Carver TLW, Robbins MP, Zeyen RJ and Dearne GA (1992) Effects of PAL-specific inhibition on suppression of activated defense and quantitative susceptibility of oats to *Erysiphe graminis*. Physiol Mol Plant Pathol 41: 149—163

Cordier C, Pozo MJ, Barea MJ, Gianinazzi S and Gianinazzi-Pearson V (1998) Cell defense and responses associated with localized and systemic resistance to *Phytophtora parasitica* induced in tomato by an arbuscular mycorrhizal fungus, MPMI 11: 1017—1028

Dangl JL, Dietrich RA and Richberg MH (1996) Death don't have no mercy: Cell death programs in plant-microbe interactions. Plant Cell 8: 1793—1807

Deising H, Nicholson RL, Haug M, Howard RJ and Mendgen K (1992) Adhesion pad formation and the involvement of cutinase and esterases in the attachment of uredospores to the host cuticle. Plant Cell 4: 1101—1111

Dickman MB, Podila GK and Kolattukudy PE (1989) Insertion of cutinase gene into a wound pathogen enables it to infect host. Nature 342: 446—448

Enkerli K, Hahn MG and Mims CW (1997) Immunogold localisations of callose and other plant cell wall components in soybean roots infected with the oomycete *Phytophtora sojae*. Can J Bot 75: 1509—1517

Fernandez MR and Heath MC (1991) Interactions of the nonhost French bean plant (*Phaseolus vulgaris*) with parasitic and saprophytic fungi. IV. Effect of preinoculation with the bean rust fungus on growth of parasitic fungi nonpathogenic on beans. Can J Bot 69: 1642—1646

Forrest RS and Lyon GD (1990) Substrate degration patterns of polygalacturonic acid lyase from *Erwinia carotovora* and *Bacillus polymixa* and release of phytoalexin-eliciting oligosaccharides from potato cell walls. J Exp Bot 41: 481—488

Freialdenhoven A, Peterhänsel C, Kurth J, Kreuzaler F. and Schulze-Lefert P (1996) Identification of genes required for the function of non-race-specific mlo resistance to powdery mildew in barley. Plant Cell 8: 5—14

Gold RE, Aist JR, Hazen BE, Stolzenburg MC, Marshal MR and Israel HW (1986) Effects of calcium nitrate and chlortetracycline on papilla formation, ml-o resistance and susceptibility of barley to powdery mildew. Physiol Mol Plant Pathol 129: 115—129

Gollote A, Gianinazzi-Pearson V, Giovannetti M, Sbrana C, Avio L and Gianinazzi S (1993) Cellular localization and cytochemical probing of resistance reactions to arbuscular mycorrhizal fungi in a 'locus a' mutant of *Pisum sativum* (L.). Planta 191: 112—122

Gow NAR (1993) Nonchemical signals used for host location and invasion by fungal pathogens. Trends Microbiol 1: 45—50

Grambow HJ and Riedel S (1977) The effect of morphogenically active factors from host and nonhost plants on the *in vitro* development of Puccinia graminis f. sp. tritici. Physiol Plant Pathol 11: 213—224.

Gross P, Julius C, Schmelzer E and Hahlbrock K (1993) Translocation of cytoplasm and nucleus to fungal penetration sites is associated with depolymerization of microtubules and defense gene activation in infected, cultured parsley cells. EMBO J 12: 1735—1744

Hächler H and Hohl HR (1984) Temporal and spatial distribution patterns of collar and papillae wall appositions in resistant and susceptible tuber tissue of *Solanum tuberosum* infected by *Phytophthora infestans*. Physiol Plant Pathol 24: 107—118

Hahn MG (1996) Microbial elicitors and their receptors in plants. Annu Rev Plantpathol 34: 387—412

Harder DE and Chong J (1991) Rust Haustoria. In: Mendgen K and Lesemann DE (eds) Electron Microscopy of Plant Pathogens, pp 235—250. Springer Verlag, Berlin

Hardham AR (1992) Cell biology of pathogenesis. Ann Rev Plant Mol Biol 43: 491—526.

Hawes C, Crooks K, Coleman J and Satiat-Jeunemaitre B (1995) Endocytosis in plants: fact or artefact? Plant Cell Environ 18: 1245—1252

Heath MC (1977) A comparative study of non-host interactions with rust fungi. Physiol Plant Pathol 10: 73—88

Heath MC (1988) Effect of fungal death or inhibition induced by oxycarboxin or polyoxin D on the interaction between resistant or susceptible bean cultivars and the bean rust fungus. Phytopathology 78: 1454—1462

Heath MC (1997) Signalling between pathogenic rust fungi and resistant or susceptible host plants. Ann. Bot 80: 713 —720

Heath MC and Stumpf MA (1986) Ultrastructural observations of penetration sites of the cowpea rust fungus in untreated and silicon-depleted French bean cells. Physiol Mol Plant Pathol 29: 27—39

Heitefuss R (1997) Cell wall modifications in relation to resistance. In: Hartleb H, Heitefuss R and Hoppe HH (eds) Resistance of Crop Plants against Fungi, pp 100—125. Fischer, Jena

Hinch JM, Wetherbee R, Mallett JE and Clarke AE (1985) Response of *Zea mays* roots to infection with *Phytophthora cinnamomi*. The epidermal layer. Protoplasma 126: 178—187

Hoch HC, Staples RC, Whitehead B, Comeau J and Wolf AD (1987) Signaling for growth orientation and cell differentiation by surface topography in *Uromyces*. Science 235: 1659—1662

Holloway PJ (1969) The effects of superficial wax on leaf wettability. Ann Appl Biol 63: 145—153

Hoson T (1991) Structure and function of plant cell walls: immunological approaches. Int Rev Cytol 130: 233-268

Howard RJ and Valent B (1996) Breaking and entering: host penetration by the fungal rice blast pathogen *Magnaporthe grisea*. Annu Rev Microbiol 50: 491—512

Hugouvieux-Cotte-Pattat N, Condemine G, Nasser W and Reverchon S (1996) Regulation of pectinolysis in *Erwinia chrysanthemi*. Annu Rev Microbiol 50: 213—257

Iiyama K, Lam TBT and Stone BA (1994) Covalent cross-links in the cell wall. Plant Physiol 104: 315—320

Jørgensen JH (1987) Three kinds of powdery mildew resistance in barley. Barley Genet 5: 583—592

Kauss H (1990) Role of the plasma membrane in host pathogen interactions. In: Larsson C and Moller IM (eds) The plant plasma membrane, pp 320—350. Springer Verlag, Heidelberg

Keegstra K, Talmadge KW, Bauer WD and Albersheim P (1973) Structure of plant cell walls. III. A model of the walls of suspension-cultured sycamore cells based on their interconnections of the macromolecular components. Plant Physiol 51: 188—197

Kerstiens G (1996) Signaling across the divide: a wider perspective of cuticular structure-function relationship. Trends Plant Sci 1: 125—129

Kobayashi I, Kobayashi Y and Hardham AR (1994) Dynamic reorganization of microtubules and microfilaments in flax cells during the resistance response to flax rust infection. Planta 195: 237—247

Kunoh H (1990) Ultrastructure and mobilization of ions near infection sites. Annu Rev Phytopathol 28: 93—111

Kunoh H, Nicholson RL, Yoshioka H, Yamaoka N and Kobayashi I (1990) Preparation of the infection court by *Erysiphe graminis*: Degradation of the host cuticle. Physiol Mol Plat Pathol 36: 397—407

Kurek B (1992) Potential applications of fungal peroxidases in the biological processing of wood, lignocellulose and related compounds. In: Penel C, Gaspar T and Greppin H (eds) Plant Peroxidases 1980-1990, Topics and Detailed Literature on Molecular, Biochemical, and Physiological Aspects, pp 139—186. University of Geneva

Lagrimini LM, Vaughn J, Erb WA, Miller SA (1993) Peroxidase overproduction in tomato: wound-induced polyphenol deposition and disease resistance. Hort Science 28: 218—221

Lamb CJ (1994) Plant disease resistance genes in signal perception and transduction. Cell 76: 419—422

Lamport DTA (1986) The primary cell wall: a new model. In: Young RA and Rowell RM (eds) Cellulose: Structure, Modification and Hydrolysis, pp 77—90. John Wiley, New York

Low PS and Chandra S (1994) Endocytosis in plants. Annu Rev Plant Physiol Plant Mol Biol 45: 609—631

Maher EA, Bate NJ, Ni W, Elkind Y and Dixon RA (1994) Increased disease susceptibility of transgenic tobacco plants with suppressed levels of preformed phenylpropanoid products. Proc. Natl. Acad. Sci. USA 91: 7802—7806

Mauch-Mani B and Slusarenko AJ (1996) Production of salicylic acid precursors is a major function of phenylalanine ammonia-lyase in the resistance of *Arabidopsis* to *Peronospora parasitica*. Plant Cell 8: 203—212

McCann MC, Roberts K, Wilson RH, Gidley MJ, Gibeaut DM, Kim JB and Carpita NC (1995) Old and new ways to probe plant cell-wall architecture. Can J Bot 73: S103—S113

Mendgen K, Hahn M and Deising H (1996) Morphogenesis and mechanisms of penetration by plant pathogenic fungi. Annu Rev Phytopathol 34: 367—386

Moerschbacher BM, Noll U, Gorrichon L and Reisener HJ (1990) Specific inhibition of lignification breaks hypersensitive resistance of wheat to stem rust. Plant Physiol 93: 465—470

Moerschbacher BM and Reisener HJ (1997) The hypersensitive resistance reaction. In: Hartleb H, Heitefuss R and Hoppe HH (eds) Resistance of Crop Plants against Fungi, pp 126—158. Fischer Verlag, Stuttgart

Mort AJ and Chen EMW (1996) Separation of 8-aminonaphthalene-1,3,6-trisulfonate (ANTS)-labeled oligomers containing galacturonic acid by capillary electrophoresis: Application to determining the substrate specificity of endopolygalacturonases. Electrophoresis 17: 379—383

Müller U, Tenberge KB, Oeser B, Tudzynski P (1997) Cel1, probably encoding a cellobiohydrolase lacking the substrate binding domain, is expressed in the initial infection phase of *Claviceps purpurea* on *Secale cereale*. MPMI 10: 268—279

Nicholson RL and Epstein L (1991) Adhesion of fungi to the plant surface: prerequisite for pathogenesis. In: Cole GT and Hoch HC (eds) The Fungal Spore and Disease Initiation in Plants and Animals, pp 3—23. Plenum Press, New York

Pearce RB and Holloway PJ (1984) Suberin in the sapwod of oak (*Quercus robur* L): its composition from a compartmentalization barrier and its occurrence in tyloses in undecayed wood. Physiol Plant Pathol 24: 71—82

Piontek K, Glumoff T, Winterhalter KH and Shoemaker HE (1993) Crystal structure of glycosylated lignin peroxidase. Are carbohydrate residues involved in the redox cycle? In: Welinder KG, Rasmussen SK, Penel C and Greppin H (eds) Plant Peroxidases: Biochemistry and Physiology, pp 9—14. University of Geneva

Podila GK, Dickman MB and Kolattukudy PE (1988) Transcriptional activation of a cutinase gene in isolated fungal nuclei by plant cutin monomers. Science 242: 922—925

Podila GK, Rosen E, San Francisco MJD, Kollattukudy PE (1995) Targeted secretion of cutinase in *Fusarium solani f. sp. pisi* and *Colletotrichum gloeosporioides*. Phytopathology 85: 238—242

Ralph J and Helm RF (1993) Lignin/hydroxy-cinnamic acid/polysaccharide complexes: synthetic models for regiochemical characterization. In: Forage Cell Wall Structure and Digestibility, pp 201—246. ASA-CSSA-SSSA, Madison

Rauscher M, Mendgen K and Deising H (1995) Extracellular proteases of the rust fungus *Uromyces viciae-fabae*. Exp Mycol 19: 26—34

Read ND, Kellock LJ, Knight H and Trewavas AJ (1992) Contact sensing during infection by fungal pathogens. In: Callow JA and Green JR (eds) Perspectives in plant cell recognition, pp 137—172. Cambridge University Press, Cambridge

Read ND, Kellock LJ, Collins TJ and Gundlach AM (1997) Role of topography sensing for infection-structure differentiation in cereal rust fungi. Planta 202: 163—170

Rice MA (1945) The cytology of host-parasite relations. II. Bot Rev 11: 288—289

Ride JP (1983) Cell walls and other structural barriers in defense. In: Callow JA (ed) Biochemical Plant Pathology, pp 215—236. Wiley, New York

Rioux D (1996) Compartmentalization in trees: New findings during the study of Dutch Elm disease. In: Nicole M and Gianinazzi-Pearson V (eds) Histology, Ultrastructure and Molecular Cytology of Plant-Microorganism Interactions, pp 211—225. Kluwer Academic Publishers, Dordrecht

Rioux D, Nicole M, Simard M and Oulette GB (1998) Immunocytochemical evidence that secretion of pectin occurs during gel (gum) and tylosis formation in trees. Phytopathology 88: 494—505

Rioux D and Oullette GB (1991) Barrier zone formation in host and nonhost trees inoculated with *Ophisostoma ulmi* I. Anatomy and histochemistry. Can J Bot 69: 2055—2073

Robb J, Brisson JD, Busch L and Lu BC (1979) Ultrastructure of wilt syndrome caused by *Verticillium dahliae*. VII. Correlated light and transmission electron microscope identification of vessel coating and tyloses. Can J Bot 57: 822—834

Robertson JG (1982) Coated and smooth vesicles in the biogenesis of cell walls, plasma membranes, infection threads and peribacteroid membranes in root hairs and nodules of white clover. J Cell Sci 58: 63—78

Rodriguez-Gálvez E and Mendgen K (1995a) The infection process of *Fusarium oxysporum* in cotton root tips. Protoplasma 189: 61—72

Rodriguez-Gálvez E and Mendgen K (1995b) Cell wall synthesis in cotton roots after infection with *Fusarium oxysporum*. Planta 197: 535—545

Rogers LM, Flaishman MA and Kolattukudy PE (1994) Cutinase gene disruption in *Fusarium solani* f sp *pisi* decreases its virulence on pea. Plant Cell 6: 935—945

Romantschuk M (1992) Attachment of plant pathogenic bacteria to plant surfaces. Annu Rev Phytopathol 30: 225—243

Rubiales D and Niks RE (1992) Low appressorium formation by rust fungi on *Hordeum chilense* lines. Phytopathology 82: 1007—1012

Russo VM and Pappelis AJ (1981) Observations of *Colletotrichum circinans* f sp *dematium* on *Allium cepa*: Halo formation and penetration of epidermal cells. Physiol Plant Pathol 19: 127—136

Russo VM and Bushnell WR (1989) Responses of barley cells to puncture by microneedles and to attempted penetration by *Erysiphe graminis* f sp *hordei*. Can J Bot 67: 2912—2921

Schäfer W (1994) Molecular mechanisms of fungal pathogenicity to plants. Annu Rev Phytopathol 32: 461—477

Scheel D (1998) Resistance response physiology and signal transduction. Curr Opinion Plant Biol 1: 305—310

Schmutz A, Jenny T, Amrhein N and Ryser U (1993) Caffeic acid and glycerol are constituents of the suberin layers in green cotton fibers. Planta 189: 453—460

Shaw BD, Kuo KC and Hoch HC (1998) Germination and appressorium development of *Phyllosticta ampelicida* pycnidiospores. Mycologia 90: 258—268

Showalter AM (1993) Structure and function of plant cell wall proteins. The Plant Cell 5: 9—23

Smart MG (1991) The plant cell wall as a barrier to fungal invasion. In: Cole GT and Hoch HC (eds) The fungal spore and disease initiation in plants and animals, pp 47—66. Plenum, New York

Stahl DJ and Schäfer W (1992) Cutinase is not required for fungal pathogenicity on pea. Plant Cell 4: 621—629

Stark-Urnau M and Mendgen K (1995) Sequential deposition of plant glycoproteins and carbohydrates into the host-parasite interface of *Uromyces vignae* and *Vigna sinensis*. Evidence for endocytosis and secretion. Protoplasma 186: 1—11

Talbot NJ, Ebbole DJ and Hamer JE (1993) Identification and characterization of *MPG1*, a gene involved in pathogenicity from the rice blast fungus *Magnaporthe grisea*. Plant Cell 5: 1575—1590

Taylor J and Mims CW (1991). Fungal development and host cell responses to the rust fungus *Puccinia substriata* var *indica* in seedlings and mature leaves of susceptible and resistant pearl millet. Can J Bot 69: 1207—1219

Tiburzy R, Noll U and Reisener HJ (1990) Resistance of wheat to *Puccinia graminis* f. sp. *tritici:* Histological investigation of resistance cuased by the *Sr5* gene. Physiol Mol Plant Pathol 36: 95—108

Umemoto N, Kakitani M, Iwamatsu A, Yoshikawa M, Yamaoka N and Ishida I (1997) The structure and function of a soybean beta-glucan-elicitor-binding protein. Proc Natl Acad Aci USA 94: 1029—1034

Vance CP, Kirk TK and Sherwood TT (1980) Lignification as a mechanism of disease resistance. Annu Rev Phytopathol 18: 259—288

von Röpenack E, Parr A and Schulze-Lefert P (1998) Structural analyses and dynamics of soluble and cell wall-bound phenolics in a broad spectrum resistance to the powdery mildew fungus in barley. J Biol Chem 273: 9013—9022

Wallace G and Fry SC (1994) Phenolic components of the plant cell wall. Int Rev Cytol 151: 229—267

Walton JD (1994) Deconstructing the cell wall. Plant Physiol 104: 1113—1118

Wessels JGH (1996) Fungal hydrophobins: proteins that function at an interface. Trends Plant Sci 1: 9—15

Whetten R and Sederoff R (1995) Lignin biosynthesis. Plant Cell 7: 1001—1013

Williams PH, Aist JR and Bhattacharya PK (1973) Host-parasite relations in cabbage clubroot. In: Byrde RJW and Cutting CV (eds) Fungal Pathogenicity and the Plant's Response, pp 141—155. Academic Press, London

Xu H and Mendgen K (1994) Endocytosis of 1,3-β-glucans by broad bean cells at the penetration site of the cowpea rust fungus (haploid stage). Planta 195: 282—290

Xu H, Mendgen K (1997) Targeted cell wall degradation at the penetration site of cowpea rust basidiosporelings Mol Plant-Microbe Interact 10: 87—94

Young PA (1926) Penetration phenomena and facultative parasitism in *Alternaria, Diplodia* and other fungi. Bot Gaz 81: 258—278

Zeyen RJ (1991) Analytical electron microscopy in plant pathology: X-ray microanalysis and energy loss spectroscopy. In: Mendgen K and Lesemann DE (eds) Electron Microscopy of Plant Pathogens, pp 59—71. Springer, Berlin

Zeyen RJ and Bushnell WR (1979) Papilla response of barley epidermal cells caused by *Erysiphe graminis*: Rate and method of deposition determined by microcinematography and transmission electron microscopy. Can J Bot 57: 898—913

Zeyen RJ and Ahlstrand GG (1993) X-ray microanalysis of frozen-hydrated, freeze-dried, and critical point dried leaf specimens: Determination of soluble and insoluble chemical elements at *Erysiphe graminis* epidermal cell papilla sites in barley isolines containing Mlo and mlo alleles. Can J Bot 71: 284—296

Zeyen RJ, Ahlstrand GG and Carver TLW (1993) X-ray microanalysis of frozen-hydrated, freeze-dried, and critical point dried leaf specimens: determination of soluble and insoluble chemical elements at Erysiphe graminis epidermal cell papilla sites in barley isolines containing Ml-o and ml-o allels. Can J Bot 71: 284—296

Zeyen RJ, Bushnell WR, Carver TLW, Robbins MP, Clark TA, Boyles DA and Vance CP (1995) Inhibiting phenylalanine ammonia lyase and cinnamyl-alcohol dehydrogenase suppresses *Mla1* (HR) but not *mlo5* (non-HR) barley powdery mildew resistances. Physiol Mol Plant Pathol 47: 119—140

CHAPTER 6

THE HYPERSENSITIVE RESPONSE

THORSTEN JABS[1] & ALAN J. SLUSARENKO[2]

[1] *Present address: BASF AG, Agricultural Research Station AP/RF,
D-67114 Limburgerhof, Germany (thorsten.jabs@msm.basf-ag.de)*
[2] *Institut für Biologie III, RWTH Aachen, D-52056 Aachen, Germany
(slusarenko@bio3.rwth-aachen.de)*

Summary

In this chapter the hypersensitive response is defined, described and analyzed. HR is described at the levels of physiology (e.g. changes in membrane potentials, ion channels etc.), biochemistry (e.g. protein phosphorylation in signal transduction, lipoxygenase and reactive oxygen intermediates in membrane damage), and genetics (e.g. studies with various classes of mutants).

An attempt has been made to put recent developments in context with the wealth of earlier data on the HR to produce a "unifying" consensus. Thus, the recently discovered role of the type III bacterial secretory pathway in delivering both virulence and avirulence gene products directly into the host cell cytoplasm has solved many incongruencies between Avr protein action, elicitor effects, and the HR induced by intact living, micro-organisms.

Emerging information on programmed cell death phenomena in the HR, analagous to some of those observed in animal systems, is discussed in the light of earlier observations and speculations, and here again the analysis of mutants is clearly of central importance.

The relationship of the HR to other defence reponses and to resistance *per se* against various different kinds of pathogens is also considered. The last section draws attention to some of the questions which, despite recent progress, remain to be answered.

Abbreviations

avr	avirulence
CAD	caspase-induced DNase
CHS	chalcone synthase
DAD	defender against apoptotic death
DFF	DNA fragmentation factor
GST	glutathione-S-transferase
HR	hypersensitive response

A. Slusarenko, R.S.S. Fraser, and L.C. van Loon (eds), Mechanisms of Resistance to Plant Diseases, 279-323.
© 2000 *Kluwer Academic Publishers. Printed in the Netherlands.*

hrp   hypersensitive response and pathogenicity
ICE   interleukin-1β converting enzyme
LOX   lipoxygenase
LRR   leucine-rich repeat
LZ   leucine zipper
MAP   mitogen-activated protein
$\Delta\Psi_m$ mitochondrial transmembrane potential
NLS   nuclear localization signal sequence
PAL   phenylalanine ammonia-lyase
PCD   programmed cell death
PR   pathogenesis-related (protein)
ROI   reactive oxygen intermediates
SA   salicylic acid
SAR   systemic acquired resistance
SOD   superoxide dismutase
TMS   trans membrane-spanning (helix)
TMV   tobacco mosaic virus
XR K^+/H^+ exchange response

I.   Introduction

In 1915 E.C. Stakman published a paper in the Journal of Agricultural Research
(U.S. Department of Agriculture) reporting on the histology of an extreme type of
resistance seen when isolates of *Puccinia graminis* from a given susceptible host
were inoculated onto a non-host plant (oats, rye, wheat or barley). The degree of
resistance was very high and in contrast to the compatible interaction between host
and pathogen, where a 'vigorous development of the pathogen without serious injury
to the host' was seen. In the incompatible combination, the fungus 'was found to kill
some of the host cells very soon after gaining entrance, and the fungus grew but little'.
Detailed descriptions of this type of resistance in *Bromus* spp. to the brown rust fungus
Puccinia dispersa had been published by Ward some years previously (Ward, 1902).
Stakmann, however, coined the descriptive term 'hypersensitiveness' for the phenomenon
which he defined as follows: 'Hypersensitiveness is used here to indicate the abnormally
rapid death of the host plant cells when attacked by rust hyphae'. With great preciseness,
and what in retrospect seems like almost prophetic foresight, he went on to add 'It is
used in this sense without any implication as to the exact physiological nature of the
phenomenon, referring, therefore, only to the facts substantiated by visual evidence'.
Nowadays one speaks of 'hypersensitivity', 'hypersensitive resistance', and most
commonly the 'hypersensitive response' or the 'hypersensitive reaction' - commonly
abbreviated to 'HR' and the controversy as what exactly constitutes the HR or simply
occurs concomitantly, still exists.

The HR occurs in plants in response to infection by plant pathogenic fungi, bacteria
and viruses. When an HR occurs the plant does not succumb to infection and damage to
the plant is limited to the cells in the HR lesion. The mechanisms involved in generating

the HR and ultimately causing resistance have been the subjects of intensive research. Much of the research has been carried out with simplified experimental systems, e.g. with fungal components called elicitors, which cause necrosis in whole plant tissues or plant cell suspension cultures. The HR caused by plant pathogenic bacteria has also been much studied. Historically, this was because of the possibility of separating prokaryotic (pathogen) metabolism from eukaryotic (host) metabolism by using selective antibiotics and more recently because of the relative ease of using molecular genetic methodology on the pathogen. The HR can also be induced by viruses and a similar phenomenon has been reported in resistance of plants to some nematodes. Information about the HR has been obtained by studying many different host plants, such as *Arabidopsis*, barley, bean, cucumber, lettuce and tomato, in response to viruses, bacteria, fungi or a whole range of different elicitor molecules. The HR has been investigated in whole plants or in cell culture systems. Thus, it is perhaps wise to treat generalizations with caution while trying at the same time to sift out a unified picture of the HR, as far as is possible, from the mass of information available from different experimental systems. It is probably fair to say that the most complete picture we have is of the HR in response to plant pathogenic bacteria and this will be the major emphasis of this chapter.

An HR occurs when cells of a plant pathogenic bacterium are introduced into tissues of a non-host plant, e.g. the pea pathogen *Pseudomonas syringae* pv. *pisi* in tobacco, or when cells of an avirulent strain of a pathogen are introduced into a host possessing a major resistance gene (see chapters 2 and 4) effective against that particular isolate, e.g. race 1 isolates of the bean pathogen *P. s.* pv. *phaseolicola* in *Phaseolus vulgaris* cv. Red Mexican bean plants.

Klement (1971; 1986) defined three phases in the HR to plant pathogenic bacteria:

1. The INDUCTION PHASE, which requires the presence of living bacteria in the intercellular spaces. Avirulence (*avr*) genes (see chapter 4) are activated in the bacteria and the *avr* gene products are delivered directly into the host cells by a special secretory mechanism (see section III).
2. The LATENT PHASE, during which living bacteria are no longer required. No macroscopic symptoms occur during this phase, but changes in the physiology of the plant cells can be detected. Host gene expression alters and cytological changes are visible in electron micrographs. During this phase the irreversible membrane damage associated with the HR occurs.
3. The PRESENTATION or COLLAPSE PHASE, during which host cells in the inoculated region collapse and desiccate, giving the leaf first a silvered, then a bronzed appearance.

The duration of these three phases depends upon the host/pathogen combination and the environmental conditions. For example, tobacco reacts very quickly and host cell collapse occurs as early as 6-8 h after infiltration of bacteria into the intercellular spaces. In contrast, in French bean inoculated with an avirulent race of *P. s.* pv. *phaseolicola*, cell collapse occurs from 18-24 h after inoculation. Historically the phases were defined by the use of antibiotics. Infiltration of streptomycin or other inhibitors of prokaryotic protein synthesis during the induction phase prevents or delays cell collapse occurring in the presentation phase. After a certain period of time streptomycin no longer prevents cell collapse and this defines the length of the induction phase. The induction period in bean inoculated with *P. s.* pv. *phaseolicola* is approximately 4 h (Roebuck *et al.*, 1978). It is now known that the requirement for live bacteria during the induction period is for the expression of bacterial avirulence *(avr)* genes and hypersensitive response and pathogenicity *(hrp)* genes (see section III in this chapter and chapter 4) to set in motion the events which lead to host cell collapse some 14-20 hours later.

The period after the induction phase is over and until cell collapse occurred has been called the latent phase because no macroscopically visible symptoms are apparent. However, ultrastructural changes can be observed (Roebuck *et al.*, 1978) and ions leak out into the intercellular spaces as the permeability of the plant cell membranes increases (Cook and Stall, 1968; Goodman, 1968). These phenomena were interpreted to be indicators of increasing membrane damage in the cells undergoing HR. Using antibiotics such as blasticidin S or cycloheximide, which inhibit eukaryotic protein synthesis, it was discovered early on that host protein synthesis during the early part of the latent phase was necessary for host cell collapse in the presentation phase (Lyon and Wood, 1977; Keen *et al.*, 1981). Thus, Keen *et al.* (1981) showed that blasticidin S, an inhibitor of protein synthesis in eucaryotes, infiltrated into soybean leaves for up to 9 h after inoculation with an avirulent isolate of *P. s.* pv. *glycinea* prevented the HR and accumulation of the phytoalexin glyceollin. It was interpreted that during the induction phase the bacteria set events in the plant cell in motion which depend on active host metabolism and which lead, ultimately, to the degeneration of cell membranes and death of the cell. A distinct set of changes in host gene expression was characterized in tissues undergoing an HR; although the functions of these genes was not known, the orderly sequence was reminiscent of and prompted comparison to, apoptosis or programmed cell death (PCD) in animal systems (Collinge and Slusarenko, 1987; Slusarenko and Longland, 1986).

II.   Physiological characteristics of cells undergoing an HR

To induce an HR lesion visible to the naked eye (so-called confluent necrosis) about 5 x 10^6 cells ml^{-1} of *Pseudomonas syringae* pv. *syringae* in tobacco (Klement and Goodman, 1967), and about 2 x 10^6 cells ml^{-1} of *P. mors-prunorum* or *P. s.* pv. *phaseolicola* in *Phaseolus vulgaris* cv. Red Mexican (Lyon and Wood, 1976) are needed. Although these titres seem very high, they correspond to about 2-5 bacterial cells for each dead leaf cell. It is important to consider this aspect because originally the HR to plant pathogenic bacteria was viewed by many as an artifact caused by inoculating plants with high bacterial titres, rather than as a resistance phenomenon relevant to disease in the field. However, 2-5 bacteria per dead plant cell does not seem so unrealistic compared to bacterial numbers which might swim into leaves through stomata or hydathodes in natural infections in the field. At inoculum concentrations below the threshold necessary to induce confluent necrosis, *P. s.* pv. *pisi* introduced into tobacco gave rise to dead leaf cells at a ratio of 1 per bacterium (Turner and Novacky, 1974). Similarly in bean, individual dead plant cells can be observed microscopically in the absence of a confluent HR (Slusarenko and Longland, 1986). Now it is known that the HR can be induced by a single bacterial cell attaching to a host cell and delivering *avr* gene products directly into it via a specialized secretory mechanism (see section III).

Early in HR research it was observed that cells undergoing HR cell collapse show changes in the membranes indicative of damage. Thus, electrolyte leakage (Cook and Stall, 1968; Goodman, 1968; Lyon and Wood, 1976) and failure of cells to plasmolyse properly (Woods *et al.*, 1988) have both been interpreted as indicating early membrane dysfunction. However, current ideas, many developed from the study of elicitor-treated cell suspension cultures, suggest that the ion fluxes seen are the result of tightly coordinated changes in the activity of ion channels and pumps rather than the product of membrane damage.

Cells undergoing the HR show changes in the electrical transmembrane potential. The electrical transmembrane potential arises because of the activities of electrogenic pumps, for example the plasmalemma H^+-ATPase, and because of the differences in the rates of passive diffusion of ions due to permeability differences in the membrane for particular ions. Thus, $\Delta E_m = \Delta E_p + \Delta E_D$, where ΔE_m is the electrical transmembrane potential and ΔE_p and ΔE_D are the components provided by electrogenic ion pumps and the diffusion potential, respectively. Pavlovkin *et al.* (1986) examined changes in the electrical membrane potential in cotyledons of cotton *(Gossypium hirsutum)* inoculated with *P. s.* pv. *tabaci* (nonhost HR) or *Xanthomonas campestris* pv. *malvacearum* (susceptible reaction). In the HR, in the first two hours after inoculation, the E_p dropped rapidly to approximately half of the control value, whereas the E_D dropped only slightly. From 2 to 10 h after inoculation E_p did not drop further, whereas E_D continued its steady decline to a level approximately half that of the control value. In contrast, in the susceptible reaction there was a total loss of E_p but the value of E_D remained unchanged (Novacky and Ullrich-Eberius, 1986). These results are important in illustrating fundamental differences in physiology between plant cells in the incompatible and compatible interactions, respectively. While in both cases the electrogenic ion pumps seem to be affected the difference lies in the effects on the passive diffusion potential (E_D) component of the E_m. Passive diffusion of ions occurs through the lipid matrix and/or protein channels of the membrane. Since an increase in the level of lipid peroxidation was observed in cucumber cotyledons during the HR it was suggested that peroxidative membrane damage might lead to the observed effects of the loss of E_D in the HR (Keppler and Novacky, 1986). The generally deleterious effects of lipid peroxidation on membrane function are well known in medicine (Halliwell and Gutteridge, 1989) and there is at least one report of specific damage to ATPase function resulting from lipid peroxidation (Mishra *et al.*, 1989). The latter authors investigated the effect of peroxidation on the function of Na^+/K^+-ATPase in brain cell membranes and showed an 80% inhibition of activity associated with changes in the affinities of the active sites. The affinity for ATP was increased whereas affinities for K^+ and Na^+ were reduced. Increased affinity for ATP would favor phosphorylation of the enzyme at low ATP concentrations whereas decreased affinity for K^+ would hinder the dephosphorylation of the enzyme-P complex resulting in the unavailability of energy for transmembrane processes. The authors concluded that their results demonstrated a specific modification of active sites in a selective manner as a result of lipid peroxidation rather than a non-specific destructive process.

The decrease in E_m in cotton cotyledons was accompanied by loss of electrolytes from the tissue during the HR (Pavlovkin and Novacky, 1986). Approximately 80% of the electrolyte loss was due to K^+; suggesting a specific increase in permeability to K^+ during the HR. In control tissue only 60% of the electrolyte loss, which was predominantly the result of non-specific electrolyte efflux from the cut surface, was due to K^+. After complete HR cell collapse (24 h), the relative amount of K^+ present in the bathing medium decreased to 30% of the total electrolytes. The authors suggested that changes in K^+ concentration resulted initially from electrogenic ion flux and that electroneutral leakage occurred later during HR cell collapse. These observations would be consistent with a rather non-specific damage to the lipid matrix or protein channels which the authors suggest might be explained by increased lipid peroxidation. Keppler and Novacky (1986) demonstrated increased lipid peroxidation during the HR in their experiments and this observation has been confirmed for the HR in several other pathosystems (reviewed in Slusarenko, 1996).

Atkinson and Baker (1987) have described a plasmalemma K^+/H^+ exchange mechanism occurring in tobacco cell suspension cultures after inoculation with avirulent bacteria. This response has been called the XR (exchange response) to identify it as a discrete part of the overall HR phenomenon (Atkinson et al., 1993). Initially, changes in the membrane lipid phase, or in protein channels were proposed to explain the XR (Keppler et al., 1988). The authors detected decreased uptake of fluorescein diacetate into plant cells during the HR, and concluded that this decrease in membrane permeability and fluidity might result from lipid peroxidation or phospholipase activity. Subsequently, Atkinson and Baker (1989) proposed that the plasmalemma H^+-ATPase activity is required for the XR. Using a variety of ATPase inhibitors, inhibitors of respiration, a protonophore and a slightly alkaline external pH, they showed that the K^+/H^+ exchange response could be inhibited to varying degrees. The K^+/H^+ exchange response was reported to be dependent upon an influx of Ca^{2+} (Atkinson et al., 1990) suggesting that it is not the result of non-specific membrane damage. However, the relationship of the increase in the extracellular pH (alkalinization) to the K^+ efflux from cells in the XR is not completely clear. Thus, depending on which counter ion accompanies the K^+, a rise in extracellular pH could occur without H^+ having to be transported into the cell. This would be due to the phenomenon of 'salt hydrolysis' (also known as 'strong ion difference') on the dissociation constant of water. This is, for example, why a solution of potassium acetate will have an alkaline pH whereas a solution of potassium chloride will be neutral and ammonium chloride acidic. Thus, it may be that there is not a stoichiometric exchange of K^+ and H^+ across the membrane at all, but rather that K^+ leakage alone is responsible for the pH rise outside, and the apparent pH drop within the cell because it is accompanied predominantly by a weak counter ion. It is to be hoped that this interesting problem will be solved as more becomes known of events leading to cell collapse; perhaps through the study of some of the mutants affected in HR-like phenomena (see section IV). Whatever the actual mechanism of the XR might turn out to be, evidence has been presented that the response may depend on a phosphoinositide-specific phospholipase C step in the signal transduction pathway (Atkinson *et al.*, 1993).

An early event in the HR which occurs very soon after contact of avirulent bacterial cells with cells of a resistant host is a biphasic generation of reactive oxygen intermediates (ROI); also called active oxygen species (AOS) or reactive oxygen species (ROS). This has come to be described as an 'oxidative burst' by analogy with that observed in macrophages encountering bacteria. The first phase of ROI production, which occurs within minutes of cell contact, is non-specific (i.e. it occurs with virulent, avirulent and saprophytic bacteria), whereas the second, more sustained phase only occurs in an incompatible combination (Baker *et al.*, 1991). It has been suggested that this second sustained burst of ROI production can in some way orchestrate the HR (Levine *et al.*, 1994). The oxidative burst in relation to the HR will be dealt with in detail in section III.

It was shown using parsley cell suspensions treated with elicitor that transient Ca^{2+} influx into the cells was mediated by an elicitor-activated ion channel (Zimmermann *et al.*, 1997). The operation of distinct Ca^{2+} channels in different elicitor/plant model systems has been inferred from inhibitor studies. Thus, piperazines inhibited elicitor-dependent Ca^{2+} fluxes in parsley (Nürnberger *et al.*, 1994a) but not in soybean cells (Ebel *et al.*, 1995) whereas La^{3+}, another Ca^{2+} channel blocker, was effective in parsley (Zimmermann *et al.*, 1997) and soybean (Stäb and Ebel, 1987). In both parsley (Jabs *et al.*, 1997) and tobacco cells (Tavernier *et al.*, 1995), elicitor-induced Ca^{2+} influx preceded the oxidative burst (see section III) and in tobacco it was shown that a protein phosphorylation event was upstream of Ca^{2+} influx (Tavernier *et al.*, 1995) which itself preceded the oxidative burst (see section III). In a very elegant study of the cowpea/cowpea rust pathosystem using confocal scanning laser microscopy and a Ca^{2+}-reporter dye in the intact tissues, a slow but prolonged increase in the intracellular Ca^{2+} concentration ($[Ca^{2+}]_i$) in epidermal cells of resistant but not susceptible plants was observed (Xu and Heath, 1998). The increase in $[Ca^{2+}]_i$ occurred as the fungus grew through the cell wall and preceded other cytoplasmic manifestations of the HR. However, $[Ca^{2+}]_i$ in epidermal cells subsequently declined as the fungus entered and grew within the cell lumen. The baseline $[Ca^{2+}]_i$ in epidermal cells could be increased above the level induced by fungus by treating the cells with kinetin and the authors concluded that raising $[Ca^{2+}]_i$ alone was not sufficient to trigger the HR. On the other hand, calcium channel inhibitors both delayed the HR and prevented the $[Ca^{2+}]_i$ increase, indicating that calcium is involved in some way in transducing an HR-inducing signal.

From the preceding paragraphs it is clear that, although we can observe several physiological changes in cells which are undergoing an HR, the mechanistic significance behind many of the well known 'HR markers' is not well understood. The relationship of various physiological changes to each other is also not always clear. How the ion fluxes and the oxidative burst relate to the action of bacterial avirulence genes which are delivered into the plant cells by a special secretion system (see section III), is also still unclear.

III.   Recognition of the pathogen and signal transduction in the activation of the HR response

The ability of plant pathogenic bacteria to induce an HR is conditioned both by *avr* and *hrp* genes (Gabriel, 1986; Lindgren *et al.*, 1986, 1988; Bogdanove *et al.*, 1996; see chapter 4). The *avr* genes confer the race-specific ability of the bacterium to induce an HR in a given resistant cultivar of the host which carries the corresponding 'classical' resistance *(R)* gene (Staskawicz *et al.*, 1984). The *hrp* gene cluster was identified in *P. s.* pv. *phaseolicola* as a chromosomal region where mutations render the bacteria unable to cause disease in bean (path⁻ phenotype) and unable to cause the HR in the nonhost, tobacco (HR⁻ phenotype). The *hrp* region in plant pathogenic bacteria contains from 20-26 genes, depending upon the organism. The transcriptional organization of the *hrp* gene cluster was studied using an ice-nucleation reporter-gene construct with transposon Tn3 (Lindgren *et al.*, 1989). It was found that some members of the *hrp* gene cluster were induced in planta within two hours of inoculation, and that some of the *hrp* genes had a regulatory function (Rahme *et al.*, 1988; Lindgren *et al.*, 1989; Grimm *et al.*, 1995). The requirement for live bacteria for induction of HR is explained because induction depends upon de novo expression of *hrp* and *avr* genes. The *hrp* gene cluster consists of genes whose products have several pathogenesis-related functions.

Some *hrp* genes are homologous to genes in bacteria pathogenic on animals. Thus, in *Yersinia*, the causal agent of bubonic plague, several homologues exist and the products of these genes form a specialized secretory system, known as a type III secretion apparatus, that delivers *Yersinia* virulence proteins or Yops (*Yersinia* outer proteins) directly into host cells (Gough *et al.*, 1993; Van Gijsegem *et al.*, 1993; Rosqvist *et al.*, 1994; Hueck, 1998). Evidence is accumulating that this is also the function of some of the genes at the *hrp* locus in plant pathogenic bacteria (Fig. 1). So far nine genes have been identified that are conserved between the *hrp* gene cluster and the Type III secretion system found in *Yersinia* and other bacteria pathogenic on humans, such as *Shigella* and *Salmonella spp*. These nine conserved genes in the *hrp* cluster of plant pathogenic bacteria have now been redesignated as *hrc* genes, for *h*ypersensitive *r*esponse *c*onserved (Bogdanove *et al.*, 1996). These *hrc* genes are conserved between *Erwinia, Ralstonia* (the novel genus containing what was formerly *Pseudomonas solanacearum*), *Pseudomonas syringae, Xanthomonas* and the animal pathogens *Yersinia* and *Salmonella*. On the basis of sequence analysis and some experimental evidence they are predicted to encode one outer-membrane protein, one outer membrane-associated lipoprotein, five inner membrane proteins, and two cytoplasmic proteins, one of which is a putative ATPase. With the exception of the outer membrane protein all of the predicted gene products are similar to proteins involved in flagellum biosynthesis (Bogdanove *et al.*, 1996).

Two classes of Hrp proteins are extracellular, namely the harpins and the pilins. The harpins are dealt with separately (see Box *Harpins*). The pilins may be involved in attachment of bacteria to the host cell or in the transfer of the Hrp proteins into the host cells. Thus, the product of the *hrpA* gene from *P. syringae* builds a special pilus-like structure which can be clearly identified on the bacterial surface and can be distinguished from other pili by immunolabeling and electron microscopy. Mutations in the *hrpA* gene eliminate bacterial attachment to host cells.

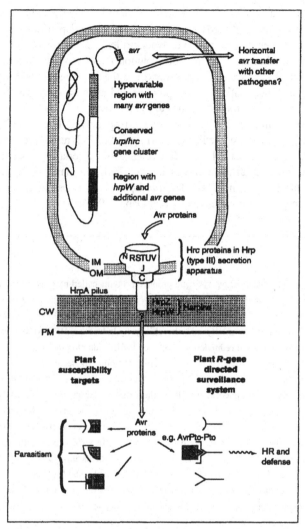

Figure 1. Proposed Model for Bacterial Pathogenicity and Coevolution with Plants Based on Injection by a Conserved Hrp (Type III) Secretion System of Bacterial Avr Proteins.

A typical *Pseudomonas syringae* strain is depicted with many *avr* genes linked to the *hrp/hrc* gene cluster in a region containing mobile genetic elements carried on plasmids. The Hrp secretion apparatus is capable of delivering the products of *avr* genes introduced from other pathovars or even other genera of plant pathogenic bacteria. Widely conserved Hrc proteins are core components of a secretion apparatus that translocates Avr proteins across the bacterial inner membrane (IM) and outer membrane (OM). Extracellular Hrp proteins such as the HrpA pilus protein and possibly the HrpZ and HrpW harpins are proposed to contribute to the subsequent transfer of Avr proteins across the plant cell wall (CW) and plasma membrane (PM). Inside plant cells, the recognition of a single Avr protein by the R-gene surveillance system triggers the HR and plant defenses that lead to resistance. Avr proteins are also proposed to interact with putative susceptibility targets that produce unknown changes in plant metabolism favoring growth of the parasite in the apoplast. The collective contribution of several Avr-like proteins appears to be necessary for parasitism, whereas a single Avr protein is sufficient for betrayal to the defense system (Collmer, 1998). (Figure reproduced by permission of the author and publishers).

The primary function of the type III secretory system in bacteria pathogenic on animals is to deliver virulence proteins directly into host cells. The *hrp*-dependent type III secretory apparatus in plant pathogenic bacteria also delivers virulence determinants directly into plant cells (Fig. 1). It seems that in the course of evolution plants have developed mechanisms to recognize some of these bacterial proteins and use them to trigger resistance responses; i.e. they then serve as functional avirulence proteins and are designated the products of *avr* genes. This is the case for the products of the *avrBs2* and *avrRpm1* genes from *Xanthomonas campestris* pv. *vesicatoria* and *P. s.* pv. *maculicola*, respectively. The *avrBs2* gene product determines avirulence on pepper cultivars carrying the *Bs2* resistance gene. However, the *avrBs2* protein is also essential for full virulence of bacteria on susceptible cultivars which lack the *Bs2* gene (Kearney and Staskawicz, 1990). Similarly, *avrRpm1* determines the HR in resistant *RPM1* genotypes of *Arabidopsis*, and is essential for virulence of the pathogen in susceptible genotypes (Ritter and Dangl, 1995).

Box 1. Harpins

Both *Erwinia amylovora* and *P. s.* pv. *syringae* produce proteins called harpins, which are glycine-rich, lack cysteine residues, and are products of the *hrpN* and *hrpZ* loci respectively. *Erwinia* Harpin$_{Ea}$ mutants gave the expected HR⁻ and path⁻ phenotypes (Kim and Beer, 1998), whereas *Pseudomonas* Harpin$_{Pss}$ mutants retain pathogenicity and the ability to induce an HR (Alfano *et al.*, 1996), suggesting a degree of functional redundancy in relation to HR induction in vivo.

Harpin$_{Ea}$ is a cell envelope-associated protein whereas Harpin$_{Pss}$ is secreted to the cell exterior. Both are dependent upon the *hrp* secretion pathway, but they are not injected by it into the host plant cell and they can both induce an HR from the extracellular milieu at 0.1 µM (Wei *et al.*, 1992; He *et al.*, 1993). In this respect they are unlike *avr* gene products which cannot induce an HR when intercellular and must be introduced into the plant cells to be active (see above). Harpins are also distinct from the *avr* gene class in that *avr* mutants gain virulence on cultivars of the host which have the corresponding *R* gene. In contrast, the action of harpins in eliciting the HR is *R* gene-independent. The HR-eliciting activity is heat stable. The biological function of harpins and the means by which they elicit cell death is unknown. However, it is interesting that the effects depend upon active host metabolism (He *et al.*, 1993) and the implication is that a cell death program is being activated in the plant (Collmer, 1998).

Some elegant experiments have shown that *avr* gene products function inside the host cells when they elicit the HR. Thus, when the *avrBs3* gene from *X. c.* pv. *vesicatoria* was put under the control of the CaMV 35S promoter, which is constitutively active in plants but inactive in bacteria, and delivered into plant cells via *Agrobacterium*, transient expression of the *avrBs3* gene resulted in an HR specifically in those plant cells harboring the *Bs3* resistance gene (Van den Ackerveken *et al.*, 1996). There is no evidence for AvrBs3 secretion into the extracellular environment and infiltration of solutions containing the AvrBs3 protein into the intercellular space of pepper leaves does not induce an HR. These experiments show clearly that the AvrBs3 protein is recognized inside the host plant cell. Additionally, the AvrBs3 sequence contains nuclear localization sequences (NLS) in the C-terminal region of the protein and these were shown to be effective in targeting the AvrBs3 protein to the plant nucleus and to be required for the HR-inducing activity (Van den Ackerveken *et al.*, 1996). A protein has been identified in pepper which interacts with AvrBs3 and which has homology with yeast, human and *Arabidopsis* importin-α, a protein involved in trafficking target proteins to the nucleus (B. Szurek, G.F.J.M. Van den Ackerveken and U. Bonas, unpublished). It is thought that AvrBs3 might act directly as a transcription factor in the plant nucleus but which genes are affected is not yet known.

Nothing is known about delivery of *avr* gene products from pathogenic fungi to plant cells. Certainly, there are no indications that there is anything equivalent to the type III secretory apparatus used by plant pathogenic bacteria. Many of the plant resistance genes so far cloned have putative membrane spanning domains (Baker *et al.*, 1997); it may be that recognition of fungal avirulence determinants occurs at the cell surface. In addition, many plant pathogenic fungi grow through cell walls and contact the plasma membrane directly, while others produce haustoria which invaginate the plasma membrane deep into the cell. Since the haustorium is a feeding organ of the fungus, there are certainly mechanisms for the exchange of macromolecules between the extrahaustorial membrane and the plant cell. It has been reported that there is continuity between the extrahaustorial membrane and the endoplasmic reticulum (ER) in rust- and downy mildew-infected plant cells (Harder and Chong, 1991).

In addition to race-specific *avr* gene products, several bacterial and fungal plant pathogens contain avirulence factors or produce elicitors that trigger defense responses and sometimes even the HR in a 'nonhost' resistant plant species (Dangl *et al.*, 1992; Nürnberger *et al.*, 1994b; Kang *et al.*, 1995; Kamoun *et al.*, 1998), suggesting that the traditional separation between 'host' and 'nonhhost' resistance may not reflect fundamentally different mechanisms of pathogen recognition and defense. Thus, a 10-kDa extracellular protein (INF1 elicitin) of *Phytophthora infestans*, the causal agent of potato and tomato late blight disease, was recently shown to be responsible for nonhost HR and resistance in *Nicotiana benthamiana*; confirming that INF1 functions as an avirulence factor (like the already described race-specific *avr* gene products) in the interaction of *N. benthamiana* and *P. infestans* (Kamoun *et al.*, 1998). The authors used a gene-silencing strategy to inhibit INF1 production in the fungus. These silenced strains retained virulence on the natural host (potato) and gained virulence on *N. benthamiana*. To our knowledge this is the first demonstration that species-specific pathogen protein elicitors function as avirulence factors in nonhost plants.

Recognition of the pathogen sets in motion the signals which lead to activation of the HR response. Recent research has focused on the central role of reactive oxygen intermediates (ROI) in HR signaling. ROI are generated rapidly and transiently after challenge at the site of inoculation. This oxidative burst (for details see Box 2) plays a key role in plant defense (reviewed by Alvarez and Lamb, 1997; Lamb and Dixon, 1997; Jabs, 1999). In plants, ROI can directly act as antimicrobial agents (Peng and Kuc, 1992), but they also contribute to three major features of the HR establishment: as co-substrates for cell wall strengthening (Brisson *et al.*, 1994), as second messengers in the activation of transcription-dependent plant defense responses (Jabs *et al.*, 1997) and as mediators of the HR-associated cell death program (Levine *et al.*, 1994; Jabs *et al.*, 1996).

***Box 2.* Oxidative Burst**

One of the earliest responses of plants to microbial pathogens is the production of ROI such as superoxide (O_2^-) and hydrogen peroxide (H_2O_2). This so-called oxidative burst is initiated immeadiately after pathogen recognition and preceeds phytoalexin production and other more delayed plant defense responses.

Although there is still some debate about the primary source of ROI produced during the oxidative burst (Bolwell *et al.*, 1995), recent findings indicate that plants contain a mechanism for O_2^- production that is homologous to the macrophage NADPH oxidase complex. Elicitor induction of ROI generation and defense responses can be blocked in soybean and parsley by diphenylene iodonium, an NADPH oxidase inhibitor (Levine *et al.*, 1994; Jabs *et al.*, 1997). Antibodies against some components of the human neutrophil NADPH oxidase complex, e.g. p22*phox*, p47*phox*, p67*phox* and the small G protein Rac2, cross-react with plant proteins of similar molecular mass in extracts from soybean, cotton, *Arabidopsis* and tomato (Tenhaken *et al.*, 1995; Desikan *et al.*, 1996; Dwyer *et al.*, 1996; Kieffer *et al.*, 1997; Xing *et al.*, 1997). However, using these heterologous antibodies to screen lambda expression libraries from *Arabidopsis* and soybean yielded sequences which were unrelated to putative plant NADPH oxidase subunits (Tenhaken and Rübel, 1998). Nevertheless, genes coding for plant homologues of the main catalytic subunit gp91*phox* have recently been cloned (Groom *et al.*, 1996; Keller *et al.*, 1998). The *rbohA* (*respiratory burst oxidase homologue A*) gene from *Arabidopsis* was shown to encode a 108-kDa protein, with a large hydrophilic N-terminal extension that is not present in human gp91*phox* (Keller *et al.*, 1998). This additional domain contains two Ca^{2+} binding motifs and has extended homology to a human GTPase-activating protein. Thus, Keller *et al.* (1998) propose that plants have a plasma membrane NADPH oxidase similar to the neutrophil homologue but with plant-specific regulatory mechanisms for Ca^{2+} and G protein stimulation of ROI production. Interestingly, earlier pharmacological studies have already implicated Ca^{2+} and G proteins in the regulation of the plant oxidative burst (Legendre *et al.*, 1992; Jabs *et al.*, 1997). For example, mastoparan, a peptide that constitutively activates G proteins, induced ROI production in the absence of elicitor in soybean (Legendre *et al.*, 1992) and parsley (Kauss and Jeblick, 1995).

In addition to a plasma membrane NADPH oxidase, peroxidases as well as other extracellular enzymes such as germin/oxalate oxidases or amine oxidases, might contribute to the generation of ROI in the oxidative burst (summarized by Bolwell and Wojtaszek, 1997). Generation of H_2O_2 directly by peroxidases is strongly dependent upon extracellular alkalinization and the presence of a reductant in the cell wall, both of which have been extensively studied in the model system of suspension-cultured cells of French bean in response to elicitor derived from the cell walls of the bean pathogen *Colletotrichum lindemuthianum* (Bolwell *et al.*, 1995). There is some evidence for the operation of a peroxidase-dependent ROI generation mechanism in plant-bacterial interactions. Inoculation of lettuce leaves with cells of wild-type *Pseudomonas syringae* pv. *phaseolicola* caused a rapid HR during which highly localized accumulation of H_2O_2 was found in plant cell walls adjacent to attached bacteria (Bestwick *et al.*, 1997). H_2O_2 production was more sensitive to inhibitors of peroxidase (cyanide, azide) than to the NADPH oxidase inhibitor, diphenylene iodonium.

One of the questions still to be resolved is which ROI are most important during the HR cell death program? While some studies suggest that H_2O_2 is sufficient to cause HR-like cell death (Levine et al., 1994), compelling evidence indicates that O_2^- is the key ROI triggering cell death in the *Arabidopsis* lesion mimic mutant *lsd1* (Jabs et al., 1996, further details see section IV). These mutant plants exhibit spontaneous HR-like lesions in the absence of pathogens. But these *lsd1*-type lesions are not definite in size; once a lesion is initated it will eventually spread over the entire leaf. The authors could visualize O_2^- generation specifically in *lsd1* leaf tissues surrounding spreading lesions using the reduction of nitrobluetetrazolium to blue-colored formazan. This 'blue border' indicates that O_2^- generation precedes *lsd1*-type cell death. Furthermore, infiltration of either superoxide dismutase (SOD), which catalyzes the dismutation of O_2^- to H_2O_2, or diphenylene iodonium, which at submicromolar concentrations is a specific inhibitor of the O_2^--generating NADPH oxidase (for details see Box 2 and Fig. 2), blocked the formation of *lsd1*-type lesions. Interestingly, HR development in wild-type *Arabidopsis* leaves (*Rpm1* genotype) inoculated with an avirulent strain of *Pseudomonas syringae* pv. *maculicola* (expressing the *avr* gene *avrRpm1*) was also blocked by the presence of 0.5 µM diphenylene iodonium (I. Kiefer, A.J. Slusarenko and T. Jabs, unpublished). Likewise, infiltration of SOD into tobacco leaves infected with tobacco mosaic virus compromised the development of the HR confirming a significant role of O_2^- during the HR (Doke and Ohashi, 1988).

The oxidative burst together with rapid ion fluxes are the earliest changes observed after pathogen recognition (i.e. interaction of *R* gene product with *avr* gene product, see Chapter 4) (Levine et al., 1994; Nürnberger et al., 1994b; May et al., 1996a; Bestwick et al., 1997; Jabs et al., 1997; Honée et al., 1998). In a study using elicitor-stimulated cultured parsley cells, it has been demonstrated that ion fluxes (Ca^{2+} and H^+ influx, K^+ and Cl^- efflux) across the plant plasma membrane causally precede O_2^- production (Jabs et al., 1997). Moreover, ion fluxes and ROI production are suggested to be equally necessary and sufficient for activation of the plant defense machinery as indicated by detailed loss- and gain-of-function experiments (Jabs et al., 1997). In contrast, using a soybean cell culture system, the reverse causal relationship was established, i.e. that ROI production induces the rapid influx of Ca^{2+} ions which then leads to HR cell death (Levine et al., 1994; Levine et al., 1996). Additionally, experiments with elicitins and tobacco cell suspension cultures have suggested that H_2O_2 from the oxidative burst was neither sufficient nor necessary for hypersensitive cell death (Dorey et al., 1999). However, results gained from model systems using elicitor-treated plant cell cultures have to be interpreted cautiously, as a recent study indicates that the induction of defense responses in a resistant Cf9 tomato cultivar by the race-specific AVR9 peptide elicitor from the tomato pathogen *Cladosporium fulvum* is developmentally regulated and is absent in undifferentiated callus tissue and suspension-cultured cells (Honée et al., 1998). The gene-for-gene interaction of tomato and *C. fulvum* is described in more detail in the case study in Chapter 1.

Furthermore, a number of reports have separated the oxidative burst from HR-associated phenomena such as phytoalexin induction (Davis et al., 1993; Deighton et al., 1994) and have presented evidence that the oxidative burst was either not sufficient

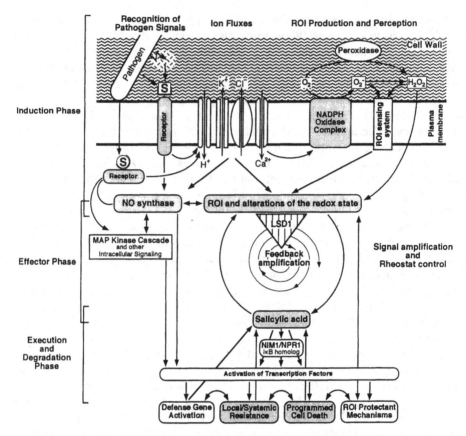

Figure 2. Speculative Model of Programmed Cell Death Regulation During the HR.
Plant receptor proteins (most likely *R* gene products) interact with pathogen interaction-derived signal molecules (S) at the plant cell surface or in the cytosol. Cytosolic pathogen recognition has been shown to take place when phytopathogenic bacteria deliver their *avr* gene products via a Hrp type III secretion system directly into the plant cell. The subsequent cell death-inducing second messenger systems are not fully understood and might depend on the plant-pathogen interaction studied. Participation of ion fluxes, NO synthesis, ROI production and cellular redox state, MAP kinase cascades, and other intracellular signaling mechanisms is discussed in the text. Strong biochemical and genetic evidence indicates an important role of the signaling molecule SA, which might form a feedback amplification cycle in concert with alterations of ROI metabolism. The putative transcription factor LSD1 may function as a negative control element of this central coordination step. Downstream signaling of SA-dependent pathways involves an I-κB homologue, NIM1/NPR1. A detailed comparison of this model to apoptosis regulation in mammals has been presented earlier (Jabs, 1999). (Figure reproduced by permission of the publishers).

(Glazener *et al.*, 1996; Jabs *et al.*, 1997) or even not necessary for HR induction (Gustine *et al.*, 1994). Moreover, several defense-related genes (e.g. *PR1, GST, SOD, LOX*) that are induced by ROI are also up-regulated by abiotic stress. This may indicate that ROI production is a part of a general oxidative stress defense machinery, rather than just an HR-specific signaling mechanism in response to pathogens. However, abiotic stress is often associated with necrotic cell death, which can also show some features of PCD. In the latter regard, two recent reports suggest that nitrogen II oxide (NO), which is well known as a signal in the immune, nervous and vascular systems of vertabrates (Schmidt and Walter, 1994), works synergistically with ROI to promote the HR (Delledonne *et al.*, 1998; Durner *et al.* 1998). Delledonne *et al.* (1998) have shown that NO potentiates the ROI-mediated induction of HR cell death in soybean cells and functions independently of ROI to induce defense-related genes. Moreover, inhibitors of NO synthesis (N^ω-nitro L-arginine, [L-NNA]; S,S'-1,3-phenylene-bis(1,2-ethanediyl)-bis-isothiourea, [PBITU]) interfered with the HR of *Arabidopsis* leaves to an avirulent strain of *P. s.* pv. *maculicola*, and promote disease development and bacterial growth. These data suggest that NO plays a critical role in disease resistance in plants and that, as in animals, NO is generated by the reaction arginine → citrulline + NO. In vertebrates, this reaction is catalyzed by NO synthase (Schmidt and Walter, 1994). NO synthase activity has also been detected in plants and fungi (Ninnemann and Maier, 1996). Now, Durner *et al.* (1998) have reported that NO synthase is induced by tobacco mosaic virus (TMV) in resistant tobacco plants (expressing *N* resistance gene) but not in susceptible tobacco plants. *N* gene-mediated resistance is discussed in detail in the TMV Case Study in Chapter 1. When animal NO synthase or NO-releasing compounds were infiltrated into tobacco leaves, defense-related genes (*PR1, PAL*) were activated. Most interestingly, this NO-dependent gene-expression is mediated by guanylate cyclase, that acts via cyclic GMP. Cyclic GMP and also cyclic ADP-ribose, two important downstream components of NO signaling in mammals, have been shown to further stimulate the NO-activated *PR1* and *PAL* gene expression. Thus, these two recent studies have highlighted a substantial role for NO as a potential 'master signal' that induces the HR cell death program and defense gene activation. This is discussed in more detail by Van Camp *et al.* (1998), Bolwell (1999) and Durner and Klessig (1999). Furthermore, NO is known to interact with O_2^-, one of the prominent ROI generated during the oxidative burst, to form peroxynitrite radicals, which are an extremely potent causes of cellular destruction and trigger of apoptotic cell death (Leist *et al.*, 1997a; Lin *et al.*, 1997).

Only a few studies of the molecular signaling events following these initial steps of the HR cell death program have been published so far (summarized by Hammond-Kosack and Jones, 1996). Recently, a mitogen-activated protein (MAP) kinase was identified that acts downstream of ion fluxes but independently or upstream of the oxidative burst (Ligterink *et al.*, 1997). Upon receptor-mediated activation, this MAP kinase is translocated to the nucleus where it might interact with transcription factors that induce defense gene activation. However, a direct participation of this or other components of MAP kinase cascades in triggering the HR remains to be demonstrated. Interestingly, phosphorylation by protein kinases is essential for the assembly of an active NADPH oxidase complex in the plasma membrane. An alternative pathway involving tyrosine phosphorylation and the MAP kinase cascade also regulates the respiratory burst in human neutrophils (Bokoch, 1994). The participation of calmodulin signaling pathways in plant defense has been postulated for a long time (Renelt *et al.*, 1993). Very recently, an elegant approach using transgenic tobacco plants confirmed this hypothesis (Harding and Roberts, 1998). Tobacco lines which expressed a hyperactive calmodulin variant displayed an enhanced oxidative burst and HR cell death in response to infection with an incompatible pathogen.

Currently, the complex interactions between ROI and NO as well as other signaling molecules, such as salicylic acid (SA, see section V and Box 4) and ethylene, are under intense investigation. Characterization of the ROI-producing NADPH oxidase and of NO synthase from plants is in progress. This, together with genetic approaches involving generation of large libraries of *Arabidopsis* lines that have mutational insertions of T-DNAs or transposons intended to cover the whole genome, should lead to isolation of mutant plants defective in these signaling pathways. The development of these genetic and cytological tools to study redox signaling processes at the subcellular level should give rise to a substantial increase in the understanding of HR signaling pathways.

IV.   Genetic regulation of cell death in plants

Genetic programs leading to animal cell death have been studied extensively during the last two decades, although they are far from being precisely understood (for details see). In plant science this field of research is just emerging. As happens in animals, plant cell death is triggered in response to pathogen infection amongst other developmental and environmental causes. It is now clear that the HR is a noteworthy example among these disease-related circumstances in which PCD or apoptosis (for definition see Box 3) has been suggested to play an important role (Greenberg, 1997; Morel and Dangl, 1997). The first mention of apoptosis in connection with the HR was by Slusarenko and Longland (1986), but little further characterization took place until cell death mutants from *Arabidopsis* and other species were described (Dietrich et al., 1994; Greenberg et al., 1994).

It is now clear that HR cell death is not directly caused by the destructive potential of the pathogen but rather results from the activation of an intrinsic plant genetic program which involves numerous factors (Slusarenko and Longland, 1986; Collinge and Slusarenko, 1987). This contention is supported by several lines of evidence:

- R gene products are required for pathogen recognition leading to the HR (Table 1; see also chapter 2).
- Protein kinases and other putative signaling proteins and transcription factors are involved (Table 2, 3 and 4; Dong, 1998; Innes, 1998).
- The HR requires active plant metabolism, including the transcription and translation machinery (Keen et al., 1981; He et al., 1994).
- Certain fungal elicitors can induce many aspects of the multicomponent defense response during disease resistance (Nürnberger et al., 1994b) and lesions which match the HR in the absence of pathogens (May et al., 1996a).
- Expression of various foreign genes sometimes activates HR-like PCD (Table 4; also summarized by Dangl et al., 1996; Mittler and Lam, 1996).
- A large class of plant mutants, so-called lesion mimics, display spontaneous or conditional cell death which in some cases resembles the HR (Table 3; summarized by Dangl et al., 1996). Interestingly, mutations within plant disease resistance genes can also lead to unregulated cell death (Table 1).

Table 1.   Plant Resistance (*R*) Genes Are Involved in Cell Death Regulation and Share Common
Structural Motifs., such as LRR[1], LZ and NBS.

R genes which have been cloned are listed with their corresponding protein structure
and their resistance function against plant pathogens from diverse origin. At least two,
mlo and *Rpl*, are directly involved in HR or other cell death programs, indicated by
the observation that mutations in these genes cause spontaneous cell death or HR lesion
mimic phenotype. The structure and function of *R* genes in plant defense are discussed
in more detail in Chapter 4.

Gene	Plant	Structure	Resistance Function	Mutant Phenotype
RPS2	*Arabidopsis*	LRR, NBS, LZ	*Pseudomonas syringae* pv. *maculicola*	
RPM1	*Arabidopsis*	LRR, NBS, LZ	*P. syringae* pv. *maculicola*	
mlo	barley	at least 6 TMS-helices	*Blumeria graminis* f.sp. *hordei*	spontaneous cell death
L6, M	flax	LRR, NBS	*Melampsora*	
Hml	maize	toxin reductase	fungal toxin	
Rpl	maize	unknown	*P. sorghi*	spontaneous cell death
Xa21	rice	LRR, protein kinase	*Xanthomonas oryzae*	
Hs¹-Pro	sugar beet		cyst nematode	
N	tobacco	LRR, NBS	TMV	
Pto	tomato	protein kinase	*P. syringae* pv. *tomato*	
I2C	tomato	LRR, NBS	*Fusarium oxysporum*	
Cf-2,4,5,9	tomato	LRR	*C. fulvum*	

[1]   Abbreviations. LRR, leucine-rich repeat; LZ, leucine zipper; NBS, nucleotide-binding site; TMS, trans-membrane spanning.

The characteristic morphology and development of HR cell death has been well
described in the interaction between the biotrophic fungus *Uromyces vignae*, which
causes cowpea rust, and cowpea. Thus, Chen and Heath (1991) observed the following
sequence of cytological events at 15 h after inoculation: (1) migration of the nucleus
to the site of attempted fungal penetration and intense cytoplasmic streaming in the
attacked plant cells, (2) cessation of cytoplasmic streaming, condensation of the nucleus
and the cytoplasm, accumulation of granules at the periphery of the cytoplasm and (3)
collapse of the protoplast and death of the infected cell. Similar cytological changes
were observed in several other incompatible plant-pathogen interactions (Bushnell,
1981; Freytag *et al.*, 1994; Mittler *et al.*, 1997). Interestingly, some of these cytological
events occurring during the HR resemble typical features of apoptotic cell death (see
Box 3). For example, Levine *et al.* (1996) detected plasma membrane blebbing, nuclear
and cytoplasmic condensation, cell shrinkage, and sometimes structures that might
constitute apoptotic bodies in soybean HR disease resistance response against avirulent
strains of *Pseudomonas syringae* pv. *glycinea*. However, they did not detect DNA
cleavage or 'laddering' (for definition see Box 3), a widely-used molecular marker
for apoptotic cell death.

Box 3. **Programmed cell death**

Programmed cell death (PCD) is an integral part of many aspects of animal (Jacobson *et al.*, 1997) and plant development (Pennell and Lamb, 1997) and selectively eliminates unwanted cells (Ellis *et al.*, 1991). Cell suicide programs may also be activated in response to biotic or abiotic stimuli such as environmental stress or pathogens. PCD is often described by the use of cytological criteria, which include chromatin aggregation, cytoplasmic and nuclear condensation and fragmentation of cytoplasm and nucleus into membrane-bound vesicles. These attributes are frequently accompanied by a phenomenon called DNA laddering, the cleavage of the chromatin at internucleosomal sites resulting in DNA fragments that are multimers of about 180 bp (Wyllie *et al.*, 1984). The term apoptosis, Greek: *apo* away from, *ptosis* falling) was introduced by J.F.R. Kerr and colleagues to distinguish this particular type of PCD from necrotic cell death in animals (Kerr *et al.*, 1972). Necrotic cell death results from high doses of cytotoxic agents or severe injury such as trauma and ischemia (injury which occurs on re-oxygenation of tissues after hypoxia) and is characterized by cell and organelle swelling and membrane rupture. Therefore, apoptosis and necrosis have been regarded as morphologically and mechanistically distinct modes of animal cell death. In contrast, in plant pathology any cell or tissue death has traditionally been termed necrosis irrespective of the mechanism leading to death.

However, an increasing body of evidence suggests that apoptosis and necrosis represent just two ends of a wide range of possible cytological and biochemical deaths (Leist and Nicotera, 1997). Diverse triggers of cell death can induce both apoptosis and necrosis. Even proteins that were thought to be highly specific to apoptosis, such as caspases (cysteine proteases similar to interleukin-1β converting enzyme [ICE]) and Bcl-2 (the prototypic regulator of mammalian cell death), seem to participate in necrosis too. Moreover, the 'shape' of cell death (apoptotic or necrotic morphology) is determined by the intracellular ATP level (reviewed by Nicotera and Leist, 1997). ATP depletion, often in conjunction with raised ROI levels and lipid peroxidation, redirects the cell death program from apoptosis to necrosis, independent of the cell-death promoting stimulus (Leist *et al.*, 1997b).

Several pathways and a repertoire of proteins induce, control, extend, and suppress apoptosis (reviewed by White, 1996; Jabs, 1999). Some of the genes that regulate PCD have been highly conserved in evolution, to the extent that at least some of them, such as *bcl-2* which codes for an intracellular inhibitor of PCD in mammals, can perform the same function in cells of a nematode, *Caenorhabditis elegans*.

The cell death program can be subdivided into three functionally different phases: a stimulus-dependent induction phase, an effector phase during which the wide range of death-stimuli are translated to a central coordinator, and a degradation phase during which the alterations commonly considered to define PCD (apoptotic morphology of the nucleus and chromatin fragmentation) become apparent (Kroemer *et al.*, 1995; Jones and Dangl, 1996). The current knowledge about HR-specific degradation mechanisms leading to plant cell collapse will be discussed in the following section 5.

There is striking new evidence for the importance of signals from mitochondria during the effector phase of apoptosis (Kroemer, 1997; Kroemer *et al.*, 1997). Cell-free apoptosis requires mitochondrial function and energy supply (Newmeyer *et al.*, 1994). Furthermore, apoptosis-inducing agents trigger uncoupling of electron transport from ATP production, leading to a decrease of mitochondrial transmembrane potential ($\Delta\Psi_m$) and subsequent ROI production (Zamzami *et al.*, 1995). These events can be attributed to a well-known phenomenon, the mitochondrial permeability transition (PT), which is beleived to facilitate the diffusion of low molecular mass componds (<1500 Da) between the intermembrane space and the cytosol (Zamzami *et al.*, 1996). Mitochondrial PT consists in opening of large conductance, cyclosporin A-inhibited channels or pores (~1.5 nS), that can be formed as multiprotein complexes at the inner-outer membrane contact sites. Functional and genetic experiments indicate that loss of $\Delta\Psi_m$ by PT and subsequent nuclear apoptosis cannot be dissociated (Kroemer *et al.*, 1997; Kuwana *et al.*, 1998). Consequently, cyclosporin A and bongkrekate, which are specific inhibitors of PT, prevent $\Delta\Psi_m$ disruption as well as post-mitochondrial apoptotic changes such as nuclear DNA fragmentation in several different cell systems (Hortelano *et al.*, 1997). Overall, these results indicate that mitochondrial PT is the central coordinator of mammalian PCD.

Several pieces of evidence suggest that the anti-apoptotic protein, Bcl-2, inhibits apoptosis by direct regulation of the mitochondrial PT pore (summarized by Petit *et al.*, 1996; Kroemer, 1997). First, Bcl-2 is anchored to the outer mitochondrial membrane (Nguyen *et al.*, 1994). Second, overexpression of the *bcl-2* gene in the outer mitochondrial membrane abolishes PT induced by various pro-apoptotic agents (Zamzami *et al.* 1996). Third, Bcl-2 co-localizes with the PT pore complex to the contact sites between outer and inner mitochondrial membrane (Riparbelli *et al.*, 1995). Fourth, Bcl-2 family members form potential ion-conducting channels (Schlesinger *et al.*, 1997).

Mitochondrial PT results not only in $\Delta\Psi_m$ disruption but also causes depletion of glutathione and NAD(P)H, matrix Ca^{2+} outflow and massive generation of ROI (Zamzami *et al.*, 1995). In addition, mitochondria undergoing PT release at least three apoptogenic proteins (Apaf-1, -2, 3), one of which (Apaf-2) is cytochrome c (Liu *et al.*, 1996). Moreover, the anti-apoptotic proteins Bcl-2 and Bcl-X$_L$ which reside in the outer mitochondrial membrane, prevent the release of cytochrome c and the activation of caspase-3 (Kluck *et al.*, 1997a), whereas the pro-apoptotic Bcl-2 family member, Bax, when expressed in yeast, induces cytochrome c release from mitochondria and cell death (Manon *et al.*, 1997). Once released into the cytosol, cytochrome c activates ICE-like cysteine proteases such as caspase-3 (Kluck *et al.*, 1997b) and caspase-8 (Kuwana *et al.*, 1998), thereby leading to downstream apoptotic events.

One other apoptogenic factor, Apaf-1, was recently identified as the first human homologue of the *C. elegans* cell death protein CED-4 (Zou *et al.*, 1997), while Apaf-3 was identified as a member of the caspase family, caspase-9 (Li *et al.*, 1997). When combined with Apaf-1, the caspase-9/Apaf-3 zymogen becomes processed to an active enzyme, but only in the presence of cytochrome c (for more detailed information see Reed, 1997).

Besides proteolysis by caspases, chromatin degradation is one of the most prominent processes during the degradation phase of apoptosis. Recently, one of the executioners of this final step of apoptotic cell death has been identified, the DNA fragmentation factor (DFF; Liu *et al.*, 1997). This endonuclease is activated by caspase-3, thereby earning its synonym, caspase-activated DNase (CAD) (Enari *et al.*, 1998). In summary, these data draw a direct signal transduction chain for the final steps of apoptosis: caspase-3 to CAD/DFF to chromatin fragmentation.

In plants, the nature of the executioner(s) or regulator(s) of cell death programs remains unclear. Bcl-2-like functions, mitochondrial PT or caspases have not been described yet in plants.

DNA cleavage is a hallmark for PCD in animals. Ryerson and Heath (1996) described this phenomenon in plant cells killed via HR or abiotic treatment. HR development in two resistant cowpea cultivars was accompanied by the cleavage of nuclear DNA into oligonucleosomal fragments (DNA laddering). Furthermore, terminal deoxynucleotidyl transferase-mediated dUTP nick end in situ labeling (TUNEL staining) of leaf sections showed that pathogen-induced DNA-fragmentation occurred only in haustorium-containing cells and was detectable early in the process of HR cell destruction. Similar observations were made concurrently by Wang et al. (1996) when they treated tomato cells with the host-selective AAL mycotoxin secreted by the fungus *Alternaria alternata* f. sp. *lycopersici*. Interestingly, DNA laddering was enhanced by Ca^{2+} and inhibited by Zn^{2+}, suggesting that specific Ca^{2+}-controlled signaling mechanisms might be involved in plant PCD.

However, a clear correlation between one particular morphology of cell death and either HR-mediated resistance or disease symptoms is still lacking. In a very few cases, apoptosis-like cell death correlates with resistance while cytologically defined necrosis (as defined in animal systems, see Box 3) correlates with susceptibility and disease development (Levine et al., 1996). In other systems, HR-mediated resistance is characterized by morphological changes reminiscent of animal necrosis (Bestwick et al., 1995). Variation in plants from the 'standard' morphological picture of PCD in animals may simply indicate divergence in the mechanisms for PCD between plants and animals in the course of evolution. In addition, pathogen-specific signals may have an influence on the development of the HR, thereby determining apoptotic or necrotic cell death morphology. Moreover, a recent theory in the animal field points out that apoptosis and necrosis might just be extremes of a continuum of multiple forms of PCD (for details see Box 3). This theory is supported by several plant and animal studies which have shown that not all forms of PCD involve all of the morphological changes associated with apoptosis.

Much emphasis has been placed on the search for mutants at loci essential for normal HR development (Table 2; Freialdenhoven et al., 1994; Hammond-Kosack and Jones, 1994; Freialdenhoven et al., 1996; Parker et al., 1996). In addition, genetic approaches have been useful to identify 'private' (i.e. stimulus-specific) as well as 'general' signal transduction pathways during both the early induction and the following effector phase of the HR (for definition see Box 3 and Fig. 2). For example, in barley the *rar1* mutation abolishes *Mla*-based hypersensitive cell death and resistance against powdery mildew (*Blumeria graminis* f. sp. *hordei*) while it does not suppress the HR due to *Mlg*-mediated resistance (Peterhänsel et al., 1997). Similarly, the *Arabidopsis* mutant, *eds-1*, suppresses HR-mediated resistance to various isolates of the fungal pathogen *Peronospora parasitica* but not to an avirulent race of the bacterial pathogen *Pseudomonas syringae* (Parker et al., 1996). In contrast, the *ndr1* mutant suppresses HR-mediated resistance to both pathogens, *P. syringae* and *P. parasitica* (Century et al., 1995). In sum, these mutations provide compelling evidence for the convergence of signals downstream of distinct recognition events into a single pathway triggering the HR. This is analogous to the convergence of signaling pathways in mammalian cell death programs (Fig. 2; Morel and Dangl, 1997; Jabs, 1999).

Table 2. Mutant Screens for Loss of Specific Plant Defense Responses or Disease Resistance as such Identified Genes Required for Disease Resistance (*RDR*).

In contrast to *R* genes listed in Table 1, *RDR* genes do not share common structural motifs, but belong to different protein families such as lipases, membrane proteins, protein kinases, transcription factors or as in the case of *Prf*, even show characteristic *R* gene elemtents such as LRR and NBS. Interestingly, mutations in *RDR* genes often suppress resistance mediated by several *R* genes and directed against various pathogen pathovars or even species as listed in column 4. This points to a plant defense signaling network with common master switches.

Gene	Plant	Structure[1]	*R* Genes Suppressed by Mutation[2] (Pathogen)	Mutant/Gene Description
EDS1	*Arabidopsis*	lipase	*RPP2, RPP4, RPP19* (*Peronospora parasitica*), *RPS4* (*Pseudomonas s.* pv. *maculicola*)	enhanced disease susceptibility; but no decreased HR and SAR
EDS5	*Arabidopsis*	partial homology to lipases	*RPP2, RPP4, RPP19* (*Peronospora parasitica*)	enhanced disease susceptibility; but no decreased HR and SAR
NDR1	*Arabidopsis*	2 TMS helices	*RPM1, RPS2, RPS5* (*Pseudomonas s.* pv. *maculicola*)	non-specific disease resistance
NPR1/NIM1	*Arabidopsis*	ankyrin-repeat, I-κB homolgy; NLS	*RPP12, RPP14* (*Peronospora parasitica*)[3]	nonexpressor of *PR* genes; no immunity; *SA* insensitive
PAD1,2,3	*Arabidopsis*	unknown	*RPP2, RPP4, RPP19* (*P. parasitica*)[4]	phytoalexin-deficient
PAD4	*Arabidopsis*	putative transcription factor	*RPP2, RPP4, RPP19* (*P. parasitica*)	phytoalexin-deficient
Rar1,2	barley	unknown	*Mla-1, Mla-13, Mla-23* (*Blumeria* graminis f.sp. *hordei*)	required for Mla-resistance
Ror1,2	barley	unknown	*mlo* (*B. g.* f.sp. *hordei*)	required for mlo-resistance
Prf	tomato	NBS, LRR	*Pto* (*Pseudomonas s.* pv. *tomato*)	required for Pto-resistance
Pti1[5]	tomato	Ser/Thr-kinase	[6]OE enhances *avrPto*-mediated HR	Pto-interacting protein
Pti4,5,6[5]	tomato	transcription factors	bind to promoter element of *PR* genes[5]	Pto-interacting protein
Rcr-1,2,3,5	tomato	unknown	*Cf-2/-9* (*Cladosporium fulvum*)	required for resistance against *C. fulvum*

[1] Includes partial sequence homologies without functional evidence. [2]Includes partial and complete suppression of pathogen resistance. [3]This mutation improves hyphal growth of avirulent *P. parasitica* strains, but does not permit fungal sporulation. [4]These mutations do not suppress resistance, except in double mutant combinations with each other. [5]These genes have been identified via yeast two-hybrid screen using Pto as the bait, no mutation has been identified so far.

[6] Abbreviations. OE, phenotype of transformants with ectopic 'sense' overexpression of the gene of interest.

An alternative approach to screening for loss of HR development or loss of resistance as such, is to screen for loss of a specific plant defense response. Mutants recovered from such approaches can be tested subsequently for effects on disease resistance. Some such mutants, e.g. those for phytoalexin accumulation in *Arabidopsis*, are listed in Table 2. Most of these mutants show only intermediate levels of compromised resistance but double mutants, generated by simple crossing, showed additive effects. Thus, five phytoalexin-deficient (*pad*) mutants have been identified in *Arabidopsis*, only one of which (*pad3*) completely lacked the *Arabidopsis* phytoalexin camalexin (Glazebrook and Ausubel, 1994; Glazebrook *et al.*, 1997). None of these single mutants was altered in resistance to avirulent strains of *Pseudomonas syringae*, and at least *pad1*, *pad2*, and *pad3* mutations were very little affected in resistance to avirulent *P. peronospora* strains. Interestingly, however, all combinations of double mutants between the 3 *pad* mutants displayed significantly enhanced decreases in resistance to *P. peronospora* (Glazebrook *et al.*, 1997). In contrast, *pad4* completely abolished resistance mediated by the *P. peronospora* R genes *RPP2* and *RPP4*. Thus PAD4, a putative transcription factor, may function in at least some signal transduction pathways conferring pathogen resistance. The relation of phytoalexin production to resistance and the HR will discussed in section VI.

Another *Arabidopsis* mutant screen, especially designed to detect loss of defense gene induction, identified the *npr1* mutation (*no PR* gene expression; allelic to *nim1* and *sai1*; Cao *et al.*, 1994). Interestingly, *npr1* mutants display no noticeable reduction in resistance to *P. syringae* strains, and only very minor decreases in resistance to *P. parasitica* strains (Delaney *et al.*, 1995). These data suggest that *PR* gene expression is not necessary for HR resistance to these pathogens. This hypothesis becomes even more evident but also more complex, when analyzing crosses between the *npr1* and the dominant *cpr6* mutant, the latter conferring constitutive *PR* gene expression and enhanced resistance to an otherwise virulent strain of *P. syringae*. Constitutive *PR* gene expression is not suppressed in this double mutant (Clarke *et al.*, 1998). Thus, constitutive *PR* gene expression does not require NPR1. However, the *cpr6*-dependent resistance to an otherwise virulent *P. syringae* strain is suppressed in the *cpr6/npr1* double mutant, despite the onset of *PR* gene expression. Clarke *et al.* (1998) conclude from these data that *cpr6*-dependent resistance to this *P. syringae* strain must be accomplished through unidentified antibacterial gene products that are regulated through NPR1. Nevertheless, *npr1* mutants appeared to be totally defective in SA perception as well as induction of systemic acquired resistance (SAR; for a discussion of this plant defense phenomenon see Chapter 4). This latter observation confirms an important defense-related role for NPR1. NPR1 seems to be localized to the nucleus (nuclear localization signal) and acts as modifier of transcription. NPR1 has homology to I-κB, an inhibitor of the mammalian redox-regulated transcription factor NF-κB (Cao *et al.*, 1997; Ryals *et al.*, 1997).

The possibility to 'hunt' for gene products which might interact with known proteins using so-called yeast two-hybrid screening is a very promising and advanced tool which has been used to identify proteins that interact with known R gene products. So far, however, only a few of these attempts have been successful. Thus, Zhou *et al.* (1995, 1997) found several proteins which interact with the Pto *R*-gene product of tomato which confers resistance to *P. syringae* pv. *tomato* strains that carry the *avrPto* avirulence gene. *Pto* itself encodes a serine/threonine kinase and unlike most other *R* genes does not contain the 'classical' leucine-rich repeat (LRR) domain or a nucleotide binding site (NBS) or a leucine zipper (LZ; Loh and Martin, 1995). *Pti1*, the first identified *Pto*-interacting gene, also encodes a serine/threonine-kinase, and is specifically phosphorylated by Pto (Zhou *et al.*, 1995). These data are the first proof for the existence of protein kinase signalling cascades in plant defense, although these have been postulated for some time, mainly based on evidence from plant cell culture studies using elicitors and protein kinase and phosphatase inhibitors (Großkopf *et al.*, 1990; Felix *et al.*, 1994)

Consistent with this hypothesis, ectopic overexpression of *Pti1* enhances *avrPto*-mediated HR cell death. Other candidates from the two hybrid screen, such as *Pti4,5,6*, appeared to be transcription factors of the ethylene-responsive element binding protein class (Zhou *et al.*, 1997). At least Pti5 and Pti6 bind to a promoter element contained in several *PR* genes in *Arabidopsis*, bean, potato, tobacco and tomato. However, a definite role for these Pti proteins in signal transduction after recognition remains to be demonstrated.

Regarding the genetic control of the HR, the following problems remain to be solved: What determines the extent of the leaf area undergoing PCD after pathogen recognition, as there is no cell differentiation-dependent determination? Do specific 'anti-cell-death' proteins or pathways exist in plants, which, for example, might inhibit the induction of PCD in the leaf tissue surrounding the area of an HR lesion? And even more intriguing, if it exists, what constitutes the central executioner (in analogy to animal PCD, see Box 3) of plant cell death in the decision to die or not to die? So far, only few similarities to the animal apoptotic signaling machinery have been elucidated in plants. All attempts to identify plant homologues to the prototypic regulator of mammalian cell death, Bcl-2, or any other members of the so-called Bcl-2 family (summarized by Kroemer, 1997), have been fruitless. Additionally, overexpression of the human antiapoptotic protein Bcl-X$_L$ in tobacco did not suppress HR lesion formation in response to viral or bacterial pathogens (Mittler *et al.*, 1996). Likewise, caspases, which play a central role in animal PCD, have yet to be described in plants. Nevertheless, molecular evidence that some mechanisms underlying PCD may indeed be conserved in animals and plants has been unraveled recently. The gene encoding the *defender against apoptotic death* protein (DAD1) has been identified from *Arabidopsis* and rice EST databases (Gallois *et al.*, 1997; Tanaka *et al.*, 1997). The *Arabidopsis* cDNA has been shown to be as efficient as human *dad1* in rescuing mutant hamster cells from apoptotic cell death (Gallois *et al.*, 1997). However, no function has been shown for DAD1 in HR plant cell death so far.

Nevertheless, anti-cell death pathways do seem to exist in plants as indicated by the existence of cell death control mutants, such as *acd2* (Greenberg *et al.*, 1994), *lsd1* (Dietrich *et al.*, 1994), and *lls1* (Gray *et al.*, 1997). These mutants exhibit impaired control of cell death in the absence of the pathogen and cannot control the spread of cell death once it is initiated (Table 3). Very recently, the *Lls1* gene was found to encode an aromatic ring-hydroxylating dioxygenase (Gray *et al.*, 1997) suggesting that its target, most likely a mediator of cell death, might be a phenol. One candidate that may fit well in this role is SA, which exhibits a 10- to 50-fold increase during the HR and some types of oxidative stress-induced PCD (Hammond-Kosack and Jones, 1996). In addition, SA is known to cause an increase of intracellular H_2O_2 (Chen *et al.*, 1993), to potentially form a cell-damaging free-radical (see Box 4; Durner and Klessig, 1996; Anderson *et al.*, 1998) and to promote cell death during the HR (Naton *et al.*, 1996) and in *Arabidopsis* cell death mutants, such as *lsd1*. In sum, this suggests that LLS1 acts as a cell death suppressor by scavenging SA or a related death-promoting phenolic compound.

Table 3.   Mutant Screens for Constitutive Pathogen Resistance and Lesion-Mimic Phenotype Identified Genes Required for Plant Defense and Cell Death Regulation.

A lot of research is directed to identify the genes involved in cell death regulation. So far, only *LSD1* and *Lls1* have been cloned, the latter of which may participate in regulation of one of the final steps during the degradation phase of the HR (see also section V).

Gene	Plant	Gene Structure[1]	Function	Mutant Description
ACD1,2	Arabidopsis	unknown	unknown	accelerated cell death
CIM2,3	Arabidopsis	unknown	unknown	constitutive immunity, but no lesion mimicry
CPR1,5,6	Arabidopsis	unknown	unknown	constitutive expressors of *PR* genes
DND	Arabidopsis	unknown	unknown	defense, no death; constitutive SAR
LSD1	Arabidopsis	zinc finger motif	transcription factor, rheostat sensor for cell death-promoting signal	lesions simulating disease resistance
LSD2-7	Arabidopsis	unknown	unknown	lesions simulating disease resistance
Lls1	maize	putative aromatic ring- hydroxylation dioxygenase	sensor for oxidants and/ or iron, degrades phenolic mediator of HR	lethal leaf spot

Another negative regulator of plant PCD is the LSD1 protein from *Arabidopsis* (Dietrich *et al.*, 1997). Mutant *lsd1* plants show spontaneous (daylength-dependent) HR-like lesions and enhanced resistance to pathogens. However, the lesion size is not limited, and a phenomenon that has been termed 'runaway cell death' develops where large areas of tissue die. The *LSD1* gene encodes a novel zinc finger protein that may regulate transcription of cell death-effectors (Dietrich *et al.*, 1997). Interestingly, O_2^- is necessary and sufficient to initiate lesion formation in *lsd1* mutants (Jabs *et al.*, 1996). It accumulates before the onset of cell death and subsequently in live cells adjacent to the spreading *lsd1* lesions. Thus, O_2^- is the critical ROI species whose perception and removal is perturbed by the *lsd1* mutation. Recent experiments indicate that the activity of several antioxidant enzymes such as Cu/Zn-SOD and catalase (CAT) is altered in *lsd1* mutant plants (T. Jabs, unpublished). In sum, these data suggest that the putative LSD1 transcription factor monitors a self-amplifying signal normally leading to cell death in plants (Fig. 2; Jabs et al., 1996). Furthermore, LSD1 may act as a rheostat, sensing signals that activate the HR cell death program. Thus, LSD1 could allow the initiation or spread of cell death by promoting transcription of cell-death effectors at high inducing-signal levels but be able to slow down or stop the cell death program as signal levels fall. This rheostat sensor might respond to alterations of the cellular redox state or increases of SA or ROI levels, as found in the close vicinity of pathogen infection sites. Alternatively, LSD1 could inhibit constitutive low level signals feeding into the ROI-generating system of the oxidative burst. The demonstration that O_2^- is a critical signal in a feedback amplifying process monitored by a putative transcription factor to regulate the spread of cell death reinforces the hypothesis that similar strategies are used to control PCD in plants and in animals (Jabs, 1999).

Thus, it is not surprising that expression of antioxidative enzymes, such as SOD, CAT, peroxidase, glutathione-*S*-transferase (GST) and glutathione peroxidase, correlates with the induction of the HR (Levine et al., 1994; Jabs et al., 1996). This enhanced antioxidative capacity may protect neighboring cells from uncontrolled diffusion of death signals, such as ROI, throughout the entire leaf. Likewise, transgenic plants with lowered antioxidative capacity often develop necrotic lesions, induce defense gene expression in the absence of pathogens and display enhanced resistance to pathogens (Table 4; Chamnongpol et al., 1996; Takahashi et al., 1997). However, *Arabidopsis* mutants which are defective in the synthesis of two major plant antioxidants, glutathione or ascorbate, respectively, show only minor or no changes in pathogen resistance (Table 4; Conklin et al., 1996; May et al., 1996b). The latter might indicate that concentration changes of a specific antioxidant may be tolerated by the plant to a certain extent or compensated readily by another antioxidant species. Nevertheless, a few critical plant components are known which when mutated cause oxidative stress followed by necrotic cell death. For example, a defect in uroporphyrinogen decarboxylase (UROD) causes by the dominant mutation *les22* in maize cannot be compensated and leads to a well described lesion mimic phenotype (Hu et al., 1998). The *les22* mutation causes heme deficiency and excessive accumulation of photo-excitable tetrapyrroles which leads to light-inducible oxidative stress and necrotic cell death. Interestingly, *UROD* mutations in humans are also dominant and cause the metabolic disorder porphyria, which manifests very much like the plant symptoms as a light-induced skin morbidity. Moreover, 'plant porphyria' caused by UROD or coporphyrinogen oxidase (CPO) deficiency is sometimes correlated with constitutive pathogen defense mechanisms such as *PR* gene expression, phytoalexin production and increased ROI and SA synthesis (Johal et al., 1995; Mock et al., 1997; Mock et al., 1998).

In summary, we have increasing evidence that reactive oxygen intermediates (ROI) serve as direct and indirect mediators of PCD in mammalian and plant cells. Overexpression of genes encoding pro- and antioxidant enzymes in transgenic animals and plants has been informative as to the function of ROI. Recent data imply a dual role of ROI in the apoptotic process, firstly as a facultative signal during the

Table 4.   Transgenics and metabolism mutants

Mutant	Plant	Gene structure[1]	Putative Function	Mutant description
Genes from the pro/antioxidant pathways				
cad2-1	*Arabidopsis*	unknown	glutathione synthesis	*cadmium*-sensitive
soz1	*Arabidopsis*	transcription factor	ascorbate synthesis	sensitive for *ozone*, reduced resistance against *P. parasitica*
Cat3	maize	Catalase	Chilling stress resistance	unknown
Les22	maize	uroporphyrinogen decarboxylase	avoidance of accumulation of photoexcitable tetrapyrroles	light-induced *lesion* mimicry via porphyria, heme-deficiency (dominant mutation)
GO	potato	fungal glucose oxidase	H_2O_2 production	OE: enhanced resistance against *e. carotovora* and *P. infestans*
CPO	tobacco	coporphyrinogen oxidase	avoidance of accumulation of photoexcitable tetrapyrroles	AS: light-inducible lesion mimicry (photobleaching herbicide-like effect)
UROD	tobacco	uroporphyrinogen decarboxylase	avoidance of accumulation of photoexcitable tetrapyrroles	AS: light-inducible lesion mimicry, heme deficiency (photobleaching herbicide-like effect)
Catl	tobacco	Catalase	detoxification of H_2O_2	AS: spontaneous cell death
LOX	tobacco	Lipoxygenase	lipid peroxidation (jasmonic acid synthesis)	AS: reduced resistance against *P. parasitica* and *R. solani*
Genes from putative signalling pathways				
NahG	*Arabidopsis*, tobacco	bacterial salicylate hydroxylase	SA degradation	OE: no SAR, reduced resistance against pathogens
NPR1[1]	*Arabidopsis*	ankyrin-repeat, I-kB homolgy: NLS	OE: enhanced resistance against *P. parasitica* and *P. parasitica*	
PDF1.2	*Arabidopsis*	thionin	antifungal activity	OE: partial resistance against *F. oxysporum*
PKC	Potato	Protein kinase C	signals PR gene expression	unknown
	Tobacco	bacterio-opsin	proton pump	
rgp1	Tobacco	small GTP binding protein	Transgenic: TMV resistant	
Invertase	Tobacco	hexose sensing	OE: SAR constitutive	OE: systemic necrosis, SAR-like resistance to several pathogens

[1] This gene is also described in Table 1. Abbreviations: AS, phenotype of transformants with ectopic antisense expression of the gene of interest; NLS, nuclear localisation signal; OE, phenotype of transformants with ectopic overexpression of the gene of interest; SAR, systemic acquired resistance.

induction phase and secondly as a common mediator leading to the final destruction of the cell (summarized by Jabs, 1999). In *Arabidopsis*, the biological relevance of all the genetic data (e.g. *R* gene products, putative interacting proteins, signaling components, transcription factors, ROI- and NO-generating protein complexes, as well as antioxidative enzymes) can now be tested via identification of insertional mutations in these genes. Huge libraries of *Arabidopsis* lines are becoming available that have insertions of T-DNAs or transposons with a theoretical density of at least one 'hit' per plant gene. This so-called 'gene machine' can be screened by polymerase chain reaction (PCR) technologies to identify the lines with an insertional mutation in the gene of interest (Wisman *et al.*, 1998). In the near future, reverse genetics will dramatically speed up the process of identification and classification of HR and plant defense signal transduction components.

V. Mechanisms leading to membrane damage and cell collapse

In animal cells undergoing genetically determined cell death programs a group of cysteine proteases called caspases (see section IV) help to dismantle the cells. In animals the need to destroy, dissemble and remove cells in an orderly fashion to prevent an undesirable inflammation response, similar to that which ensues after tissue damage, is clear. However, plants tend to be more tolerant of diseased or necrotic tissues and cells than animals (e.g. huge stem cankers in trees are essentially 'ignored' and can be seen in trees which attain a very old age). Plants do not have an immune system with circulating antibodies and thus do not show inflammation as we know it in animals. Thus, the cell dismantling process during PCD in plants might be expected to show some differences to that seen in animals. Up to now a group of proteases equivalent to caspases has not yet been identified in plants and how the cells die during the HR is not known with certainty. A recent report (Solomon *et al.*, 1999) provides evidence that cysteine proteases might be involved in regulating PCD in plants but unlike caspases, these proteases do not cleave after an aspartic acid residue and thus have different substrate specificity to the caspases involved in PCD in animals.

One idea put forward to explain HR cell death is that the vacuole functions like a lysosome in animal cells. The plant cell vacuole has many hydrolytic enzymes (Boller and Kende, 1979) which, after liberation due to destruction of the tonoplast, could be responsible for degrading the cytoplasmic contents. The emphasis in plant systems would then be on the programmed initiation of the cell death under the required circumstances, whereas the cell destruction process itself might be less coordinated than in animal cells. In this scenario membranes are the primary targets of the genetically programmed events and lysosomal enzymes, already present in the plant cells, finish off the cell destruction process. This attractive hypothesis for the HR would fit in nicely with the well known lysosomal concept of the plant cell vacuole (Matile, 1975; 1987). Membrane damage is one of the prominent features associated with cells undergoing HR (see section II). Much research into the HR has concentrated on events at or in membranes to elucidate mechanisms which might lead to membrane destruction.

However, in other cases some of the typical features of animal cell apoptosis have been observed in plant cells undergoing an HR. Thus, in cells of cowpea (*Vignia unguiculata*) undergoing an HR with *Uromyces vignae*, cleavage of nuclear DNA into oligonucleosomal fragments has been reported (Ryerson and Heath, 1996). This cleavage gave rise to DNA-ladders on DNA gels which is also a typical characteristic for animal cells undergoing apoptosis (see Box 3). Although cyanide also caused DNA-laddering in cowpea, physical damage did not (Ryerson and Heath, 1996). This latter observation would rather speak against the mechanism of hypersensitive cell death being due to the simple liberation of hydrolytic enzymes from the vacuole after the tonoplast becomes damaged. Nevertheless, membrane damage remains one of the earliest characteristics of HR cells and it is important to consider mechanisms which might be activated to bring about that damage.

During the HR, membrane lipid peroxidation occurs and leads to the liberation from the leaves of several volatile substances synthesized ultimately from fatty acid hydroperoxides (Croft *et al.*, 1993). These are easily measured and some, e.g. *E*-2-hexenal, are potently antimicrobial (ibid), while others, e.g. ethane, are regarded as a specific index for membrane damage via peroxidation (Riely *et al.*, 1974). Lipid peroxidation can occur by both enzymic, via lipoxygenase (LOX), and non-enzymic means via the action of selected ROI and organic free radicals (Thompson *et al.*, 1987). Lipoxygenase (EC 1.13.11.12) will oxidize unsaturated fatty acids that have a cis-1,4-pentadiene moiety, e.g. linoleic (18:2) and linolenic (18:3) acids, both of which are common in plant membranes. Because free fatty acids are better substrates for some LOX isoforms than fatty acids esterified in membrane lipids, LOX might work in conjunction with lipolytic acyl hydrolase, or a more specific phospholipase activity to cause membrane damage. However, recently there have been several reports of specialized LOX isoforms acting directly on fatty acids esterified in membranes (e.g. Maccarrone *et al.*, 1994; Vianello *et al.*, 1995) and an elicitor-induced LOX isoform was reported which shows preferential activity against membrane phospholipids rather than free fatty acids (Kondo *et al.*, 1993). Indeed, it has been postulated that LOX may play an important role in the 'remodeling' and degradation of plant membranes in situ as circumstances, including pathogen and stress, dictate (Macré *et al.*, 1994). LOX activity has been shown to increase in resistant tissues for several host-pathogen combinations and the potential role of LOX in causing HR-associated membrane damage was recently reviewed (Slusarenko, 1996). Essentially, one hypothesis is that the requirement for active host metabolism during the induction phase of the HR is, at least in part, to provide the membrane degrading potential due to LOX. In this regard, recently published experiments with antisense *LOX* constructs transformed into tobacco are perhaps very significant (Rancé *et al.*, 1998; see also Table 4). The antisense plants showed reduced levels of pathogen-induced LOX activity and a normally avirulent race 0 isolate of *Phytophthora infestans* was able to colonize the plants and cause disease. Although resistance is normally associated with HR in this pathosystem the authors do not go as far as categorically stating that HR cell collapse is prevented in the antisense plants and unfortunately do not show any microscopical details of infection sites. Thus, although transformed from resistance to susceptibility, it remains unclear what effect the antisense *LOX* expression has on the HR phenotype.

Alternatively, the lipid peroxidation observed during HR could result from ROI or organic free radicals (Slusarenko *et al.*, 1991; Slusarenko, 1996). At this point it is important to consider the chemistry of ROI produced during the oxidative burst because neither H_2O_2 nor O_2^- are reactive enough to peroxidize fatty acids in their own right (Aikens and Dix, 1993). Since these are the two most important ROI produced in the pathogen-induced oxidative burst there is a requirement to explain how H_2O_2 or O_2^- might be invoked to promote lipid peroxidation. Superoxide anion production has been linked with membrane damage and *in vivo* toxicity in numerous instances (Wolff *et al.*, 1986). It has often been postulated that the deleterious effects of O_2^- result from its conversion to OH$^\bullet$ radicals in the presence of iron as a catalyst, via H_2O_2 resulting from dismutation of O_2^-. Whether iron is available at the site of O_2^- production in the oxidative burst is not known but experiments with DMSO could not detect any OH$^\bullet$ radical during a bacterially-induced HR in cucumber (Popham and Novacky, 1991). The availability of iron in the plant cell wall and at the plasmalemma under normal physiological conditions is a factor which needs investigating. However, protonation of O_2^- to give its conjugate acid, the perhydroxyl radical (HO$_2^\bullet$, pK 4.8), can occur in cellular microenvironments where there are locally high concentrations of protons, for example at the negatively charged surfaces of membranes (Bielski *et al.*, 1983; Fridovich, 1988). The perhydroxyl radical is a strong oxidizing agent and will react with polyunsaturated fatty acids directly (Bielski *et al.*, 1983; Sutherland, 1991). Thus, the perhydroxyl radical is perhaps the most biologically relevant ROI potentially involved in membrane lipid peroxidation.

Regarding a potential role for organic free radicals, a very interesting hypothesis has recently been advanced suggesting that SA radicals might cause lipid peroxidation in vivo. SA accumulates in tissues undergoing resistance responses. SA can donate an electron to catalase or peroxidase and in doing so becomes a free radical because it is left with an unpaired electron. SA treatment of plants leads to lipid peroxidation and lipid peroxides were found to activate *PR* genes in tobacco cells (Anderson *et al.*, 1998). The activation of *PR* genes by SA was inhibited by diethyldithiocarbamic acid, a compound that converts lipid peroxides into their hydroxyl derivatives. The authors suggested that the lipid peroxidation effects due to SA radicals might be involved in causing HR necrosis (Anderson *et al.*, 1998). The potential roles of SA in contributing to cell death are discussed in more detail in Box 4.

VI.   Relation of the HR to resistance and to other plant defense responses

It is quite clear that resistance to disease can occur without an HR, even where the pathogen is an obligate biotroph. For example, the *mlo*-determined resistance of barley to *Blumeria graminis* pv. *hordei* is determined by the speed of papilla formation at fungal penetration sites (see chapter 1), even though the *mlo* gene is associated with a lesion mimic phenotype in older leaf tissue (see section IV; Table 1). Similarly, resistance of soybean cv. Clark 63 to pathogenic strains of *Xanthomonas campestris* pv. *glycinea* is not associated with an HR. Virulent and avirulent bacterial isolates multiply at the same rate in the tissues, but symptoms of infection do not develop in the 'incompatible' combination (Fett, 1984).

Box 4.

SA and the HR

Salicylic acid (SA) is a simple phenolic acid which has recently become the focus of much research into cell signaling in response to pathogen attack. It was first postulated that SA might be the systemic signal which moves out of the stimulated leaf and is transported in the phloem to co-ordinate systemic acquired resistance (SAR) in the rest of the plant (Malamy *et al.*, 1990; Métraux *et al.*, 1990). Subsequently it was shown that, although essential for the expression of SAR, SA is not the systemic signal (Rasmussen *et al.*, 1991). Transgenic *Arabidopsis* plants, which contained a gene (*NahG*; see also Table 4) for SA catabolism from a *Pseudomonas* bacterium and were thus unable to accumulate SA, were also rendered highly susceptible to genetically avirulent isolates of *Peronospora parasitica* (Delaney *et al.*, 1994). This showed, for the first time, that SA also played a role in genetically determined incompatibility and not just in SAR-related phenomena. The hypersensitive response, which is normally associated with incompatibility was completely suppressed, at least in the *avr/R* gene combinations studied (Delaney *et al.*, 1994). Additionally, it was shown that the spontaneous lesion-building phenotype of certain *lsd* mutants of *Arabidopsis* (e.g. *lsd6*) depended upon SA (Weymann *et al.*, 1995). This has led to the suggestion of an 'amplification' or 'feed-back loop' role for SA in the cell death signaling pathway (Fig. 2; Dangl *et al.*, 1996). The inference that SA might be necessary for HR cell death is supported by studies with the *Arabidopsis/ Peronospora* pathosystem using 2-aminoindan-2-phosphonic acid (AIP), an inhibitor of PAL activity. AIP treatment caused a transition from HR-associated resistance to complete susceptibility shown by substantial conidiophore and oospore production (Mauch-Mani and Slusarenko, 1996). This susceptible condition could be reverted to the resistant state by complementation with SA. PAL catalyses the deaminiation of phenylalanine to cinnamic acid which in turn can be converted via β-oxidation to benzoic acid, the direct precursor of SA (Yalpani *et al.*, 1993; Ribnicky *et al.*, 1998). Recent work suggests that the induction of SA synthesis during the first non-specific oxidative burst (see section II and III) might serve to potentiate the HR, at least in suspension-cultured cells (Draper, 1997; Shirasu *et al.*, 1997).

In addition to a role in cell signaling it has been suggested that SA-radical-induced lipid peroxidation might participate in producing the cell death associated with the HR, at least in tobacco (Anderson *et al.*, 1998). However, the role of SA in regulating HR-associated cell death is still unclear and requires further investigation.

Another interesting example is the behavior of a new class of *Arabidopsis* mutants, designated *dnd* (defense, no death). Mutant *dnd1* plants, carrying the *R* genes *Rpm1* or *Rps2*, do not mount an HR after inoculation with strains of *P. syringae* expressing the corresponding *avr* genes *AvrRpm1* or *avrRpt2*, respectively, but retain characteristic responses to avirulent *P. syringae* such as induction of *PR* gene expression and strong restriction of the pathogen (Yu *et al.*, 1998). Interestingly, *dnd1* mutants also exhibit enhanced resistance against a broad spectrum of normally virulent fungal, bacterial and viral plant pathogens, but this remained distinguishable from the gene-for-gene resistance mediated by *RPM1* and *RPS2*. This mutant phenotype may imply that the HR is not always essential for *avr/R* gene-specific resistance. However, the constitutive PR gene expression and enhanced levels of SA in the *dnd1* mutant, two signs of SAR activation, might effectively substitute for HR cell death and lead to pathogen resistance. Therefore, one has to interpret the physiological and pathological behavior of mutant plants very cautiously and precisely before drawing conclusions.

Additionally, even when an HR has occurred resistance is not necessarily guaranteed. Thus, when Samsun-NN tobacco plants (homozygous for the *N* resistance gene) are inoculated with an avirulent strain of tobacco mosaic virus (TMV) at 22°C, an HR occurs and the virus is normally limited to the local lesion. However, if the plants are shifted to 28°C the virus breaks out of the local HR lesions and a systemic infection ensues. In spite of these observations, the HR is a very common plant response accompanying the unsuccessful attempt of a pathogen to colonize a plant. There are two important questions to be considered:

1) *How does the HR contribute to resistance as such?*

and

2) *What is the relation of the HR to other defense responses?*

In considering the first question it is important to remember that the plant cells die as a result of the HR. For obligate biotrophs, which need to establish a highly regulated nutritional relationship with living host cells in order to survive, it is quite clear that the rapid death of host cells in an HR could be sufficient to explain resistance. Interestingly, necrotrophs, for example soft-rot-causing *Erwinia* spp. tend not to induce an HR.

An interesting category of pathogens is the facultative biotrophs, i.e. pathogens that enter into a biotrophic relationship with the host for part of the disease cycle but are also capable of existing in necrotic tissue at later stages of the infection e.g. *Phytophthora infestans*. With these pathogens the answer to the first question leads directly into the second question because it is not immediately clear that death of host cells as such would be a hinderence to colonisation. Thus, the contribution of cell death to resistance would be expected to be minor. However, even if the end result of necrosis in the HR is not important in stopping facultative biotrophs, it may be that the programmed sequence of events occuring in the HR is important in coordinating or setting in motion other defense responses which are effective. What can certainly be said is that the HR, where it occurs, is generally a very good marker for resistance and thus the assumption that it is causally linked with resistance is not improbable.

The HR occurs in parallel with a number of other changes in the host's biochemistry which are viewed as defense responses (see chapters 5, 7 and 8) and the second question relates to the positioning of defense responses in chains of dependence or as largely independent mechanisms. Thus, it has been proposed that the HR, and here is meant the process of the genetically determined PCD rather than the dead cell as an end product, might be tied to the activation of other defense pathways such as, for example, phytoalexin biosynthesis. Bailey (1982) observed that host cell injury, often leading to host cell death, was a common denominator in the action of biotic and abiotic elicitors, and pathogen challenge, which preceded accumulation of phytoalexins. In compatible interactions the rapid accumulation of phytoalexins was not observed, at least in the initial biotrophic phase where gross cell injury was avoided. Bailey (1982) postulated that injured cells released constitutive elicitors into the surrounding healthy cells, which responded by synthesizing phytoalexins. Dead cells also act as a sink for phytoalexin accumulation (Hargreaves and Bailey, 1978).

In some model systems, for example bean cell-suspension cultures treated with elicitor, transcriptional activation of the genes involved in phytoalexin biosynthesis was very rapid and occurred within 5-10 min of elicitor treatment (Templeton and Lamb, 1988). Maximum steady state transcript levels occurred at 3-4 h after treatment with elicitor. This would seem to argue against Bailey's hypothesis. However, in more natural infections involving whole plant tissues, events proceed at a more leisurely pace. Thus, when French bean hypocotyls are inoculated with conidia of *Colletotrichum lindemuthianum*, a 30-40 h period is required for spore germination, infection peg production, and penetration of the host to occur. Accumulation of phenylalanine ammonia-lyase (PAL) and chalcone synthase (CHS) transcripts were first detected 39 h after inoculation, with maximum steady state levels of mRNA observed between 70 and 75 h after inoculation (Cramer *et al.*, 1985; Lamb *et al.*, 1986). Additionally, in bean leaves infiltrated with *Pseudomonas syringae* pv. *phaseolicola*, accumulation of PAL and CHS transcripts was observed in the directly inoculated region within 3 h and chitinase transcripts within six hours in the incompatible combination (Meier *et al.*, 1993). HR cell collapse occurred approximately 18 h after inoculation and thus the increased steady state levels of some defense gene transcripts clearly precedes cell collapse. However, phytoalexins accumulated in the leaves from 24 to 48 h after inoculation (ibid) and the relation of the HR genetic program as such to the activation of genes leading to phytoalexin synthesis remains unclear.

Experiments with *hrp* mutants of phytopathogenic bacteria have shown that certain defense responses, which are normally induced in association with an HR, can be induced without HR occurring. Thus, in bean plants (*Phaseolus vulgaris*) transcripts of genes involved in phytoalexin biosynthesis and transcripts for the hydrolytic enzyme chitinase accumulated after inoculation with either *Pseudomonas syringae* pv. *tabaci* (HR), heat- or antibiotic-killed wild-type cells or an *hrp* mutant derived from the wild-type (Jakobek and Lindgren, 1993). In addition the authors showed that phytoalexins accumulated by 8 hours after inoculation with the wild-type, the *hrp* mutant, antibiotic-killed cells of the wild-type and even cells of the saprophyte *Pseudomonas fluorescens*. As described earlier in this chapter, dead bacteria, saprophytic bacteria and *hrp* mutants do not induce an HR, thus suggesting that the HR can effectively be uncoupled from these defense responses. However, using the same host, Meier *et al.* (1993, see also preceding paragraph) found no induction of phytoalexin accumulation and only weak accumulation of transcripts for CHS and chitinase after inoculation with *Pseudomonas fluorescens* in contrast to the HR induced by an avirulent isolate of *P. s.* pv. *phaseolicola*. In this context it also interesting that in soybean inoculated with an avirulent race of *Xanthomonas campestris* pv. *glycinea* at an inoculum concentration just below that required for confluent HR cell death, no phytoalexin accumulation was observed, whereas with slightly higher inoculum levels and confluent HR, phytoalexins were easily detectable (Wyman and Van Etten, 1982). Thus, the separation of appreciable amounts of phytoalexin synthesis from the HR is not completely unequivocal.

Other defense phenomena, such as papilla synthesis, and the localized generation of H_2O_2 in lettuce cell walls have also been induced by *hrp* mutants, but membrane damage associated with electrolyte leakage and phytoalexin accumulation were not observed (Bestwick *et al.*, 1995, 1997; Brown *et al.*, 1998).

It is clear that not all defense responses are dependent upon the initiation of the HR signaling pathway and are independently regulated. This also makes evolutionary sense by avoiding putting all the defensive 'eggs' in one basket. However, the relationship of the HR to other plant defense responses is at present only poorly understood and hopefully more light will be thrown on this question by studies with mutants in the cell death pathway and other mutants with lesions in specific defense pathways for example the phytoalexin-deficient *pad* mutants in *Arabidopsis*.

VII.   Open questions

When reading the preceding sections it becomes clear that, while we know a lot about what is going on in plant cells during the HR, there are still a lot of open questions and black boxes. For bacterially induced HR, the recent discovery of type III secretion of Avr proteins directly into host cells was an enormous leap which solved many previous riddles about Avr function. How recognition occurs with fungal and viral avirulence determinants can be speculated upon but is not yet known. Signal transduction subsequent to recognition, in terms of ion fluxes, oxidative burst, cyclic nucleotides, phosphorylation of target proteins, NO and SA synthesis, is also becoming clearer. That the HR is a form of PCD is now widely accepted. Exactly which genes, in terms of positive and negative regulators, are involved, and what the mechanism of cell death is, are not yet known. In earlier years irreversible membrane damage was considered of central importance in HR cell death. Whether the emphasis will shift to searches for caspase-like proteases and DNases associated with HR cell collapse is unclear. With respect to mechanisms causing membrane damage it is still unclear whether an enzymic-, ROI- or an organic radical-dependent process, or some combination of these, is most important. In the case of ROI involvement, it would be useful to know the availability of iron in the cell wall or plasmalemma. How the genetic program leading ultimately to host cell collapse in the HR is integrated with the regulation of other defense mechanisms is also still unclear.

Acknowledgments

Studies by the authors discussed in the present chapter were supported by the Deutsche Forschungsgemeinschaft (DFG Grants JA-830/2-1 and Sl 30/1-1). Thanks are due to Profs. Ulla Bonas and John Mansfield for providing us with material prior to publication.

References

Aikens J and Dix TA (1993) Hydrodioxyl (perhydroxyl), peroxyl, and hydroxyl radical-Initiated lipid peroxidation of large unilamellar vesicles (liposomes): comparative and mechanistic studies. Arch Biochem Biophys 305: 516—525

Alfano JR, Bauer DW, Milos TM and Collmer A (1996) Analysis of the role of the *Pseudomonas syringae* pv. *syringae* HrpZ harpin in elicitation of the hypersensitive response in tobacco using functionally nonpolar deletion mutations, truncated HrpZ fragments and *hrmA* mutations. Mol Microbiol 19: 715—728

Alvarez ME and Lamb C (1997) Oxidative burst-mediated defense responses in plant disease resistance. In: Scandelios JG (ed) Oxidative Stress and the Molecular Biology of Antioxidant Defenses, pp 815—839. Cold Spring Harbor Laboratory Press, New York

Anderson MD, Chen Z and Klessig DF (1998) Possible involvement of lipid peroxidation in salicylic acid-mediated induction of *PR*-1 gene expression. Phytochemistry 47: 555—566

Atkinson MM and Baker CJ (1987) Association of host plasma membrane K^+/H^+ exchange with multiplication of *Pseudomonas syringae* pv. *syringae* in *Phaseolus vulgaris.* Phytopathology 77: 1273—1279

Atkinson MM and Baker CJ (1989) Role of plasmalemma H^+-ATPase in *Pseudomonas syringae*-induced pv. *syringae* in K^+/H^+ exchange in suspension-cultured tobacco cells. Plant Physiol 91: 298—303

Atkinson MM, Keppler LD, Orlandi EW, Baker CJ and Mischke CF (1990) Involvement of plasma membrane calcium influx in bacterial induction of the K^+/H^+ and hypersensitive responses in tobacco. Plant Physiol 92: 215—221

Atkinson M, Bina J and Sequeira L (1993) Phosphoinositide breakdown during the K^+/H^+ exchange response of tobacco to *Pseudomonas syringae* pv. *syringae.* Mol Plant-Microbe Interact 6: 253—260

Bailey JA (1982) Physiological and biochemical events associated with the expression of resistance to disease. In: Wood RKS (ed) Active Defence Mechanisms in Plants, NATO ASI Series, pp 39—65. Plenum Press, New York

Baker CJ, O'Neill N, Keppler D and Orlandi EW (1991) Early responses during plant-bacteria interactions in tobacco cell suspensions. Phytopathology 81: 1504—1507

Baker B, Zambryski P, Staskawicz B and Dinesh-Kumar SP (1997) Signaling in plant-microbe interactions. Science 276: 726—733

Bestwick CS, Bennett MH and Mansfield JW (1995) Hrp mutant of *Pseudomonas syringae* pv. *phaseolicola* induces cell wall alterations but not membrane damage leading to the hypersensitive reaction in lettuce. Plant Physiol 108: 503—516

Bestwick CS, Brown IR, Bennett MH and Mansfield, JW (1997) Localization of hydrogen peroxide accumulation during the hypersensitive reaction of lettuce cells to *Pseudomonas syringae* pv *phaseolicola.* Plant Cell 9: 209—221

Bielski BHJ, Arudi RL and Sutherland MW (1983) A study of the reactivity of $HO_2^{\bullet}/O_2^{-}$ with unsaturated fatty acids. J Biol Chem 258: 4759—4761

Bogdanove AJ, Beer SV, Bonas U, Boucher CA, Collmer A, Coplin DL, Cornelis GR, Huang, H-C, Hutcheson, SW, Panopoulos, NJ and Van Gijsegem, F (1996) Unified nomenclature for broadly conserved *hrp* genes of phytopathogenic bacteria. Mol Microbiol 20: 681—683

Bokoch GM (1994) Regulation of the human neutrophil NADPH oxidase by the Rac GTP-binding proteins. Curr Opin Cell Biol 6: 212—218

Boller T and Kende H (1979) Hydrolytic enzymes in the central vacuole of plant cells. Plant Physiol

63: 1123—1132

Bolwell GP, (1999) Role of active axygen species and NO in plant defence responses. Curr Op Plant Bio 2: 287—294

Bolwell GP and Wojtaszek P (1997) Mechanisms for the generation of reactive oxygen species in plant defence - a broad perspective. Physiol Mol Plant Pathol 51: 347—366

Bolwell GP, Butt VS, Davies DR and Zimmerlin A (1995) The origin of the oxidative burst in plants. Free Rad Res 23: 517—532

Brisson LF, Tenhaken R and Lamb C (1994) Function of oxidative cross-linking of cell wall structural proteins in plant disease resistance. Plant Cell 6: 1703—1712

Brown IR, Trethowan J, Kerry M, Mansfield J and Bolwell GP (1998) Localization of components of the oxidative cross-linking of glycoproteins and of callose synthesis in papillae formed during the interaction between non-pathogenic strains of Xanthomonas campestris and French bean mesophyll cells. Plant J 15: 333—343

Bushnell WR (1981) Incompatibility conditioned by the Mla gene in powdery mildew of barley: the halt of cytoplasmic streaming. Phytopathology 71: 1062—1066

Cao H, Bowling SA, Gordon S and Dong X (1994) Characterization of an Arabidopsis mutant that is nonresponsive to inducers of systemic acquired resistance. Plant Cell 6: 1583—1592

Cao H, Glazebrook J, Clarke JD, Volko S and Dong X (1997) The Arabidopsis NPR1 gene that controls systemic acquired resistance encodes a novel protein containing ankyrin repeats. Cell 88: 57—63

Century KS, Holub EB and Staskawicz BJ (1995) NDR1, a locus of Arabidopsis thaliana that is required for disease resistance to both a bacterial and a fungal pathogen. Proc Natl Acad Sci USA 92: 6597—6601

Chamnongpol S, Willekens H, Langebartels C, Van Montagu M, Inzé D and Van Camp W (1996) Transgenic tobacco with a reduced catalase activity develops necrotic lesions and induces pathogenesis-related expression under high light. Plant J 10: 491—503

Chen CY and Heath MC (1991) Cytological studies of the hypersensitive death of cowpea epidermal cells induced by basidiospores-derived infection by the cowpea rust fungus. Can J Bot 69: 1199—1206

Chen Z, Silva H and Klessig DF (1993) Active oxygen species in the induction of plant systemic acquired resistance by salicylic acid. Science 262: 1883—1886

Clarke JD, Liu Y, Klessig DF and Dong X (1998) Uncoupling PR gene expression from NPR1 and bacterial resistance: characterization of the dominant Arabidopsis cpr6-1 mutant. Plant Cell 10: 557—569

Collinge DB and Slusarenko A (1987) Plant gene expression in response to pathogens. Plant Mol Biol 9: 389—410

Collmer A (1998) Determinants of pathogenicity and avirulence in plant pathogenic bacteria. Curr Opin Plant Biol 1: 329—335

Conklin PL, Williams EH and Last RL (1996) Environmental stress sensitivity of an ascorbate-deficient Arabidopsis mutant. Proc Natl Acad Sci USA 93: 9970—9974

Cook AA and Stall RE (1968) Effect of Xanthomonas vesicatoria on loss of electrolytes from leaves of Capsicunum annuum. Phytopathology 58: 617—619

Cramer CL, Bell JN, Ryder TB, Bailey JA, Schuch W, Bolwell GP, Robbins MP, Dixon RA and Lamb CJ (1985) Co-ordinated synthesis of phytoalexin biosynthetic enzymes in biologically-stressed cells of bean (Phaseolus vulgaris L.). EMBO J 4: 285—289

Croft KPC, Jüttner F and Slusarenko AJ (1993) Volatile products of the lipoxygenase pathway evolved from Phaseolus vulgaris (L.) leaves inoculated with Pseudomonas syringae pv. phaseolicola. Plant Physiol 101: 13—24

Dangl JL, Ritter C, Gibbon MJ, Mur LAJ, Wood JR, Goss S, Mansfield J, Taylor JD and Vivian A (1992) Functional homologs of the Arabidopsis *RPM1* disease resistance gene in bean and pea. Plant Cell 4: 1359—1369

Dangl JL, Dietrich RA and Richberg MH (1996) Death don't have no mercy: cell death programs in plant-microbe interactions. Plant Cell 8: 1793—1807

Davis D, Merida J, Legendre L, Low PS and Heinstein P (1993) Independent elicitation of the oxidative burst and phytoalexin formation in cultured plant cells. Phytochemistry 32: 607—611

Deighton N, Lyon GD, Johnston D, Glidwell SM and Goodman BA (1994) Are free radical generation and phytoalexin biosynthesis coupled? Proc Roy Soc Edin 102B: 253—255

Delaney TP, Friedrich L and Ryals JA (1995) *Arabidopsis* signal transduction mutant defective in chemically and biologically induced disease resistance. Proc Natl Acad Sci USA 92: 6602—6606

Desikan R, Hancock JT, Coffey MJ and Neill SJ (1996) Generation of active oxygen in elicited cells of *Arabidopsis thaliana* is mediated by a NADPH oxidase-like enzyme. FEBS Lett 382: 213—217

Delaney TP, Uknes S, Vernooij B, Friedrich L, Weymann K, Negrotto D, Gaffney T, Gut-Rella M, Kessmann H, Ward ER and Ryals J (1994) A central role of salicylic acid in plant disease resistance. Science 266: 1247—1249

Delledonne M, Xia Y, Dixon RA and Lamb CJ (1998) Nitric oxide functions as a signal in plant disease resistance. Nature 394: 585—588

Dietrich RA, Delaney TP, Uknes SJ, Ward ER, Ryals JA and Dangl JL (1994) *Arabidopsis* mutants simulating disease resistance response. Cell 77: 565—577

Dietrich RA, Richberg MH, Schmidt R, Dean C and Dangl JL (1997) A novel zinc finger protein is encoded by the *Arabidopsis LSD1* gene and functions as a negative regulator of plant cell death. Cell 88: 685—694

Doke N and Ohashi Y (1988) Involvement of an O_2—generating system in the induction of necrotic lesions on tobacco leaves infected with tobacco mosaic virus. Physiol Mol Plant Pathol 32: 163—175

Dong X (1998) SA, JA, ethylene, and disease resistance in plants. Curr Opin Plant Biol 1: 316—323

Dorey S, Kopp M, Geoffroy P, Fritig B and Kauffmann S (1999) Hydrogen peroxide from the oxidative barst is neither necessary nor sufficient for hypersensitive cell death induction, phenyl alanine ammonia lyase stimulation, salicylic acid accumulation, or scopoletin consumption in cultured tobacco cells treated with elicitin. Plant Physiol 121: 163—171

Draper J (1997) Salicylate, superoxide synthesis and cell suicide in plant defence. Trends Plant Sci 2: 162—165

Durner J and Klessig DF (1996) Salicylic acid is a modulator of tobacco and mammalian catalases. J Biol Chem 271: 28492—28501

Durner J and Klessing DF (1999) Nitric oxide as a signal in plants. Curr Op Plant Biol 2: 369—374

Durner J, Wendehenne D and Klessig DF (1998) Defense gene induction in tobacco by nitric oxide, cyclic GMP, and cyclic ADP-ribose. Proc Natl Acad Sci USA 95: 10328—10333

Dwyer SC, Legendre L, Low PS and Leto TL (1996) Plant and human neutrophil oxidative burst complexes contain immunologically related proteins. Biochim Biophys Acta 1289: 231—237

Ebel J, Bhagwat AA, Cosio EG, Feger M, Kissel U, Mith fer A and Waldmüller T (1995) Elicitor-binding proteins and signal transduction in the activation of a phytoalexin defense response. Can J Bot 73: S506—S510

Ellis RE, Yuan J and Horwitz HR (1991) Mechanisms and functions of cell death. Annu Rev Cell Biol 7: 663—698

Enari M, Sakahira H, Yokoyama H, Okawa K, Iwamatsu A and Nagata S (1998) A caspase-activated DNase that degrades DNA during apoptosis, and its inhibitor ICAD. Nature 391: 43—50

Felix G, Regenass M, Spanu P and Boller T (1994) The protein phosphatase inhibitor calyculin A mimics elicitor action in plant cells and induces rapid hyperphosphorylation of specific proteins as revealed by pulse labeling with [^{33}P]phosphate. Proc Natl Acad Sci USA 91: 952—956

Fett WF (1984) Accumulation of isoflavonoids and isoflavone glucosides after inoculation of soybean leaves with *Xanthomonas campestris* pv. *glycinea* and pv. *campestris* and a study of their role in resistance. Physiol Mol Plant Pathol 28: 67—77

Freialdenhoven A, Scherag B, Hollricher K, Collinge DB, Thordal-Christensen H and Schulze-Lefert P (1994) *Nar-1* and *Nar-2*, two loci required for *Mla$_{12}$*-specified race-specific resistance to powdery mildew in barley. Plant Cell 6: 983—994

Freialdenhoven A, Peterhänsel C, Kurth J, Kreuzaler F and Schulze-Lefert P (1996) Identification of genes required for the function of non-race-specific *mlo* resistance to powdery mildew in barley. Plant Cell 8: 5—14

Freytag S, Arabatzis N, Hahlbrock K and Schmelzer E (1994) Reversible cytoplasmic rearrangements precede cell wall apposition, hypersensitive cell death and defense-related gene activation in potato/*Phytophthora infestans* interactions. Planta 194: 123—135

Fridovich I (1988) The biology of oxygen radicals. In: Halliwell B (ed) Oxygen Radicals and Tissue Injury, Proceedings of a Brook Lodge Symposium, pp 1—5. The Upjohn Company, Bethesda, Maryland

Gabriel DW (1986) Specificity and gene function in plant-pathogen interactions. American Soc Microbiol News 52: 19—25

Gallois P, Makishima T, Hecht V, Despres B, Laudié M, Nishimotot T and Cooke R (1997) An *Arabidopsis thaliana* cDNA complementing a hamster apoptosis suppressor mutant. Plant J 11: 1325—1331

Glazebrook J and Ausubel FM (1994) Isolation of phytoalexin-deficient mutants of *Arabidopsis thaliana* and characterization of their interactions with bacterial pathogens. Proc Natl Acad Sci USA 91: 8955—8959

Glazebrook J, Zook M, Mert F, Kagan I, Rogers EE, Crute IR, Holub EB, Hammerschmidt R and Ausubel FM (1997) Phytoalexin-deficient mutants of *Arabidopsis* reveal that *PAD4* encodes a regulatory factor and that four *PAD* genes contribute to downy mildew resistance. Genetics 146: 381—392

Glazener JA Orlandi, EW and Baker CJ (1996) The active oxygen response of cell suspensions to incompatible bacteria is not sufficient to cause hypersensitive cell death. Plant Physiol 110: 759—763

Goodman RN (1968) The hypersensitive reaction in tobacco: a reflection of changes in host cell permeability. Phytopathology 58: 872—873

Gough CL, Genin S, Lopes V and Boucher CA (1993) Homology between the HrpO protein of *Pseudomonas solanacearum* and bacterial proteins implicated in a signal peptide-independent secretion mechanism. Mol Gen Genet 239: 378—392

Gray J, Close PS, Briggs SP and Johal GS (1997) A novel suppressor of cell death in plants encoded by the *Lls1* gene of maize. Cell 89: 25—31

Greenberg JT (1997) Programmed cell death in plant-pathogen interactions. Annu Rev Plant Physiol 48: 525—545

Greenberg JT, Guo A, Klessig DF and Ausubel FM (1994) Programmed cell death in plants: a pathogen-triggered response activated coordinately with multiple defense functions. Cell 77: 551—563

Grimm C, Aufsatz W and Panopoulos NJ (1995) The *hrpRS* locus of *Pseudomonas syringae* pv. *phaseolicola* constitutes a complex regulatory unit. Mol Microbiol 15: 155—165

Groom QJ, Torres MA, Fordham-Skelton AP, Hammond-Kosack KE, Robinson NJ and Jones JDG (1996) *rbohA*, a rice homologue of the mammalian *gp91phox* respiratory burst oxidase gene. Plant J

10: 515—522

Großkopf DG, Felix G, and Boller T (1990). K-252a inhibits the response of tomato cells to fungal elicitors *in vivo* and their microsomal protein kinase *in vitro*. FEBS Lett 275: 177—180

Gustine DL, Sherwood RT, Lukezic FL, Moyer BG and Devlin WS (1994) Metabolites of *Pseudomonas corrugata* that elicit plant defense reactions. In: Hedin PA (ed) Bioregulators for Crop Protection and Pest Control, ACS Symposium Series 557, pp 169—181. American Chemical Society, Washington

Halliwell B and Gutteridge JMC (1989) Free Radicals in Biology and Medicine. Clarendon Press, Oxford

Hammond-Kosack KE and Jones JDG (1994) Identification of two genes required in tomato for full *Cf-9*-dependent resistance to *Cladosporium fulvum*. Plant Cell 6: 361—374

Hammond-Kosack KE and Jones JDG (1996) Resistance gene-dependent plant defense responses. Plant Cell 8: 1773—1791

Harder DE and Chong J (1991) Rust haustoria. In: Mendgen K and Leseman DE (eds) Electron Microscopy of Plant Pathogens, pp 235—250. Springer, Berlin

Harding SA and Roberts DM (1998) Incompatible pathogen infection results in enhanced reactive oxygen and cell death responses in transgenic tobacco expressing a hyperactive mutant calmodulin. Planta 206: 253—258

Hargreaves JA and Bailey JA (1978) Phytoalexin production by hypocotyls of *Phaseolus vulgaris* in response to constitutive metabolites released by damaged bean cells. Physiol Plant Pathol 13: 89—100

He SY, Huang H-C and Collmer A (1993) *Pseudomonas syringae* pv. *syringae* harpin$_{Pss}$: a protein that is secreted via the Hrp pathway and elicits the hypersensitive response in plants. Cell 73: 1255—1266

He SY, Bauer DW, Collmer A and Beer SV (1994) Hypersensitive response elicited by *Erwinia amylovora* harpin requires active plant metabolism. Mol Plant-Microbe Interact 7: 289—292

Honée G, Buitink J, Jabs T, De Kloe J, Sijbolts F, Apotheker M, Weide R, Sijen T, Stuiver M and de Wit PJGM (1998) Induction of defense-related responses in Cf9 tomato cells by the AVR9 elicitor peptide of *Cladosporium fulvum* is developmentally regulated. Plant Physiol 117: 809—820

Hortelano S, Dallaporta B, Zamzami N, Hirsch T, Susin SA, Marzo I, Bosca L, Kroemer G (1997) Nitric oxide induces apoptosis via triggering mitochondrial permeability transition. FEBS Lett 410: 373—377

Hu G, Yalpani N, Briggs SP and Johal GS (1998) A porphyrin pathway impairment is responsible for the phenotype of a dominant disease lesion mimic mutant of maize. Plant Cell 10: 1095—1105

Hueck CJ (1998) Type III secretion systems in bacterial pathogens of animals and plants. Microbiol Mol Biol Rev 62: 379—433

Innes RW (1998) Genetic dissection of R gene signal transduction pathways. Curr Opin Plant Biol 1: 229—304

Jabs T (1999) Reactive oxygen intermediates as mediators of programmed cell death in plants and animals. Biochem Pharmacol 57: 231—245

Jabs T, Dietrich RA and Dangl JL (1996) Initiation of runaway cell death in an *Arabidopsis* mutant by extracellular superoxide. Science 273: 1853—1856

Jabs T, Colling C, Tschöpe M, Hahlbrock K and Scheel D (1997) Elicitor-stimulated ion fluxes and O_2^- from the oxidative burst are essential components in triggering defense gene activation and phytoalexin synthesis in parsley. Proc Natl Acad Sci USA 94: 4800—4805

Jacobson MD, Weil M and Raff MC (1997) Programmed cell death in animal development. Cell 88: 347—354

Jakobek JL and Lindgren PB (1993) Generalized induction of defence responses in bean is not correlated with the induction of the hypersensitive reaction. Plant Cell 5: 49—56

Johal GS, Hulbert SH and Briggs SP (1995) Disease lesion mimics of maize: a model for cell death in plants. BioEssays 17: 685—692

Jones AM and Dangl JL (1996) Logjam at the Styx: programmed cell death in plants. Trends Plant Sci 1: 114—119

Kamoun S, van West P, Vleeshouwers VGAA, de Groot KE and Govers F (1998) Resistance of *Nicotiana benthamiana* to *Phytophthora infestans* is mediated by the recognition of the elicitor protein INF1. Plant Cell 10: 1413—1425

Kang S, Sweigard JA and Valent B (1995) The *PWL* host-specificity gene family in the blast fungus *Magnaporthe grisea*. Mol Plant-Microbe Interact 8: 939—948

Kauss H and Jeblick W (1995) Pretreatment of parsley suspension cultures with salicylic acid enhances spontaneous and elicited production of H_2O_2. Plant Physiol 108: 1171—1178

Kearney B and Staskawicz B (1990) Widespread distribution and fitness contribution of *Xanthomonas campestris* avirulence gene *avrBs2*. Nature 346: 385—386

Keen NT, Ersek T, Long M, Bruegger B and Holliday M (1981) Inhibition of the hypersensitive reaction of soybean leaves to incompatible *Pseudomonas* spp. by blastocidin S, streptomycin or elevated temperature. Physiol Plant Pathol 18: 325—337

Keller T, Damude HG, Werner D, Doerner P, Dixon RA and Lamb C (1998) A plant homolog of the neutrophil NADPH oxidase gp91^{phox} subunit gene encodes a plasma membrane protein with Ca^{2+} binding motifs. Plant Cell 10: 255—266

Keppler LD and Novacky A (1986) Involvement of lipid peroxidation in the development of a bacterially induced hypersensitive reaction. Phytopathology 76: 104—108

Keppler LD, Atkinson MM and Baker CJ (1988) Plasma membrane alteration during bacteria-induced hypersensitive reaction in tobacco suspension cells as monitored by intracellular accumulation of fluorescein. Physiol Mol Plant Pathol 32: 209—219

Kerr JFR, Wyllie AH and Currie AR (1972) Apoptosis: a basic biological phenomenon with wide ranging implication in tissue kinetics. Br J Cancer 26: 239—257

Kieffer F, Simon-Plas F, Maume BF and Blein JP (1997) Tobacco cells contain a protein, immunologically related to the neutrophil small G protein Rac2 and involved in elicitor-induced oxidative burst. FEBS Lett 403: 149—153

Kim JF and Beer SV (1998) HrpW of *Erwinia amylovora*, a harpin that contains a domain homologous to pectate lyases of a distinct class. J Bacteriol 180: 5203—5210

Klement Z (1971) Development of the hypersensitivity reaction induced by plant pathogenic bacteria. Proceedings of the 3rd International Conference on Plant Pathogenic Bacteria, Wageningen, pp 157—164

Klement Z (1986) Hypersensitivity. In: Mount MS and Lacy GH (eds) Phytopathogenic Prokaryotes, pp 149—177. Academic Press, New York

Klement Z and Goodman RN (1967) The hypersensitive reaction to infection by bacterial pathogens. Annu Rev Phytopathol 5:17—44

Kluck RM, Bossy-Wetzel E, Green DR and Newmeyer DD (1997a) The release of cytochrome c from mitochondria: a primary site for Bcl-2 regulation of apoptosis. Science 275: 1132—1136

Kluck RM, Martin SJ, Hoffman BM, Zhou JS, Green DR and Newmeyer DD (1997b) Cytochrome c activation of CPP32-like proteolysis plays a critical role in a *Xenopus* cell-free apoptosis system. EMBO J 16: 4639—4649

Kondo Y, Kawai Y, Hayashi T, Ohnishi M, Miyazawa T, Itoh S and Mizutani J (1993) Lipoxygenase in soybean seedlings catalyzes the oxygenation of phospholipid and such activity changes after treatment with fungal elicitor. Biochim Biophys Acta 1170: 301—306

Kroemer G (1997) The proto-oncogene Bcl-2 and its role in regulating apoptosis. Nature Med 3: 614—620

Kroemer G, Petit P, Zamzami N, Vayssiere JL and Mignotte B (1995) The biochemistry of programmed cell death. Faseb J 9: 1277—1287

Kroemer G, Zamzani N and Susin SA (1997) Mitochondrial control of apoptosis. Immunol Today 18: 44—51

Kuwana T, Smith JJ, Muzio M, Dixit V, Newmeyer DD and Kornbluth S (1998) Apoptosis induction by caspase-8 is amplified through the mitochondrial release of cytochrome c. J Biol Chem 273: 16589—16594

Lamb CJ and Dixon RA (1997) The oxidative burst in plant disease resistance. Annu Rev Plant Physiol Plant Mol Biol 76: 419—422

Lamb CJ, Corbin DR, Lawton M, Sauer N and Wingate VPM (1986) Recognition and response in plant-pathogen interactions. In: Lugtenberg B (ed) Recognition in Microbe-Plant Symbiotic and Pathogenic Interactions, NATO ASI Series, pp 333—344. Springer, Berlin

Legendre L, Heinstein PF and Low PS (1992) Evidence for participation of GTP-binding proteins in elicitation of the rapid oxidative burst in cultured soybean cells. J Biol Chem 267: 20140—20147

Leist M and Nicotera P (1997) The shape of cell death. Biochem Biophys Res Commun 236: 1—9

Leist M, Fava E, Montecucco C and Nicotera P (1997a) Peroxynitrite and nitric oxide donors induce neuronal apoptosis by eliciting autocrine excitotoxicity. Eur J Neurosci 9: 1488—1498

Leist M, Single B, Castoldi AF, Kuhnle S and Nicotera P (1997b) Intracellular adenosine triphosphate (ATP) concentration: a switch in the decision between apoptosis and necrosis. J Exp Med 185: 1481—1486

Levine A, Tenhaken R, Dixon RA and Lamb C (1994) H_2O_2 from the oxidative burst orchestrates the plant hypersensitive disease resistance response. Cell 79: 583—593

Levine A, Pennell RI, Alvarez ME, Palmer R and Lamb C (1996) Calcium-mediated apoptosis in a plant hypersensitive disease resistance response. Curr Biol 6: 427—437

Li P, Nijhawan D, Budihardjo I, Srinivasula SM, Ahmad M, Alnemri ES and Wang X (1997) Cytochrome c and dATP-dependent formation of Apaf-1/caspase-9 complex initiates an apoptotic protease cascade. Cell 91: 479—489

Ligterink W, Kroj T, zur Nieden U, Hirt H and Scheel D (1997) Receptor-mediated activation of a MAP kinase in pathogen defense of plants. Science 276: 2054—2057

Lin KT, Xue JY, Sun FF and Wong PYK (1997) Reactive oxygen species participate in peroxynitrite-induced apoptosis in HL-60 cells. Biochem Biophys Res Commun 230: 115—119

Lindgren, PB, Peet RC and Panopoulos NJ (1986) Gene cluster of *Pseudomonas syringae* pv. *phaseolicola* controls pathogenicity of bean plants and hypersensitivity on nonhost plants. J. Bacteriol 168: 512—22

Lindgren PB, Panopoulos NJ, Staskawicz BJ and Dahlbeck D (1988) Genes required for pathogenicity and hypersensitivity are conserved and interchangeable among pathovars of *Pseudomonas syringae*. Mol Gen Genet 211: 499—506

Lindgren PB, Frederick R, Govindarajan AG, Panopoulos NJ, Staskawicz BJ and Lindow SW (1989). An ice nucleation reporter gene system: Identification of inducible pathogenicity genes in *Pseudomonas syringae* pv. *phaseolicola*. EMBO J 8: 1291—301

Liu X, Kim CN, Yang J, Jemmerson R and Wang X (1996) Induction of apoptotic program in cell-free extracts: requirement for dATP and cytochrome c. Cell 86: 147—157

Liu X, Zou H, Slaughter C and Wang X (1997) DFF, a heterodimeric protein that functions downstream of caspase-3 to trigger DNA fragmentation during apoptosis. Cell 89: 175—184

Loh Y-T and Martin GB (1995) The *Pto* bacterial resistance gene and the *Fen* insecticide sensitivity gene encode functional protein kinases with serine/threonine specificity. Plant Physiol 108: 1735—1739

Lyon F and Wood RKS (1976) The hypersensitive reaction and other responses of bean leaves to bacteria. Ann Bot 40: 479—491

Lyon F and Wood RKS (1977) Alterations of response of bean leaves to compatible and incompatible bacteria. Ann Bot 41: 359—367

Maccarrone M, van Arle PGM, Veldink GA and Vliegenthart JFG (1994) In vitro oxygenation of soybean biomembranes by lipoxygenase-2. Biochim Biophys Acta 1190: 164—169

Macré F, Braidot E, Petrussa E and Vianello A (1994) Lipoxygenase activity associated to isolated soybean plasma membranes. Biochim Biophys Acta 1215: 109—114

Malamy J, Carr JP, Klessig DF and Raskin I (1990) Salicylic acid: a likely endogenous signal in the resistance response of tobacco to viral infection. Science 250: 1002—1004

Manon S, Chaudhuri B and Guerin M (1997) Release of cytochrome c and decrease of cytochrome c oxidase in Bax-expressing yeast cells, and prevention of these effects by coexpression of Bcl-xL. FEBS Lett 415: 29—32

Matile P (1975) The Lytic Compartment of Plant Cells. Springer, Wien and New York

Matile P (1987) The sap of plant cells. New Phytol 105: 1—26

Mauch-Mani B and Slusarenko AJ (1996) Production of salicylic acid precursors is a major function of phenylalanine ammonia lyase in the resistance of *Arabidopsis* to *Peronospora parasitica*. Plant Cell 8: 203—212

May MJ, Hammond-Kosack KE and Jones JDG (1996a) Involvement of reactive oxygen species, glutathione metabolism, and lipid peroxidation in the *Cf*-gene-dependent defense response of tomato cotyledons induced by race-specific elicitors of *Cladosporium fulvum*. Plant Physiol 110: 1367—1379

May MJ, Parker JE, Daniels MJ, Leaver CJ and Cobbett CS (1996b) An *Arabidopsis* mutant depleted in glutathione shows unaltered responses to fungal and bacterial pathogens. Mol Plant-Microbe Interact 9: 349—356

Meier BM, Shaw N and Slusarenko AJ (1993) Spatial and temporal accumulation of defense gene transcripts in bean (*Phaseolus vulgaris*) leaves in relation to bacteria-induced hypersensitive cell death. Mol Plant-Microbe Interact 6: 453—466

Métraux JP, Signer H, Ryals J, Ward E, Wyss-Benz M, Gaudin J, Raschdorf K, Schmid E, Blum W and Inverardi B (1990) Increase in salicylic acid at the onset of systemic acquired resistance in cucumber. Science 250: 1004—1006

Mishra OP, Delivoria-Papadopoulos M, Cahillane G and Wagerle LC (1989) Lipid peroxidation as the mechanism of modification of the affinity of the Na^+,K^+-ATPase active sites for ATP, K^+, Na^+, and strophanthidin in vitro. Neurochem Res 14: 845—851

Mittler R and Lam E (1996) Sacrifice in the face of foes: pathogen-induced programmed cell death in plants. Trends Microbiol 4: 10—15

Mittler R, Shulaev V, Seskar M and Lam E (1996) Inhibition of programmed cell death in tobacco plants during a pathogen-induced hypersensitive response at low oxygen pressure. Plant Cell 8: 1991—2001

Mittler R, Simon L and Lam E (1997) Pathogen-induced programmed cell death in tobacco. J Cell Sci 110: 1333—1344

Mock H-P, Keetman U, Kruse E, Rank B and Grimm B (1997) Defense responses to tetrapyrrole-induced oxidative stress in transgenic plants with reduced uroporphyrinogen decarboxylase or coporphyrinogen oxidase activity. Plant Physiol 116: 107—116

Mock H-P, Keetman U and Grimm B (1998) Photosensitizing tetrapyrroles induce anti-oxidative and pathogen defence responses in plants, Proceedings of the 7th International Congress of Plant Pathology, Abstract 1.2.16. International Society for Plant Pathology, Edinburgh

Morel J-B and Dangl JL (1997) The hypersensitive response and induction of cell death in plants. Cell Death Differ 4: 671—683

Naton B, Hahlbrock K and Schmelzer E (1996) Correlation of rapid cell death with metabolic changes in fungus-infected, cultured parsley cells. Plant Physiol 112: 433—444

Newmeyer DD, Farschon DM and Reed JC (1994) Cell-free apoptosis in *Xenopus* egg extracts: inhibition by Bcl-2 and requirement for an organelle fraction enriched in mitochondria. Cell 79: 353—364

Nguyen M, Branton PE, Walton PA, Oltvai ZN, Korsmeyer SJ and Shore GC (1994) Role of membrane anchor domain of Bcl-2 in suppression of apoptosis caused by E1B-defective adenovirus. J Biol Chem 269: 16521—16524

Nicotera P and Leist M (1997) Energy supply and the shape of death in neurons and lymphoid cells. Cell Death Differ 4: 435—442

Ninnemann H and Maier J (1996) Indications for the occurrence of nitric oxide synthases in fungi and plants and the involvement in photoconidiation of *Neurospora crassa*. Photochem Photobiol 64: 393—398

Novacky A and Ullrich-Eberius CI (1982) Relationship between membrane potential and ATP level in *Xanthomonas campestris* pv. *malvacearum* infected cotton cotyledons. Physiol Plant Pathol 21:237—249

Nürnberger T, Colling C, Hahlbrock K, Jabs T, Renelt A, Sacks WR and Scheel D (1994a) Perception and transduction of an elicitor signal in cultured parsley cells. Biochem Soc Symp 60: 173—182

Nürnberger T, Nennstiel D, Jabs T, Sacks WR, Hahlbrock K and Scheel D (1994b) High-affinity binding of a fungal oligopeptide elicitor to parsley plasma membranes triggers multiple defense responses. Cell 78: 449—460

Parker JE, Holub EB, Frost LM, Falk A, Gunn N and Daniels MJ (1996) Characterization of *eds1*, a mutation in *Arabidopsis* suppressing resistance to *Peronospora parasitica* specified by several different *Rpp* genes. Plant Cell 8: 2033—2046

Pavlovkin J, Novacky A and Ullrich-Eberius CI (1986) Membrane potential changes during bacteria-induced hypersensitive reaction. Physiol Mol Plant Pathol 28: 125—135

Peng M and Kuc J (1992) Peroxidase-generated hydrogen peroxide as a source of antifungal activity *in vitro* and on tobacco leaf disks. Phytopathology 82: 696—699

Pennell RI and Lamb C (1997) Programmed cell death in plants. Plant Cell 9: 1157—1168

Peterhänsel C, Freialdenhoven A, Kurth J, Kolsch R and Schulze-Lefert P (1997) Interaction analyses of genes required for resistance responses to powdery mildew in barley reveal distinct pathways leading to leaf cell death. Plant Cell 9: 1397—1409

Petit PX, Susin SA, Zamzami N, Mignotte B and Kroemer G (1996) Mitochondria and programmed cell death: back to the future. FEBS Lett 396: 7—13

Popham PL and Novacky A (1991) Use of dimethyl sulphoxide to detect hydroxyl radical during bacteria-induced hypersensitive reaction. Plant Physiol 96: 1157—1160

Rahme L, Frederick RT, Grim C, Minderinos M and Panopoulos NJ (1988) Transcriptional organization of pathogenicity/hypersensitivity controlling genes (*hrp*) in *Pseudomonas syringae* pv. *phaseolicola* (Abstract). Fifth International Congress of Plant Pathology, Japan

Rancé I, Fournier J, Esquerré-Tugayé M-T (1998) The incompatible interaction between *Phytophthora parasitica* var. *nicotianae* race 0 and tobacco is suppressed in transgenic plants expressing antisense lipoxygenase sequences. Proc Natl Acad Sci USA 95: 6554—6559

Rasmussen JB, Hammerschmidt R and Zook MN (1991) Systemic induction of salicylic acid accumulation in cucumber after inoculation with *Pseudomonas syringae* pv. *syringae*. Plant Physiol 97: 1342—1347

Reed JC (1997) Cytochrome c: can't live with it - can't live without it. Cell 91: 559—562

Renelt A, Colling C, Hahlbrock K, Nürnberger T, Parker JE, Sacks WR and Scheel D (1993) Studies on elicitor recognition and signal transduction in plant defence. J Exp Bot 44S: 257—268

Ribnicky DM, Shulaev V and Raskin I (1998) Intermediates of salicylic acid biosynthesis in tobacco. Plant Physiol 118: 565—572

Riely CA, Cohen G and Liebermann M (1974) Ethane evolution: a new index of lipid peroxidation. Science 183: 208—210

Riparbelli MG, Callaini G, Tripodi SA, Cintorino M, Tosi P and Dallai R (1995) Localization of the Bcl-2 protein to the outer mitochondrial membrane by electron microscopy. Exp Cell Res 221: 363—369

Ritter C and Dangl JL (1995) The *avrRpm1* gene of *Pseudomonas syringae* pv. *maculicola* is required for virulence on *Arabidopsis*. Mol Plant-Microbe Interact 8: 444—453

Roebuck P, Sexton R and Mansfield JW (1978) Ultrastructural observations on the development of the hypersensitive reaction in leaves of *Phaseolus vulgaris* cv. red Mexican inoculated with *Pseudomonas phaseolicola* (Race 1). Physiol Plant Path 12: 151—157

Rosqvist R, Magnusson K-E and Wolf-Watz H (1994) Target cell contact triggers expression and polarized transfer of *Yersinia* YopE cytotoxin into mammalian cells. EMBO J 13: 964—972

Ryals J, Weymann K, Lawton K, Friedrich L, Ellis D, Steiner HY, Johnson J, Delaney TP, Jesse T, Vos P and Uknes S (1997) The Arabidopsis NIM1 protein shows homology to the mammalian transcription factor inhibitor I kappa B. Plant Cell 9: 425—439

Ryerson DE and Heath MC (1996) Cleavage of nuclear DNA into oligonucleosomal fragments during cell death induced by fungal infection or by abiotic treatments. Plant Cell 8: 393—402

Schlesinger PH, Gross A, Yin XM, Yamamoto K, Saito M, Waksman G and Korsmeyer SJ (1997) Comparison of the ion channel characteristics of proapoptotic BAX and antiapoptotic BCL-2. Proc Natl Acad Sci USA 94: 11357—11362

Schmidt HHHW and Walter U (1994) NO at work. Cell 78: 919—925

Shirasu K, Nakajima H, Rajasekhar VK, Dixon RA and Lamb C (1997) Salicylic acid potentiates an agonist-dependent gain control that amplifies pathogen signals in the activation of defense mechanisms. Plant Cell 9: 261—270

Solomon M, Belenghi B, Delledonne M, Menachem E and Levine A (1999) The involvemnet of cysteine proteases and protease inhibitor genes in the regulation of programmed cell death in plants. Plant Cell 11: 431—434

Slusarenko AJ (1996) The role of lipoxygenase in resistance to infection. In: Piazza GJ (ed) Lipoxygenase and Lipoxygenase Pathway Enzymes, pp 176—197. American Oil Chemists Society Press, Champaign, Illinois

Slusarenko AJ and Longland A (1986) Changes in gene activity during expression of the hypersensitive response in *Phaseolus vulgaris* cv. Red Mexican to an avirulent race 1 isolate of *Pseudomonas syringae* pv. *phaseolicola*. Physiol Mol Plant Pathol 29: 79—84

Slusarenko AJ, Croft KPC and Voisey CR (1991) Biochemical and molecular events in the hypersensitive response of bean to *Pseudomonas syringae* pv. *phaseolicola*. In: Smith C (ed) Biochemistry and Molecular Biology of Plant:Pathogen Interactions, pp 126—143. Oxford University Press, Oxford

Stäb MR and Ebel J (1987) Effects of Ca^{2+} on phytoalexin induction by fungal elicitor in soybean cells. Arch Biochem Biophys 257: 416—423

Stakmann EC (1915) Relation between *Puccinia graminis* and plants highly resistant to its attack. J Agric Res 4: 193—199

Staskawicz BJ, Dahlbeck D and Keen NT (1984) Cloned avirulence gene of *Pseudomonas syringae* pv. *glycinea* determines race-specific incompatibility on *Glycine max* (L). Merr. Proc Natl Acad Sci USA 81: 6024—6028

Sutherland MW (1991) The generation of oxygen radicals during host plant responses to infection. Physiol

Mol Plant Pathol 39: 79—93

Takahashi H, Chen ZX, Du H, Liu YD and Klessig DF (1997) Development of necrosis and activation of disease resistance in transgenic tobacco plants with severely reduced catalase levels. Plant J 11: 993—1005

Tanaka Y, Makishima T, Sasabe M, Ichinose Y, Shiraishi T, Nishimoto T and Yamada T (1997) *dad-1*, a putative programmed cell death suppressor gene in rice. Plant Cell Physiol 38: 379—383

Tavernier E, Wendehenne D, Blein J-P and Pugin A (1995) Involvement of free calcium in action of cryptogein, a proteinaceous elicitor of hypersensitive reaction in tobacco cells. Plant Physiol 109: 1025—1031

Templeton MD and Lamb CJ (1988) Elicitors and defence gene activation. Plant Cell Environ 11: 395—401

Tenhaken R, Levine A, Brisson LF, Dixon RA and Lamb C (1995) Function of the oxidative burst in hypersensitive disease resistance. Proc Natl Acad Sci USA 92: 4158—4163

Tenhaken R, Rübel C (1998) Cloning of putative subunits of the soybean plasma membrane NADPH-oxidase involved in the oxidative burst by antibody expression screening. Protoplasma, in press

Thompson JE, Legge RL and Barber RF (1987) The role of free radicals in senescence and wounding. New Phytol 105: 317—344

Turner JG and Novacky A (1974) The quantitative relation between plant and bacterial cells involved in the hypersensitive reaction. Phytopathology 64: 885—890

Van den Ackerveken, G, Marois, E and Bonas, U (1996) Recognition of the bacterial avirulence protein AvrBs3 occurs inside the host plant cell. Cell 87: 1307—1316

Van Gijsegem F, Genin S and Boucher C (1993) Conservation of secretion pathways for pathogenicity determinants of plant and animal bacteria. Trends Microbiol 1: 175—180

Van Camp W, Van Montagu M and Inzé D (1998) H_2O_2 and NO: redox signals in disease resistance. Trends Plant Sci 3: 330—334

Vianello A, Braidot E, Bassi G and Macré F (1995) Lipoxygenase activity on the plasmalemma of sunflower protoplasts and its modulation. Biochim Biophys Acta 1255: 57—62

Wang H, Li J, Bostock RM and Gilchrist DG (1996) Apoptosis: a functional paradigm for programmed plant cell death induced by a host-selective phytotoxin and invoked during development. Plant Cell 8: 375—391

Ward HM (1902) On the relations between host and parasite in the bromes and their brown rust, *Puccinia dispersa* (Erikss). Ann Bot 16: 233—315

Wei Z-M, Laby RJ, Zumoff CH, Bauer DW, He SY, Collmer A and Beer SV (1992) Harpin, elicitor of the hypersensitive response produced by the plant pathogen *Erwinia amylovora*. Science 257: 85—88

Weymann K, Hunt M, Uknes S, Neuenschwander U, Lawton K, Steiner H-Y and Ryals J (1995) Suppression and restoration of lesion formation in *Arabidopsis lsd* mutants. Plant Cell 7: 2013—2022

White E (1996) Life, death, and the pursuit of apoptosis. Genes Dev 10: 1—15

Wisman E, Cardon GH, Fransz P and Saedler H (1998) The behaviour of the autonomous maize transposable element En/Spm in *Arabidopsis thaliana* allows efficient mutagenesis. Plant Mol Biol 37: 989—999

Wolff SP, Garner A and Dean RT (1986) Free radicals, lipids and protein degradation. Trends Biochem Sci 11: 27—31

Woods AM, Fagg J and Mansfield J (1988) Fungal development and irreversible membrane damage in cells of *Lactuca sativa* undergoing the hypersensitive reaction to the downy mildew fungus *Bremia lactucae*. Physiol Mol Plant Pathol 32: 483—497

Wyllie AH, Morris RG, Smith AL and Dunlop D (1984) Chromatin cleavage in apoptosis: association with condensed chromatin morphology and dependence on macromolecular synthesis. J Pathol 142: 67—77

Wyman JG and Van Etten HD (1982) Isoflavonoid phytoalexins and nonhypersensitive resistance of beans to *Xanthomonas campestris* pv. *phaseoli*. Phytopathology 72: 1419—1424

Xing T, Higgins VJ and Blumwald E (1997) Race-specific elicitors of *Cladosporium fulvum* promote translocation of cytosolic components of NADPH oxidase to the plasma membrane of tomato cells. Plant Cell 9: 249—259

Xu H and Heath MC (1998) Role of calcium in signal transduction during the hypersensitive response caused by basidiospore-derived infection of the cowpea rust fungus. Plant Cell 10: 585—597

Yalpani N, Leon J, Lawton MA and Raskin I (1993) Pathway of salicylic acid biosynthesis in healthy and virus-inoculated tobacco. Plant Physiol 103: 315—321

Yu IC, Parker J and Bent AF (1998) Gene-for-gene disease resistance without the hypersensitive response in Arabidopsis *dnd1* mutant. Proc Natl Acad Sci USA 95: 7819—7824

Zamzami N, Marchetti P, Castedo M, Decaudin D, Macho A, Hirsch T, Susin SA, Petit PX, Mignotte B and Kroemer G (1995) Sequential reduction of mitochondrial transmembrane potential and generation of reactive oxygen species in early programmed cell death. J Exp Med 182: 367—377

Zamzami N, Susin SA, Marchetti P, Hirsch T, Gomez-Monterrey I, Castedo M and Kroemer G (1996) Mitochondrial control of nuclear apoptosis. J Exp Med 183: 1533—1544

Zimmermann S, Nürnberger T, Frachisse J-M, Wirtz W, Guern J, Hedrich R and Scheel D (1997) Receptor-mediated activation of a plant Ca^{2+}-permeable ion channel involved in pathogen defense. Proc Natl Acad Sci USA 94: 2751—2755

Zhou J, Loh Y-T, Bressan RA and Martin GB (1995) The tomato gene *Pti1* encodes a serine/threonine kinase that is phosphorylated by Pto and is involved in the hypersensitive response. Cell 83: 925—935

Zhou JM, Tang XY and Martin GB (1997) The Pto kinase conferring resistance to tomato bacterial speck disease interacts with proteins that bind a cis-element of pathogenesis-related genes. EMBO J 16: 3207—3218

Zou H, Henzel WJ, Liu X, Lutschg A and Wang X (1997) Apaf-1, a human protein homologous to *C. elegans* CED-4, participates in cytochrome c-dependent activation of caspase-3. Cell 90: 405—413

CHAPTER 7

ANTIMICROBIAL COMPOUNDS AND RESISTANCE

THE ROLE OF PHYTOALEXINS AND PHYTOANTICIPINS

J.W. MANSFIELD

Department of Biological Sciences, Wye College,
University of London, Ashford, Kent, TN25 5AH, UK
(J.Mansfield@wye.ac.uk)

Summary

Many antimicrobial compounds produced by plants have important roles in their resistance to infection by bacteria, fungi and nematodes. The defensive compounds may be broadly classified into phytoanticipins which are constitutive, and phytoalexins which are synthesized in response to challenge by microorganisms. The two groups of secondary metabolites include a wide range of chemical families produced by different biosynthetic pathways. Phytoanticipins are primarily involved in non-host rather than varietal resistance. Many pathogens have evolved mechanisms of tolerance to their host's phytoanticipins and phytoalexins. Tolerance often involves enzymatic degradation, and transfer of genes for such detoxification can extend host range. Activation of phytoalexin biosynthesis is associated with the hypersensitive reaction or the formation of necrotic lesions. Accumulation of phytoalexins is highly localized. Two strategies appear to have evolved in pathogens to overcome the phytoalexin response, either the defence is not activated or the pathogen tolerates the presence of the inhibitor. Antimicrobial compounds from plants have potential as fungicides themselves or in providing classes of compounds suitable for development as biocides. The biosynthetic pathways to phytoanticipins and phytoalexins offer targets for genetic manipulation to engineer resistance. Progress in the development of our understanding of the role of antimicrobial compounds in resistance is critically discussed, with particular emphasis on the application of molecular genetics to unravel plant-microbe interactions.

I.   Introduction

There is no doubt that antimicrobial compounds can be recovered from both healthy and infected plants and that many of these compounds have been demonstrated to have striking activity *in vitro* against potential pathogens including, bacteria, fungi and nematodes. Indeed, the presence of toxic compounds within plant tissues has been known and utilized by man for centuries. What is less certain is the role such

A. Slusarenko, R.S.S. Fraser, and L.C. van Loon (eds), *Mechanisms of Resistance to Plant Diseases*, 325-370.
© 2000 *Kluwer Academic Publishers. Printed in the Netherlands.*

compounds have in disease resistance. Wood (1967) and Deverall (1977) emphasized that proof of a role requires demonstration that the substance is present, (1) at the right time; (2) in the right place and (3) in sufficient quantity to cause any observed inhibition of microbial colonization of plant tissue. Fulfilment of these criteria requires a detailed knowledge of the microenvironment within which the fungus or bacterium is restricted, and the timing of restriction of microbial growth. Although the routes taken by many pathogens have been described by microscopical studies, surprisingly little is known about the conditions prevailing around fungal hyphae which, for example like those of *Botrytis fabae*, may be creating their own environment as they grows through the degraded plant cell wall, or around bacterial cells which may be embedded in their own extracellular polysaccharides in the intercellular spaces between mesophyll cells (Mansfield and Richardson, 1980; Mansfield, 1990; Brown *et al.*, 1993).

Antimicrobial compounds from plants are broadly classified into two categories: *phytoanticipins*, which are recognised as, "...low molecular weight, antimicrobial compounds that are present in plants before challenge by microorganisms, or are produced after infection solely from pre-existing constituents," and *phytoalexins* which are defined as "low molecular weight, antimicrobial compounds that are both synthesized by and accumulated in plants after exposure to microorganisms," (Paxton, 1980; VanEtten *et al.*, 1994). The aim of this chapter is to review the roles of phytoanticipins and phytoalexins in disease resistance. Emphasis is placed on case studies and more recent genetical approaches, but the reader is urged not to ignore elegant earlier work and the historical perspectives as discussed in detail by Ingham (1973), Deverall (1982) and Mansfield (1983), which have led to some of the generalisations presented here. A very thorough survey of the chemical diversity of antifungal compounds from higher plants was recently published by Grayer and Harborne (1994). When scanning the literature the reader is advised to view statements about antimicrobial activity with caution. As a general rule any compound which lacks appreciable activity (e.g. 50% inhibition of fungal germ-tubes) at more than $10^{-4}M$ is probably unlikely to have a major role in resistance.

II.   Phytoanticipins

Several examples of compounds which are either present in their active form and others which are converted to activity by a simple biochemical reaction are given in Table 1 and selected chemical structures illustrated in Fig. 1. Difficulties inherent in any precise demarcation of what is or is not a phytoanticipin are apparent from Table 1, which includes some compounds which may also be classified as phytoalexins. For example the flavanone sakuranetin, is constitutive in blackcurrant leaves but induced in rice leaves (Kodama *et al.* 1988). Sakuranetin is present in glands on the adaxial but not the abaxial leaf surface of blackcurrant. The presence or absence of the glands and therefore the phytoanticipin, has been shown to be associated with the relative failure and success, respectively, of *Botrytis cinerea* to germinate on the two surfaces (Atkinson and Blakeman, 1982). Some compounds may also be phytoanticipins in one organ and

Table 1.   Phytoanticipins are found in many different plant families and chemical classes

Plant family	Species	Compounds	Chemical class	Active form	Organ most studied	Microorganism studied	Reference
Alliaceae	Allium cepa (onion)	Catechol	Phenol	-	Bulb	Colletotrichum circinans	Walker and Stahmann, 1955
		Protocatechuic acid	Phenol	-			
Cruciferae	Brassica spp (including oilseed rape)	Glucosinolates	Thioglucoside	Isothiocyanates, nitriles and thiocyanates	Leaves	Leptosphaeria maculans	Mithen, 1992
Gramineae	Avena sativa (oats)	Avenacins	Triterpenoid saponins	-	Roots	Gaeumannomyces graminis	Crombie et al., 1984
		Avenacosides	Steroidal saponins	Deglucoavenacosides	Leaves	Septoria avenae	Kesselmeier & Urban, 1983
	Sorghum species	Dhurrin	Cyanogenic glucoside	HCN	Leaves	Gleocercospora sorghi	Wang and VanEtten, 1992
Grossulariaceae	Ribes nigrum (currant)	Sakuranetin	Flavanone	-	Leaf glands	Botrytis cinerea	Atkinson & Blakeman, 1982
Lauraceae	Persea americano (avocado)	1-acetoxy-2-hydroxy-4-oxo-heneicosa-12,15-diene	Long chain alcohol	-	Fruit peel	Colletotrichum gloeosporioides	Prusky et al., 1982
Leguminoseae	Lupinus albus (lupin)	Luteone	Isoflavone	-	Leaf surface	Helminthosporium carbonum	Ingham et al., 1983
		Wighteone	Isoflavone	-			

Table 1.   Continued

Plant family	Species	Compounds	Chemical class	Active form	Organ most studied	Microorganism studied	Reference
Liliaceae	Tulipa gesneriana (tulip)	Tuliposides	Butyrolactone	Tulipalins	Bulb	Fusarium oxysporum	Bergman & Beijersbergen, 1968
Pinaceae	Pinus radicata	ω-hydroxyhexa-decanoic acid	Long chain fatty acid	-	Needle surface	Dothistroma pini	Franich et al., 1983
Rutaceae	Glycosmis cyanocarpa	Sinharine	Sulphur containing amide	-	Leaves	Cladosporium cladosporioides	Greger et al., 1992
Solanaceae	Lycopersicon esculentum (tomato)	α-Tomatine	Steroidal glycoalkaloid	-	Green fruit	Nectria haematococca	Défago and Kern, 1983

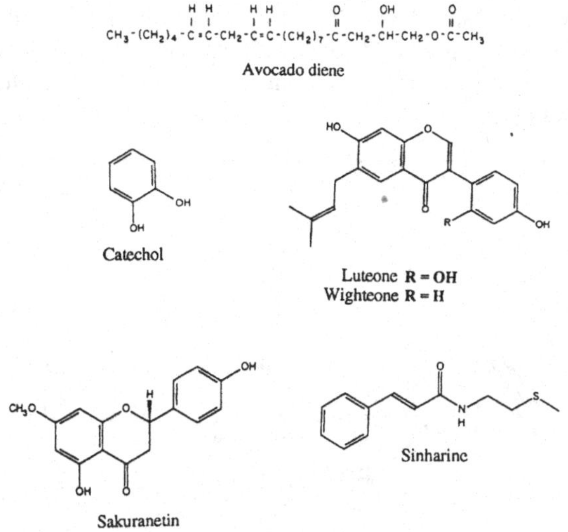

$$CH_3-(CH_2)_4-\overset{\overset{H}{|}}{C}-\overset{\overset{H}{|}}{C}-CH_2-\overset{\overset{H}{|}}{C}-\overset{\overset{H}{|}}{C}-(CH_2)_7-\overset{\overset{O}{\|}}{C}-CH_2-\overset{\overset{OH}{|}}{CH}-CH_2-O-\overset{\overset{O}{\|}}{C}-CH_3$$

Avocado diene

Catechol

Luteone **R = OH**
Wighteone **R = H**

Sinharine

Sakuranetin

Figure 1. Phytoanticipins from different chemical families.

phytoalexins in another of the same plant, an example being momilactone A which occurs constitutively in rice seed but is induced in leaves (Cartwright *et al.*, 1980).

Perhaps surprisingly, there are few well characterised examples of antimicrobial compounds present on plant surfaces. Inhibition of microbial development in the phylloplane or rhizosphere has more often been attributed to competitive effects of the resident microflora (Blakeman, 1973). Some exceptions are the epicuticular fatty acids from pine needles (Franich *et al.*, 1983), isoflavones from lupin (Ingham *et al.*, 1983; Tahara *et al.*, 1984) and the simple phenolics catechol and protocatechuic acid, which diffuse into water droplets on the surface of red onion bulb scales, and were some of the earliest compounds to be implicated in disease resistance (Walker and Stahmann, 1955).

A.   QUIESCENT INFECTIONS

The phenomenon of quiescent infection (sometimes referred to as latency) has considerable impact on the development of post harvest diseases (Verhoeff, 1974; Swinburne, 1983). Infections established during the maturation of fruits and bulbs often remain symptomless until after harvest. Phytoanticipins have often been implicated in the restriction of pathogens after the initial infection - subsequent colonization following a reduction in the concentration of the phytoanticipin within the mature tissue. Major causes of such quiescent infections in fruits are species of *Colletotrichum* (teliomorph *Glomerella*). The quiescence of *Colletotrichum* often occurs following penetration into the cuticle by an infection peg which remains dormant until the fruit ripens. Prusky and Plumbley (1992) described quiescent infections of avocado, banana, mango and papaya and assessed the possible role for phytoanticipins. The more convincing data come from work on avocado and mango and the phytoanticipins, 1-acetoxy-2-hydroxy-4-oxo-heneicosa-12,15-diene (Fig. 1) and a group of

5-substituted resorcinols respectively. In unripe fruits, the phytoanticipins are found in high concentration in the peel. The apparently causal relationship between changes in diene levels in avocado and the development of lesions by *C. gloeosporioides* is clearly illustrated in Fig. 2.

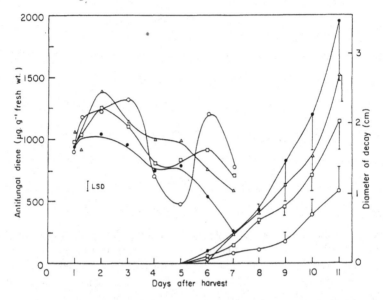

Figure 2. Changes in antifungal diene levels in the peel of unripe avocado fruits and lesion formation by *Colletotrichum gleosporioides*. Fruits were exposed to different concentrations of CO_2 for 24 h at 20°C; Δ, 11%; □, 16%; 0,30% and •, untreated controls (adapted from Prusky and Plumley, 1992).

Maintenance of a high level of phytoanticipins after harvesting is a good target for possible manipulation of the plant to increase resistance to infection. Attempts to achieve elevated diene levels in avocado have involved various post-harvest treatments. Some success has been achieved using exposure to an atmosphere of 30% CO_2 for 24h; which caused both increased diene levels and reduced disease incidence (Fig. 2). Understanding the reason for the decline in phytoanticipin concentrations might also provide an insight into potentially beneficial manipulation. Decline in diene levels in avocado has been linked to increased lipoxygenase (LOX) activity during maturation (Prusky *et al.*, 1985). The interaction between LOX and the diene is further complicated by the presence of the flavan-3-ol, epicatechin, which acts as a potent inhibitor of LOX in the peel and whose concentration declines from 500 to 8µg g^{-1} fresh weight, in unripe and ripe fruits respectively. Development of the fungus is therefore occurring amidst changes in three components regulating antifungal activity; the diene, LOX and epicatechin. Awareness of the possible interactions affecting quiescence provides a background for disease control strategies.

The healthy tulip plant contains very potent phytoanticipins, the tuliposides. The active principles in extract have been identified as the butyrolactones tulipalin A and B (Tscheche *et al.*, 1968). The active lactones are formed from tuliposides A and B which are glucosides thought to be stored in plant cell vacuoles. Conversion to tulipalins occurs via the unstable intermediate acids (Fig. 3). The concentrations and proportions of each tuliposide vary between tissues, tuliposides A and B occurring in particularly high concentrations in the white skin of tulip bulbs and in pistils respectively. Unsaturated lactones occur as glucosides in many plants notably members of the Liliaceae, Ranunculaceae and Rosaceae (Slob *et al.*, 1975; Schönbeck and Schlösser, 1976).

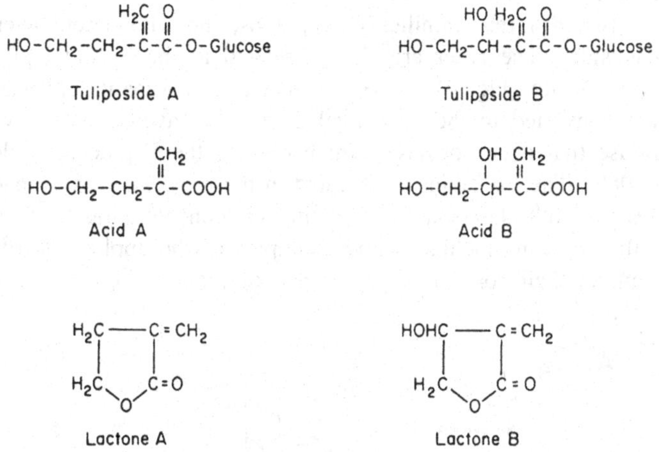

$$HO-CH_2-CH_2-\underset{\underset{Tuliposide\ A}{}}{\overset{\overset{H_2C\ \ O}{\|\ \ \|}}{C-C}}-O-Glucose$$

$$HO-CH_2-\underset{\underset{Tuliposide\ B}{}}{\overset{\overset{HO\ H_2C\ \ O}{\|\ \ \ \|\ \ \|}}{CH-C-C}}-O-Glucose$$

$$HO-CH_2-CH_2-\underset{\underset{Acid\ A}{}}{\overset{\overset{CH_2}{\|}}{C}}-COOH$$

$$HO-CH_2-\underset{\underset{Acid\ B}{}}{\overset{\overset{OH\ CH_2}{\|\ \ \|}}{CH-C}}-COOH$$

Lactone A

Lactone B

Figure 3. Generation of antimicrobial lactones (tulipalins) from tuliposides, phytoanticipins in tulip tissues.

Fusarium oxysporum causes a rot of tulip bulbs, but infection only develops during the last few weeks before harvest, even if the mother bulb planted in the autumn was heavily infected or covered with conidia. The quiescence of infection in the outer bulb scales of the developing bulb is associated with high levels of tuliposide A, several times more than that which is fungistatic to mycelium of *F. oxysporium*. During the short period of susceptibility before lifting, the outer scale desiccates and turns brown and loses its high concentration of tuliposides. The underlying white scales which are temporarily susceptible to colonization also have a low tuliposide content. In the absence of the preformed barrier of tuliposides *F. oxysporium* is able to cause damaging lesions (Bergman and Beijersbergen, 1968). As with the tropical fruits, the mechanisms for and physiological significance of the rapid changes occurring in phytoanticipin concentrations during plant development remain largely unknown.

B.   AVENACIN, AVENACOSIDES AND OTHER SAPONINS

Saponins are glycosylated compounds that are found in many plant families (Osbourn, 1996a). They are divided into three major groups depending on the structure of the aglycone which may be a triterpenoid, a steroid or a steroidal alkaloid; examples being avenacin and avenacoside (both from oats), and α-tomatine (from tomato), respectively. The toxic action of saponins is associated with the ability of these compounds to complex with plasma membrane sterols and cause pore formation (Fenwick *et al.*, 1992).

Plants may protect themselves from their own saponins by compartmentalizing them in the cell vacuole or in other organelles, the membranes of which may avoid lysis due to low or altered sterol composition (Steel and Drysdale, 1988; Keukens *et al.*, 1995). Bacteria and the oomycetes, including the plant pathogenic genera *Phytophthora* and *Pythium*, contain low levels of sterols in their membranes and are largely insensitive to saponins. These pathogens must, therefore, be restricted by other methods of defence.

Oat contains two different families of saponins, the triterpenoid avenacins and steroidal avenacosides. The latter are closely related to the steroidal glycoalkaloid, α-tomatine, from tomato (Fig. 4). Unlike avenacins and tomatine, avenacosides are inactive but are converted to the antifungal 26-desglucoavenacocides by a specific glucosyl hydrolase (oat avenacosidase) which cleaves the D-glucose molecule from C-26 (Nisius, 1988). The avenacins are located in the root, whereas avenacosides are found in the leaves; both classes are at their highest concentrations in the epidermis. Research into the saponins includes good examples of the application of molecular genetics to examine their role in the resistance of plants to colonization by fungi (Osbourn 1996b).

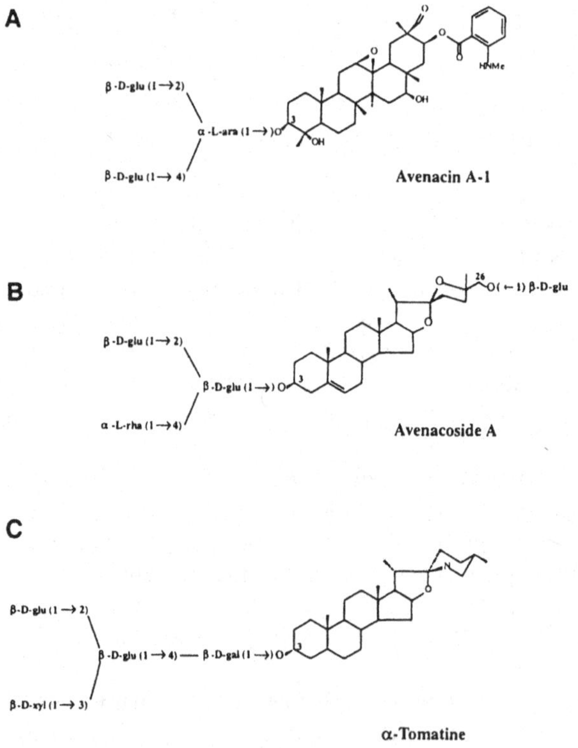

Figure 4. Saponins from oats and tomato. Note that avenacin A-1 and a tomatine exist in plants in their active forms but avenacoside A is converted to activity by cleavage of the glucose molecule at C-26 (adapted from Osbourn, 1996b).

Gaeumannomyces graminis var. *tritici* invades roots of wheat and barley causing the economically important take-all disease, but does not cause a lasting infection of oats. Oat roots are, however, susceptible to a different form of the fungus, *G. graminis* var. *avenae*. In elegant early work, Turner (1953; 1961) demonstrated that *G.g.* var. *avenae* was tolerant of avenacin, an inhibitor she detected in extracts of oat roots. Some indication of the capacity of phytoanticipins to confer resistance to oat roots came from her demonstration that the extract from one root tip in 2 cm³ of water was sufficient to cause a 50% reduction in growth of mycelium of *G.g.* var. *tritici*. She also found that the tolerance of *G.g.* var. *avenae* was associated with production of an extracellular glycosidase which removed the terminal sugars from the carbohydrate chain (Fig. 4). Five other fungi tested, including *G.g* var. *tritici*, failed to produce the detoxifying enzyme named avenacinase. Subsequent analysis has confirmed that avenacinase removes the β,1-2 - and β,1-4 -linked terminal D-glucose molecules from avenacin A-1 (Crombie *et al.*, 1984; 1986; Osbourn *et al.*, 1991).

The importance of the enzymatic detoxification of avenacin A-1 for the pathogenicity of *G.g.* var. *avenae* towards oats and conversely, the potential significance of the phytoanticipin in disease resistance, has been clearly demonstrated by the genetical analysis carried out by Bowyer *et al* (1995). This important work merits discussion in some detail to illustrate the route taken to achieve targeted disruption of the avenacinase gene. Avenacinase was first purified from culture filtrates of *G.g.* var. *avenae*. Polyclonal antisera raised to avenacinase were used to select immunoreactive phage plaques from a λ cDNA expression library, prepared from *G.g.* var. *avenae* mRNA collected during active secretion of the enzyme. A genomic DNA clone containing the putative avenacinase gene was then isolated by homology to the cDNA clone and tested for its ability to encode the enzyme by heterologous expression in the genetically tractable saprophytic fungus *Neurospora crassa*.

The next step in analysis of the role of avenacinase was to disrupt the gene in *G.g.* var. *avenae*. Transformation-mediated targeted disruption of *G.g.* var. *avenae* allowed the isolation of avenacinase-minus mutants which were about 8x more sensitive to avenacin A-1, no longer produced a protein reacting with anti-avenacinase antisera and had no detectable avenacinase activity. The targeted disruption of the avenacinase gene rendered *G.g.* var. *avenae* unable to cause disease on oats; the mutant fungus always failed to penetrate beyond the epidermal tissues of the oat root. As concluded by Bowyer *et al.* (1995), these experiments demonstrated that avenacin A-1 detoxification is an absolute requirement for the pathogenicity of *G.g.* var. *avenae* to oats. Interestingly, however, the avenacinase-minus mutant retained wild-type levels of pathogenicity to wheat; there was no hidden role for the glucosidase in basic parasitic ability in the absence of its primary substrate, avenacin A-1.

A genetical approach has also been adopted with the plant side of the avenacin A-1/avenacinase model. Osbourn *et al.* (1994) reported that one diploid oat species, *A. longiglumis* which was found to lack avenacin A-1, was significantly more susceptible to infection by *G.g.* var. *tritici*. Recently, mutants of the diploid oat *A. strigosa* which lack the normally high levels of avenacins in this species have been isolated after sodium azide mutagenesis (Osbourn pers.comm.). Crosses between the mutant and wild type *A. strigosa* have produced progeny in which there is co-segregation for the presence of avenacin A-1 and resistance to a range of fungi including *G.g.* var. *tritici*, and the root infecting *F. avenacearum, F. culmorum* and *F. graminearum*. In the absence of avenacin A-1 there was also increased susceptibility to *G.g.* var. *avenae*. So far, the close correlation between the presence of avenacin A-1 and resistance has been confirmed. An interesting experiment would be to transfer the plasmid allowing expression of avenacinase to the wheat pathogen, *G. g.* var. *tritici*, and to test for increased ability to invade oats.

Other saponins are also detoxified by pathogens of their plants of origin (Arneson and Durbin, 1967; 1968). For example α–tomatine from tomato (Fig. 4) is detoxified by *Septoria lycopersici*, a tomato leaf spotting pathogen, but the relationship between specific enzymic degradation and tolerance is less clear than has proved to be the case with avenacin A-1 (Sandrock *et al.*, 1995). Part of the difficulty may relate to the requirement for more than one glucosidase to carry out complete detoxification. For example α–tomatine is converted to β_2-tomatine by a β-1,2-D glucosidase produced by *Septoria lycopersici* that removes the terminal glucose from the tetrasaccharide chain (Fig. 4). However β_2-tomatine remains toxic to many fungi and additional glucosidases are probably required for removal of other sugars thereby leading to more complete detoxification. Determination of the role of enzymes such as β_2-tomatinase in the pathogenicity of fungi towards tomato requires gene disruption experiments as described for avenacinase. Although β_2-tomatinase has now been successfully cloned from *S. lycopersici* (Osbourn *et al.*, 1995), the next stage in analysis is currently hindered by the lack of suitable transformation technology for the tomato pathogen (Sandrock *et al.*, 1995). VanEtten *et al.* (1995) have also proposed that factors in addition to the production of "tomatinases" may confer tolerance of *S. lycopersici* to α tomatine. Avenacosides, the oat leaf saponins, are detoxified by specific glucosidases that do not act on avenacin. The role of the avenacosidases in pathogenicity has not yet been critically assessed (Wubben *et al.*, 1996).

Overall, a common feature of plant pathogenic fungi is their tolerance to phytoanticipins present in their host plants; typically this involves an ability to degrade the inhibitors to less toxic products. As we have discussed, with avenacin A-1 such a tolerance has often been associated with enzymic detoxification, additional examples of degradative tolerance are the interactions between *Stemphylium loti* and HCN produced from the cyanogenic glycoside in *Lotus corniculatus* (Fry and Myers, 1981), and *Botrytis tulipae* and tulipalins (Schönbeck and Schroeder, 1972).

C.   GLUCOSINOLATES

Cruciferous plants have long been known to contain a family of secondary metabolites termed the glucosinolates, which are sulphur containing glucosides (Fenwick *et al.*, 1983). The glucosinolate molecule comprises a common glycone moiety with a remarkably variable side chain derived from amino acids. Three major classes of glucosinolates occur; aliphatic, derived from methionine; indolyl, from tryptophan and aralkyl from phenylalanine. Following cellular damage, glucosinolates undergo hydrolysis catalysed by the enzyme myrosinase (thioglucoside glucohydrolase), to produce glucose, sulphate and a variety of low molecular weight products possessing diverse chemical and biological properties (Fig. 5). Some of the degradation products, notably isothiocyanates and oxazolidine-2-thiones, have been shown to possess antifungal activity as well as acting as stimuli for feeding and egg deposition in insects (Mithen *et al.*, 1986; Chew, 1988; Mithen 1992; Mari *et al.*, 1993). Analysis of the potential role of glucosinolates is complicated by the diverse structural effects that pH may have on the products generated by enzymic hydrolysis. The availability of myrosinase and its substrates also varies between tissues.

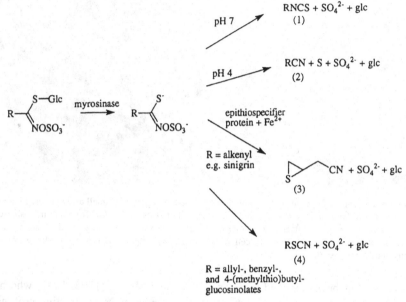

Figure 5. Generation of antimicrobial products from glucosinolates, the thioglucosides from crucifers (adapted from Bones and Rossiter, 1996). Note that the enzyme myrosinase gives rise to an unstable thiohydroximate-sulphonate which degrades to various products.

Recent success has been achieved in demonstrating the localization of myrosinase and sinigrin (propenyl glucosinolate) in cells of *Brassica juncea* cotyledons, using immunocytochemistry at the electron microscope level (P. Kelly and J. Rossiter, per.comm.). Antisera to the enzyme were raised following conventional purification, but the low molecular weight sinigrin had first to be conjugated to bovine serum albumin as discussed by Hassan *et al.* (1988). As illustrated in Fig. 6, the enzyme was found in myrosin granules and also in protein bodies in which it was co-localized with its substrate. Similar immunocytochemical approaches should be adopted with other secondary metabolites, either phytoanticipins or phytoalexins, to provide greater understanding of the spatial organisation of cellular defence reactions in plants.

Several reviewers suggest that glucosinolates and their products play a major role in resistance to economically important pathogens (Fenwick *et al.*, 1983; Chew, 1988; Bones and Rossiter, 1996). However, more recent genetical analysis indicates that they may only function as factors in non-host resistance, pathogens of the crucifers being tolerant of their potential toxicity in the plant. Critical appraisal of the role of

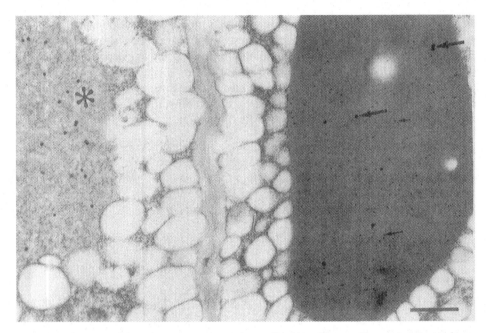

Figure 6. Immunocytochemical localization of sinigrin (small gold particles, small arrows), and myrosinase (large gold particles, large arrow), within cells in a cotyledon of *Brassica juncea*. Note that myrosinase and its substrate sinigrin are co-localized within myrosin grains, myrosinase is also located separately within the cytoplasm of an adjacent aleurone-like cell (asterisk). Bar = 0.5 μm. Micrograph kindly provided by Peter Kelly and John Rossiter.

glucosinolates has been achieved by Giamoustaris and Mithen (1995, 1997) who have developed a series of near isogenic lines of *Brassica napus* with contrasting profiles of leaf glucosinolates. The lines vary both in the total amount of aliphatic glucosinolates and in the ratio of different side chain structures. They were used in field experiments to assess interactions between *Brassica*, pathogens and also generalist and specialist pests. Analysis of the lines has indicated the likely success to be achieved in breeding for resistance to pests and diseases by targeted increase in quantity and/or quality of glucosinolates. The unequivocal conclusion reached for the necrotrophic pathogens *Leptosphaeria maculans* and *Alternaria* spp. was that resistance was *not* correlated with glucosinolate concentrations. By contrast, generalist pests appeared to be deterred by the glucosinolates, whereas the specialist pests such as flea beetle were positively attracted to lines rich in the thioglucosides.

D.   PASSIVE OR ACTIVE DEFENCE?

The work on glucosinolates illustrates the potential complexity in allocating functions to secondary metabolites in plants. Specific pathogens and pests may have become adapted to take advantage of constitutive secondary metabolites which may nonetheless have a genuine role in non-host resistance. Varietal resistance does not appear to be based on the presence of different levels of phytoanticipins, but compounds which are present in an inactive form might be activated during expression of resistance involving plant cell death as occurs during the hypersensitive reaction (HR). For example, the resistance of tomato to the biotrophic leaf mould fungus, *Cladosporium fulvum* is expressed by the HR which restricts fungal development soon after stomatal penetration (de Wit, 1995). *Cladosporium fulvum* does not produce tomatinase but if it is engineered to do so, the transformed strain produces significantly more mycelium than wild-type in resistant varieties, although they do not become fully susceptible (Melton *et al.*, 1998). The conclusion to be drawn from this experiment is that the release of tomatine during the *R*-gene specific response of tomato, has a role in restricting fungal growth.

A crude but effective experiment to examine the presence of antimicrobial concentrations of phytoanticipins in plants is simply to freeze and thaw the tissue and then either to challenge with a known pathogen or simply watch the saprophytes appear. Most killed plant tissues do not resist microbial colonization but there are exceptions, for example Schönbeck and Schroeder (1972) showed that frozen and thawed tulip pistils remained highly resistant to *Botrytis* and other microorganisms for several days. Plant tissues with particularly high levels of phytoanticipins are exceptional, in most plants there is no doubt that resistance to colonization requires active metabolism involving mRNA and protein synthesis and respiratory activity. In several different plant-pathogen interactions, the active responses have been prevented by application of very specific inhibitors of these plant processes (Mansfield, 1983).

III. Phytoalexins

Phytoalexins represent one component of a battery of induced defence mechanisms used by plants, including the important formation of physical barriers to invasion by alterations to the plant cell wall, the transient generation of antimicrobial active oxygen species (such as H_2O_2) which are generated by the oxidative burst, and release of biologically active lipids as a result of lipid peroxidation (Ride, 1986; Graham and Graham, 1991; Croft *et al.*, 1993; Levine *et al.*, 1994; Low and Merida, 1996; Bestwick *et al.*, 1997). It is important to recognize that phytoalexin accumulation may be part of a co-ordinated defence strategy (including phytoanticipins), in which any one factor may alone be unable to account for restriction of the potential pathogen. There are some interactions in which the speed of accumulation of the inhibitors and their high levels of toxicity argue strongly that they are the principal cause of restriction of microbial growth. Such examples are the accumulation of phaseollin in *Phaseolus vulgaris* hypocotyls and wyerone derivatives in cotyledons of *Vicia faba* (Mansfield *et al.*, 1980; Bailey, 1982).

The involvement of phytoalexins in disease resistance is not restricted to a particular interaction between genotypes, for example race-specific or non-host resistance, but is more closely associated with morphologically similar types of response. Most data supporting a role for phytoalexin accumulation as the cause of the inhibition of microbial growth come from interactions in which resistance is expressed following penetration into the plant and is associated with the necrosis of plant cells. Such a local lesion response, often described as a hypersensitive reaction (HR), can occur during the expression of non-host, race-specific and race-non-specific resistance (see Jabs and Slusarenko, this volume).

The phytoalexin concept was developed by Müller and Borger (1940) in a classic paper describing the resistance of potato tuber tissue to *Phytophthora infestans*. Prior inoculation with an avirulent race, which caused the HR, protected tissue against invasion following a second inoculation with a virulent race of *P. infestans* or other pathogenic fungi. The concept that plant tissue was able to generate persistently antimicrobial conditions by the production of phytoalexins (from the Greek phyton = plant, and alexin = protecting substance), was next extended by Müller (1958) and Jerome and Müller (1958) using bean pod inoculations. Suspensions of fungal spores were inoculated onto endocarp exposed by opening the pod and removing seeds; the seed cavities formed a useful receptacle for inoculum droplets. The fungi used included *P. infestans* which elicited a non-host HR in pod tissue. Droplets recovered from infection sites became highly antifungal whereas droplets from healthy pods were highly stimulatory. The major antifungal compound within the droplets was subsequently identified as phaseollin (Cruickshank and Perrin, 1971). The importance of these experiments is not only the identification of the phytoalexin, but also the clear demonstration that the inhibitor can increase to reach fungitoxic concentrations within conditions which must closely mimic those within the infected tissue. The essential criteria stated to prove a role for phytoalexins were, in fact, largely fulfilled by some of the earliest experiments using the drop diffusate technique in legume pods. Similarly convincing antifungal activity has been found using pods of pea (*Pisum sativum*) and broad bean (*Vicia faba*) in which pisatin and wyerone acid are the major phytoalexins (Cruickshank and Perrin, 1960; Perrin and Bottomley, 1962; Deverall 1967, 1977).

A.   CHEMOTAXONOMY

The realization that plants might produce a range of novel secondary metabolites following microbial challenge has led to a remarkably sustained burst of activity amongst chemists studying natural products. In 1982, Bailey and Mansfield recorded that phytoalexins had been identified in 15 plant families; this number was increased to 31 in the review by Grayer and Harborne (1994). Two main conclusions may be drawn from studies on the chemical diversity of phytoalexins: (1) members of plant families usually produce chemically similar types of phytoalexin, for example isoflavonoids from the Leguminoseae and sesquiterpenoids from the Solanaceae; and (2) although accumulation of one compound may predominate in a particular species, most plants produce several closely related phytoalexins. *Vicia faba* provides a notable exception to these general rules; like most other legumes it produces an isoflavonoid phytoalexin, in this case medicarpin, but the principal induced antimicrobial compounds are furanoacetylenic wyerone derivatives. Nine furanoacetylenes have been recognized as phytoalexins in *Vicia faba* (Fig. 7).

The multicomponent phytoalexin response that is found in many plants is clearly demonstrated by bioassays carried out on thin layer chromatograms. Fig. 8 illustrates such a bioassay carried out on an extract of *Vicia faba* leaf tissue that had been

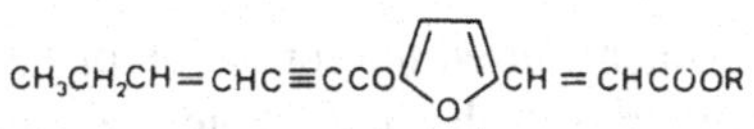

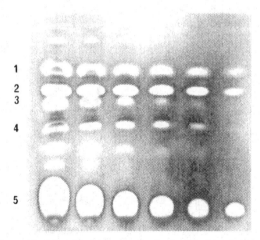

CH$_3$CH$_2$CH=CHC≡CCO⟨O⟩CH=CHCOOR

Wyerone acid R = H
Wyerone R = CH$_3$

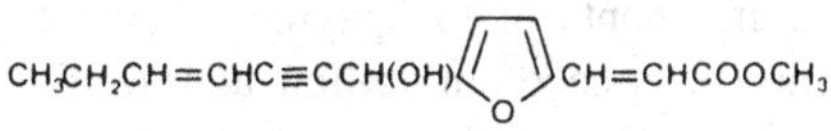

CH$_3$CH$_2$CH=CHC≡CCH(OH)⟨O⟩CH=CHCOOCH$_3$

Wyerol

CH$_3$CH$_2$CH$_2$CH$_2$C≡CCO⟨O⟩CH=CHCOOR

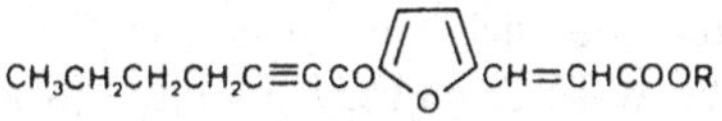

Dihydrowyerone acid R = H
Dihydrowyerone R = CH$_3$

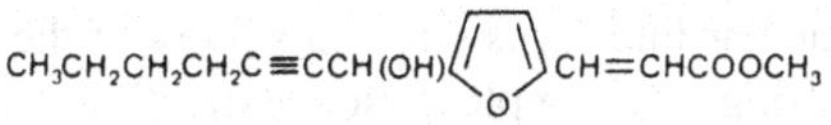

CH$_3$CH$_2$CH$_2$CH$_2$C≡CCH(OH)⟨O⟩CH=CHCOOCH$_3$

Dihydrowyerol

CH$_3$CH$_2$CH–CHC≡CCO⟨O⟩CH=CHCOOCH$_3$

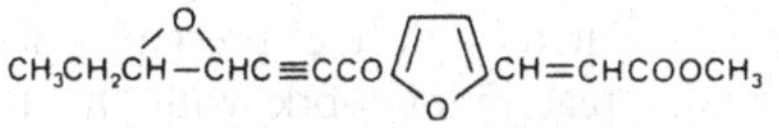

24 Wyerone epoxide

Figure 7. Furanocetylenic wyerone derivatives, part of the multicomponent phytoalexin response in *Vicia faba.*

Figure 8. Thin layer chromatography plate bioassay of extracts from inoculation sites in leaves of *Vicia faba* challenged by *Botrytis cinerea.* Tracks on the chromatogram from left to right were from diethyl ether extracts of 0.4, 0.2, 0.1, 0.05, 0.025, and 0.0125g of fresh tissue. Extracts were applied to 1cm origins and after development, the chromatogram was sprayed with a suspension of spores of *Cladosporium herbarum* in nutrient solution. The white zones of inhibition correspond to the presence of the phytoalexins:1, wyerone; 2, wyerone epoxide; 3, wyerol; 4, medicarpin and hydrohydroxyketo-wyerone (unresolved) and 5, wyerone acid. Note that no antifungal activity was detected in equivalent extracts from uninoculated tissue. Thin layer chromatography does not resolve the unsaturated and dihydro- forms of the wyerone derivatives (Fig. 7). Figure reproduced from Hargreaves *et al.* (1977).

challenged with *B. cinerea* which caused the formation of limited lesions (Hargreaves *et al.*, 1977). After chromatography, the TLC plate was sprayed with a suspension of spores of *Cladosporium herbarum* in nutrient solution. White zones of inhibition occurred where the dark green fungus failed to grow. Fig. 8 also illustrates the relative contribution of different components of the phytoalexin response. Wyerol is weakly antifungal and fails to cause a zone of inhibition at low dilution of the extract. By contrast wyerone epoxide, although present in comparatively low levels, is highly toxic and maintains activity in all samples. No activity was detected in healthy leaves using this assay technique. Dihydrowyerone derivatives (see Fig. 7) are not resolved from their unsaturated homologues by TLC; HPLC is required (Mansfield *et al.*, 1980).

The diversity of chemical classes and plant families in which phytoalexins have been identified is outlined in Table 2, examples of structures are given in Fig. 9 and in later sections. For a more detailed chemotaxonomic analysis, see Grayer and Harborne (1994). Of particular interest is a recent survey of the Rosaceae by Kokubun and Harborne (1994). Using the simple tissue extraction technique as applied to bean leaves (Fig. 8) phytoalexin production was detected in only about 15% of the species tested. Grayer and Harborne (1994) indicate that failure to give a positive phytoalexin response was correlated with the presence of phytoanticipins within the tissues. Another interesting feature of work with the Rosaceae is the striking variability in the phytoalexin response between different tissues. For example in *Malus pumila* (apple) the fruit and sapwood produce benzoic acid, and the biphenyls

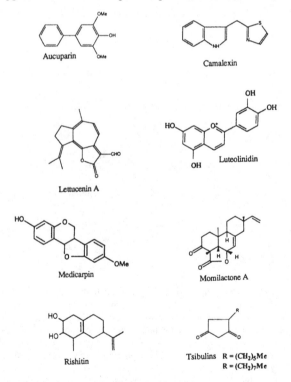

Figure 9. Phytoalexins from different chemical familites.

Table 2. Phytoalexins accumulate in many different plant families and are produced by the activation of diverse biosynthetic pathways

Plant family	Species	Compound(s)	Chemical class	Reference
Alliaceae	Allium cepa	Tsibulins	Cyclic dione	Tverskoy et al., 1991
Compositae	Lactuca sativa (lettuce)	Lettucenin A	Sesquiterpene lactone	Bennett et al., 1994
Cruciferae	Arabidopsis thaliana	Camalexin	Indole	Tsuji et al., 1992
Gramineae	Sorghum bicolor	Luteolinidin	Anthocyanidin	Snyder et al., 1991
	Oryza sativa	Momilactone A	Diterpene	Cartwright et al., 1981
Leguminosae	Medicago sativa (alfalfa)	Medicarpin	Isoflavonoid	Ingham, 1982
	Phaseolus vulgaris (French bean)	Phaseollin	Isoflavonoid	Ingham, 1982
Malvaceae	Gossypium spp (cotton)	Lacinilene C	Sesquiterpene	Essenberg et al., 1992b
Rosaceae	Malus pumila (apple)	Aucuparin	Biphenyl	Kokubun et al., 1994
		Benzoic acid	Phenolic acid	
Solanaceae	Solanum tuberosum (potato)	Rishitin	Sesquiterpene	Kuc, 1982
	Lycopersicon esculentum (tomato)			
Sterculiaceae	Theobroma cacao (cacao)	Arjunolic acid	Triterpene	Cooper et al., 1996
		3,4 dihydroxy-acetophenone	Acetophenone	
		Cyclooctasulphur	Elemental sulphur	
Umbelliferae	Petrostelium crispum (parsley)	Bergapten	Furanocoumarin	Hahlbrock et al., 1995
Vitaceae	Vitis vinifera (grapevine)	Viniferins	Stilbenes	Stoessl, 1982

aucuparin and 2'-methoxyaucuparin respectively. Such clear variation between tissues was not apparent in studies of the Leguminoseae (Ingham, 1981; 1982). There are also examples of members of different families producing the same phytoalexins. For example, the stilbene resveratrol has been found as a component of the phytoalexin response in *Arachis* (Leguminoseae), *Broussonetia* (Moraceae), *Festuca* and *Saccharum* (Gramineae), and *Veratrum*, a member of the Liliaceae (Grayer and Harborne, 1994). Although itself weakly antifungal (Langcake and Pryce, 1977), resveratrol probably polymerizes in the presence of peroxidase and H_2O_2 to produce more toxic oligomers such as the viniferins found in grapevine (*Vitis vinifera*) as shown in Fig. 10.

B.   BIOSYNTHESIS

A fundamental aspect of the definition of phytoalexins is that they are synthesized from remote precursors. Thus, simple labelling studies have demonstrated that the amino acid phenylalanine is used for the synthesis of a complex isoflavonoid such as phaseollin (Stoessl, 1982). Activation of defence responses usually leads to a massive, albeit transient and local diversion of normal metabolism into synthesis of groups of secondary products. The interactions between primary and secondary metabolism leading to biosynthesis of chemically diverse phytoalexins and the sources of essential substrates such as phenylalanine, acetyl-CoA, malonyl-CoA and mevalonic acid are outlined in Fig. 11. A much more detailed diagram of the current understanding of phenylpropanoid biosynthesis is given in Fig. 12 which relies heavily on the work of Dixon and colleagues, dealing in particular with alfalfa. The long term goal of their research is to define targets for genetic manipulation to enhance disease resistance in plants. Phenylpropanoids include several classes of phytoalexin including furanocoumarins, flavonoids, isoflavonoids and stilbenes, all of which are ultimately derived from *p*-coumarate. For certain compounds such as the stilbene resveratrol, a single enzymic step is required to produce the phytoalexin from common metabolites, whereas synthesis of glyceollin requires the co-ordinated activity of a multienzyme pathway. It is also apparent from Figs 11 and 12, that phytoalexin synthesis may generate a degree of competition between substrates, for example those used in lignin formation, or non-defensive activities such as the production of flavanoid pigments (Dixon *et al.*, 1992; Nicholson and Hammerschmidt, 1992).

In order to examine the control of biosynthetic pathways leading to phytoalexin production in detail, it has been necessary to use artificial model systems and in particular the use of cell cultures which produce phytoalexins after treatment with various elicitors. Experiments with cell cultures have shown that induction of phenylpropanoid synthesis is typically the result of increased transcription of genes encoding the corresponding biosynthetic enzymes (Kneusel *et al.*, 1989; Dixon and Paiva, 1995). Increased transcription rates for enzymes of both the central phenylpropanoid pathway and specific branch pathways of isoflavonoid or furanocoumarin synthesis are observed at the onset of the phytoalexin response in elicitor-treated cell suspensions of alfalfa and parsley, respectively (Oommen *et al.*, 1994; Dixon and Paiva, 1995; Hahlbrock *et al.*, 1995). The transcriptional changes which occur have implications for the signal transduction mechanisms involved.

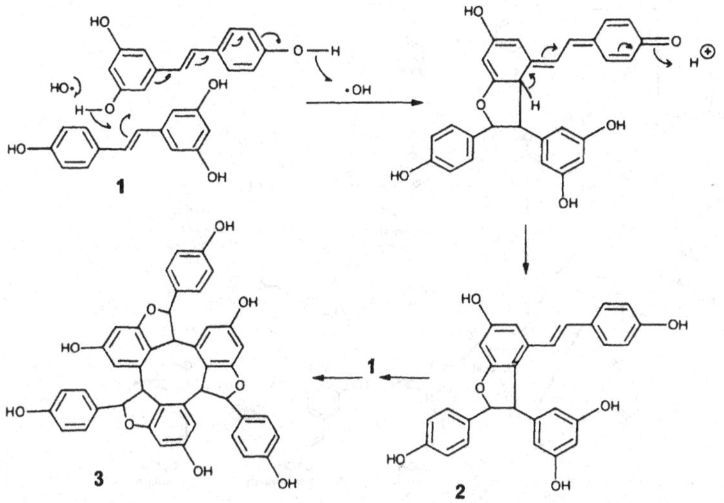

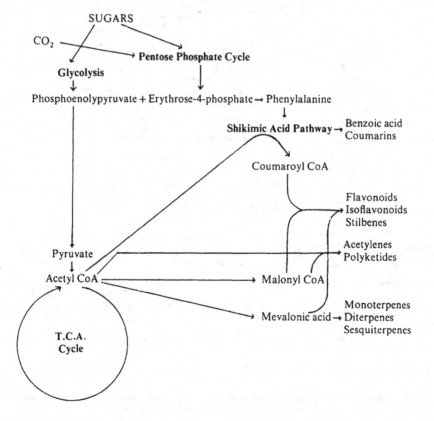

Figure 10. Possible routes to the synthesis of viniferins from resveratrol in grapevine, 1, resveratrol; 2, ε-viniferin; 3, α-viniferin (from Stoessl, 1982).

Figure 11. Interrelationships between primary metabolism and synthesis of phytoalexins (adapted from Bailey, 1982).

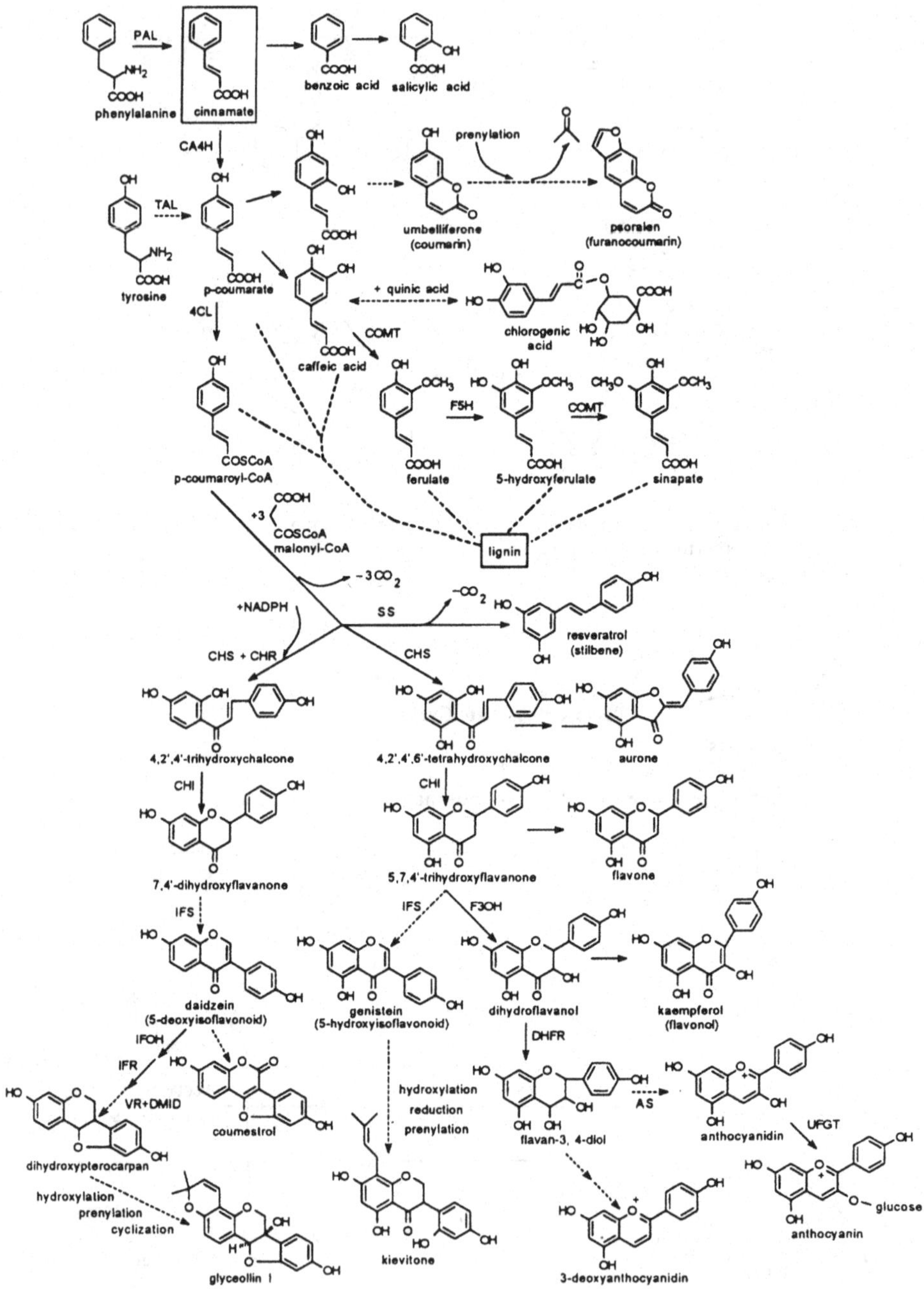

Figure 12. Biosynthesis of phenylpropanoids including the phytoalexins benzoic acid, psoralen, resveratrol, glyceollin 1, kievitone and 3-deoxyanthocyanidin (from Dixon and Paiva, 1995).

For example, transcription of phenylalanine ammonia lyase (PAL) and chalcone synthase (CHS) genes in bean and alfalfa is extremely rapid and co-ordinated whereas transcription of some branch pathway enzymes, such as the bergaptol -methyltransferase required for furanocoumarin synthesis in parsley, may be delayed (Hahlbrock *et al.*, 1995). These findings imply the involvement of multiple signals for activation of complex pathways as a whole. Consistent with this picture is the existence of common sequence motifs in the promoters of PAL and CHS genes from a number of sources (Dixon and Harrison, 1990) whereas genes encoding later branch pathway enzymes, such as the isoflavone reductase (IFR) of pterocarpan phytoalexin biosynthesis in alfalfa may lack exact copies of these motifs (Oommen *et al.*, 1994). Further analysis of transcriptional regulators is needed to identify the factors involved in orchestrating phytoalexin biosynthesis (Dixon *et al.*, 1995).

An elicitor used with great success in biochemical analysis of alfalfa is derived by heat treatment of yeast cell walls; plants are unlikely to be exposed to the complex product under natural conditions (Dixon *et al.*, 1995). Having used such elicitors to unravel the fascinating biochemistry underlying biosynthesis it will, therefore, be necessary to extend experiments to live pathogens and whole plants to determine if the same controls operate within infected tissue.

C. *THEOBROMA CACAO, VERTICILLIUM DAHLIAE* AND PHYTOALEXIN LOCALIZATION

Perhaps the simplest phytoalexin discovered to date is elemental sulphur as cycloocta-sulphur (S8) which was detected with other phytoalexins in *Theobroma cacao* (cocoa) stems challenged by the vascular wilt fungus *Verticillium dahliae* (Cooper *et al.*, 1996). In addition to sulphur, two acetophenone derivatives and the triterpenoid arjunolic acid (Fig. 13) contributed to the phytoalexin response. Sulphur-containing phytoalexins are otherwise known only in the indole-based compounds of crucifers, such as camalexin (Fig. 9). The concentrations of the phytoalexins found in stems of a resistant cocoa genotype and their toxicities are summarized in Table 3; note that no significant antifungal activity was found in the susceptible genotype.

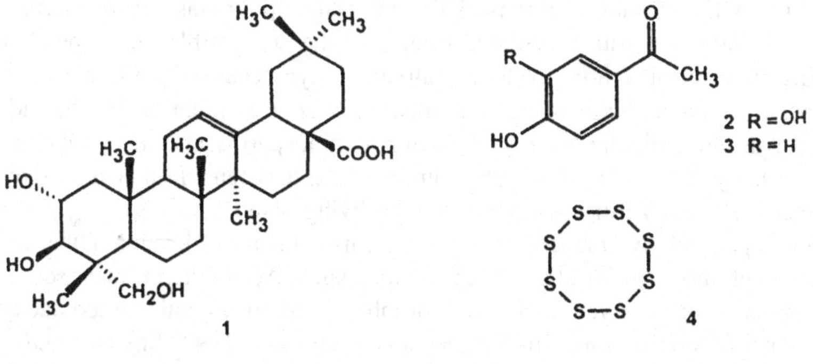

Figure 13. Phytoalexins from cocoa represent three chemical families; 1) arjunolic acid; 2) 3,4-di-hydroxyacetophenone; 3) 4-hydroxyacetophenone;4) sulphur.

Table 3.   Toxicities and levels of phytoalexins from stems of cocoa resistant to *Verticillium dahliae* (data from Cooper *et al.*, 1996)

Phytoalexin	ED5O ($\mu g\ ml^{-1}$) germination [a]	Concentration in inoculated stems ($\mu g\ g^{-1}\ fr.\ wt$) [b]
Arjunolic acid	12.8	168
3,4-Dihydroxyacetophenone	92.5	4.9
4-Hydroxyacetophenone	7.2	2
Sulphur	3.6	51

[a] Conidia were incubated in distilled water for 15h at 25°C

[b] Tissues were extracted 15 days after inoculation of cv. Pound-7. No fungitoxicity was detected in extracts of a susceptible genotype

Verticillium dahliae is a vascular wilt fungus which preferentially invades and multiplies in xylem vessels. The fungus either penetrates through vessel end walls as mycelia or is dispersed as microconidia in the transpiration stream (Durrands and Cooper, 1988). In resistant genotypes, inhibition of growth within vessels is often associated with the formation of tyloses and secretion of gels into the vessel lumen providing, potentially, both physical and chemical barriers to colonization (Cooper, 1981). The importance of the localization of any phytoalexins at the sites of restriction of fungal growth is, therefore, particularly relevant with the vascular wilt fungi. Identification of sulphur as a phytoalexin allowed its localization by the highly specific and sensitive technique of coupled scanning electron microscopy and energy dispersive X-ray microanalysis. Results obtained are illustrated in Fig. 14.

High concentrations of sulphur were found in scattered xylem parenchyma cells in direct contact with xylem vessels, within vessel walls and in gels occluding xylem vessels. The remarkable localization of sulphur indicates that it is present close to the invading fungus at concentrations much higher than calculated in terms of tissue fresh weight (Table 3). The distribution of sulphur within xylem parenchyma was consistent with its accumulation in cells observed to undergo the HR during attempted colonization by the resistant genotype. No elemental sulphur was recovered from intact or wounded plants or from inoculated tissues of the susceptible genotype. It will be fascinating to determine how cyclooctasulphur is synthesized in cocoa and whether it may act as a phytoalexin in other plants; sulphur accumulation has already been reported in the epicuticular waxes of several gymnosperms and angiosperms (Kylin *et al.*, 1994). As Cooper *et al.* (1996) comment, "....it is tenable that man's probably oldest pesticide already functions in this role in living plants".

Sulphur is particularly amenable to analysis of its cellular localization. Other attributes have also been used to localize other phytoalexins. At the light microscope level, localization has been achieved on the basis of inherent colour as with the red deoxyanthocyanidin phytoalexins of sorghum (Snyder and Nicholson, 1990; Snyder *et al.*, 1991). Autofluorescence has also been used with some success, notably with wyerone derivatives in *Vicia faba*, and lacinilenes in cotton (Mansfield, 1982; Essenberg *et al.*, 1992a).

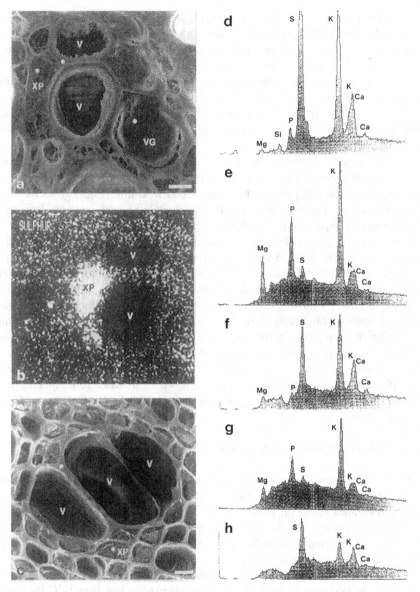

Figure 14. Localization of sulphur in cocoa plants challenged by *Verticillium dahliae*. Accumulation of elemental sulphur as a phytoalexin is located using energy dispersive X-ray analysis of transverse sections of stems.

XP, xylem parenchyma cell; V, xylem vessel; VG, vascular gel. The X-ray map (*b*) was produced from the equivalent area in (*a*); spot analyses (*d-h*) were made from *a* and *c* at the points marked *; the bar on *a*, *b* represents 10µm. In uninoculated stems (*c*) localized accumulation of S was never observed and spot analyses of XP cells revealed S as a minor peak with K the predominant element (*e*). Inoculated stems (*a*) contained scattered XP cells with a high accumulation of S evident from the X-ray dot maps (*b*); X-ray spectra revealed S peaks often equivalent to or greater than K (*d*). Adjacent xylem vessel cell walls also contained an unusually high proportion of S (*f*) compared with walls in control plants (*g*). In vascular gels, S was sometimes the main element (*h*); the non-cytoplasmic nature of these structures is apparent from the low K content. Figure reproduced from Cooper *et al.* (1996) and kindly provided by Richard Cooper.

Histochemistry is generally applicable to groups of related compounds and therefore lacks specificity (Mace *et al.*, 1978; O'Brien and McCully, 1981; Dai *et al.*, 1996).

A very specific method, but one which is not generally available, is laser microprobe mass analaysis. Moesta *et al.* (1982) have used this technique to localize glyceollin to groups of cells in soybean. The fine beam of the laser is focused on sites within single cells and ionized and volatilized phytoalexin are detected and identified by mass spectroscopy. This approach is substance specific but does not produce quantitative data, also the laser tends to damage more tissue than would be ideal, limiting localization to, at best, the level of a single cell.

In a valuable combination of microscopy and phytoalexin analysis, fluorescence-activated cell sorting has been used to isolate cells containing high concentrations of the fluorescent sesquiterpene phytoalexins found in cotton (Fig. 15). Cells which have undergone the HR in cotton cotyledons inoculated with an avirulent strain of *Xanthomonas campestris* pv. *malvacearum* display a yellow/green fluorescence characteristic of lacinilene derivatives (Pierce and Essenberg, 1987). Responding mesophyll tissues were digested with macerating enzymes and suspensions containing mixtures of responding and unaffected cells sorted into two groups, either cells with bright fluorescence or those with at most a weak response (Fig. 15). High concentrations of lacinilene C and 2,7-dihydroxycadalene were found in cells which had undergone the HR and were highly fluorescent. This study of cotton is one of few in which an attempt has been made to correlate precise localization with quantitative analysis. The overall conclusion was that more than 90% of the most active phytoalexins recovered from whole cotyledons are concentrated in dead, fluorescent cells at infection sites. Average phytoalexin concentrations in these cells were calculated to be much greater than those required to inhibit bacterial growth *in vitro* (Essenberg *et al.*, 1992a; Pierce *et al.*, 1996).

D.   *COLLETOTRICHUM*, BEAN AND THE HR

The interaction between *Coletotrichum lindemuthianum* and bean (*Phaseolus vulgaris*) has provided a very good model for studies of the role of phytoalexins in the expression of varietal resistance based on gene-for-gene interactions. The more detailed studies have examined hypocotyl infections. *Colletotrichum lindemuthianum* is an example of a fungus which has a temporarily biotrophic relationship with its host (Bailey *et al.*, 1992). Following penetration into epidermal cells large primary hyphae are produced which grow between the plant plasma membrane and cell walls. These intracellular hyphae penetrate into surrounding cells and at temperatures less than 17°C the fungus may develop in this biotrophic state for several days. The production of narrow, secondary hyphae is associated with a shift to necrotrophy, the pathogen causing more extensive plant cell wall degradation and killing plant cells in advance of colonization as the black anthracnose lesion develops. Varietal resistance is expressed by the HR of penetrated cells: the timing of plant cell collapse and restriction of fungal growth depending on the resistance gene in the plant. Resistance to colonization within genetically susceptible tissue also occurs as lesion limitation at higher temperatures (Bailey *et al.*, 1980).

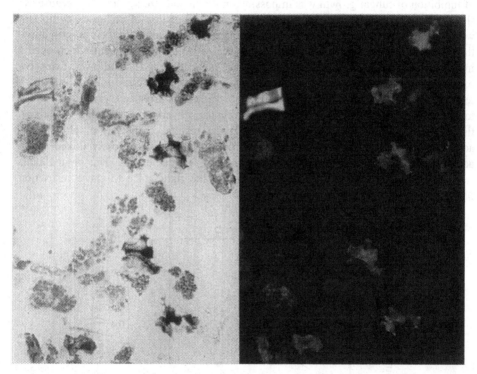

Figure 15. Structures of the phytoalexins from cotton; 1) dihydroxy cadalene and 2) lacinilene C. A, transmitted white light and B, fluorescence (excitation, 460-485nm) micrographs of mesophyll cells isolated from a cotyledon of the bacterial blight-resistant cotton line OK1.2 inoculated with *Xanthomonas campestris* pv. *malvacearum,* in preparation for fluorescence activated cell sorting. The yellow/green autofluorescent cells which have undergone the HR, contain high concentrations of the lacinilene phytoalexins. Red autofluorescence in the undamaged cells is due to chlorophyll. Micrographs kindly provided by Margaret Pierce.

Phaseolus vulgaris produces several isoflavonoid phytoalexins including phaseollin, phaseollinisoflavan, phaseollidin and the isoflavanone, kievitone. Phaseollin accumulates to greatest concentrations and has considerable activity against germ-tubes of *C. lindemuthianum* being fungicidal at 3µg ml^{-1} (9.4 x 10^{-6}M). Activity against the greater biomass of mycelium is much less, ED50 concentrations against growth in liquid culture being about 30µg ml^{-1}. In susceptible tissues challenged by *C. lindemuthianum*, no phytoalexins accumulate during the initial biotrophic phase of colonization. By contrast, the onset of the HR is associated with rapid phaseollin production. As reported in several interactions, the bean phytoalexins were found to be remarkably localized to cells which had undergone the HR. Dissection experiments indicated that intracellular hyphae are probably exposed to concentrations of more than 3 mg g^{-1} fr.wt. within epidermal cells, many times the levels fungicidal to germ-tubes *in vitro* (Bailey and Deverall, 1971).

Despite the positive correlation between phytoalexin accumulation and the restriction of *Colletotrichum*, Bailey *et al.* (1980) pointed out that a critical analysis of the timing of inhibition of fungal growth was impossible in tissue undergoing the HR, because the short hyphae produced were obscured by brown pigmentation of collapsed cytoplasm within the hypersensitive cell. In order to overcome this limitation to their investigation, they utilized the temperature-sensitive resistance to lesion development which occurs in this interaction. Using a susceptible cultivar, incubation at 16°C allowed extensive biotrophic growth for three days. Hypocotyls were then transferred to 25°C at which temperature a shift to necrotrophy occurred but the lesions produced were soon limited to the site of inoculation. The growth of hyphae observed within the inoculated tissue is recorded in Fig. 16 which also includes results of analysis of phytoalexin accumulation. The slowing down in growth rate of hyphae was closely linked to the accumulation of phaseollin and phaseollinisoflavin within lesions. The phytoalexins were, therefore, present at the right time and in the right place to cause restriction of fungal growth.

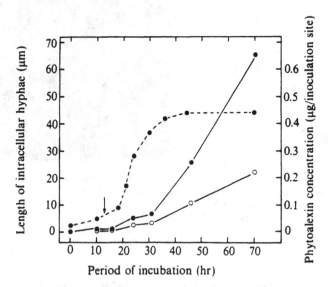

Figure 16. Relationships between phytoalexin accumulation and the restriction of growth of *Colletotrichum lindemuthianum* in *Phaseolus vulgaris*. Hypocotyls were inoculated, incubated at 16°C for 72h (= 0 on the x axis) and transferred to 25°C to induce a resistant response. The arrow indicates when cellular necrosis became visible. Restriction of hyphal growth (•- - - - - •) followed accumulation of phytoalexins, phaseollin (•—•) and phaseollinisoflavan (o —— o); adapted from Bailey *et al.* (1980).

The close association between plant cell death during the HR or limited lesion formation, prompted Bailey and colleagues to examine the accumulation of phytoalexins following treatments which killed some cells within a largely healthy tissue - to some extent mimicking the occurrence of the HR. Treatment with chemicals to kill surface layers and freezing and thawing of regions of bean hypocotyls, were found to lead to phytoalexin accumulation. Hargreaves and Bailey (1978) designed an interesting series of experiments using whole or part freezing of hypocotyls, and bringing frozen (and thawed) and live tissues together before incubation and subsequent extraction. They found that phytoalexin accumulation took place at the interface between living and dead tissue; live cells were essential for synthesis of phytoalexins but dead cells were the sites of their accumulation. Additional experiments in which the metabolism of phaseollin by living bean tissue led Bailey (1982) to propose the operation of the scheme outlined in Fig. 17 to explain the observed accumulation of phytoalexins during the HR in bean.

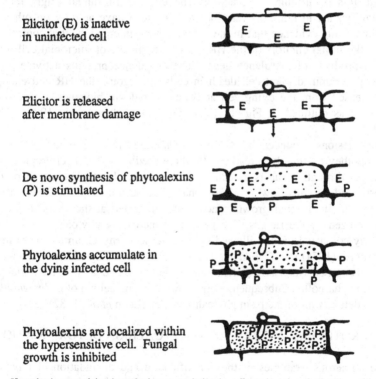

Elicitor (E) is inactive in uninfected cell

Elicitor is released after membrane damage

De novo synthesis of phytoalexins (P) is stimulated

Phytoalexins accumulate in the dying infected cell

Phytoalexins are localized within the hypersensitive cell. Fungal growth is inhibited

Figure 17. Hypothesis to explain phytoalexin accumulation in cells undergoing the hypersensitive reaction (HR). This scheme was initially proposed for the interaction between *Colletotrichum lindemuthianum* and French bean (adapted from Bailey, 1982).

A key component of the proposal is that phytoalexin synthesis is induced in living cells by factors termed endogenous elicitors which are released from the cells undergoing hypersensitive cell death. The phytoalexins accumulate within the dead cells to inhibit fungal growth and any excess in living tissue is metabolized by the plant when lesion development has been restricted. Such a simple hypothesis satisfactorily explains the highly localized accumulation of phytoalexins in tissues undergoing the HR as has been observed in all plant-microbe interactions studied in detail. It seems probable that cell suspension cultures, after treatment with elicitors, respond in a manner which mimics that of cells surrounding those undergoing the HR.

The production of elicitors of phytoalexin accumulation by *C. lindemuthianum* has also been examined. Several different components of the fungal cell wall, including glucans and glycoproteins (Tepper *et al.*, 1989; Coleman *et al.* 1992) have been found to elicit phaseollin accumulation but they do so in a pattern of activity which is not highly cultivar specific. Thus, the glucan elicitor from one isolate will induce phytoalexin production in tissue from both resistant and susceptible cultivars. As susceptible cells are directly exposed to the fungal cell wall during biotrophic infection but do not produce phytoalexins, it is difficult to reconcile a genuine role for the cell wall derived elicitors in this plant/fungus interaction. The elicitors of fungal origin may be of greater significance in non-host reactions as they also have activity on other plants. There is increasing evidence that in the gene-for-gene interactions that control varietal resistance, the initial elicitors of the HR are not fragments of microbial cell walls but the protein products of avirulence genes. The avirulence proteins activate processes leading to programmed plant cell death in cells undergoing the HR and endogenous elicitors released from the dying cell activate phytoalexin biogenesis in surrounding living tissues as outlined in Fig. 17 (Van den Ackerveken *et al.*, 1996; Mansfield *et al.*, 1997).

Although lesions produced by *C. lindemuthianum* are limited at 25°C, at lower temperatures they eventually spread to rot whole seedlings. The necrotrophic progress of the fungus through tissue is associated with low levels of phytoalexin production at the lesion edge. Only low levels of phaseollin accumulate, much less than that required to inhibit mycelial growth, and none is detected at the lesion centre which is fully colonized by the fungus. The pattern of increasing phytoalexin concentration within newly colonized tissue followed by a decrease as mycelium advances indicated to Burden *et al.* (1974) that *C. lindemuthianum* was able to degrade the phytoalexins to which it was exposed. Such a detoxification was demonstrated *in vitro*, the initial product being hydroxyphaseollin. Subsequent work showed that mycelium of *C. lindemuthianum* was able to detoxify all of the bean phytoalexins (VanEtten *et al.*, 1982).

E.   NECROTROPHS, *NECTRIA* AND PHYTOALEXIN DETOXIFICATION

There are numerous examples of the detoxification and degradation of phytoalexins by fungi (VanEtten *et al.*, 1982; 1995). As described for phytoanticipins, it has often been argued that ability to detoxify the host's phytoalexin is an important determinant of pathogenicity. VanEtten *et al.* (1982) listed 22 phytoalexins (including sesquiterpenoids, furanoterpenoids, furanoacetylenes and isoflavonoids) which were degraded by pathogens of their plant of origin.

The formation of a necrotic lesion within plant tissue will inevitably lead to the activation of some phytoalexin biosynthesis following the release of endogenous elicitors. The amount of phytoalexin which accumulates depends on the rate and duration of phytoalexin synthesis which will be greatly affected by the speed at which cells are killed by the advancing pathogen, and the ability of the fungus or bacterium to tolerate and detoxify the phytoalexins to which it may be exposed. Consideration of these factors has led to the concept of a balance between phytoalexin production by the plant and phytoalexin degradation by the pathogen occurring at microsites within infected tissue.

The operation of such a balance in the interaction between *Botrytis* spp. and *Vicia faba* is illustrated in Fig. 18. *Botrytis fabae* is a totally necrotrophic pathogen of bean leaves, the chocolate spot fungus penetrates rapidly into and degrades epidermal cell walls killing numerous cells around penetration points. Areas of dead tissue soon coalesce and the fungus ramifies through the leaf producing dark brown lesions which rapidly spread from the inoculation site. Species that are not pathogens of been, such as *B. cinerea*, *B. elliptica* and *B. tulipae* also penetrate into the epidermis but kill far fewer cells than *B. fabae* during the early stages of infection and their hyphae stop growing within epidermal cell walls by 12 h after inoculation. Lesions which develop at sites inoculated with the non-pathogens are confined to the tissue beneath inoculum droplets (Mansfield and Hutson, 1980).

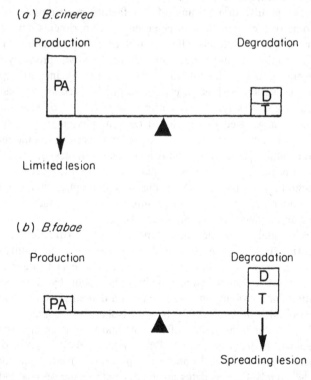

Figure 18. Lesion development and the balance between phytoalexin production by the plant and degradation by the pathogen. The development of *B. cinerea* and *B. fabae* in *Vicia faba* are compared. The degradation weighting incorporates both metabolic capacity to detoxify the inhibitors (D) and a tolerance factor (T) which accounts for the differential sensitivity of the species to wyerone derivatives.

Restriction of *B. cinerea* was associated with the rapid accumulation of the phytoalexins wyerone and wyerone acid (and their dihydro-homologues, see Fig. 7) to concentrations which were fungicidal to *B. cinerea in vitro* (Mansfield, 1982). When spreading lesions were examined, an initial increase in phytoalexin concentrations was followed by a decrease as tissue became completely necrotic and colonized by *B. fabae*. In applying the balance hypothesis to the interaction between *B. cinerea* and *B. fabae* as shown in Fig. 18, the weightings applied to each side of the fulcrum have been chosen on the basis of the following results. 1) Both *B. cinerea* and *B. fabae* possess the enzymic capacity to detoxify (D) wyerone derivatives. 2) *B. fabae* is more tolerant (T) of the inhibitors and therefore able to grow and detoxify higher concentrations. 3) Bean cells are killed much more rapidly by *B. fabae* than by *B. cinerea*, thereby establishing hyphae within a lesion containing low phytoalexin (PA) concentration at the start of the interaction (Rossall *et al.*, 1978; Mansfield, 1980, 1982). Mutational analysis of *B. fabae* has revealed that pathogenicity is more closely correlated to cell killing ability than to wyerone acid tolerance, but none of the mutants obtained were as sensitive to the phytoalexin as the wild-type isolate of *B. cinerea* (Hutson and Mansfield, 1980). The mechanisms underlying the non-degradative tolerance of *B.fabae* are not understood but may be associated with a reduced frequency of sensitive sites in the fungal cell membrane (Rossall *et al.*, 1977).

In bean leaves, modification of any of the factors contributing to the balance affects the outcome of the interaction in bean leaves. Experiments have demonstrated the expected shifts occurring to spreading or limited lesion formation. For example, decreasing spore numbers reduces development of *B. fabae* probably by reducing the numbers of cells killed at inoculation sites and thereby increasing the capacity for phytoalexin production and accumulation around invading hyphae (Mansfield and Hutson, 1980). Conversely, the increasing susceptibility of senescing leaves to *B. cinerea* has been associated with reduced rates of phytoalexin biosynthesis such that wyerone acid does not accumulate to levels which restrict even the more sensitive species (Abu-Blan and Mansfield, unpublished data). Studies of the *Botrytis/Vicia* interaction highlight the many factors which determine the success or failure of necrotrophic fungi to colonize their host plants and emphasize the importance of analysis of the rate of plant cell death at infection sites, a factor which is often overlooked in the absence of microscopical studies.

The most detailed analyses of the role of phytoalexin detoxification in pathogenicity is the work carried out by VanEtten and colleagues on the ability of the pea pathogen *Nectria haematococca* (anamorph *Fusarium solani*) to degrade the major pea phytoalexin, pisatin. Detoxification is brought about by a cytochrome P_{450}-mediated demethylation involving the enzyme pisatin demethylase (PDA), as shown in Fig. 19 (Matthews and VanEtten, 1983). Natural isolates of *N. haematococca* mating population (MP)VI, occur in a number of habitats and have been classified into three phenotypes: inability to detoxify pisatin (Pda⁻); low levels of pisatin detoxification after long exposure to pisatin (Pda^L) and rapid pisatin-induced production of pisatin demethylase (Pda^H). Only Pda^H isolates are highly pathogenic on pea and the genetic linkage between pathogenicity and the presence of Pda^H has not been broken in crosses (VanEtten *et al.*, 1980; Tegtmeier and VanEtten, 1982). Conventional genetic analysis has identified three Pda[H] genes (*PDA1, 4* and *5*) and four Pda[L] genes (*PDA2, 3, 6-1*and *6-2*)as discussed by VanEtten *et al.* (1995).

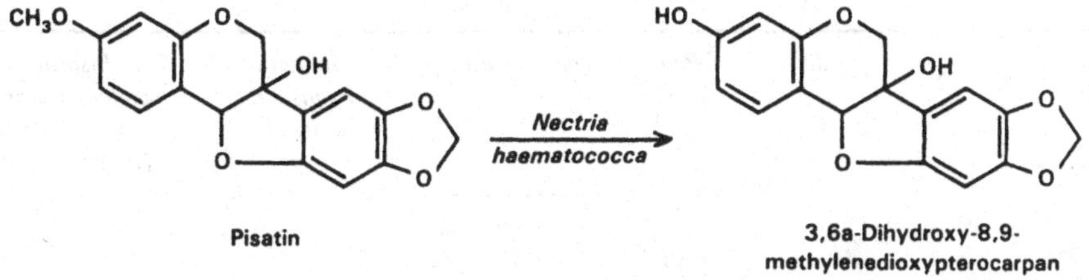

Pisatin

3,6a-Dihydroxy-8,9-methylenedioxypterocarpan

Nectria haematococca

Figure 19. Detoxification of pisatin by the enzyme pisatin demethylase which is a cytochrome P450 mono-oxygenase.

The cloning of the *PDA1* gene by Weltring *et al.* (1988) has allowed the role of pisatin demethylation in pathogenicity to be examined with more precision. Wherever a gene has been cloned, single gene changes can be made to fungi by either the transformation of the cloned gene into a recipient that lacks that gene, or specific mutation of the wild type isolate by transformation mediated gene disruption. When the *PDA[H]* gene was transformed into the maize pathogen *Cochliobulus heterostrophus*, it greatly increased the ability of the fungus to cause lesions on pea, indicating that PDA enhanced pathogenicity and also that pisatin accumulation had a major role in restricting growth of the untransformed strain (Schäfer *et al.*, 1989). Similar increases in pathogenicity to pea were found in *Ascochyta rabiei* (chickpea pathogen) or a Pda⁻ isolate of *N. haematococca* after transformation with *PDA[H]* (Weltring *et al.*, 1995; Ciufetti and VanEtten, 1996; Barz and Welle, 1997).

Surprisingly, the targeted disruption of *PDA[H]* to construct Pda⁻ mutants of three Pda[H] isolates of *N. haematococca* reduced but did not completely eliminate pathogenicity to pea. Some data obtained for pathogenicity tests of Pda⁻ mutants and wild-type strains by Wasmann and VanEtten (1996) are summarized in Table 4. Targeted gene disruption has apparently broken the absolute link between Pda[H] and pathogenicity which had been established by conventional genetic analysis.

An explanation of how loss of pisatin demethylating ability had always been associated with complete loss of pathogenicity in crosses, has come from observations of isolate karyotypes. The *PDA[H]* genes have been located on small dispensable chromosomes or dispensable regions of the *PDA[H]* containing chromosomes (Miao *et al.*, 1991; Miao and VanEtten, 1992). Loss of the dispensable chromatin occurs during crossing and therefore *PDA⁻* progeny have probably also lost other genes which may also have an important role in pathogenicity. Wasmann and VanEtten (1996) coined the name *PEP[D]* (pea pathogenicity dispensable chromosome) genes for these potentially important determinants of pathogenicity.

As discussed by Wasmann and VanEtten (1996) the demonstration that Pda⁻ mutants generated by gene disruption can produce significant lesions on pea, even in the presence of large amounts of pisatin (Table 4), raises questions regarding the relative importance of pisatin demethylase for ability to grow in the presence of this phytoalexin. It is clear that pisatin degradation is a tolerance mechanism because when *PDA* genes are transformed into Pda⁻ fungi their sensitivity to pisatin decreases

Table 4.   Pisatin demethylase and pathogenicity to pea: characteristics of parental isolate, transformants
with disrupted *PDA*-gene and ascospore isolates with Pda⁻ phenotype (data compiled from
Wasmann and VanEtten, 1996).

Isolate	Pda[a]	Pea pathogenicity test		% inhibition by pisatin $(160\ \mu g\ ml^{-1})^d$	Pisatin concentration in lesions $(\mu g\ g^{-1}\ fr\ wt)^e$
		Lesion length[b]	% reduction of wild-type[c]		
Pathogenic wild-type					
77-13-5	+	13.9	-	36	NT[f]
Transformants of 77-13-5					
Tr 6.2	-	8.2*	41.0	59	NT
Tr 7.8	-	8.6*	38.1	59	NT
Tr 53.1	-	8.4*	40.0	58	NT
Pathogenic wild-type					
77-13-7	+	11.8	-	27	76
Transformants of 77-13-7					
Tr 18.2	-	9.0*	23.7	64	7,644
Tr 18.5	-	9.4*	20.3	65	4,192
Pda- ascopore isolates					
156-2-1	-	4.0	-	NT	6,138
44-100	-	3.4	-	58	NT

[a] Pda phenotype
[b] Pathogenicity assays on pea stems. Values differing significantly P (< 0.05) from that of the corresponding
recipient are denoted by an asterisk.
[c] Entries are the % reduction in the mean lesion size of the transformant relative to the corresponding
recipient isolate
[d] Assessed by radial growth of mycelium on agar
[e] Pisatin recovered from lesions six days after inoculation
[f] NT, not tested

(Weltring *et al.*, 1995; Ciuffetti and VanEtten, 1996). Denny *et al.* (1987) have shown,
however, that Pda⁻ isolates of *N. haematococca* MPVI, whether naturally occúrring or
engineered mutants, are substantially more tolerant of pisatin than related fungi such as
N. haematococca MPI. This 'non-degradative tolerance' of *N. haematococca* MPVI
to pisatin is an induced property which may involve a change in membrane structure
or function that decreases the net intracellular concentration of pisatin (Denny *et al.*,
1987). Thus, *N. haematococca* MPVI apparently utilizes at least two types of pisatin
tolerance mechanism to circumvent the toxic environment created by the synthesis
of pisatin in infected pea tissue. Such a redundancy of tolerance mechanisms may
explain why certain Pda⁻ mutants, though reduced in their vigour, remain pathogenic
on pea. The balance as outlined in Fig. 18, probably operates in the interaction between

N. haematococca and pea. Although pisatin metabolism has been examined in great detail, other factors of potential significance contributing to the balance remain to be studied.

F. *ARABIDOPSIS* AND CAMALEXIN

The mutational analysis of the oat plant adopted by Osbourn (1996b) has provided evidence that avenacin has a key role in resistance to *G. graminis* var *tritici*. By contrast, a similar approach using *Arabidopsis* has provided strong evidence *against* the role of the phytoalexin camalexin in restricting multiplication of pathovars of the bacterium *Pseudomonas syringae* (Glazebrook and Ausubel, 1994; Glazebrook *et al.*, 1996). The indole phytoalexin fluoresces under UV radiation and this property was used in a rapid screen for camalexin production following challenge with the virulent bacterium *P.s.* pv. *maculicola*. Three phytoalexin deficient mutants were isolated and their phenotype found to be determined by recessive alleles of three different genes, *pad1*, *pad2* and *pad3*. Following challenge by an avirulent strain of *Pseudomonas*, very low levels of the phytoalexin were found in *pad1* and *pad2* mutants whereas no camalexin could be detected in *pad3* plants.

Infection of *pad* mutant plants with strains carrying cloned avirulence genes, and which caused a rapid HR, revealed that the *pad* mutation did not greatly affect the plant's ability to restrict the growth of these strains, but the *pad1* mutation did allow more rapid growth of avirulent strains before their restriction. Both *pad1* and *pad2* allowed significantly more rapid growth of virulent strains of *P.s.* pv. *tomato* and *P.s.* pv *maculicola*, whereas in *pad3* mutant plants (which produced no detectable camalexin), growth of each strain was the same as in the wild-type. These results, particularly the lack of effect of the *pad3* mutation on infection development, argue that camalexin does not have any role in the expression of resistance. Glazebrook and Ausubel (1994) suggested that the *pad1* and *pad2* mutations may exert pleiotropic effects on other defence responses which are required for limiting pathogen growth. For example, the *pad1* and *pad2* mutations may be in genes encoding components of the signal transduction pathway leading to activation of defence responses. They explained the phenotype of the *pad3* mutant by suggesting that this mutation may block the camalexin biosynthetic pathway at a point at which a precursor accumulates to antimicrobial concentrations, i.e. acts as a phytoalexin.

The elegant genetical analyses of camalexin production have generated numerous areas for further work, not least the molecular characterization of the *pad* mutations. Even without the *pad* mutants, however, the evidence available demonstrating the low concentrations of camalexin reached during interactions with *P. syringae* and the weak activity of the phytoalexin against the bacterium, argue against a major role for camalexin in disease resistance (Tsuji *et al.*, 1992; Rogers *et al.*, 1996). Camalexin may be one component of a defence response in which phytoalexin accumulation has variable importance depending on the challenging microbe. In this respect it will be of great interest to examine the responses of *pad* mutants to other pathogens such as the downy mildew fungus, *Peronospora parasitica*. It is unfortunate that it has not yet been possible to carry out similar genetical analyses using a system in which the biochemical and physiological evidence argues strongly in favour of a key role for phytoalexins in resistance, for example in the French bean/*Colletotrichum* interaction.

IV.   Phytoanticipins and phytoalexins in disease control

A.   AS FUNGICIDES

Perhaps surprisingly, there have been few critical assessments of the use of phyto-anticipins or phytoalexins as fungicides (Kuc, 1992). One exception is the study of seven isoflavonoid phytoalexins from bean carried out by Rathmell and Smith (1980); their results were not very encouraging. Although kievitone, in particular, had some protectant activity it did not compare with established fungicides, benomyl and mancozeb. Three features argue against the use of the plant's own antibiotics; 1) comparative lack of activity compared with highly toxic fungicides, at 10^{-5} compared with 10^{-7} M in many cases; 2) their high mammalian toxicity, and 3) lack of systemic activity - they do not move throughout the plant but are concentrated often at sites of tissue damage. Such a localization is advantageous for their role in the plant but not as an applied chemotherapeutic agent.

There is one fungicide, sulphur, which we now know is a phytoalexin in cocoa. Given the enormous diversity of chemical families containing phytoanticipins and phytoalexins, it would seem probable that some additional compounds should prove valuable either as fungicides or bacteriocides themselves or as important lead compounds which, with minor structural modification, might prove useful in disease control. Such a scenario has led to the development of a new group of fungicides, the methoxyacrylates based on the antifungal strobilurins produced by the toadstools *Oudemansiella mucida* and *Strobilurus tenacellus* (Godwin *et al.*, 1992).

B.   GENETICAL MANIPULATION TO ENGINEER RESISTANCE

Consideration of the interactions between *G. graminis* and its hosts wheat and oats, raises the simple idea that engineering wheat to produce avenacin A-1 would confer resistance to the very damaging take-all disease. Such an approach is, however, impractical because of the complex biosynthetic pathway leading to avenacin accumulation. The wheat plant would have to be engineered to produce not one new enzyme but several which would allow controlled diversion of metabolites from their normal functions in wheat cells. Even if production of avenacin were to be achieved, there may be a penalty to be paid by the disruption of normal metabolism which might reduce plant growth.

Fortunately, not all potentially useful antimicrobial compounds require complex biogenesis and there has already been one pioneering report (Hain *et al.*, 1993) of the successful introduction of a new phytoalexin, the stilbene resveratrol, into tobacco which normally produces the sesquiterpenoid phytoalexins capsidiol and phytuberin (Kuc, 1982). The genetic manipulation achieved by Hain *et al.* (1993) allowed induced resveratrol synthesis and conferred significant resistance to the necrotroph *Botrytis cinerea*.

The key to the success with resveratrol is that it is the product of one enzyme, stilbene synthase which catalyses the formation of hydroxy-stilbenes from malonyl-CoA and *p*-coumaroyl-CoA (Fig. 12) both of which are common plant metabolites.

Transgenic plants, engineered to express the stilbene synthase gene (which was isolated from a genomic library of grapevine), accumulated high concentrations of resveratrol. The amount of heterologous phytoalexin produced was positively correlated with resistance to *B. cinerea*, as shown in Fig. 20 which summarizes data for all leaves tested. The difference in level of resistance between transgenic and wild-type tobacco was more pronounced on younger leaves. *Botrytis cinerea* was perhaps not a good choice of pathogen for this type of study because its ability to develop lesions is greatly influenced by behaviour of conidia on the leaf surface, which is probably not influenced by phytoalexin production (Mansfield, 1980). The inherent variability in pathogenicity of *B. cinerea* probably explains the large number of points with no disease incidence at low resveratrol concentrations in Fig. 20. Overall, the results of Hain *et al.* (1993) are, therefore, very encouraging for the likely success of strategies of disease control based on heterologous expression of phytoalexins or phytoanticipins. In addition to the direct effect of the introduced compound it is likely that synergistic interactions occur with the host's normal defence mechanisms. It would be interesting to assess the accumulation of the tobacco phytoalexins in the stilbene synthase transformed plants. Further experiments are also needed to test the reaction of the resveratrol-producing transgenic tobacco to other pathogens. It is debatable whether modification of the phytoalexin response would limit growth of a biotrophic pathogen such as *Peronospora tabacina*, the cause of blue mould. During colonisation by biotrophic fungi and bacteria, the signalling pathways leading to the HR and phytoalexin accumulation are usually not activated (Bailey, 1982; Bennett *et al.*, 1994) - so although the plant has a potentially enhanced defensive "weapon" it may not be "fired". When engineering resistance, it may be necessary to link new defence responses to promoters which have been shown to be activated during normally susceptible interactions, a strategy emphasized by Dixon *et al.*(1995).

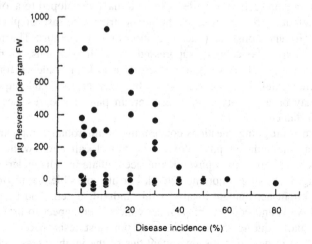

Figure 20. Relationship between resveratrol accumulation and resistance to *Botrytis cinerea* in transgenic tobacco expressing stilbene synthase from grapevine (adapted from Hain *et al.*, 1993).

Other examples where introduction of a single new enzyme might enhance resistance by altering secondary metabolites include sesquiterpene cyclase and prenyl transferase. The structure of sesquiterpenoids depends on the products generated from the common precursor farnesyl pyrophosphate by sesquiterpene cyclases which catalyse different patterns of cyclization (Chappell, 1995). Transfer of the enzyme from one plant to another may generate novel profiles of sesquiterpenoids. A good candidate plant for such an approach is lettuce which contains a range of constitutive sesquiterpene lactones (e.g. lactucopicrin) and also produces the related, highly toxic phytoalexin lettucenin A (Fig. 9, Bennett et al., 1994). Prenyl transferases also have potential for modifying the structure, and thereby activity, of isoflavonoids (Biggs et al., 1990; Blount et al., 1993). Many prenylated isoflavonoids, such as glyceollin, are more strongly antimicrobial than their non-prenylated precursors. Production of suitable prenyl transferases should therefore enhance the efficiency of the reponse of a plant such as alfalfa which produces the non-prenylated phyoalexin medicarpin (Fig. 12).

Genetic manipulation of secondary metabolites is an area of molecular biology that remains in its infancy. More detailed understanding of the enzymology of phytoalexin synthesis is needed. Dixon et al.(1996) point out that attempts to engineer metabolic pathways may give unexpected results, revealing fundamentally redundant enzyme systems, metabolic channelling, or unexpected translational control mechanisms. A continued research focus on phytoalexins is, therefore, likely to yield data on fundamental aspects of plant metabolism as well as being of direct benefit to the development of disease control strategies.

V.   Concluding remarks

The interactions occurring between antimicrobial compounds and invading potential pathogens are summarized in Fig. 21. It is clear from earlier sections that not all branches of the scheme are present in all plants. Biotrophic pathogens, notably the economically damaging rusts and mildews seem to have developed to avoid activation of the plant's defences, for example by failing to trigger release of phytoanticipins, or to elicit the HR and consequent accumulation of phytoalexins. The idea that the defence processes may be actively suppressed by such biotrophs, as noted by an alternative route in Fig. 21, is intriguing. Despite a lack of evidence for production of suppressor molecules it remains a possibility that an essential requirement for pathogenicity may be an ability to suppress certain plant responses (see discussions by Heath and Skalamera, 1997).

One of the most intriguing questions concerning antimicrobial compounds in plants, and in particular the induced phytoalexins, is their localization within challenged tissues. The mechanism by which phytoalexins accumulate in cells undergoing the HR as outlined in Fig. 17, is simply not understood. It is also possible that phytoalexins and phytoanticipins might accumulate locally within individual cells and be incorporated into altered cell walls and deposits of papillae which often appear to prevent invasion by fungi, particularly during the expression of non-host resistance. Because of the difficulties inherent in detecting the low quantities of the inhibitors at such microsites

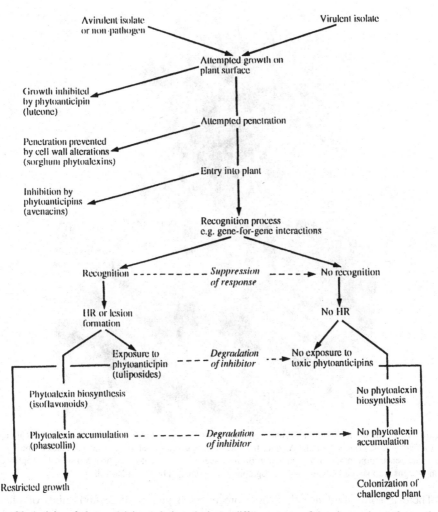

Figure 21. Activity of phytoanticipins and phytoalexins at different stages of the plant pathogen interaction. Note that the routes taken by virulent pathogens to avoid or overcome the effects of antimicrobial compounds are indicated by the dashed lines. Examples of some compounds known to be involved at the key stages are given in parentheses.

there is little experimental evidence to support this proposal. However, the coloured phytoalexins from sorghum have been found to accumulate at such sites, as illustrated in Fig. 22. The highly localized response that does not involve the HR is, therefore, also included in the overall scheme (Fig. 21). The application of immunocytochemical approaches, as outlined for glucosinolates (see Fig. 6), may reveal changes in phytoalexin and phytoanticipin concentrations and distribution occurring at the sub-cellular level which may be critical for co-ordinated plant defence. A close link has been established between the HR and major synthesis of phytoalexins, but we may be missing more subtle activation and activities because of insensitive analytical techniques.

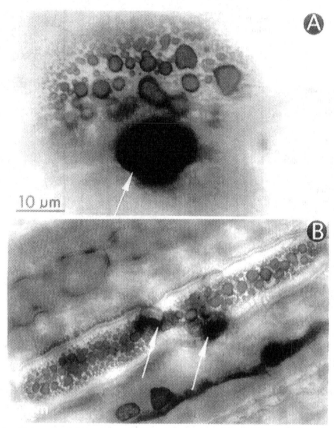

Figure 22. Accumulation of red deoxyanthocyanidin phytoalexins within sorghum epidermal cells challenged by *Colletotrichum graminicola*. In A, note the localized aggregation of pigmented granules at the site of attempted penetration into a living cell and in B, the dispersal of pigment throughout a dead cell. Appressoria of the fungus are marked with arrows. Micrographs kindly donated by Ralph Nicholson.

There is some evidence that the more evolved genera of certain tribes of plants may produce more complex and potentially more toxic phytoalexins. For example, among the legumes, in the tribe Phaseoleae compounds with prenyl substitution occur in the more advanced systematic groupings (Ingham 1981). There are some plants for example cucumber, from which neither phytoanticipins nor phytoalexins have been identified, and antimicrobial activity has not consistently been reported in extracts from challenged tissues (Mansfield and Bailey, 1982). Although the active compounds may not be recovered by the extraction methods which have been applied, it is probable that in certain plants other mechanisms of resistance may be more important, for example the rapid lignification of challenged tissue as proposed for cucumber by Hammerschmidt and Kuc (1982). Phytoanticipins and phytoalexins should always be considered as components of a co-ordinated battery of defences available to plants. It seems, however, highly probable that an ability to produce the antimicrobial secondary metabolites has had a major role in shaping the evolution of flowering plants since their emergence during the Devonian period.

VI. Acknowledgements

I wish to thank numerous colleagues for invaluable discussion; Anne Osbourn, Margaret Essenberg, Mark Bennett, Richard Dixon, Peter Kelly and John Rossiter deserve special mention. Margaret Critchley deserves a prize for spotting references I have missed (as usual) and the U.K. Biotechnology and Biological Sciences Research Council must be acknowledged for supporting our research on antimicrobial compounds.

Reference

Arneson PA and Durbin RD (1967) Hydrolysis of tomatine by *Septoria lycopersici*, a detoxification . mechanism. Phytopathology 57: 1358—1360

Arneson PA and Durbin RD (1968) The sensitivity of fungi to α–tomatine. Phytopathology 58: 536—537

Atkinson P and Blakeman JP (1982) Seasonal occurrence of an antimicrobial flavonone, sakuranetin, associated with glands on leaves of *Ribes nigrum*, New Phytol 92: 63-74

Bailey JA (1982) Physiological and biochemical events associated with the expression of resistance to diseases. In: Wood RKS (ed) Active Defense Mechanisms in Plants, pp 39—65. Plenum Press, New York, London

Bailey JA and Deverall BJ (1971) Formation and activity of phaseollin in the interaction between bean hypocotyls (*Phaseolus vulgaris*) and physiological races of *Collectotrichum lindemuthianum*. Physiol Plant Pathol 1: 435-449

Bailey JA and Mansfield JW (1982) Phytoalexins. Blackie, Glasgow

Bailey JA, Rowell PM and Arnold GM (1980) The temporal relationship between infected cell death, phytoalexin accumulation and the inhibition of hyphal development during resistance of *Phaseolus vulgaris* to *Colletotrichum lindemuthianum*. Physiol Plant Pathol 17: 329—339

Bailey JA, O'Connell RJ, Pring RJ and Nash C (1992) Infection strategies of *Colletotrichum* species. In: JA Bailey and MJ Jeger (eds) *Colletotrichum*: Biology, pathology and control, pp 88-121. CAB International, Wallingford.

Barz W and Welle R (1992) Biosynthesis and metabolism of isoflavones and pterocarpan phytoalexins in chickpea, soybean and phytopathogenic fungi. In: Stafford HA and Ibrahim RK (eds) Phenolic metabolism in plants, pp 134—164. Plenum Press, New York

Bennett MH, Gallagher MDS, Bestwick CS, Rossiter JT and Mansfield JW (1994) The phytoalexin response of lettuce to challenge by *Botrytis cinerea*, *Bremia lactucae* and *Pseudomonas syringae* pv. *phaseolicola*. Physiol Mol Plant Pathol 44: 321—333

Bergman BHH and Beigersbergen JCM (1968) A fungitoxic substance extracted from tulips and its possible role as a protectant against disease. Neth J Plant Pathol 74: 157—162

Bestwick CS, Brown IR, Bennett MHR and Mansfield J (1997) Localization of hydrogen peroxide accumulation during the hypersensitive reaction of lettuce cells to *Pseudomonas syringae* pv. *phaseolicola*. Plant Cell 9: 209—221

Biggs DR, Welle R and Grisebach H (1990) Intracellular localization of prenyltransferases of isoflavonoid phytoalexin biosynthesis in bean and soybean. Planta 181: 244—48

Blakeman JP (1973) The chemical environment of leaf surfaces with special reference to germination of pathogenic fungi. Pestic Sci 4: 575—588

Blount JW, Dixon RA and Paiva NL (1993) Stress responses in alfalfa (*Medicago sativa* L.) XVI. Antifungal activity of medicarpin and its biosynthetic precursors: implications for the genetic manipulation of stress metabolites. Physiol Mol Plant Pathol 41: 333—349

Bones AM and Rossiter JT (1996) The myrosinase-glucosinolate systems, its organisation and biochemistry. Phyiol Plant 97: 194—208

Bowyer P, Clarke BR, Lunness P, Daniels MJ and Osbourne AE (1995) Host range of a plant pathogenic fungus determined by a saponin detoxifying enzyme. Science 267: 371—374.

Brown I, Mansfield J, Irlam I, Conrads-Strauch J and Bonas U (1993) Ultrastructure of interactions between *Xanthomonas campestris* pv. *vesicatoria* and pepper, including immunocytochemical localization of extracellular polysaccharides and the AvrBs3 protein. Mol Plant-Microbe Interact 6: 376—386

Burden RS, Bailey JA and Vincent GG (1974) Metabolism of phaseollin by *Colletotrichum lindemuthianum*. Phytochemistry 13: 1789—1791

Cartwright DW, Langcake P and Ride JP (1981) Phytoalexin production in rice and its enhancement by a dichlorocyclopropane fungicide. Physiol Plant Pathol 17: 259—267

Chappell J (1995) The biochemistry and molecular biology of isoprenoid metabolism. Plant Physiol 107: 1—6

Chew FS (1988) Biological effects of glucosinolates. In: Cutler HG (ed) Biologically active natural products: potential use in agriculture, pp 151-81. American Chemical Society, Washington DC, USA

Ciuffetti LM and VanEtten HD (1996) Virulence of a pisatin-deficient *Nectria haematococca* MPVI isolate is increased by transformation with a pisatin demethylase gene. Molec Plant-Microbe Interact 9: 787—792

Coleman MJ, Mainzer J and Dickerson AG (1992) Characterization of a fungal glycoprotein that elicits a defence response in French bean. Physiol Mol Plant Pathol 40: 333-351

Cooper RM (1981) Pathogen-induced changes in host ultrastructure. In: Plant disease control: resistance and susceptibility (eds RC Staples and GH Toenniesen) Wiley, New York. pp 105-142

Cooper RM, Resende MLV, Flood J, Rowan MG, Beale MH and Potter U (1996) Detection and cellular localization of elemental sulphur in disease-resistant genotypes of *Theobroma cacao*. Nature 379: 159-162

Croft KPC, Jütter F and Slusarenko AJ (1993) Volatile products of the lipoxygense pathway evolved from *Phaseolus vulgaris* (L.) leaves inoculated with *Pseudomonas syringae* pv. *phaseolicola*. Plant Physiol 101: 13-24

Crombie L, Crombie WML and Whiting DA (1984) Structures of the four avenacins, oat root resistance factors to take-all disease. J Chem Soc Chem Comm 4: 246—248

Crombie WML, Crombie L, Green JB and Lucas JA (1986) Pathogenicity of the take-all fungus to oats: its relationship to the concentration and detoxification of the four avenacins. Phytochemistry 25: 207—2083

Cruickshank IAM and Perrin DR (1960) Isolation of a phytoalexin from *Pisum sativum* L. Nature 187: 799-800

Cruickshank IAM and Perrin DR (1971) Studies on phytoalexins. XI. The induction, anti-microbial spectrum and chemical assay of phaseollin. Phytopath Z 70: 209-229

Dai GH, Nicole M, Andary C, Martinez C, Bresson E, Boher B, Daniel JF and Geiger JP (1996) Flavonoids accumulate in cell walls, middle lamellae and callose-rich papillae during an incompatible interaction between *Xanthomonas campestris* pv. *malvacearum* (Race 18) and cotton. Physiol Mol Plant Pathol 49: 285-306

Défago G and Kern H (1983) Induction of *Fusarium solani* mutants insensitive to tomatine, their pathogenicity and aggressiveness to tomato fruits and pea plants. Physiol Mol Plant Pathol 22: 29—37

Denny TP, Matthews PS and VanEtten HD (1987) A possible mechanism of nondegradative tolerance of pisatin in *Nectria haematococca* MP VI. Physiol Mol Plant Pathol 30: 93—107

Deverall BJ (1967) Biochemical changes in infection dropletscontaining spores of *Botrytis* spp. incubated in the seed cavities of pods of bean (Vicia faba L.). Ann appl Biol 59: 375—387

Deverall BJ (1977) Defence mechanisms of plants. Cambridge University Press, Cambridge

Deverall BJ (1982) Introduction. In: Phytoalexins (eds JA Bailey and JW Mansfield). pp 1-16, Blackie, Glasgow.

de Wit PJGM (1995) Fungal avirulence genes and plant resistance genes: unravelling the molecular basis of gene-for-gene interactions. Advances in Botanical Research 21: 148—185

Dixon RA and Harrison MJ (1990) Activation, structure, and organization of genes involved in microbial defense in plants. Adv Genet 28: 165-217

Dixon RA and Paiva NL (1995) Stress induced phenylpropanoid metabolism. Plant Cell 7: 1085—1097

Dixon RA, Choudhary AD, Dalkin D, Edwards R, Fahrendorf T, Gowri G, Harrison MJ, Lamb CJ, Loake GJ, Maxwell CA, Orr J and Paiva NL (1992) Molecular biology of stress-induced phenylpropanoid and isoflavonoid biosynthesis in alfalfa. In: Phenolic metabolism in plants (eds HA Stafford and RK Ibrahim) Plenum Press, New York, pp 91-138

Dixon RA, Paiva NL and Bhattacharyya MK (1995) Engineering disease resistance in plants: An overview. In: Molecular methods in plant pathology (eds RP Singh and US Singh) pp. 249-270 CRC Press, Boca Raton

Dixon RA, Lamb CJ, Masoud S, Sewalt VJH and Paiva NL (1996) Metabolic engineering: prospects for crop improvement through the genetic manipulation of phenylpropanoid biosynthesis and defense responses - a review. Gene 179: 61-71

Durrands PK and Cooper RM (1988) Selection and characterization of pectinase-deficient mutants of the vascular wilt pathogen *Verticillium albo-atrium*. Physiol Mol Plant Pathol 32: 343-362

Essenberg M, Pierce ML, Cover EC, Hamilton B, Richardson PE and Scholes VE (1992a) A method for determining phytoalexin concentrations in fluorescent, hypersensitively necrotic cells in cotton leaves. Physiol Mol Plant Pathol 41: 101—109

Essenberg M, Pierce ML, Hamilton B, Cover EC, Scholes VE and Richardson PE (1992b) Development of fluorescent, hypersensitively necrotic cells containing phytoalexins adjacent to colonies of *Xanthomonas campestris* pv. *malvacearum* in cotton leaves. Physiol Mol Plant Pathol 41: 85—99

Fenwick GR, Heaney RK and Mullin WJ (1983) Glucosinolates and their breakdown products in food and food plants. CRC Critical Reviews in Food Science and Nutrition 18: 123—201

Fenwick GR, Price KR, Tsukamota C and Okubo K (1992) Saponins. In: Toxic substances in crop plants (eds JP D'Mello, CM Duffus and JH Duffus) Royal Society of Chemistry, Cambridge, UK. pp. 285-327

Franich RA, Gadgil PD and Shain L (1983) Fungistatic effects of *Pinus radiata* needle epicuticular fatty and acid resins on *Dothiostroma pini*. Physiol Plant Path 23: 183-195

Fry WE and Myers DF (1981) Hydrogen cyanide metabolism by fungal pathogens of cyanogenic plants In: Vennesland B, Knowles CJ, Conn EE, Westley J and Wissing F (eds) Cyanide in Biology, pp. 321-334. Academic Press, London

Giamoustaris A and Mithen RF (1995) The effect of modifying the glucosinolate content of leaves of oilseed rape (*Brassica napus* ssp. *oleifera*) on its interaction with specialist and generalist pests. Ann. appl. Biol. 126: 347—363

Giamoustaris A and Mithen R (1997) Glucosinolates and disease resistance in oilseed rape (*Brassica napus* ssp. *oleifera*). Pl Pathol 46: 271-275

Glazebrook J and Ausubel FM (1994) Isolation of phytoalexin-deficient mutants of *Arabidopsis thaliana* and characterization of their interactions with bacterial pathogens. Proc Natl Acad Sci USA 91: 8955—8959

Glazebrook J, Rogers EE and Ausubel FM (1996) Isolation of *Arabidopsis* mutants with enhanced disease susceptibility by direct screening. Genetics 143: 973-982

Godwin J, Anthony VM, Clough JM and Godfrey CRA (1992) ICIA5504: a novel, broad spectrum, systemic β-methoxyacrylate fungicide. Proc 1992 Brighton Crop Protection Conference - Pests and Diseases 1: 435-442.

Graham MY and Graham TL (1991) Rapid accumulation of anionic peroxidases and phenolic polymers in soybean cotyledon tissues following treatment with *Phytophthora megasperma* f.sp *glycinea* wall glucan. Plant Physiol 97: 1445—1455

Grayer RJ and Harborne JJ (1994) A survey of antifungal compounds from higher plants. Phytochemistry 37: 19—42

Greger H, Hofer O, Kählig H and Wurz G (1992) Sulfur containing cinnamides with antifungal activity from *Glycosmis cyanocarpa*.Tetrahedron 48: 1209-1218

Hahlbrock K, Scheel D, Logemann E, Nürnberger T, Parniske M, Reinold S, Sacks WR and Schmelzer E (1995) Oligopeptide elicitor-mediated defense gene activation in cultured parsley cells. Proc Natl Acad Sci USA 92: 4150—4157

Hain R, Reif H-J, Krause E, Langebartels R, Kindl H, Vornam B, Wiese W, Schmeizer E, Schreier, PH, Stocker RH. and Stenzel K (1993) Disease resistance results from foreign phytoalexin expression in a novel plant. Nature 361: 153—156

Hammerschmidt R and Kuc J (1982) Lignification as a mechanism for induced systemic resistance in cucumber. Physiol Mol Plant Path 20: 61-71

Hargreaves JA and Bailey JA (1978) Phytoalexin production by hypocotyls of *Phaseolus vulgaris* in response to constitutive metabolites released by damaged cells. Physiol Plant Pathol 13: 89—100

Hargreaves JA, Mansfield JW and Rossall S (1977) Changes in phytoalexin concentration in tissues of the broad bean plant (*Vicia faba* L.) following inoculation with species of *Botrytis*. Physiol Plant Pathol 11: 227—242

Hassan F, Rothnia NE, Yeung SP and Palmer MV (1988) Enzyme-linked immunosorbent assays for alkenyl glucosinolates. J Agric Food Chem 36: 398—403

Heath MC and Skalamera D (1997) Cellular interactions between plants and biotrophic fungal parasites. Adv Bot Res 24: 196-225

Hutson RA and Mansfield JW (1980) A genetical approach to the analysis of mechanisms of pathogenicity in *Botrytis/Vicia* interactions. Physiol Plant Pathol 17: 309—317

Ingham JL (1973) Disease resistance in higher plants. The concept of pre-infectional and post-infectional resistance. Phytopathol Z 78: 314—335

Ingham J (1981) Phytoalexin induction and its taxonomic significance in the Leguminosae. In: Advances in legume systematics (eds. RM Polhill and PH Raven) HMSO, London. pp 599-626

Ingham JL (1982) Phytoalexins from the Leguminosae. In: Bailey J and Mansfield JW (eds) Phytoalexins, pp 21—80. Blackie, Glasgow

Ingham JL, Tahara S and Harborne JB (1983) Fungitoxic isoflavones from *Lupinus albus* and other *Lupinus* species. Z Naturforsch 38c: 194-200

Jerome SMR and Müller KO (1958) Studies on phytoalexins II Influence of temperature on resistance of *Phaseolus vulgaris* towards *Sclerotinia fructicola* with reference to phytoalexin output. Aust J Biol Sci 11: 301—314

Kesselmeier J and Urban B (1983) Subcellular localization of saponins in green and etiolated leaves and green protoplasts of oat (*Avena sativa* L.). Protoplasma 114: 133-140

Keukens EAJ, de Vrije T, van den Boom C, de Waard P, Plasmna HH, Thielm F, Chupin V, Jongen WMF and de Kruijff B (1995) Molecular basis of glycoalkaloid induced membrane disruption. Biochim Biophys Acta 1240: 216—228

Kneusel RE, Matern U and Nicolay K (1989) Formation of *trans*-caffeoyl-CoA from *trans*-4-coumaroyl-CoA by Zn^{2+}-dependent enzymes in cultured plant cells and its activation by an elicitor-induced pH shift. Arch Biochem Biophys 269: 455—462.

Kodama O, Suzuki T, Miyakawa J and Akatsuka T (1988) Ultraviolet-induced accumulation of phytoalexins in rice leaves. Agric Biol Chem 52: 2469-2473

Kokubun T and Harborne JB (1994) A survey of phytoalexin induction in leaves of the Rosaceae by copper ions. Z Naturforsch 49c: 628-634

Kuc J (1982) Phytoalexins from the Solanaceae In: Bailey JA and Mansfield JW (eds) Phytoalexins, pp 81—100. Blackie, Glasgow

Kuc J (1992) Antifungal compounds in plants. In: Nigg HN and Seigler D (eds) Phytochemical resources for medicine and agriculture, pp. 159-184. Plenum Press, New York

Kylin H, Atuma S, Hovander L and Jensen S (1994) Elemental sulphur (S$_8$) in higher plants - biogenic of anthropogenic origin? Experientia 50: 80-85

Langcake P and Pryce RJ (1977) A new class of phytoalexins from grapevines. Experientia 33: 151—152

Levine A, Tenhaken R, Dixon R and Lamb C (1994) H$_2$O$_2$ from the oxidative burst orchestrates the plant hypersensitive disease resistance response. Cell 79: 583—593

Low PS and Merida JR (1996) The oxidative burst in plant defense: function and signal transduction. Physiol Plant 96: 533—542

Mace ME, Bell AA and Stipanovic RD (1978) Histochemistry and identification of flavanols in *Verticillium* wilt-resistant and -susceptible cottons. Physiol Plant Pathol 13: 143—149

Mansfield JW (1980) Mechanisms of resistance to *Botrytis* In: Coley-Smith JR, Verhoeff K and Jarvis WR (eds) The biology of *Botrytis*, pp 181—218. Academic Press, London

Mansfield JW (1982) Role of phytoalexins in disease resistance. In: Bailey J and Mansfield JW (eds) Phytoalexins, pp 253-288. Blackie, Glasgow

Mansfield JW (1983) Antimicrobial compounds. In: Callow JA (ed) Biochemical plant pathology, pp.155—164. John Wiley & Sons, Chichester

Mansfield JW (1990) Recognition and response in plant-fungus interactions. In: Fraser RSS (ed) Recognition and response in plant-virus interaction. pp. 31—52. Springer-Verlag, Berlin

Mansfield JW and Bailey JA (1982) Phytoalexins: current problems and future prospects. In: JA Bailey and JW Mansfield (eds) pp. 319-322. Phytoalexins. Blackie, Glasgow

Mansfield JW and Hutson RA (1980) Microscopical studies on fungal development and host responses in broad bean and tulip leaves inoculated with five species of *Botrytis*. Physiol Plant Pathol 17: 131—145

Mansfield JW and Richardson A (1981) The ultrastructure of interactions between *Botrytis* species and broad bean leaves. Physiol Plant Pathol 19: 41—48

Mansfield JW, Porter AEA and Smallman RV (1980) Dihydrowyerone derivatives as components of the furanoacetylenic phytoalexin response of tissues of *Vicia faba*. Phytochemistry 9: 1057—1061

Mansfield JW, Bennett MH, Bestwick CS and Woods-Tor AM (1997) Phenotypic variation in gene-for-gene interactions: variation from recognition to response. In: Crute IR, Burden JJ and Holub EB (eds) pp. 265—291 The gene-for-gene relationship in host-parasite interactions. CAB International, London UK

Mari M, Iori R, Leoni O and Marchi A (1993) *In vitro* activity of glucosinolate-derived isothiocyanates against postharvest fruit pathogens. Ann appl Biol 123: 155—164

Matthews DE and VanEtten HD (1983) Detoxification of the phytoalexin pisatin by a fungal cytochrome P-450. Arch Biochem Biophys 224: 494—505

Melton RE, Flegg LM, Brown JKM, Oliver RP, Daniels MJ and Osbourn AE (1998) Heterologous expression of *Septoria lycopersicae* tomatinase in *Cladosporium fulvum*: Effects on compatible and incompatible interactions with tomato seedlings. Molec Plant-Microbe Interact, 11: 228—235.

Miao VPW, Covert SF and VanEtten HD (1991) A fungal gene for antibiotic resistance on a dispensable ('B') chromosome. Science 254: 1773—1776

Miao VPW and VanEtten HD (1992) Three genes for metabolism of the phytoalexin maackiain in the plant pathogen *Nectria haematococca*: Meiotic instability and relationship to a new gene for pisatin demethylase. Appl Environ Microbiol 58: 801—808

Mithen RF (1992) Leaf glucosinolate profiles and their relationship to pest and disease resistance in oilseed rape. Euphytica 63: 71—83

Mithen RF, Lewis BG and Fenwick GR (1986) *In vitro* activity of glucosinolates and their products against *Leptosphaeria maculans*. Trans Brit. Mycol. Soc. 87: 433—440

Moesta P, Soyedl U, Lindner B and Grisebach H (1982) Detection of glyceollin at the cellular level in infected soybean by laser microprobe mass analysis. Zeit Naturforsch 37c: 748—751

Müller KO (1958) Studies on phytoalexins. 1. The formation and immunological significance of phytoalexin produced by *Phaseolus vulgaris* in response to infections with *Sclerotinia fructicola* and *Phytophthora infestans*. Aust J Biol Sci 11: 275-300

Müller KO and Borger H (1940) Experimentelle Untersuchungen uber die *Phytophthora* - Resistenz der Kartoffel - zugleich ein Beitrag zum Problem der 'erworbenen Resistenz' im Pflanzenreich. Arb Biol Anst Reichsanst (Berl) 23: 189—231

Nicholson RL and Hammerschmidt R (1992) Phenolic compounds and their role in disease resistance. Ann Rev Phytopathol 30: 369—380

Nisius H (1988) The stromacentre in plastids: An aggregation of β-glucosidase responsible for the activation of oat-leaf saponins. Planta 173: 474—481

O'Brien TP and McCully ME (1981) The study of plant structure: principles and selected methods. Termarcarphi Pty Ltd., Melbourne, Australia

Oommen A, Dixon RA and Paiva NKL (1994) The elicitor-inducible alfalfa isoflavone reductase promoter confers different patterns of developmental expression in homologous and heterologous transgenic plants. Plant Cell 6: 1789—1803

Osbourn AE (1996a) Saponins and plant defence - a soap story. Trends Plant Sci 1: 4—9.

Osbourn AE (1996b) Preformed antimicrobial compounds and plant defense against fungal attack. Plant Cell 8: 1821-1831

Osbourn AE, Clarke BR, Dow JM and Daniels MJ (1991) Partial characterization of avenacinase from *Gaeumannomyces graminis* var. *avenae*. Physiol Mol Plant Pathol 38: 301—312

Osbourn AE, Clarke BR, Lunness P, Scott PR and Daniels MJ (1994) An oat species lacking avenacin is susceptible to infection by *Gaeumannomyces graminis* var. *tritici*. Physiol Mol Plant Pathol. 45: 457—467

Osbourn AE, Bowyer P, Lunness P, Clarke B and Daniels M (1995) Fungal pathogens of oat roots and tomato leaves employ closely related enzymes to detoxify host plant saponins. Mol Plant-Microbe Interact 8: 971—978

Paxton JD (1981) Phytoalexins - a working redefinition. Phytopathol Z 101: 106-109.

Perrin DR and Bottomley W (1962) Studies on phytoalexins. V. The structure of pisatin from *Pisum sativum* L. J Am Chem Soc 84: 1919—1922

Pierce M and Essenberg M (1987) Localization of phytoalexins in fluorescent mesophyll cells isolated from bacterial blight-infected cotton cotyledons and separated from other cells by fluorescence-activated cell sorting. Physiol Mol Plant Pathol 31: 273—290

Pierce ML, Cover EC, Richardson PE, Scholes VE and Essenberg M (1996) Adequacy of cellular phytoalexin concentrations in hypersensitively responding cotton leaves. Physiol Mol Plant Pathol 48: 305—324

Prusky D and Plumbley RA (1992) Quiescent infections of *Colletotrichum* in tropical and subtropical fruits. In: *Colletotrichum*: biology, pathology and control (eds JA Bailey and MJ Jeger) pp 289-308, CAB International, Wallingford.

Prusky D, Keen NT, Sims JJ and Midland SL (1982) Possible involvement of an antifungal compound in latency of *Colletotrichum gloeosporoides* in unripe avocado fruits. Phytopathology 72: 1578—1582

Prusky D, Kobiler I, Jacoby B, Sims JJ and Midland SL (1985) Effect of inhibitors of lipoxygenase activity and its possible relation to latency of *Colletotrichum gloeosporioides* on avocado fruits. Physiol Mol Plant Pathol 27: 269—279

Prusky D, Plumbley RA and Kobiler I (1991) Effect of CO_2 treatment on the induction of antifungal diene in unripe avocado fruits. Physiol Mol Plant Pathol 39: 325—334

Rathmell WG and Smith DA (1980) Lack of activity of selected isoflavonoid phytoalexins as protectant fungicides. Pestic Sci 11: 568-572

Ride JP (1986) Induced structural defences in plants. In: Gould GW, Rhodes-Roberts ME, Charnley AK, Cooper RM and Board RG (eds) Natural antimicrobial systems in plants and animals, pp 159—165. University Press, Bath

Rogers EE, Glazebrook J and Ausubel FM (1996) Mode of action of the *Arabidopsis thaliana* phytoalexin camalexin and its role in *Arabidopsis*-pathogen interactions. Mol Plant-Microbe Interact 9: 748-757

Rossall S and Mansfield JW (1978) The activity of wyerone acid against *Botrytis* Ann appl Biol 89: 359-362

Rossall S, Mansfield JW and Price NC (1977) The effect of reduced wyerone acid on the antifungal activity of the phytoalexin wyerone acid against *Botrytis fabae*. J Gen Microbiol 102: 203-205

Sandrock RW, Della-Penna D and VanEtten HD (1995) Purification and characterization of β_2-tomatinase, an enzyme involved in the degradation of α-tomatine and isolation of the gene encoding β_2-tomatinase from *Septoria lycopersici*. Mol Plant-Microbe Interact 8: 960—970

Schäfer W, Straney D, Ciuffetti L, VanEtten HD and Yoder OC (1989) One enzyme makes a fungal pathogen but not a saprophyte, virulent on a new host plant. Science 246: 247—249

Schönbeck F and Schlosser E (1976) Preformed substances as potential protectants. In Heitefuss R and Williams PH (eds) Encyclopedia of plant physiology Vol 4, pp 653—678. Springer-Verlag, Berlin

Schönbeck F and Schroeder C (1972) Role of antimicrobial substances (tuliposides) in tulips attacked by *Botrytis* spp. Physiol Plant Pathol 2: 91—99

Slob A, Jekel B, de Jong B and Schlatmann, E (1975) On the occurrence of tuliposides in the Liliiflorae. Phytochemistry 14: 1997—2005

Snyder BA and Nicholson RL (1990) Synthesis of phytoalexins in sorghum as a site specific response to fungal ingress. Science 248: 1637—1639

Snyder B A, Hipskind J, Butler LJ and Nicholson RL (1991) Accumulation of sorghum phytoalexins induced by *Colletotrichum graminocola* at the infection site. Physiol Mol Plant Pathol 39: 463—470

Steel CS and Drysdale RB (1988) Electrolyte leakage from plant and fungal tissues and disruption of liposome membranes by α-tomatine. Phytochemistry 27: 1025—1030

Stoessl A (1982) Biosynthesis of phytoalexins. In: Bailey JA and Mansfield JW (eds) Phytoalexins, pp 133—180. Blackie, Glasgow

Swinburne TR (1983) Quiescent infections in post-harvest disease. In: Dennis C (ed) Post-harvest pathology of fruits and vegetables, pp 1—21. Academic Press, London

Tahara S, Ingham JC, Nakahara S, Mizutani S, Mizutani J and Harborne JB (1984) Fungitoxic dihydro-furanoisoflavones and related compounds in white lupin, *Lupinus albus* Phytochemistry 23: 1889-1900

Tegtmeier KJ and VanEtten HD (1982) The role of pisatin tolerance and degradation in the virulence of *Nectria haematococca* on peas: a genetic analysis. Phytopathology 72: 608—612

Tepper CS, Albert FG and Anderson AJ (1989) Differential mRNA accumulation in three cultivars of bean in response to elicitors from *Colletotrichum lindemuthianum*. Physiol Mol Plant Pathol 34: 85-89

Tscheche R, Kammerer FJ, Wulff G and Schönbeck F (1978) Uber die antibiotisch wirksamen substanzen der Tulpe (*Tulipa geseriana*). Tetrahedron Lett 6: 701—706

Tsuji J, Jackson EP, Gage DA, Hammerschmidt R and Somerville SC (1992) Phytoalexin accumulation in *Arabidopsis thaliana* during the hypersensitive reaction to *Pseudomonas syringae* pv. *syringae*. Plant Physiol 98: 1304—1309

Turner EM (1953) The nature of resistance of oats to the take-all fungus. J Exp Bot 4: 264—271.

Turner EM (1961) An enzymic basis for pathogen specificity in *Ophiobolus graminis* J Exp Bot 12: 169—175

Tverskoy L, Dmitriev A, Kozlovsky A and Gradzinsky D (1991) Two phytoalexins from *Allium cepa* bulbs. Phytochemistry 30: 799

Van den Ackerveken G, Marois E and Bonas U (1996) Recognition of the bacterial avirulence protein AvrBs3 occurs inside the host plant cell. Cell 87: 1307—1316

VanEtten HD, Matthews PS, Tegtmeier KJ, Dietert MF and Stein JI (1980) The association of pisatin tolerance and demethylation with virulence on pea in *Nectria heamatococca*. Physiol Plant Pathol 16: 257—268

VanEtten HD, Matthews DE and Smith DA (1982) Metabolism of phytoalexins. In: Bailey JA and Mansfield JW (eds) Phytoalexins, pp 181—217. Blackie, Glasgow

VanEtten HD, Mansfield JW, Bailey JW and Farmer EE (1994) Letter to the editor: two classes of plant antibiotics: phytoalexins versus "phytoanticipins". Plant Cell 6: 1191—1192

VanEtten HD, Sandrock RW, Wasmann CC, Soby SD, McCluskey K and Wang P (1995) Detoxification of phytoanticipins and phytoalexins by phytopathogenic fungi. Can J Bot 73: S518—S525

Verhoeff K (1974) Latent infections by fungi. Ann Rev Phytopathol 12: 99—110

Walker JC and Stahmann MA (1955) Chemical nature of disease resistance in plants. Ann Rev Plant Physiol 6: 351—366

Wang P and VanEtten HD (1992) Cloning and properties of a cyanide hydratase gene from the phytopathogenic fungus *Gloeocercospora sorghi*. Biochem Biophys Res Commun 187: 1048-1054

Wasmann CC and VanEtten HD (1996) Transformation-mediated chromosome loss and disruption of a gene for pisatin demethylase decrease the virulence of *Nectria haematococca* on pea. Molec Plant-Microbe Interact 9: 793-803

Welle R, Schroeder G, Schiltz E, Grisebach H and Schröder J (1991) Induced plant responses to pathogen attack: analysis and heterologous expression of the key enzyme in the biosynthesis of phytoalexins in soybean (*Glycine max*. L.). Eur J Biochem 196: 423—430

Weltring KM, Turgeon BG, Yoder OC and VanEtten HD (1988) Isolation of a phytoalexin-detoxification gene from the plant pathogenic fungus *Nectria haematococca* by detecting its expression in *Aspergillus nidulans*. Gene 68: 335—344

Weltring KM, Schaub HP and Barz W (1995) Metabolism of pisatin stereoisomers by *Ascochyta rabiei* strains transformed with the pisatin demethylase of *Nectria haematococca* MPVI. Molec Plant-Microbe Interact 8: 499—505

Wood RKS (1967) Physiological Plant Pathology. Blackwell Scientific, Oxford.

Wubben JP, Price KR, Daniels MJ and Osbourn AE (1996) Detoxification of oat leaf saponins by *Septoria avenae*. Phytopathology 86: 986—992

CHAPTER 8

INDUCED AND PREFORMED ANTIMICROBIAL PROTEINS

W.F. BROEKAERT[1], F.R.G. TERRAS[1] & B.P.A. CAMMUE[1,2]

[1] *F.A. Janssens Laboratory of Genetics, Katholieke Universiteit Leuven,*
Kardinaal Mercierlaan 92, B-3001 Heverlee - Leuven - Belgium
[2] *Flanders Interuniversity Institute for Biotechnology,*
Kardinaal Mercierlaan 92, B-3001 Heverlee - Leuven - Belgium
(willem.broekaert@agr.kuleuven.ac.be)

Summary

To protect themselves from invasion by pathogenic microorganisms, plants produce a wide array of proteins that exert direct antimicrobial activity. Based on homology at the level of the amino acid sequence and/or three-dimensional folding pattern, these antimicrobial proteins (AMPs) can be classified into over 18 distinct protein families. Some information is emerging on how these proteins interfere with the growth of microorganisms, fungi in particular. Hydrolases such as PR-3-type chitinases and PR-2-type glucanases, and possibly also PR-4-type proteins, affect fungal growth by disturbing the structural integrity of their cell wall. Thionins, 2S albumins, lipid transfer proteins and puroindolines have been demonstrated to partially lyse artificial phospholipid vesicles and are therefore believed to interfere with the phospholipid bilayer of the microbial plasma membrane. Other AMPs, such as PR-5-type proteins and plant defensins are proposed to affect plasma membrane receptors, although the evidence is still circumstantial. For many different types of AMPs it has been shown that particular combinations by two result in synergistic antimicrobial effects, suggesting that maximal antimicrobial potency of AMPs is achieved when they act in concert. Most plant tissues express simultaneously a number of AMP genes. In uninfected vegetative tissues expression is predominant in gateways for microbial invasion such as epidermal cells, stomata, hydathodes and cells in vascular strands. Infection of vegetative tissues by viruses, bacteria or fungi results in the coordinate activation of sets of AMP genes via multiple signalling pathways. These genes are in some but not all cases the same as those that confer basal expression in microbial gateway cells. At the subcellular level, AMPs are usually either deposited in the extracellular space or stored in vacuoles, in which case they are released when host cell lysis occurs. Vacuolar and extracellular AMP isoforms are usually products of different genes. The contribution of AMP genes to the resistance of plants has been studied extensively by overexpression or antisense down-regulation of AMP genes in transgenic plants, or by using mutant or transgenic plants showing either up- or down-regulation of signalling pathways leading to coordinate AMP gene expression.

A. Slusarenko, R.S.S. Fraser, and L.C. van Loon (eds), Mechanisms of Resistance to Plant Diseases, 371-477.

Many of these studies prove the concept that particular AMPs contribute to resistance to particular groups of microbial pathogens.

Abbreviations

AFP	antifungal protein
AMP	antimicrobial protein
nsLTP	nonspecific lipid transfer protein
PR	pathogenesis-related
RIP	ribosome-inactivating protein
TMV	tobacco mosaic virus

I.   Introduction

Prevalent among the chemical weapons used by plants to defend themselves against microorganisms are antimicrobial proteins (AMPs). AMPs can be defined as polypeptides that directly interfere with the growth, differentiation, multiplication and/or spread of microbial organisms (mostly viruses, bacteria or fungi) that are present on or inside a plant tissue. Hence, defense-related proteins involved in the synthesis or transport of metabolites that are inhibitory to microorganisms, are not considered as AMPs, although expression of such genes often plays a very similar role in host defense.

Proteins are usually identified as being antimicrobial by the use of a bioassay in which growth of a microorganism is assessed both in the presence and the absence of those proteins. It should be noted that it is technically much easier to detect antimicrobial activity against a microorganism that can grow in a medium in vitro as compared to an obligate biotrophic microorganism that requires living host cells for growth. This explains why most antimicrobial activity tests are performed in vitro and use either perthotrophic or non-obligate biotrophic microorganisms.

Historically, AMPs have been discovered via three separate search strategies. The first approach is an undirected one and consists of testing the antimicrobial properties of a protein that was purified for a different purpose based on a property different from the antimicrobial activity *per se*. A second approach is based on the use of antimicrobial activity tests to guide the fractionation and purification processes, thus leading directly to a purified antimicrobial protein. A third approach, finally, makes use of biochemical techniques to identify proteins that are more abundant in pathogen-challenged plant tissues than in healthy non-treated plant tissues, whereafter the purified pathogen-induced proteins are subjected to antimicrobial activity tests. The pioneering work of the latter approach was carried out on tobacco leaves infected with tobacco mosaic virus (TMV) (Gianinazzi *et al.*, 1970; Van Loon and Van Kammen, 1970) but was later extended to numerous other plant-pathogen combinations. Proteins found to be induced in pathogen-challenged tissues were called pathogenesis-related proteins or PR-proteins. As this approach initially resulted in the identification of proteins with unknown biological activity, the PR-proteins were categorized into families of structurally homologous proteins via a numbering system, e.g. PR-1 proteins, PR-2 proteins, etc. (Van Loon, 1990; Van Loon *et al.*, 1994). Several members of well-characterized PR-protein families have now been demonstrated to exert antimicrobial effects (e.g. PR-proteins 1 to 5 and PR-11 proteins), while for other PR-protein families no intrinsic antimicrobial activity has yet been found, suggesting that those may fulfill a different function. For instance, proteinase inhibitors of the PR-6 family are believed to be involved mainly in defense against insect herbivores (Johnson *et al.*, 1989; Duan *et al.*, 1996).

It is not surprising that each of these different strategies has generated its own nomenclature, resulting in babylonic confusion in the many cases in which the same types of proteins were uncovered via parallel routes. Because the PR-protein nomenclature is currently widely accepted and used in the literature on antimicrobial proteins, we will adopt this nomenclature in the present chapter where appropriate. The inconvenience of the PR-protein nomenclature, however, is that it only applies to those members of a protein family that are induced upon pathogen stress, whereas highly homologous members encoded by constitutively expressed genes are, strictly speaking, excluded from the group. The latter proteins are sometimes designated as "PR-like proteins" (Van Loon *et al.*, 1994). In this chapter we will use the broader term "PR-type proteins" (e.g. PR-1-type proteins) to designate a family of structurally related proteins, some members of which may be inducible while others may be preformed. In a number of cases no PR-protein category has yet been ascribed to a family of antimicrobial proteins, even though particular members have been shown to be pathogen-inducible. In those cases we will use a generic name for the protein family, pending extension of the PR-protein nomenclature.

II. Shortkey to the Main Families of Antimicrobial Proteins

Proteins or peptides exerting antimicrobial activity can be classified into distinct families based on primary structure homology. Table 1 summarizes the main properties of 12 families of well characterized antimicrobial proteins or peptides. Among the antimicrobial proteins are the PR-1-, PR-2-, PR-3-, PR-4-, PR-5-, PR-8-, PR-11-type proteins and the ribosome inactivating proteins (RIPs). The antimicrobial peptides (having less than 100 amino acids) include: non-specific lipid transfer proteins (nsLTPs) and structurally related proteins such as 2S albumins and puroindolines; hevein-type proteins; thionins; plant defensins; and miscellaneous peptides of which an insufficient number has been isolated to allow definitive classification. Each of the AMP families can be further divided into various subtypes which frequently need to be revisited or extended due to the ongoing discovery of members with deviating features. PR-type proteins are often discriminated into acidic and basic subtypes according to whether their isoelectric point is below or above 7.0, respectively. For at least one member of each AMP family mentioned in table 1, detailed information on the 3D structure is available. As shown in Fig. 1, the folds adopted by the distinct AMP families vary widely. Remarkably, however, the PR-2-type proteins and PR-8-type proteins both have loosely similar $(\beta/\alpha)_8$-barrel structures despite the fact that they do not share any

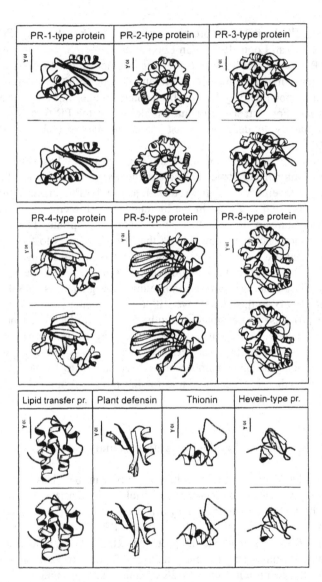

Figure 1. **Three-dimensional structures of different types of antimicrobial proteins from plants.**
Figures represent stereoviews of the following representatives of antimicrobial protein families: **PR-1-type proteins**, P14a from tomato leaves (Fernandez *et al.*, 1997); **PR-2-type proteins**, β-1,3-glucanase isoenzyme GII from barley seeds (Varghese *et al.*, 1994); **PR-3-type proteins**, class II chitinase from barley seeds (Hart *et al.*, 1993); **PR-4-type proteins**, class II PR-4-type protein (barwin) from barley seeds (Ludvigsen and Poulsen, 1992); **PR-5-type proteins**, PR-5-type protein (thaumatin isoform I) from *Thaumatococcus danielli* fruits (Ogata *et al.*, 1992); **PR-8-type proteins**, class III chitinase/lysozyme (hevamine) from rubber tree latex (Terwisscha van Scheltinga *et al.*, 1994, 1996); **Nonspecific lipid transfer proteins**, LTP from wheat seeds (Gincel *et al.*, 1994); **Plant defensins**, Rs-AFP1 from radish seeds (Fant *et al.*, 1998); **Thionins**, β-purothionin from wheat seeds (Stec *et al.*, 1995); **Hevein-type antimicrobial peptides**, hevein from rubber tree latex (Andersen *et al.*, 1993). Peptide backbones are shown in molscript presentation with helices representing α-helices and arrows representing β-strands. Only major secondary structure elements are shown. Bars represent 10 Å on the Y-axis to allow for size comparison.

homology at the amino acid sequence level. The structural aspects of the different AMP families will be treated in substantial detail in section V of this chapter.

Table 1. Key properties of the main families of AMPs

AMP family	Typical size (kDa)	Enzymatic properties	Proposed microbial target
PR-1-type proteins	15	N.E.A.R. [a]	U. [b]
PR-2-type proteins	30	β-1,3-glucanase	β-1,3-1,6-glucan
PR-3-type proteins	25-30	chitinase	chitin
PR-4-type proteins	15-20	chitinase (?) [c]	chitin (?)
PR-5-type proteins	25	N.E.A.R.	membrane (membrane protein?)
PR-8-type proteins	28	chitinase	chitin (?)
PR-11-type proteins	40	chitinase	chitin (?)
Ribosome inactivating proteins	30	polynucleotide adenosine glycosidase	polynucleotides
Lipid transfer proteins (nsLTPs)	9	N.E.A.R.	membrane (phospholipids)
Hevein-type AMPs	3-5	N.E.A.R.	U.
Thionins	5	N.E.A.R.	membrane (phospholipids)
Plant defensins	5	N.E.A.R.	membrane (nonphospholipid component?)

[a] No enzymatic activity reported
[b] Unknown
[c] Question mark indicates lack of direct evidence

A number of AMPs listed in table 1 have been ascribed enzymatic properties. PR-2-type proteins, for instance, are β-1,3-glucanases; PR-3-, PR-8-, and PR-11- and possibly also PR-4-type proteins hydrolyse chitin or related substrates; and ribosome-inactivating proteins have polynucleotide adenosine glycosidase activity. In the case of PR-2- and PR-3-type proteins the antifungal properties have been linked to their ability to hydrolyse fungal cell wall polysaccharides (Arlorio *et al.*, 1992; Benhamou *et al.*, 1993a). Likewise, the enzymatic activity of ribosome-inactivating proteins has been shown to be required for their antiviral activity (Tumer *et al.*, 1997). For many AMP families devoid of enzymatic activity the target appears to be the microbial plasma membrane. Both thionins and non-specific lipid transfer proteins are known to interact directly with phospholipid membranes thus causing structural disturbance of membranes (Caaveiro *et al.*, 1997; Gasanov *et al.*, 1993; Suribade *et al.*, 1995; Thevissen *et al.*, 1996; Wall *et al.*, 1995). Plant defensins and PR-5-type proteins, on the other hand, have been proposed to affect particular binding sites on membranes (Thevissen *et al.*, 1997; Yun *et al.*, 1997b), although a direct proof for this assumption is still lacking. For a more elaborate discussion on the antimicrobial properties of the different AMP families, the reader is referred to section V of this chapter. A schematic overview of the putative targets of plant AMPs on a fungal pathogen is presented in Fig. 2.

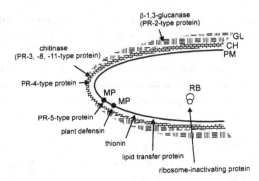

Figure 2.   **Schematic representation of the proposed targets on a fungal hypha for different families of AMPs.**
Many of the proposed targets, except those of chitinases, β-1,3-glucanases and thionins, have not yet been fully validated. CH: chitin; GL: β-linked glucan; MP: membrane binding site; PM: plasma membrane; RB: ribosome.

III.   Expression and Localization of Antimicrobial Proteins

AMPs have been discovered in every plant species that has been looked at so far and can be present in virtually any organ. In most cases AMPs are encoded by small gene families usually counting from 5 to over 50 members (see references cited in Linthorst, 1991 for PR-1 to PR-5; see Kader, 1997 for nsLTPs; see Epple *et al.*, 1997a for plant defensins; see Bohlmann, 1994 for thionins) and different members of the gene family can be differentially expressed in terms of spatial distribution, developmental regulation and regulation by external stimuli (e.g. Holtorf *et al.*, 1995; Thomma and Broekaert, 1998; Tornero *et al.*, 1997). In keeping with the model systems presented in chapter 1 of this book, we will discuss in this section general aspects of AMP gene expression in leaves, roots, flowers and seeds of three model plants: the dicots tobacco (*Nicotiana tabacum*, Solanaceae) and Arabidopsis (*Arabidopsis thaliana*, Brassicaceae) and the monocot barley (*Hordeum vulgare*, Poaceae). Where relevant, we will also mention complementary data from studies performed on other plant species. A limited literature guide on the expression of AMP genes in leaves of tobacco, Arabidopsis and barley is given in Table 2. Table 3 provides a limited key to references on AMP expression in roots, flowers and seeds of these three plant species.

A.   EXPRESSION IN LEAVES

By far the best studied plant organs for what concerns expression of AMP genes are the leaves. A general property of leaf-expressed AMP genes is that many of them are upregulated during and following invasion attempts by microorganisms. As can be seen in Table 2, most of the antimicrobial protein families mentioned in section II have pathogen-inducible members in tobacco, Arabidopsis and barley. A pathogen-inducible RIP has so far not been described in these plant species but it has been reported that viruses can induce accumulation of two type 1 RIPs in sugar beet leaves (Girbés *et al.*, 1996).

Table 2. Examples of AMPs induced by various forms of stress in leaves of tobacco, Arabidopsis or barley. References relating to tobacco, Arabidopsis or barley are indicated in normal, *italic* or **bold** face, respectively.

AMP family gen	Pathogen stress	Wounding	Cold stress	Osmotic stress	UV light or active oxy-species
PR-1	Ward et al., 1991 Uknes et al., 1992 **Reiss and Bryngelsson, 1996**				Brederode et al., 1991 Ernst et al., 1992 *Sharma et al., 1996*
PR-2	Ward et al. 1991 Uknes et al., 1992 **Reiss and Bryngelsson, 1996**	Brederode et al., 1991			Brederode et al., 1991 Ernst et al., 1992
PR-3	Ward et al., 1991 Dempsey et al., 1993 **Reiss and Bryngelsson, 1996**	Brederode et al., 1991	**Tronsmo et al., 1993**	Yun et al., 1996	Brederode et al., 1991 Ernst et al., 1992
PR-4	Ward et al., 1991 Ponstein et al., 1994 Potter et al., 1993 **Hejgaard et al., 1992**	Ponstein et al., 1994			Brederode et al., 1991 Ponstein et al., 1994
PR-5	Ward et al., 1991 Uknes et al., 1992 **Bryngelsson and Green, 1989**	Brederode et al., 1991		King et al., 1988	Brederode et al., 1991
PR-8	Lawton et al., 1992 *Samac and Shah, 1991*				

Table 2. Continued

AMP family	Pathogen stress	Wounding	Cold stress	Osmotic stress	UV light or active oxygen species
PR-11	Melchers et al., 1994	Melchers et al., 1994			Melchers et al., 1994
nsLTP	**Molina and Garcia-Olmedo, 1993**		**White et al., 1994**	**White et al., 1994**	
Thionin	*Epple et al., 1995* **Bohlmann et al., 1988**			**Garcia-Olmedo et al., 1995**	
Plant defensin	*Penninckx et al., 1996*				*Penninckx et al., 1996*
RIP		**Reinbothe et al., 1994**		**Reinbothe et al., 1994**	

Table 3. Examples of AMPs occurring in roots, flowers and seeds of tobacco, Arabidopsis or barley
References relating to tobacco, Arabidopsis or barley are indicated in normal, *italic* or **bold face**, respectively

AMP family	Roots	Flowers	Seeds
PR-1	Tahiri-Alaoui *et al.*, 1993	Lotan *et al.*, 1989	
PR-2	Memelink *et al.*, 1990	Lotan *et al.*, 1989 Bucciaglia and Smith, 1994	**Leah *et al.*, 1991**
PR-3	Memelink *et al.*, 1990 *Samac et al., 1990*	Neale *et al.*, 1990 Lotan *et al.*, 1989	**Leah *et al.*, 1991**
PR-4	*Potter et al., 1993*		Hejgaard *et al.*, 1991
PR-5	King *et al.*, 1988	Kononowicz *et al.*, 1992	Vigers *et al.*, 1991
PR-8	*Samac and Shah, 1991*		
PR-11		Heitz *et al.*, 1994b	
nsLTP	Canevascini *et al.*, 1996	Koltunow *et al.*, 1990 *Thoma et al., 1994* **Molina and Garcia-Olmedo, 1993**	Mundy and Rogers, 1986
Thionin	**Steinmuller *et al.*, 1986**	Epple *et al.*, 1995	Ponz *et al.*, 1986
Plant defensin	*Thomma and Broekaert, 1998*	Gu *et al.*, 1992 *Epple et al., 1997a*	Mendez *et al.*, 1990 *Penninckx et al., 1996*
RIP			Roberts and Selitrennikoff, 1986

Induction of sets of AMP genes has been observed upon infection of plants with various types of pathogens. In tobacco, for instance, inoculation with TMV induces a series of acidic and basic antimicrobial PR-proteins, including those belonging to PR-types 1 to 5, 8 and 11 (Brederode *et al.*, 1991, Gianinazzi *et al.*, 1970; Melchers *et al.*, 1994; Lawton *et al.*, 1992; Ponstein *et al.*, 1994; van Loon and Van Kammen, 1970; Ward *et al.*, 1991). At least some of these proteins have been shown to accumulate upon infection of tobacco by bacteria (Ahl *et al.*, 1981; Meins and Ahl, 1989), true fungi (Gianinazzi *et al.*, 1980) and Oomycetous fungi (Meins and Ahl, 1989; Tuzun *et al.*, 1989; Ye *et al.*, 1989). The same PR-1, PR-2 and PR-5 genes are induced in Arabidopsis leaves upon infection with turnip crinkle virus (Dempsey *et al.*, 1993; Uknes *et al.*, 1993), the bacterium *Pseudomonas syringae* pv. *tomato* (Uknes *et al.*, 1992) and the fungus *Fusarium oxysporum* (Epple *et al.*, 1995; Mauch-Mani and Slusarenko, 1994). However, the entire repertoire of pathogen-responsive AMP genes is not always induced by a particular pathogen. For instance in barley, inoculation with the fungus *Erysiphe graminis* induces particular PR-1, thionin and nsLTP genes, while inoculation with the bacterium *Pseudomonas syringae* induces the former two genes but suppresses nsLTP genes below basal expression level (Garcia-Olmedo *et al.*, 1995; Molina *et al.*, 1996). This indicates that different signalling pathways must exist for the induction of AMP genes in leaves. In addition, the extent by which a particular pathogen induces pathogen-responsive AMP genes often depends on its virulence on a particular host. In many cases, avirulent races or pathovars of a pathogen have been found to induce AMP genes earlier and/or to a higher level in comparison to their virulent counterparts (e.g. Conrads-Strauch *et al.*, 1990; Fink *et al.*, 1990; Vöisey and Slusarenko, 1989). In these cases, the virulent pathogen is only poorly detected by the pathogen recognition system of the host. Virulence of the pathogen then relies mainly on its ability to avoid the triggering of some inducible defense responses. As AMP expression correlates with avirulence it can be conceived that the induced AMPs contribute at least in part to the control of the invading avirulent pathogen. On the other hand, other examples have been described where AMP expression was more intense upon inoculation with virulent than with avirulent pathogens (e.g. Dong *et al.*, 1991; Garcia-Olmedo *et al.*, 1995). In such cases, the virulent and avirulent pathogens most probably trigger similar AMP gene expression levels per mass of pathogen but because the virulent pathogen is able to grow and spread in the host tissues it will induce a stronger response. Clearly, in these cases the avirulent pathogens must be controlled by mechanisms which do not involve AMPs but rather other mechanisms such as for instance the hypersensitive response or induced cell wall fortification.

The primary signal for induction of pathogen-responsive AMP genes are the elicitors secreted by or released from microbial invaders (see Chapter 4; Coté and Hahn, 1994; Ebel and Cosio, 1994). Each pathogen produces a particular mixture of elicitors, sometimes accompanied by suppressors, and these molecules interact with receptors on the host cells that further translate the primary signal into particular events at the level of the plasma membrane, the cytosol and/or the nucleus (see Chapter 4). The induction of AMP genes generally requires that the primary signals (elicitors) give rise to the production of secondary endogenous signal molecules (stress hormones) by the challenged cells in the infection site. These secondary signal molecules in turn set in motion signal transduction cascades in receiving cells, eventually leading to activation of, amongst others, pathogen-responsive AMP genes. Secondary signal molecules thus serve to amplify and spread out the response of the host following initial recognition of the pathogen. Several secondary signal molecules whose synthesis is increased in response to elicitor recognition and which are involved in the activation of AMP genes have been identified. These include H_2O_2 or other active oxygen species (reviewed by Lamb and Dixon, 1997; Mehdy, 1994), salicylic acid (reviewed by Dürner *et al.*, 1997), ethylene (reviewed by Boller, 1991), jasmonic acid (reviewed by Creelman and Mullet, 1997) and abscisic acid (reviewed by Zeevaart, 1988), but it is likely that still others remain to be discovered. Some of these, such as H_2O_2, have a high turnover rate and can only activate genes in a limited area immediately surrounding the infection site (Levine *et al.*, 1994). Others, such as salicylic acid, are transported over a longer distance and are even spread via the vascular system to leaves or other organs that are distant from the infection site (Mölders *et al.*, 1996; Shualev *et al.*, 1995). Many aspects of these mobile signals are not fully understood. For instance in the case of salicylic acid it is still unclear whether it is transported as such, as a derivative (e.g. as methyl salicylate; Shualev *et al.*, 1997) or accompanied by other unknown mobile signal molecules (see chapter 10 and reviews by Dürner *et al.*, 1997; Ryals *et al.*, 1996). AMP genes responding to signal molecules with limited mobility are said to be locally induced, while those responding to mobile signals are said to be systemically induced. To which secondary signal molecules a particular pathogen-responsive AMP gene responds depends primarily on the presence and relative position of binding sites for transcription factors in its promoter (for a review on *cis*-regulatory elements in panthogen-responsive AMP genes, see Somssich, 1994).

In view of the number of different secondary signal molecules it is not surprising that the different AMP genes do not respond in the same way to a particular pathogen. In tobacco, for instance, it has been observed that transcripts encoding acidic PR-1, PR-2, PR-3 and PR-5 proteins accumulate coordinately both in TMV-inoculated leaves and non-inoculated leaves of the same plant, while transcripts encoding their basic counterparts accumulate in TMV-inoculated leaves but not or very little in non-inoculated leaves (Brederode *et al.*, 1991; Ward *et al.*, 1991). This clearly indicates that at least two separate pathways for induction of pathogen-responsive AMP genes exist in tobacco. Despite the fact that there is less transcription of basic PR-protein genes versus acidic PR-protein genes in tobacco leaves distant from TMV-inoculated leaves, basic PR-proteins clearly accumulate in these leaves as detected by Western blot analysis, suggesting either a low transcript or limited protein turnover rate for basic PR-protein genes in tobacco (Heitz *et al.*, 1994a). Hence, post-transcriptional events can also influence the expression pattern of pathogen-responsive AMP genes. The occurrence of at least two separate AMP gene induction pathways in tobacco can also be inferred from the observation that acidic PR-protein genes are more responsive to exogenous application of salicylic acid than ethylene, while the reverse is true for basic PR-protein genes (Beffa *et al.*, 1995; Brederode *et al.*, 1991). Induction of the acidic PR-protein genes in tobacco has convincingly been demonstrated to require salicylic acid as a secondary signal molecule. Transgenic plants engineered to express a bacterial salicylate hydroxylase gene (*nahG*) failed to accumulate salicylate in response to pathogen attack, which made them unable to induce acidic PR-1, PR-2 and PR-3 genes (Friedrich *et al.*, 1995; Gaffney *et al.*, 1993). Pathogen-induced expression of the basic PR-proteins, on the other hand, appears to involve the secondary signal molecule ethylene. This was inferred from experiments using transgenic tobacco plants transformed with a dominant-negative mutant allele of the Arabidopsis ethylene receptor gene ETR1. The transgenic plants with a disrupted ethylene response pathway failed to induce genes encoding basis PR-1 and PR-5 upon infection with TMV, while the expression of a gene encoding an acidic PR-1 was unaffected (Knoester *et al.*, 1998). Evidence has also been presented for the occurrence of two separate pathways for AMP gene induction in Arabidopsis, one of which is salicylate-dependent while the other is dependent on the stress hormones jasmonate and ethylene. Transgenic Arabidopsis expressing the *nahG* salicylate hydroxylase gene lost their ability to induce at least the PR-1 gene upon microbial infection (Delaney *et al.*, 1994; Lawton *et al.*, 1995; Penninckx *et al.*, 1996). On the other hand, nahG-expressing plants still induced systemically a plant defensin gene upon inoculation with the fungus *Alternaria brassicicola* (Penninckx *et al.*, 1996). Pathogen-induced expression of the plant defensin gene, on the other hand, was abolished in a mutant affected in its response to ethylene (*ein2*) as well as in another mutant affected in the jasmonate response (*coil*). Both ethylene and jasmonic acid, when applied exogenously, induce the plant defensin gene but not the PR-1 gene, and vice versa for salicylic acid treatment (Penninckx *et al.*, 1996). In order to activate the plant defensin gene, both the ethylene and the jasmonate signal pathways must be activated concomitantly (Penninckx *et al.*, 1998). Hence, in this case, pathogen-induced expression does require the concerted action of at least two secondary signal molecules. Interactions between different secondary signal molecules has also been inferred from the synergistic enhancement of AMP gene expression by a combination of two signal molecules. This has been observed with ethylene and methyl jasmonate for basic PR-5 gene expression in tobacco (Xu *et al.*, 1994) and for plant defensin gene expression in Arabidopsis (Penninckx *et al.*, 1998), and with ethylene and salicylic acid for PR-1 gene expression in Arabidopsis (Lawton *et al.*, 1994). The ability of particular AMP genes to respond to combinations of secondary signal molecules further enhances the plasticity of the host defense response. Whereas signalling pathways for pathogen-induced gene expression appear to be conserved in most plant species, the set of AMP genes that is induced along a particular pathway may vary from plant to plant. For instance, PR-8 genes in tobacco are systemically induced upon pathogen stress and respond to salicylic acid (Ward *et al.*, 1991), while the Arabidopsis PR-8 gene is only induced in the vicinity of pathogen lesions and is not induced by salicylic acid (Samac *et al.*, 1990; Samac and Shah, 1991). Also, quantitative differences in expression levels of AMP genes have been observed among different plants. In pathogen-stressed tobacco plants the most abundant AMPs are PR-1 proteins (van Loon, 1985) while in infected cucumber plants the prominent AMPs are PR-8 proteins (Schneider *et al.*, 1996).

It has been shown that plant cells often respond to primary signal molecules from pathogens by rapidly generating active oxygen species such as H_2O_2 and superoxide anion (reviewed by Lamb and Dixon, 1997). These reactive oxygen species may in turn activate via largely unknown signal transduction events the production of secondary signal molecules such as salicylic acid (Yalpani *et al.*, 1994), ethylene (see Schraudner *et al.*, 1996) and jasmonic acid (Conconi *et al.*, 1996). However, pathogen attack is not the only cause of stimulated formation of reactive oxygen species in plants. Indeed, several other forms of stress including cold stress, heat stress, drought stress, salinity stress, wounding and UV light exposure are known to trigger either transient or sustained generation of active oxygen species (reviewed by Bowler *et al.*, 1992; Foyer *et al.*, 1994). This may explain why many pathogen-responsive AMP genes have also been found to be induced when plants are subjected to these different forms of stress (see Table 2). Again, not all AMP genes respond similarly to these stimuli. In tobacco, for instance, basic PR-2 and PR-3 genes respond to a lower ozone exposure level than the acidic PR-1 and PR-3 genes (Ernst *et al.*, 1992), while in Arabidopsis a plant defensin gene but not a PR-1 gene is induced by sub-herbicidal levels of the superoxide anion-generating compound paraquat (Penninckx *et al.*, 1996). In some cases the stress-induced AMP genes may be different from the pathogen-induced genes. However, at least in the case of the basic PR-5 gene from tobacco, expression of a reporter gene driven by the promoter of this gene is induced by salinity stress (La Rosa *et al.*, 1992). It should further be noted that many AMP genes known to be induced by either drought or salinity stress can be activated by exogenous application of abscisic acid. This applies for either PR-3 type proteins (Chen *et al.*, 1994), PR-5 type proteins (La Rosa *et al.*, 1992; Singh *et al.*, 1989; Zhu *et al.*, 1995) or nsLTPs (Soufleri *et al.*, 1996; Torres-Schumann *et al.*, 1992; White *et al.*, 1994). This is consistent with the observation that abscisic acid accumulates in many plants in response to drought stress (Zeevaart, 1988).

Not all leaf-expressed AMPs are induced upon pathogen attack. For instance, out of 8 electrophoretically separable acidic chitinase isozymes found in extracts from healthy tobacco leaves, only two are present at higher levels in extracts from leaves infected with the Oomycete *Peronospora tabacina* (Pan *et al.*, 1992). Similarly, Arabidopsis has at least two leaf-expressed genes for thionins and at least two more leaf-expressed genes for plant defensins, but in both cases only one of them is pathogen-inducible (Epple *et al.*, 1995, 1997a). In addition, pathogen-inducible AMP genes have a basal expression level which in some cases varies with the developmental stage of the plant. In tobacco, basic PR-2, PR-3, PR-5 and PR-11 genes are more abundantly expressed in old leaves than in young leaves (Heitz *et al.*, 1994a; Kononowicz *et al.*, 1992; Neale *et al.*, 1990; Shinshi *et al.*, 1987; Vögeli-Lange *et al.*, 1994b), while particular constitutive acidic chitinase isoforms are present in young apical leaves but not in mature leaves (Trudel *et al.*, 1989). This all means that AMPs are present in healthy non-infected leaves, albeit usually at lower levels as compared to infected leaves. In many cases the cellular distribution of AMPs in healthy leaves is uneven and reaches highest levels in any of the following cell types: epidermal cells (e.g. basic PR-2 and PR-3 in tobacco, Keefe *et al.*, 1990; basic PR-3 in potato, Garcia-Garcia *et al.*, 1994; thionin in barley, Reimann-Philipp *et al.*; 1989, nsLTP in barley, Molina and Garcia-Olmedo, 1993, tobacco, Canevascini *et al.*, 1996, and Arabidopsis, Thoma *et al.*, 1994) stomatal cells and hydathode cells (e.g. PR-3 in bean, Mauch *et al.*, 1992, and sugar beet, Nielsen *et al.*, 1996a; nsLTP in tobacco, Canevascini *et al.*, 1996, and sugar beet, Nielsen *et al.*, 1996b; PR-8 in Arabidopsis, Samac and Shah, 1991; plant defensin in sugar beet,

Kragh *et al.*, 1995), cells in vascular strands (e.g. basic PR-1 in tobacco, Eyal *et al.*, 1993; PR-2 and PR-3 in bean, Del Campillo and Lewis, 1992; PR-8 in Arabidopsis, Samac and Shah, 1991; nsLTP in Arabidopsis, Thoma *et al.*, 1994, barley, Molina and Garcia-Olmedo, 1993, and sugar beet, Nielsen *et al.*, 1996a; plant defensin in sugar beet, Kragh *et al.*, 1995) and cells in the leaf abscision zone (e.g. basic PR-1 in tobacco, Eyal *et al.*, 1993; nsLTP in Arabidopsis, Thoma *et al.*, 1994). It is noteworthy that all leaf cell types in which basally expressed AMPs accumulate preferentially represent entry routes for microbial pathogens. Indeed, all pathogens initially enter unwounded leaves either via the epidermis, stomata or hydathodes. The vascular tissues, on the other hand, are used by several pathogens as a route for systemic spread in plants. Hence, the constitutive presence of AMPs at these gateways warrants a first line of defense against invading pathogens. Once the pathogen has breached this first line of defense, further accumulation of inducible AMPs in other cell layers may occur. Indeed, while basic class I PR-2 and PR-3 proteins are mainly restricted to the epidermis in healthy tobacco leaves, they accumulate in all leaf cells upon treatment with ethylene, a treatment that simulates pathogen stress (Keefe *et al.*, 1990). Similarly, a basic PR-3 protein was confined to epidermal cells in healthy potato leaves, but was homogeneously distributed through mesophyll and epidermis in the area around infection sites by the Oomycete *Phytophthora infestans* (Garcia-Garcia *et al.*, 1994). The preferential basal expression of AMP genes in protective cell layers of unchallenged plant organs bears strong resemblance to the situation in animals where genes encoding AMPs such as defensins are predominantly expressed in several epithelial tissues (Diamond *et al.*, 1996; Schonwetter *et al.*, 1995; Zhao *et al.*, 1996).

At the subcellular level, AMPs are generally either deposited in the vacuole or in the extracellular space. Evidence for an extracellular location has been presented for acidic PR-1 to PR-5 in tobacco (Dore *et al.*, 1991; Parent and Asselin, 1984; Van Loon and Gerritsen, 1989), basic PR-1 in tobacco (Niderman *et al.*, 1995), acidic PR-8 in tobacco (Lawton *et al.*, 1992), nsLTP in Arabidopsis (Thoma *et al.*, 1993), thionin in barley (Bohlmann *et al.*, 1988; Ebrahim-Nesbat *et al.*, 1989, 1993), plant defensin in sugar beet (Kragh *et al.*, 1995) and RIP in pokeweed (Ready *et al.*, 1986). On the other hand, vacuolar deposition was demonstrated for basic PR-2, PR-3 and PR-5 in tobacco (Keefe *et al.*, 1990; Singh *et al.*, 1987; Van den Bulcke *et al.*, 1989) and thionin in barley (Reimann-Philipp *et al.*, 1989). Thus, AMPs belonging to the same family can be deposited either in the vacuole or in the extracellular space in the same plant. Vacuolar and extracellular AMPs are, however, usually products of different genes and vacuolar deposition seems generally to be correlated with the presence of a C-terminal propeptide in the precursors of vacuolar isoforms which acts as a vacuolar targetting determinant (Neuhaus *et al.*, 1991b; Shinshi *et al.*, 1988; Sticher *et al.*, 1992a; Van den Bulcke *et al.*, 1989). One exception appears to be the thionins in barley leaves of which vacuolar isoforms are identical to extracellular isoforms (Reimann-Philipp *et al.*, 1989). As in this case over 95% of the thionin content in leaves is deposited in the vacuoles, extracellular deposition might be the result of an overflow from the main trafficking route.

The presence of AMPs at these two distinct subcellular locations is believed to have a physiological meaning. Most pathogens grow extracellularly during at least the early stage of the infection process. During this stage they will encounter extracellular AMPs that will interfere with their growth and development. On the other hand, plant cells can be lysed following pathogen invasion or due to the hypersensitive response. Lysed cells will release their vacuolar content including the vacuolar AMPs. This sudden release of AMPs can result in high local AMP concentrations and preclude adaptation of the pathogen to low constitutive doses of AMPs. In animals it is well documented that lysosomes, the organelles equivalent to vacuoles in plants, can fuse with the plasma membrane and thus release their contents extracellularly by Ca^{2+}-regulated stimuli (Rodriguez *et al.*, 1997), but it is not known whether a similar process occurs during plant-pathogen interactions.

In a number of cases the concentration of AMPs in leaves has been estimated and compared to the concentration required for microbial growth inhibition in vitro. In infected pea pods, basic vacuolar chitinases and glucanases accumulate each at about 40 to 120 µg per g fresh weight. Assuming that the leaf is in a liquid state, this would correspond to a concentration of about 40 to 120 µg/ml, which is about four-fold above the concentration required to inhibit fungal growth by a combination of the purified enzymes (Mauch *et al.*, 1988b). In infected *Arabidopsis* leaves, plant defensins accumulate to about 10 µg per g fresh weight (Penninckx *et al.*, 1996). Given the fact that plant defensins are deposited extracellularly (Terras *et al.*, 1995; Terras FRG, unpublished results), the concentration in the extracellular space should be at least 10-fold higher, thus reaching at least 100 µg/ml, which is sufficient to inhibit growth of most fungi in vitro (Terras *et al.*, 1992b). The concentration of nsLTPs in healthy barley leaves has been estimated to be about 100 µg per g fresh weight (Molina and Garcia-Olmedo, 1993). As the major part of the nsLTPs are present at the tissue surface, the local concentration should be at least one order of magnitude higher and thus well above the concentrations needed to inhibit microbial growth in vitro (Molina *et al.*, 1993b). Hence, in many cases studied the accumulation of AMPs appears to be sufficiently high to interfere with microbial growth *in planta*.

B.   EXPRESSION IN ROOTS

Plant roots generally contain high constitutive levels of at least some AMPs (see Table 3). In tobacco, the genes for basic vacuolar PR-2, PR-3 and PR-5 proteins are constitutively expressed in roots while genes encoding acidic extracellular PR-1 to PR-5 have a much lower constitutive expression level in this organ (Nelson *et al.*, 1992; Memelink *et al.*, 1990). Basic PR-3 proteins, for instance, can constitute as much as 4% of the total soluble protein content in healthy tobacco roots (Shinshi *et al.*, 1987). Such high constitutive expression levels even in absence of pathogens may indicate a need for protection against the continuous presence of microorganisms in the soil. At least for the basic PR-2 and PR-5 genes from tobacco it was shown by promoter-reporter gene fusions that constitutive expression in roots and pathogen-inducible expression in leaves are conferred by the same genes (Nelson *et al.*, 1992; van de Rhee *et al.*, 1993; Vögeli-Lange *et al.*, 1994b). Some AMPs in roots have low basal expression levels and are upregulated upon pathogen infection as in the case of many leaf-expressed AMPs. For instance, tobacco roots contain a protein that is immunoreactive with antibodies raised against an acidic PR-1 protein from tobacco leaves. This protein is extracellularly located and is strongly enriched in roots infected with the fungus *Chalara elegans* (Tahiri-Alaoui *et al.*, 1993). In tomato, extracellular PR-1, PR-2 and PR-3 proteins have been identified which accumulate in response to infection by *Fusarium oxysporum* f.sp. *lycopersici* (Benhamou *et al.*, 1989, 1990, 1991). It is not known, however, whether or not these proteins are products of pathogen-responsive leaf-expressed PR-genes.

C.   EXPRESSION IN FLOWERS

Expression of AMPs in flowers has so far best been studied in species of the family Solanaceae. As shown in Table 3, members of most AMP protein families have been detected in tobacco flower organs. These flower-expressed AMPs can be products of flower-specific genes, such as the potato *SK2* gene encoding a PR-3-type chitinase that is apparently only expressed in the transmitting tract of the styles (Wemmer *et al.*, 1994) and the tobacco *Tag1* gene encoding an anther-specific PR-2-type glucanase (Bucciaglia and Smith, 1994). In other cases, AMPs present in flower tissues are products of pathogen-responsive leaf-expressed genes. For instance, the promoter from a gene for an acidic PR-2 expressed in tobacco leaves infected with TMV, also confers expression in flower organs (Hennig *et al.*, 1993). Using antisera raised against acidic PR-1, PR-2 and PR-3 proteins from tobacco leaves, Lotan *et al.* (1989) detected a PR-1-type protein in sepals, a PR-2-type protein in pistils and a PR-3-type protein in pedicels, sepals, anthers and ovaries of tobacco flowers. In addition, a PR-11-type protein was found in pedicels, sepals and petals (Heitz *et al.*, 1994b), a plant defensin in petals, anthers and pistils (Gu *et al.*, 1992) and an nsLTP in anthers (Koltunow *et al.*, 1990) of tobacco flowers. The reason for this diversity in expression patterns among different flower organs is not known. Some of these AMPs may play a role in a developmental process. For instance, an anther-specific tobacco β-1,3-glucanase (Tag1) may participate in the separation of microspore tetrads and thus be involved in pollen formation (Bucciaglia and Smith, 1994), while a plant defensin-like protein present in the coat of pollen from *Brassica oleracea* may have a function in pollen-pistil interactions (Stanchev *et al.*, 1996). On the other hand, it has also been speculated that AMPs in flower organs may help to protect these vulnerable tissues. For instance, the abundant presence of an acidic glucanase in the transmitting tract of tobacco pistil styles, which represents up to 12% of the extracellular proteins, may help to control ingress of opportunistic fungi which follow the track of invading pollen tubes (Ori *et al.*, 1990). In addition, the predominant occurrence of a plant defensin in the epidermis of the adaxial surface of petals and in the peripheral cell layers of the style, the ovary, the stamen filaments and the anthers, is also consistent with a role in protection of these organs against microbial invaders (Gu *et al.*, 1992).

D.   EXPRESSION IN SEEDS

Plant seeds are generally a very rich source of AMPs. The presence of AMPs in seeds has been studied most exhaustively in barley and nearly all types of AMPs have been purified from seeds of this species (see Table 3). The most abundant AMPs probably act primarily as storage proteins that provide a source of nitrogen and sulfur during germination.

Typical for seed storage proteins is that they are stored in vacuole-derived protein bodies and are degraded during germination (Shewry *et al.*, 1995). 2S albumins from Brassicaceae seeds certainly can be considered as storage proteins (Higgins, 1984), but also thionins occurring in the protein bodies of the endosperm of Poaceae seeds (Bohlmann, 1994; Carmona *et al.*, 1993a) and basic chitinases in cucumber seeds (Majeau *et al.*, 1990) have been proposed to fulfill a role as storage proteins. In this context it is also worth to mention that one of the major storage proteins in *Canavalia ensiformis* seeds, concanavalin B, shows sequence and 3D-structure homology to PR-8-type chitinases whilst having lost chitinolytic activity due to mutations in the active site (Hennig *et al.*, 1995a). It is not excluded that storage proteins exihibiting antimicrobial activity have a dual function and protect seeds from molding during the dormant period.

Other seed AMPs have a localization profile that is not consistent with a storage role. Plant defensins from radish seeds, for instance, are located extracellularly in peripheral cells surrounding all seed organs (Terras *et al.*, 1995). It has been proposed that these plant defensins play an important role in the protection of seeds or seedlings against invasion by soil-borne fungi (Terras *et al.*, 1995). This hypothesis is based on the observation that radish seeds germinating on a medium supporting the growth of a fungal colony, cause a growth inhibition halo which can be mimicked by application on the medium of drops containing as little as 1 µg of the purified radish plant defensins. Seeds that were kept dormant by external application of abscisic acid did not produce the inhibition halo unless their seed coats were mechanically perforated. Analysis of the imbibition solution of seeds with a mechanically incised seed coat, revealed that plant defensins accounted for 30% of released proteins, although they are minor proteins in the seed (0.5% of total seed proteins). The amount of plant defensins released from a single seed was estimated to be at least 1 µg, which is the amount of peptide required to mimic the inhibition halo formed around a germinating seed. All of these experiments indicate that plant defensins are released from radish seeds when the seed coat is perforated (either by the radicle of the germinating embryo under natural conditions or artificially with the aid of a scalpel), and moreover, that the released amounts are sufficient to create a zone around the seeds in which fungal growth is suppressed. Hence, these findings strongly suggest that plant defensins play a role in the protection of seedling tissues during the early stage of emergence and thus may contribute to the enhancement of seedling survival rates. The simple fact that chemical fungicides are commonly used for the coating of crop seeds to increase seedling stand illustrates that soil-borne or seed-borne fungi form a considerable threat to germinating seeds. Release of AMPs during seed imbibition has also been described for barley seeds. Of five different chitinase isoforms found in barley seeds, two embryo-specific ones were released preferentially during imbibition (Swegle *et al.*, 1992). However, the released amounts were only 70 ng per seed after 10 h of imbibition and it is uncertain whether this would be sufficient to protect the germinating seeds against soil-borne pathogens. In sorghum seeds, imbibition caused PR-5-type proteins and chitinases to leach out of the endosperm and to accumulate in the pericarp (Seetharamam *et al.*, 1996). Retention of AMPs in the pericarp may help to protect the seed surface from fungal invasion during radicle emergence, although this hypothesis requires substantiation by experimental evidence.

AMPs including PR-1-, PR-2-, PR-3- and PR-5-type proteins have also been shown to be synthesized *de novo* during germination of seeds including those of tobacco, barley, maize and sorghum (Casacuberta *et al.*, 1991; Hennig *et al.*, 1993; Leah *et al.*, 1991; Seetharamam *et al.*, 1996; Vögeli-Lange *et al.*, 1994a). A PR-1-type gene from maize was found to be upregulated upon infection of seedlings with the fungus *Fusarium moniliforme* (Casacuberta *et al.*, 1991), suggesting that at least some of these AMPs produced by germinating seedlings have a similar protective function as their counterparts in leaves or roots. Promoter-reporter gene fusions have shown that both a basic and an acidic PR-2-type protein found in tobacco seedlings are products of pathogen-responsive leaf-expressed PR-genes (Hennig *et al.*, 1993; Vögeli-Lange *et al.*, 1994a). The tobacco gene for the basic PR-2 protein is expressed in the germinating seeds exclusively in the micropylar region of the endosperm and it has been proposed that it plays a role in the weakening of this tissue to allow radicle emergence (Leubner-Metzger *et al.*, 1995). However, it is not excluded that this protein also plays a role in the protection of the ruptured seed tissues against microbial infection during germination.

IV.   Contribution of Antimicrobial Proteins to Resistance

The fact that some plant proteins are termed "pathogenesis-related" or "antimicrobial" relies on their property to be induced *in planta* after pathogen-attack or to exert antibiotic activity against pathogens in vitro, respectively (see section I). These findings, however, do not necessarily imply and certainly do not prove a role in warding off microbial invaders. More evidence for such a role *in planta* can be obtained by different genetic strategies. These can include the overexpression of AMP genes in either homologous or heterologous plants, followed by assessment of the resistance to microbial pathogens of the transgenic plants. Alternatively, one can analyse whether a plant's susceptibility to invasion by microorganisms is increased by downregulating the expression of endogenous AMP genes (e.g. through antisense RNA techniques). A third powerfull approach to unravel the potential in vivo role of AMPs in plant defense is the analysis of mutant plants with up- or downregulated signalling pathways leading to induced AMP synthesis.

A.   OVEREXPRESSION OF AMP GENES

During the last decade over 25 different reports dealing with the overexpression of AMP genes in transgenic plants have appeared. This information is comprehensively summarized in Table 4 and some of the reports will be highlighted below.

PR-1 proteins, being among the first PR-proteins characterized, were also the first to be analysed in transgenic plants. Based on earlier observations that (i) PR-1 proteins are the most abundant of the PR-proteins accumulating upon TMV-infection of tobacco, (ii) the synthesis of PR-1 proteins in tobacco correlates with reduced expansion of lesions caused by TMV, (iii) interspecific tobacco hybrids (*Nicotiana glutinosa* X *N. debneyi*) constitutively producing PR-1 protein are highly resistant to TMV (Ahl and Gianinazzi, 1982), a role for these proteins in viral resistance was postulated. Two independent research groups transformed tobacco with the gene encoding the acidic PR-1a from tobacco, which is the most abundant of the PR-1 set of proteins accumulating in tobacco cultivar Xanthi-nc in response to TMV infection. In both studies it was found that constitutive expression of the PR-1a gene under the control of the strong constitutive CaMV35S promoter did not affect the susceptibility of tobacco plants to infection by either TMV, alfalfa mosaic virus or potato virus Y (Table 4; Linthorst *et al.*, 1989; Alexander *et al.*, 1993). Similar results were obtained for the acidic PR-1b, an isoform of PR-1a (Cutt *et al.*, 1989). These data indicate that the PR-1 proteins are not sufficient as independent factors for viral resistance. It is, however, not excluded that they may function as such in combination with other induced (PR-) proteins. Tobacco plants overexpressing PR-1a were also assessed for resistance against different bacterial and fungal pathogens (Table 3; Alexander *et al.*, 1993). Only in the cases of infection by Oomycetous fungi could a significant reduction of disease severity be observed in the transgenic plants. Homozygous tobacco lines constitutively expressing the PR-1a gene produced 42% less blue mold symptoms than untransformed plants (7 days after infection with *Peronospora tabacina*). In the same plants, development of the Black Shank disease caused by *Phytophthora parasitica* var. *nicotianae* was reduced by 32% at the end of the test (14 days post infection). The level of other PR-proteins (PR-2, PR-3, PR-4 or PR-5) was verified and found not to be induced as a consequence of expressing PR-1a as a constitutive transgene, hence indicating that PR-1a can act on its own to protect tobacco against oomycetes (Alexander *et al.*, 1993). Whether the observed disease resistance was due to a direct antimicrobial effect of the constitutively expressed PR-1a is not proven, but the results are consistent with the observation that PR-1-type proteins exhibit antimicrobial activity against at least one Oomycete, *Phytophthora infestans* (Niderman *et al.*, 1995; see further in section V.A.2.).

As can be deduced from Table 4, most of the studies on overexpression of AMPs were so far focussed on the overexpression of PR-2 (β-1,3-glucanases) and PR-3 (chitinases) proteins. Broglie and coworkers (1991) were the first to report on transformation of plants with a chitinase-encoding transcription unit. This construct consisted of a basic vacuolar chitinase gene from bean driven by the constitutive CaMV35S promoter. Transgenic lines of both tobacco and canola were found to exhibit enhanced resistance towards the soil-borne pathogenic fungus *Rhizoctonia solani*. In tobacco, homozygous transgenic lines showed an increase of chitinase activity up to 4- and 44-fold in roots and leaves, respectively. The mortality of tobacco seedlings caused by *R. solani* infection decreased from 53% (control plants) to 23% in some homozygous transgenic lines. The extent of disease was also evaluated under lower infection pressure conditions by monitoring the loss of root fresh weight in infected plants. Roots of chitinase-overexpressing plants lost on average 5 to 15% of their fresh weight, compared to 46% for control plants. The observed resistance in the transgenic tobacco or canola plants varied with the amount of fungal inoculum, a property characteristic of quantitative resistance. All these findings are consistent with the observed in vitro antifungal activity of the bean chitinase towards *R. solani* (Broglie *et al.*, 1991). The fungitoxic effect of the expressed chitinase was later demonstrated by electron microscopy to result, at least in part, from the direct lytic effect of the enzyme on fungal cell walls (Benhamou *et al.*, 1993b).

Table 4. Examples of plants genetically engineered to overexpress plant antimicrobial protein genes for enhanced disease resistance

Expressed transgene[a,b]	Transgene source	Target plant	Target pathogen	Effect on disease resistance[c]	Reference
PR-1a	tobacco (Nicotiana tabacum)	tobacco (Nicotiana tabacum)	Tobacco Mosaic Virus (TMV)	=	Linthorst et al., 1989
			Alfalfa Mosaic Virus (AlMV)	=	
PR-1a	tobacco (Nicotiana tabacum)	tobacco (Nicotiana tabacum)	Peronospora tabacina	↗	Alexander et al., 1993
			Phytophthora parasitica var. nicotianae	↗	
			Cercospora nicotianae	=	
			Pseudomonas syringae pv tabaci	=	
			Pseudomonas tabacina	=	
			Tobacco Mosaic Virus (TMV)	=	
			Potato Virus Y (PVY)	=	
PR-1b	tobacco (Nicotiana tabacum)	tobacco (Nicotiana tabacum)	Tobacco Mosaic Virus (TMV)	=	Cutt et al., 1989
PR-2 (basic β-1,3-glucanase)	soybean (Glycine max)	tobacco (Nicotiana tabacum)	Phytophthora parasitica var. nicotianae	↗	Yoshikawa et al., 1993
			Alternaria alternata (syn. A. longipes)	↗	
PR-2 (acidic β-1,3-glucanase, PR-N)	tobacco (Nicotiana tabacum)	tobacco (Nicotiana tabacum)	Peronospora tabacina	↗	Lusso and Kuc, 1996
			Phytophthora parasitica var. nicotianae	↗	
			Tobacco Mosaic Virus (TMV)	=	
			Tobacco Etch Virus (TEV)	=	
			Tobacco Vein Mottling Virus (TVMV)	=	
PR-2 (acidic β-1,3-glucanase)	alfalfa (Medicago sativa)	tobacco (Nicotiana tabacum)	Cercospora nicotianae	↗	Zhu et al., 1994
PR-3 (basic chitinase)	rice (Oryza sativa)	tobacco (Nicotiana tabacum)	Cercospora nicotianae	↗	

Table 4. Continued.

Expressed transgene[a,b]	Transgene source	Target plant	Target pathogen	Effect on disease resistance[c]	Reference
[PR-2 + PR-3]	alfalfa (*Medicago sativa*) and rice (*Oryza sativa*)	tobacco (*Nicotiana tabacum*)	*Cercospora nicotianae* *Rhizoctonia solani*	↗↗ ↗↗	
PR-2 (β-1,3-glucanase)	barley (*Hordeum vulgare*)	tobacco (*Nicotiana tabacum*)	*Rhizoctonia solani*	↗	Jach *et al.*, 1995
PR-3 (class II chitinase)	barley (*Hordeum vulgare*)	tobacco (*Nicotiana tabacum*)	*Rhizoctonia solani*	↗	
ribosome-inactivating protein (Type-I, RIP)	barley (*Hordeum vulgare*)	tobacco (*Nicotiana tabacum*)	*Rhizoctonia solani*	↗	
[PR-2 + PR-3]	cfr above	tobacco (*Nicotiana tabacum*)	*Rhizoctonia solani*	↗↗	
[PR-3 + RIP]	cfr above	tobacco (*Nicotiana tabacum*)	*Rhizoctonia solani*	↗↗	
PR-3 (class I chitinase)	bean (*Phaseolus vulgaris*)	tobacco (*Nicotiana tabacum*) canola (*Brassica napus*)	*Rhizoctonia solani* *Pythium aphanidermatum* *Rhizoctonia solani*	↗ = ↗	Broglie *et al.*, 1991
PR-3 (class I chitinase)	tobacco (*Nicotiana tabacum*)	tobacco (*Nicotiana tabacum*)	*Rhizoctonia solani*	↗	Lawton *et al.*, 1993
PR-8 (class III chitinase)	tobacco (*Nicotiana tabacum*)	tobacco (*Nicotiana tabacum*)	*Rhizoctonia solani*	↗	

Table 4.   Continued.

Expressed transgene[a,b]	Transgene source	Target plant	Target pathogen	Effect on disease resistance[c]	Reference
PR-8 (class III chitinase)	cucumber (Cucumis sativus)	tobacco (Nicotiana tabacum)	Rhizoctonia solani	↗	
PR-8 (acidic class III chitinase)	sugar beet (Beta vulgaris)	tobacco (Nicotiana benthamiana)	Cercospora nicotianae	=	Nielsen et al., 1993
PR-2 (basic β-1,3-glucanase)	tobacco (Nicotiana tabacum)	tomato (Lycopersicon esculentum)	Fusarium oxysporum f.sp. lycopersici race 1	=	Van den Elzen et al., 1993
PR-3 (class I chitinase)	tobacco (Nicotiana tabacum)	tomato (Lycopersicon esculentum)	Fusarium oxysporum f.sp. lycopersici race 1	=	
[PR2 + PR3]	tobacco (Nicotiana tabacum)	tomato (Lycopersicon esculentum)	Fusarium oxysporum f.sp. lycopersici race 1	↗	
PR-3 (class I chitinase)	tobacco (Nicotiana tabacum)	Nicotiana sylvestris	Cercospora nicotianae	= (↗)	Neuhaus et al., 1991a
			Rhizoctonia solani	↗	Vierheilig et al., 1993
PR-2 (acidic β-1,3-glucanase)	alfalfa (Medicago sativa)	alfalfa (Medicago sativa)	Stemphylium alfalfae	=	Masoud et al., 1996
			Colletotrichum trifolii	=	
			Phoma medicaginis	↗	
			Phytophthora megasperma f.sp. medicaginis	↗	
PR-3 (basic chitinase)	rice (Oryza sativa)	alfalfa (Medicago sativa)	Stemphylium alfalfae	=	
			Colletotrichum trifolii	=	
			Phoma medicaginis	= (↗)	
			Phytophthora megasperma f.sp. medicaginis	=	

Table 4. Continued.

Expressed transgene[a,b]	Transgene source	Target plant	Target pathogen	Effect on disease resistance[c]	Reference
[PR-2 + PR-3]	cfr above	alfalfa (*Medicago sativa*)	*Stemphylium alfalfae*	=	
			Colletotrichum trifolii	=	
			Phoma medicaginis	= (↗)	
			Phytophthora megasperma f.sp. *medicaginis*	=	
PR-8 (acidic chitinase, class III)	sugar beet (*Beta vulgaris*)	tobacco (*Nicotiana tabacum*)	*Cercospora nicotianae*	=	Nielsen *et al.*, 1993
PR-3 (acidic chitinase)	petunia (*Petunia X hybrida*)	cucumber (*Cucumis sativus*)	*Alternaria cucumeris*	=	Punja and Raharjo 1996
			Botrytis cinerea	=	
			Colletotrichum lagenarium	=	
			Rhizoctonia solani	=	
		carrot (*Daucus carota*)	*Alternaria radicini*	=	
			Botrytis cinerea	=	
			Rhizoctonia solani	=	
			Sclerotium rolfsii	=	
			Thielaviopsis basicola	=	
PR-3 (basic chitinase)	tobacco (*Nicotiana tabacum*)	cucumber (*Cucumis sativus*)	*Alternaria cucumeris*	=	
			Botrytis cinerea	=	
			Colletotrichum lagenarium	=	
			Rhizoctonia solani	=	
		carrot (*Daucus carota*)	*Alternaria radicini*	=	
			Botrytis cinerea	↗	
			Rhizoctonia solani	↗	
			Sclerotium rolfsii	↗	
			Thielaviopsis basicola	=	

Table 4. Continued.

Expressed transgene[a,b]	Transgene source	Target plant	Target pathogen	Effect on disease resistance[c]	Reference
PR-3 (basic chitinase)	bean (*Phaseolus vulgaris*)	cucumber (*Cucumis sativus*)	*Alternaria cucumeris* *Botrytis cinerea* *Colletotrichum lagenarium* *Rhizoctonia solani*	= = = =	
PR-3 (chitinase)	tomato (*Lycopersicon esculentum*)	oilseed rape (*Brassica napus* var. *oleifera*)	*Cylindrosporium concentricum* *Phoma lingam* *Sclerotinia sclerotiorum*	↗ ↗ ↗	Grison *et al.*, 1996
PR-3 (class I chitinase)	rice (*Oryza sativa*)	rice (*Oryza sativa*)	*Rhizoctonia solani*	↗	Lin *et al.*, 1995
PR-4 (class II, prohevein)	rubber tree (*Hevea brasiliensis*)	tomato (*Lycopersicon esculentum*)	*Trichoderma hamatum* *Botrytis cinerea* *Rhizoctonia solani*	↗ = =	Lee and Raikhel, 1995
PR-5 (osmotin)	tobacco (*Nicotiana tabacum*)	tobacco (*Nicotiana tabacum*) potato (*Solanum tuberosum*)	*Phytophthora parasitica* var. *nicotianae* *Phytophthora infestans*	= ↗	Liu *et al.*, 1994
PR-5 (osmotin)	potato (*Solanum tuberosum*)	potato (*Solanum commersonii*)	*Phytophthora infestans*	↗	Zhu *et al.*, 1996
PR-5 (acidic)	tobacco (*Nicotiana tabacum*)	tobacco (*Nicotiana tabacum*)	Tobacco Mosaic Virus (TMV)	=	Linthorst *et al.*, 1989
ribosome-inactivating protein (PAP)	pokeweed (*Phytolacca americana*)	tobacco (*Nicotiana tabacum*)	Potato Virus X (PVX) Potato Virus Y (PVY) Cucumber Mosaic Virus (CMV)	↗ ↗ ↗	Lodge *et al.*, 1993 Tumer *et al.*, 1997

Table 4. Continued.

Expressed transgene[a,b]	Transgene source	Target plant	Target pathogen	Effect on disease resistance[c]	Reference
ribosome-inactivating protein (trichosantin)	*Trichosanthes kirilowii*	*Nicotiana benthamiana* potato (*Solanum tuberosum*)	Potato Virus Y (PVY) Potato Virus X (PVX) Potato Virus Y (PVY) Turnip Mosaic Virus (TuMV)	↗ ↗ ↗ ↗	Lam *et al.*, 1996
ribosome-inactivating protein (RIP)	barley (*Hordeum vulgare*)	tobacco (*Nicotiana tabacum*)	*Rhizoctonia solani*	↗	Logemann *et al.*, 1992
α-thionin (type I)	barley (*Hordeum vulgare*) wheat (*Triticum aestivum*)	tobacco (*Nicotiana tabacum*) tobacco (*Nicotiana tabacum*)	*Pseudomonas syringae* pv. *tabaci* *Pseudomonas syringae* pv. *syringae* *Pseudomonas syringae* pv. *tabaci* *Pseudomonas syringae* pv. *syringae*	↗ ↗ = =	Carmona *et al.*, 1993b
α-thionin (type I)	barley (*Hordeum vulgare*)	tobacco (*Nicotiana tabacum*) tomato (*Lycopersicon esculentum*)	*Pseudomonas syringae* pv. *tabaci* *Pseudomonas solanacearum* *Clavibacter michiganense* sp. *michiganense* *Pseudomonas solanacearum* *Xanthomonas campestris* pv. *vesicatoria*	= = = = =	Florack and Stiekema, 1994
thionin (Thi2.1; type III)	*Arabidopsis thaliana*	*Arabidopsis thaliana*	*Fusarium oxysporum* f.sp. *matthiolae*	↗	Epple *et al.*, 1997b
lipid transfer protein (LPT2)	barley (*Hordeum vulgare*)	tobacco (*Nicotiana tabacum*) *Arabidopsis thaliana*	*Pseudomonas syringae* pv. *tabaci* *Pseudomonas syringae* pv. *tomato*	↗ ↗	Molina and Garcia-Olmedo 1997

Table 4. Continued.

Expressed transgene[a,b]	Transgene source	Target plant	Target pathogen	Effect on disease resistance[c]	Reference
plant defensin (Rs-AFP2)	radish (*Raphanus sativus*)	tobacco (*Nicotiana tabacum*)	*Alternaria longipes*	↗	Terras *et al.*, 1995
knottin-type AMPs (Mj-AMPs)	four o'clock plant (*Mirabilis jalapa*)	tobacco (*Nicotiana tabacum*)	*Alternaria longipes* *Botrytis cinerea*	= =	De Bolle *et al.*, 1996
hevein-type AMPs (Ac-AMPs)	amaranth (*Amaranthus caudatus*)	tobacco (*Nicotiana tabacum*)	*Alternaria longipes* *Botrytis cinerea*	= =	De Bolle *et al.*, 1996

a   all transgenes were expressed under the control of the Cauliflower Mosaic Virus 35S (CaMV35S) promoter or derived sequences, except in Lodge *et al.* (1993) (both CaMV35S and Figwort Mosaic Virus 35S promoter) and in Logemann *et al.* (1992) (wound-inducible promoter of the potato *wun1* gene)

b   [*+*] indicates coexpression of two different antimicrobial protein transgenes

c   =   : no significant difference in disease tolerance as compared to control plants
    ↗   : increased disease tolerance as compared to control plants
    ↗↗   : increased disease tolerance (of plants expressing a combination of transgenes) as compared to plants expressing the individual transgenes

The potential of constitutive expression of chitinases to increase disease resistance was also investigated in other crops (Table 4), but success was not achieved in all cases. Consistent with the results of Broglie *et al.* (1991), Vierheilig and coworkers (1993) reported that overexpression of a basic vacuolar PR-3-type tobacco chitinase (that is 76% homologous to the bean chitinase mentioned above) in *Nicotiana sylvestris* plants resulted in increased resistance to *R. solani*. However, *N. sylvestris* transformed with the same chitinase construct lacking the carboxy-terminal propeptide were as susceptible to *R. solani* as control plants (Vierheilig *et al.*, 1993). In the first case, the chitinase was targeted to the vacuole, whereas in the latter the chitinase was destined to the extracellular space (Neuhaus *et al.*, 1991b). These results indicate that *R. solani*, which invades plant tissues via intercellularly growing hyphae that eventually kill host cells, is halted more efficiently by an AMP packaged in the vacuole than by an extracellular AMP. On the other hand, *N. sylvestris* plants overexpressing the vacuolar form of the basic tobacco chitinase did not show enhanced resistance to the fungal leaf pathogen *Cercospora nicotianae* (Neuhaus *et al.*, 1991a). This negative result may be either due to non-susceptibility of this fungus to the chitinase *in planta* or to a lack of contact between the vacuolar chitinase and the hyphae growing in the apoplast. Unfortunately, the susceptibility to *C. nicotianae* of transgenic plants expressing the chitinase in the extracellular space has not been investigated.

Not only the subcellular location of the chitinase may influence the disease resistance of chitinase-overexpressing transgenic plants, but also the intrinsic properties of the expressed chitinase itself, as well as the plant in which it is expressed. Punja and Raharjo (1996) found that expression of a basic PR-3-type chitinase from tobacco, but not an acidic PR-3-type chitinase from petunia, resulted in enhanced resistance of transgenic carrot plants to *R. solani* and *Botrytis cinerea*. Introduction of the same constructs in cucumber, however, did not increase resistance against any of these fungi. It should be noted that each plant species synthesizes its own characteristic set of antimicrobial compounds in response to pathogen attack, and depending on the composition and properties of this cocktail, it may or may not be reinforced by the AMP expressed from the transgene.

Overexpression of a PR-2 (β-1,3-glucanase) gene has also been shown to result in increased resistance of transgenic plants to fungal pathogens (Yoshikawa *et al.*, 1993; Lusso and Kuc, 1996). In one study , the resistance against both fungi and viruses was investigated (Lusso and Kuc, 1996). While constitutively expressed endogenous acidic β-1,3-glucanase in tobacco exerted a protective effect towards fungi, the same transgenic lines did not show any resistance to either tobacco etch virus, tobacco mosaic virus or tobacco vein mottling virus (Lusso and Kuc, 1996). Interestingly, Yoshikawa *et al.* (1993) observed that expression of a soybean β-1,3-glucanase in tobacco resulted in enhanced resistance to both the Oomycete *Phytophthora parasitica* and the Ascomycetous fungus *Alternaria alternata*, while the purified enzyme did not inhibit growth of these organisms in vitro. It was proposed that the protective effect was in this case due to the potential of the enzyme to release elicitor-active β-glucan oligosaccharides from fungal cell walls, which are known inducers of defense responses. In support of this assumption it was found that the β-1,3-glucanase-overexpressing plants produced more phytoalexin-synthesis enzymes when challenged with *Phytophthora parasitica* compared to similarly challenged control plants (Yoshikawa *et al.*, 1993).

A number of studies have also reported on the overexpression of small cysteine-rich antimicrobial peptides in plants (De Bolle *et al.*, 1996; Epple *et al.*, 1997b; Terras *et al.*, 1995). Terras *et al.* (1995) found that overexpression of the radish plant defensin Rs-AFP2 in tobacco plants conferred significantly increased resistance to infection by the fungus *Alternaria longipes*. On the other hand, no increased resistance to this fungus was observed in plants expressing similar levels of either the hevein-type antimicrobial peptide Ac-AMP2 or the knottin-type antimicrobial peptide Mj-AMP2. In in vitro assays, Rs-AFP2, Ac-AMP2 and Mj-AMP2 inhibit a broad range of fungi at very similar doses (Cammue *et al.*, 1992; Broekaert *et al.*, 1992; Terras *et al.*, 1992b) and both Mj-AMP2 and Rs-AFP2 are equally effective in suppressing the development of disease caused by *Cercospora beticola* when sprayed on sugar beet plants (De Bolle *et al.*, 1993b). The main difference in terms of in vitro antifungal activity is that Rs-AFP2 is substantially less affected by inorganic cations than both Ac-AMP2 and Mj-AMP2 (Broekaert *et al.*, 1992; Cammue *et al.*, 1992; Terras *et al.*, 1992b). Hence, it has been pointed out by De Bolle *et al.* (1996) that the lack of disease resistance in plants overexpressing either Mj-AMP2 or Ac-AMP2 may be due to counteraction of their activity by inorganic cations present in plant tissues.

Besides single-gene expression, several research groups have focussed on the combinatorial expression of two different AMP genes. This strategy is based on earlier observations that combinations of certain AMPs, such as for instance chitinases and β-1,3-glucanases have synergistic antifungal effects when tested in vitro (see further in section V.B.5.). Transgenic tomato plants expressing either a basic tobacco chitinase or a basic tobacco β-1,3-glucanase alone did not show increased resistance when challenged with the fungus *Fusarium oxysporum* f.sp. *lycopersici*. However, concurrent expression of both genes resulted in a decrease of disease severity which correlated with the expression level of the transgenes (van den Elzen *et al.*, 1993). Synergistic effects of co-expressed chitinases and β-1,3-glucanases were later confirmed by others (Jach *et al.*, 1995; Zhu *et al.*, 1994). Another example of constitutive co-expression of different AMPs leading to increased disease resistance is the combination between a basic chitinase and a type I RIP (Jach *et al.*, 1995). Whereas tobacco plants expressing the individual genes were significantly more resistant to the soilborne pathogen *R. solani*, a synergistically increased protection was observed in plants expressing both the chitinase and the RIP gene. Preliminary infection assays with other pathogenic fungi such as *Alternaria alternata* and *Botrytis cinerea* confirmed the synergy between chitinase and RIP (Jach *et al.*, 1995). It should be noted that synergy between chitinase and RIP was also observed in in vitro antifungal activity assays (Leah *et al.*, 1991).

Only one type of AMPs, namely ribosome-inactivating proteins (RIPs) have so far been demonstrated to confer resistance to viruses when overexpressed in transgenic plants. Both transgenic tobacco and potato expressing pokeweed antiviral protein (PAP), a type I RIP occurring in cell walls of *Phytolacca americana*, were protected from infection by different viruses (Lodge *et al.*, 1993). Resistance was effective against both mechanically and aphid-transmitted viruses and appeared to occur at an early stage of viral infection. In contradiction to some experiments described above, no correlation could be observed between the level of expression of PAP and the level of resistance to viral infection. The basis of the protective effect of PAP against viruses is still unclear but it has been shown that it relies on its adenine:glycosidase activity, as a catalytic site mutant of PAP was unable to confer resistance to viruses when expressed constitutively in tobacco (Tumer *et al.*, 1997).

Only very few studies have so far investigated the role of AMPs in resistance to bacterial diseases. Already from the early 1970s such a role has been proposed for thionins, based on their in vitro antibacterial activity (Fernandez de Caleya *et al.*, 1972). Evidence for such a function *in planta* has been obtained by overexpressing a barley thionin (hordothionin) in tobacco, which resulted in enhanced resistance against two different pathovars of *Pseudomonas syringae* (Carmona *et al.*, 1993b). In a transgenic line expressing the highest level of hordothionin, less than 10% of the leaf inoculation points developed necrotic lesions 90 h after infection with *P. syringae* pv. *tabaci*, as compared to more than 70% in control plants. The level of disease resistance appeared to be correlated with the expression level of the transgene. Similarly, overexpression in tobacco and *Arabidopsis* of a barley leaf nsLTP, which also exerts antibacterial activity in vitro, conferred significant resistance to *P. syringae* pv. *tabaci* and *P. syringae* pv. *tomato*, respectively (Molina and Garcia-Olmedo, 1997). On the other hand, Florack and Stiekema (1994) were unable to observe enhanced disease resistance to *P. syringae* pv. *tabaci* in tobacco plants overexpressing a hordothionin construct similar to the one used by Carmona *et al.* (1993b). The discrepancy between the results obtained by both research groups might be explained by a difference in the *P. syringae* pv. *tabaci* strain used.

From the experiments summarized in Table 4 it can be concluded that overexpression of different AMPs can lead to an increased resistance of the transgenic plants to different types of microbial pathogens including viruses, bacteria and fungi, supporting the hypothesis that these proteins play a protective role in vivo. Some considerations should, however, be made when extrapolating this function to the plant of origin.

First, in the majority of the studies summarized in Table 4, the transgene was expressed in an heterologous plant species. In the experiments of Liu and coworkers (1994), a tobacco PR-5 gene, encoding osmotin, was expressed in both tobacco and potato. They observed that constitutive expression of osmotin resulted in increased tolerance against *Phytophthora infestans* in potato, but not in tobacco against *Phytophthora parasitica*. Based on their results and on experiments dealing with overexpression of other PR-proteins in homologous plants (Cutt *et al.*, 1989; Neuhaus *et al.*, 1991a) they postulated a specific activity of PR-proteins against pathogens that are not pathogenic to the plant species of origin, as a result of host-pathogen co-evolution. This postulation has somehow been countered by later findings that resistance to *P. infestans* could also be obtained in potato by overexpression of its endogenous PR-5 gene (Zhu *et al.*, 1996) and by other successful experiments involving overexpression of endogenous genes (Alexander *et al.*, 1993; Lawton *et al.*, 1993; Lin *et al.*, 1995; Lusso and Kuc, 1996; Masoud *et al.*, 1996).

Further, when overviewing Table 4, it is clear that in all cases, except one (Logemann *et al.*, 1992), the transgene was expressed under the control of the constitutive CaMV35S promoter or derivatives from it. It should be kept in mind that many AMPs are produced in their plant of origin in a pathogen-inducible way. Hence, under natural conditions the effect of AMPs on invading microorganisms depends on the timing of the induction of AMP genes. As mentioned before in section III, AMP production is often faster and more intense in many incompatible plant-pathogen interactions versus compatible interactions. Hence, failure to contain pathogen ingress is not always due to inefficacy of the AMPs but often to the slowness of the pathogen detection and signal transduction events. When AMPs are expressed constitutively in transgenic plants, matters of pathogen perception and timing of the defense response become irrelevant, allowing to better study their direct impact on microorganisms.

When further overviewing Table 4 it can be seen that in more than half of the studies the overexpression of a certain AMP did not result in increased disease tolerance of the target plant. This does not necessarily imply that the specific AMP has no defense-related function in its plant of origin. Much depends, for example, on the infection conditions applied when assaying the disease resistance of the transgenic plants. Transgenic potato plants constitutively expressing the endogenous PR-5 protein were found to be significantly more resistant than control plants to *P. infestans* in infection assays using 50 zoospores per inoculum (Zhu *et al.*, 1996). However, the same transgenic plants were as susceptible to *P. infestans* as control plants when 100 zoospores per inoculum were applied in the infection assays. This clearly indicates that resistance conferred by AMP expression is conditional to the level of infection pressure.

All but one of the reports listed in Table 4 have described the performance of AMP-overexpressing plants as assessed in a greenhouse or growth chamber. Grison *et al.* (1996) have evaluated the disease resistance of transgenic oilseed rape constitutively expressing a tomato chitinase gene in field trials at two different geographical locations in France. The transgenic genotypes exhibited on average from 23% to 79% less disease symptoms as compared to control plants after being challenged with either of the fungal pathogens *Cylindrosporium concentricum, Phoma lingam* or *Sclerotinia sclerotiorum,* which are responsible for the most severe damages to oilseed rape under European culture conditions. This clearly illustrates the agronomic potential of practical applications derived from the basic study of plant AMPs (see also chapter 11).

Plants are not only invaded by pathogens but can also be colonized by similar types of microorganisms with benefial effects on plant growth. In order to evaluate a possible role of AMPs in the plant's defense it is important to address the question whether the observed antimicrobial activity is specific to pathogenic invaders. Vierheilig *et al.* (1995) have therefore investigated the effect of different AMPs, overexpressed *in planta,* on beneficial vesicular-arbuscular mycorrhizal (VAM) fungi. The latter colonize most herbaceous plants and enhance the uptake of mineral nutrients in exchange for assimilates provided by the plant (Koide and Schreiner, 1992). These endophytic fungi belong to the order Glomales and possess chitin- and β-1,3-glucan-containing cell walls like most higher fungi. In addition, plant colonization by VAM occurs via similar processes as infection by phytopathogenic fungi. *Nicotiana sylvestris* plants constitutively expressing either one of different tobacco PR-proteins, including PR-1a, PR-2, PR-3, PR-4 and PR-5, were evaluated for their colonization ability by the VAM fungus *Glomus mosseae* (Vierheilig *et al.*, 1995). Most of the expressed transgenes did not affect either the level or time course of colonization by *G. mosseae*. Only high levels of the acidic β-1,3-glucanase had a slightly negative effect on the symbiotic potential of the VAM fungus (Vierheilig *et al.*, 1995). The reason for the apparent insensivity of VAM fungi to most PR-proteins is unclear. It is conceivable, however, that the mutualistic VAM fungi, like highly successful pathogenic fungi, have evolved mechanisms to passively or actively avoid the damaging effect of host AMPs.

B. DOWN-REGULATION OF AMP GENES

The expression of endogenous genes in plants can effectively be blocked by genetic transformation with (parts of) the transcribed region of these genes in antisense orientation (reviewed by Mol *et al.*, 1994). Although not as numerous as in the previous section, several groups have applied this strategy to unravel the biological role of certain plant AMPs (summarized in Table 5).

The most convincing data obtained so far on the role of individual AMPs in disease resistance were presented by Lusso and Kuc (1996). These authors found that antisense inhibition of an acidic β-1,3-glucanase in tobacco resulted in transgenic plants that were more susceptible to the Oomycetes *Peronospora tabacina* and *Phytophthora parasitica* (see Table 5). These results were fully in line with their observation that overexpression of this glucanase in tobacco yielded plants with increased resistance to these same pathogens (see Table 4). On the other hand, both the sense and antisense plants were equally sensitive to infection by three different viruses. Taken together, these results convincingly support the idea that the acidic β-1,3-glucanase of tobacco plays a role in defense against at least oomycetous fungi but not against the viruses tested.

On the other hand, Beffa and Meins (1996) found, quite surprisingly, that antisense down-regulation of a basic vacuolar β-1,3-glucanase in tobacco and *Nicotiana sylvestris* increased the resistance to TMV and tobacco necrosis virus, respectively. The increased resistance to viruses was correlated with increased callose deposition around virally infected cells. Callose (β-1,3-glucan) is believed to serve as a physical barrier to the spread of viruses or other pathogens from the infection site to distant cells. β-1,3-glucanase may play a role in the breakdown of the callose layer once the pathogen is inactivated. As suggested by Beffa *et al.* (1996) this process might be impaired in the β-1,3-glucanase deficient plants, leading to more and more prolonged callose deposition. Interpretation of the data from these experiments is, however, complicated by the observation that the antisense transgenic plants impaired in basic vacuolar β-1,3-glucanase apparently compensate in part this deficiency by upregulating the expression of another type of pathogen-inducible β-1,3-glucanase gene, called an "ersatz" (i.e. compensatory) gene (Beffa *et al.*, 1993). However, the fact that partial compensation of the β-1,3-glucanase enzyme deficiency occurs in these transgenic plants in response to pathogen infection strongly speaks for the importance of this enzyme in pathogenesis.

In two other studies, antisense inhibition of antimicrobial pathogenesis-related protein genes did not result in the expected significant decrease in disease resistance level (Samac and Shah, 1994; Zhu *et al.*, 1996). However, Samac and Shah (1994) observed that antisense inhibition of the basic PR-3 gene in *Arabidopsis* had resulted in severe down-regulation of basal expression levels but only to a two-fold reduction of expression levels in pathogen-infected tissues. The incomplete suppression of gene expression during pathogenesis may account for these negative results.

Table 5.   Examples of plants genetically engineered with antisense constructs of plant antimicrobial protein genes.

Expressed transgene in antisense orientation[a]	Transgene source	Target plant	Target pathogen	Effect on disease resistance[b]	Reference
PR-2 (basic β-1,3-glucanase)	tobacco (Nicotiana tabacum)	Nicotiana sylvestris	Cercospora nicotianae	=	Neuhaus et al., 1992
PR-2 (basic β-1,3-glucanase)	tobacco (Nicotiana tabacum)	tobacco (Nicotiana tabacum)	Tobacco Mosaic Virus (TMV)	↗	Beffa et al., 1996
		Nicotiana sylvestris	Tobacco Necrosis Virus (TNV)	↗	
PR-2 (acidic β-1,3-glucanase)	tobacco (Nicotiana tabacum)	tobacco (Nicotiana tabacum)	Peronospora tabacina	↗	Lusso and Kuc, 1996
			Phytophthora parasitica	↗	
			Tobacco Mosaic Virus (TMV)	=	
			Tobacco Etch Virus (TEV)	=	
			Tobacco Vein Mottling Virus (TVMV)	=	
PR-3 (class I basic chitinase)	Arabidopsis thaliana	Arabidopsis thaliana	Botrytis cinerea	(↗)	Samac and Shah, 1994
PR-5 (osmotin)	potato (Solanum commersonii)	potato (Solanum commersonii)	Phytophthora infestans	=	Zhu et al., 1996

a   all transgenes were expressed under the control of the Cauliflower Mosaic Virus 35S (CaMV35S) promoter or derived sequences

b   = : no significant difference in disease tolerance as compared to control plants

　　↗ : increased disease tolerance as compared to control plants

　　↗ : decreased disease tolerance as compared to control plants

　　(↗) : tendency to increased susceptibility but differences not significant due to high variability

C.   TRANSGENIC OR MUTANT PLANTS AFFECTED IN SIGNAL PATHWAYS FOR AMP GENE EXPRESSION

During the last five years considerable research efforts have been dedicated to the elucidation, via genetic approaches, of signal pathways leading to induced defense responses (see also Chapter 4 of this book). This has resulted in the identification of several mutants affected in their defense mechanisms mounted against pathogens. In other cases, transgenic plants have been created aimed at either blocking or overactivating a signal pathway required for induction of pathogenesis-related genes. Table 6 provides some examples of mutant or transgenic plants that show either impairment in their ability to induce a set of PR-genes or show constitutive expression of several PR-genes in vegetative tissues. In most cases examined so far, mutants or transgenic lines which are blocked in their PR-response show increased susceptibility to viral, bacterial and fungal pathogens, whereas mutants or transgenic lines with a constitutitve PR-response are more resistant than wild-type plants.

At first sight, these data provide very strong support for the supposed function of antimicrobial proteins in general disease resistance. However, some care should be taken in the interpretation of the data as resistance or susceptibility need not necessarily in all cases be due to the antimicrobial effect of the AMPs on the pathogens tested. Indeed, many of the mutants and transgenic plants presented in Table 6 have either elevated or suppressed levels of salicylic acid correlating with elevated or suppressed PR-gene expression (reviewed in Dürner *et al.*, 1997). Salicylic acid is not only involved in regulating the expression of some PR-genes but also in potentiating the hypersensitive response (Chapter 6; Shirasu *et al.*, 1997) and possibly also in regulating the alternative oxidase (reviewed in Vanlerberghe and McIntosh, 1997). The function of the alternative oxidase in defense is still unclear but experiments with chemical inhibitors of this enzyme have shown that it is implicated in resistance to viruses but not to fungi and bacteria (Chivasa *et al.*, 1997). Indeed, inhibition of the alternative oxidase in tobacco abolished salicylate-induced resistance to TMV but not its ability to induce PR-genes or to mount resistance to the necrotrophic pathogens *Erwinia carotovora* or *Botrytis cinerea* following salicylate treatment. This observation strenghtens a concluding view that, with the exception of ribosome-inactivating proteins, most PR/AMP genes characterized to date are not involved in resistance to viruses but rather in resistance to at least certain types of bacteria, true fungi and Oomycetous fungi. This notion is consistent with the overall data on overexpression of AMP genes discussed in section IV.A., and summarized in Table 4. These data showed that in most cases analysed so far, except those involving ribosome-inactivating proteins, overexpression of an AMP gene did not result in improved protection against viral diseases, while in many cases increased resistance to fungal and bacterial diseases was observed.

Table 6.   Examples of genetically tranformed or mutant plants affected in their disease signalling pathways

Mutant or transformant specification	Plant species	AMPs demonstrated to be affected in their expression[e]	Target pathogen	Effect on disease tolerance[f]	Reference
nahG[a]	tobacco (Nicotiana tabacum)	PR-1: ↓; PR-2: ↓; PR-3: ↓	Tobacco Mosaic Virus (TMV)	↗	Gaffney et al., 1993
			Pseudomonas syringae pv. tabaci	↗	Friedrich et al., 1995
			Phytophthora parasitica	↗	
			Cercospora nicotianae	↑	
Cholera toxin A1[b]	tobacco (Nicotiana tabacum)	PR-1: ↑; PR-2: ↑; PR-3: ↑	Pseudomonas syringae pv. tabaci	↑	Beffa et al., 1995
etr1-1[c]	tobacco (Nicotiana tabacum)	basic PR-1: ↓; basic PR-5: ↓	Tobacco Mosaic Virus (TMV)	=	Knoester et al., 1998
			Pythium sulvaticum	↗	
Bacterio-opsin[d]	potato (Solanum tuberosum)	PR-2: ↑; PR-3: ↑; PR-5: ↑	Phytophthora infestans strain US1	↑	Abad et al., 1997
			Phytophthora infestans strain US8	=	
			Erwinia carotovora	=	
			Potato Virus X	↗	
nahG[a]	Arabidopsis thaliana	PR-1: ↓	Pseudomonas syringae pv. tomato	↗	Delaney et al., 1994
			Pseudomonas syringae pv. maculicola	↗	Lawton et al., 1995
			Peronospora parasitica	↗	
npr1	Arabidopsis thaliana	PR-1: ↓; PR-2: ↓; PR-5: ↓	Pseudomonas syringae pv. maculicola	↗	Cao et al., 1994
			Peronospora parasitica	↗	
cpr1	Arabidopsis thaliana	PR-1: ↑; PR-2: ↑; PR-5: ↑	Pseudomonas syringae pv. maculicola	↑	Bowling et al., 1994
			Peronospora parasitica	↑	

Table 6. Examples of genetically tranformed or mutant plants affected in their disease signalling pathways

Mutant or transformant specification	Plant species	AMPs demonstrated to be affected in their expression[e]	Target pathogen	Effect on disease tolerance[f]	Reference
acd2	*Arabidopsis thaliana*	PR-1: ↑; PR-2: ↑; PR-5: ↑	*Pseudomonas syringae* pv. *tomato* *Pseudomonas syringae* pv. *maculicola*	↗ ↗	Greenberg *et al.*, 1994
lsd1, lsd2, lsd4	*Arabidopsis thaliana*	PR-1: ↑; PR-2: ↑; PR-5: ↑	*Pseudomonas syringae* pv. *maculicola* *Peronospora parasitica*	↗ ↗	Dietrich *et al.*, 1994 Hunt *et al.*, 1997
coi1	*Arabidopsis thaliana*	Plant defensin: ↓; basic PR-3: ↓; basic PR-4: ↓	*Botrytis cinerea*	↗	Thomma *et al.*, 1998

[a] transgene = *Pseudomonas putida* salicylate hydroxylase gene under the control of the CaMV35S promoter
[b] transgene = Cholera toxin subunit A1 gene under the control of the wheat cab-1 promoter
[c] transgene = Dominant-negative mutant allele of *Arabidopsis thaliana* ethylene receptor
[d] transgene = *Halobacterium halobium* bacterio-opsin proton pump
[e] ↓ : expression down-regulated as compared to wild-type plant
↑ : expression up-regulated as compared to wild-type plant
[f] = : no significant difference in disease tolerance as compared to wild-type plants
↗ : increased disease tolerance as compared to wild-type plants
↘ : decreased disease tolerance as compared to wild-type plants

V.   Structural Aspects and Antimicrobial Properties of The Main Families of Antimicrobial Proteins from Plants

In the section below a detailed description will be given of the structural properties for each of the different antimicrobial protein families. In addition, information will be presented on their antimicrobial properties and mode of action.

A.   PR-1-TYPE PROTEINS

1.   *Structural features*
PR-1-type proteins are monomeric proteins of about 15 kDa with no known catalytic activity. They were first detected and purified from tobacco leaves infected with tobacco mosaic virus (Antoniw and Pierpoint, 1978; Antoniw et al., 1980). Tobacco PR-1-type proteins are encoded by a small gene family comprising at least 8 members, as determined by Southern blot analysis (Cornelissen et al., 1987). Proteins that are highly homologous to the tobacco PR-1-type proteins have either been purified or deduced from cloned cDNAs or genes in several monocot and dicot species, including amongst others Arabidopsis, barley, maize, parsley, and tomato (see Yun et al., 1997a).

Comparison of the determined amino-terminus of a mature acidic PR-1 protein from tobacco (PR-1a) and a basic PR-1 protein from tomato (P14a) (Lucas et al., 1985) with the amino acid sequences derived from their respective cDNA sequences (Pfitzner and Goodman, 1987; Tornero et al., 1993) reveals that both proteins, as well as other PR-1-type proteins, are synthesized as preproteins consisting of an amino-terminal hydrophobic signal peptide followed by the mature PR-1 domain. Some, but not all, basic PR-1-type proteins feature a short (6 to 19 amino acids) carboxy-terminal segment that is absent in most other PR-1-type proteins (see Fig. 3). This carboxy-terminal extension may function as a vacuolar targetting determinant in analogy to the proven role of the carboxy-terminal propeptide of some basic PR-2, PR-3 and PR-5 proteins (*vide infra*; Neuhaus et al., 1991b; Melchers et al., 1993). In the case of the basic PR-1 proteins with   a carboxy-terminal extension, it remains to be demonstrated that the extension is truly cleaved off and plays a role in subcellular targetting.

A representative set of PR-1 amino acid sequences from different plant species is aligned in Fig. 3. The alignment of multiple PR-1 proteins reveals that, taking the 138-amino acidic tobacco PR-1a as a reference, 26% of the residues are strictly conserved in this set of PR-1 proteins. Among these are six cysteines. In at least 12 of the 16 aligned sequences, an additional 37 of the positions are occupied by amino acids belonging to a same amino acid homology group. The PR-1 protein sequences thus show a high overall conservation.

Left largely uncovered in plant PR-1 literature is the fact that plant PR-1-type proteins show significant homology to some proteins of nonplant origin. Fang et al. (1988) were the first to notice considerable homology between the amino acid sequence of antigen 5, an allergen present in hornet venom, and plant PR-1 proteins. Hornet antigen 5 is related to mandarotoxin, a neurotoxin that reduces sodium currents in presynaptic neuromuscular junctions of a lobster walking leg (Abe et al., 1982; King, 1990). A second nonplant protein found to resemble PR-1-type proteins through computer searches in protein sequence databases belongs to the family of mammalian sperm-coating glycoproteins (Kasahara *et al.*, 1989). These proteins are secreted by testis epididymal epithelium to become associated with the sperm surface (Kasahara *et al.*, 1989). Similar proteins, called cysteine-rich secretory proteins (CRISPs) are secreted by mouse B-lymphocytes and the salivary gland (Haendler *et al.*, 1993; Pfisterer *et al.*, 1996). A last nonplant protein similar to plant PR-1s is the human glioma pathogenesis-related protein (GliPR), a protein specifically expressed in astrocytes-derived human brain tumors (astrocytomas or gliomas) but not in normal or fetal brain (Murphy *et al.*, 1995). In analogy to virally-induced plant PR-1 proteins, accumulation of GliPR can be induced in cultured human astrocytoma cell lines by infection with the cytomegalovirus. High-level *GliPR* expression can also be induced by activation of a human macrophage cell line. It should also be noted that astrocytes function as immune cells in the central nervous system besides being neuron-supporting cells (Murphy *et al.* 1995). In Fig. 2, the PR-1-like domains of three different nonplant PR-1-homologs are aligned with the tobacco PR-1a and PR-1g sequences. Although the cysteine-patterning is different between the plant PR-1 and the nonplant PR-1-like proteins, there is still a considerable sequence similarity. Seventeen % of the positions are perfectly conserved in the five aligned sequences. Especially noteworthy is the conserved heptapeptide sequence HYTQ (hydrophobic residue) VW at position 124-130 (numbering as in Fig. 3). When homologous amino acid substitutions are allowed, an additional 24% of the residues is homologous in at least four of the five proteins. Apart from the association of plant PR-1 proteins and human GliPR-proteins with disease, there is no known functional similarity between plant PR-1 proteins and their homologs in animals.

A three-dimensional fold supposedly shared by all members of the PR-1-type protein family has been determined by NMR for the tomato P14a protein (see Fig. 1; Fernandez *et al.*, 1997). This protein features an α-β-α sandwich structure in which α-helices are tightly packed on both sides of a central β-sheet. The P14a residues that are exposed to the surface mostly occur in regions with higher sequence divergence among different PR-1-type proteins. The heptapeptide sequence HYTQ (hydrophobic residue) VW, that is conserved among plant and animal PR-1-like proteins, constitutes an α-helix that is largely buried inside the protein core and thus appears to play a structural role, possibly as a folding nucleus.

2. Antimicrobial Properties

So far only one report has appeared on *in vitro* antimicrobial properties of PR-1-type proteins. Niderman *et al.* (1995) found that several PR-1-type proteins inhibit zoospore germination of the Oomycete *Phytophthora infestans*, the causal agent of potato late blight. In comparative tests, basic PR-1 proteins from either tomato or tobacco were significantly more inhibitory than acidic isoforms. For instance, the basic tobacco protein PR-1g inhibits *P. infestans* zoospore germination by more than 90% at a concentration of 20 µg/ml, whereas 200 µg/ml is required of the acidic isoform PR-1a from tobacco to obtain the same inhibition rate (Niderman *et al.*, 1995). The activity of PR-1-type proteins on other microorganisms has not yet been investigated, neither is anything known on how they may affect growth processes of Oomycetes.

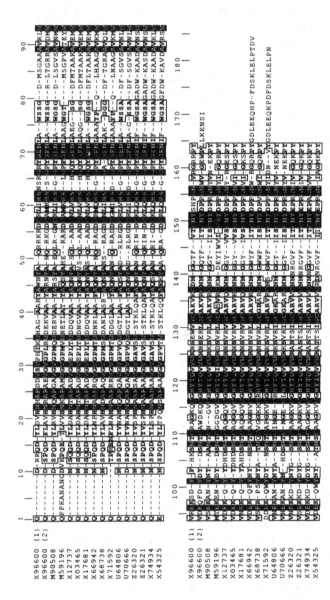

Figure 3. **Alignment of plant PR-1-type protein sequences.**

Aligned are the amino acid sequences of a number of plant PR-1 proteins as deduced from cDNAs or genes with GenBank accession numbers indicated on the left. In all cases have the putative signal peptides been omitted. The cDNAs/genes were isolated from the following plants: *Arabidopsis thaliana*: X96600 (1) and X96600 (2), the first and second of tandemly repeated *PR-1* genes (Kloska S and Schuster W, unpublished), M90508 (*PR-1* cDNA; Uknes *et al.*, 1992) and M59196 (*PR-1* gene; Metzler *et al.*, 1991). *Nicotiana tabacum*: X12737 (*PR-1a* gene; Payne *et al.*, 1988), X03465 (*PR-1b* gene; Cornelissen *et al.*, 1986b), X17681 (*PR-1c* gene; Ohshima *et al.*, 1990) and X66942 (basic *PR-1g* gene; Eyal *et al.*, 1992). *Lycopersicon esculentum*: X68738 (*P14a* cDNA; Tornero *et al.*, 1993) and X71592 (*PR-1a* cDNA; Tornero *et al.*, 1994). *Brassica napus*: U64806 (*Ypr1* cDNA; Zhang P and Fristensky BW, unpublished) and U70666 (*Ypr1* cDNA; Zhang P and Fristensky BW, unpublished). *Hordeum vulgare*: Z26320 (basic *pbr1-2* cDNA; Mouradov *et al.*, 1994), Z26321 (basic *pbr1-3* cDNA; Mouradov *et al.*, 1994) and X74934 (basic *PR-1* gene; Bryngelsson *et al.*, 1994). *Zea mays*: X54325 (basic *PRms* gene; Casacuberta *et al.*, 1991).

Gaps introduced to obtain an optimal alignment are indicated by hyphens. Residues occurring in all depicted sequences are present in a black box. Residues belonging to the same amino acid homology group (FWY; MILV; RKH; NQDE; PAGST) and occurring in at least 12 out of the 16 aligned PR-1 sequences are bold-faced and boxed.

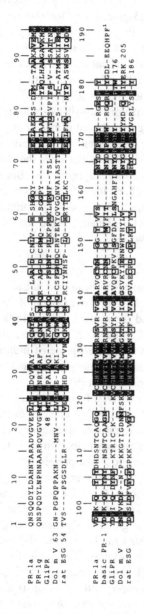

Figure 4. **Alignment of plant PR-1 protein sequences with non-plant PR-1-like protein sequences.**

Aligned are the sequences of the mature tobacco PR-1a and basic PR-1g proteins (see also Fig. 1) and the PR-1-like domains of the human glioma pathogenesis-related protein (GliPR, mature protein contains 219 amino acids; Murphy *et al.*, 1995), the venom allergen antigen 5 of the white-faced hornet (Dol m V, mature protein contains 205 amino acids; Lu *et al.*, 1993), and the rat epididymal secretory glycoprotein (rat ESG, mature protein contains 227 amino acids; Brooks *et al.*, 1986). Gaps introduced for optimal alignment are denoted by hyphens. Residues perfectly conserved in all sequences are present in a black box. Residues belonging to the same amino acid homology group (FWY; MILV; RKH; NQDE; PAGST) and occurring in at least 4 out of the 5 aligned sequences are bold-faced and boxed. Numbers of the first and last residues of the aligned parts of the nonplant PR-1-like proteins are given immediately in front of and after the depicted sequences. The 10 amino acid carboxy-terminal extension of the tobacco basic PR-1 protein is omitted.

B.   B-1,3-GLUCANASES (PR-2-TYPE PROTEINS) AND CHITINASES
      (PR-3-, PR-8- AND PR-11-TYPE PROTEINS)

Plants contain several types of enzymes that have the ability to hydrolyze polysaccharides from microbial cell walls. Endo-β-1,3-glucanases, for instance, are enzymes that specifically hydrolyze β-1,3 ether linkages in β-1,3-glucans and β-1,3-1,6-glucans, resulting in a range of oligosaccharides down to disaccharides (Hrmova and Fincher, 1993; Moore and Stone, 1972b; Wong and Maclachan, 1979; Keen and Yoshikawa, 1983). Most endo-β-1,3-glucanases are structurally related and form a family known as PR-2-type proteins. Plants also contain enzymes that hydolyze the β-1,4 ether linkage in either poly-β-1,4-N-acetylglucosamine (chitin) or poly-[β-1,4-N-acetylglucosamine-β-1,4-N-acetylmuramic acid] (the carbohydrate component of bacterial peptidoglycan). These enzymes are called either chitinases or lysozymes, respectively. Many purified plant chitinases exhibit lysozyme activity and vice versa, but since the chitinase activity is usually predominant the name chitinase is more appropriate (Boller, 1988). As hundreds of sequences of plant chitinases have been identified it has become clear that chitinases can be separated in three main groups, called PR-3-type proteins, PR-8-type proteins and PR-11-type proteins. The endo-β-1,3-glucanases and the three main types of chitinases will be discussed separately below.

1.   *PR-2-Type Proteins*
β-1,3-glucanases were first purified from leaves of bean and tobacco in the early 1970s (Abeles *et al.*, 1971; Moore and Stone, 1972a). Comparison of data from β-1,3-glucanases purified from various monocot and dicot plant species indicates that they are all monomers with a molecular weight of about 30 kDa (Boller, 1988). In 1987, Kauffmann *et al.* reported that four pathogenesis-related proteins induced in tobacco upon TMV infection were in fact β-1,3-glucanases. These proteins and their homologs are now known as PR-2-type proteins (van Loon *et al.*, 1994).

Phylogenetic analysis of PR-2-type proteins derived from cDNAs or genes isolated from several plants yields a dendrogram with at least five clusters of related proteins (Bucciaglia and Smith, 1994; Hird *et al.*, 1993). Representative sequences of five β-1,3-glucanase classes are given in Fig. 5. Basic glucanases belong to class I, whereas class II accommodates acidic glucanases as well as a PR-2-like protein from tobacco styles. The highly abundant stylar PR-2-like protein is, however, heavily glycosylated and not induced by elicitor treatment (Ori *et al.*, 1990). Class III β-1,3-glucanases harbors an acidic tobacco enzyme slightly different from the class II enzyme. The products of three tandemly arranged Arabidopsis β-1,3-glucanase genes (Dong *et al.*, 1991) also fall into this class. Barley β-1,3- and structurally related β-1,3-1,4-glucanases are classified in a fourth class of PR-2-type proteins. Tag1, the PR-2-like protein from tobacco anthers forms a fifth class of β-1,3-glucanases (Bucciaglia and Smith, 1994). Anther-specific proteins from *Brassica napus* and Arabidopsis similar to β-1,3-glucanases but with a 114-amino acid carboxy-terminal extension (relative to Tag1) and 37% identity to Tag1 might form a sixth class of β-1,3-glucanases (Hird *et al.*, 1993). The only nonplant proteins reported so far to display similarity to plant PR-2-type proteins are fungal β-1,3-glucanases (Hird *et al.*, 1993).

In tobacco, acidic class II PR-2 proteins are synthesized as preproteins whereas the mature basic class I PR-2 proteins originate from a preproprotein (Melchers *et al.*, 1993; Shinshi *et al.*, 1988; Sticher *et al.*, 1992a; Van den Bulcke *et al.*, 1989). The amino-terminal residue of basic tobacco PR-2 most likely is a glutamine, an observation consistent with the presence of a 21-residue signal peptide in the preproprotein (Shinshi *et al.*, 1988). A similar signal peptide is also present in all other PR-2 pre(pro)proteins aligned in Fig. 5. After entry into the ER-lumen, at which point the signal peptide is removed, the carboxy-terminal extension of the tobacco basic β-1,3-glucanase is *N*-glycosylated with a ~2 kDa glycan. The glycosylated 22-residue carboxy-terminal extension is subsequently removed during transport to or in the vacuole as this propeptide is absent from the mature protein (Shinshi *et al.*, 1988).

The three-dimensional structure of the barley β-1,3-glucanase shown in Fig. 1 has been determined by X-ray crystallography (Varghese *et al.*, 1994). The 3-D structure is dominated by a series of eight consecutive β-sheet/α-helix folds, thus forming a so-called $(\alpha/\beta)_8$-barrel structure (see Fig. 1). The most prominent structural feature of a space-filling model of the β-1,3-glucanase is a ~40 Å long deep cleft able to accommodate a linear stretch of 8 β-1,3-linked glycosyl residues (Hrmova *et al.*, 1995; Varghese *et al.*, 1994). The β-glycosyl residues are not bound equally tight to the individual subsites and a slight repulsion rather than a binding is observed for the interaction between the third and fourth subsite (relative to the nonreducing terminus of the substrate) and the β-1,3-glucan substrate. This might facilitate the hydrolytic cleavage catalyzed by amino acids located between the same third and fourth β-glycosyl-interacting sites (Hrmova *et al.*, 1995). Hydrolysis of the β-1,3-glycosidic bond is thought to be initiated by its protonation by the catalytic acid E309 (numbering as in Fig. 5). The pK_a of the E309 side-chain carboxyl group might be elevated by the neighbouring E300, whereas the negative charge of the deprotonated E309 can be stabilized by residue K303. Protonation of the glycosidic oxygen leads to bond cleavage via an positively charged oxycarbonium intermediate that is stabilized by the catalytic nucleophile E246 (Høj and Fincher, 1995). The catalytic nucleophile E246 is held in position by hydrogen bonds to residues Y35, N96, E97, N180 and Y182. A hydrophobic surface interacting with the substrate and probably partially determining substrate specificity is formed by Y182, F185, F295 and F312 (Varghese *et al.*, 1994). The importance of residues E246, E309, E300 and K303 was proven by mutagenesis of these amino acids. Depending on the substituted amino acid, a 240- to 20000-fold reduction of the β-1,3-glucanase activity was recorded (Chen *et al.*, 1995).

2. *PR-3-Type Proteins*

PR-3-type chitinases are monomeric enzymes ranging from about 25 to 35 kDa in size (Boller, 1988; Graham and Sticklen, 1994). They act as endochitinases and release chitooligosaccharides of varying degrees of polymerisation, of which the smallest are usually chitobiose and chitotriose (Molano *et al.*, 1979; Boller *et al.*, 1983; Shinshi *et al.*, 1987; Mauch *et al.*, 1988a; Broekaert *et al.*, 1988). A PR-3-type chitinase was first isolated from bean leaves (Abeles *et al.*, 1971; Boller *et al.*, 1983) but similar enzymes have now been purified from about 20 different monocot and dicot plants (reviewed in Graham and Sticklen, 1994). The link between PR-proteins and

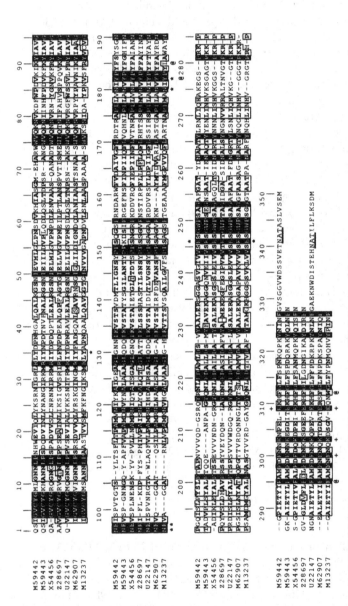

Figure 5. **Alignment of plant ß-1,3-glucanases and ß-1,3-1,4-glucanases (PR-2-type proteins).**

Aligned are the amino acid sequences of the (pro)protein parts of ß-1,3- and ß-1,3-1,4-glucanases as deduced from cDNAs or genes with GenBank accession numbers indicated on the left. In all cases have the putative signal peptides been omitted. The cDNAs/genes were isolated from the following plants: *Nicotiana tabacum*: M59442 (basic class I ß-1,3-glucanase gene; Linthorst *et al.*, 1990a), M59443 (acidic class II ß-1,3-glucanase gene; Linthorst *et al.*, 1990a), X54456 (acidic class III ß-1,3-glucanase cDNA; Payne *et al.*, 1990) and Z28697 (anther-specific ß-1,3-glucanase *Tag1* cDNA; Bucciaglia and Smith, 1994). *Hevea brasiliensis*: U22147 (latex ß-1,3-glucanase HGN1 cDNA; Chye and Cheung, 1995). *Hordeum vulgare*: M62907 (ß-1,3-glucanase GII cDNA; Leah *et al.*, 1991) and M13237 (ß-1,3-1,4-glucanase EII gene; Fincher *et al.*, 1986; Høj and Fincher, 1995).

Gaps introduced to obtain an optimal alignment are indicated by hyphens. Residues occurring in all depicted sequences are present in a black box. Residues belonging to the same amino acid homology group (FWY; MILV; RKH; NQDE; PAGST) and occurring in at least 6 out of the 7 aligned PR-2 and PR-2-like sequences are bold-faced and boxed. The catalytic proton-donating nucleophile E246 is indicated by an asterisk above and under the sequence, and the reaction intermediate-stabilizing residue E309 is indicated by plus signs above and under the sequences. Residues associated with E246 are marked by an asterisk under the sequences; residues associated with E309 are marked by a plus sign under the sequences. Residues hydrophobically interacting with the substrate are marked by '@' under the sequences. *N*-Glycosylation sites in the carboxy-terminal propeptides of class I ß-1,3-glucanases from tobacco and rubber tree are double underlined.

chitinases was first made by Legrand *et al.* (1987) who observed that four related PR-proteins from TMV-infected tobacco leaves, now known as PR-3 proteins, exhibited chitinase activity.

Inspection of the amino acid sequences deduced from cDNAs or genes encoding PR-3-type proteins from different plants reveals that they all have in common a core domain harboring the catalytic site and hence called the chitinolytic or catalytic domain (see Fig. 6). In addition, all PR-3-type proteins have an N-terminal hydrophobic signal peptide. The PR-3-type proteins can be further divided in three subgroups, which are historically designated as class I, class II and class IV chitinases (Collinge *et al.*, 1993). The catalytic domains of class I, II and IV PR-3-type proteins share about 25% identical residues and an additional 53% homologous residues (Fig. 6). Class II chitinase precursors consist of a signal peptide and a chitinolytic domain. Class I chitinase precursors have a more complex structure and are multidomain proteins. From amino-terminus to carboxy-terminus they feature a signal peptide, a cysteine/glycine-rich chitin-binding lectin domain (also called the hevein-like domain), a hinge region usually rich in proline residues (some of which can be post-translationally hydroxylated, Sticher *et al.*, 1992b), the chitinolytic domain, and at least in the case of most basic class I chitinases, a carboxy-terminal propeptide. Mature class I and class II chitinases were originally defined as being basic and acidic, respectively (Shinshi *et al.*, 1990). However, many acidic chitinases with and basic chitinases without a hevein-like domain have been described (Graham and Sticklen, 1994 and references therein; Yun *et al.*, 1996) and hence it seems more appropriate to define class I and class II chitinases as above, irrespective of their isoelectric point. Most basic class I chitinases are synthesized with a carboxy-terminal propeptide that is removed during transport to or in the vacuole (Neuhaus *et al.*, 1991b; Sticher *et al.*, 1993). The tobacco basic class I chitinase propeptide has been shown to be necessary and sufficient for transport to the vacuole (Neuhaus *et al.*, 1991b; Sticher *et al.*, 1993). A peculiar class I chitinase from stinging nettle rhizomes contains a doubled chitin-binding lectin domain which is apparently cleaved off during processing and exists as the chitin-binding lectin UDA (*Urtica dioica* agglutinin; Beintema and Peumans, 1992; Lerner and Raikhel, 1992). Class IV chitinases mainly differ from the class I chitinases by four deletions, one occurring in the chitin-binding lectin domain, the other three in the catalytic domain (Collinge *et al.*, 1993; see Fig. 6). They also lack a carboxy-terminal propeptide. Class IV chitinases accommodate basic as well as acidic proteins (Herget *et al.*, 1990; Margis-Pinheiro *et al.*, 1991; Nielsen *et al.*, 1994; Rasmussen *et al.*, 1992). Some proline residues in the hinge region of sugar beet and bean class IV chitinases are also hydroxylated (Lange *et al.*, 1996; Nielsen *et al.*, 1994).

A class II chitinase from barley endosperm (Fig. 6) has been crystallized and its 3-D structure determined by X-ray crystallography (Hart *et al.*, 1993; see Fig. 1). Almost half of the protein (47%) is compacted in ten helical segments. Similar to the 3-D structure of the basic class II β-1,3-glucanase, an ~40 Å long cleft putatively housing the chitin substrate is running through the molecule (Hart *et al.*, 1993). Although the primary sequence of the crystallized barley chitinase is not significantly homologous to the primary sequences of animal or T4 phage lysozymes, the 3-D structures of these enzymes can be superimposed (Holm and Sander, 1994). This comparison also reveals that the conserved amino acid E125 (numbering as in Fig. 6) may be involved in the

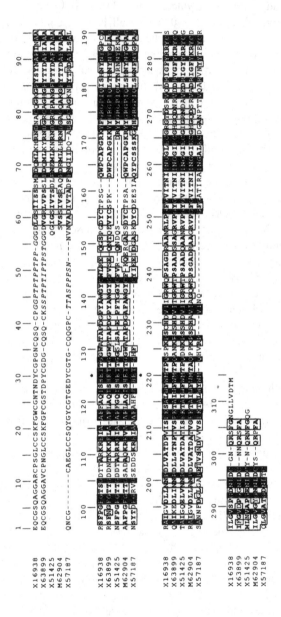

Figure 6. **Alignment of the amino acid sequences of chitinases of the PR-3-type protein family.**

Aligned are the amino acid sequences of the (pro)protein parts of PR-3-type chitinases cloned as cDNAs or genes of which the GenBank accession numbers are indicated on the left. In all cases have the putative signal peptides been omitted. The cDNAs/genes were isolated from the following plants: *Nicotiana tabacum:* X16938 (basic class I chitinase gene; Shinshi *et al.,* 1990) and X51425 (acidic class II chitinase cDNA; Linthorst *et al.,* 1990b); *Pisum sativum:* X63899 (acidic class I chitinase cDNA; Vad *et al.,* 1993); *Hordeum vulgare:* M62904 (basic class II chitinase cDNA; Leah *et al.,* 1991); *Phaseolus vulgaris:* X57187 (acidic class IV chitinase cDNA; Margis-Pinheiro *et al.,* 1991). Gaps introduced to obtain an optimal alignment are indicated by hyphens. Residues occurring in all depicted sequences are present in a black box. Residues belonging to the same amino acid homology group (FWY; MILV; RKH; NQDE; PAGST) and occurring in at least 4 out of the 5 aligned PR-3 family chitinases are bold-faced and boxed. Hinge regions between the amino-terminal chitin-binding lectin domain and the carboxy-terminal chitinolytic domain are italicized. The putative catalytic proton-donating amino acid is indicated by an asterisk above and under the sequences.

catalytic action of plant PR-3-type chitinases, as this residue is located in the 3-D structure at the position of the catalytic glutamate residue of animal and T4 phage lysozymes (Holm and Sander, 1994).

3. PR-8-Type Proteins

The PR-8 protein family accommodates the chitinases which were previously designated as class III chitinases (van Loon *et al.*, 1994). This new nomenclature reflects the fact that PR-8-type chitinases are structurally unrelated to PR-3-type chitinases. PR-8-type chitinases were first isolated as lysozymes from the latex of fig (Glazer *et al.*, 1969) and papaya (Howard and Glazer, 1969). Mature PR-8-type chitinases are monomeric proteins with a molecular weight of about 28 kDa and can be either acidic or basic. In most cases, PR-8-type chitinases have been reported to exhibit both chitinase and lysozyme activities (Bernasconi *et al.*, 1987; Glazer *et al.*, 1969; Howard *et al.*, 1969; Métraux *et al.*, 1989; Tata *et al.*, 1983) but at least a sugar beet PR-8 chitinase exhibits endo- and exochitinase activity but no lysozyme activity (Nielsen *et al.*, 1993), indicating that bifunctional chitinase/lysozyme activity is not a general property of this enzyme family.

Several PR-8-type chitinase cDNAs and genes have been cloned from various plant species and the deduced amino acid sequences of four representative members is shown in Fig. 7. All PR-8-type chitinases contain an N-terminal signal peptide. As shown in Fig. 7, PR-8-type chitinases display homology to concanavalin B, an enzymatically inactive protein from jack bean (*Canavalia ensiformis*). More intriguingly, plant PR-8-type proteins show sequence homology with a chitinase from baker's yeast (*Saccharomyces cerevisiae*). The overall identity between these proteins is 18%, including 6 invariant cysteines, with an additional 43% of homologous amino acids in at least five out of the six aligned sequences.

The 3-D structure of hevamine, the rubber tree PR-8-type chitinase/lysozyme, was determined by X-ray crystallography (Terwisscha van Scheltinga *et al.*, 1994, 1996). As could be expected based on the lack of primary sequence homology between PR-3- and PR-8-type chitinases, the 3-D structure of hevamine is completely different from that of the PR-3-type barley chitinase. Hevamine displays a $(\beta/\alpha)_8$-barrel topology (Fig. 1) similar to that determined for the barley β-1,3-glucanase (Terwisscha van Scheltinga *et al.*, 1996; Varghese *et al.*, 1994). The cleft in hevamine accommodates six *N*-acetylglucosamine residues and the catalytic acid (proton donor) was identified as being E142 (numbering as in Fig. 7; Terwisscha van Scheltinga *et al.*, 1994, 1995). No catalytic nucleophile was identified but the positively charged oxazoline reaction intermediate might be stabilized by D140 (or N140 in the *Arabidopsis* PR-8-type chitinase; Samac *et al.*, 1990; Terwisscha van Scheltinga *et al.*, 1995). Amino acid residues important for substrate recognition in hevamine and related PR-8-type chitinases are thought to be Y12, Y201 and W279 (numbering as in Fig. 7; Terwisscha van Scheltinga *et al.*, 1995).

4. PR-11-Type Proteins

A single endochitinase from tobacco constitutes the class V chitinases, recently reclassified as PR-11 proteins (Heitz *et al.*, 1994b; Melchers *et al.*, 1994; van Loon *et al.*, 1994).

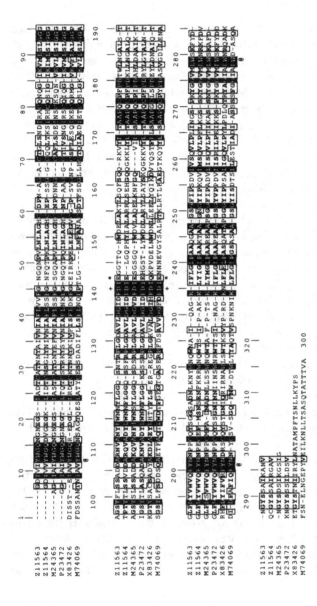

Figure 7. **Alignment of the amino acid sequences of the chitinases of the PR-8-type protein family and of plant and nonplant PR-8-like proteins.**
Aligned are the amino acid sequences of the (pro)protein parts of the plant PR-8 and nonplant PR-8-like proteins deduced from cDNAs or genes retrieved from the GenBank database. The hevamine sequence was retrieved from the SwissProt database. All accession numbers are indicated on the left. In all cases have the putative signal peptides been omitted. The following sequences are aligned: *Nicotiana tabacum*: Z11563 (acidic class III chitinase cDNA; Lawton *et al.*, 1992) and Z11564 (basic class III chitinase cDNA; Lawton *et al.*, 1992); *Cucumis sativus*: M24365 (class III chitinase/lysozyme cDNA; Métraux *et al.*, 1989); *Hevea brasiliensis*: P23472 (hevamine; Jekel *et al.*, 1991); *Canavalia ensiformis*: X83426 (concanavalin B cDNA; Schlesier *et al.*, 1996) and *Saccharomyces cerevisiae*: M74069 (chitinase gene; Kuranda and Robbins, 1991). Gaps introduced to obtain an optimal alignment are indicated by hyphens. Residues occurring in all depicted sequences are present in a black box. Residues belonging to the same amino acid homology group (FWY; MILV; RKH; NQDE; PAGST) and occurring in at least 5 out of the 6 aligned PR-8 family proteins are bold-faced and boxed. The last 232 amino acids of the yeast chitinase have been omitted. The catalytic proton-donating amino acid at position 142 is marked by asterisks above and under the sequences. The residue putatively stabilizing the reaction intermediate is likewise marked with plusses. Residues thought to be important for substrate recognition are marked with '@' under the sequences.

This enzyme is synthesized as a preprotein with an amino-terminal prepeptide and a carboxy-terminal propeptide loosely homologous to the tobacco basic β-1,3-glucanase carboxy-terminal extensions, although an N-glycosylation site is missing. The tobacco PR-11 chitinase shows significant homology to bacterial and nematode chitinases (Melchers *et al.*, 1994) and to a lesser extent also to PR-8-type chitinases. Eight percent identical and an additional 24% homologous amino acids can be found in an alignment of 5 fragments covering 193 residues of the *Serratia marcescens* chitinase A (Jones *et al.*, 1986) with PR-8-type chitinases; with PR-11 chitinase, 36% identical and an additional 21% homologous amino acids are tractable in 4 fragments covering 280 residues of the *S. marcescens* chitinase A (alignments not shown). Crystal structures of several bacterial chitinases have been determined. Similar to β-1,3-glucanases and PR-8-type chitinases, these bacterial enzymes adopt a $(β/α)_8$-barrel fold (Perrakis *et al.*, 1994; Rao *et al.*, 1995b; Van Roey *et al.*, 1994). The class V chitinase can thus be expected to be folded in the same way. The active site amino acids of bacterial chitinases (Schmidt *et al.*, 1994; Watanabe *et al.*, 1993) reside in the crystals at similar positions as the catalytic acid E140 and the stabilizing residue D138 in the 3-D structure of hevamine (see above; Terwisscha van Scheltinga *et al.*, 1996). A highly conserved stretch of 11 amino acids containing the two residues involved in chitinolysis occurs in plant PR-11 and PR-8 chitinases and in bacterial, nematode, viral and mammalian chitinases, as well as in chitinase-like proteins such as human gp39 and mouse Yml. This observation is represented in Fig. 8. Also remarkable is the fact that some fungal chitinases are more homologous to PR-8-type chitinases (e.g. Kuranda and Robbins, 1991), while other fungal chitinases are more related to PR-11-type chitinases (e.g. Hayes *et al.*, 1994). The catalytic glutamate is not conserved in the human gp39, mouse Yml and a *Drosophila* secreted glycoprotein, which are all devoid of chitinolytic activity (Hakala *et al.*, 1993; Kirkpatrick *et al.*, 1995). Furthermore, the aspartate (or asparagine in the *Arabidopsis* PR-8-type chitinase) stabilizing the reaction intermediate is substituted for an alanine in the human gp39 and the *Drosophila* secreted glycoprotein. The biological role of the animal chitinase-like proteins lacking chitinolytic activity is not clear at present. The human gp39 is, however, secreted by cells associated with the inflammatory or degenerative processes of rheumatoid arthritis (Hakala *et al.*, 1993). Interestingly, high levels of a PR-11-like chitinase are found in plasma of patients showing symptoms of the Gaucher disease that is caused by a defective glucocerebrosidase in lysosomes of macrophages (Boot *et al.*, 1995; Hollak *et al.*, 1994). This chitinase is also produced by activated macrophages and it was speculated that the enzyme might function in warding off chitin-containing pathogens (Boot *et al.*, 1995).

5. Antimicrobial Properties

The cell wall of most fungi and Oomycetes contains β-1,3-glucans, with β-1,6-linked glucosyl residues (Gooday, 1994). These β-1,3-1,6-glucans form the major fibrous cell wall polymers together with either chitin (in most true fungi) or cellulose (in most Oomycetes). Based on the fact that β-1,3-1,6-glucan and chitin are substrates of plant β-1,3-glucanases and chitinases, respectively, Abeles *et al.* (1971) originally proposed that these enzymes may act as antifungal compounds. However, it was not until the late 1980s, that the presumed antifungal properties of chitinases and glucanases

Accession Number	Organism	Active Site Region	Reference
		+ *	
M34107	*Arabidopsis* class III chitinase	L D G I D F N I E L G	Samac et al., 1990
M24365	cucumber class III chitinase	L D G V D F D I E S G	Métraux et al., 1989
P23472	(SwissProt) hevamine	L D G I D F D I E H G	Jekel et al., 1991
M74069	yeast chitinase	V D G F D F D I E N N	Kuranda and Robbins, 1991
X77111	tobacco class V chitinase	F H G L D L D W E Y P	Melchers et al., 1994
L22858	*Autographa californica* nuclear polyhedrosis virus, ORF 126	F D G V D I D W E F P	Ayres et al., 1994
U02270	tobacco hornworm chitinase	F D G L D L D W E Y P	Kramer et al., 1993
M73689	*Brugia malayi* chitinase (nematode)	F D G F D L D W E Y P	Fuhrman et al., 1992
M80927	secreted human glycoprotein gp39	F D G L D L A W L Y P	Hakala et al., 1993
S27879	(PIR) secreted mouse macrophage protein Ym1	F D G L N L D W Q Y P	Jin et al., 1998
U29615	human macrophage chitotriosidase	F D G L D L D W E Y P	Boot et al., 1995
P32470	(SwissProt) *Aphanocladium album* chitinase (fungus)	F D G I D I D W E Y P	Blaiseau et al., 1992
L14614	*Trichoderma harzianum* chitinase (fungus)	F D G I D V D W E Y P	Hayes et al., 1994
P20533	(SwissProt) *Bacillus circulans* chitinase (bacterium)	F D G V D L D W E Y P	Watanabe et al., 1993
M82804	*Streptomyces plicatus* chitinase (bacterium)	F D G I D L D W E Y P	Robbins et al., 1992
Z36294	*Serratia marcescens* chitinase A (bacterium)	F D G V D I D W E F P	Brurberg et al., 1994
P09805	(SwissProt) *Kluyveromyces lactis* killer toxin α-subunit/chitinase	L D G I D L D W E Y P	Stark and Boyd, 1986
U13825	*Drosophila melanogaster* chitinase	F D G L D V A W Q F P	Kirkpatrick et al., 1995
		+ *	

Figure 8. **Comparison of the active site regions of plant PR-11-type chitinases and nonplant chitinases and chitinase-like proteins.**

The 11-residue region containing the catalytic glutamate and the reaction intermediate stabilizing aspartate/asparagine (indicated by an asterisk and a plus above and under the sequences, respectively) from a plant PR-11 chitinase (tobacco class V chitinase) and related chitinases and chitinase-like proteins from different sources are aligned. For comparison, the corresponding regions of four PR-8-type chitinases are also given. Amino acids occurring in the majority of the active site regions of the different chitinases are boxed. Unless indicated otherwise, the accession numbers confer to the GenBank database.

were actually demonstrated by Boller and coworkers (Schlumbaum *et al.*, 1986; Mauch *et al.*, 1988b).

The most striking feature of the antifungal properties of chitinases and glucanases is that their efficacy is much higher when they act in combination. This synergistic antifungal activity of chitinase/glucanase combinations was first observed by Mauch *et al.* (1988b) who reported that all of 15 different fungi were inhibited by a combination of a basic glucanase and a basic class I chitinase isolated from infected pea pods. Of these 15 fungi only one was inhibited by the glucanase alone and another one by the chitinase alone, at least at the doses used in the tests with the chitinase/glucanase combinations. Of three different Oomycetes tested, none was inhibited by either the chitinase, glucanase or the enzyme mixture. Synergistic activities of chitinase/glucanase combinations have later been confirmed with enzymes purified from other plant sources, including infected tobacco leaves (Sela-Buurlage *et al.*, 1993), infected tomato leaves (Joosten *et al.*, 1995; Lawrence *et al.*, 1996) and infected cucumber leaves (Ji and Kuc, 1996).

Purified chitinases can also exert antifungal activity on their own against some but not all fungi (reviewed in Yun *et al.*, 1997a). When the antifungal potency of basic and acidic PR-2 and PR-3 proteins of either tobacco (Sela-Buurlage *et al.*, 1993) or tomato (Joosten *et al.*, 1995; Lawrence *et al.*, 1996) was compared, the basic isoforms were found to be significantly more active. This difference in antifungal activity may be related to a higher specific hydrolytic activity on their substrates, as was shown at least for different tobacco PR-2 (Kauffmann *et al.*, 1987) and PR-3 (Legrand *et al.*, 1987) enzymes. The observation that basic glucanase and chitinase isoforms from tobacco and tomato are more active than their acidic counterparts must not be generalized, as at least one acidic glucanase from cucumber has been identified that exerts significant antifungal activity on its own (Ji and Kuc, 1996).

The best studied chitinases in terms of their antifungal activity are the class I PR-3-type chitinases. Iseli *et al.* (1993) demonstrated that a recombinant class I PR-3 protein from which the hevein-like domain had been deleted still inhibited fungal growth, although both its antifungal activity and chitinolytic activity were about three-fold lower compared to those of the enzyme with a hevein-like domain. Although not essential for antifungal activity, the hevein-like domain with its intrinsic chitin-binding properties may potentiate the activity of the chitinase by delivering it more efficiently to its target substrate. The fact that the hevein-like domain is not an absolute requirement for antifungal activity is also illustrated by the observation that some basic class II PR-3-type proteins also exhibit substantial antifungal activity (Yun *et al.*, 1996). Little information is available on the antifungal activity of class III PR-8-type chitinases. At least one class III PR-8-type chitinase from chickpea was reported to be devoid of antifungal activity (Vogelsang and Barz, 1993). However, it is highly unlikely that none of the class III PR-8-type chitinases would affect fungi, as this family of enzymes is structurally related to the class V PR-11-type chitinases, of which at least one member with antifungal properties is known (Melchers *et al.*, 1994). Many class III PR-8-type proteins also exhibit significant lysozyme activity (see V.B.3.) suggesting that they may also interfere with bacteria, especially Gram-positive bacteria. In a comparative test including ten different tobacco chitinases, two basic class III PR-8-type proteins were found to be much more efficient in lysing cell walls of the Gram-positive bacterium *Micrococcus lysodeikticus* as compared to the class I or class II PR-3-type proteins, or the class V PR-11-type proteins (Brunner *et al.*, 1998). Unfortunately, no data are yet available on the bactericidal or bacteriostatic activity of plant chitinases.

The antifungal effects of chitinases and glucanases are believed to be due to breakdown of either formed or nascent chitin and β-linked glucan polymers in fungal cell walls, respectively. Microscopic observation of fungal hyphae treated with various PR-3-type chitinases revealed swelling of the tips and in some cases lysis of swollen tips in hypotonic but not in isotonic media (Mauch *et al.*, 1988b; Broekaert *et al.*, 1988; Arlorio *et al.*, 1992). Analysis of glucanase-treated *Trichoderma longibrachatum* by electron microscopy revealed substantial removal of an electron-dense outer cell wall layer containing β-linked glucans, whereas treatment of either *T. longibrachatum* or *Rhizoctonia solani* with chitinase caused loosening and detaching of fibrils from the cell wall, especially at the tip (Arlorio *et al.*, 1992; Benhamou *et al.*, 1993a). Moreover, glucanase-gold complexes were shown by electron microscopy to be present in patches along the hyphal shaft, whereas chitinase-gold complexes were located mainly at the hyphal tip (Arlorio *et al.*, 1992). This pattern is consistent with the current models for hyphal cell wall architecture which propose that chitin and β-glucans are synthesized at the hyphal tip, while this loose primary layer becomes more fibrillar, more intensively crosslinked and covered with a secundary layer of β-glucans and additional polymers in the lateral cell walls (Gooday, 1994). Hence, the experimental data of Arlorio *et al.* (1992) and Benhamou *et al.* (1993a) demonstrate that glucanases and chitinases indeed reach their target substrates in hyphal walls and cause physical damage to the cell wall ultrastructure. The observed synergistic action of glucanases and chitinases can be easily explained by assuming that glucanases provide a better access for chitinases to their substrate and vice versa.

In addition to playing a role as antimicrobial agents, β-1,3-glucanases and chitinases have also been proposed to contribute to the release of elicitors from fungal cell walls. Indeed, it is well documented that particular β-linked oligoglucosides and chitooligosaccharides, can act as inducers of defense genes, including PR-genes themselves (see also chapter 4 and a review by Coté and Hahn, 1994). Receptors for β-linked oligoglucosides and chitooligosaccharides have been shown to occur in the plasma membrane of plant cells and are believed to play an important role in the signalling of defense responses (see chapter 4). Some fungi produce inhibitors of particular β-1,3-glucanase isoforms and can thus interfere with the function of these enzymes in the induction of defense reponses or with their antimicrobial effect (Ham *et al.*, 1997). In addition, not all chitinases and glucanases in plants are implicated in defense, as some particular forms appear to be involved in developmental processes such as embryogenesis (De Jong *et al.*, 1992; Kragh *et al.*, 1996), microsporogenesis (Bucciaglia and Smith, 1994), seed germination (Fincher, 1989; Leubner-Metzger *et al.*, 1995; Vögeli-Lange *et al.*, 1994a) or in freeze tolerance (Hincha *et al.*, 1997; Hon *et al.*, 1995).

C.   PR-4-TYPE PROTEINS

1.   *Structural Features*
PR-4-type proteins form a family of proteins ranging from 15 to 20 kDa in size. The first PR-4 proteins to be purified were isolated from TMV-infected tobacco leaves (Friedrich *et al.*, 1991; Linthorst *et al.*, 1991) and *Cladosporium fulvum* infected tomato leaves (Linthorst *et al.*, 1991), although cDNA sequences encoding PR-4-type proteins had been characterized before as wound-induced (*win*) transcripts in potato (Stanford *et al.*, 1989) and rubber tree (Broekaert *et al.*, 1990).

Comparison of the proteins predicted from various cDNAs and genes encoding PR-4-type proteins reveals that all have signal peptides at their amino terminus and share a domain of about 120 amino acids that is located at the carboxy-terminus and called the PR-4 domain. Alignment of the cDNA deduced sequences of five different PR-4-type proteins from four different plant species reveals that the overall identity in their PR-4 domains is 50% with an additional 28% homology (Fig. 9). Among the fully conserved residues are 6 cysteines. PR-4-type proteins can be subdivided in two major classes. Class II PR-4-type proteins are encoded as preproteins and their precursors consist of a signal peptide followed by a PR-4 domain. Examples are the tobacco TMV-induced PR-4a (Friedrich *et al.*, 1991) and a barley seed protein called barwin (Svensson *et al.*, 1992). Class I PR-4-type proteins have a more complex structure and, in analogy to class I PR-3-type proteins, contain a chitin-binding lectin domain with a short hinge region. The chitin-binding lectin domain is located between the signal peptide and the PR-4 domain. In addition, class I PR-4-type precursors have a short carboxy-terminal propeptide which is absent from the mature protein and most likely functions in vacuolar targetting (Ponstein *et al.*, 1994). Examples of class I PR-4-type proteins are a TMV-induced chitin-binding protein CBP20 from tobacco (Ponstein *et al.*, 1994) and preprohevein from rubber tree latex. In addition to removal of the N-terminal and C-terminal propeptides, the rubber tree preprohevein precursor undergoes further processing resulting in mature hevein, a 5 kDa protein corresponding to the chitin-binding lecting domain (Archer *et al.*, 1969; Walujono *et al.*, 1975) and a 15 kDa protein corresponding to the PR-4 domain (Lee *et al.*, 1991). The latter protein may be partially degraded as hevein occurs in a 30-fold molar excess relative to the 15 kDa PR-4 domain protein (Soedjanaatmadja *et al.*, 1995). Such processing beyond removal of the N-terminal prepeptide and C-terminal propeptide apparently does not occur in the case of the tobacco CBP20 class I PR-4 protein (Ponstein *et al.*, 1994).

It has been reported that class II PR-4-type proteins are able to bind chitin and chitin oligomers (Hejgaard *et al.*, 1992; Ludvigsen and Poulsen, 1992). In contrast, the 15 kDa PR-4 domain processing product of prohevein did not interact with chitin, while hevein itself and unprocessed prohevein, both containing the chitin-binding lectin domain, bound reversibly to chitin (Lee *et al.*, 1991). Hence, chitin-binding capacity is associated with the chitin-binding domain and also with the PR-4 domain of some but not all PR-4 type proteins. Ponstein *et al.* (1994) have reported that a class I PR-4 protein from tobacco (CBP20) exhibits a very low chitinase activity (about 100-fold lower compared to class I plant chitinases) and a very low lysozyme activity (about 1000-fold lower compared to hen egg white lysozyme but in the same range as for

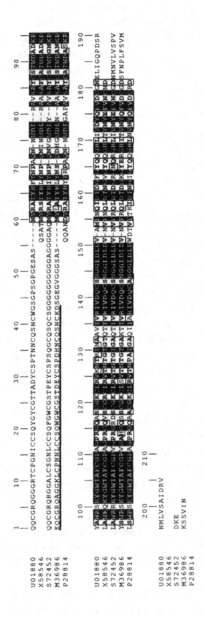

Figure 9. **Alignment of plant PR-4-type protein sequences.**
Aligned are the sequences of PR-4 proteins as deduced from cDNAs retrieved from the GenBank database. The barwin sequence was retrieved from the SwissProt database. All accession numbers are indicated on the left. In all cases have the (putative) signal peptides been omitted. The following sequences are aligned: *Arabidopsis thaliana*: U01880 (hevein-like PR-4; Potter *et al.*, 1993); *Nicotiana tabacum*: X58546 (PR-4a; Friedrich *et al.*, 1991) and S72452 (CBP20; Ponstein *et al.*, 1994); *Hevea brasiliensis*: M36986 (prohevein; Broekaert *et al.*, 1990); *Hordeum vulgare*: P28814 (barwin; Svensson *et al.*, 1992). The mature hevein amino acid sequence is double underlined. Gaps introduced to obtain an optimal alignment are indicated by hyphens. Residues belonging to the same amino acid homology group (FWY; MILV; RKH; NQDE; PAGST) and occurring in at least 4 out of the 5 aligned PR-4 sequences are bold-faced and boxed.

class I plant chitinases). These data were later confirmed by Brunner *et al.* (1998). However, with such low specific activities it is difficult to exclude contribution from contaminating chitinase/lysozymes. Class II PR-4-type proteins from barley, on the other hand, were reported not to exhibit chitinase, lysozyme nor chitosanase activity, although no detection limits of the assays were given (Hejgaard *et al.*, 1992). Hence, the possible enzymatic activity of PR-4-type proteins warrants further investigation.

The 3D-structure of barwin was analyzed by ^{1}H nuclear magnetic resonance spectroscopy (see Fig. 1; Ludvigsen and Poulsen, 1992). An opened-shell-like structure is formed by two β-sheets each consisting of 4 antiparallel β-strands. Thirty-nine percent of the barwin amino acids are involved in these β-sheets, while about 23% of the residues form four short α-helices. Amino acid substitutions in PR-4 proteins relative to barwin occur most often in loop regions P70-P81, P110-G120 and in the helix region W151-G168 (numbering as in Fig. 9; Ludvigsen and Poulsen, 1992). Chitotetraose binds relatively weakly to barwin and interactions with residues Y65, H66, Y67, D148, D150, W151, Y165 and H169 were identified (Ludvigsen and Poulsen, 1992).

2.   *Antimicrobial Properties*

In vitro antifungal activity has so far been reported for four different PR-4-type proteins. One class II PR-4-type protein from barley seeds and two class II PR-4 proteins from infected barley leaves inhibit growth of the fungus *Trichoderma harzianum* (Hejgaard *et al.*, 1992). The antifungal activity of these proteins was synergistically enhanced when mixed with either a PR-3-type chitinase or a PR-5-type protein from barley seeds (Hejgaard *et al.*, 1992). A class I PR-4 protein purified from infected tobacco leaves inhibited two out of three tested fungi (Ponstein *et al.*, 1994). Growth inhibition was accompanied by swelling of hyphal tips. Synergistic enhancement of the antifungal activity of tobacco class I PR-4 protein was observed in the presence of either a tobacco basic class I PR-3 chitinase or a tobacco basic class I glucanase (Ponstein *et al.*, 1994). The target site of PR-4-type proteins is unknown but based on some reports that these proteins can bind chitin and exert chitinolytic activity (Ponstein *et al.*, 1994; Heitz *et al.*, 1994b) it may be speculated to be the chitin layer in fungal walls. Activities of PR-4-type proteins against microorganisms other than fungi have not yet been investigated.

D.   PR-5-TYPE PROTEINS (THAUMATIN-LIKE PROTEINS)

1.   *Structural Features*

PR-5-type proteins form a family of related proteins which exist as monomers of about 25 kDa. Although PR-5-type proteins have no known enzymatic activity, several but not all of them have a specific binding affinity for β-1,3-glucans (Trudel *et al.*, 1998). The first PR-5-type protein to be purified was thaumatin, isolated because of its sweet tasting properties from the fruits of the West-African shrub *Thaumatococcus danielli* (van der Wel and Loeve, 1972). Upon analyzing cDNAs corresponding to transcripts that are highly induced in TMV-infected tobacco plants, Cornelissen *et al.* (1986a) discovered a cDNA encoding a PR-protein, later designated as PR-5 (van Loon, 1990; van Loon *et al.*, 1994), that showed significant homology to thaumatin. Hence, the name thaumatin-like proteins was suggested to denote this class of proteins (Cornelissen *et al.*, 1986a). Some confusion in the nomenclature of PR-5-type proteins arises from the fact that two other generic names have been proposed. The first one is osmotins, a name given to basic PR-5 proteins first discovered in osmotically-stressed cultured tobacco cells (Singh *et al.*, 1987); the other one is permatins, a name proposed for thaumatin-like proteins isolated from monocot grains which have the property to permeabilize fungal membranes (Vigers *et al.*, 1991). To further add to confusion, an antifungal permatin from maize seeds, called zeamatin, is in fact identical or nearly identical to a protein previously described as being a bifunctional trypsin/α-amylase inhibitor (Richardson *et al.*, 1987; Huynh *et al.*, 1992; Malehorn *et al.*, 1994; Roberts and Selitrennikoff, 1990).

In tobacco, three subclasses of PR-5 proteins are distinguished: basic PR-5 (called osmotin), neutral PR-5 (called osmotin-like protein) and acidic PR-5 proteins (Koiwa *et al.*, 1994). Acidic PR-5 proteins are synthesized as preproteins whereas the others also contain a carboxy-terminal propeptide in their precursor (see Fig. 10). This carboxy-terminal extension has been shown to be necessary for targeting the mature protein to the vacuoles (Melchers *et al.*, 1993). Thaumatin itself is also derived from a preproprotein (Edens *et al.*, 1982).

A peculiar mosaic protein showing homology to PR-5-type proteins is the *Arabidopsis* PR5K, which consists of an extracellular PR-5-like receptor connected via a membrane-spanning domain to an intracellular serine/threonine kinase (Wang *et al.*, 1996). The *Arabidopsis* PR5K is unique in the sense that it is so far the only protein for which it is shown that a PR-like domain might function as a receptor connected to a kinase most likely involved in intracellular signalling (Wang *et al.*, 1996). The PR5K ligand has not yet been identified. The kinase domain belongs to the same group of kinases as the tomato Pto (Martin *et al.*, 1993) interacting with the *Pseudomonas syringae* pv. *tomato* avirulence gene product AvrPto (Scofield *et al.*, 1996; Tang *et al.*, 1996), and the *Brassica oleracea* self-incompatibility associated SRK (Stein and Nasrallah, 1993). The extracellular domain of SRK is also present as a highly homologous secreted protein, SLG (Nasrallah *et al.*, 1987), a similar situation as noted for PR-5 and PR5K (Wang *et al.*, 1996). Interestingly, proteins with a higher molecular weight than the PR-5 proteins have been detected in tobacco flowers along with genuine PR-5-type proteins (Neale *et al.*, 1990; Richard *et al.*, 1992). The amino acid sequences of genuine PR-5 proteins, thaumatin, the maize trypsin/amylase inhibitor and the PR-5-like domain of the *Arabidopsis* receptor protein kinase PR5K have been aligned in Fig. 10. Identical residues are found in 22% of the positions and 16 conserved positions are occupied by cysteines, which are all pairwise disulfide linked (Batalia *et al.*, 1996; Ogata *et al.*, 1992). An additional 44% of homologous amino acids occur in at least 6 out of the 8 aligned proteins.

One of the thaumatin isoforms, thaumatin I, has been crystallized and its 3-D structure determined via X-ray crystallography (see Fig. 1; Ogata *et al.*, 1992). Over one-third of the residues form a flattened β-sandwich consisting of 11 β-strands. The strands in the top and bottom sheets of the sandwich are lying parallel to each other. Two additional β-sheets are involved in a second sandwich oriented back-to-back with the major β-sandwich. The third main domain comprises four short α-helical

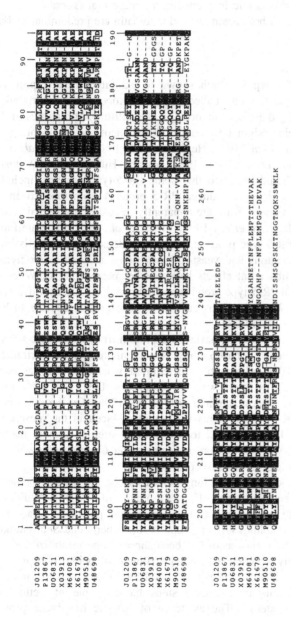

Figure 10. **Alignment of plant PR-5-type protein sequences.**

Aligned are the sequences of PR-5 (pro)proteins as deduced from cDNAs retrieved from the GenBank database. The maize bifunctional inhibitor sequence was retrieved from the SwissProt database. All accession numbers are indicated on the left. In all cases have the (putative) signal peptides been omitted. The following sequences are aligned: *Thaumatococcus daniellii*: J01209 (thaumatin 2; Edens *et al.*, *1982*); *Zea mays*: P13867 (trypsin/-amylase inhibitor; Richardson et al., 1987) and U06831 (zeamatin-like protein; Malehorn et al., *1994*); *Nicotiana tabacum*: X03913 (acidic PR-5; Cornelissen et al., 1986a), M64081 (neutral osmotin-like protein; *Takeda et al., 1991*) and X61679 (basic osmotin; Kumar and Spencer, *1992*); *Arabidopsis* thaliana: M90510 (thaumatin-like protein; Uknes et al., 1992) and U48698 (PR-5-like domain of PR5K receptor kinase; Wang et al., 1996). Gaps introduced to obtain an optimal alignment are indicated by hyphens. Residues belonging to the same amino acid homology group (FWY; MILV; RKH; NQDE; PAGST) and occurring in at least 6 out of the 8 aligned PR-5 sequences are bold-faced and boxed.

segments held together by four disulfides (Ogata *et al.*, 1992). The 3-D crystal structure of zeamatin has also been determined (Batalia *et al.*, 1996). The overall structure of zeamatin is very similar to the thaumatin structure. One main difference is, however, that the sandwich-cleft is highly acidic in zeamatin, a feature that is less pronounced for thaumatin. The back sides of both zeamatin and thaumatin are predominantly basic (Batalia *et al.*, 1996).

2.   Antimicrobial Properties

The first thaumatin-like (PR-5-type) protein shown to exhibit antifungal properties was a protein isolated from maize seeds, called zeamatin (Roberts and Selitrennikoff, 1990). Subsequently, several groups have reported on the antifungal properties of over 10 different PR-5-type proteins originating from different dicot and monocot species (Abad *et al.*, 1996; Borgmeyer *et al.*, 1992; Hejgaard *et al.*, 1991; Huynh *et al.*, 1992; Malehorn *et al.*, 1994; Vigers *et al.*, 1991, 1992; Vu and Huynh, 1994; Woloshuk *et al.*, 1991). However, it should be noted that some PR-5-type proteins, including thaumatin itself, some acidic pathogen-induced PR-5 proteins from tobacco and an acidic PR-5-type protein from cherry fruit, are apparently devoid of antifungal activity (Fils-Lycaon *et al.*, 1996; Malehorn *et al.*, 1994; Vigers *et al.*, 1992). The antifungal activity spectrum of active PR-5-type proteins, when considered as a group, is broad and includes numerous filamentous and yeast-like true fungi as well as Oomycetes (see Yun *et al.*, 1997a). Antibacterial effects have not yet been reported. When the activity spectrum of individual PR-5-type proteins are compared, significant differences can be observed (Vigers *et al.*, 1992). Thus each PR-5-type protein appears to have a characteristic specificity to certain fungi.

The antifungal potency of osmotin, the basic PR-5 protein from tobacco, is strongly potentiated in the presence of a tobacco chitinase. To inhibit spore germination of *Botrytis cinerea* by 50%, 10 µg/ml osmotin was required, whereas in the presence of 32 µg/ml of chitinase, which by itself was not inhibitory, 0.3 µg/ml of osmotin caused 50% inhibition of spore germination (Lorito *et al.*, 1996). On the other hand, the antifungal potency of PR-5-type proteins is strongly reduced in the presence of inorganic salts. Addition of 100 mM NaCl to a low ionic strength growth medium raised the minimum inhibitory concentration of zeamatin on *Candida albicans* from 5 µg/ml to 80 µg/ml (Roberts and Selitrennikoff, 1990).

The target site of PR-5-type proteins on fungi appears to be the plasma membrane (Abad *et al.*, 1996; Roberts and Selitrennikoff, 1990; Yun *et al.*, 1997b). Addition of these proteins to fungi results in dissipation of the membrane potential (Abad *et al.*, 1996) and release of cytosolic solutes (Abad *et al.*, 1996; Roberts and Selitrennikoff, 1990). A PR-5-type protein from flax seeds has been shown to permeabilize artificial membrane vesicles, especially vesicles with a high content of acidic phospholipids and sterols (Anzlovar *et al.*, 1998). However, it has been argued that the mode of action of PR-5-type proteins must involve more steps than a simple interaction with membrane phospholipids and sterols. The existence of specific membrane proteins that facilitate insertion in fungal membranes has been postulated but not yet proven (Yun *et al.*, 1997b). On the other hand, fungal cell wall components appear to play a role as well, as inferred from the observation that baker's yeast (*Saccharomyces cerevisiae*) can acquire resistance to osmotin by overexpressing particular cell wall proteins (Yun *et al.*, 1997b). It has also been shown that several PR-5-type proteins exhibiting antifungal activity can bind to β-1,3-glucans, including glucans isolated from fungal cell walls (Trudel *et al.*, 1998). However, it is not known at present whether this binding activity is a prerequisite for the antifungal activity of PR-5-type proteins.

E.   RIBOSOME-INACTIVATING PROTEINS (RIPS)

1.   *Structural Features*

Ribosome-inactivating proteins or RIPs are divided into two groups (Stirpe *et al.*, 1992). Type 1 RIPs are single A-chain molecules (~30 kDa) exhibiting the ribosome-inactivating activity. Type 2 RIPs additionally contain a lectin B-chain (~30 kDa) connected to the active A-chain via a single disulfide bond. The lectins of most type 2 RIPs specifically recognize galactose or *N*-acetylglucosamine (reviewed in Barbieri *et al.*, 1993; Hartley *et al.*, 1996). The mechanism of the ribosome-inactivating activity was clarified by Endo *et al.* (1987) who showed that the A-chain of the type 2 castor bean RIP ricin is an *N*-glycosidase specifically removing an adenine from a conserved loop of the 28S ribosomal RNA. This mechanism was confirmed later for genuine type 1 RIPs (Taylor and Irvin; 1990). In vitro, dicot RIPs depurinate ribosomal RNA from different sources including endogenous ribosomes (Irvin and Uckun, 1992; Taylor and Irvin; 1990), whereas at least some monocot (cereal) RIPs do not inactivate either endogenous or tobacco ribosomes (Bass *et al.*, 1992; Taylor and Irvin, 1990). Some, but not all, type 1 RIPs remove more than one adenosine per ribosome and can even remove adenosine from non-ribosomal RNA and DNA (Olivieri *et al.*, 1996). Hence, the term polynucleotide:adenosine glycosidase is more appropriate to denote the enzymatic activity of plant RIPs (Olivieri *et al.*, 1996). Another enzymatic activity associated with some RIPs is endonuclease activity specifically exerted on supercoiled DNA (Ling *et al.*, 1995).

Type 1 RIPs from dicots are synthesized either as pre- or preproproteins (Carzaniga *et al.*, 1994; Kung *et al.*, 1990; Lin *et al.*, 1991; Ready *et al.*, 1986). Vacuolar targeting of at least one type 1 RIP, dianthin from carnation, is predicted from its cDNA-derived amino acid sequence which contains an 18-residue carboxy-terminal propeptide homologous to the barley lectin propeptide (Legname *et al.*, 1991). The latter was shown to be necessary and sufficient for vacuolar targeting of the barley lectin (Bednarek and Raikhel, 1991). Whereas dicot type 1 RIPs are derived from pre- or preproproteins, monocot type 1 RIPs apparently lack a signal peptide and are presumably located in the cytoplasm, a location compatible with their inactivity towards endogenous ribosomal RNA (Bass *et al.*, 1992; Habuka *et al.*, 1993; Leah *et al.*, 1991). Type 2 RIPs are produced as preproproteins. After entry into the ER with co-translational removal of the signal peptide, the precursor of the type 2 RIP ricin is *N*-glycosylated at four positions and intra- and inter-chain disulfides are established. This modified precursor undergoes a last post-translational maturation step in the seed protein bodies where a 12-amino acid linker is removed between the A- and B-chain (Lord *et al.*, 1994).

2.   *Antimicrobial Properties*

A type 1 ribosome-inactivating protein from barley seeds is the only RIP reported so far to exert antifungal activity (Leah *et al.*, 1991; Roberts and Selitrennikoff, 1986). The antifungal activity of this protein was synergistically enhanced in combination with either a basic class IV glucanase or a basic class II chitinase from barley seeds (Leah *et al.*, 1991). The basis of the antifungal activity of this RIP may rely on its inhibitory effect on fungal ribosomes, although it has not been demonstrated that this occurs in vivo. If the RIP would act on fungal ribosomes, then it would have to somehow cross the plasma membrane barrier, which also remains to be proven. The synergism between the barley RIP and the cell wall hydrolases is most likely explained by assuming that the hydrolases facilitate access for the RIP to reach the plasma membrane or any other target site.

Better documented is the antiviral activity of several type 1 RIPs (reviewed in Verma *et al.*, 1995). When purified RIPs are applied on plants together with viruses, they drastically suppress virus multiplication and symptom development (Irvin, 1975; Oliveiri *et al.*, 1996; Stirpe *et al.*, 1981; Wyatt and Shepherd, 1969). The adenosine glycosidase activity of RIPs appears to be essential for exerting antiviral properties, as a variant of a RIP from pokeweed (*Phytolacca americana*) with an amino acid substitution in its catalytic site was not antiviral against TMV when applied on tobacco (Tumer *et al.*, 1997). It is supposed that RIPs enter cells together with the viruses and exert their adenosine glycosidase activity in the cytosol to affect either host ribosomes (via 28S rRNA) or possibly viral RNA. A direct proof for this assumption has, however, not yet been presented. A striking observation is that at least some RIPs not only inhibit virus multiplication locally in RIP-treated leaves but also systemically in non-treated leaves (Smirnov *et al.*, 1997; Verma *et al.*, 1984; Verma *et al.*, 1996). Hence, RIPs release an unknown signal that induces systemic resistance to viruses and, again, the adenosine glycosidase activity of RIPs is required for release of the signal (Smirnov *et al.*, 1997). The factors contributing to RIP-induced systemic resistance are not yet known but appear not to involve pathogenesis-related proteins, at least not PR-1, PR-2 and PR-3 types (Smirnov *et al.*, 1997).

F.   HEVEIN-TYPE ANTIMICROBIAL PEPTIDES

1.   *Structural Features*

In the foregoing sections it has already been mentioned that class I and class IV PR-3-type chitinases as well as class I PR-4 type proteins are multidomain proteins of which one domain corresponds to a cysteine/glycine-rich domain of about 40 residues, known as the chitin-binding lectin domain or hevein-like domain. In at least one PR-4 type protein, namely preprohevein from rubber tree, processing leads to the release of a 5 kDa peptide corresponding to the hevein-like domain, in this case called hevein (Archer *et al.*, 1969; Walujono *et al.*, 1975; Broekaert *et al.*, 1990). Another example is that of the PR-3-type protein preproUDA from stinging nettle (*Urtica dioica*) which is processed to release the haemagglutinating lectin UDA (Lerner and Raikhel, 1992). UDA binds to chitin (Peumans *et al.*, 1983; Shibuya *et al.*, 1986) and consists of two tandem repeats of the hevein-like domain (Beintema and Peumans, 1992). Tandem repeats of hevein-like domains, four in total, also occur in WGA, the haemagglutinating and chitin-binding lectin from wheat embryos (Raikhel and Wilkins, 1987). A separate class of chitin-binding hevein-type antimicrobial peptides, tentatively called single domain hevein-type antimicrobial peptides, is exemplified by Ac-AMP1 and Ac-AMP2 from *Amaranthus caudatus* seeds (Broekaert *et al.*, 1992). The Ac-AMPs lack a carboxy-terminal portion of the canonical hevein-like domain such that they contain only 6 disulfide-linked cysteines as opposed to the 8 disulfide-linked cysteines present in the hevein-like domains of hevein, UDA and WGA. The Ac-AMPs are made as preproteins with an amino-terminal signal sequence and a carboxy-terminal propeptide of 31 residues (De Bolle *et al.*, 1993a). Other proteins that fall into this class of single domain hevein-type antimicrobial peptides are IWF4, a protein isolated from the intercellular fluid of sugar beet leaves (Nielsen *et al.*, 1997); An1, a protein occurring in *Atriplex nummularia* seeds (Last and Llewellyn, 1997); Pn-AFP1 and Pn-AFP2, proteins isolated from seeds of Japanese morning glory (Koo *et al.*, 1998). The precursor of the latter protein has a preproprotein structure similar to that of Ac-AMP2 but the mature peptide has eight cysteines instead of six for Ac-AMP2. A non-plant protein showing significant homology to the hevein-like domain is the killer toxin α-subunit of the yeast *Kluyveromyces lactis* (Butler *et al.*, 1991). This protein has an internal domain which aligns well to hevein. The amino acid sequences of the above-mentioned proteins, WGA excepted, have been aligned in Fig. 11.

The fact that the hevein-like domain occurs within the coding regions of totally unrelated mosaic proteins might be due to transposition events involving hevein-encoding DNA. In support of this hypothesis, Shinshi *et al.* (1990) have found imperfect repeats flanking the nucleotide sequence encoding the hevein-like domain of class I PR-3-type chitinases. Plant transposons often create sequence duplication upon insertion into a target site.

The three-dimensional structure of WGA has been obtained by X-ray crystallography (Wright, 1987). All four hevein domains of WGA adopt a similar fold, a relatively flat saucer-like molecule with three antiparallel β-strands and a short α-helical segment organized around four disulfide bonds (Wright, 1987). By ^{1}H NMR, similar foldings were determined in solution for hevein itself (Andersen *et al.*, 1993; see Fig. 1), for Ac-AMP2 (Martins *et al.*, 1996) and for a ragweed pollen antigen (Metzler *et al.*, 1992). Chitin interacts with WGA subdomains via the aromatic residues Y21 (W in hevein), Y23 (W in hevein), Y30 and via S19 and E/D29 (Wright, 1992). The corresponding aromatic residues of Ac-AMP2, as well as its amino-terminal three amino acids interact with chitotriose (Verheyden *et al.*, 1995).

2. Antimicrobial properties

Antifungal properties of hevein-type antimicrobial peptides were first reported for UDA, the chitin-binding lectin with a duplicated hevein-like domain isolated from stinging nettle rhizomes (Broekaert *et al.*, 1989). Later studies have confirmed the antimicrobial activity of this class of proteins, notably in the case of hevein, although the activity of this protein is rather weak (Van Parijs *et al.*, 1991), and in the case of the Ac-AMP chitin-binding peptides from amaranth seeds (Broekaert *et al.*, 1992), IWF4 from sugar beet leaves (Nielsen *et al.*, 1997), An1 from *Atriplex nummularia* seeds (Last and Llewellyn, 1997), and Pn-AFPs from Japanese morning glory seeds (Koo *et al.*, 1998) WGA, the wheat lectin with a quadruple hevein-like domain, lacks antimicrobial activity (Schlumbaum *et al.*, 1986) but instead has insecticidal activity (Heusing *et al.*, 1991). The antimicrobial spectrum of Ac-AMP1 and Ac-AMP2 is relatively broad and includes all of 7 tested fungal species, both of two tested Gram-positive bacteria, but none of two tested Gram-negative bacteria (Broekaert *et al.*, 1992). The antimicrobial activity of these peptides is antagonised when the concentration of divalent cations in the medium is raised above 1 mM or when the concentration of monovalent cations is raised above 50 mM (Broekaert *et al.*, 1992; Cammue *et al.*, 1995). In media containing 1 mM $CaCl_2$ and 50 mM KCl, these peptides do not affect fungal growth at concentrations below 100 µg/ml, whereas they generally inhibit fungi at concentrations below 10 µg/ml in low ionic strength media (Broekaert *et al.*, 1992).

Most of the above-mentioned antimicrobial peptides have been shown to bind to chitin (Broekaert *et al.*, 1989, 1992; Nielsen *et al.*, 1997; Van Parijs *et al.*, 1991). However, it is not known whether their antifungal activity is really due to interference with chitin metabolism in fungi. The fact that their activity is antagonised by cations rather suggest that they interfere with the plasma membrane, as such antagonism is often observed for peptides that are active on membranes (Cociancich *et al.*, 1993; Lehrer *et al.*, 1988).

G.   LIPID TRANSFER PROTEINS, STORAGE 2S ALBUMINS AND PUROINDOLINES

1.   *Structural Features*

The ~9 kDa plant lipid transfer proteins are capable of in vitro shuffling of different types of phospholipids, fatty acids or acyl-CoAs between artificial membranes or from liposomes to mitochondria. These proteins were originally purified based on this activity (Kader *et al.*, 1996; Kader, 1997). Due to lack of specificity for the phospholipids they can transport, these proteins are referred to as nonspecific lipid transfer proteins (nsLTPs). It was originally proposed that nsLTPs play a role in the intracellular trafficking of phospholipids from the endoplasmic reticulum, the site of phospholipid biosynthesis, to other cellular organelles. However, this hypothesis has now been largely abandoned because it is not consistent with a number of features of nsLTPs, e.g. the fact that they are not located in the cytosol (see also below). A number of nsLTPs possess antimicrobial activity (Cammue *et al.*, 1995; Molina *et al.*, 1993b; Nielsen *et al.*, 1996a; Segura *et al.*, 1993; Terras *et al.*, 1992a) while others inhibit animal α-amylases (Bernhard and Somerville, 1989; Campos and Richardson, 1984), and hence a role for these proteins in plant defense appears to be more likely (Garcia-Olmedo *et al.*, 1995).

All plant nsLTP cDNA/genes characterized so far encode preproteins (Kader *et al.*, 1996), thus excluding deposition in the cytosol. The sole exception is a cDNA encoding Ace-AMP1, an antimicrobial protein from onion seeds which highly resembles nsLTPs and which is synthesized as a preproprotein with a carboxy-terminal propeptide (Cammue *et al.*, 1995). Although Ace-AMP1 shares many of the consensus amino acids with nsLTPs, it is unable to shuffle phospholipids in the classical lipid transfer activity assays (Cammue *et al.*, 1995).

The 2S albumins are a class of highly abundant dicot seed storage proteins consisting of a small (~4 kDa) subunit linked by two disulfides to a large (~8 kDa) subunit (Shewry *et al.*, 1995). Loosely homologous monomeric proteins also exist in monocots, where they act as inhibitors of trypsin (Odani *et al.*, 1983), α-amylase (Gautier *et al.*, 1990) or both (Yu *et al.*, 1988).

Puroindolines are basic cysteine-rich proteins (~13 kDa) isolated from wheat endosperm and carry a unique tryptophan-rich domain. Complementary DNA cloning revealed that there are at least two isoforms which are 55% similar (Blochet *et al.*, 1993; Gautier *et al.*, 1994).

The 2S albumins and puroindolines are synthesized as preproproteins. After entry in the ER, both are proteolytically processed at their amino-termini as well as at their carboxy-termini. An internal propeptide present in a solvent-exposed loop (Rico *et al.*, 1996) is additionally removed from the 2S albumin precursor, thus generating the two subunits (Gautier *et al.*, 1994; Krebbers *et al.*, 1988). Processing of the 2S albumin propeptides is believed to be the result of cooperation between several vacuolar proteases, one of which is an aspartic endoprotease (D'Hondt *et al.*, 1993; Hiraiwa *et al.*, 1997).

No significant amino acid sequence homology exists between nsLTPs, 2S albumins and puroindolines. When only taking the cysteine pattern into account, however, a remarkable similarity is observed (see Fig. 12). Also included in Fig. 12 are a number of other proteins displaying a similar cysteine pattern. One of these is a hydrophobic protein from soybean (Odani *et al.*, 1987). The maize hybrid proline-rich protein (HyPRP; Josè-Estanyol *et al.*, 1992) is a mosaic protein consisting of an amino-terminal proline-rich domain with sequence motifs common to the soybean proline-rich protein (Hong *et al.*, 1989) and the maize extensin (Stiefel *et al.*, 1990). The carboxy-terminal domain of this protein has an nsLTP-like cysteine-patterning (see Fig. 12; Josè-Estanyol *et al.*, 1992). Similar hybrid proteins exist in tomato (Salts *et al.*, 1991) and tobacco (Wu *et al.*, 1993). A family of anther-/tapetum-specific proteins can also be classified as nsLTP-like based on their cysteine topology, although the amino acid stretches interspersing several cysteine residues are significantly shorter (see Fig. 12; Aguirre and Smith, 1993; Crossley *et al.*, 1995; Nacken *et al.*, 1991; Paul *et al.*, 1992).

Three-dimensional structures have been determined by X-ray diffraction for the soybean hydrophobic protein (Baud *et al.*, 1993) and the maize nsLTP (Shin *et al.*, 1995) and by proton NMR for the wheat nsLTP (Gincel *et al.*, 1994; see Fig. 1) and a 2S albumin (Rico *et al.*, 1996). Although the number of cysteines and/or the cysteine-interspacing and/or the cystine connections are different in the three proteins (see Fig. 12), they adopt a similar folding dominated by a four -α-helix bundle. In the 2S albumin, this motif is formed by the carboxy-terminal α-helix of the small subunit and the three α-helices of the large subunit. An additional fifth α-helix is formed by the amino-terminal half of the small subunit (Rico *et al.*, 1996). In the nsLTPs, the α-helices surround a hydrophobic tunnel (diameter ~5 Å, length ~15 Å) in which fit 12 carbon atoms of a single fatty acid acyl chain (Gomar *et al.*, 1996; Shin *et al.*, 1995). The hydrophobic tunnel also seems to be present in the soybean

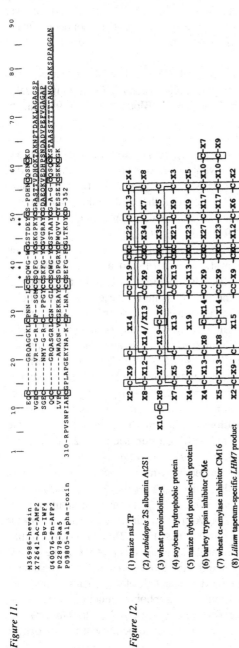

Figure 11.

```
                    1          10         20         30         40         50         60         70         80         90
                    |    |     |    |     |    |     |    |     |    |     |    |     |    |     |    |     |    |     |

M36986-hevein                  EQC----GRQAGGKLC--LCCSQWG-WGSTDEYC--SPDHNC--QSNC-KD
X72641-Ac-AMP2                 VGEC----VR--G--R--CP--SGMCCSQFG-YCGKGPKYCGRASTVDHQKTAKNPTDAKAGAGSP
-Bv-IWF4                       SGEC----NMY-G-R--C--PPGYCCSKFG-YCGVGRAYCDAEQKVEDHFSNDADVPFVGAGAP
U40076-Pn-AFP2                 QQC----GRQASGRLCGCN-GLCCSQWG-YCGSTAAYC-A-GQSQCKSTAASSTTTTANQSTAKSDPRAGGAN
P02878-Ra5                     LVEC----AWAGN-VCSEKRAYCCSDPGRCPWQVV--CYESSEIGSKKGEK
P09805-alpha-toxin             310-RPVSNPIAECGPLAPGEKYNA-K-CP-LNA-CSEFG-YCGLTKDYC-352
```

Figure 12.

<table>
<tr><td>(1) maize nsLTP</td></tr>
<tr><td>(2) Arabidopsis 2S albumin At2S1</td></tr>
<tr><td>(3) wheat puroindoline-a</td></tr>
<tr><td>(4) soybean hydrophobic protein</td></tr>
<tr><td>(5) maize hybrid proline-rich protein</td></tr>
<tr><td>(6) barley trypsin inhibitor CMe</td></tr>
<tr><td>(7) wheat α-amylase inhibitor CM16</td></tr>
<tr><td>(8) Lilium tapetum-specific LHM7 product</td></tr>
</table>

Figure 11. **Alignment of hevein-like protein sequences.**

Aligned are the mature protein sequences of hevein (*Hevea brasiliensis*, GenBank M36986; Broekaert *et al.*, 1990), Ac-AMP2 with its carboxy-terminal propeptide (*Amaranthus caudatus* antimicrobial protein 2, GenBank X72641; De Bolle *et al.*, 1993a), Bv-IWF4 with its carboxyterminal propeptide (*Beta vulgaris* intercellular washing fluid protein 4: Nielsen *et al.*, 1997), Pn-AFP2 with its carboxy-terminal propeptide (*Pharbitis nil* antifungal protein 2, GenBank U40076: Koo *et al.*, 1998), mature Ra5 (*Ambrosia artemisiifolia* var. *elatior*, ragweed, pollen allergen 5, SwissProt P02878: Mole *et al.*, 1975), and an internal part of alpha-toxin (*Kluyveromyces lactis* killer toxin α-subunit chitin-binding domain, SwissProt P09805; Stark and Boyd, 1986). Residue numbers of the first and last residue of the Killer α-subunit portion are shown before and behind the sequence. Gaps introduced to obtain an optimal alignment are indicated by hyphens. Cysteines are bold-faced and boxed. Hevein-residues important for chitin-binding are indicated by asterisks above the hevein sequence. The Ac-AMP2, Bv-IWF4 and Pn-AFP2 carboxy-terminal propeptides are double underlined.

Figure 12. **Cysteine-patterning in nsLTPs, 2S albumins, puroindolines and related proteins.**

The cysteine-spacing and, where known, cysteine interconnections of an nsLTP, a 2S albumin, puroindoline-a and a set of related proteins are shown. The number of amino acids ('X') in between two consecutive cysteines is indicated after the 'X'. Cysteines are boxed. Only the cysteine-rich region of the maize hybrid proline-rich protein is represented. References: (1) Tchang *et al.*, 1988; (2) Krebbers *et al.*, 1988; (3) Blochet *et al.*, 1993. (4) Odani *et al.*, 1987; (5) José-Estanyol *et al.*, 1992; (6) Odani *et al.* 1983; (7) Gautier *et al.*, 1990; (8) Crossley *et al.*, 1995.

hydrophobic protein but the carboxy-terminal tyrosine essential for binding the fatty acid or phospholipid carboxyl- or carbonyl-group in nsLTPs is exposed to the solvent in the soybean protein and a polar cluster of charged residues involved in the stabilization of the lipid-protein complex is missing (Gomar *et al.*, 1996). Both features most likely exclude a lipid carrier function for the soybean hydrophobic protein (Gomar *et al.*, 1996). Interestingly, Ace-AMP1, the nsLTP-like antimicrobial protein which lacks lipid transfer activity, has a very narrow hydrophobic cavity that is obstructed by bulky aromatic residues (Tassin *et al.*, 1998). Similarly, no hydrophobic space is present in the core of the 2S albumin structure and this protein also lacks the carboxy-terminal tyrosine that is essential for lipid-binding (Rico *et al.*, 1996). Secondary structure determination of the puroindolines indicates that these proteins as well exhibit the four-α-helix bundle folding motif (Le Bihan *et al.*, 1996). Interestingly, this motif is also found in some non-homologous non-plant pore-forming or membranolytic proteins: amoebapore produced by *Entamoeba histolytica* (Dandekar and Leippe, 1997; Leippe and Müller-Eberhard, 1994), NK-lysin from porcine lymphocytes (Andersson *et al.*, 1995; Dandekar and Leippe, 1997) and ectatomin, a venom component of the selva ant (Arseniev *et al.*, 1994).

2. Antimicrobial Properties

Antimicrobial properties of nsLTPs were first reported by Terras *et al.* (1992a). Meanwhile, several reports have described the antimicrobial activity of nsLTPs from different plant sources (Cammue *et al.*, 1995; Molina *et al.*, 1993b; Nielsen *et al.*, 1996b; Segura *et al.*, 1993). The antifungal activity of nsLTPs can differ significantly between different members. For instance, an onion seed LTP (Ace-AMP1) is highly active against a broad range of fungi (with IC_{50} values ranging from 1 to 6 µg/ml), whereas a radish seed nsLTP is only moderately active (IC_{50} values ranging from 7 to 100 µg/ml) and maize and wheat seed nsLTPs are virtually inactive against most fungi (Cammue *et al.*, 1995). Also, the sensitivity of their antifungal activity to the presence of cations varies with the nsLTP type. The onion seed nsLTP was almost as potent in a low ionic strength medium as in the same medium supplemented with 1 mM Ca^{2+} and 50 mM K^+, whereas the radish nsLTP showed drastically reduced antifungal activity in the presence of cations (Cammue *et al.*, 1995). nsLTPs have also been shown to possess antibacterial activity. Those from barley and spinach leaves were highly active against the Gram-positive pathogen *Clavibacter michiganense* and the Gram-negative pathogen *Ralstonia (Pseudomonas) solanacearum* (Molina *et al.*, 1993b; Segura *et al.*, 1993). The most active LTP from barley leaves had an IC_{50} value of 0.6 µg/ml on *Clavibacter michiganense*. The onion seed LTP, which exhibits strong antifungal activity, was also shown to be inhibitory to Gram-positive bacteria but not to a range of Gram-negative bacteria. In contrast, none of the nsLTPs from the seed of onions, radishes, maize, or wheat caused lysis of human erythrocytes, nor did they affect the viability of cultured human fibroblasts (Cammue *et al.*, 1995). 2S-albumins, cereal trypsin or α-amylase inhibitors and puroindolines, which are all structurally related to nsLTPs (see above), have also been demonstrated to exert antifungal and antibacterial activities (Dubreil *et al.*, 1998; Halim *et al.*, 1973; Terras *et al.*, 1992b, 1993a, b). However, their antimicrobial potency is relatively weak and is extremely sensitive to the presence of cations, especially divalent cations.

One striking property shared by nsLTPs, 2S-albumins, cereal trypsin/α-amylase inhibitors and puroindolines is that they all strongly potentiate the antimicrobial activity of thionins (Molina et al., 1993b; Terras et al., 1993a; Dubreil et al., 1998), even in the presence of physiological concentrations of cations. On the fungus Alternaria brassicicola, for instance, a thionin from wheat seeds was inhibitory down to 0.32 µg/ml in the presence of 50 µg/ml of rapeseed 2S albumin, whereas the same thionin was only active down to 11 µg/ml in the absence of the 2S albumin, which corresponds to a potentiation by a factor of 34 (Terras et al., 1993a). Even the wheat seed nsLTP displaying little or no antifungal activity on its own, was able to potentiate the antifungal potency of thionins (Dubreil et al., 1998). Interestingly, the nsLTP from onion seeds, which is a very potent fungal growth inhibitor despite its lack of lipid transfer activity (Cammue et al., 1995), did not act synergistically with thionins, indicating that its mode of action may be different from that of the other nsLTPs and related proteins.

The most plausible site of interaction of nsLTPs and related proteins is the phospholipid membrane. The lipid transfer activity of nsLTPs implies that these proteins can at least transiently interact with phospholipid membranes. It was shown that the wheat seed nsLTP can interact with phospholipid monolayers in a surface-pressure dependent way (Suribade et al., 1995). However, wheat seed nsLTP was unable to permeabilise phospholipid bilayer vesicles whereas barley leaf nsLTP, wheat seed puroindoline and mustard seed 2S albumin clearly caused permeabilisation in such assays (Caaveiro et al., 1997; Onaderra et al., 1994; Tassin et al., 1998). This correlates with the observation that puroindoline, barley leaf nsLTP and mustard 2S albumins are antifungal agents whereas wheat seed nsLTP is almost inactive (Cammue et al., 1995; Dubreil et al., 1998; Molina et al., 1993b; Terras et al., 1993b). Interestingly, the ability of barley leaf nsLTP to permeabilise phospholipid vesicles was strongly antagonised by cations, which again is consistent with the antagonistic effect of cations on the antimicrobial activity of nsLTPs (Caaveiro et al., 1997). When the ability of the onion nsLTP to permeabilise phospholipid bilayer vesicles was compared to that of puroindoline, it was found that it was less active in this assay, although it is far superior in its antimicrobial activity (Tassin et al., 1998). This may again indicate that the onion nsLTP acts on a different target, perhaps a membrane-located protein or a nonphospholipid membrane lipid.

H.   THIONINS

1.   Structural Features

Thionins are small (~5 kDa) cysteine-rich peptides first identified in wheat flour (Balls et al., 1942). Since then, several thionins have been identified in various monocot and dicot species. All cysteines present in thionins are involved in disulfide bridges (Bohlmann, 1994). According to the number of cysteines, thionins can be separated in two main classes: the eight-cysteine type thionins and the six-cysteine type thionins (Broekaert et al., 1997). A third class, also known as type V thionins, is represented by a wheat thionin which, compared to eight-cysteine and six-cysteine type thionins, lacks a carboxy-terminal segment and has a different cysteine arrangement (Castagnero et al., 1992, 1995). Eight-cysteine type thionins have been purified from seeds and leaves of various monocots (Békés and Lásztity, 1981; Bohlmann *et al.*, 1988; Ebrahim-Nesbat *et al.*, 1989; 1991; 1993; Hernandez-Lucas *et al.*, 1978; Jones and Cooper, 1980; Reimann-Philipp *et al.*, 1989; Redman and Fisher, 1968, 1969) and dicots (Vernon *et al.*, 1985; Schrader-Fisher and Apel, 1993). Six-cysteine type have so far only been discovered in dicot plants (Epple *et al.*, 1995; Mellstrand and Samuelsson, 1974; Samuelsson, 1974; Samuelsson and Petterson, 1977; Schrader-Fisher and Apel, 1993, 1994; Terras *et al.*, 1996; Thunberg and Samuelsson, 1982). The amino acid sequences of several thionins is presented in Fig. 13. Overall, there is relatively little sequence conservation among the different thionins as fully conserved residues are restricted to four cysteines, a basic residue at position 10 and an aromatic residue at position 14 (numbering as in Fig. 13).

Irrespective of thionin type or source, cDNAs and genes predict that the mature thionins are derived from preproprotein precursors (~18 kDa) with an amino-terminal signal peptide and an acidic carboxy-terminal propeptide (Bohlmann *et al.*, 1988; Castagnaro *et al.*, 1992; Epple *et al.*, 1995; Ponz *et al.*, 1986; Schrader and Apel, 1991; Schrader-Fischer and Apel, 1993, 1994). Within a species, the carboxy-terminal propeptides are highly conserved, even when attached to thionin domains of a different type (Castagnaro *et al.*, 1992; Schrader-Fischer and Apel, 1993, 1994). Homology also exists between the acidic propeptides of thionins from different species but is less pronounced. The removal of the carboxy-terminal propeptide is a post-translational process (Ponz *et al.*, 1983) and a vacuolar proteinase capable of processing thionin precursors of different species has been isolated from barley leaves (Romero *et al.*, 1997). The intact propeptide could not be detected whereas the mature thionin was found in the vacuole (Romero *et al.*, 1997). The acidic propeptide is generally thought to neutralize the basic thionin domain and, as such, to protect cells from the cytotoxic effect of thionins until they are sequestered extracellularly or in the vacuole (Bohlmann, 1994). A consensus sequence of the acidic propeptide is also given in Fig. 13.

Three-dimensional structures have been determined for six-cysteine type thionin from *Crambe abyssinica* seeds (Brünger *et al.*, 1987) and from mistletoe leaves (Lecomte *et al.*, 1987), and for eight-cysteine type thionins from wheat and barley grains (Lecomte *et al.*, 1982; Rao *et al.*, 1995a; Stec *et al.*, 1995). In all cases, an L-shaped amphipathic molecule is drawn in which the long arms are formed by two antiparallel α-helices and the short arms by two antiparallel β-strands. A loop is present at the carboxy-terminus of the thionins (see Fig. 1 for the wheat α_1-purothionin).

2.   Antimicrobial Properties

It has been known for over 50 years that thionins inhibit the growth of bacteria and fungi in vitro (Stuart and Harris, 1942). Fernandez de Caleya *et al.* (1972) first demonstrated that thionins are inhibitory to a number of plant pathogenic bacteria, thus suggesting that thionins may fulfill a protective role *in planta*. In subsequent studies, the antimicrobial activity spectrum of thionins has been analyzed more extensively and shown to include several Gram-positive and Gram-negative plant pathogenic bacteria (Cammue *et al.*, 1992; Florack *et al.*, 1993; Molina *et al.*, 1993a), as well as several phytopathogenic fungi and Oomycetes, with IC_{50} values generally ranging from 1 to 15 µg/ml (Cammue *et al.*, 1992; Molina *et al.*, 1993a). The antifungal activity of

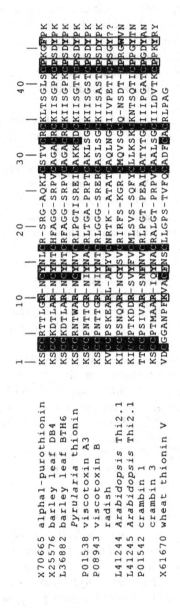

```
                              1              10            20            30            40
X70665 alpha1-purothionin     KSCCRTTLGR-NCYNLCR-SRG-AQKLCSTVCRCKLTSGLSCPKGFPK
X25576 barley leaf DB4        KSCCKDTLAR-NCYNTCHFAGG-SRPVCAGACRCKIISGPKCPSDYPK
L36882 barley leaf BTH6       KSCCKDTLAR-NCYNTCRFAGG-SRPVCAGACRCKIISGPKCPSDYPK
       Pyrularia thionin      KSCCRNTWAR-NCYNVCRLPGTISREICAKKCDCKIISGTTCPSDYPK
P01538 viscotoxin A3          KSCCPNTTGR-NIYNACRLTGA-PRPTCAKLSGCKIISGSTCPSDYPK
P08943 viscotoxin B           KSCCPNTTGR-NIYNTCRLGGG-SRERCASLSGCKIISASTCPSDYPK
       radish                 KVCCPSKEARL-AFYVCNRTK--ATATCAQLNGCIIVPETICPSGY??
L41244 Arabidopsis Thi2.1     KICCPSNQAR-NGYSVCRIRFS-KGR-CMQVSGQ-NSDT-CPRGWVN
L41245 Arabidopsis Thi2.1     KICCPTKDDR-SVYFVCMLSVS-SQFYCLLKSKNTSQTICPPGYTN
P01542 crambin 1              TTCCPSIVAR-SNFNVCRLPGT-PEAICATYTGCIIIPGATCPGDYAN
       crambin 3              KSCCPTMAAR-IQYNACRALGT-PRPVCAALSGCIILDVTKCDPKRY
X61670 wheat thionin V        VDCGGANPFKVACENSCLLGPS-TVFQCADFCACRLPAG
```

consensus acidic carboxy-terminal propeptide

X(2-15)-C-X2-G-C-X2-S-X-C-X(12-21)-C-X6-C-X(6-14)

Figure 13. **Alignment of thionin protein sequences.**
Aligned are the amino acid sequences of the mature thionins as deduced from cDNAs or genes or as experimentally determined. Available accession numbers (GenBank or SwissProt) are indicated on the left. Gaps introduced for optimal alignment are indicated by hyphens. Cysteines are shown in black boxes. Identical/homologous amino acids occurring in all sequences are bold-faced and boxed. The carboxy-terminal amino acids of the radish thionins are not known. The consensus sequence of the acidic carboxy-terminal propeptide is also given: X denotes any amino acid, the number following the X indicates the number of amino acids between two consecutive cysteines. All three thionin subgroups are represented: eight-cysteine type thionins: α_1-purothionin (Castagnaro *et al.*, 1994), barley leaf DB4 and BTH6 (Bohlmann and Apel, 1987; Holtorf *et al.*, 1995), and *Pyrularia* thionin (Vernon *et al.*, 1985); six-cysteine type thionins: viscotoxins A3 and B (Schrader and Apel, 1991), radish thionin (Terras *et al.*, 1996), the *Arabidopsis* thionins (Epple *et al.*, 1995), crambin 1 and crambin 3 (Schrader-Fischer and Apel, 1994); type V thionins: thionin wheat thionin V (Castagnaro *et al.*, 1992).

thionins is inhibited by Ca^{2+} at concentrations above 5 mM, but not by Mg^{2+} or Ba^{2+} at concentrations up to 10 mM, nor by monovalent cations at concentrations up to 50 mM (Okada *et al.*, 1970; Terras *et al.*, 1992b; Cammue *et al.*, 1995). As mentioned in section V.G.2., the antifungal potency of thionins is strongly enhanced in the presence of nsLTPs and structurally related proteins.

Some Gram-negative bacteria such as a number of *Pseudomonas*, *Ralstonia* and *Erwinia* species are apparently insensitive to thionins (Fernandez de Caleya *et al.*, 1972; Cammue *et al.*, 1992; Florack *et al.*, 1993; Pineiro *et al.*, 1995). Titarenko *et al.* (1997) have demonstrated that, at least in the case of *Ralstonia* (*Pseudomonas*) *solanacearum*, insensitivity to thionins is due to the protective effect of particular lipopolysaccharides.

In addition to their effects on microorganisms, thionins have also been shown to exert adverse effects, often involving permeabilization, on various cultured mammalian and insect cells and plant protoplasts (reviewed in Florack and Stiekema, 1994). Moreover, thionins are toxic to insects (Kramer *et al.*, 1979) and mammals (Coulson *et al.*, 1942; Rosell and Samuelsson, 1966; Evett *et al.*, 1986; Mellstrand and Samuelsson, 1973) when injected in their body fluids, but not when administered orally (Coulson *et al.*, 1942; Samuelsson, 1974).

The mechanism underlying the antimicrobial activity of thionins was first investigated in the yeast *Saccharomyces cerevisiae* (Okada and Yoshizumi, 1973). It was found that a wheat seed thionin causes permeabilization of the yeast cells as evidenced by leakage into the culture medium of K^+, PO_4^{3-}, and cellular components absorbing at 265 and 280 nm. A drop in cytoplasmic PO_4^{3-} content is known to result in adenosine triphosphate (ATP) hydrolysis, which in turn entails a block of energy-requiring processes (Guihard *et al.*, 1993). Also for the filamentous fungus *Neurospora crassa* it was found that a barley seed thionin can cause a substantial release of ^{14}C-isoaminobutyric acid from hyphae preloaded with this nonmetabolizable compound (Thevissen *et al.*, 1996). Moreover, the dose-response curve of the isoaminobutyric acid release correlated well with that of fungal growth inhibition, suggesting that growth arrest may directly or indirectly result from the permeabilization process. Thionin was also shown to mediate a transient influx of Ca^{2+} into *Neurospora crassa* hyphae, a steady efflux of K^+, and a steady alkalinization of the culture medium, the latter possibly due to increased H^+ influx. All these ion fluxes could be observed within 1 min after addition of the protein (Thevissen *et al.*, 1996). It was also shown that a combination of thionins and 2S albumins that exerted synergistic antifungal effects also caused a synergistic increase of K^+ efflux from hyphae of *Fusarium culmorum* (Terras *et al.*, 1993a). The changes in the permeability of the fungal membrane toward ions and other small solutes may be the result of a direct interaction between membrane phospholipids and thionins. Indeed, using artificial planar lipid bilayers it was shown that thionins can alter electrical and physicochemical properties of the membrane at relatively low concentrations (1 µg/ml). This process does not seem to involve the formation of bona fide ion channels by the thionins since current-time plots showed highly irregular current spikes, but not square-like fluctuations as usually observed for bona fide channel-forming proteins (Thevissen *et al.*, 1996). The irregular current spikes may result from the formation of transient pores, as also suggested by kinetic measurements of the release of fluorochromes from thionin-treated phospholipid vesicles (Caaveiro *et al.*, 1998). The interaction between thionins and membranes may require specific phospholipid domains, most likely consisting of patches with acidic phospholipids (Vernon, 1992). A radiolabeled thionin from *Pyrularia pubera* was shown to bind efficiently to artificial liposomes composed of phosphatidylserine (Vernon, 1992). In addition, physicochemical analysis of the incubation of this thionin with phospholipid bilayers at varying lipid compositions has shown that inclusion of either phosphatidylserine or phosphatidylglycerol in the bilayers makes them more susceptible to thionin-induced structural disturbance and permeabilisation (Caaveiro *et al.*, 1997; Gasanov *et al.*, 1993). As both thionins and the nsLTP/2S albumin/puroindoline superfamily proteins interact with phospholipid bilayers, it is likely that the synergy between these proteins is based on cooperativity in the binding to or the disturbance of the membranes.

I.   PLANT DEFENSINS

1.   Structural Features
Plant defensins are highly basic ~5kDa peptides containing 8 cysteine-residues involved in 4 disulfide bridges (Broekaert *et al.*, 1995). The first purified plant-defensin-like proteins originate from barley and wheat grains and were called γ-thionins (Colilla *et al.*, 1990; Mendez *et al.*, 1990) despite the low similarity to true thionins. Peptides homologous to γ-thionins were soon thereafter isolated from sorghum seeds and characterized as inhibitors of insect α-amylases (Bloch and Richardson, 1991; Nitti *et al.*, 1995). Gamma-thionins also partially inhibit human saliva α-amylase (Méndez *et al.*, 1996). The first peptides of this protein family that were shown to exert antimicrobial activity were isolated from radish seeds and were termed Rs-AFP1 and Rs-AFP2 (*Raphanus sativus* antifungal proteins; Terras *et al.*, 1992b). The radish seed Rs-AFPs are readily released into the surrounding medium during germination and can protect the young seedlings against fungal infection (Terras *et al.*, 1995). Based on their protective effect and their homology to insect defensins (see below), the name plant defensins was proposed to denote this class of peptides (Terras *et al.*, 1995; Broekaert *et al.*, 1995). Plant defensins that exhibit antimicrobial activity have now been isolated from numerous dicot plants (Kragh *et al.*, 1995; Moreno *et al.*, 1994; Osborn *et al.*, 1995; Penninckx *et al.*, 1996; Segura *et al.*, 1998; Terras *et al.*, 1993b). In addition, many more plant defensins with unknown biological activity were identified via cDNA cloning (Albani *et al.*, 1990; Brandstädter *et al.*, 1996; Chiang and Hadwiger, 1991; Choi *et al.*, 1995; Domon *et al.*, 1990; Epple *et al.*, 1997a; Gu *et al.*, 1992; Ishibashi *et al.*, 1990; Meyer *et al.*, 1996; Milligan and Gasser, 1995; Karunanandaa *et al.*, 1994; Sharma and Lönneborg, 1996; Stanchev *et al.*, 1996).

Based on an analysis of the available plant defensin cDNAs, the vast majority of plant defensins are made as preproteins. Only a plant defensin occurring in tobacco flowers (Gu *et al.*, 1992) and a plant defensin from sunflower anthers (Domon *et al.*, 1990) feature a carboxy-terminal domain that may or may not act as a propeptide.

Comparison of a series of sequences of plant defensins from various plant species reveals that there is very little sequence conservation outside the cysteine residues and some glycine residues (Fig. 14). The structural hallmark of plant defensins is the cysteine-stabilized αβ motif characterized by the CXXXC, GXC and CXC blueprint (with X being any amino acid) (see Fig. 14; Cornet *et al.*, 1995). The motif is indicative for a three-dimensional structure consisting of an α-helix (containing CXXXC) connected by two disulfides to the second (containing CXC) of two consecutive β-strands after the α-helix. The first of these consecutive β-strands contains the GXC submotif and its cysteine is involved in a disulfide bond with a cysteine-residue in the amino-terminal loop. A mammalian peptide with vasoconstricting activity, endothelin (Yanagisawa *et al.*, 1988), was the first protein in which this motif was recognized (Kobayashi *et al.*, 1991). This motif was subsequently found in insect defensin A, a 40-residue cysteine-rich antibacterial peptide accumulating in the haemolymph of the fleshfly after bacterial infection (Cornet *et al.*, 1995; Lambert *et al.*, 1989). A highly homologous insect defensin was later isolated from bacterially challenged *Drosophila* (Dimarcq *et al.*, 1994). Another protein that is induced in bacterially infected *Drosophila* and which contains the cysteine-stabilized αβ motif is drosomycin. Drosomycin is a potent antifungal peptide and is remarkably homologous (38%) to the antifungal plant defensin Rs-AFP1 (Fehlbaum *et al.*, 1994; Fig. 13). Interestingly, *Drosophila* mutants affected in their capacity to induce the antifungal drosomycin were found to be highly sensitive to fungal infection, whereas another mutant that was unable to synthesize the antibacterial insect defensin as well as other antibacterial peptides in response to bacterial challenge, was highly susceptible to bacterial infection (Lemaitre *et al.*, 1995, 1996). This clearly illustrates the importance of antimicrobial peptides in insect immunity.

The 3D-structure has been determined of a number of plant defensins including γ1-purothionin, γ1-hordothionin, ω-hordothionin and Rs-AFP1 (Bruix *et al.*, 1993, 1995; Fant *et al.*, 1998; see Fig. 1). All these plant defensins share a similar fold which is dominated by an α-helix packed against a triple-stranded β-sheet. As predicted by the primary structure homology, this structure is very similar to the one adopted by drosomycin (Landon *et al.*, 1997). The plant defensin fold is also similar to that of insect defensin (Bonmatin *et al.*, 1992), except that the latter features a double-stranded β-sheet instead of a triple-stranded one. In addition, the polypeptide backbone fold of plant defensins is also comparable to that of some scorpion toxins (e.g. Darbon *et al.*, 1991; Fontecilla-Camps *et al.*, 1988; Landon *et al.*, 1996).

2. Antimicrobial Properties

The first plant defensins that were demonstrated to possess antifungal activity were the two plant defensin isoforms Rs-AFP1 and Rs-AFP2 isolated from radish seeds (Terras *et al.*, 1992b). To date, a whole range of plant defensins have been thoroughly analyzed to determine their antimicrobial activity spectrum. Based on the antimicrobial effects observed on fungi, at least two groups of plant defensins can be distinguished. The "morphogenic" plant defensins cause reduced hyphal elongation with a concomitant increase in hyphal branching, whereas the "nonmorphogenic" plant defensins only slow down hyphal elongation but do not induce marked morphological distortions. The antifungal activity of plant defensins, whether morphogenic or not, is reduced by increasing the ionic strength of the fungal growth assay medium. This antagonism was found to be due to the cations, with divalent cations being at least one order of magnitude more potent than monovalent cations (Terras *et al.*, 1992b, 1993b). Ca^{2+}, Mg^{2+}, and Ba^{2+} are equally effective at inhibiting the antifungal activity of plant

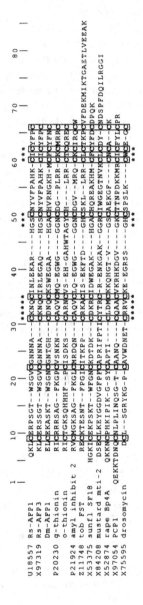

Figure 14. **Alignment of plant defensin and plant defensin-like protein sequences.**

Aligned are the amino acid sequences of the mature or proprotein domains of the plant defensin (-like) proteins as deduced from cDNAs or genes or as experimentally determined. Available accession numbers are indicated on the left. Gaps introduced for optimal alignment are indicated by hyphens. Cysteines are boxed. Cysteine-stabilized α-helix footprints are indicated by asterisks above and under the sequences. Rs-AFP1 and 3: *Raphanus sativus* antifungal proteins 1 and 3 (Terras *et al.*, 1995); Dm-AFP1: *Dahlia merckii* antifungal protein 1 (Osborn *et al.*, 1995); g-thionin: barley γ-thionin (Méndez *et al.*, 1990); o-thionin: barley ω-thionin (Méndez *et al.*, 1996); amyl inhibit 2: sorghum α-amylase inhibitor 2 (Bloch and Richardson, 1991); tob FST: tobacco flower-specific thionin (Gu *et al.*, 1992); sunfl SF18: sunflower cDNA SF18 (Domon *et al.*, 1990); mustard mti2: *Sinapis alba* trypsin inhibitor 2 (Ceci *et al.*, 1995); rape Bp4C: *Brassica napus* pollen-coating protein (Albani *et al.*, 1990); PCP1: *B. oleracea* pollen-coating protein (Stanchev *et al.*, 1996) and drosomycin: *Drosophila* antifungal protein (Fehlbaum *et al.*, 1994).

defensins (Terras *et al.*, 1993b; Broekaert WF unpublished results). The antagonistic effect of cations is strongly dependent on the fungus and on the plant defensin type (Osborn *et al.*, 1995). When comparing different plant defensins for their relative activity on various fungi, marked differences in activity spectrum can be observed, and such differences are even accentuated after increasing the ionic strength of the medium. For instance, in a potato dextrose broth medium supplemented with 1 mM $CaCl_2$ and 50 mM KCl, the plant defensin from horse chestnut (Ah-AMP1) is active against *Leptosphaeria maculans* with an IC_{50} value of 6 µg/ml, but is basically inactive against *Fusarium culmorum*. However, a plant defensin from *Heuchera sanguinea* (Hs-AFP1) is inactive against the former fungus, but inhibits the latter with an IC_{50} value of 3 µg/ml (Osborn *et al.*, 1995).

Some plant defensins also exert antibacterial activity. For instance a *Clitoria ternatea* plant defensin (Ct-AMP1) is active against *Bacillus subtilus* (Osborn *et al.*, 1995) and plant defensins from potato tubers and spinach leaves inhibit *Ralstonia* (*Pseudomonas*) *solanacearum* and *Clavibacter michiganense* (Moreno *et al.*, 1994; Segura *et al.*, 1998). So far, none of the plant defensins has been found to cause detrimental effects on cultured human or plant cells (Terras *et al.*, 1992b; Broekaert WF and Cammue BPA, unpublished results).

It is not known at present how exactly plant defensins inhibit fungal growth. It was shown that the plant defensins from radish seed (Rs-AFP2) and dahlia seed (Dm-AMP1) had no effect on electrical current measured in artificial phospholipid membrane systems (Thevissen *et al.*, 1996) nor did a plant defensin from potato tubers cause permeabilisation of artificial phospholipid vesicles (Caaveiro *et al.*, 1997). Hence, permeabilization through direct protein-phospholipid interactions does not seem to be the primary cause of fungal growth inhibition by plant defensins, unlike in the case of thionins. On the other hand, plant defensins did cause a marked and sustained increased Ca^{2+} influx when added to *Neurospora crassa* hyphae, reaching up to tenfold the level of Ca^{2+} influx in untreated hyphae (Thevissen *et al.*, 1996). Concomitant with the Ca^{2+} influx, the plant defensins caused efflux of K^+ and alkalinization of the external medium (Thevissen *et al.*, 1996). Such ion fluxes were not observed with a recombinant Rs-AFP2 variant in which the tyrosine at position 38 was replaced by a glycine and which lacked substantial antifungal activity (De Samblanx *et al.*, 1997). Two other single amino substitution variants with increased antifungal potency caused enhanced Ca^{2+} influx (De Samblanx *et al.*, 1997). Our current view is that the ion fluxes observed upon treatment of fungi with plant defensins are indicative of membrane permeabilisation initiated after the binding of plant defensins to a particular plasma membrane nonphospholipid component. High affinity binding sites for Hs-AFP1, the plant defensin from *Heuchera sanguinea* seeds, have recently been demonstrated to reside on the plasma membrane of *Neurospora crassa* hyphae (Thevissen *et al.*, 1997). Interestingly, Rs-AFP2 was able to compete with radiolabeled Hs-AFP1, whereas the inactive Rs-AFP2 variant with the substitution at position 38 was unable to displace Hs-AFPs from the binding sites. The presence of particular binding sites for plant defensins in fungal membranes is consistent with the observation that these proteins exert specificity in their antimicrobial activity.

J.   MISCELLANOUS CYSTEINE-RICH ANTIMICROBIAL PEPTIDES

1.   *Mirabilis jalapa Antimicrobial Peptides (Mj-AMPs)*

Two basic cysteine-rich peptides that are 90% identical to one another were isolated
from seeds of the four o' clock plant (*M. jalapa* L.) using an assay for detecting in vitro
antifungal activity. These peptides, termed Mj-AMP1 and Mj-AMP2, are about 4 kDa
in size and contain 6 disulfide linked cysteines (Cammue *et al.*, 1992).

Mj-AMPs were reported to share homology with μ-agatoxins, spider venom peptides
neurotoxic to insects (Cammue *et al.*, 1992; Skinner *et al.*, 1989). This homology can now
be extended to ω-conotoxins from marine snails (Yanagawa *et al.*, 1988), scorpion venom
peptides (Fazal *et al.*, 1989), an amaranth seed α-amylase inhibitor (Chagolla-Lopez
et al., 1994), a sweet-taste-suppressing peptide from leaves of *Gymnema sylvestra* (Kamei
et al., 1992), squash-type trypsin inhibitors (Heitz *et al.*, 1989; Wilusz *et al.*, 1983), potato
carboxypeptidase inhibitor (Clore *et al.*, 1987) and, the race-specific elicitor AVR9 of
the fungal plant pathogen *Cladosporium fulvum* (de Wit, 1992). The overall homology
between these peptides with very different biological activities is low except for the
arrangement of their six cysteines (Cammue *et al.*, 1992; Chagolla-Lopez *et al.*, 1994;
Vervoort *et al.*, 1997) and where determined, their three-dimensional folding (Bode
et al., 1989; Chagolla-Lopez *et al.*, 1994; Clore *et al.*, 1987; Heitz *et al.*, 1989; Nemoto
et al., 1995; Omecinsky *et al.*, 1996; Sevilla *et al.*, 1993; Vervoort *et al.*, 1997).
The above-mentioned peptides indeed all contain the so-called cystine knot motif in
which two disulfides form a ring through which the third disulfide runs. This cystine
knot motif holds three antiparallel β-strands rigidly together (Le-Nguyen *et al.*, 1990).
As outlined by Chagolla-Lopez *et al.* (1994), the cellulose-binding domain of fungal
cellobiohydrolases also contains the cystine-knot motif and these enzymes are indeed
folded as knottin-type proteins (Kraulis *et al.*, 1989). The chitin-binding hevein-type
proteins (see section V.F.1.) have their (first) six cysteines also arranged in a knottin-type
motif (Chagolla-Lopez *et al.*, 1994). However, their structures deviate from genuine
knottin-type structures in that a short α-helical segment is introduced in between the second
and third β-strand (Andersen *et al.*, 1993; Le-Nguyen *et al.*, 1990; Martins *et al.*, 1996).

2.   *Macadamia integrifolia Antimicrobial Peptide (MiAMP1)*

From the nut kernels of the Australian native plant *Macadamia integrifolia,* an ~8
kDa peptide was isolated on the basis of its in vitro antifungal activity (Marcus *et
al.*, 1997). The peptide inhibits growth of several fungi, and Gram-positive, but not
Gram-negative, bacteria. The mature protein is derived from a signal-peptide containing
preprotein and no significant homology was found to any protein sequence present in
databases (Marcus *et al.*, 1997).

3.   *Zea mays Basic Peptide (MBP-1)*

A ~4 kDa arginine-rich peptide, called MBP-1, was isolated from maize kernels and
shown to contain four disulfide-linked cysteines (Duvick *et al.*, 1992). No homology
of MBP-1 with other proteins was reported but when we compared the MBP-1 protein
sequence with a series of animal defense peptides (listed in Lee and Prasad, 1995),
we found some limited homology with antimicrobial and antiviral peptides from
horseshoe crabs, the tachyplesins and polyphemusins (see Fig. 15; Miyata *et al.*,

1989; Morimoto *et al.*, 1991; Murakami *et al.*, 1991; Muta *et al.*, 1990; Nakamura *et al.*, 1988; Tamamura *et al.*, 1993). Although amino acid sequence homology exists between MBP-1 and these animal peptides, some important differences should also be noticed. Tachyplesins and polyphemusins are carboxy-terminally amidated and have an antiparallel β-sheet structure connecting cysteines 2 with 3 and 1 with 4 (Kawano *et al.*, 1990). MBP-1, however, has a free carboxy-terminus and is predominantly α-helical as determined by circular dichroism (Duvick *et al.*, 1992).

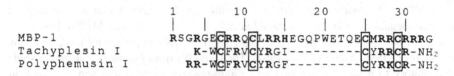

Figure 15. **Alignment of MBP-1 with animal MBP-1-like protein sequences.**
Aligned are the experimentally determined amino acid sequences of the maize basic protein 1 (MBP-1; Duvick *et al.*, 1992), and the horseshoe crab antimicrobial proteins tachyplesin 1 and polyphemusin 1 (Miyata *et al.*, 1989; Nakamura *et al.*, 1988). Gaps introduced to obtain optimal alignment are indicated

4. Impatiens balsamina Antimicrobial Peptides (Ib-AMPs)

The smallest (20 amino acids) plant antimicrobial peptides known to date were recently isolated from *Impatiens balsamina* seeds (Tailor *et al.*, 1997). Four isoforms, called Ib-AMP1 to 4, were identified which are highly basic and contain four disulfide-linked cysteines.

Weak sequence homology between the Ib-AMPs and the α-conotoxins was reported (Patel *et al.*, 1998) and it was also noted that Ib-AMPs have the same cysteine-pattern as the amino-terminal part of mature eight-cysteine type thionins (Tailor *et al.*, 1997). Three-dimensional structure determination of Ib-AMP1 revealed that this peptide is U-shaped with the bend formed by three consecutive β-turns (Patel *et al.*, 1998).

A cDNA was obtained encoding a complex Ib-AMP preproprotein. The proprotein consists of an amino-terminal propeptide (35 amino acids) followed by six Ib-AMP domains (Ib-AMP3, three times Ib-AMP1, Ib-AMP2 and Ib-AMP4) each of them interspersed by a 16- to 28-residue linker. The Ib-AMP4 domain is followed by a carboxy-terminal 35-residue extension (Tailor *et al.*, 1997). All propeptides contain di-acidic motifs (DD, EE, DE or ED) which are also found in the amino-terminal and internal propeptides of 2S storage albumins (Krebbers *et al.*, 1988) and in *Nicotiana alata* proteinase inhibitor precursor propeptides (Atkinson *et al.*, 1993).

The Ib-AMP precursor is so far the only example in plants of a single precursor delivering multiple antimicrobial peptide isoforms. However, a very similar multipeptide structure was also reported for the precursor of apidaecins, pathogen-inducible antimicrobial peptides from the honey bee (Casteels-Josson *et al.*, 1993). In the case of the apidaecin precursor, processing results in a set of isoforms with a slightly divergent antimicrobial activity spectrum (Casteels *et al.*, 1994). Multipeptide precursors can hence be regarded as a means to diversify and broaden the activity spectrum of defensive peptides.

5.   Antimicrobial Properties

Mj-AMP1 and Mj-AMP2 isolated from *Mirabilis jalapa* seeds have been shown to inhibit a broad range of fungi and Gram-positive bacteria but not two tested Gram-negative bacteria (Cammue *et al.*, 1992). The antimicrobial activity of these peptides is strongly antagonized by inorganic cations, especially divalent cations (Terras *et al.*, 1992b; Broekaert WF and Cammue BPA, unpublished results). The broad spectrum antimicrobial properties of the Mj-AMPs are also shared by the unrelated peptides from *Macadamia integrifolia* seeds (MiAMP1, Marcus *et al.*, 1997) and maize seeds (MBP-1, Duvick *et al.*, 1992). The maize seed MBP-1 in addition is also bactericidal against at least one Gram-negative bacterium, *Escherichia coli*. Both MiAMP1 and MBP-1 have been reported to be strongly antagonized in their antimicrobial properties by inorganic salts (Duvick *et al.*, 1992; Marcus *et al.*, 1997). The Ib-AMPs from *Impatiens balsamina* seeds are inhibitory to a broad range of fungi, Gram-positive bacteria and some but not all Gram-negative bacteria (Tailor *et al.*, 1997). Their antifungal activity is also antagonized by inorganic salts but the most basic isoform, Ib-AMP4, is still able to inhibit several fungi at concentrations between 6 and 50 µg/ml in a medium containing 1 mM $CaCl_2$ and 50 mM KCl. The Ib-AMPs induce marked morphological alterations to fungi, including increased branching and sometimes twisting of the hyphae, whereas such alterations are not observed in fungi treated with either Mj-AMPs or MiAMP1 (Tailor *et al.*, 1997; Cammue BPA and Broekaert WF, unpublished results). Hence, it is likely that the mode of action of Ib-AMPs differs from that of Mj-AMPs and MiAMP1.

Ackmowledgement

The authors acknowledge the financial support of research in their laboratory by the Commission of the European Community (AIR2-CT94-1356) and by the Fonds voor Wetenschappelijk Onderzoek - Vlaanderen (G.0218.97). FRG Terras is the recipient of a fellowship from the 'Collen Research Foundation'. The authors wish to thank Anita Vermassen for preparation of the manuscript and Dr. Franky Fant for his help in preparation of the figure showing three-dimensional structures of AMPs.

References

Abad LR, D'Urzo MP, Liu D, Narasimhan ML, Reweni M, Zhu JK, Niu X, Singh NK, Hasegawa PM and Bressan RA (1996) Antifungal activity of tobacco osmotin has specificity and involves plasma membrane permeabilization. Plant Sci 118: 11—23

Abad MS, Hakimi SM, Kaniewski WK, Rommens CMT, Shualev V, Lam E and Shah DM (1997) Characterization of acquired resistance in lesion-mimic transgenic potato expressing bacterio-opsin. Mol Plant-Microbe Interact 5: 635—645

Abe T, Kawai N and Niwa A (1982) Purification and properties of a presynaptically acting neurotoxin, mandarotoxin, from hornet (*Vespa mandarinia*). Biochemistry 21: 1693—1697

Abeles FB, Bosshart RP, Forrence LE and Habig WH (1971) Preparation and purification of glucanase and chitinase from bean leaves. Plant Physiol 47: 129—134

Aguirre PJ and Smith AG (1993) Molecular characterization of a gene encoding a cysteine-rich protein preferentially expressed in anthers of *Lycopersicon esculentum*. Plant Mol Biol 23: 477—487

Ahl P and Gianinazzi S (1982) b-proteins as a constitutive component in highly (TMV) resistant interspecific hybrids of *Nicotiana glutinosa* x *Nicotiana debneyi*. Plant Sci Lett 26: 173-181

Ahl P, Benjama A, Samson R and Gianinazzi S (1981) Induction chez le tabac par *Pseudomonas syringae* de nouvelles protéines (protéines "b") associées au développement d'une résistance non spécifique à une deuxième infection. Phytopathol Z 102: 201—212

Albani D, Robert LS, Donaldson PA, Altosaar I, Arnison PG and Fabijanski SF (1990) Characterization of a pollen-specific gene family from *Brassica napus* which is activated during early microspore development. Plant Mol Biol 15: 605—622

Alexander D, Goodman RM, Gut-Rella M, Glascock C, Weymann K, Friedrich L, Maddox D, Ahl-Goy P, Luntz T, Ward E and Ryals J (1993) Increased tolerance to two oomycete pathogens in transgenic tobacco expressing pathogenesis-related protein 1a. Proc Natl Acad Sci USA 90: 7327—7331

Andersen NH, Cao B, Rodriguez-Romero A and Arreguin B (1993) Hevein: NMR assignment and assessment of solution-state folding for the agglutinin-toxin motif. Biochemistry 32: 1407—1422

Andersson M, Gunne H, Agerbeth B, Boman A, Bergman T, Sillard R, Jörnvall H, Mutt V, Olsson B, Wigzell H, Dagerlind Å, Boman HG and Gudmundsson GH (1995) NK-lysin, a novel effector peptide of cytotoxic T and NK cells. Structure and cDNA cloning of the porcine form, induction by interleukin 2, antibacterial and antitumour activity. EMBO J 14: 1615—1625

Antoniw JF and Pierpoint WS (1978) The purification and properties of one of the "b" proteins from virus-infected tobacco plants. J Gen Virol 39: 343—350

Antoniw JF, Ritter CE, Pierpoint WS and Van Loon LC (1980) Comparison of three pathogenesis-related proteins from plants of two cultivars of tobacco infected with TMV. J Gen Virol 47: 79—87

Anzlovar S, Dalla Serra M, Dermastia M and Menestrina G (1998) Mol Plant-Microbe Interact 7: 610—617

Archer BL, Audley BG, Sweeney GP and Hong TC (1969) Studies on composition of latex serum and 'bottom fraction' particles. J Rubber Res Inst Malaysia 21: 560—569

Arlorio M, Ludwig A, Boller T and Bonfante P (1992) Inhibition of fungal growth by plant chitinases and β-1,3-glucanases. A morphological study. Protoplasma 171: 34—43

Arseniev AS, Pluzhnikov KA, Nolde DE, Sobol AG, Torgov MY, Sukhanov SV and Grishin EV (1994) Toxic principle of selva ant venom is a pore-forming protein transformer. FEBS Lett 347: 112—116

Atkinson AH, Heath RL, Simpson RJ, Clarke AE and Andersson MA (1993) Proteinase inhibitors in *Nicotiana alata* stigmas are derived from a precursor protein which is processed into five homologous inhibitors. Plant Cell 5: 203—213

Audy P, Trudel J and Asselin A (1988) Purification and characterization of a lysozyme from wheat germ. Plant Sci 58: 43—50

Ayres MD, Howard SC, Kuzio J, Lopez-Ferber M and Possee RP (1994) The complete DNA sequence of *Autographa californica* nuclear polyhedrosis virus. Virology 202: 586—605

Balls AK, Hale WS and Harris TH (1942) A crystalline protein from a lipoprotein of wheat flour. Cereal Chem 58: 360—361

Barbieri L, Battelli MG and Stirpe F (1993) Ribosome-inactivating proteins from plants. Biochim Biophys Acta 1154: 237—282

Bass HB, Webster C, O'Brian CR, Roberts JKM and Boston RS (1992) A maize ribosome-inactivating protein is controlled by the transcriptional activator Opaque-2. Plant Cell 4: 225—234

Batalia MA, Monzingo AF, Ernst S, Roberts W and Robertus JD (1996) The crystal structure of the antifungal protein zeamatin, a member of the thaumatin-like, PR-5 protein family. Nature Struct

Biol 3: 19—23

Baud F, Pebay-Peyroula E, Cohen-Addad C, Odani S and Lehmann MS (1993) Crystal structure of hydrophobic protein from soybean; a member of a new cysteine-rich family. J Mol Biol 231: 877—887

Bednarek SY and Raikhel NV (1991) The barley lectin carboxyl-terminal propeptide is a vacuolar protein sorting determinant in plants. Plant Cell 3: 1195—1206

Beffa R and Meins F Jr (1996) Pathogenesis-related functions of plant β-1,3-glucanases investigated by antisense transformation - a review. Gene 179: 97—103

Beffa R, Szell M, Meuwly P, Pay A, Vögeli-Lange R, Métraux J-P, Meins F and Nagy F (1995) Cholera toxin elevates pathogen resistance and induces defense reactions in transgenic tobacco plants. EMBO J 14: 5753—5761

Beffa RS, Hofer R, Thomas M and Meins F Jr (1996) Decreased susceptibility to viral disease of β-1,3-glucanase-deficient plants generated by antisense transformation. Plant Cell 8: 1001—1011

Beffa RS, Neuhaus J-M and Meins F Jr (1993) Physiological compensation in antisense transformants: Specific induction of an "ersatz" glucan endo-1,3-β-glucosidase in plants infected with necrotizing viruses. Proc Natl Acad Sci USA 90: 8792—8796

Beintema JJ and Peumans WJ (1992) The primary structure of stinging nettle (*Urtica dioica*) agglutinin: a two domain member of the hevein family. FEBS Lett 299: 131—134

Békés F and Lásztity R (1981) Isolation and determination of amino acid sequences of avenothionin, a new purothionin analogue form oat. Cereal Chem 58: 360—361

Benhamou N, Broglie K, Broglie R and Chet I (1993a) Antifungal effect of bean endochitinase on *Rhizoctonia solani*: ultrastructural changes and cytochemical aspects of chitin breakdown. Can J Microbiol 39: 318—328

Benhamou N, Broglie K, Chet I and Broglie R (1993b) Cytology of infection of 35S bean chitinase transgenic canola plants by *Rhizoctonia solani*: cytochemical aspects of chitin breakdown *in vivo*. Plant J 4: 295—305

Benhamou N, Grenier J and Asselin A (1991) Immunogold localization of pathogenesis-related protein P14 in tomato root cells infected by *Fusarium oxysporum* f.sp. *radicis-lycopersici*. Physiol Mol Plant Pathol 38: 237—253

Benhamou N, Grenier J, Asselin A and Legrand M (1989) Immunogold localization of β-1,3-glucanases in two plants infected by vascular wilt fungi. Plant Cell 1: 1209—1221

Benhamou N, Joosten MHAJ and De Wit PJGM (1990) Subcellular localization of chitinase and of its potential substrate in tomato root tissues infected by *Fusarium oxysporum* f.sp. *radicis-lycopersici*. Plant Physiol 92: 1108—1120

Bernasconi P, Locher R, Pilet PE, Jollès J and Jollès P (1987) Purification and N-terminal amino acid sequence of a basic lysozyme from *Parthenocissus quinquifolia* cultured *in vitro*. Biochim Biophys Acta 915: 254—260

Bernhard WR and Somerville CR (1989) Coidentity of putative amylase inhibitors from barley and finger millet with phospholipid transfer proteins inferred from amino acid sequence homology. Arch Biochem Biophys 269: 695—697

Blaak H and Schrempf H (1995) Binding and substrate specificities of a *Streptomyces olivaceoviridis* chitinase in comparison with its proteolytically processed form. Eur J Biochem 229: 132—139

Blaiseau PL, Kunz C, Grison R, Bertheau Y and Brygoo Y (1992) Cloning and expression of a chitinase gene from the hyperparasitic fungus *Aphanocladium album*. Curr Genet 21: 61—66

Bloch C and Richardson M (1991) A new family of small (5 kDa) protein inhibitors of insect α-amylases from seeds of sorghum (*Sorghum bicolor* (L.) Moench) have sequence homologies with wheat γ-purothionins.

FEBS Lett 279: 101—104

Blochet J-E, Chevalier C, Forest E, Pebay-Peyroula E, Gautier M-F, Joudrier P, Pézolet M and Marion D (1993) Complete amino acid sequence of puroindoline, a new basic and cystine-rich protein with a unique tryptophan-rich domain, isolated from wheat endosperm by Triton X-114 phase partitioning. FEBS Lett 329: 336—340

Bode W, Greyling HJ, Huber R, Otlewski J and Wilusz T (1989) The refined 2.0 Å X-ray crystal structure of the complex formed between bovine α-trypsin and CMTI-I, a trypsin inhibitor from squash seeds (*Cucurbita maxima*). Topological similarity of the squash seed inhibitors with the carboxypeptidase A inhibitor from potatoes. FEBS Lett 242: 285—292

Bohlmann H (1994) The role of thionins in plant protection. Crit Rev Plant Sci 13: 1—16

Bohlmann H and Apel K (1987) Isolation and characterization of cDNAs coding for leaf-specific thionins closely related to the endosperm-specific hordothionin of barley *Hordeum vulgare* L. Mol Gen Genet 207: 446—454

Bohlmann H, Clausen S, Behnke S, Giese H, Hiller C, Reimann-Philipp U, Schrader G, Barkholt C and Apel K (1988) Leaf-specific thionins of barley: a novel class of cell wall proteins toxic to plant-pathogenic fungi and possibly involved in the defense mechanism of plants. EMBO J 7: 1559—1565

Boller T (1988) Ethylene and the regulation of antifungal hydrolases in plants. In: Miflin BJ and Miflin HF (eds) Oxford Surveys of Plant Molecular and Cell Biology, Vol 5, pp 145—174. Oxford University Press, Oxford

Boller T (1991) Ethylene in pathogenesis and disease resistance. In: Mattoo AK and Suttle JC (eds) The Plant Hormone Ethylene, pp 293—314. CRC Press, Boca Raton

Boller T, Gehri A, Mauch F, Vögeli U (1983) Chitinase in bean leaves: induction by ethylene, purification, properties, and possible function. Planta 157: 22—31

Bonmatin J-M, Bonnat J-L, Gallet X, Vovelle F, Ptak M, Reichhart J-M, Hoffmann JA, Keppi E, Legrain M and Achstetter T (1992) Two-dimensional ^{1}H-NMR study of recombinant insect defensin A in water: Resonance assignments, secondary structure and global folding. J Biomol NMR 2: 235—256

Boot RG, Renkema GH, Strijland A, van Zonneveld AJ and Aerts JMFG (1995) Cloning of a cDNA encoding chitotriosidase, a human chitinase produced by macrophages. J Biol Chem 270: 26252—26256

Borgmeyer JR, Smith CE and Huynh QK (1992) Isolation and characterization of a 25 kDa antifungal protein from flax seeds. Biochem Biophys Res Comm 187: 480—487

Bowler C, Van Montagu M and Inzé D (1992) Superoxide dismutase and stress tolerance. Annu Rev Plant Physiol Plant Mol Biol 43: 83—116

Bowling SA, Guo A, Cao H, Gordon AS, Klessig DF and Dong X (1994) A mutation in Arabidopsis that leads to constitutive expression of systemic acquired resistance. Plant Cell 6: 1845—1857

Brandstädter J, Rosslach C and Theres K (1996) Expression of genes for a defensin and a proteinase inhibitor in specific areas of the shoot apex and the developing flower in tomato. Mol Gen Genet 252: 146—154

Brandt A and Thomsen KK (1986) Primary structure of the (1-3,1-4)-β-D-glucan 4-glucohydrolase from barley aleurone. Proc Natl Acad Sci USA 83: 2081—2085

Brederode F Th, Linthorst HJM and Bol JF (1991) Differential induction of acquired resistance and PR gene expression in tobacco by virus infection, ethephon treatment, UV light and wounding. Plant Mol Biol 17: 1117—1125

Breu V, Guerbette F, Kader J-C, Kannagara CG, Svensson B and Von Wettstein-Knowles P (1989) A 10 kD barley basic protein transfers phosphatidylcholine from liposomes to mitochondria. Carlsberg Res Comm 54: 81—84

Broekaert W, Lee H-I, Kush A, Chua N-H and Raikhel N (1990) Wound-induced accumulation of mRNA containing a hevein sequence in laticifers of rubber tree (*Hevea brasiliensis*). Proc Natl Acad Sci USA 87: 7633-7637

Broekaert WF, Cammue BPA, De Bolle MFC, Thevissen K, De Samblanx GW and Osborn RW (1997) Antimicrobial peptides from plants. Crit Rev Plant Sci 16: 297—323

Broekaert WF, Mari'n W, Terras FRG, De Bolle MFC, Proost P, Van Damme J, Dillen L, Claeys M, Rees SB, Vanderleyden J and Cammue BPA (1992) Antimicrobial peptides from *Amaranthus caudatus* seeds with sequence homology to the cysteine/glycine-rich domain of chitin-binding proteins. Biochemistry 31: 4308—4314

Broekaert WF, Terras FRG, Cammue BPA and Osborn RW (1995) Plant defensins: Novel antimicrobial peptides as components of the host defense system. Plant Physiol 108: 1353—1358

Broekaert WF, Van Parijs J, Allen AK and Peumans WJ (1988) Comparison of some molecular, enzymatic and antifungal properties of chitinases from thorn-apple, tobacco and wheat. Physiol Mol Plant Pathol 33: 319—331

Broekaert WF, Van Parijs J, Leyns F, Joos H and Peumans WJ (1989) A chitin-binding lectin from stinging nettle rhizomes with antifungal properties. Science 245: 1100—1102

Broglie K, Chet I, Holliday M, Cressman R, Biddle P, Knowlton S, Mauvais CJ and Broglie R (1991) Transgenic plants with enhanced resistance to the fungal pathogen *Rhizoctonia solani*. Science 254: 1194—1197

Brooks DE, Means AR, Wright EJ, Singh SP and Tiver KK (1986) Molecular cloning of the cDNA for androgen-dependent sperm-coating glycoproteins secreted by the rat epididymis. Eur J Biochem 161: 13—18

Bruix M, González C, Santoro J, Soriano F, Rocher A, Méndez E and Rico M (1995) ^{1}H-NMR studies on the structure of a new thionin from barley endosperm. Biopolymers 36: 751—763

Bruix M, Jiménez MA, Santoro J, González C, Colilla FJ, Méndez E and Rico M (1993) Solution structure of γ1-H and γ1-P thionins from barley and wheat endosperm determined by ^{1}H-NMR: a structural motif common to toxic arthropod proteins. Biochemistry 32: 715—724

Brünger AT, Campbell RL, Clore GM, Gronenborn AM, Karplus M, Petsko GA and Teeter MM (1987) Solution of a protein crystal structure with a model obtained from NMR interproton distance restraints. Science 235: 1049—1053

Brunner F, Stintzi A, Fritig B and Legrand M (1998) Substrate specificities of tobacco chitinases. Plant J 14: 225—234

Brurberg MB, Eijsink VG and Nes IF (1994) Characterization of a chitinase gene (chiA) from *Serratia marcescens* BJL200 and one-step purification of the gene product. FEMS Microbiol Lett 124: 399—404

Bryngelsson T and Green B (1989) Characterization of a pathogenesis-related, thaumatin-like protein isolated from barley challenged with an incompatible race of mildew. Physiol Mol Plant Pathol 35: 45—52

Bryngelsson T, Sommer-Knudsen J, Gregersen PL, Collinge DB, Ek B and Thordal-Christensen H (1994) Purification, characterization, and molecular cloning of basic PR-1-type pathogenesis-related proteins from barley. Mol Plant-Microbe Interact 7: 267—275

Bucciaglia PA and Smith AG (1994) Cloning and characterization of *Tag1*, a tobacco anther β-1,3-glucanase expressed during tetrad dissolution. Plant Mol Biol 24: 903—914

Butler AR, O'Donnell RW, Martin VJ, Gooday GW and Stark MJR (1991) *Kluyveromyces lactis* toxin has an essential chitinase activity. Eur J Biochem 199: 483—488

Caaveiro JMM, Molina A, Gonzales-Manas JM, Rodriguez-Palenzuela P, Garcia-Olmedo F, Goni FM

(1997) Differential effects of five types of antipathogenic plant peptides on model membranes. FEBS Lett 410: 338—342

Caaveiro JMM, Molina A, Rodriguez-Palenzuela P, Goni FM and Gonzales-Manas JM (1998) Interaction of wheat α-thionin with large unilamellar vesicles. Protein Sci 7: 2567—2577

Cammue BPA, De Bolle MFC, Terras FRG, Proost P, Van Damme J, Rees SB, Vanderleyden J and Broekaert WF (1992) Isolation and characterization of a novel class of plant antimicrobial peptides from *Mirabilis jalapa* L. seeds. J Biol Chem 267: 2228—2233

Cammue BPA, Thevissen K, Hendriks M, Eggermont K, Goderis IJ, Proost P, Van Damme J, Osborn RW, Guerbette F, Kader J-C and Broekaert WF (1995) A potent antimicrobial protein from onion seeds showing sequence homology to plant lipid transfer proteins. Plant Physiol 109: 445—455

Campos FAP and Richardson M (1984) The complete amino acid sequence of the α-amylase inhibitor I-2 from seeds of ragi (Indian finger millet, *Eleusine coracana* Gaertn.). FEBS Lett 167: 221—225

Canevascini S, Caderas D, Mandel T, Fleming AJ, Dupuis I and Kuhlemeier C (1996) Tissue-specific expression and promoter analysis of the tobacco *ltp1* gene. Plant Physiol 112: 513—524

Cao H, Bowling SA, Gordon AS and Dong X (1994) Characterization of an Arabidopsis mutant that is nonresponsive to inducers of systemic acquired resistance. Plant Cell 6: 1583—1592

Carmona MJ, Hernandez-Lucas C, San Martin C, Gonzales P and García-Olmedo F (1993a) Subcellular localization of type I thionins in the endosperm of wheat and barley. Protoplasma 173: 1—7

Carmona MJ, Molina A, Fernández JA, López-Fando JJ and García-Olmedo F (1993b) Expression of the α-thionin gene from barley in tobacco confers enhanced resistance to bacterial pathogens. Plant J 3: 457—462

Carrasco L, Vásquez D, Hernández-Lucas C, Carbonero P and García-Olmedo F (1981) Thionins: plant peptides that modify membrane permeability in cultured mammalian cells. Eur J Biochem 116: 185—189

Carzaniga R, Sinclair L, Fordham-Skelton AP, Harris N and Croy RRD (1994) Cellular and subcellular distribution of saporins, type-1 ribosome-inactivating proteins, in soapwort (*Saponaria officinalis* L.). Planta 194: 461—470

Casacuberta JM, Puigdomènech P and San-Segundo B (1991) A gene coding for a basic pathogenesis-related (PR-like) protein from *Zea mays*. Molecular cloning and induction by a fungus (*Fusarium moniliforme*) in germinating maize seeds. Plant Mol Biol 16: 527—536

Castagnaro A, Maraña C, Carbonero P and García-Olmedo (1994) cDNA cloning and nucleotide sequences of α1 and α2 thionins from hexaploid wheat endosperm. Plant Physiol 106: 1221-1222

Castagnaro A, Maraña C, Carbonero P and García-Olmedo F (1992) Extreme divergence of a novel wheat thionin generated by a mutational burst specifically affecting the mature protein domain of the precursor. J Mol Biol 224: 1003—1009

Castagnaro A, Segura A and Garcia-Olmedo F (1995) High conservation among sequences encoding type-V thionins in wheat and Aegilops. Plant Physiol 107: 1475-1476

Casteels P, Romagnolo J, Castle M, Casteels-Josson K and Tempst PD (1994) Biodiversity of apidaecin-type peptide antibiotics. Prospects of manipulating the antibacterial spectrum and combating acquired resistance. J Biol Chem 269: 26107—26115

Casteels-Josson K, Capaci T, Casteels P and Tempst PD (1993) Apidaecin multipeptide precursor structure: a putative mechanism for amplification of the insect antibacterial response. EMBO J 12: 1569—1578

Ceci LR, Spoto N, De Virgilio M and Gallerani R (1995) The gene coding for the mustard trypsin inhibitor is discontinuous and wound-inducible. FEBS Lett 364: 179—181

Chagolla-Lopez A, Blanco-Labra A, Patthy A, Sánchez R and Pongor S (1994) A novel α-amylase inhibitor from amaranth (*Amaranthus hypocondriacus*) seeds. J Biol Chem 269: 23675—23680

Chen L, Garrett TPJ, Fincher GB and Høj PB (1995) A tetrad of ionizable amino acids is important for catalysis in barley β-glucanases. J Biol Chem 270: 8093—8101

Chen R, Wang F and Smith AG (1996) A flower-specific gene encoding an osmotin-like protein from *Lycopersicon esculentum*. Gene 179: 301—302

Chen R-D, Yu L-X, Greer AF, Cheriti H and Tabaeizadeh Z (1994) Isolation of an osmotic stress- and absiscic acid-induced gene encoding an acidic endochitinase from *Lycopersicon chilense*. Mol Gen Genet 245: 195—202

Chiang CC and Hadwiger LA (1991) The *Fusarium solani*-induced expression of a pea gene family encoding high cysteine content proteins. Mol Plant-Microbe Interact 4: 324—331

Chivasa S, Murphy AM, Naylor M and Carr JP (1997) Salicylic acid interferes with tobacco mosaic virus replication via a novel salicylhydroxamic acid-sensitive mechanism. Plant Cell 9: 547—557

Choi Y, Ahn JH, Choi YD and Lee JS (1995) Tissue-specific and developmental regulation of a gene encoding a low molecular weight sulfur-rich protein in soybean seeds. Mol Gen Genet 246: 266—268

Chye ML and Cheung KY (1995) β-1,3-glucanase is highly expressed in laticifers of *Hevea brasiliensis*. Plant Mol Biol 29: 397—402

Clore GM, Gronenborn AM, Nilges M and Ryan CA (1987) Three-dimensional structure of potato carboxypeptidase inhibitor in solution. A study using nuclear magnetic resonance, distance geometry, and restrained molecular dynamics. Biochemistry 26: 8012—8023

Cociancich S, Ghazi A, Hétru C, Hoffmann JA and Letellier L (1993) Insect defensin, an inducible antibacterial peptide forms voltage-dependent channels in *Micrococcus luteus*. J Biol Chem 268: 19239—19245

Colilla FJ, Rocher A and Mendez E (1990) γ-Purothionins: amino acid sequences of two polypeptides of a new family of thionins from wheat endosperm. FEBS Lett 270: 191—194

Collinge DB, Kragh KM, Mikkelsen JD, Nielsen KK, Rasmussen U and Vad K (1993) Plant chitinases. Plant J 3: 31—40

Conconi A, Smerdon MJ, Howe GA and Ryan CA (1996) The octadecanoid signalling pathway in plants mediates a response to ultraviolet radiation. Nature 383: 827—829

Conrads-Strauch J, Dow JM, Milligan DE, Parra R and Daniels MJ (1990) Induction of hydrolytic enzymes in *Brassica campestris* in response to pathovars of *Xanthomonas campestris*. Plant Physiol 93: 238—243

Cornelissen BJC, Hooft van Huijsduijnen RAM and Bol JF (1986a) A tobacco mosaic virus-induced tobacco protein is homologous to the sweet-tasting protein thaumatin. Nature 321: 531—532

Cornelissen BJC, Hooft van Huijsduijnen RAM, Van Loon LC and Bol JF (1986b) Molecular characterization of messenger RNAs for 'pathogenesis-related' proteins 1a, 1b and 1c, induced by TMV infection of tobacco. EMBO J 5: 37—40

Cornelissen BJC, Horowitz J, Van Kan JAL, Goldberg RB and Bol JF (1987) Structure of tobacco genes encoding pathogenesis-related proteins from the PR-1 group. Nucleic Acids Res 15: 6799—6811

Cornet B, Bonmatin J-M, Hétru C, Hoffmann JA, Ptak M and Vovelle F (1995) Refined three-dimensional solution structure of insect defensin A. Structure 3: 435—448

Coté F and Hahn MG (1994) Oligosaccharins: structure and signal transduction. Plant Mol Biol 26: 1379—1411

Coulson EJ, Harris TH and Axelrod B (1942) Effect on small laboratory animals of the injection of the crystalline hydrochloride of a sulfur protein from wheat four. Cereal Chem 19: 301—307

Creelman RA and Mullet JE (1997) Biosynthesis and action of jasmonates in plants. Annu Rev Plant

Physiol Plant Mol Biol 48: 355—381

Crossley SJ, Greenland AJ and Dickinson HG (1995) The characterisation of tapetum-specific cDNAs isolated from a *Lilium henryi* L. meiocyte subtractive cDNA library. Planta 196: 523—529

Cutt JR, Harpster MH, Dixon DC, Carr JP, Dunsmuir P and Klessig DF (1989) Disease response to Tobacco Mosaic Virus in transgenic tobacco plants that constitutively express the pathogenesis-related PR1b gene. Virology 173: 89—97

D'Hondt K, Bosch D, Van Damme J, Goethals M, Vandekerckhove J and Krebbers E (1993) An aspartic proteinase in seeds cleaves *Arabidopsis* 2S albumin precursors *in vitro*. J Biol Chem 268: 20884—20891

Dandekar T and Leippe M (1997) Molecular modeling of amoebapore and NK-lysin: a four-α-helix bundle motif of cytolytic peptides from distantly related organisms. Folding Design 2: 47—52

Darbon H, Weber C and Braun W (1991) Two-dimensional ¹H nuclear magnetic resonance study of AaH IT, an anti-insect toxin from the scorpion *Androctonus australis* Hector. Sequential resonance assignments and folding of the polypeptide chain. Biochemistry 30: 1836—1845

De Bolle MFC, David KMM, Rees SB, Vanderleyden J, Cammue BPA and Broekaert WF (1993a) Cloning and characterization of a cDNA encoding an antimicrobial chitin-binding protein from amaranth, *Amaranthus caudatus*. Plant Mol Biol 22: 1187—1190

De Bolle MFC, Eggermont K, Duncan RE, Osborn RW, Terras FRG and Broekaert WF (1995) Cloning and characterization of two cDNA clones encoding seed-specific antimicrobial peptides from *Mirabilis jalapa* L. Plant Mol Biol 28: 713—721

De Bolle MFC, Osborn RW, Goderis IJ, Noe L, Acland D, Hart CA, Torrekens S, Van Leuven F and Broekaert WF (1996) Antimicrobial peptides from *Mirabilis jalapa* and *Amaranthus caudatus*: expression, processing, localization and biological activity in transgenic tobacco. Plant Mol Biol 31: 993—1008

De Bolle MFC, Terras FRG, Cammue BPA, Rees SB and Broekaert WF (1993b) *Mirabilis jalapa* antimicrobial peptides and *Raphanus sativus* antifungal proteins: a comparative study of their structure and biological activities. In: Fritig B and Legrand M (eds) Mechanisms of Plant Defense Responses, pp 433-436, Kluwer Academic Publishers

De Jong AJ, Cordewener J, Lo Schiavo F, Terzi M, Vandekerckhove J, Van Kammen A and De Vries SC (1992) A carrot somatic embryo mutant is rescued by chitinase. Plant Cell 4: 425—433

Del Campillo E, Lewis NL (1992) Identification and kinetics of accumulation of proteins induced by ethylene in bean abscission zones. Plant Physiol 98: 955—961

Delaney T, Uknes S, Vernooij B, Friedrich L, Weymann K, Negrotto D, Gaffney T, Gut-Rella M, Kessmann H, Ward E and Ryals J (1994) A central role of salicylic acid in plant disease resistance. Science 266: 1247—1250

De Samblanx, GW, Goderis IJ, Thevissen K, Raemaeker SR, Fant F, Borremans F, Acland D, Osborn RW, Patel, S and Broekaert WF (1997) Mutational analysis of a plant defensin from radish (*Raphanus sativus* L.) reveals two adjacent sites important for antifungal activity. J Biol Chem 272: 1171—1179

Dempsey DMA, Wobbe KK and Klessig DF (1993) Resistance and susceptible responses of *Arabidopsis thaliana* to turnip crinkle virus. Phytopathology 83: 1021—1029

De Wit PJGM (1992) Molecular characterization of gene-for-gene systems in plant-fungus interactions and the application of avirulence genes in control of plant pathogens. Annu Rev Phytopathol 30: 391—418

Diamond G, Russell JP and Bevins CL (1996) Inducible expression of an antibiotic peptide gene in lipopolysaccharide-challenged tracheal epithelial cell. Proc Natl Acad Sci USA 93: 5156—5160

Dietrich RA, Delany TP, Uknes SJ, Ward ER, Ryals JA and Dangl JL (1994) Arabidopsis mutants simulating disease resistance response. Cell 77: 565—577

Dimarcq JL, Hoffmann D, Meister M, Bulet P, Lanot R, Reichhart J-M and Hoffmann JA (1994) Characterization and transcriptional profiles of a *Drosophila* gene encoding an insect defensin. A study in insect immunity. Eur J Biochem 221: 201—209

Domon C, Evrard J-L, Herdenberger F, Pillay DTN and Steinmetz A (1990) Nucleotide sequence of two anther-specific cDNAs from sunflower (*Helianthus annuus* L.). Plant Mol Biol 15: 643—646

Dong X, Mindrinos M, Davis KR and Ausubel FM (1991) Induction of Arabidopsis defense genes by virulent and avirulent *Pseudomonas syringae* strains and by a cloned avirulence gene. Plant Cell 3: 61—72

Dore I, Legrand M, Cornelissen BJC and Bol JF (1991) Subcellular localization of acidic and basic PR proteins in tobacco mosaic virus-infected tobacco. Arch Virol 120: 97—107

Duan X, Li X, Xue Q, Abo-El-Saad M, Xu D and Wu R (1996) Transgenic rice plants harboring an introduced potato proteinase inhibitor II gene are insect resistant. Nature Biotechnol 14: 494—498

Dubreil L, Gaborit, T, Bouchet B, Gallant DJ, Broekaert WF, Quillien L and Marion D (1998) Spatial and temporal distribution of the major isoforms of puroindolines (puroindoline-a and puroindoline-b) and non specific lipid transfer protein (ns-LTPe$_1$) in *Triticum aestivum* seeds. Relationships with their *in vitro* antifungal properties. Plant Sci 138: 121—135

Düring K, Porsch P, Fladung M and Lörz H (1993) Transgenic potato plants resistant to the phytopathogenic bacterium *Erwinia carotovora*. Plant J 3: 587—598

Dürner J, Shah J and Klessig DF (1997) Salicylic acid and disease resistance in plants. Trends Plant Sci 2: 266—274

Duvick JP, Rood T, Rao AG and Marshak DR (1992) Purification and characterization of a novel antimicrobial peptide from maize (*Zea mays* L.) kernels. J Biol Chem 267: 18814—18820

Ebel J and Cosio E (1994) Elicitors of plant defense responses. Int Rev Cytol 148: 1—26

Ebrahim-Nesbat F, Apel K and Heitefuss R (1991) Immunological evidence for the presence of barley thionins in rachis and stalk of wheat: decrease of antibody binding at *Fusarium culmorum* penetration sites. J Phytopathol 131: 259—264

Ebrahim-Nesbat F, Behnke S, Kleinhofs A and Apel K (1989) Cultivar-related differences in the distribution of cell-wall-bound thionins in compatible and incompatible interactions between barley and powdery mildew. Planta 179: 203—210

Ebrahim-Nesbat F, Bohl S, Heitefuss R and Apel K (1993) Thionin in cell walls and papillae of barley in compatible and incompatible interactions with *Erisyphe graminis* f.sp. *hordei*. Physiol Mol Plant Pathol 43: 343—352

Edens L, Hesling L, Klock R, Ledeboer AM, Maat J, Toonen MY, Visser C and Verrips CT (1982) Cloning of cDNA encoding the sweet-tasting plant protein thaumatin and its expression in *Escherichia coli*. Gene 18: 1—12

Endo Y, Mitsui K, Motizuki M and Tsurugi K (1987) The mechanism of action of ricin and related toxin lectins on eukaryotic ribosomes. The site and the characteristics of the modification in 28S ribosomal RNA caused by toxins. J Biol Chem 262: 5908—5912

Epple P, Apel K and Bohlmann H (1995) An *Arabidopsis thaliana* thionin gene is inducible via a signal transduction pathway different from that for pathogenesis-related proteins. Plant Physiol 109: 813—820

Epple P, Apel K and Bohlmann H (1997a) ESTs reveal a multigene family for plant defensins in *Arabidopsis thaliana*. FEBS Lett 400: 168—172

Epple P, Apel K and Bohlmann H (1997b) Overexpression of an endogenous thionin enhances resistance of

Arabidopsis against *Fusarium oxysporum*. Plant Cell 9: 509—520

Ernst D, Schraudner M, Langebartels C and Sandermann H (1992) Ozone-induced changes of mRNA levels of β-1,3-glucanase, chitinase and pathogenesis-related protein 1b in tobacco plants. Plant Mol Biol 20: 673—682

Evans J, Wang Y, Shaw K-P and Vernon LP (1989) Cellular responses to *Pyrularia* thionin are mediated by Ca^{2+} influx and phospholipase A_2 activation and are inhibited by thionin tyrosine iodination. Proc Natl Acad Sci USA 86: 5849—5853

Evett GE, Donaldson DM and Vernon LP (1986) Biological properties of *Pyrularia* thionin prepared from nuts of *Pyrularia pubera*. Toxicon 24: 622—625

Eyal Y, Meyer Y, Lev-Yadun S and Fluhr R (1993) A basic-type PR-1 promoter directs ethylene-responsiveness, vascular and abscission-specific expression. Plant J 4: 225—234

Eyal Y, Sagee O and Fluhr R (1992) Dark-induced accumulation of a basic pathogenesis-related (PR-1) transcript and a light requirement for its induction by ethylene. Plant Mol Biol 19: 589—599

Fang KSY, Vitale M, Fehlner P and King TP (1988) cDNA cloning and primary structure of a white-face hornet venom allergen, antigen 5. Proc Natl Acad Sci USA 85: 895-899

Fant F, Vranken WF, Broekaert WF and Borremans FAM (1998) Determination of the three-dimensional solution structure of *Raphanus sativus* antifungal protein 1 by ^{1}H NMR. J Mol Biol 279, 257—270.

Fazal A, Beg OU, Shafqat J, Zaidi ZH and Jornvall H (1989) Characterization of two different peptides from the venom of the scorpion *Buthus sindicus*. FEBS Lett 257: 260—262

Fehlbaum P, Bulet P, Michaut L, Lageux M, Broekaert WF, Hétru C and Hoffmann JA (1994) Insect immunity. Septic injury of *Drosophila* induces the synthesis of a potent antifungal peptide with sequence homology to plant antifungal proteins. J Biol Chem 269: 33159—33163

Fernandez C, Szyperski T, Bruyère T, Ramage P, Mösinger E and Wüthrich (1997) NMR solution structure of the pathogenesis-related protein P14a. J Mol Biol 266: 576—593

Fernández de Caleya R, Gonzalez-Pascual B, García-Olmedo F and Carbonero P (1972) Susceptibility of phytopathogenic bacteria to wheat purothionins *in vitro*. Appl Microbiol 23: 998—1000

Fernández de Caleya R, Hernández-Lucas C, Carbonero P and García-Olmedo F (1976) Gene expression in alloploids: genetic control of lipopurothionins in wheat. Genetics 83: 687—699

Fernández-Puentes C and Larrasco L (1980) Viral infection permeabilizes mammalian cells to toxin proteins. Cell 20: 769—775

Fils-Lycaon BR, Wiersma PA, Eastwell KC and Sautiere P (1996) A cherry protein and its gene, abundantly expressed in ripening fruit, have been identified as thaumatin-like. Plant Physiol 111: 269—273

Fincher GB (1989) Molecular and cellular biology associated with endosperm mobilization in germinating cereal grains. Annu Rev Plant Physiol Plant Mol Biol 40: 305—346

Fincher GB, Lock PA, Morgan MM, Lingelbach K, Wettenhall REH, Mercer JFB, Brandt A and Thomsen KK (1986) Primary structure of the (1-3,1-4)- β-D-glucan 4-glucohydrolase from barley aleurone. Proc Natl Acad Sci USA 83: 2081—2085

Fink W, Liefland M and Mengden K (1990) Comparison of various stress responses in oat in compatible and nonhost resistant interactions with rust fungi. Physiol Mol Plant Pathol 37: 309—321

Fleming AJ, Mandel T, Hofmann S, Sterk P, de Vries SC and Kuhlemeier C (1992) Expression pattern of a tobacco lipid transfer protein gene within the shoot apex. Plant J 2: 855—862

Florack DEA and Stiekema WJ (1994) Thionins: properties, possible biological roles and mechanisms of action. Plant Mol Biol 26: 25—37

Florack DEA, Visser B, de Vries PM, Van Vuurde JWL and Stiekema WJ (1993) Analysis of the toxicity of purothionins and hordothionins for plant pathogenic bacteria. Neth J Plant Pathol 99: 259—268

Fontecilla-Camps JC, Habersetzer-Rochat C and Rochat H (1988) Orthorombic crystals and three-dimensional structure of the potent toxin II from the scorpion *Androctonus australis* Hector. Proc Natl Acad Sci USA 85: 7443—7447

Foyer CH, Descourvières P and Kumert KJ (1994) Protection against oxygen radicals: an important defence mechanism studied in transgenic plants. Plant Cell Environ 17: 507—523

Frendo P, Didierjean L, Passelegue E and Burkard G (1992) Abiotic stresses induce a thaumatin-like protein in maize; cDNA isolation and sequence analysis. Plant Sci 85: 61—69

Friedrich L, Lawton K, Dincher S, Winter A, Staub T, Uknes S, Kessmann H and Ryals J (1996) Benzothiadiazole induces systemic acquired resistance in tobacco. Plant J 10, 61-70

Friedrich L, Moyer M, Ward E and Ryals J (1991) Pathogenesis-related protein 4 is structurally homologous to the carboxy-terminal domains of hevein, Win-1 and Win-2. Mol Gen Genet 230: 113—119

Friedrich L, Vernooij B, Gaffney T, Morse A and Ryals J (1995) Characterization of tobacco plants expressing a bacterial salicylate hydroxylase gene. Plant Mol Biol 29: 959—968

Fuhrman JA, Lane WS, Smith RF, Piessens WF and Perler FB (1992) Transmission-blocking antibodies recognize microfilarial chitinase in brugian lymphatic filariasis. Proc Natl Acad Sci USA 89: 1548—1552

Gaffney T, Friedrich L, Vernooij B, Negrotto D, Nye G, Uknes S, Ward E, Kessmann H and Ryals J (1993) Requirement of salicylic acid for the induction of systemic acquired resistance. Science 261: 754—756

Ganz T, Metcalf JA, Gallin JI, Boxer LA and Lehrer RI (1988) Microbiocidal/cytotoxic proteins in neutrophils are deficient in two disorders: Chediak-Higashi syndrome and "specific" granule deficiency. J Clin Invest 82: 552—556

Garcia-Garcia F, Schmelzer E, Hahlbrock K and Roxby R (1994) Differential expression of chitinase and β-1,3-glucanase genes in various tissues of potato plants. Z Naturforschung 49c: 195—203

García-Olmedo F, Carbonero P, Hernández-Lucas C, Paz-Ares J, Ponz F, Vicente O and Sierra JM (1983) Inhibition of eukaryotic cell-free protein synthesis by thionins from wheat endosperm. Biochim Biophys Acta 740: 52—56

García-Olmedo F, Molina A, Segura A and Moreno M (1995) The defensive role of non-specific lipid transfer proteins in plants. Trends Microbiol 3: 72—74

Gasanov SE, Vernon LP and Aripov TF (1993) Modification of phospholipid membrane structure by the plant toxic peptide *Pyrularia* thionin. Arch Biochem Biophys 301: 367—374

Gatehouse AMR, Barbieri L, Stirpe F and Croy RRD (1990) Effects of ribosome inactivating proteins on insect development - differences between Lepidoptera and Coleoptera. Entomol Exp Appl 54: 43—51

Gausing K (1987) Thionin genes specifically expressed in barley leaves. Planta 171: 241—246

Gausing K (1994) Lipid transfer protein genes specifically expressed in barley leaves and coleoptiles. Planta 192: 574—580

Gautier M-F, Alany R and Joudrier P (1990) Cloning and characterization of a cDNA encoding the wheat (*Triticum durum* Desf.) CM16 protein. Plant Mol Biol 14: 313—322

Gautier M-F, Aleman M-E, Guirao A, Marion D and Joudrier P (1994) *Triticum aestivum* puroindolines, two basic cystine-rich seed proteins: cDNA sequence analysis and developmental gene expression. Plant Mol Biol 25: 43—57

Gehrz RC, Wilson C, Eckhardt J, Meyers D, Irvin JD and Uckun FM (1991) Treatment of human cytomegalovirus (HCMV) with novel antiviral immunoconjugates. In: Landin MP (ed) Progress in Cytomegalovirus Research, pp 353—356. Elsevier Science Publishers BV, Amsterdam, The

Netherlands

Gianinazzi S, Ahl P, Cornu A and Scalla R (1980) First report of host b-protein appearance in response to a fungal infection in tobacco. Physiol Plant Pathol 16: 337—342

Gianinazzi S, Martin C and Vallée JC (1970) Hypersensibilité aux virus, température et protéines solubles chez le *Nicotiana* Xanthi nc. Apparition de nouvelles macromolécules lors de la répression de la synthèse virale. CR Acad Sci Paris 270D: 2883—2886

Gincel E, Simorre J-P, Caille A, Marion D, Ptak M and Vovelle F (1994) Three-dimensional structure in solution of a wheat lipid-transfer protein from multidimensional ^{1}H-NMR data. A new folding for lipid carriers. Eur J Biochem 226: 413—422

Girbés T, De Torre C, Iglesias R, Ferreras JM and Méndez E (1996) RIP for viruses. Nature 379: 777—778

Glazer AN, Barel AO, Howard JB and Brown DM (1969) Isolation and characterization of fig lysozyme. J Biol Chem 244: 3583—3589

Gomar J, Petit M-C, Sodano P, Sy D, Marion D, Kader J-C, Vovelle F and Ptak M (1996) Solution structure and lipid binding of a nonspecific lipid transfer protein extracted from maize seeds. Protein Sci 5: 565—577

Gooday GW (1994) Cell walls. In: Gow NAR and Gadd GM (eds) The Growing Fungus, pp 43—62. Chapman and Hall, London

Graham LS and Sticklen MB (1994) Plant chitinases. Can J Bot 72: 1057—1083

Green R and Fluhr R (1995) UV-B-Induced PR-1 accumulation is mediated by active oxygen species. Plant Cell 7: 203—212

Greenberg JT, Guo A, Klessig DF and Ausubel FM (1994) Programmed cell death in plants - A pathogen-triggered response activated coordinately with multiple defense functions. Cell 77: 551—563

Grison R, Grezes-Besset B, Schneider M, Lucante N, Olsen L, Leguay J-J and Toppan A (1996) Field tolerance to fungal pathogens of *Brassica napus* constitutively expressing a chimeric chitinase gene. Nature Biotechnol. 14: 643—646

Grosbois M, Guerbette F, Jolliot A, Quintin F and Kader J-C (1993) Control of maize lipid transfer protein activity by oxido-reducing conditions. Biochim Biophys Acta 1170: 197—203

Gu Q, Kanata EE, Morse MJ, Wu HM and Cheung AY (1992) A flower-specific cDNA encoding a novel thionin in tobacco. Mol Gen Genet 234: 89—96

Guihard G, Bénédetti H, Besnard M and Letellier L (1993) Phosphate efflux through the channels formed by colicins and phage T5 in *Escherichia coli* cells is responsible for the fall in cytoplasmic ATP. J Biol Chem 268: 17775—17780

Habuka N, Kataoka J, Miyano M, Tsuge H, Ago H and Noma M (1993) Nucleotide sequence of a genomic clone encoding tritin, a ribosome-inactivating protein from *Triticum aestivum*. Plant Mol Biol 22: 171—176

Haendler B, Krätzschmar J, Theuring F and Schleuning W-D (1993) Transcripts for cysteine-rich secretory protein-1 (CRISP-1; DE/AEG) and the novel related CRISP-3 are expressed under androgen control in the mouse salivary gland. Endocrinology 133: 192—198

Hakala BE, White C and Recklies AD (1993) Human cartilage gp-39, a major secretory product of articular chondrocytes and synovial cells, is a mammalian member of a chitinase protein family. J Biol Chem 268: 25803—25810

Hakuba N, Murakami Y, Noma M, Kudo T and Horikoshi K (1989) Amino acid sequence of *Mirabilis* antiviral protein, total synthesis of its gene and expression in *Escherichia coli*. J Biol Chem 264: 6629—6637

Halim AH, Wassom CE, Mitchell HL and Edmunds LK (1973) Suppression of fungal growth by isolated trypsin inhibitors of corn grain. J Agr Food Chem 21: 1118—1119

Ham K-S, Wu S-C, Darvill AG and Albersheim P (1997) Fungal pathogens secrete an inhibitor protein that distinguishes isoforms of plant pathogenesis-related endo-β-1,3-glucanases. Plant J 11: 169—179

Hart PJ, Monzingo AF, Ready MP, Ernest SR and Robertus JD (1993) Crystal structure of an endochitinase from *Hordeum vulgare* L. seeds. J Mol Biol 229: 189—193

Hartley MR, Chaddock JA and Bonness MS (1996) The structure and function of ribosome-inactivating proteins. Trends Plant Sci 1: 254—260

Hase T, Matsubara H and Yoshizumi H (1978) Disulfide bonds of purothionin, a lethal toxin for yeasts. J Biochem 83: 1671—1678

Hause B, Zur Nieden U, Lehmann J, Wasternack C and Parthier B (1994) Intracellular localization of jasmonate-induced proteins in barley leaves. Bot Acta 107: 333-341

Hayes CK, Klemsdal S, Lorito M, Di Pietro A, Peterbauer C, Nakas JP, Tronsmo A and Harman GE (1994) Isolation and sequence of an endochitinase-encoding gene from a cDNA library of *Trichoderma harzianum*. Gene 138: 143—148

Heitz A, Chiche L, Le-Nguyen D and Castro B (1989) ^{1}H 2D NMR and distance geometry study of the folding of *Ecballium elaterium* trypsin inhibitor, a member of the squash inhibitors family. Biochemistry 28: 2392—2398

Heitz T, Fritig B and Legrand M (1994a) Local and systemic accumulation of pathogenesis-related proteins in tobacco plants infected with tobacco mosaic virus. Mol Plant-Microbe Interact 7: 776—779

Heitz T, Segond S, Kauffmann S, Geoffroy P, Prasad V, Brunner F, Fritig B and Legrand M (1994b) Molecular characterization of a novel pathogenesis-related (PR) protein: a new plant chitinase/lysozyme. Mol Gen Genet 245: 246—254

Hejgaard J, Jacobsen S and Svendsen I (1991) Two antifungal thaumatin-like proteins from barley grain. FEBS Lett 291: 127—131

Hejgaard J, Jacobsen S, Bjørn SE and Kragh KM (1992) Antifungal activity of chitin-binding PR-4 type proteins from barley grain and stressed leaf. FEBS Lett 307: 389—392

Hennig J, Deweg RE, Cutt JR and Klessig DF (1993) Pathogen, salicylic acid and developmental dependent expression of a β-1,3-glucanase/GUS gene fusion in transgenic tobacco plants. Plant J 4: 481—493

Hennig M, Jansonius JN, Terwissscha van Scheltinga AC, Dijkstra BW and Schlesier B (1995a) Crystal structure of concanavalin B at 1.65 Å resolution. An "inactivated" chitinase from seeds of *Canavalia ensiformis*. J Mol Biol 254: 237—246

Hennig M, Schlesier B, Pfeffer-Hennig S, Dauter Z, Nong VH and Wilson KS (1995b) Crystal structure of narbonin at 1.8 Å resolution. Acta Crystallog D 51: 177—189

Herget T, Schell J and Schreier PH (1990) Elicitor-specific induction of one member of the chitinase gene family in *Arachis hypogea*. Mol Gen Genet 224: 469—476

Hernández-Lucas C, Carbonero P and García-Olmedo F (1978) Identification and purification of a purothionin homologue from rye (*Secale cereale* L.). J Agric Food Chem 26: 794—796

Higgins TJV (1984) Synthesis and regulation of major proteins in seeds. Annu Rev Plant Physiol 35: 191—221

Hincha DK, Meins F Jr and Schmitt JM (1997) β-1,3-glucanase is cryoprotective in vitro and is accumulated in leaves during cold acclimation. Plant Physiol 114: 1077—1083

Hiraiwa N, Kondo M, Nishimura M and Hara-Nishimura I (1997) An aspartic endopeptidase is involved in the breakdown of storage proteins in protein-storage vacuoles of plants. Eur J Biochem 246: 133—141

Hird DL, Worrall D, Hodge R, Smartt S, Paul W and Scott R (1993) The anther-specific protein encoded by the *Brassica napus* and *Arabidopsis thaliana* A6 gene diplays similarity to β-1,3-glucanases. Plant J 4: 1023—1033

Hiscock SJ, Doughty J, Willis AC and Dickinson HG (1995) A 7-kDa pollen coating-borne peptide from *Brassica napus* interacts with S-locus glycoprotein and S-locus-related glycoprotein. Planta 196: 367—374

Høj PB and Fincher GB (1995) Molecular evolution of plant β-glucan endohydrolases. Plant J 7: 367—379

Hollak CEM, Van Weely S, van Oers MHJ and Aerts JMFG (1994) Marked elevation of plasma chitotriosidase activity. A novel hallmark of Gaucher disease. J Clin Invest 93: 1288—1292

Holm L and Sander C (1994) Structural similarity of plant chitinases and lysozymes from animals and phage. An evolutionary connection. FEBS Lett 340: 129—132

Holtorf S, Apel K and Bohlmann K (1995) Specific and different expression patterns of two members of the leaf thionin multigene family of barley in transgenic tobacco. Plant Sci 111: 27—37

Hon W-C, Griffith M, Mlymarz A, Kwok YC and Young DSC (1995) Antifreeze proteins in winter rye are similar to pathogenesis-related proteins. Plant Physiol 109: 879—889

Hong JC, Nagao RT and Key JL (1989) Developmentally regulated expression of soybean proline-rich cell wall protein genes. Plant Cell 1: 937—943

Hong Y, Saunders K, Hartley MR and Stanley J (1996) Resistance to geminivirus infection by virus-induced expression of dianthin in transgenic plants. Virology 220: 119—127

Howard JB and Glazer AN (1969) Papaya lysozyme. Terminal sequences and enzymatic properties. J Biol Chem 244: 1399—1409

Howard JB and Glazer AN (1969) Papaya lysozyme Terminal sequences and enzymatic properties. J Biol Chem 244: 1399—1409

Hrmova M and Fincher GB (1993) Purification and properties of three (1-3)-β-D-glucanase isoenzymes from young leaves of barley (*Hordeum vulgare*). Biochem J 289: 453—461

Hrmova M, Garrett TPJ and Fincher GB (1995) Subsite affinities and disposition of catalytic amino acids in the subsite-binding region of barley 1,3-β-glucanases. J Biol Chem 270: 14556—14563

Huesing JE, Murdock LL and Shade RE (1991) Effect of wheat germ isolectins on development of cowpea weevils. Phytochemistry 30: 785—788

Hunt MD, Delaney TP, Dietrich RA, Weymann KB, Dangl JL and Ryals JA (1997) Salicylate-independent lesion formation in Arabidopsis *lsd* mutants. Mol Plant-Microbe Interact 10: 531—536

Huynh QK, Borgmeyer JR and Zobel JF (1992) Isolation and characterization of a 22 kDa protein with antifungal properties from maize seeds. Biochem Biophys Res Comm 182: 1—5

Irvin JD (1975) Purification and partial characterization of the antiviral protein from *Phytolacca americana* which inhibits eukaryotic protein synthesis. Arch Biochem Biophys 169: 522—528

Irvin JD and Uckun FM (1992) Pokeweed antiviral protein: ribosome inactivation and therapeutic applications. Pharmac Ther 55: 279—302

Iseli B, Boller T and Neuhaus J-M (1993) The N-terminal cysteine-rich domain of tobacco class I chitinase is essential for chitin binding but not for catalytic or antifungal activity. Plant Physiol 103: 221—226

Ishibashi N, Yamauchi D and Minamikawa T (1990) Stored mRNA in cotyledons of *Vigna unguiculata* seeds: nucleotide sequence of cloned cDNA for a stored mRNA and induction of its synthesis by precocious germination. Plant Mol Biol 15: 59—64

Jach G, Gornhardt B, Mundy J, Logemann J, Pinsdorf E, Leah R, Schell J and Maas C (1995) Enhanced quantitative resistance against fungal diseases by combinatorial expression of different barley antifungal proteins in transgenic tobacco. Plant J 8: 97—109

Jekel PA, Hartmann BH and Beintema JJ (1991) The primary structure of hevamine, an enzyme with lysozyme/chitinase activity from *Hevea brasiliensis* latex. Eur J Biochem 200: 123—130

Ji C and Kuc J (1996) Antifungal activity of cucumber β-1,3-glucanase and chitinase. Physiol Mol Plant Pathol 49: 257—265

Jin HM, Copeland NG, Gilbert DJ, Jenkins NA, Kirkpatrick RB and Rosenberg M (1998) Genetic characterization of the murine Yml gene and identification of a cluster of highly homologous genes. Genomics 54: 316—322

Johnson R, Narvaez J, An G and Ryan C (1989) Expression of proteinase inhibitors I and II in transgenic tobacco plants: Effects on natural defense against *Manduca sexta* larvae. Proc Natl Acad Sci USA 86: 9871—9875

Jones BL and Cooper DB (1980) Purification and characterization of a corn (*Zea mays*) protein similar to purothionins. J Agric Food Chem 28: 904—928

Jones JDG, Grady KL, Suslow TV and Bedbrook JR (1986) Isolation and characterization of genes encoding two chitinase enzymes from *Serratia marcescens*. EMBO J 5: 467—473

Joosten MHAJ, Bergmans CJB, Meulenhoff EJS, Cornelissen BJC and De Wit PJGM (1990) Purification and serological characterization of three basic 15-kilodalton pathogenesis-related proteins from tomato. Plant Physiol 94: 585—591

Joosten MHAJ, Verbakel HM, Nettekoven ME, Van Leeuwen J, Van der Vossen RTM and De Wit PJGM (1995) The phytopathogenic fungus *Cladosporium fulvum* is not sensitive to the chitinase and β-1,3-glucanase defence proteins of its host, tomato. Physiol Mol Plant Pathol 46: 45—59

Josè-Estanyol M, Ruiz-Avila L and Puigdomènech P (1992) A maize embryo-specific gene encodes a proline-rich and hydrophobic protein. Plant Cell 4: 413—423

Kader J-C (1997) Lipid-transfer proteins: a puzzling family of plant proteins. Trends Plant Sci 2: 66—70

Kader J-C, Grosbois M, Guerbette F, Jolliot A and Oursel A (1996) Lipid transfer proteins: structure, function and gene expression. In: Smallwood M, Knox JP and Bowles DJ (eds) Membranes: Specialized functions in plants, pp 165—178. Bios Scientific Publishers Ltd, Oxford, UK

Kamei K, Takano R, Miyasaka A, Imoto T and Hara S (1992) Amino acid sequence of sweet-taste-suppressing peptide (gurmarin) from the leaves of *Gymnema sylvestre*. J Biochem 111: 109—112

Karunanandaa B, Singh A and Kao T-H (1994) Characterization of a predominantly pistil-expressed gene encoding a γ-thionin-like protein of *Petunia inflata*. Plant Mol Biol 26: 459—464

Kasahara M, Gutknecht J, Brew K, Spurr N and Goodfellow PN (1989) Cloning and mapping of a testis-specific gene with sequence similarity to a sperm-coating glycoprotein gene. Genomics 5: 527—534

Kauffmann S, Legrand M, Geoffroy P and Fritig B (1987) Biological function of "pathogenesis-related" proteins. Four PR-proteins of tobacco have 1,3-β-glucanase activity. EMBO J 6: 3209—3212

Kawano K, Yonega T, Miyata T, Yoshikawa K, Tokunaga F, Terada Y and Iwanaga S (1990) Antimicrobial peptide, tachyplesin I, isolated from hemocytes of the horseshoe crab (*Tachypleus tridentatus*). NMR determination of the β-sheet structure. J Biol Chem 265: 15365—15367

Keefe D, Hinz U and Meins F (1990) The effect of ethylene on the cell-type specific and intracellular localization of β-1,3-glucanase and chitinase in tobacco leaves. Planta 182: 43—51

Keen NT and Yoshikawa M (1983) β-1,3-endoglucanase from soybean releases elicitor-active carbohydrates from fungus cell walls. Plant Physiol 71: 460—465

King GJ, Turner VA, Hussey CE, Wurtele ES and Lee SM (1988) Isolation and characterization of tomato cDNA clone which codes for a salt-induced protein. Plant Mol Biol 10: 401—412

King TP (1990) Insect venom allergens. Monogr Allergy 28: 84—100

Kirkpatrick RB, Matico RE, McNulty DE, Strickler JE and Rosenberg M (1995) An abundantly secreted glycoprotein from *Drosophila melanogaster* is related to mammalian secretory proteins produced in rheumatoid tissues and by activated macrophages. Gene 153: 147—154

Knoester M, Van Loon, LC, Vanden Heuvel J, Hennig J, Bol JF and Linthorst HJM (1998) Ethylene-insensitive tobacco lacks nonhost resistance against soil-borne fungi. Proc Natl Acad Sci USA 95: 1933—1937

Kobayashi Y, Sato A, Takashima H, Tamaoki H, Nishimura S, Kyogoku Y, Ikenaka K, Kondo I, Mikoshika K, Hojo H, Aimoto S and Moroder L (1991) A new α-helical motif in membrane active peptides. Neurochem Int 18: 523—534

Koide RT and Schreiner RP (1992) Regulation of the vesicular-arbuscular mycorrhizal symbiosis. Annu Rev Plant Physiol Plant Mol Biol 43: 557—581

Koiwa H, Sato F and Yamada Y (1994) Characterization of accumulation of tobacco PR-5 proteins by IEF-immunoblot analysis. Plant Cell Physiol 35: 821—827

Koltunow AM, Truettner J, Cox KH, Wallroth M and Goldberg RB (1990) Different temporal and spatial expression patterns occur during anther development. Plant Cell 2: 1201—1224

Kononowicz AK, Nelson DE, Singh NK, Hasegawa PM and Bressan RA (1992) Regulation of the osmotin gene promoter. Plant Cell 4: 513—524

Koo JC, Lee SY, Chun HJ, Cheong YH, Choi JS, Kawabata S, Miyagi M, Tsunasawa S, Ha KS, Bae DW, Han CD, Lee BL and Cho MJ (1998) Two hevein homologs isolated from the seed of *Pharbitis nil* L. exhibit potent antifungal activity. Biochim Biophys Acta 1382: 80—90

Kragh KM, Hendriks T, De Jong A, Lo Schiavo F, Bucherna N, Hojrup P, Mikkelsen JD and De Vries SC (1996) Characterization of chitinases able to rescue somatic embryos of the temperature-sensitive carrot variant *ts11*. Plant Mol Biol 31: 631—645

Kragh KM, Nielsen JE, Nielsen KK, Dreboldt S and Mikkelsen JD (1995) Characterization and localization of new antifungal cysteine-rich proteins from *Beta vulgaris*. Mol Plant-Microbe Interact 8: 424—434

Kramer KJ, Corpuz L, Choi HK and Muthukrishnan S (1993) Sequence of a cDNA and expression of the gene encoding epidermal and gut chitinases of *Manduca sexta*. Insect Biochem Mol Biol 23: 691—701

Kramer KJ, Klassen LW, Jones BL, Speirs RD and Kammer AE (1979) Toxicity of purothionin and its homologues to the tobacco hornworm, *Manduca sexta* (L.) (Lepidoptera: Sphingidae). Toxicol Appl Pharmacol 48: 179—183

Kraulis PJ, Clore GM, Nilges M, Jones TA, Petterson G, Knowles J and Gronenborn AM (1989) Determination of the three-dimensional solution structure of the C-terminal domain of cellobiohydrolase I from *Trichoderma reesei*. A study using nuclear magnetic resonance and hybrid distance geometry-dynamic simulated annealing. Biochemistry 28: 7241—7257

Krebbers E, Herdies L, De Clercq A, Seurinck J, Leemans J, Van Damme J, Segura M, Gheysen G, Van Montagu M and Vandekerckhove J (1988) Determination of the processing sites of an *Arabidopsis* 2S albumin and characterization of the complete gene family. Plant Physiol 87: 859—866

Kumar V and Spencer ME (1992) Nucleotide sequence of an osmotin cDNA from the *Nicotiana tabacum* cv. White burley generated by the polymerase chain reaction. Plant Mol Biol 18: 621—622

Kung S, Kimura M and Funatsu G (1990) The complete amino acid sequence of antiviral protein from the seeds of pokeweed (*Phytolacca americana*). Agric Biol Chem 34: 3301—3318

Kuranda MJ and Robbins PW (1991) Chitinase is required for cell separation during growth of *Saccharomyces cerevisiae*. J Biol Chem 266: 19758—19767

Lam Y-H, Wong Y-S, Wang B, Wong RN-S, Yeung H-W, Shaw P-C (1996) Use of trichosanthin to reduce infection by turnip mosaic virus. Plant Sci 114: 111—117

Lamb C and Dixon RA (1997) The oxidative burst in plant disease resistance. Annu Rev Plant Physiol Plant Mol Biol 48: 251—275

Lambert J, Keppi E, Dimarcq J-L, Wicker C, Reichhart J-M, Dunbar B, Lepage P, Van Dorsselaer A, Hoffmann J, Fothergill J and Hoffmann D (1989) Insect immunity: Isolation from immune blood of the dipteran *Phormia terranovae* of two insect antibacterial peptides with sequence homology to rabbit lung macrophage bactericidal peptides. Proc Natl Acad Sci USA 89: 262—266

Landon C, Cornet B, Bonmatin J-M, Kopeyan C, Rochat H, Vovelle F and Ptak M (1996) ^{1}H-NMR-derived secondary structure and the overall fold of the potent anti-mammal and anti-insect toxin III from the scorpion *Leiurus quinquestriatus quinquestriatus*. Eur J Biochem 236: 395—404

Landon C, Sodano P, Hétru C, Hoffmann C and Ptak M (1997) Solution structure of drosomycin, the first inducible antifungal protein from insects. Protein Sci 6: 1878—1884

Lange J, Mohr U, Wiemken A, Boller T and Vögeli-Lange R (1996) Proteolytic processing of class IV chitinase in the compatible interaction of bean roots with *Fusarium solani*. Plant Physiol 111: 1135—1144

LaRosa PC, Chen Z, Nelson DE, Singh NK, Hasegawa PM and Bressan RA (1992) Osmotin gene expression is posttranscriptionally regulated. Plant Physiol 100: 409—415

Last DI and Llewellyn DJ (1997) Antifungal proteins from seeds of australian native plants and isolation of an antifungal peptide from *Atriplex nummularia*. New Zealand J Bot 35: 385—394

Lawrence CB, Joosten MHAJ and Tuzun S (1996) Differential induction of pathogenesis-related proteins in tomato by *Alternaria solani* and the association of a basic chitinase isozyme with resistance. Physiol Mol Plant Pathol 48: 377—383

Lawton K, Uknes S, Friedrich L, Gaffney T, Alexander D, Goodman R, Métraux JP, Kessmann H, Ahl-Goy P, Gut Rella M, Ward E and Ryals J (1993) The molecular biology of systemic acquired resistance. In: Fritig B and Legrand M (eds) Mechanisms of Plant Defense Responses, pp 422-432, Kluwer Academic Publishers

Lawton K, Ward E, Payne G, Mayer M and Ryals J (1992) Acidic and basic class III chitinase mRNA accumulation in response to TMV infection in tobacco. Plant Mol Biol 19: 735—743

Lawton K, Weymann K, Friedrich L, Vernooij B, Uknes S and Ryals J (1995) Systemic acquired resistance in Arabidopsis requires salicylic acid but not ethylene. Mol Plant Microbe Interact 8: 863—870

Lawton KA, Potter SL, Uknes S and Ryals J (1994) Acquired resistance signal transduction in Arabidopsis is ethylene independent. Plant Cell 6: 581—588

Le Bihan T, Blochet J-E, Désormaux A, Marion D and Pézolet M (1996) Determination of the secondary structure and conformation of puroïndolines by infrared and Raman spectroscopy. Biochemistry 35: 12712—12722

Leah R, Tommerup H, Svendsen I and Mundy J (1991) Biochemical and molecular characterization of three barley seed proteins with antifungal properties. J Biol Chem 266: 1564—1573

Lecomte JTJ, Jones BL and Llinas M (1982) Proton magnetic resonance studies of barley and wheat thionins: structural homology with crambin. Biochemistry 21: 4843—4849

Lecomte JTJ, Kaplan D, Llinas M, Thunberg E and Samuelsson G (1987) Proton magnetic resonance characterization of phoratoxins and homologous proteins related to crambin. Biochemistry 26: 1187—1194

Lee H-I and Raikhel NV (1995) Prohevein is poorly processed but shows enhanced resistance to a chitin-binding fungus in transgenic tomato plants. Braz J Med Biol Res 28: 743—750

Lee H-I, Broekaert WF and Raikhel NV (1991) Co- and post-translational processing of the hevein preproprotein of latex of the rubber tree (*Hevea brasiliensis*). J Biol Chem 266: 15944—15948

Lee T, Crowell M, Shearer MH, Aron GH and Irvin JD (1990) Poliovirus-mediated entry of pokeweed antiviral protein. Antimicrob Agents Chemother 34: 2034—2037

Lee WM and Prasad UK (1995) Structure-activity studies on magainins and other host defense peptides. Biopolymers 37: 105—122

Legname G, Bellosta P, Gromo G, Modena D, Keen JN, Roberts LM and Lord JM (1991) Nucleotide sequence of cDNA coding for dianthin 30, a ribosome-inactivating protein from *Dianthus caryophyllus*. Biochim Biophys Acta 1090: 119—122

Legrand M, Kauffmann S, Geoffroy P and Fritig B (1987) Biological function of "pathogenesis-related" proteins: four tobacco PR-proteins are chitinases. Proc Natl Acad Sci USA 84: 6750—6754

Lehrer RI, Ganz T and Selsted ME (1991) Defensins: Endogenous antibiotic peptides of animal cells. Cell 64: 229—230

Lehrer RI, Ganz T, Szklarek D and Selsted ME (1988) Modulation of the in vitro candicidal activity of human neutrophil defensin by target cell metabolism and divalent cations. J Clin Invest 81: 1829—1835

Leippe M and Müller-Eberhard HJ (1994) The pore-forming peptide of *Entamoeba histolytica*, the protozoan parasite causing human amoebiasis. Toxicology 87: 5—18

Lemaitre B, Kramer-Metzger E, Michaut L, Nicolas E, Meister M, George P, Reichhart J-M and Hoffmann JA (1995) A recessive mutation, immune deficiency (*imd*), defines two distinct control pathways in the *Drosophila* host defense. Proc Natl Acad Sci USA 92: 9465—9469

Lemaitre B, Nicolas E, Michaut L, Reichhart J-M and Hoffmann JA (1996) The dorsoventral regulatory gene cassette *spätzle/Toll/cactus* controls the potent antifungal response in Drosophila adults. Cell 86: 973—983

Le-Nguyen DL, Heitz A, Chiche L, Castro B, Baigegrain F, Favel A and Coletti-Previero (1990) Molecular recognition between serine proteases and new bioactive microproteins with a knotted structure. Biochimie 72: 431—435

Lerner DR and Raikhel NV (1992) The gene for stinging nettle lectin (*Urtica dioica* agglutinin) encodes both a lectin and a chitinase. J Biol Chem 267: 11085—11091

Leubner-Metzger G, Fründt C, Vögeli-Lange F and Meins F Jr (1995) Class I β-1,3-glucanases in the endosperm of tobacco during germination. Plant Physiol 109: 751—759

Levine A, Tenhaken R, Dixon R and Lamb C (1994) H_2O_2 from the oxidative burst orchestrates the plant hypersensitive disease resistance response. Cell 79: 583—593

Liao YC, Kreuzaler F, Fisher R, Reisener H-J and Tiburzy R (1994) Characterization of a wheat class Ib chitinase gene differentially induced in isogenic lines by infection with *Puccinia graminis*. Plant Sci 103: 177—187

Lin Q, Chen ZC, Antoniw JF and White RF (1991) Isolation and characterization of a cDNA clone encoding the antiviral protein from *Phytolacca americana*. Plant Mol Biol 17: 609—614

Lin W, Anuratha CS, Datta K, Potrykus I, Muthukrishnan S and Datta SK (1995) Genetic engineering of rice for resistance to sheath blight. Bio/Technol 13: 686—691

Ling J, Li X-D, Wu X-H and Liu W-Y (1995) Topological requirements for recognition and cleavage of DNA by ribosome-inactivating proteins. Biol Chem Hoppe-Seyler 376: 637—641

Linthorst HJM, Melchers LS, Mayer A, Van Roekel JS, Cornelissen BJ and Bol JF (1990a) Analysis of gene families encoding acidic and basic β-1,3-glucanases of tobacco. Proc Natl Acad Sci USA 87: 8756—8760

Linthorst HJM, Van Loon LC, van Rossum CM, Mayer A, Bol JF, Van Roekel JS, Meulenhoff EJ and Cornelissen BJ (1990b) Analysis of acidic and basic chitinases from tobacco and petunia and their constitutive expression in transgenic tobacco. Mol Plant-Microbe Interact 3: 252—258

Linthorst HJM (1991) Pathogenesis-related proteins of plants. Crit Rev Plant Sci 10: 123—150

Linthorst HJM, Danhash N, Brederode FT, Van Kan JAL, De Wit PJGM and Bol JF (1991) Tobacco and tomato PR proteins homologous to *win* and pro-hevein lack the "hevein" domain. Mol Plant-Microbe Interact 4: 585—592

Linthorst HJM, Meuwissen RLJ, Kauffmann S and Bol JF (1989) Constitutive expression of pathogenesis-related proteins PR-1, GRP, and PR-S in tobacco has no effect on virus infection. Plant Cell 1: 285—291

Liu D, Raghothama KG, Hasegawa PM and Bressan RA (1994) Osmotin overexpression in potato delays development of disease symptoms. Proc Natl Acad Sci USA 91: 1888-1892

Lodge JK, Kaniewski WK and Turner NE (1993) Broad-spectrum virus resistance in transgenic plants expressing pokeweed antiviral protein. Proc Natl Acad Sci USA 90: 7089—7093

Logemann J, Jach G, Tommerup H, Mundy J and Schell J (1992) Expression of a barley ribosome-inactivating protein leads to increased fungal protection in transgenic tobacco plants. Biotechnology 10: 305—308

Lord JM, Roberts LM and Robertus JD (1994) Ricin: structure, mode of action and some current applications. FASEB J 8: 201—208

Lorito M, Woo SL, Díambrosio M, Harman GE, Hayes CK, Kubicek CP and Scala F (1996) Synergistic enhancement between cell wall degrading enzymes and membrane affecting compounds. Mol Plant-Microbe Interact 9: 206—213

Lotan T, Ori N, Fluhr R (1989) Pathogenesis-related proteins are developmentally regulated in tobacco flowers. Plant Cell 1: 881—887

Lu G, Villalba M, Coscia MR, Hoffman DR and King TP (1993) Sequence analysis and antigenic cross-reactivity of a venom allergen, antigen 5, from hornets, wasps, and yellow jackets. J Immunol 150: 2823—2830

Lucas J, Camacho Henriquez A, Lottspeich F, Henschen A and Sänger HL (1985) Amino acid sequence of the 'pathogenesis-related' leaf protein p14 from viroid-infected tomato reveals a new type of structurally unfamiliar proteins. EMBO J 4: 2745—2749

Ludvigsen S and Poulsen FM (1992) Three-dimensional structure in solution of barwin, a protein from barley seed. Biochemistry 31: 8783—8789

Lusso M and Kuc J (1996) The effect of sense and antisense expression of the PR-N gene for β-1,3-glucanase on disease resistance of tobacco to fungi and viruses. Physiol Mol Plant Pathol 49: 267—283

Ma D-P, Liu H-C, Tan H, Creech RG, Jenkins JN and Chang Y-F (1997) Cloning and characterization of a cotton lipid transfer protein gene specifically expressed in fiber cells. Biochim Biophys Acta 1344: 111—114

Majeau N, Trudel J and Asselin A (1990) Diversity of cucumber isoforms and characterization of one seed basic chitinase with lysozyme activity. Plant Sci 68: 9—16

Malehorn DE, Borgmeyer JR, Smith CE and Shah DM (1994) Characterization and expression of an antifungal zeamatin-like protein (*Zlp*) gene from *Zea mays*. Plant Physiol 106: 1471—1481

Marcus JP, Goulter KC, Green JL, Harrison SJ and Manners JM (1997) Purification, characterisation and cDNA cloning of an antimicrobial peptide from *Macadamia integrifolia*. Eur J Biochem 244: 743—749

Margis-Pinheiro M, Metz-Boutigue MH, Awada A, De Tapia M, Le Ret M and Burkard G (1991) Isolation of a complementary DNA encoding the bean PR4 chitinase: an acidic enzyme with an amino-terminus cysteine-rich domain. Plant Mol Biol 17: 243—253

Martin GB, Brommonschenkel SH, Chunwongse J, Frary A, Ganal MW, Spivey R, Wu T, Earle ED and Tanksley SD (1993) Map-based cloning of a protein kinase gene conferring disease resistance in tomato. Science 262: 1432—1436

Martins JC, Maes D, Loris R, Pepermans HAM, Wyns L, Willem R and Verheyden P (1996) ¹H NMR study of the solution structure of Ac-AMP2, a sugar binding antimicrobial protein isolated from *Amaranthus caudatus*. J Mol Biol 258: 322—333

Masoud SA, Zhu Q, Lamb C and Dixon RA (1996) Constitutive expression of an inducible β-1,3-glucanase in alfalfa reduces disease severity caused by the oomycete pathogen *Phytophtora megasperma* f.sp. *medicaginis*, but does not reduce disease severity of chitin-containing fungi. Transgenic Res. 5: 313—323

Mauch F, Hadwiger LA and Boller T (1988a) Antifungal hydrolases in pea tissue. I. Purification and characterization of two chitinases and two β-1,3-glucanases differentially regulated during development and in response to fungal infection. Plant Physiol 87: 325-333

Mauch F, Mauch-Mani B and Boller T (1988b) Antifungal hydrolases in pea tissue. II. Inhibition of fungal growth by combinations of chitinase and β-1,3-glucanase. Plant Physiol 88: 936—942

Mauch F, Meehl JB and Staehelin LA (1992) Ethylene-induced chitinase and β-1,3-glucanase accumulate specifically in the lower epidermis and along vascular strands of bean leaves. Planta 186: 367—375

Mauch-Mani B and Slusarenko AJ (1994) Systemic acquired resistance in *Arabidopsis thaliana* induced by a predisposing infection with a pathogenic isolate of *Fusarium oxysporum*. Mol Plant-Microbe Interact 7: 378—383

Mehdy MC (1994) Active oxygen species in plant defense against pathogens. Plant Physiol 105: 467—472

Meijer EA, De Vries SC, Sterk P, Gadella DWJ Jr, Wirtz KWA and Hendriks T (1993) Characterization of the non-specific lipid transfer protein EP2 from carrot (*Daucus carota* L.). Mol Cell Biochem 123: 159—166

Meins F JR and Ahl P (1989) Induction of chitinase and β-1,3-glucanase in tobacco plants infected with *Pseudomonas tabaci* and *Phytophthora parasitica* var. *nicotianae*. Plant Sci 61: 155-161

Melchers LS, Sela-Buurlage MB, Vloemans SA, Woloshuk CP, Van Roekel JSC, Pen J, Van den Elzen PJM and Cornelissen BJC (1993) Extracellular targeting of the vacuolar tobacco proteins AP24, chitinase and β-1,3-glucanase in transgenic plants. Plant Mol Biol 21: 583—593

Melchers LS, Apotheker-de Groot M, Van der Knaap JA, Ponstein AS, Sela-Buurlage MB, Bol JF, Cornelissen BJC, Van den Elzen PJM and Linthorst HJM (1994) A new class of tobacco chitinases homologous to bacterial exo-chitinases display antifungal activity. Plant J 5: 469—480

Mellstrand ST and Samuelsson G (1974) Phoratoxin, a toxic protein from the mistletoe *Phoradendron tomentosum* subsp. *macrophyllum* (Loranthaceae). The amino acid sequence. Acta Pharm Suec 11: 347—360

Mellstrand ST and Samuelsson, G (1973) Phoratoxin, a toxic protein from the mistletoe *Phoradendron tomentosum* subsp. *macrophyllum* (Loranthaceae). Improvements in the isolation procedure and further studies on the properties. Eur J Biochem 32: 143—147

Memelink JH, Linthorst HJM, Schilperoort RA and Hoge JHC (1990) Tobacco genes encoding acidic and basic isoforms of pathogenesis-related proteins display different expression patterns. Plant Mol Biol 14: 119—126

Méndez E, Moreno A, Colilla F, Pelaez F, Lìmas GG, Mendez R, Soriano F, Salinas M and de Haro C (1990) Primary structure and inhibition of protein synthesis in eukaryotic cell-free system of a novel thionin, γ-hordothionin, from barley endosperm. Eur J Biochem 194: 533—539

Méndez E, Rocher A, Calero M, Girbés T, Citores L and Soriano F (1996) Primary structure of ω-hordothionin, a member of a novel family of thionins from barley endosperm, and its inhibition of protein synthesis in eukaryotic and prokaryotic cell-free systems. Eur J Biochem 239: 67—73

Métraux J-P, Burkhart W, Moyer M, Dincher S, Middlesteadt W, Williams S, Payne G, Carnes M and Ryals J (1989) Isolation of a complementary DNA encoding a chitinase with structural homology to a bifunctional lysozyme/chitinase. Proc Natl Acad Sci USA 86: 896—900

Metzler MC, Cutt JR and Klessig DF (1991) Isolation and characterization of a gene encoding a PR-1-like protein from *Arabidopsis thaliana*. Plant Physiol 96: 346—348

Metzler WJ, Valentine K, Roebber M, Marsh DG and Mueller L (1992) Proton resonance assignments and three-dimensional solution structure of the ragweed allergen Amb a V by nuclear magnetic resonance spectroscopy. Biochemistry 31: 8697-8705

Meyer B, Houlné G, Pozueta-Romero J, Schantz M-L and Schantz R (1996) Fruit-specific expression of a defensin-type gene family in bell pepper. Upregulation during ripening and upon wounding. Plant Physiol 112: 615—622

Milligan SB and Gasser CS (1995) Nature and regulation of pistil-expressed genes in tomato. Plant Mol Biol 28: 691—711

Miyata T, Tokunaga F, Yoneya T, Yoshikawa K, Iwanaga S, Niwa M, Takao T and Shimonishi Y (1989) Antimicrobial peptides, isolated from horseshoe crab hemocytes, tachyplesin II, and polyphemusins I and II: chemical structures and biological activity. J Biochem 106: 663—668

Mol JNM, Van Blockland R, De Lange P, Stam M and Kooter JM (1994) Post-transcriptional inhibition of gene expression: sense and antisense genes. In: Paszkowski J (ed), Gene Inactivation and Homologous Recombination in Plants. pp 309—334, Kluwer, Dordrecht

Molano J, Polacheck I, Duran A and Cabib E (1979) An endochitinase from wheat germ. J Biol Chem 254: 4901—4907

Mölders W, Buchala A and Métraux J-P (1996) Transport of salicylic acid in tobacco necrosis virus-infected cucumber plants. Plant Physiol 112: 787—792

Mole LE, Goodfriend L, Lapkoff CB, Kehoe JM and Capra JD (1975) The amino acid sequence of ragweed pollen allergen Ra5. Biochemistry 14: 1216—1220

Molina A and García-Olmedo F (1993) Developmental and pathogen-induced expression of three barley genes encoding lipid transfer proteins. Plant J 4: 983—991

Molina A and Garcia-Olmedo F (1997) Enhanced tolerance to bacterial pathogens caused by the transgenic expression of barley lipid transfer protein LTP2. Plant J 12: 669—675.

Molina A, Ahl-Goy P, Fraile A, Sánchez-Monge R and García-Olmedo F (1993a) Inhibition of bacterial and fungal plant pathogens by thionins of types I and II. Plant Sci 92: 169—177

Molina A, Diaz I, Vasil IK, Carbonero P and Garcia-Olmedo F (1996) Two cold-inducible genes encoding lipid transfer protein LTP4 from barley show differential responses to bacterial pathogens. Mol Gen Genet 252: 162—168

Molina A, Segura A and García-Olmedo F (1993b) Lipid transfer proteins (nsLTPs) from barley and maize leaves are potent inhibitors of bacterial and fungal plant pathogens. FEBS Lett 316: 119—122

Moore AE and Stone BA (1972a) A β-1,3-glucan hydrolase from *Nicotiana glutinosa*. I. Extraction, purification and physical properties. Biochim Biophys Acta 258: 238—247

Moore AE and Stone BA (1972b) A β-1,3-glucan hydrolase from *Nicotiana glutinosa*. II. Specificity, action pattern and inhibitor studies. Biochem Biophys Acta 258: 248—264

Moreno M, Segura A and García-Olmedo F (1994) Pseudothionin-St1, a potato peptide active against potato pathogens. Eur J Biochem 223: 135—139

Morimoto M, Mori H, Otake T, Ueba N, Kumita N, Niwa M, Murakami T and Iwanaga S (1991) Inhibitory effects of tachyplesin I on the proliferation of human immunodeficiency virus in vitro.

Chemotherapy 37: 206—211

Mouradov A, Mouradova E and Scott KJ (1994) Gene family encoding basic pathogenesis-related 1 proteins in barley. Plant Mol Biol 26: 503—507

Mundy J and Rogers JC (1986) Selective expression of a probable α-amylase/protease inhibitor in barley aleurone cells: comparison to the barley α-amylase/subtilisin inhibitor. Planta 169: 51—63

Murakami T, Niwa M, Tokunaga F, Miyota T and Iwanaga S (1991) Direct virus inactivation of tachyplesin I and its isopeptides from horseshoe crab hemocytes. Chemotherapy 37: 327—334

Murphy EV, Zhang Y, Zhu W and Biggs J (1995) The human glioma pathogenesis-related protein is structurally related to plant pathogenesis-related proteins and its gene is expressed specifically in brain tumors. Gene 159: 131—135

Muta T, Fujimoto T, Nakajima H and Iwanaga S (1990) Tachyplesins isolated from hemocytes of Southeast Asian horseshoe crabs (*Carcinoscorpius rotundicauda* and *Tachypleus gigas*): identification of a new tachyplesin, tachyplesin III, and a processing intermediate of its precursor. J Biochem 108: 261—266

Nacken WKF, Huijsen P, Beltran J-P, Saedler H and Sommer H (1991) Molecular characterization of two stamen-specific genes, *tap1* and *fil1*, that are expressed in the wild type, but not in the *deficiens* mutant of *Antirrhinum majus*. Mol Gen Genet 229: 129—136

Nakamura T, Furunaka F, Miyata T, Tokunaga F, Muta T, Iwanaga S, Niwa M, Takao T and Shimonishi Y (1988) Tachyplesin, a class of antimicrobial peptide from the hemocytes of the horseshoe crab (*Tachypleus tridentatus*). Isolation and chemical structure. J Biol Chem 263: 16709—16713

Nasrallah JB, Kao T-H, Chen C-H, Goldberg ML and Nasrallah ME (1987) Amino-acid sequence of glycoproteins encoded by three alleles of the *S* locus of *Brassica oleraceae*. Nature 326: 617—619

Neale AD, Wahleithner JA, Lund M, Bonnett HT, Kelly A, Meeks-Wagner DR, Peacock WJ and Dennis ES (1990) Chitinase, β-1,3-glucanase, osmotin and extensin are expressed in tobacco explants during flower formation. Plant Cell 2: 673-684

Nelson DE, Raghothama KG, Singh NK, Hasegawa PM and Bressan RA (1992) Analysis of structure and transcriptional activation of an osmotin gene. Plant Mol Biol 19: 577—588

Nemoto N, Kubo S, Yoshida T, Chino N, Kimura T, Sakakibara S, Kyogoku Y and Kobayashi Y (1995) Solution structure of ω-conotoxin MVIIC determined by NMR. Biochem Biophys Res Comm 207: 695—700

Neuhaus J-M, Ahl-Goy P, Hinz U, Flores S and Meins F (1991a) High-level expression of a tobacco chitinase gene in *Nicotiana sylvestris*. Susceptibility of transgenic plants to *Cercospora nicotianae* infection. Plant Mol Biol 16: 141—151

Neuhaus J-M, Sticher L, Meins F Jr and Boller T (1991b) A short C-terminal sequence is necessary and sufficient for the tarteting of chitinases to the plant vacuole. Proc Natl Acad Sci USA 88: 10362—10366

Neuhaus JM, Flores S, Keefe D, Ahl-Goy P and Meins F Jr (1992) The function of vacuolar β-1,3-glucanase investigated by antisense transformation. Susceptibility of transgenic *Nicotiana sylvestris* plant to *Cercospora nicotianae* infection. Plant Mol Biol 19: 803—813

Niderman T, Genetet I, Bruyère T, Gees R, Stintzi A, Legrand M, Fritig B and Mösinger E (1995) Pathogenesis-related PR-1 proteins are antifungal. Isolation and characterization of three 14-kilodalton proteins of tomato and of a basic PR-1 of tobacco with inhibitory activity against *Phytophthora infestans*. Plant Physiol 108: 17—27

Nielsen KK, Bojsen K, Roepstorff P and Mikkelsen JD (1994) A hydroxyproline-containing class IV chitinase of sugar beet is glycosylated with xylose. Plant Mol Biol 25: 241-257

Nielsen KK, Mikkelsen JD, Kragh KM and Bojsen K (1993) An acidic class III chitinase in sugar beet:

induction by *Cercospora beticola*, characterization, and expression in transgenic tobacco plants. Mol Plant-Microbe Interact 6: 495—506

Nielsen JE, Nielsen KK and Mikkelsen JD (1996a) Immunological localization of a basic class IV chitinase in *Beta vulgaris* leaves after infection with *Cercospora beticola*. Plant Sci 119: 191—202

Nielsen KK, Nielsen JE, Madrid SM and Mikkelsen JD (1996b) New antifungal proteins from sugar beet (*Beta vulgaris* L.) showing homology to non-specific lipid transfer proteins. Plant Mol Biol 31: 539—552

Nielsen KK, Nielsen JE, Madrid SM and Mikkelsen JD (1997) Characterization of a new antifungal chitin-binding peptide from sugar beet leaves. Plant Physiol 113: 83—91

Nitti G, Orrù S, Bloch C Jr, Morhy L, Marino G and Pucci P (1995) Amino acid sequence and disulphide-bridge pattern of three γ-thionins from *Sorghum bicolor*. Eur J Biochem 228: 250—256

Odani S, Koide T and Ono T (1983) The complete amino acid sequence of barley trypsin inhibitor. J Biol Chem 258: 7998—8003

Odani S, Koide T, Ono T, Seto Y and Tanaka T (1987) Soybean hydrophobic protein. Isolation, partial characterization and the complete primary structure. Eur J Biochem 162: 485—491

Ogata CM, Gordon PF, de Vos AM and Kim S-H (1992) Crystal structure of a sweet tasting protein thaumatin I, at 1.65 Å resolution. J Mol Biol 228: 893—908

Ohshima M, Harada N, Matsuoka M and Ohashi Y (1990) The nucleotide sequence of pathogenesis-related (PR) 1c protein gene of tobacco. Nucleic Acids Res 18: 182

Okada T and Yoshizumi H (1973) The mode of action of toxic protein in wheat and barley on brewing yeast. Agr Biol Chem 37: 2289—2294

Okada T, Yoshizumi H and Terashima Y (1970) A lethal toxic substance for brewing yeast in wheat and barley. Agric Biol Chem 34: 1084—1088

Olivieri F, Prasad V, Valbonesi P, Srivastava S, Ghosal-Chowdhury P, Barbieri L, Bolognesi A and Stirpe F (1996) A systemic antiviral resistance-inducing protein isolated from *Clerodendrum inerme* Gaertn. is a polynucleotide:adenosine glycosidase (ribosome-inactivating protein). FEBS Lett 396: 132—134

Omecinsky DO, Holub KE, Adams ME and Reily MD (1996) Three-dimensional structure analysis of μ-agatoxins: further evidence for common motifs among neurotoxins with diverse ion channel specificities. Biochemistry 35: 2836—2844

Onaderra M, Monsalve RI, Manche–o JM, Villalba M, Martinez del Pozo A, Gavilanes JG and Rodriguez R (1994) Food mustard allergen interaction with phospholipid vesicles. Eur J Biochem 225: 609—615

Ordentlich A, Elad Y and Chet I (1988) The role of chitinase of *Serratia marcescens* in biocontrol of *Sclerotium rolfsii*. Phytopathology 78: 84—88

Ori N, Sessa G, Lotan T, Himmelhoch S and Fluhr R (1990) A major stylar matrix polypeptide (sp41) is a member of the pathogenesis-related proteins superclass. EMBO J 9: 3429-3436

Osborn RW, De Samblanx GW, Thevissen K, Goderis I, Torrekens S, Van Leuven F, Attenborough S, Rees SB and Broekaert WF (1995) Isolation and characterisation of plant defensins from seeds of Asteraceae, Fabaceae, Hippocastanaceae and Saxifragaceae. FEBS Lett 368: 257—262

Pan SQ, Ye XS and Kuc J (1992) Induction of chitinases in tobacco plants systemically protected against blue mold by *Peronospora tabacina* or tobacco mosaic virus. Phytopathology 82: 119—123

Parent J-G and Asselin A (1984) Detection of pathogenesis-related proteins (PR or b) and of other proteins in the intercellular fluid of hypersensitive plants infected with tobacco mosaic virus. Can J Bot 62: 564—569

Patel SU, Osborn R, Rees S and Thornton JM (1998) Structural studies of *Impatiens balsamina* antimicrobial protein (Ib-AMP1). Biochemistry 37: 983—990

Paul W, Hodge R, Smartt S, Draper J and Scott R (1992) The isolation and characterisation of the tapetum-specific *Arabidopsis thaliana* A9 gene. Plant Mol Biol 19: 611—622

Payne G, Middlesteadt W, Desai N, Williams S, Dincher S, Carnes M and Ryals J (1989) Isolation and sequence of a genomic clone encoding the basic form of pathogenesis-related protein 1 from *Nicotiana tabacum*. Plant Mol Biol 12: 595—596

Payne G, Parks TD, Burkhart W, Dincher S, Ahl P, Métraux J-P and Ryals J (1988) Isolation of the genomic clone for pathogenesis-related protein 1a from *Nicotiana tabacum* cv Xanthi-nc. Plant Mol Biol 11: 89—94

Payne G, Ward E, Gaffney T, Goy PA, Moyer M, Harper A, Meins F Jr and Ryals J (1990) Evidence for a third structural class of β-1,3-glucanase in tobacco. Plant Mol Biol 15: 797—808

Penninckx IAMA, Eggermont K, Terras FRG, Thomma BPHJ, De Samblanx GW, Buchala A, Métraux J-P, Manners JM and Broekaert WF (1996) Pathogen-induced systemic activation of a plant defensin gene in Arabidopsis follows a salicylic acid-independent pathway. Plant Cell 8: 2309—2323

Penninckx IAMA, Thomma BPHJ, Buchala A, Métraux J-P and Broekaert WF (1998) Concomitant activation of jasmonate and ethylene response pathways is required for induction of a plant defensin gene in Arabidopsis. Plant Cell 10: 2103—2114

Perrakis A, Tews I, Dauter Z, Oppenheim AB, Chet I, Wilson KS and Vorgias CE (1994) Crystal structure of a bacterial chitinase at 2.3 Å resolution. Structure 2: 1169—1180

Peumans WJ, De Ley M and Broekaert WF (1983) An unusual lectin from stinging nettle (*Urtica dioica*) rhizomes. FEBS Lett 177: 99—103

Pfisterer P, König H, Hess J, Lipowsky G, Haendler B, Schleuning W-D and Wirth T (1996) CRISP-3, a protein with homology to plant defense proteins, is expressed in mouse B cells under the control of Oct2. Mol Cell Biol 16: 6160—6168

Pfitzner UM and Goodman HM (1987) Isolation and characterization of cDNA clones encoding pathogenesis-related proteins from tobacco mosaic virus infected tobacco plants. Nucl Acids Res 15: 4449—4465

Pieterse CMJ, Van Wees SCM, Hoffland E, Van Pelt JA and Van Loon LC (1996) Systemic resistance in Arabidopsis induced by biocontrol bacteria is independent of salicylic acid accumulation and pathogenesis-related gene expression. Plant Cell 8: 1225—1237

Pineiro M, Diaz I, Rodriguez-Palanzuela P, Titarenko E and Garcia-Olmedo F (1995) Selective disulphide linkage of plant thionins with other proteins. FEBS Lett 369: 239—242

Ponstein AS, Bres-Vloemans SA, Sela-Buurlage MB, Van den Elzen PJM, Melchers LS and Cornelissen BJC (1994) A novel pathogen- and wound-inducible tobacco (*Nicotiana tabacum*) protein with antifungal activity. Plant Physiol 104: 109—118

Ponz F, Par-Ares J, Hernandez-Lucas C, Garcia-Olmedo F and Carbonero P (1986) Cloning and nucleotide sequence of a cDNA encoding the precursor of the barley toxin α hordothionin. Eur J Biochem 156: 131—135

Ponz F, Paz-Ares J, Hernández-Lucas C, Carbonero P and García-Olmedo F (1983) Synthesis and processing of thionin precursors in developing endosperm from barley (*Hordeum vulgare* L.). EMBO J 2: 1035—1040

Potter S, Uknes S, Lawton, K, Winter AM, Chandler D, DiMaio J, Novitzky R, Ward E and Ryals J (1993) Regulation of a hevein-like gene in *Arabidopsis*. Mol Plant-Microbe Interact 6: 680—685

Punja ZK and Raharjo SHT (1996) Response of transgenic cucumber and carrot plants expressing different chitinase enzymes to inoculation with fungal pathogens. Plant Disease 80: 999—1005

Pyee J and Kolattukudy PE (1994) Identification of a lipid transfer protein as the major protein in the surface wax of broccoli (*Brassica oleracea*) leaves. Arch Biochem Biophys 311: 460—468

Raikhel NV and Wilkins TA (1987) Isolation and characterization of a cDNA clone encoding wheat germ agglutinin. Proc Natl Acad Sci USA 84: 6745—6749

Rao U, Stec B and Teeter MM (1995a) Refinement of purothionins reveals solute particles important for lattice formation and toxicity. Part 1: α_1-purothionin revisited. Acta Cryst D51: 904—913

Rao VH, Guan C and Van Roey P (1995b) Crystal structure of endo-β-N-acetylglucosaminidase H at 1.9 Å resolution: active site geometry and substrate recognition. Structure 3: 449—457

Rasmussen U, Bojsen K and Collinge DB (1992) Cloning and characterization of a pathogen induced chitinase in *Brassica napus*. Plant Mol Biol 20: 277—287

Ready MP, Brown DT and Robertus JD (1986) Extracellular location of pokeweed antiviral protein. Proc Natl Acad Sci USA 83: 5053—5056

Redman DG and Fisher N (1968) Fractionation and comparison of purothionin and globulin components of wheat. J Sci Food Agric 19: 651—655

Redman DG and Fisher N (1969) Purothionin analogues from barley flour. J Sci Food Agric 20: 427—432

Reimann-Philipp U, Schrader G, Martinoia E, Barkholt V and Apel K (1989) Intracellular thionins of barley. A second group of leaf thionins closely related to but distinct from cell wall-bound thionins. J Biol Chem 264: 8978—8984

Reinbothe S, Reinbothe C, Lehmann J, Becker W, Apel K and Parthier B (1994) JIP-60, a methyl jasmonate-induced ribosome inactivating protein involved in plant stress reactions. Proc Natl Acad Sci USA 91: 7012—7016

Reiss E and Bryngelsson T (1996) Pathogenesis-related proteins in barley leaves, induced by infection with *Drechslera teres* (Sacc.) Shoem. and by treatment with other biotic agents. Physiol Mol Plant Pathol 49: 331—341

Richard L, Arró M, Hoebeke J, Meeks-Wagner DR and Van KTT (1992) Immunological evidence of thaumatin-like proteins during tobacco floral differentiation. Plant Physiol 98: 337—342

Richardson M, Valdes-Rodriguez S and Blanco-Labra A (1987) A possible function for thaumatin and TMV-induced protein suggested by homology to a maize inhibitor. Nature 327: 432—434

Rico M, Bruix M, González C, Monsalve RI and Rodr'guez R (1996) ^{1}H NMR assignment and global fold of napin BnIb, a representative 2S albumin seed protein. Biochemistry 35: 15672—15682

Robbins PW, Overbye K, Albright CF, Benfield B and Pero J (1992) Cloning and high-level expression of chitinase-encoding genes of *Streptomyces plicatus*. Gene 111: 69—76

Roberts WK and Selitrennikoff CP (1986) Isolation and partial characterization of two antifungal proteins from barley. Biochim Biophys Acta 880: 161—170

Roberts WK and Selitrennikoff CP (1990) Zeamatin, an antifungal protein from maize with membrane-permeabilizing activity. J Gen Microbiol 136: 1771—1778

Rodriguez A, Webster P, Ortego J and Andrews NW (1997) Lysosomes behave as Ca^{2+}-regulated exocytic vesicles in fibroblasts and epithelial cells. J Cell Biol 137: 93—104

Romero A, Alamillo JM and García-Olmedo F (1997) Processing of thionin precursors in barley leaves by a vacuolar proteinase. Eur J Biochem 243: 202—208

Rosell S and Samuelsson G (1966) Effect of mistletoe viscotoxin and phoratoxin on blood circulation. Toxicon 4: 107—110

Salts Y, Wachs R, Gruissem W and Barg R (1991) Sequence coding for a novel proline-rich protein preferentially expressed in young tomato fruit. Plant Mol Biol 17: 149—150

Samac DA and Shah DM (1991) Developmental and pathogen-induced activation of the Arabidopsis acidic chitinase promoter. Plant Cell 3: 1063—1072

Samac DA and Shah DM (1994) Effect of chitinase antisense RNA expression on disease susceptibility of

Arabidopsis plants. Plant Mol Biol 25: 587—596

Samac DA, Hironaka CM, Yallaly PE and Shah DM (1990) Isolation and characterization of the genes encoding basic and acidic chitinase in *Arabidopsis thaliana*. Plant Physiol 93: 907—914

Samuelsson G (1974) Mistletoe toxins. Syst Zool 22: 566—569

Samuelsson G and Pettersson B (1977) Toxic proteins from the mistletoe *Dendrophtora clavata*. II. The amino acid sequence of denclatoxin B. Acta Pharm Suec 14: 245—254

Schlesier B, Nong V, Horstmann C and Hennig M (1996) Sequence analysis of concanavalin B from *Canavalia ensiformis* reveals homology to chitinases. J Plant Physiol 147: 665—674

Schlumbaum A, Mauch F, Vögeli U and Boller T (1986) Plant chitinases are potent inhibitors of fungal growth. Nature 324: 365—367

Schmidt BF, Ashizawa E, Jarnagin AS, Lynn S, Noto G, Woodhouse L, Estell DA and Lad P (1994) Identification of two aspartates and a glutamate essential for the activity of endo-β-N-acetylglucosaminidase H from *Streptomyces plicatus*. Arch Biochem Biophys 311: 350—353

Schneider M, Schweizer P, Meuwly P and Métraux JP (1986) Systemic acquired resistance in plants. Int Rev Cytol 168: 303—340

Schonwetter BS, Stolzenberg ED and Zasloff MA (1995) Epithelial antibiotics induced at sites of inflammation. Science 267: 1645—1648

Schrader G and Apel K (1991) Isolation and characterization of cDNAs encoding viscotoxins of mistletoe (*Viscum album*). Eur J Biochem 198: 549—553

Schrader-Fischer G and Apel K (1993) cDNA-derived identification of novel thionin precursors in *Viscum album* that contain highly divergent thionin domains but conserved signal and acidic polypeptide domains. Plant Mol Biol 23: 1233—1242

Schrader-Fischer G and Apel K (1994) Organ-specific expression of highly divergent thionin variants that are distinct from the seed-specific crambin in the crucifer *Crambe abyssinica*. Mol Gen Genet 245: 380—389

Schraudner M, Langebartels C and Sandermann HJ (1996) Plant defense systems and ozone. Biochem Soc Trans 24: 456—461

Scofield SR, Tobias CM, Rathjen JP, Chang JH, Lavelle DT, Michelmore RW and Staskawicz BJ (1996) Molecular basis of gene-for-gene specificity in bacterial speck disease of tomato. Science 274: 2063—2065

Seetharaman K, Waniska RD and Rooney LW (1996) Physiological changes in sorghum antifungal proteins. J Agric Food Chem 44: 2435—2441

Segura A, Moreno M and García-Olmedo F (1993) Purification and antipathogenic activity of lipid transfer proteins (LTPs) from the leaves of *Arabidopsis* and spinach. FEBS Lett 332: 243—246

Segura A, Moreno M, Molina A and Garcia-Olmedo F (1998) Novel defensin subfamily from spinach (*Spinacia oleracea*). FEBS Let 435: 159—162

Sela-Buurlage MB, Ponstein AS, Bres-Vloemans SA, Melchers LS, Van den Elzen PJM and Cornelissen BJC (1993) Only specific tobacco (*Nicotiana tabacum*) chitinases and β-1,3-glucanases exhibit antifungal activity. Plant Physiol 101: 857—863

Sevilla P, Bruix M, Santoro J, Gago F, García AG and Rico M (1993) Three-dimensional structure of ω-conotoxin GVIA determined by 1H NMR. Biochem Biophys Res Comm 192: 1238—1244

Sharma P and Lönneborg A (1996) Isolation and characterization of a cDNA encoding a plant defensin-like protein from roots of Norway spruce. Plant Mol Biol 31: 707—712

Sharma YK, Leon J, Raskin I and Davis KR (1996) Ozone-induced responses in *Arabidopsis thaliana*: the role of salicylic acid in the accumulation of defense-related transcripts and induced resistance. Proc

Natl Acad Sci USA 93: 5099—5104

Shewry PR, Napier JA and Tatham AS (1995) Seed storage proteins: structures and biosynthesis. Plant Cell 7: 945—956

Shibuya N, Goldstein IJ, Shaper JA, Peumans WJ and Broekaert WF (1986) Carbohydrate binding properties of the stinging nettle (*Urtica dioica*) rhizome lectin. Arch Biochem Biophys 249: 215—224

Shin DH, Lee JY, Hwang KY, Kim KK and Suh SW (1995) High-resolution crystal structure of the non-specific lipid-transfer protein from maize seedlings. Structure 3: 189—199

Shinshi H, Mohnen D and Meins F Jr (1987) Regulation of a plant pathogenesis-related enzyme: inhibition of chitinase and chitinase mRNA accumulation in cultured tobacco tissues by auxin and cytokinin. Proc Natl Acad Sci USA 84: 89—93

Shinshi H, Neuhaus J-M, Ryals J and Meins F Jr (1990) Structure of a tobacco endochitinase gene: evidence that different chitinase genes can arise by transcription of sequences encoding a cysteine-rich domain. Plant Mol Biol 14: 357—368

Shinshi H, Wenzler H, Neuhaus J-M, Felix G, Hofsteenge J and Meins F Jr (1988) Evidence for N- and C-terminal processing of a plant defense-related enzyme: Primary structure of tobacco prepro-β-1,3-glucanase. Proc Natl Acad Sci USA 85: 5541—5545

Shirasu K, Nakajima H, Rajasekhar VK, Dixon RA and Lamb C (1997) Salicylic acid potentiates an agonist-dependent gain control that amplifies pathogen signals in the activation of defense mechanisms. Plant Cell 9: 261—270

Shualev V, Léon J and Raskin I (1995) Is salicylic acid a translocated signal of systemic acquired resistance in plants? Plant Cell 7: 1691—1701

Shualev V, Silverman P and Raskin I (1997) Airborne signaling by methyl salicylate in plant pathogen resistance. Nature 385: 718—721

Singh NK, Bracker CA, Hasegawa PM, Handa AK, Buckel S, Hermodsom MA, Pfankoch E, Regnier FE and Bressan RA (1987) Characterization of osmotin. A thaumatin-like protein associated with osmotic adaptation in plant cells. Plant Physiol 85: 529—536

Singh NK, Nelson DE, Kuhn D, Hasegawa PM and Bressan RA (1989) Molecular cloning of osmotin and regulation of its expression by ABA and adaptation to low water potential. Plant Physiol 90: 1096—1101

Skinner WS, Adams ME, Quistad GB, Kataoka H, Cesarin BJ, Enderlin FE and Schooley DA (1989) Purification and characterization of two classes of neurotoxins from the funnel web spider *Agelenopsis aperta*. J Biol Chem 264: 2150—2155

Smirnov S, Shualev V and Tumer NE (1997) Expression of pokeweed antiviral protein in transgenic plants induces virus resistance in grafted wild-type plants independently of salicylic acid accumulation and pathogenesis-related protein synthesis. Plant Physiol 114: 1113—1121

Soedjanaatmadja UMS, Subroto T and Beintema JJ (1995) Processed products of the hevein precursor in the latex of the rubber tree (*Hevea brasiliensis*). FEBS Lett 363: 211—213

Somssich IE (1994) Regulatory elements governing pathogenesis-related (PR) gene expression. In: Nover L (ed) Plant Promoters and Transcription Factors, pp 163—179. Springer Verlag, Berlin

Sossountzov L, Ruiz-Avilla L, Vignols F, Jolliot A, Arondel V, Tchang F, Grosbois M, Guerbette F, Miginiac E, Delseny M, Puigdomènech P and Kader J-C (1991) Spatial and temporal expression of a maize lipid transfer protein gene. Plant Cell 3: 923—933

Soufleri IA, Vergnolle C, Miginiac E and Kader J-C (1996) Germination-specific lipid transfer protein cDNAs in *Brassica napus* L. Planta 199: 229—237

Staehelin C, Granado J, Müller J, Wiemken A, Mellor RB, Felix G, Regenass M, Broughton WJ and Boller

T (1994) Perception of *Rhizobium* nodulation factors by tomato cells and inactivation by root chitinases. Proc Natl Acad Sci USA 91: 2196—2200

Stanchev BS, Doughty J, Scutt CP, Dickinson H and Croy RRD (1996) Cloning of *PCP1*, a member of a family of pollen coat protein (*PCP*) genes from *Brassica oleracea* encoding novel cysteine-rich proteins involved in pollen-stigma interactions. Plant J 10: 303—313

Stanford A, Bevan M and Northcote D (1989) Differential expression within a family of novel wound-induced genes in potato. Mol Gen Genet 215: 200—208

Stark MJR and Boyd A (1986) The killer toxin of *Kluyveromyces lactis*: characterization of the toxin subunits and identification of the genes which encode them. EMBO J 5: 1995—2002

Stec B, Rao U and Teeter MM (1995) Refinement of purothionins reveals solute particles important for lattice formation and toxicity. II. structure of β-purothionin at 1.7 Å resolution. Acta Crystallogr D 51: 914—924

Stein JC and Nasrallah JB (1993) A plant receptor-like gene, the *S*-locus receptor kinase of *Brassica oleracea* L., encodes a functional serine/threonine kinase. Plant Physiol 101: 1103—1106

Steinmüller K, Batschauer A and Apel K (1986) Tissue-specific and light-dependent changes of chromatin organization in barley (*Hordeum vulgare*). Eur J Biochem 158: 519—525

Sterk P, Booij H, Schellekens GA, Van Kammen A and De Vries SC (1991) Cell-specific expression of the carrot EP2 lipid transfer protein gene. Plant Cell 3: 907—921

Sticher L, Hinz U, Meyer AD and Meins F Jr (1992a) Intracellular transport and processing of a tobacco vacuolar β-1,3-glucanase. Planta 188: 559-565

Sticher L, Hofsteenge J, Milani A, Neuhaus J-M and Meins F Jr (1992b) Vacuolar chitinases of tobacco: a new class of hydroxyproline-containing proteins. Science 257: 655-657

Sticher L, Hofsteenge J, Neuhaus J-M, Boller T and Meins F Jr (1993) Posttranslational processing of a new class of hydroxyproline-containing proteins. Prolyl hydroxylation and C-terminal cleavage of tobacco (*Nicotiana tabacum*) vacuolar chitinase. Plant Physiol 101: 1239—1247

Stiefel V, Ruiz-Avila L, Raz R, Vallés MD, G—mez J, Pagés M, Mart'nez-Izquierdo JA, Ludevid MD, Longdale JA, Nelson T and Puigdomènech P (1990) Expression of a maize cell wall hydroxyproline-rich glycoprotein gene in early leaf and root vascular differentiation. Plant Cell 2: 785—793

Stintzi A, Heitz T, Prasad V, Wiedermann-Merdinoglu S, Kauffmann S, Geoffroy P, Legrand M and Fritig B (1993) Plant 'pathogenesis-related' proteins and their role in defense against pathogens. Biochimie 75: 687—706

Stirpe F, Barbieri L, Battelli MG, Soria M and Douglas AL (1992) Ribosome-inactivating proteins from plants: present status and future prospects. Biotechnology 10: 405—412

Stirpe F, Williams DG, Onyon LJ, Legg RF and Stevens WA (1981) Dianthins, ribosome damaging proteins with antiviral properties from *Dianthus caryophyllus* L. (carnation). Biochem J 195: 399—405

Stuart LS and Harris TH (1942) Bactericidal and fungicidal properties of a crystalline protein from unbleached wheat flour. Cereal Chem 19: 288—300

Suribade M, Salesse C, Marion D and Pézolet M (1995) Interaction of a nonspecific wheat lipid transfer protein with phospholipid monolayers imaged by fluorescence microscopy and studied by infrared spectroscopy. Biophys J 69: 974—988

Suzuki K, Fukuda Y and Shinshi H (1995) Studies on elicitor-signal transduction leading to differential expression of defense genes in cultured tobacco cells. Plant Cell Physiol 36: 281—289

Svensson B, Svendsen I, Højrup P, Roepstorff P, Ludvigsen S and Poulsen FM (1992) Primary structure of barwin: a barley seed protein closely related to the C-terminal domain of proteins encoded by wound-induced plant genes. Biochemistry 31: 8767—8770

Swegle M, Kramer KJ and Muthukrishnan S (1992) Properties of barley seed chitinases and release of embryo-associated isoforms during early stages of imbibition. Plant Physiol 99: 1009—1014

Tahiri-Alaoui A, Dumas-Gaudot E and Gianinazzi S (1993) Immunocytochemical localization of pathogenesis-related PR-1 proteins in tobacco root tissues infected *in vitro* by the black root rot fungus *Chalara elegans*. Physiol Mol Plant Pathol 42: 69—82

Tailor R, Acland DP, Attenborough S, Cammue BPA, Evans IJ, Osborn RW, Ray J, Rees SB and Broekaert WF (1997) A novel family of small cysteine-rich antimicrobial peptides from seed of *Impatiens balsamina* is derived from a single precursor protein. J Biol Chem (in press)

Takeda S, Sato F, Ida K and Yamada Y (1991) Nucleotide sequence of a cDNA for osmotin-like protein from cultured tobacco cells. Plant Physiol. 97: 844-846

Tamamura H, Kuroda M, Masuda M, Otaka A, Funakoshi S, Nakashima H, Yamamoto N, Waki M, Matsumoto A, Lancelin JM, Kohda D, Tate S, Inagaki F and Fujii N (1993) A comparative study of the solution structures of tachyplesin I and a novel anti-HIV synthetic peptide, T22([Tyr 5, 12, Lys 7]-polyphemusin II), determined by nuclear magnetic resonance. Biochim Biophys Acta 1163: 209—216

Tang X, Frederick RD, Zhou J, Halterman DA, Jia Y and Martin GB (1996) Initiation of plant disease resistance by physical interaction of AvrPto and Pto kinase. Science 274: 2060—2063

Tassin S, Broekaert WF, Marion D, Acland DP, Ptak M, Vovelle F, Sodano, P (1998) Solution structure of Ace-AMP1, a potent antimicrobial protein extracted form onion seeds. Structural analogies with non specific lipid transfer proteins. Biochemistry 37: 3623—3637

Tata SJ, Beintema JJ and Balabaskaran S (1983) The lysozyme of *Hevea brasiliensis* latex: isolation, purification, enzyme kinetics and a partial amino-acid sequence. J Rubber Res Inst Malaysia 31: 35—48

Taylor BE and Irvin JD (1990) Depurination of plant ribosomes by pokeweed antiviral protein. FEBS Lett 273: 144—146

Taylor S, Massiah A, Lomonossoff G, Roberts LM, Lord JM and Hartley M (1994) Correlation between the activities of five ribosome-inactivating proteins in depurination of tobacco ribosomes and inhibition of tobacco mosaic virus infection. Plant J 5: 827—835

Tchang F, This P, Stiefel V, Arondel V, Morch M-D, Pagés M, Puigdomènech P, Grellet F, Delseny M, Bouillon P, Huet J-C, Guerbette F, Beauvais-Cante F, Duranton H, Pernollet J-C and Kader J-C (1988) Phospholipid transfer protein: full length cDNA and amino acid sequence in maize. Amino acid sequence homologies between plant phospholipid transfer proteins. J Biol Chem 263: 16849—16855

Terras FRG, Eggermont K, Kovaleva V, Raikhel NV, Osborn RW, Kester A, Rees SB, Torrekens S, Van Leuven F, Vanderleyden J, Cammue BPA and Broekaert WF (1995) Small cysteine-rich antifungal proteins from radish: Their role in host defense. Plant Cell 7: 573—588

Terras FRG, Goderis IJ, Van Leuven F, Vanderleyden J, Cammue BPA and Broekaert WF (1992a) *In vitro* antifungal activity of a radish (*Raphanus sativus* L.) seed protein homologous to nonspecific lipid transfer proteins. Plant Physiol 100: 1055—1058

Terras FRG, Schoofs HME, De Bolle MFC, Van Leuven F, Rees SB, Vanderleyden J, Cammue BPA and Broekaert WF (1992b) Analysis of two novel classes of antifungal proteins from radish (*Raphanus sativus* L.) seeds. J Biol Chem 267: 15301—15309

Terras FRG, Schoofs HME, Thevissen K, Osborn RW, Vanderleyden J, Cammue BPA and Broekaert WF (1993a) Synergistic enhancement of the antifungal activity of wheat and barley thionins by radish and oilseed rape 2S albumins and by barley trypsin inhibitors. Plant Physiol 103: 1311—1319

Terras FRG, Torrekens S, Van Leuven F and Broekaert WF (1996) A six-cyteine type thionin from the radish storage organ displays weak *in vitro* antifungal activity against *Fusarium culmorum*. Plant Physiol Biochem 34: 599—603

Terras FRG, Torrekens S, Van Leuven F, Osborn RW, Vanderleyden J, Cammue BPA and Broekaert WF (1993b) A new family of basic cysteine-rich plant antifungal proteins from Brassicaceae-species. FEBS Lett 316: 233—240

Terwisscha van Scheltinga AC, Armand S, Kalk KH, Isogai A, Henrissat B and Dijkstra BW (1995) Stereochemistry of chitin hydrolysis by a plant chitinase/lysozyme and X-ray structure of a complex with allosamidin: evidence for substrate assisted catalysis. Biochemistry 34: 15619—15623

Terwisscha van Scheltinga AC, Hennig M and Dijkstra BW (1996) The 1.8 Å resolution structure of hevamine, a plant chitinase/lysozyme, and analysis of the conserved sequence and structure motifs of glycosyl hydrolase family 18. J Mol Biol 262: 243—257

Terwisscha van Scheltinga AC, Kalk KH, Beintema JJ and Dijkstra BW (1994) Crystal structure of hevamine, a plant defence protein with chitinase and lysozyme activity, and its complex with an inhibitor. Structure 2: 1181—1189

Thevissen K, Ghazi A, De Samblanx GW, Brownlee C, Osborn RW and Broekaert WF (1996) Fungal membrane responses induced by plant defensins and thionins. J Biol Chem 271: 15018—15025

Thevissen K, Osborn RW, Acland DP and Broekaert WF (1997) Specific, high affinity binding sites for an antifungal plant defensin on *Neurospora crassa* hyphae and microsomal membranes. J Biol Chem 272: 32176—32181

Thoma S, Hecht U, Kippers A, Botella J, De Vries S and Somerville C (1994) Tissue-specific expression of a gene encoding a cell wall-localized lipid transfer protein from *Arabidopsis*. Plant Physiol 105: 35—45

Thoma S, Kaneko Y and Somerville C (1993) A non-specific lipid transfer protein from *Arabidopsis* is a cell wall protein. Plant J 3: 427—436

Thomma BPHJ and Broekaert WF (1998) Tissue-specific expression of plant defensin genes *PDF2.1* and *PDF2.2* in *Arabidopsis thaliana*. Plant Physiol Biochem 36: 533—537

Thomma BPHJ, Eggermont K, Penninckx IAMA, Mauch-Mani B, Vogelsang R, Cammue BPA and Broekaert WF (1998) Separate jasmonate-dependent and salicylate-dependent defense response pathways in Arabidopsis are essential for resistance to distinct microbial pathogens. Proc Natl Acad Sci USA 95: 15107—15111

Thunberg E and Samuelsson G (1974) The amino acid sequence of ligatoxin A from the mistletoe *Phoradendron liga*. Acta Pharm Suec 19: 447—456

Titarenko E, Hargreaves J, Kean J and Gurr SJ (1993) Defence-related expression in barley coleoptile cells following infection by *Septoria nodorum*. In Fritig B, Legrand M (eds) Mechanisms of plant defense responses, pp 308—311. Kluwer Academic Publishers, Dordrecht, The Netherlands

Titarenko E, Lopez-Solanilla E, García-Olmedo F and Rodriguez-Palenzuela P (1997) Mutants of *Ralstonia* (*Pseudomonas solanacearum*) sensitive to antimicrobial peptides are altered in their LPS structure and are avirulent. J Bacteriol 179: 6699—6704

Tornero P, Conejero V and Vera P (1994) A gene encoding a novel isoform of the PR-1 protein family from tomato is induced upon viroid infection. Mol Gen Genet 243: 47—53

Tornero P, Gadea J, Conejero V and Vera P (1997) Two *PR-1* genes from tomato are differentially regulated and reveal a novel mode of expression for a pathogenesis-related gene during the hypersensitive response and development. Mol Plant Microbe Interact 10: 624—634

Tornero P, Rodrigo I, Conejero V and Vera P (1993) Nucleotide sequence of a cDNA encoding a pathogenesis-related protein, p1-p14 from tomato (*Lycopersicon esculentum*). Plant Physiol 102: 325

Torres-Schumann S, Godoy JA and Pintor-Toro JA (1992) A probable lipid transfer protein gene is induced by NaCl in stems of tomato plants. Plant Mol Biol 18: 749—757

Tronsmo AM, Gregersen P, Hjeljord L, Sandal T, Bryngelsson T and Collinge DB (1993) Cold-induced disease resistance. In: Fritig B and Legrand M (eds) Mechanisms of Plant Defense Responses, pp. 369. Kluwer Academic Press, Dordrecht

Trudel J, Andy P and Asselin A (1989) Electrophoretic forms of chitinase activity in Xanthi-nc tobacco, healthy and infected with tobacco mosaic virus. Mol Plant-Microbe Interact 2: 315—324

Trudel J, Grenier J, Potvin C and Asselin A (1998) Several thaumatin-like proteins bind to β-1,3-glucans. Plant Physiol 118: 1431—1438

Tumer NE, Hwang DJ and Bonness M (1997) A nontoxic C-terminal deletion mutant of pokeweed antiviral protein inhibits viral infection. Proc Natl Acad Sci USA 94: 3866—3871

Tuzun S, Rao MN, Vogeli U, Schlardl CL and Kuc J (1989) Induced systemic resistance to Blue Mold: early induction and accumulation of β-1,3-glucanases, chitinases, and other pathogenesis-related proteins (b-proteins) in immunized tobacco. Phytophatoglogy 79: 979—983

Uknes S, Mauch-Mani B, Moyer M, Potter S, Williams S, Dincher S, Chandler D, Slusarenko A, Ward ER and Ryals JA (1992) Acquired resistance in Arabidopsis. Plant Cell 4: 645—656

Uknes S, Winter A, Delaney T, Vernooij B, Morse A, Friedrich L, Nye G, Potter S, Ward E and Ryals J (1993) Biological induction of systemic acquired resistance in Arabidopsis. Mol Plant-Microbe Interact 6: 692—698

Vad K, De Neergaard E, Madriz-Ordenana K, Mikkelsen JD and Collinge DB (1993) Accumulation of defence-related transcripts and cloning of a chitinase mRNA from pea leaves (*Pisum sativum* L.) inoculated with *Ascochyta pisi* Lib. Plant Sci 92: 69—79

Van de Rhee M, Lemmers R and Bol JF (1993) Analysis of regulatory elements involved in stress-induced and organ-specific expression of tobacco acidic and basic β-1,3-glucanase genes. Plant Mol Biol 21: 451—461

Van den Bulcke M, Bauw G, Castresana C, Van Montagu M and Vandekerckhove J (1989) Characterization of vacuolar and extracellular β-(1,3)-glucanases of tobacco: evidence for a strictly compartmentalized plant defense system. Proc Natl Acad Sci USA 86: 2673—2677

Van den Elzen PJM, Jongedijk E, Melchers LS and Cornelissen BJC (1993) Virus and fungal resistance: from laboratory to field. Phil. Trans. R. Soc. London B 342: 271—278

Van der Wel H and Loeve K (1972) Isolation of thaumatin I and II, the sweet-tasting proteins from *Thaumatococcus danielli*. Eur J Biochem 31: 221—225

Van Loon LC (1985) Pathogenesis-related proteins. Plant Mol Biol 4: 111—116

Van Loon LC and Gerritsen YAM (1989) Localization of pathogenesis-related proteins in infected and non-infected leaves of Samsun NN tobacco during the hypersensitive reaction to tobacco mosaic virus. Plant Sci 63: 131—139

Van Loon LC and van Kammen A (1970) Polyacrylamide disc electrophoresis of the soluble leaf proteins from *Nicotiana tabacum* var "Samsun" and "Samsun NN" II. Changes in protein constitution after infection with tobacco mosaic virus. Virology 40: 199—211

Van Loon LC (1990) The nomenclature of pathogenesis-related proteins. Physiol Mol Plant Pathol 37: 229—230

Van Loon LC, Pierpoint WS, Boller T and Conejero V (1994) Recommendations for naming plant pathogenesis-related proteins. Plant Mol Biol Rep 12: 245—264

Van Parijs J, Broekaert WF, Goldstein IJ and Peumans WJ (1991) Hevein: an antifungal protein from rubber-tree (*Hevea brasiliensis*) latex. Planta 183: 258—264

Van Roey P, Rao V, Plummer TH Jr and Tarentino AL (1994) Crystal structure of endo-β-*N*-acetylglucosaminidase F1, an α/β-barrel enzyme adapted for a complex substrate. Biochemistry 33: 13989—13996

VanEtten CH, Nielsen HC and Peters JE (1965) A crystalline polypeptide from the seed of *Crambe abyssinica*. Phytochemistry 4: 467—473

Vanlerberghe GC and McIntosh L (1997) Alternative oxydase: from gene to function. Annu Rev Plant Physiol Plant Mol Biol 48: 703—734

Varghese JN, Garrett TPJ, Colman PM, Chen L, Høj PB and Fincher GB (1994) Three-dimensional structures of two plant β-glucan endohydrolases with distinct substrate specificities. Proc Natl Acad Sci USA 91: 2785—2789

Verheyden P, Pletinckx J, Maes D, Pepermans HAM, Wyns L, Willem R and Martins JC (1995) ^{1}H NMR study of the interaction of *N, N',N''*-triacetyl chitotriose with Ac-AMP2, a sugar binding antimicrobial protein isolated from *Amaranthus caudatus*. FEBS Lett 370: 245—249

Verma HN, Chowdhury B and Rastogi P (1984) Antiviral activity in leaf extracts of different *Clerodendrum* L. species. Z Pflanzenkr Pflanzenschutz 91: 34—41

Verma HN, Srivastava S, Varsha and Kumar D (1996) Induction of systemic resistance in plants against viruses by a basic protein from *Clerodendrum aculeatum* leaves. Phytopathology 86: 485—492

Verma NH, Varsha and Baranwal VK (1995) Agricultural role of endogenous antiviral substances of plant origin. In: Chessin M, Deborde D and Zipf A (eds.) Antiviral Proteins in Higher Plants, pp 23—37. CRC Press, Boca Raton

Vernon LP (1992) *Pyrularia* thionin: physical properties, biological responses and comparison to other thionins and cardiotoxin. J Toxicol 11: 169—191

Vernon LP and Rogers A (1992a) Effect of calcium and phosphate ions on hemolysis induced by *Pyrularia* thionin and *Naja naja kaouthia* cardiotoxin. Toxicon 30: 701—709

Vernon LP and Rogers A (1992b) Binding properties of *Pyrularia* thionin and *Naja naja kaouthia* cardiotoxin to human and animal erythrocytes and to murine P388 cells. Toxicon 30: 711—721

Vernon LP, Evett GE, Zeikus RD and Gray WR (1985) A toxic thionin from *Pyrularia pubera*: purification, properties and amino acid sequence. Arch Biochem Biophys 238: 18—19

Vervoort J, Van den Hooven HW, Berg A, Vossen P, Vogelsang R, Joosten MHAJ and De Wit PJGM (1997) The race-specific elicitor AVR9 of the tomato pathogen *Cladosporium fulvum*: a cystine knot protein. Sequence-specific ^{1}H NMR assignments, secondary structure and global fold of the protein. FEBS Lett 404: 153—158

Vierheilig H, Alt M, Neuhaus J-M, Boller T and Wiemken A (1993) Colonization of transgenic *Nicotiana sylvestris* plants, expressing different forms of *Nicotiana tabacum* chitinase, by the root pathogen *Rhizoctonia solani* and by the mycorrhizal symbiont *Glomus mosseae*. Mol Plant-Microbe Interact 6: 261—264.

Vierheilig H, Alt M, Lange J, Gut-Rella M, Wiemken A and Boller T (1995) Colonization of transgenic tobacco constitutively expressing pathogenesis-related proteins by the vesicular-arbuscular mycorrhizal fungus *Glomus mosseae*. Appl Environ Microbiol 61: 3031—3034

Vigers AJ, Roberts WK and Selitrennikoff CP (1991) A new family of plant antifungal proteins. Mol Plant-Microbe Interact 4: 315—323

Vigers AJ, Wiedemann S, Roberts WK, Legrand M, Selitrennikoff CP and Fritig B (1992) Thaumatin-like pathogenesis-related proteins are antifungal. Plant Sci 83: 155—161

Vögeli-Lange R, Fründt C, Hart CM, Beffa R, Nagy F and Meins F Jr (1994a) Evidence for a role of β-1,3-glucanase in dicot seed germination. Plant J 5: 273—278

Vögeli-Lange R, Fründt C, Hart CM, Nagy F and Meins F Jr (1994b) Developmental, hormonal and pathogenesis-related regulation of the tobacco class I β-1,3-glucanase B promoter. Plant Mol Biol 25: 299—311

Vogelsang R and Barz W (1993) Purification, characterization and differential hormonal regulation of a β-1,3-glucanase and two chitinases from chickpea (*Cicer arictinum* L.). Planta 189: 60—69

Vöisey CR and Slusarenko (1989) Chitinase mRNA and enzyme activity in *Phaseolus vulgaris* (L) increase more rapidly in response to avirulent than to virulent cells of *Pseudomonas syringae* pv *phaseolicola*. Physiol Mol Plant Pathol 35: 403—412

Vu L and Huynh QK (1994) Isolation and characterization of a 27 kDa antifungal protein from the fruits of *Diospyros texana*. Biochem Biophys Res Comm 202: 666—672

Wada K, Ozaki Y, Matsubara H and Yoshizumi H (1982) Studies on purothionin by chemical modifications. J Biochem 91: 257—263

Walujono K, Scholma RA, Beintema JJ, Mariono A and Hahn AM (1975) Amino acid sequence of hevein. Proc Int Rubber Conf, Kuala Lumpur, 2: 518—531

Wang X, Zafian P, Choudhary M and Lawton M (1996) The PR5K receptor protein kinase from *Arabidopsis thaliana* is structurally related to a family of plant defense proteins. Proc Natl Acad Sci USA 93, 2598—2602

Ward ER, Uknes, SJ, Williams SC, Dincher SS, Wiederhold DL, Alexander DC, Ahl-Goy P, Métraux J-P and Ryals JA (1991) Coordinate gene activity in response to agents that induce systemic acquired resistance. Plant Cell 3: 1085—1094

Watanabe T, Kobori K, Miyashita K, Fujii T, Sakai H, Uchida M and Tanaka H (1993) Identification of glutamic acid 204 and aspartic acid 200 in chitinase A1 of *Bacillus circulans* WL-12 as essential residues for chitinase activity. J Biol Chem 268: 18567—18572

Wemmer T, Kaufmann H, Kirch H-H, Schneider K, Lottspeich F and Thompson RD (1994) The most abundant soluble basic protein of the stylar transmitting tract in potato (*Solanum tuberosoum* L.) is an endochitinase. Planta 194: 264—273

White AJ, Dunn MA, Brown K and Hughes MA (1994) Comparative analysis of genomic sequence and expression of a lipid transfer protein gene family in winter barley. J Exp Bot 45: 1885—1892

Wilusz T, Wieczorek M, Polanowski A, Denton A, Cook J and Laskowski M Jr (1983) Amino-acid sequence of two trypsin inhibitors, ITD I and ITD III from squash seeds (*Cucurbita maxima*). Hoppe-Seyler's Z Physiol Chem 364: 93—95

Woloshuk CP, Meulenhoff JS, Sela-Buurlage M, Van den Elzen PJM and Cornelissen BJC (1991) Pathogen-induced proteins with inhibitory activity against *Phytophthora infestans*. Plant Cell 3: 619—628

Wong Y and Maclachan GA (1979) 1,3-β-glucanases from *Pisum sativum* seedlings. II. Substrate specificities and enzymatic patterns. Biochem Biophys Acta 571: 256—259

Worrall D, Hird DL, Hodge R, Paul W, Draper J and Scott R (1992) Premature dissolution of the microsporocyte callose wall causes male sterility in transgenic tobacco. Plant Cell 4: 759—771

Wright CS (1987) Refinement of the crystal structure of wheat germ agglutinin isolectin 2 at 1.8 Å resolution. J Mol Biol 194: 501—529

Wright CS (1992) Crystal structure of a wheat germ agglutinin/glycophorin-sialoglycopeptide receptor complex. J Biol Chem 267: 14345—14352

Wu H-M, Zou J, May B, Gu Q and Cheung AY (1993) A tobacco gene family for flower cell wall proteins with a proline-rich domain and a cysteine-rich domain. Proc Natl Acad Sci USA 90: 6829—6833

Wyatt SD and Shepherd RJ (1969) Isolation and characterization of a virus inhibitor from *Phytolacca americana*. Phytopathology 59: 1787—1793

Xu Y, Chang P-FL, Liu D, Narasimhan ML, Raghothama KG, Hasegawa PM and Bressan RA (1994)

Plant defense genes are synergistically induced by ethylene and methyl jasmonate. Plant Cell 6: 1077—1085

Yalpani N, Enyedi AJ, Leon J and Raskin I (1994) Ultraviolet light and ozone stimulate accumulation of salicylic acid pathogenesis-related proteins and virus resistance in tobacco. Planta 193: 372—376

Yanagawa Y, Abe T, Satake M, Odani S, Suzuki J and Ishikawa K (1988) A novel sodium channel inhibitor from *Conus geographicus*: purification, structure and pharmacological properties. Biochemistry 27: 6256—6262

Yanagisawa M, Kurihara H, Kimura S, Tomobe Y, Kobayashi M, Mitsui Y, Yazaki Y, Goto K and Masaki T (1988) A novel potent vasoconstrictor peptide produced by vascular endothelial cells. Nature 332: 411—415

Ye XS, Pan SQ and Kuc J (1989) Pathogenesis-related proteins and systemic resistance to blue mold and tobacco mosaic virus induced by tobacco mosaic virus, *Peronospora tabacina* and aspirin. Physiol Mol Plant Pathol 35: 161—167

Yoshikawa M, Tsuda M and Takeuchi Y (1993) Resistance to fungal diseases in transgenic tobacco plants expressing the phytoalexin elicitor-releasing factor, β-1,3-endoglucanase, from soybean. Naturwissenschaften 80: 417—420

Youle RJ and Huang AHC (1978) Evidence that the castor bean allergens are the albumin storage proteins in the protein bodies of castor bean. Plant Physiol 61: 1040—1042

Youle RJ and Huang AHC (1979) Albumin storage proteins and allergens in cotton seeds. J Agric Food Chem 27: 500—503

Yu YG, Chang CH, Fowler A and Suh SW (1988) Amino acid sequence of a probable amylase/protease inhibitor from rice seeds. Arch Biochem Biophys 265: 466—475

Yun D-J, Bressan RA and Hasegawa (1997a) Plant antifungal proteins. Plant Breeding Rev 14: 39—88

Yun D-J, D'Urzo MP, Abad L, Takeda S, Salzman R, Chen Z, Lee H, Hasegawa PM and Bressan RA (1996) Novel osmotically induced antifungal chitinases and bacterial expression of an active recombinant isoform. Plant Physiol 111: 1219—1225

Yun D-J, Zhao Y, Pardo JM, Narasimhan ML, Damsz B, Lee H, Abad LR, Paino D'Urzo MP, Hasegawa PM and Bressan RA (1997b) Stress proteins on the yeast cell surface determine resistance to osmotin, a plant antifungal protein. Proc Natl Acad Sci USA 94: 7082—7087

Zeevaart JAD (1988) Metabolism and physiology of abscisic acid. Annu Rev Plant Physiol Plant Mol Biol 39: 439—473

Zhao C, Wang I and Lehrer RI (1996) Widespread expression of beta-defensin hBD-1 in human secretory glands and epithelial cells. FEBS Lett 396: 319—322

Zhu B, Chen THH and Li PH (1993) Expression of an ABA-responsive osmotin-like gene during the induction of freezing tolerance in *Solanum commersonii*. Plant Mol Biol 21: 729—735

Zhu B, Chen THH and Li PH (1995) Activation of two osmotin-like protein genes by abiotic stimuli and fungal pathogen in transgenic potato plates. Plant Physiol 108: 929-937

Zhu B, Chen THH and Li PH (1996) Analysis of late-blight disease resistance and freezing tolerance in transgenic potato plants expressing sense and antisense genes for an osmotin-like protein. Planta 198: 70—77

Zhu Q, Maher EA, Masoud S, Dixon RA and Lamb CJ (1994) Enhanced protection against fungal attack by constitutive co-expression of chitinase and glucanase genes in transgenic tobacco. Bio/technology 12: 807—812

CHAPTER 9

SPECIAL ASPECTS OF RESISTANCE TO VIRUSES

R.S.S. FRASER

Society for General Microbiology,
Marlborough House, Basingstoke Road, Spencers Wood
Reading RG7 1AE, UK (r.fraser@socgenmicrobiol.org.uk)

Summary

Plant viruses cause serious diseases and economic losses in numerous crops. The viruses show a great diversity of particle structure, and of genome organisation and expression, which are reflected in the diversity of interactions with the host and the number of types of resistance mechanisms that plants have evolved to combat them, or that man has invented. Resistance may operate at a number of points in the virus infection and replication cycle. Different types of resistance mechanisms may operate at the species, cultivar and individual plant level. These are considered using a variety of conceptual models, as well as observation and experimental evidence. Non-host resistance involves restriction of virus spread from the initially-infected cell in some cases, but other models are also considered. A study of the genetics of cultivar resistance (as used by plant breeders) indicates that resistance can be variously dominant, semi-dominant or recessive to susceptibility, and that this may be correlated with different types of biochemical mechanisms of resistance. A number of examples of these are described. Many of the resistances considered have been overcome by virulent isolates of the virus, and some quite complex gene-for-gene interactions have been established. Recombinant DNA methods have allowed the mapping of the determinants of virulence/avirulence to a number of different viral functions. Resistance may also be conferred on normally susceptible plants by a variety of treatments, or occur in particular plant parts or at particular times after infection. Examples of mechanisms considered include virus-free green islands, satellite protection, cross protection, genetic transformation and gene silencing. Transgenes derived from various parts of the viral genome, and from numerous other sources, have been shown to confer resistance by diverse mechanisms. The future prospects for exploitation of different types of resistance in crop protection are considered.

Abbreviations

BCMV	bean common mosaic virus
CMV	cucumber mosaic virus

A. Slusarenko, R.S.S. Fraser, and L.C. van Loon (eds), Mechanisms of Resistance to Plant Diseases, 479-520.

CPMV   cowpea mosaic virus
CTV   citrus tristeza virus
LMV   lettuce mosaic virus
ORF   open reading frame
PAP   pokeweed antiviral protein
PR   pathogenesis-related protein
PVX   potato virus X
PVY   potato virus Y
RFLP   restriction fragment length polymorphism
RIP   ribosome-inactivating protein
TMGMV   tobacco mild green mottle virus
TMV   tobacco mosaic virus
ToMV   tomato mosaic virus

I.   Introduction

A.   LOSSES CAUSED BY PLANT VIRUS DISEASES

Viruses can cause disease in a very wide range of crop plants, giving rise to serious economic losses, second amongst the pathogens only to those caused by fungi (Matthews, 1991; Barnett, 1995). A comprehensive review of losses caused by viruses in particular crops is given by Waterworth and Hadidi (1998). This includes some apposite comments on the difficulties of measuring loss, and the potential for mis- or over-interpretation of data. In this introduction we are concerned with the mechanisms by which economic losses occur.

The damage caused by plant viruses can take many forms (Fraser, 1992a). Loss of yield biomass as a result of decreased plant vigour and growth is common. For example, the complex of related viruses known as barley yellow dwarf virus causes serious yield losses in all major cereal crops, including bread and durum wheats, corn, oat, barley, rye, rice and sorghum (Burnett and Plumb, 1998). Such losses are normally the consequence of the highly visible effects of virus infection on plant growth and development, but may even occur without visible disease symptoms. Thus the so-called cryptic virus infections of sugar beet can cause severe reduction in output of refined sugar without the plants showing outward symptoms of virus infection (Antoniw *et al.*, 1990).

Visible symptoms of disease such as mosaic or distorted growth also result in loss of marketable quality, whether of ornamentals or food crops, where appearance is important to the purchaser. For example, zucchini yellow mosaic virus causes a loss of as much as 50% of yield biomass in courgettes and other cucurbits, but because the fruits are blistered and discoloured the loss of marketable yield in an infested crop can be 100% (Walkey, 1992).

Further economic costs are incurred if eradication or avoidance programmes are enforced, as a pre-requirement for growing a particular crop in a given area. Examples include the costs of production and certification of virus-tested seed potatoes, which in the absence of controls would carry several viruses forward into the ware crop (Slack and Singh, 1998), and the ultimately unsuccessful attempt to eradicate cacao swollen shoot virus in Ghana by removing infected trees (Hughes and Ollennu, 1994; Barnett, 1995).

These overall losses make control of virus diseases a necessity for many crops, but the problem is exacerbated by two factors. Firstly, and unlike fungal pathogens or insect pests, there are virtually no chemical weapons in the armoury. Partly, this is because viruses rely so heavily on host components for their multiplication, that any compound with antiviral activity would be likely to have antihost activity as well. Antiviral chemicals with some degree of selective effectiveness against plant viruses are indeed known, but their use has been restricted to the special case of creating virus-free lines of crops such as garlic for further vegetative propagations (Stace-Smith, 1990). Otherwise, the use of antiviral chemicals in field crop protection is ruled out for reasons of cost, human or plant toxicity, incomplete efficacy and the cost and complexity of pesticide registration procedures in many countries.

Secondly, the great diversity of types of plant viruses and other sub-cellular pathogens, and of their mechanisms of replication, pathogenesis and transmission, makes then a complex and widespread target. It is therefore appropriate to describe this diversity as a background to the consideration of resistance mechanisms in this chapter.

B.   THE DIVERSITY OF PLANT VIRUSES

The definitive list of plant viruses with descriptions of their properties and behaviour is Brunt *et al.* (1996), which is also available on the Internet at http://biology.anu.edu.au/Groups/MES/vide. A useful concise guide to the current taxonomy of viruses is provided by Mayo and Pringle (1998). At the time of writing, Matthews (1991) remains the most detailed general text on plant virus structure and biology; more recent but less detailed information is in Webster and Granoff (1994).

Before considering the diversity of plant viruses in detail, it may be useful to list the structural and behavioural features that define a virus, and by extension other sub-cellular pathogens such as viroids and various forms of satellites which share some or all of these features. This is presented in Box I, and further details of viral structures are given in Box II.

Box I. **Definition of a plant virus**

- a macromolecular structure
- made up of a small number of *types* of molecules and ions (less than 10^2)
- which is unable to replicate on its own
- but contains information as DNA or RNA sequences which can command its own replication within a living cell
- and enable onward transmission to further opportunities for replication

This definition can also be applied to viroids.

Satellite viruses and satellite RNAs fall within the overall definition, but also require a helper virus.

Box II.   Composition and structure of some plant viruses and similar pathogens

Property	Examples
Genome	
Circular or linear single-stranded RNA, extensively base-paired, no ORFs	viroid
Single-stranded RNA, messenger sense	
monopartite	tobamovirus
multipartite	cucumovirus
Single-stranded RNA, negative sense	rhabdovirus
Double stranded RNA, multipartite	phytoreovirus
Single-stranded DNA	geminivirus
Double-stranded DNA	caulimovirus
Other particle components	
Protein	
none	viroid
one coat protein type	tobamovirus
two coat protein types	comoviruses
RNA polymerase	rhabdovirus
DNA polymerase	caulimovirus
Polyamines	tymovirus
Lipid membranes	tospovirus
Cations (Ca^{2+}, Zn^{2+}), water	numerous/all
Particle architecture	
Naked nucleic acid	viroid
Rod	
rigid	tobamovirus
multicomponent rigid	hordeivirus
flexuous	potyvirus
Isometric	
single component	luteovirus
multicomponent	alfamovirus
Enveloped	rhabdovirus

It is clear that plant viruses and the other sub-cellular pathogens have a great diversity of type and arrangement of genetic material, covering most possible combinations of RNA or DNA, single-stranded or double-stranded, positive (messenger-sense) or negative, single component or multicomponent, linear or circular. If the viroids are included, the genome size stretches from the 125 kDa (359 nucleotide residues) of potato spindle tuber viroid to 12.5 mDa (phytoreovirus), a range of 1:100.

The different forms of viruses, viroids etc. have also evolved different ways of expressing the genetic material to command the various aspects of replication, spread within the plant, and transmission from plant to plant. Viroid RNAs do not code for any proteins: pathogenesis must therefore be controlled purely at the RNA level. Virus genomes may code for one or more structural proteins which go to make up the virus particle(s), one or more proteins involved in the polymerase/replicase activities responsible for multiplying and expressing the genome, one or more proteins involved in facilitating cell-to-cell spread of infection, and various other proteins with specialist functions such as facilitating transmission by a specific vector. Not all viruses code for all of these functions: the 'basic set' seems to be replicase component(s), movement protein and virus particle structural protein, as described for TMV in chapter 1.

The most common types of plant viruses are those with positive (messenger)-sense single-stranded RNA genomes coding for a number of proteins. Different groups of these have evolved different ways of solving the '80S ribosome problem'. The plant cytoplasmic ribosome will only translate the 5'-proximal open reading frame (ORF) of an mRNA coding for more than one protein. The TMV double strategy, of synthesising subgenomic mRNAs for proteins coded downstream of the first ORF, and of synthesising two protein products from the first start codon by read-through of a 'leaky' termination codon at the end of the shorter, is described in some detail in chapter 1.

The strategy adopted by potyviruses is to have all the information coded by the single virus genomic messenger RNA within a single ORF, and expressed as a single translation product or polyprotein. This contains three proteinase activities which excise various functional proteins, such as the replicase, coat protein, cell-to-cell movement and insect transmission determinant, from it. Note that despite the elegant complexity of this strategy, there is an apparent inefficiency in that it does not provide a mechanism for differential expression of those proteins needed in larger amounts. It might be assumed that structural proteins, such as the viral coat protein, would be required in larger amounts than the catalytic proteins, such as the replicase. This may explain the low concentrations reached by potyviruses in plants, together with the accumulation of comparatively high concentrations of non-structural proteins. This contrasts with the TMV situation, where production of different amounts of subgenomic messengers, and control of translational activity through the presence or absence of the 7-methyl guanosine cap, allows differential production of large amounts of the coat proteins (see chapter 1).

A different type of strategy is to fragment the genome into a number of components, thus allowing more proteins to be synthesised from 5'-proximal start codons. This is taken to an extreme by the comparatively rare phytoreoviruses, which have genomes of 12 double stranded RNAs, giving rise to mRNAs for the 12 known virus proteins.

A more common approach in the virus groups with single-stranded messenger sense RNA genomes is to have the genome fragmented into two or three components. In some cases one or more of these fragments code for single proteins; the remaining genomic fragment(s) may code for two or more proteins, and expression of the end products from these is via the subgenomic mRNA (e.g. bromovirus, alfalfa mosaic virus) or polyprotein (e.g. comovirus) strategies. Note that a corollary of the multipartite genome strategy for these ssRNA viruses is that the virus particles come in various forms and sizes, depending on which genomic fragments are packaged in individual particles. Infection obviously depends on delivery of the full genomic set. An indication of the range of types of plant virus particles is presented in Box II. This further emphasises the diversity of plant viruses as disease-causing agents, although there is no clear correlation between any particular type of structure, and any particular mechanism of resistance.

Examples of plant viruses with DNA genomes are much fewer in number than those with RNA genomes, but nevertheless also show considerable diversity in particle structure and means of replicating and expressing the genome. The caulimoviruses have a particularly complex replicative cycle, involving an intermediate RNA transcript of the DNA genome, from which the DNA for progeny viruses is produced by a reverse transcriptase activity. The caulimoviruses have similarities to animal retroviruses such as human immunodeficiency virus (HIV).

Unlike many animal viruses and bacteriophages, plant viruses do not attach to specific receptors on the outer surfaces of host cells; external receptor specificity is therefore not a possible basis for resistance mechanisms. Plant viruses require damage to penetrate plant cells, an undoubted consequence of the presence of the plant cell wall. Damage can occur in a casual or deliberate mechanical manner, or be caused by organisms that feed on or from plants. Viruses have proved adept at exploiting a large number of types of plant-feeding agents as vectors to carry them from plant to plant, and crop to crop. These include beetles, aphids, leafminers, whitefly, fungi, nematodes and parasitic plants such as dodder. In many cases, the interaction between the virus and vector is highly specific. Some viruses can replicate in both the host plant and the insect vector, and can be transmitted transovarially to the insect progeny. Luteoviruses, which are transmitted by phloem-feeding aphids, do not invade plant tissues other than the phloem.

Those viruses with no specific vector infect chance wounds occurring by contact between the plant and the soil or animals. Spread of these viruses would clearly be aided by multiplication to high levels in each infected host, and stable particles well-fitted for mechanical transmission, or long-term persistence in the environment. These features are seen in TMV, a mechanically transmitted virus with no specific vector, as described in chapter 1.

It will be clear from this brief summary that plant viruses are very diverse in structure, means of replication and means of transmission. It would therefore be expected that plants would have evolved a number of different types of resistance mechanisms against them, with the possibility of operating at various stages in the virus replicative and transmission cycles.

Plant viruses are, of all plant pathogens, the best understood in terms of their genetic make-up, and how the products of these genes contribute to the viral replicative cycle. The number of plant virus genomes that have been completely sequenced greatly outnumbers that for the totality of bacteria and fungi, let alone those pathogenic to plants. This makes it useful to approach the analysis of mechanisms of host resistance to viruses by considering some theoretical models, in the hope of shedding light on the more complex issues of plant-virus interactions including pathogenesis and host resistance responses. Sections II and III of this chapter consider two types of conceptual model, and then review knowledge of a number of different types of resistance mechanism against this background framework.

II. Resistance Mechanisms may Operate at a Number of Different Points in the Virus Replication and Infection Cycle

Despite the diversity of structure, mode of replication and means of transmission described above, the different replicative and infection cycles of different viruses share certain basic features, and these may be examined for possible targets where resistance mechanisms might interfere. These are summarised in Fig. 1.

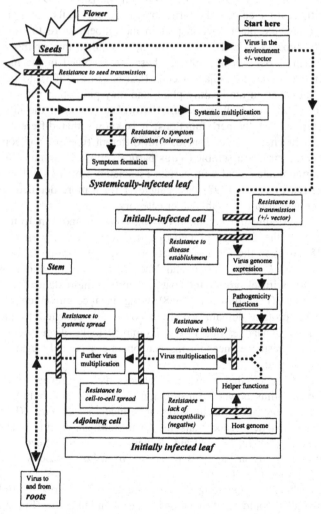

Figure 1. Possible stages in the infection and multiplication cycle of a plant virus, where different types of resistance mechanism might operate. Modified from Fraser (1990).

A.   RESISTANCE TO MULTIPLICATION AND MOVEMENT

The infection commences when the virus penetrates a suitable host plant - through a wound, or with the assistance of a plant-feeding vector. Within the infected plant cell it must expose its genetic material for translation, leading to the production of replicase activity, structural (coat) proteins and other virus-coded factors. These events require a number of host-provided functions. Some of these are general metabolic activities which are not specific to virus multiplication: supplies of energy, amino acids and nucleoside triphosphate precursors; the general machinery of protein synthesis such as ribosomes and transfer RNAs. Others may be host helper functions which are more specific to the virus: the use of one or more host-coded proteins as subunits of a multicomponent viral replicase for example (Osman and Buck, 1997).

From the initially infected cell, the virus spreads to other cells, and may then spread systemically through the plant by transport in the phloem sieve tubes (Dawson and Hilf, 1992). The process of cell-to-cell movement requires one or more virus-coded 'movement proteins' (Beck et al., 1994). These are involved in the modification of the plasmodesmata connecting adjacent cells (Lucas et al., 1993), to permit passage of larger molecules than in the healthy plant, and may play other roles in moving the infectious entity (Godefroy-Colburn et al., 1990). The spread of the infectious entity may involve further host 'helper' functions; it is known that different viruses have evolved different mechanisms for spread, with different types of nucleoprotein being transferred. In some viruses assembled virus particles move between cells; in others it can be the genomic nucleic acid with certain associated proteins (Hull 1989; Dawson and Hilf, 1992; Lartey et al., 1997). See chapter 1 for a more detailed description of TMV movement, which falls into this latter category.

Long distance systemic movement, involving entry to and exit from the phloem, may share some of the features of cell-to-cell spread, but may involve additional factors and differences in the form of infection being translocated (Dawson and Hilf, 1992; Carrington et al., 1996; Séron and Haenni, 1996; Tang and Leisner, 1997). For example, tobamoviruses require the coat protein for long-distance transport, but not for cell-to-cell spread (Dawson et al., 1988). Long distance movement of potyviruses involves the VPg protein domain which is covalently attached to the 5' end of the genomic RNA (Schaad et al., 1997). Examples to be discussed in section III suggest that plant resistance mechanisms can operate separately against cell-to-cell and long-distance spread.

There is also evidence that resistance can operate to prevent infection of the developing seed and thus onward transmission to the progeny: lettuce varieties can differ in the percentage seed transmission of lettuce mosaic virus (LMV) (Couch, 1955) although the extent of seed transmission in cultivars carrying genes for LMV resistance is also affected by virus strain (Revers et al., 1997). Seed transmission of barley stripe mosaic virus in barley is prevented by a host property controlled by a recessive gene (Carroll et al., 1979).

Resistance mechanisms operating at the basic levels of virus multiplication and spread within the plant could be of two types, *positive* and *negative* (Fig. 2). These may also be referred to as *active* and *passive*, respectively (Dawson and Hilf, 1992).

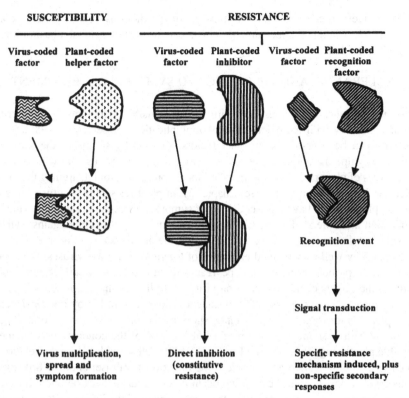

SUSCEPTIBILITY | RESISTANCE

Virus-coded factor | Plant-coded helper factor | Virus-coded factor | Plant-coded inhibitor | Virus-coded factor | Plant-coded recognition factor

Recognition event

Signal transduction

Virus multiplication, spread and symptom formation | Direct inhibition (constitutive resistance) | Specific resistance mechanism induced, plus non-specific secondary responses

Figure 2. Models for plant-virus interactions, involving recognition events between host- and pathogen-coded molecules, and the consequent responses of susceptibility or resistance.

In the *positive* model, resistance is expressed because the plant contains a molecule which inhibits some aspect of virus multiplication or spread, or which recognises some virus-coded molecule and switches on an active resistance mechanism through a signal transduction pathway. The former mechanism is constitutive, the latter induced after infection. In this model, susceptibility is lack of the inhibitor or recognition/signal transduction molecules. Genetically, resistance in this model should be dominant over susceptibility, or the phenotypic expression of resistance could show gene-dosage dependence, in which resistance would be more effective in the homozygous form (RR) than in heterozygous form (R+), because of the higher concentration of gene product. Dominance is more likely to be associated with an all-or-nothing recognition event switching on an induced response; gene-dosage dependence is more likely to be associated with the direct inhibitor model.

In the *negative* model, the resistant plant is such because it lacks some host helper function required by the virus, or contains the helper function in a form unable to interact with the virus-coded component. Thus a plant which produced a defective host-coded replicase subunit, or lacked the protein with which the virus movement protein interacted, would inhibit multiplication or cell-to-cell spread. This type of resistance is constitutive, and should be recessive to susceptibility, because plants heterozygous for resistance would contain enough host-coded factor to support pathogenesis.

These models therefore make a number of predictions about genetic control and possible biochemical mechanisms which can be tested against observation and experimental evidence.

B.   'TOLERANCE' AND RESISTANCE TO SYMPTOM DEVELOPMENT

During the multiplication phase, most viruses cause visible symptoms. The severity of symptoms can vary with the extent of virus multiplication (Fraser *et al.*, 1986), but this relationship can be over-ridden by other factors, such as by strains of the same virus which cause symptoms of different severity. 'Tolerance' is a term used in various ways in different areas of plant pathology: some practitioners prefer to avoid it, arguing that it is too imprecise to have useful meaning (see chapter 3). In plant resistance to viruses, it is used to indicate plants which have a lesser degree of symptom expression, while allowing virus multiplication and spread. The possibility that resistance might operate against symptom expression, but disconnected from the replicative cycle, is indicated in Fig. 1.

There are few well-documented examples of tolerance for plant viruses; in supposed cases where a particular host genotype reacts to infection with milder than normal symptoms, the effects of the tolerance on virus multiplication have not necessarily been accurately measured. In the case of resistance to ToMV controlled by the *Tm-1* gene in tomato, plants heterozygous for resistance show no visible mosaic symptoms (Fig. 3). The virus spreads systemically, and reaches about 30% of the concentration attained in susceptible plants (Fig. 3; Fraser and Loughlin, 1980). Other evidence on the relationship between virus multiplication and mosaic symptom severity (Fraser *et al.*, 1986) suggests that this level of virus multiplication would in other circumstances be enough to cause mosaic symptoms, and therefore that the *Tm-1* gene might have a direct effect on symptom expression as well as its inhibitory effect on virus multiplication.

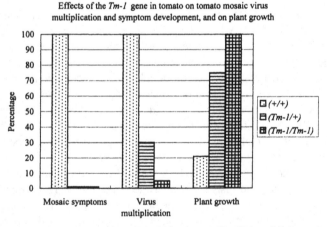

Figure 3.   Effects of the *Tm-1* gene in tomato, in heterozygous (*Tm-1/+*) and homozygous (*Tm-1/Tm*-1) form, on multiplication and symptoms caused by an avirulent (strain 0) isolate of tomato mosaic virus, and on plant growth, compared with susceptible (+/+) plants. For each characteristic, the 100% level was the maximum observed in any of the three genotypes. Mosaic symptoms were scored visually on a scale of 0-3. Virus multiplication was measured by extraction and ultraviolet spectrophotometry, and plant growth as fresh weight. Based on data in Fraser *et al.*, 1986).

C.   RESISTANCE OPERATING OUTSIDE THE PLANT OR AT THE PLANT SURFACE

Plants may differ in their attractiveness to insects and other pests, or have evolved various defence mechanisms against them, and these can indirectly provide components of resistance against viruses which use these pests as vectors. The mechanisms include differences in plant colour which affect host selection by insects, chemical barriers such as secretion of volatile insect alarm pheromones or the presence of antifeedant chemicals in the sap, and physical barriers such as dense or specialised leaf hairs and thick cuticles. A good example is the leaf hairs of the wild potato species *Solanum berthaultii*. These exude the aphid alarm pheromone (E)-β-farnesene, which causes aphids to avoid the plant and reduces transmission of potato virus Y and beet yellows virus (Gibson and Pickett, 1983). Clearly, resistance mechanisms of this type which depend on the three-way interaction of host, vector and virus are likely to be specific for the particular viruses carried by the vector. Others, such as prevention of infection by thick cuticles, have been shown to be generally effective against a number of viruses. Non-specific resistance to initial infection of a tobacco line by a number of viruses has been correlated with a reduced number of ectodesmata - thought to be one initial channel of infection (Thomas and Fulton, 1968)

III.   Different Types of Resistance Mechanism May Operate at Different Levels of Population Complexity

The model considered above referred to the case of interaction of a virus with an individual plant. Another way of looking at models of possible resistance mechanisms is to consider different levels of plant population complexity. There, separate types of mechanism may operate at the species, variety/cultivar and individual plant level. At the species level the concept is non-host resistance. At cultivar/variety level it refers to resistant varieties of normally susceptible species, and at the individual plant level it refers to the acquisition of various types of resistance by normally susceptible plants. The different types of interaction are shown in Fig. 4.

A.   NON-HOST RESISTANCE

One of the most intriguing questions in plant virology today is what precisely is non-host resistance? Most plant species do not appear to be infected by most plant viruses. Some viruses - such as cucumber mosaic virus and tomato spotted wilt virus - are found in a very wide range of host species (Brunt *et al.*, 1996). Others have a very restricted host range: bean common mosaic virus will only infect the bean *Phaseolus vulgaris* and a few other leguminous species (Frencel and Pospieszny, 1979). Sometimes the host ranges of two viruses will overlap, so that there are species which are hosts for both viruses, and species which host only one of the two. It is important to distinguish between natural infections (in wild or cultivated plants), and experimental infections in the laboratory. The latter can produce extensions to the host range which would

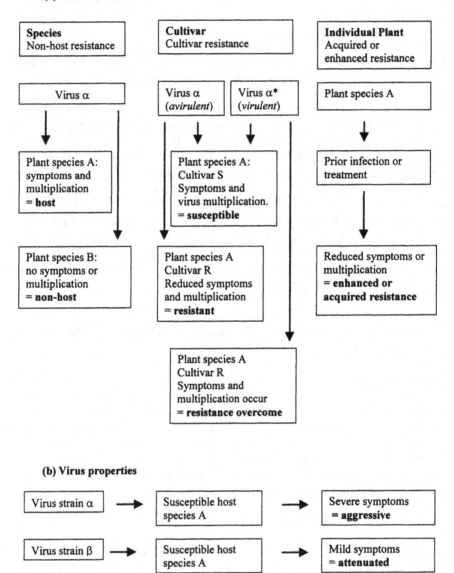

Figure 4. Different types of interaction between plants and viruses, at different levels of plant population complexity.

not be encountered in the field. Nevertheless, the question about the mechanism of host range control remains valid.

A series of classical papers by Bald and Tinsley (1967a, b, c) set out a number of theoretical models, from which some of the considerations of positive and negative mechanisms of resistance in section II.A are drawn. They postulated that the virus may produce various 'pathogenicity factors' required for multiplication and spread, which are complemented by host 'susceptibility factors'. Successful pathogenesis would require compatibility between the host- and virus-coded factors. Non-host species would not produce susceptibility factors required by the virus, or would possess them in a form incompatible with the virus pathogenicity factors. There is now an increasing body of experimental evidence that such mechanisms may be involved in host range determination.

Factors determining pathogenicity to particular species have been mapped on the viral genome, using the 'artificial recombinant' approach described in Chapter 1 (Zheng and Edwards, 1990; Weiland and Edwards, 1994). Molecular evidence of interaction of host- and virus-coded factors has also come from studies of the host- and virus-coded subunits of viral replicases (Hayes and Buck, 1990; Quadt *et al.*, 1993; Osman and Buck, 1997). A promising recent development in the study of 'host' susceptibility factors comes from the interesting discovery that yeast systems can carry out some of the stages of replication of certain plant viruses. Study of the function of the host components involved in replication, such as replicase subunits, should be aided by the more rapid mutational techniques available for yeast than for higher plants (Ishikawa *et al.*, 1997).

Earlier studies of virus host ranges involved inoculation, then observation of whether the plants developed symptoms or not. This type of experimental approach was shown to be inadequate with the demonstration of 'subliminal' infection: sensitive methods of detection showed that in some plants adjudged non-hosts at the whole plant level, virus multiplication could be detected in a small number of cells, presumably those infected in the initial inoculation (Cheo and Garard, 1971; Sulzinski and Zaitlin, 1982). There have also been numerous reports that virus multiplication could be supported when protoplasts from 'non-host' species were inoculated (Furusawa and Okono, 1978; Van Loon, 1987). These lines of evidence suggest that one mechanism in non-host resistance could be prevention of cell-to-cell spread after an initially successful establishment of an infection 'beachhead', perhaps at the single cell level. Support for this idea comes from experiments with plants infected with two viruses, where a spreading ('helper') virus can make it possible for normally non-hosted viruses to spread (Atabekov and Taliansky, 1990). A useful review by Dawson and Hilf (1992) covers cases and experimental treatments where the division between 'host' and 'non-host' becomes further blurred.

It certainly appears that the original use of the term non-host resistance in the context of complete immunity, i.e. inoculation produces no detectable symptoms of infection or virus multiplication whatsoever, is an oversimplification, although the examples showing limited infection do not exclude the possibility that complete immunity could occur in other host-virus combinations. As Dawson and Hilf (1992) point out, the literature is likely to be skewed towards positive effects; reports of failure to detect virus multiplication in protoplasts of non-host species would be difficult to publish.

Two other possible mechanisms of non-host resistance require comment. Holmes (1955) thought that non-host plants might contain a large number of genes for resistance to a particular virus, which acted in an additive manner to give complete resistance, impossible for the virus to break down by mutation. This could represent an extreme form of horizontal resistance in the Vanderplank (1984) sense, and would be very difficult to demonstrate experimentally. The second is that some plants or groups of plants might be highly resistant to some or all viruses for non-specific reasons such as unsuitable cell pH (Fraser, 1985). For example, TMV requires a local pH of 8 and a low Ca^{2+} concentration at the cellular site where it initiates the infection by beginning to uncoat its genomic RNA. Any species which did not have cytoplasmic regions with such pH and ionic conditions might well be a non-host. Alternatively, non-host behaviour might be conferred by the presence of hostile chemicals such as phenolics and numerous others. This mechanism may be behind the lack of reports of viruses from groups such as the Gymnosperms and Pteridophytes.

It should also be noted that there is evidence that the host range of a virus in nature and in a particular geographical area may be influenced by competition between viruses. A study of the occurrence of two tobamoviruses, tobacco mosaic virus and tobacco mild green mottle virus (TMGMV), in living Australian populations of *Nicotiana glauca* and herbarium specimens from the preceding century suggested that competition from TMGMV had replaced TMV in this host (Fraile *et al.*, 1997).

B.   CULTIVAR RESISTANCE

This type of resistance was first recognised as early as 1925 and characterised genetically in 1938 (see chapter 1; Holmes, 1938). Within a cultivated species normally *susceptible* to a particular virus, it involves cultivars discovered by chance to possess a mechanism of *resistance* to that virus, or those to which resistance has been deliberately transferred from other cultivars or related wild species. Resistance is effective against *avirulent* isolates of the virus; resistance-breaking isolates are referred to as *virulent* (Fig. 4). These terms parallel the use of resistance/avirulence in the classical and widespread gene-for-gene model of recognition and signalling in resistance to bacterial and fungal pathogens (Flor, 1956; Lugtenberg, 1986), which in the case of these types of pathogens is generally expressed as the hypersensitive response. However, the contention of this section is that while gene-for-gene resistance to viruses does in many cases induce the hypersensitive response, it involves overall a wider range of mechanisms and underlying types of genetic controls.

One further aspect of the terminology requires explanation. The term *virulent* is often used in virology - especially in the context of human and animal viruses - to refer to isolates causing particularly severe disease. As in 'a new and particularly virulent strain of influenza virus'. In this chapter, the terms virulence/avirulence are used in the same sense as for bacteria and fungi. Different isolates of a plant virus can differ markedly in the severity of disease symptoms caused on (susceptible) plants (e.g. Fraser, 1969) or extent of multiplication (e.g. Fraser *et al.*, 1986b). They are described as differing in *aggressiveness*, and that difference need have no relationship to the specific pattern of interaction with host resistance genes (Fig. 4).

1.   *Genetics of Resistance*

Since the pioneering work of Holmes (1938) on the *N* gene transferred from *Nicotiana glutinosa* to cultivated tobacco (see chapter 1), the genetics of resistance to viruses has been studied in a vast number of combinations of plants and viruses. For economic reasons, the plants studied have generally been cultivated varieties, or closely related wild species being investigated as potential sources of transferable resistance genes for use in breeding programmes.

In earlier reviews (Fraser, 1986a; 1990; 1992b), I reported the results of an ongoing survey of the genetics of resistance, together with a number of related factors such as whether resistance genes had been reported as being overcome by virulent strains of the virus, and the nature of the expression of the resistance phenotype. The objectives were to gain an overview of the diversity of genetic interactions and possible correlations with mechanisms, and to test the more theoretical models discussed earlier in this chapter. Table 1 shows an updated version. Another comprehensive study of the genetics of resistance to viruses has recently been published by Khetarpal *et al.* (1998).

The review sample for Table 1 was chosen randomly from the literature, but was inevitably heavily skewed to cultivated species. Studies of the genetics of resistance in wild species have concentrated on detecting major gene resistance which could easily be transferred to cultivated relatives, and co-evolution of resistance and virulence in cultivated species is heavily influenced by agricultural practices. The full picture of co-evolution of resistance and virulence in wild species might well be much more complex. By the same token, the conclusions from the survey may not be fully applicable to the interaction of viruses with wild species. However, recent extensive research into virus resistance genes in the molecular biologist's favourite weed species, *Arabidopsis thaliana*, has shown resistance to a number of viruses, each controlled at a single locus, and involving a diversity of mechanisms including HR (Lee *et al.*, 1994; Callaway *et al.*, 1996; Dempsey *et al.*, 1997).

Most resistance mechanisms in the survey sample appeared to be under simple genetic control, with resistance inherited at a single locus. Most hosts had only a single allele for resistance at the locus, although in a few cases two or more resistance alleles had been reported. A small number of hosts had separate mechanisms of resistance against one virus mapping to two or three separate loci. This limited complexity of resistance loci and alleles is exemplified by the resistances to tomato mosaic virus (ToMV) in tomato shown in Box III.

Interestingly, only 53% of the resistance alleles in the sample were dominant; 24% were recessive and the remaining 23% showed incomplete dominance (gene-dosage dependence). The distribution contrasts with plant resistance to fungal and bacterial pathogens, which tends more strongly to be dominant over susceptibility, and where recessive examples such as the *mlo* mutants of the barley *Mlo* locus (Buschges *et al.*, 1997) are infrequent. It has also to be said that Avr/R-gene interactions for fungal and bacterial pathogens operate on a much grander scale of genetic complexity than for viruses.

The diversity of dominance, incomplete dominance and recessiveness of virus resistance alleles may well reflect the diversity of resistance mechanisms. In contrast, resistance to bacteria and fungi appears to be much more focused on the hypersensitive

Table 1. Summary of the genetic basis of resistance to plant viruses (updated from Fraser 1990)

Genetic basis of resistance	Number of host-virus combinations
Resistance controlled at a single locus	
Dominant	40
Incompletely dominant	17
(gene-dosage dependent) recessive	18
Sub-total:	75
Possibly oligogenic resistance, or monogenic with effects of modifier genes or host background	21
Total number in sample	96

Resistance phenotype[1]	Immunity or subliminal infection	Local lesions (HR)	Resistance to long distance transport	Systemically effective resistance to multiplication; no inhibition of movement	Not reported
Resistance genotype					
Dominant	7	23	1	4	8
Incompletely dominant	0	0	5	13	0
Recessive	6	0	3	9	4

Virulent isolates reported[1]	Yes	No[2]	Not known or not tested
Resistance genotype			
Dominant	20	4	16
Incompletely dominant	10	3	3
Recessive	9	4	8
Totals	39	11	27

[1]Data are tabulated for all resistances operating at a single locus, and for those cases of oligogenic resistance or modifier genes where the characters of the individual major genes can be isolated.
[1]Means that no breakdown of resistance has been found after the genes were extensively tested in the field.

Box III.	**The gene-for-gene relationship between resistance in tomato and virulence in tomato mosaic virus (ToMV)**							
ToMV strain	0	1	2	2^2	1.2	1.2^2	2.2^2 (rare)	$1.2.2^2$ (rare)
Host genotype								
+/+	S	S	S	S	S	S	S	S
Tm-1	R	S	R	R	S	S	R	S
Tm-2	R	R	S	R	S	R	S	S
Tm-2^2	R	R	R	S	R	S	S	S
Tm-1/Tm-2	R	R	R	R	S	R	R	S
Tm-1/Tm-2^2	R	R	R	R	R	S	R	S
Tm-1/Tm-2/ Tm-2^2	R	R	R	R	R	R	R	S

R = resistant, no symptoms; S = susceptible, systemic mosaic and stunting; + = wild type (non-resistant) host genotype

response and secondary events dependent on this, such as the synthesis of pathogenesis-related proteins and phytoalexin accumulation (see chapters 10 and 11).

Some examples of resistance to viruses appeared to be under more complex genetic control, showing effects of genes at more than one locus. This ranges from general variation in effectiveness of a single resistance gene in hosts with different genetic backgrounds (e.g. Jones and Catherall, 1970) to oligogenic resistance systems involving additive or complementary effects between different loci.

One well characterised example of a complex oligogenic resistance system is that to bean common mosaic virus (BCMV) in *Phaseolus vulgaris* (Drijfhout, 1978). This involves four separate loci, *bc-u*, *bc-1*, *bc-2* and *bc-3*. Resistance at each locus does not involve HR, and is recessive to susceptibility. Effective resistance requires the minimum presence of resistance alleles at the *bc-u* locus and at one of the other three. Resistance conferred by the *bc-u* locus is effective against all virus strains if the plant is also homozygous for an effective *bc-1*, *bc-2* or *bc-3* allele. The *bc-u* gene is not effective on its own. Two different resistance alleles are known at each of the *bc-1* and *bc-2* loci, and these resistances are virus-strain specific, in that they can be overcome by strains containing appropriate determinants of virulence, even in the presence of homozygous *bc-u*. The single allele at the *bc-3* locus does not appear to have been overcome. Note that the resistance conferred by the complex of *bc* loci is separate from the hypersensitive resistance to BCMV and other potyviruses conferred by the dominant *I* gene, which is discussed below.

In the earlier reviews, I drew attention to cases where resistance had originally been thought to be under complex genetic control, but later work showed it to be simple at the genetic level, but complicated at the level of phenotypic expression because of genotype x environment interactions. It is undoubtedly still the case that, applying the principle of Occam's razor, the simpler explanation is to be preferred. However, application of modern techniques such as molecular genetic mapping have thrown up a number of examples of interesting complexities in the genetics of plant virus interactions, which indicate a need for broad-mindedness. A few examples are considered here.

The *I* locus in *Phaseolus vulgaris* was originally identified as conferring resistance to the potyvirus BCMV (Drijfhout, 1978), but is now known to control a form of HR resistance to nine different viruses within the potyvirus group, presumably by recognising some strongly conserved structural element common to all (Fisher and Kyle, 1994). Interestingly, this gene has not been overcome by virulent strains of the viruses, suggesting that the putative conserved virus sequence has limits on the amount of mutation it can accommodate without losing its function in the viral pathogenic cycle. Fisher and Kyle (1994) refer to this as a 'pathogenic bottleneck'. Further work by the same authors (Fisher and Kyle 1996) showed that a separate mechanism of resistance to systemic infection (as opposed to HR) by two potyviruses also appeared to be controlled from a single locus, although resistance alleles were also present at further unlinked loci.

In soybean, Yu *et al.* (1996) used degenerate oligonucleotide primers derived from the conserved nucleotide-binding site of the tobacco *N* and *Arabidopsis RPS2* resistance genes against TMV and *Pseudomonas syringae* pv *tomato* respectively to identify a superfamily of 11 classes of putative resistance genes. Two of these were shown by RFLP positional analysis to map in the vicinity of two genes for potyvirus resistance: others mapped close to genes for *Phytophthora* root rot and powdery mildew resistance. Caranta *et al.* (1996) investigated potyvirus resistance in *Capsicum* doubled haploid lines derived from the F_1 hybrid between two susceptible parent lines, and found a doubled haploid line containing a new and complete resistance to pepper veinal mottle virus. This resistance was shown to depend on complementation of recessive genes at two loci, one of which was already known as a locus for potato virus Y (PVY) resistance. This is therefore a further example of one locus being involved in resistance to more than one potyvirus.

In lettuce, resistance to LMV is controlled at the recessive *mo* locus. The effects of this have long been recognised as somewhat variable: the amount of virus accumulation permitted in resistant plants seemed to depend on the host genetic background. Molecular genetic mapping showed only weak links to *mo*, suggesting that it might involve a number of dispersed genetic components (Montesclaros *et al.*, 1997).

The *Rx* gene in potato confers hypersensitive resistance to potato virus X (PVX); the determinant of the hypersensitive response was shown to be a threonine residue at position 121 of the coat protein (Goulden and Baulcombe, 1993). Interestingly, this residue was also shown to be involved in the elicitation of hypersensitivity against PVX in *Gomphrena globosa*, suggesting that there are similarities in the host components in the two species in the way that they recognise the viral determinant of avirulence. In an intriguing further complication, it was shown that the resistance induced by PVX in *Rx* plants was also effective against the unrelated cucumber mosaic virus (CMV) (Köhm *et al.*, 1993).

2.   *Genetics of Virulence*

In the survey of resistance to plant viruses (Table 1), more than half of the examples of resistance genes had been reported to be overcome by virulent (resistance-breaking) isolates of the virus involved. Virulent isolates may have arisen by mutation of the virus in the face of a selection pressure when a resistance gene was deployed in plant breeding, or naturally present in a species. Virulent isolates may also have been present naturally in the virus population as a result of random mutation, or selection pressures operating on other functions of the virus gene which incidentally also selected virulence. Finally, there is also evidence that virulence may arise or be modified by recombination (Ding *et al.*, 1996; Macfarlane, 1997).

Some virus resistance genes have proved to have very low durability, being quickly overcome by resistance-breaking isolates after introduction in commercial cultivars. A prime example is resistance to ToMV controlled at the *Tm-1* locus in tomato, which was overcome within a year of introduction (Pelham *et al.*, 1970). Resistance at a separate locus controlled by the *Tm-2* allele was also quickly overcome, but the *Tm-2²* allele at the same locus has proved much more durable, and strains overcoming this form of resistance have rarely become established in commercial tomato crops (Hall, 1980; Fletcher, 1992).

Other resistance genes have also proved to be outstandingly durable. The *N* gene for TMV resistance in tobacco, used as a case history in chapter 1, is perhaps the outstanding example, with a gap of half a century between its genetic characterisation, and the first and to date only report of a resistance-breaking isolate (Csillery *et al.*, 1983; Ikeda *et al.*, 1993; Padgett and Beachy, 1993; Sanfaçon *et al.*, 1993). Other examples include the *bc-3* gene for resistance to BCMV in *Phaseolus vulgaris* (Drijfhout, 1978); the *I* gene for resistance to the group of nine potyviruses discussed above (Fisher and Kyle, 1994), and the *Ry* gene for PVY resistance in potato (Jellis, 1992).

In earlier reviews (Fraser 1986a; 1990; 1992b), I suggested that recessive resistance mechanisms, based on the absence of a host factor required for the development of the viral replicative and pathogenic cycle, might prove more difficult for the virus to overcome than dominant inhibitor-based or recognition-event models. The argument was that it would be more difficult for the virus to mutate to cope with the absence of an essential host co-factor, than to modify a virus-coded factor so that it continued to function, but did not interact with the host inhibitor or recognition factor (see Fig. 2). The sample of highly durable resistances is too small to test these suggestions: certainly the recessive *bc-3* resistance would fall within the prediction, but the recessive alleles at the *bc-1* and *bc-2* loci in the same host have been overcome (Drijfhout, 1978), as have numerous other recessive genes (Table 1). The other examples of highly durable resistance quoted involve dominant genes giving a hypersensitive response; clearly there must be some constraint on the ability of viruses to mutate to overcome some, but not all, mechanisms of this kind.

As described in chapter 1, production of artificial recombinants between virulent and avirulent isolates of the same virus, followed by sequencing, has allowed mapping of the determinant(s) of virulence to particular virus genes. This has now been carried out for a number of combinations of viruses and resistance genes (Box IV). It is clear that the determinant of virulence/avirulence can map to diverse viral functions. Sometimes these can be related to the apparent mode of action of the resistance mechanism. Thus virulence against *Tm-1* in tomato, which inhibits ToMV multiplication, maps to the viral replicase gene. Virulence against *Tm-2* and *Tm-2²* in the same host, which prevent virus movement from the initially-infected cells, maps to the 30 kDa movement protein (see chapter 1 and section III.B.3, this chapter). Finally, it should be noted that the coat protein is the location of the determinant of virulence/avirulence for many resistance genes.

Box IV.   **Location of the determinants of virulence/avirulence on plant virus genomes**

Virus	Host	Resistance gene or cultivar reference	Function of viral gene containing determinant of virulence/avirulence	References
Tobamoviruses:				
Pepper mild mottle	*Capsicum*	*L2*	coat protein	De la Cruz *et al.*, 1997
		L3	coat protein	Berzal-Herranz *et al.*, 1995
Tomato mosaic	tomato	*Tm-1*	replicase	Hamamoto *et al.*, 1997
		Tm-2	movement protein	Calder and Palukaitis, 1992
		Tm-2²	movement protein	Weber *et al.*, 1993
Tobacco mosaic	tobacco	*N*	replicase	Ikeda *et al.*, 1993; Padgett and Beachy, 1993
		N'	coat protein	Culver and Dawson, 1989; Pfitzner and Pfitzner, 1992
Potexviruses:				
Potato virus X	potato	*Nx*	coat protein	Santa Cruz and Baulcombe, 1995
		Rx-1	coat protein	Bendahmane *et al.*, 1995
Potyviruses:				
Tobacco etch	tobacco	V20	VPg	Schaad *et al.*, 1997
Tobacco vein mottling	tobacco	*va*	VPg	Nicolas *et al.*, 1996, 1997

The determinant of virulence/avirulence against particular host resistance genes can map to a number of different functions on viral genomes. For a single virus, the determinant of virulence/avirulence against different host resistance genes can map to different viral functions.

It is clear from the data in Box IV, and from knowledge of the genetic structure of a large number of plant viruses, that they cannot contain genes with the sole function of determination of virulence/avirulence: each such determinant must have another function in the viral replicative or pathogenic cycle, and there may be constraints on the extent to which mutation to virulence can be accommodated while retaining pathogenic function. In a study of natural and artificial mutants of ToMV capable of overcoming the *Tm-1* gene in tomato, Fraser and Gerwitz (1987) showed that the acquisition of virulence by mutation was accompanied by a loss of pathogenic fitness, as defined by ability to multiply and cause visible symptoms; a finding consistent with Vanderplank's (1984) suggestion that virulence may have a 'price' in terms of loss of efficiency of the mutated function. However, a well established ToMV strain from commercial tomato crops, able to overcome *Tm-1*, had an unimpaired ability to multiply and cause severe symptoms, suggesting that these functions were not ultimately incompatible with virulence.

How many virulence determinants may a virus genome contain? Because of the small size of plant virus genomes, there is a possibility that virulence determinants against different resistance genes, which map to the same functional region of the virus genome, could be mutually incompatible, because they involve alterations in the viral function which could not be simultaneously accommodated without loss of function. One example might explain the paucity of ToMV strains displaying virulence against both *Tm-2* and *Tm-2²* in tomato (Calder and Palukaitis, 1992). In contrast, it is possible that the mutation to virulence against one resistance gene or allele might automatically confer virulence against another: an example is at the *bc-1* locus for BCMV resistance in *Phaseolus vulgaris*, where all isolates capable of overcoming the *bc-1²* allele also appear capable of overcoming the *bc-1* allele (Drijfhout, 1978).

3.   *Some Examples of Mechanisms of Resistance*

Examination of the phenotypic expression of resistance caused by different genes in different hosts suggest that there are many classes of mechanisms. Some are better understood than others, and many as yet have no complete biochemical or physiological explanation. Some examples have been chosen to illustrate the diversity of mechanisms involved, and to try to draw out some common principles.

It is clear that one of the commonest mechanisms is the local lesion (HR) response, an example of which is described in detail in the case history of the tobacco *N* gene in chapter 1. The HR is common in resistance against other viruses (Table 1; Fraser 1986a, 1990, 1992b and references therein), and indeed in resistance to bacteria and fungi, as described elsewhere in this book. HR against viruses appears to be strongly associated with dominance, consistent with a model in which the resistance alleles code for a product which recognises a viral determinant and induces a cascade of events, one or several of which lead to localisation of the virus in or around the lesion. It is likely that many of the events associated with necrotic lesion formation, such as cell wall thickening and increased lignification, and accumulation of compounds such as phenolics and quinones with the potential to damage virus structure or function, could contribute to the localisation in a non-specific manner. Some which specifically affect the plasmodesmata, such as the deposition of callose on plasmodesmatal pores (Beffa *et al.*, 1996), could well be more directly involved in the primary mechanism of localization.

An extreme case of localization is exemplified by the *Tm-2* locus in tomato. Both alleles at that locus (*Tm-2* and *Tm-2²*) localise avirulent ToMV to the initially infected cell.

This was demonstrated by staining with fluorescent antibody against the virus coat protein after leaf inoculation, when a small number of individual cells fluoresced, presumably those directly infected in the inoculation (Nishiguchi and Motoyoshi, 1987). When protoplasts isolated from *Tm-2* or susceptible plants were inoculated with ToMV, virus multiplication was detected equally in both, confirming that the resistance mechanism did not inhibit at the level of the individual cell (Motoyoshi and Oshima, 1977). The clear implication is that in the leaf the virus multiplied in those cells which were directly infected in the initial abrasive inoculation, and that the first cell-to-cell movement was inhibited by the resistance mechanism. It is entirely consistent with this mechanism that the determinants of virulence against these two resistance alleles map to the viral movement protein. The movement proteins of the virulent isolates presumably differ in their interactions with the products of the host resistance genes so that cell-to-cell movement is not inhibited. What is not clear is whether this is a constitutive mechanism, or an extremely rapid and effective case of an induced mechanism otherwise similar to the eventual restriction of virus spread in the hypersensitive response. The former explanation would appear more likely.

As noted earlier, this type of mechanism involving restriction of the virus to the initially inoculated cells, or to a very small number of cells, is likely to be quite widespread in plants. It may be involved in subliminal infection (Sulzinski and Zaitlin, 1982) and as one possible explanation of non-host resistance (Van Loon, 1987; Dawson and Hilf, 1992)

Another group of resistance mechanisms operating against virus movement appears to be effective at the point of phloem loading, in that there are several examples where virus multiplication and spread can be detected in the initially-infected leaf, but no systemic spread is evident, or systemic spread is inhibited in comparison with susceptible plants. Examples include resistance to BCMV in *Phaseolus vulgaris* controlled by homozygous recessive *bc-u* plus *bc-1* genes (Day, 1984), resistance to cowpea chlorotic mottle virus in cowpeas (Wyatt and Kuhn, 1980), and resistance to tobacco etch virus in tobacco (Schaad *et al.*, 1997).

The biochemical processes of viral multiplication are the target of another diverse group of resistance mechanisms. In tomato, the *Tm-1* gene permits virus spread throughout the plant, but inhibits multiplication (measured by the accumulation of viral RNA and coat protein) by 70% in plants with the gene in heterozygous form (*Tm-1/+*) and by 95% in homozygotes (*Tm-1/Tm-1*) (Fig. 3; Fraser and Loughlin 1980). This clear example of gene-dosage dependence is consistent with the gene product being an incompletely effective inhibitor of virus multiplication: two copies of the resistance gene should lead to a higher concentration of inhibitor, with greater effect. It is also consistent with this model that virulence maps to the viral replicase gene, presumably resulting in an altered or non-interaction with the inhibitor.

In cowpeas of the cultivar 'Arlington', resistance to cowpea mosaic virus is controlled by a single dominant gene, which segregates with a specific protease inhibitor only found in resistant plants (Ponz *et al.*, 1988). As described in section I.B, the genetic information of comoviruses such as CPMV is expressed initially as two polyproteins translated from the two viral genomic RNAs, which are then cleaved by virus-coded protease activity. This is thought to be the target of the host resistance mechanism.

Ribosome inactivating proteins (RIPs) are a group of proteins present in numerous plant species, which inhibit the translocation step of translation. They do this by catalytically removing a specific adenine base from the larger ribosomal RNA: ribosomes so treated are unable to bind the EF-2/GTP complex required for transloca- tion (reviewed by Hartley *et al.*, 1996). Many RIPs, such as that from *Phytolacca americana*, (pokeweed antiviral protein, PAP), have been shown to have broad spectrum antiviral activity: when mixed with viruses before inoculation they protect plants from infection, and transgenic plants expressing RIPs have been shown to be resistant to infection by RNA and DNA viruses (Lodge *et al.*, 1993; Hong *et al.*, 1996; Smirnov *et al.*, 1997).

PAP is normally sequestered in the cell walls in healthy plants, and thus does not inactivate their ribosomes. This led to the 'local suicide' hypothesis, whereby PAP entered the cytosol as a result of damage caused by mechanical inoculation or vector transmission of a virus, and thereby inactivated the ribosomes and prevented virus translation (Bonness *et al.*, 1994). However, in virus-infected animal cells, PAP was shown to inhibit viral protein synthesis to a much greater extent than it inhibited cell protein synthesis. Also, experiments with mutant forms of PAP have suggested that the antiviral activity can be dissociated from the ability to depurinate host ribosomes (Tumer *et al.*, 1997). These experiments therefore suggest that some further explanation of the antiviral activity may be required. A virus-specific inhibitory effect of PAP has recently been demonstrated with yeast viruses (Tumer *et al.*, 1998), but may not be widely relevant to plant viruses.

C.   RESISTANCE MECHANISMS CONFERRED ON OR OCCURRING IN THE NORMALLY SUSCEPTIBLE PLANT

1.   *Dark Green Islands*
Many viruses which spread systemically give rise to a characteristic light green/dark green mosaic in leaves which are invaded at an early stage of their development. It has long been recognised that while the light green areas are infected, the dark green areas contain limited amounts of virus or are virus free, and are resistant to challenge inoculation with the same virus but not with unrelated viruses. The early literature is reviewed in Fraser (1986b). The resistance of dark green islands to invasion by virus present in adjoining parts of the leaf, and the resistance to challenge inoculation, might involve the same or separate mechanisms. Murakishi and Carlson (1976) found a residual level of resistance in plants regenerated from protoplasts prepared from dark green areas.

The nature of the resistance mechanism in dark green areas is not understood, but may in part involve alterations in plant hormone metabolism. The chlorophyll concentration in the dark green areas is normally higher than in healthy tissue (Whenham *et al.*, 1986). The concentration of the plant hormone abscisic acid (ABA) was shown to be higher in tobacco mosaic virus-infected tobacco leaves than in healthy leaves, and within infected leaves was higher in dark green areas than in light green infected areas. Treatment of healthy plants with exogenous ABA in a manner which caused a similar increase in ABA concentration to that caused by TMV infection increased chlorophyll concentration and resistance to challenge infection with the virus.

Cytokinins have also been implicated in the formation of dark green tissue in diseased plants and in possible resistance to infection. Whenham (1989) showed that the effect of TMV infection of tobacco leaves was to cause little change in the concentration of zeatin and its metabolites zeatin-O-glucoside, zeatin riboside and zeatin riboside-O-glucoside in the light green areas. In the dark green areas, however, the concentration of zeatin was lower than in healthy leaves, while the concentrations of zeatin-O-glucoside and zeatin riboside-O-glucoside were significantly higher than in healthy leaves. Interestingly, treatment of healthy or infected plants with ABA at physiological concentrations also imposed or reinforced these changes in the concentrations of zeatin and its metabolites. It may be therefore that ABA and cytokinins form part of an integrated mechanism involved in control of development of the light green/dark green mosaic symptoms, and the resistance of dark green tissue to virus infection, but the mechanism of resistance is not understood. One suggestion has been that ABA and cytokinins may maintain tissue in some juvenile and resistant state, akin to the resistance of plant meristems to invasion by viruses (Fraser, 1986b). A possibility that deserves further investigation is the nature of the plasmodesmata in the dark green islands. The plasmodesmata of young tissues (primary plasmodesmata) differ structurally from those of older tissues (secondary plasmodesmata) (Lucas and Gilbertson, 1994). The latter appear to be more responsive to modification after virus infection (Ding *et al.*, 1992; Lartey *et al.*, 1997).

As mentioned in section III.C.4 below, an alternative suggestion is that dark green islands may owe their resistance to some form of virus gene silencing, as a result of signals transmitted from infected parts of the plant (Voinnet and Baulcombe, 1997).

2.   *Satellite Resistance*

Viruses in four of the currently recognised forty or so genera of plant viruses (Mayo and Pringle, 1998) may be associated with additional RNA sequences known as satellites. These fall into two groups: linear sequences which share some sequence similarity at their ends with the replicase-binding regions at either end of the helper virus genomic and negative strand RNAs, and are replicated by the viral replicase via a complementary intermediate, and circular sequences similar to viroids, replicated by a rolling-circle mechanism (Matthews, 1991). Satellites depend on their helper virus for their replication. In turn, they may increase or decrease the overall severity of disease symptoms caused by the combined infection, and may reduce multiplication of the helper virus. Cases where disease severity is reduced by a satellite form a special example of attenuation phenotypically resembling a resistance mechanism.

This has been most extensively studied with CMV, which causes extensive damage in a wide range of crops (Palukaitis *et al.*, 1992). A satellite RNA which attenuates the normal symptoms of CMV has been inoculated onto tens of thousands of hectares of tobacco, pepper and tomato plants in the Peoples' Republic of China, and has had beneficial effects on yield and quality (Tien and Wu, 1991). Rigorous quality control during production of the protecting satellite RNA was necessary to avoid contamination with variants causing more severe symptoms. More limited trials in the USA and

Europe have used benign satellites to protect against severe symptoms caused by other forms of the satellites, rather than the helper virus alone (Gallitelli *et al.*, 1991; Montasser *et al.*, 1991). Further work is needed to understand the mechanisms of interaction between the host plant, helper virus and different forms of the satellite, in changing the severity of symptoms produced. One demonstrated effect involves inhibition of virus movement by the satellite (Kong *et al.*, 1997). Use of transgene forms of satellite protection is considered in section III.C.4.

3. *Cross Protection*

This mechanism is mentioned briefly in Chapter 10 and covered in more detail here. It refers to a phenomenon whereby a susceptible plant systemically infected by one isolate of a virus becomes to varying degrees resistant to disease caused by another isolate of the same virus, and was first described by McKinney (1929). In experimental virology, the effect has been used as a diagnostic test of relatedness between virus isolates (e.g. Taiwo *et al.*, 1982), but several examples of inconsistency have been reported (reviewed in Fraser, 1998). The method has largely been superseded by nucleic acid sequencing or advanced serological techniques, which have in some cases prompted re-evaluation of relationships drawn from cross protection studies (e.g. Shukla and Ward, 1989).

In practical defence of crops against virus diseases, cross protection has been used in the particular form where plants are deliberately inoculated with a mild or attenuated form of the virus, causing barely detectable visible symptoms and minimal loss of yield. In successful examples of the effect, the inoculated plants are then protected against infection by strains of the virus normally causing more severe symptoms and loss of yield. Examples of crops where the technique has been widely employed are given in Box V, and more comprehensive details are given in recent reviews by Fuchs *et al.* (1997) and Fraser (1998).

Box V. **Examples of effective use of cross protection in agriculture**

Plant host	Virus	References
Tomato	Tomato mosaic	Rast, 1975; Channon *et al.*, 1978; Fletcher, 1992
Courgette, squash, melon, cucumber	Zucchini yellow mosaic	Wang *et al.*, 1991; Lecoq *et al.*, 1991; Walkey, 1992
Sweet and sour orange, grapefruit, lime	Citrus tristeza	Müller and Costa, 1977
Papaya	Papaya ringspot	Yeh *et al.*, 1988

Cross protection of tomato against tomato mosaic virus was widely used in the USA and Europe, but has been largely superseded by the use of resistant varieties. Protection of cucurbits against zucchini yellow mosaic is increasing. Tens of millions of orange trees in Brazil are cross-protected against citrus tristeza, and the method has been used in India, the Middle East, South Africa and Florida. Cross protection of papaya against papaya ringspot gives useful increases in yield if combined with removal of plants infected by the severe strain. See Fraser (1998) for detailed references.

In all, cross protection has only been used on a few crops, for a number of reasons, although it has had some striking successes. It has been applied where no host resistance genes are available against an important virus of the crop - as in protection of courgettes (zucchini) and other cucurbits against zucchini yellow mosaic virus, or of citrus trees against citrus tristeza virus (CTV). It is also used to protect cultivars with particularly desirable agronomic or quality characters, to which available resistance genes have not yet been transferred. An example is the use of attenuated strains of ToMV to protect cherry tomato cultivars, many of which do not contain the highly effective Tm-2^2 resistance gene (Fletcher, 1992).

On the negative side, the selection of a suitable and stable attenuated mutant or naturally occurring isolate can be a protracted process (e.g. Rast, 1975). In certain cases the effectiveness of protection has been found to be dependent on the host genotype: in citrus a three-way match between CTV isolate, rootstock and scion genotypes is necessary (Müller and Costa, 1977). Mild strains have also been shown to cause a detectable loss of yield, with figures of 5-10% commonly quoted (Channon et $al.$, 1978; Walkey, 1992). This can be acceptable if the alternative of establishment of a full infection by a severe isolate could be a loss of 50% of yield biomass, or blemishing leading to complete loss of marketability. The production of mild strain inoculum, quality control for the absence of severe strain contaminants, and effective inoculation of the plants to be protected, are further costs to be set against the potential benefits.

Concern has also been expressed about the deliberate release of viruses into the cropping environment, especially in areas where that virus is not endemic, or for viruses with very wide host ranges. On the other hand, fears that protected crops might develop severe symptoms as a result of mutations or contaminants in the protecting isolates do not appear to have been realised, and there do not appear to have been reports of undesirable synergistic effects between the protecting virus and other infections by related or unrelated viruses.

Another complication is that in countries of the European Union (EU), mild strain viruses are classified as microbiological biocontrol agents, and as such require registration as pesticides. In some countries of the EU the procedures and costs of registration are a sufficient barrier to prevent or delay introduction of otherwise effective controls (Fraser, 1993). Non-EU countries may also have regulatory controls over cross-protecting agents.

What is the mechanism of cross protection? A number of theories have been proposed, and there has been a spirited debate in the literature (e.g. De Zoeten and Fulton, 1975; Zaitlin, 1976; Urban et $al.$, 1990). Partly, this may have been because cross protection probably involves a variety of mechanisms at different levels of virus-virus and virus-host interaction: conclusions drawn from one set of experimental circumstances cannot necessarily be used to invalidate a mechanism proposed for different circumstances. A specific example of this will be considered later in this section.

Early theories suggested that the cross protecting virus used up certain host metabolites required for virus multiplication. This would appear to be incompatible with the fact that mild cross-protecting strains multiply to much lower amounts than their severe counterparts, and the fact that cross protection is observed with viruses such as the potyviruses, which multiply to very low concentrations in normal infections. An alternative was that the cross protecting virus blocked host receptor sites or factors specifically involved in an aspect of virus replication, to the exclusion of the later-arriving severe strain. There is no specific evidence for this mechanism, although now that the host factors involved in virus multiplication are beginning to be unravelled (e.g. Osman and Buck, 1997), more direct experimental tests may be possible. Palukaitis and Zaitlin (1984) suggested that the negative strand of the challenging virus RNA could become sequestered in an inactive double stranded form by the excess positive-sense progeny RNA of the protecting strain. This model could work, and might also be a possible explanation of the limited multiplication of attenuated, protecting strains, but would require clear scheduling of appropriate concentrations of the different RNA forms, and possible gross differences in stability of the double-stranded RNA forms of different composition.

Most evidence has favoured an involvement of the viral coat protein in a major - although most probably not the only - role in cross protection. The mechanism for which there is the most compelling evidence is prevention by mild strain coat protein of the initial uncoating of the invading severe strain RNA. In TMV, for example, the 5' end of the RNA is exposed at one end of the virus particle in the early stage of infection, by loss of the coat protein subunits. This forms the binding site for host ribosomes, which then proceed to translate the TMV RNA, stripping further coat protein subunits from it as they progress. Wilson and Watkins (1986) showed that addition of exogenous viral coat protein inhibited this co-translational disassembly of TMV in vitro.

Further compelling evidence for a major role of coat protein in cross protection comes from numerous reports of cross protection-like effects in transgenic plants expressing viral coat protein genes (see section III.C.4; Fuchs *et al.*, 1997; Miller and Hemenway, 1998). Crucially, Bendahmane *et al.* (1997) showed that the ability of TMV coat protein transgenes to cross-protect depended on subunit-subunit interactions. Mutant genes producing coat proteins which had lost the ability to assemble conferred no protection against challenging virus: other mutants with stronger subunit-subunit interactions conferred enhanced protection.

Against this background, it is also clear that there are further contributing mechanisms in different cases. A cross protection-like effect has been observed from initial infection by a coat protein-less mutant of TMV (Gerber and Sarkar, 1989), and between viroids which lack any proteins (Niblett *et al.*, 1978). Some of these effects may be akin to the 'green island' phenomenon considered in section III.C.1, and the specificity of some of the results has been questioned (Sherwood, 1987; Urban *et al.*, 1990). There are also indications that viral coat protein itself can interfere with the challenge virus at stages other than uncoating, such as long distance transport (reviewed in Fraser, 1998). Finally, recent evidence that gene silencing may be involved in the mechanism of some types of engineered resistance, and possibly some types of natural resistance (considered in section III.C.4) suggests that this mechanism may also be of value in explaining some of the aspects of cross protection which are difficult to account for by the hypothesis involving coat protein alone.

4.   Transformation

In chapter 11, some genetic engineering approaches to control of epidemic spread of virus diseases are considered; these concentrate on the use of virus coat protein genes, and a mention of 'plantibodies'. In this chapter I will concentrate more on the range of mechanisms by which virus resistance in plants may be genetically engineered, and consider whether the results indicate any common ground between genetically-engineered resistance and natural mechanisms.

The first report of genetically engineered resistance was by Powell-Abel *et al.* (1986). TMV-susceptible tobacco plants were transformed with the coat protein gene of TMV. Expression of the gene in regenerant transformants conferred a type of resistance to virus, in the form of a marked delay in appearance of symptoms, and reduction of virus accumulation. The parallels with cross protection have been noted earlier.

Since then, ten different host species, including potato, tomato, cucurbits, rice and sugarbeet, have been transformed with coat protein (or nucleocapsid) genes from 14 plant virus groups (reviewed in Miller and Hemenway, 1998). Generally, effective resistance has been found, and in many cases is now in field trials. This type of resistance tends to be quite specific to virus isolates with coat protein sequences highly similar to those of the transgene, although plants expressing potyvirus coat proteins appeared to have resistance against a broad spectrum of other potyviruses (Namba *et al.*, 1992). It has also been possible to confer wider-spectrum resistance by transforming a single host plant to express the coat proteins of a number of viruses (Prins *et al.*, 1995).

Initially, it was thought that the mechanism of coat protein-mediated resistance was similar to that proposed for cross protection, in that the coat protein inhibited challenge virus multiplication by sequestering its RNA or preventing uncoating. The latter explanation is supported by evidence that in some cases, coat-protein-mediated resistance was substantially more effective against inoculation with virions than with viral RNA (Register and Beachy, 1988). The results of Bendahmane *et al.* (1997), demonstrating a strong correlation between assembly characteristics of mutant TMV coat proteins, and the degree of resistance conferred on plants transformed with them, also provide strong support for this mechanism of coat protein-mediated resistance. Also, in some host-virus combinations there is evidence that translationally-defective coat protein genes do not confer resistance (Van Dun *et al.*, 1988; Powell-Abel *et al.*, 1990), whereas genes which produce functional coat protein do. In general, the viruses where there is good evidence for coat protein-mediated resistance dependent on expression of the coat protein itself all belong to the alpha virus-like supergroup of virus genera related by sequence and pattern of genomic organisation (De Haan, 1998).

The experience of workers producing transgenic plants with coat protein-mediated resistance has been that it is necessary to produce many different transformant lines, to get some that have effective resistance. There has not always been a good correlation between the number of copies of the coat protein gene inserted into the plant chromosomes and the level of resistance. Insertion of coat protein genes in transcriptionally-inactive regions of the genome might offer an explanation for this effect. Others have found a poor correlation between the level of expression of the coat protein gene, and the level of resistance (reviewed in Miller and Hemenway, 1998). Conversely, expression of a transgene for an assembly-defective coat protein was shown to confer resistance to TMV (Clark *et al.*, 1995).

These inconsistencies led to the suspicion that there might be mechanisms of transgenic resistance, involving coat protein transgenes, which did not depend on the production of functional coat protein (Hammond and Dienelt, 1997). This was confirmed by observations that transgenic resistance to potyviruses could be conferred by coat protein gene constructs which produced untranslateable messenger sense RNAs, and even those producing antisense RNAs (e.g. Lindbo and Dougherty, 1992a, b; Smith *et al.*, 1994). This led to the suggestion that resistance in these transgenic plants might involve a cellular pathway involving targeted degradation of viral RNAs (Dougherty *et al.*, 1994).

In parallel with the mounting evidence from coat protein transgenic plants that resistance might involve mechanisms over and above those dependent on coat protein expression, there were reports that transformation of plants with sequences from other regions of viral genomes could confer resistance. Some examples are shown in Box VI. It is clear that sequences representing diverse viral functions, such as the replicase or movement protein, defective or truncated versions of these genes, and expression in sense and antisense configuration, can confer resistance, often to a highly effective level. Generally, these types of resistance have been found to operate at the RNA level, rather than by producing any functional virus-derived protein sequence. They are therefore referred to collectively as RNA-mediated transgenic virus resistances (reviewed in De Haan, 1998), although some examples of resistance involving transgenes producing defective virus functions might indeed operate at the protein level (e.g. Malyshenko *et al.*, 1993; Seppanen *et al.*, 1997).

What are the mechanisms involved in RNA-mediated resistance? So far, the indications are that the effects may be analogous to or directly involve the phenomenon of 'gene-silencing', an RNA-surveillance mechanism for selective removal of targeted RNAs in the plant cell (Baulcombe, 1996; Dawson, 1996). The mechanisms of RNA-mediated resistance remain to be fully elucidated, but it has been suggested that host-coded RNA-dependent RNA polymerase may be involved in synthesis of double-stranded RNA which is then degraded by a dsRNA-specific ribonuclease (De Haan, 1998). Angell and Baulcombe (1997) have suggested an interaction between viral RNA and homologous sequences in the host DNA, leading to sequence-specific DNA methylation and post-transcriptional gene silencing (English *et al.*, 1996).

Recently, it has been reported that the 'recovery' phenomenon, whereby plants which had previously shown symptoms of severe virus disease produce new growth with less severe or no symptoms, shows evidence of virus-specific gene silencing (Covey *et al.*, 1997; Ratcliffe *et al.*, 1997). The latter authors suggested that the effect might also underly the resistance of meristematic tissue and dark green islands to virus, and hypothesised that the biological significance of the effect could be in prevention of seed transmission of virus from the infected plant to the progeny. Voinnet and Baulcombe (1997) suggested that a systemic signal, probably nucleic acid in nature, moved ahead of the virus in plants, inducing gene silencing and so delaying the spread of the infection front. It will be interesting to see whether these suggestions are supported by further experimental evidence over the next few years.

Virus-resistant transgenic plants have also been produced using a variety of other constructs not derived from the viral genome, and in many cases having no connection with natural virus resistance mechanisms in plants. The use of genes for antibodies raised against plant viruses ('plantibodies') is discussed in chapter 11. Other examples

Box VI.   Some examples of resistance to viruses in transgenic plants transformed with virus-derived sequences other than coat protein

Virus donating sequence	Virus gene expressed	Host	Viruses against which resistance conferred, if different from gene donor virus	References
Tobacco mosaic	defective movement protein	tobacco		Malyshenko *et al.*, 1993
			several unrelated viruses	Cooper *et al.*, 1995
Potato virus X	mutant 12 kDa movement protein	potato	potexviruses and carlaviruses	Seppanen *et al.*, 1997
Potato virus X	24 kDa movement protein	tobacco	tobacco mosaic and tobamovirus Ob	Ares *et al.*, 1998
White clover mosaic virus (potexvirus)	mutant 13 kDa movement protein	*Nicotiana benthamiana*	potexviruses and carlaviruses	Beck *et al.*, 1994
Cucumber mosaic virus	replicase genes in various constructs	tobacco		Carr *et al.*, 1994; Hellwald and Palukaitis, 1995; Suzuki *et al.*, 1996; Wintermantel *et al.*, 1997
Tomato yellow leaf curl virus (DNA geminivirus)	truncated replicase gene	*Nicotiana benthamiana*		Noris *et al.*, 1996
Pepper mild mottle virus(tobamovirus)	truncated 54 kDa replicase segment	*Nicotiana benthamiana*		Tenllado *et al.*, 1996

include ribozymes: RNA enzymes which cleave specific (viral) target RNA sequences and inhibit translation (Edington and Nelson, 1992; De Feyter *et al.*, 1996; Yang *et al.*, 1997). In mammals, a mechanism of resistance to virus infection involves catalysis of viral RNA decay by an interferon-regulated system, involving a specific RNase dependent on a 5'-phosphorylated 2',5'linked oligoadenylate (2-5A). Mitra *et al.* (1996) showed that plants expressing the genes for the human enzymes RNase

L and 2-5A synthase formed necrotic lesions after inoculation with tobacco mosaic, alfalfa mosaic or tobacco etch viruses, whereas plants expressing only one of the two human enzymes became systemically infected. Watanabe *et al.* (1995) showed that plants transformed with and expressing a gene for a double-stranded RNA-specific RNase from yeast were less susceptible to three viruses tested.

Ribosome-inhibiting proteins have been mentioned earlier as natural resistance mechanisms (section III.B.3). Transgenic plants expressing PAP have been shown to have broad-spectrum virus resistance, to both mechanical and aphid transmission (Lodge *et al.*, 1993). In a refinement, Hong *et al.* (1996) placed the gene for another RIP, dianthin, under the control of a transactivated geminivirus promoter from African cassava mosaic virus (ACMV). The effect of this was that dianthin was only synthesised when plants were inoculated with ACMV, and was localized to virus-infected tissue, where it inhibited virus multiplication.

Now that natural plant genes for resistance to plant viruses are beginning to be isolated (Whitham *et al.*, 1994), it is proving possible to transfer resistance to other species to which the gene could not be moved by conventional plant breeding. Thus the *N* gene for TMV resistance in tobacco has been shown to confer an effective form of HR when expressed in tomato (Whitham *et al.* 1996). Disease-attenuating forms of CMV satellite RNA have also been used to make DNA transgenes, which, when expressed from the nuclear genome, caused amelioration of symptoms (Baulcombe *et al.*, 1986).

D.   RESISTANCE ENHANCED BY ADDITIONAL TREATMENTS

Acquired resistance to viruses, and to other pathogens, is considered in detail in chapter 10, where the varied terminology is also elucidated. It is mentioned here partly for completeness, but mainly in the context of how it relates to the other types of virus resistance mechanisms covered in this chapter. The first point to make is that acquired resistance, unlike the four types of mechanism considered in section C, represents a modulation of an existing resistance response to the virus, rather than starting from full susceptibility.

As explained in chapter 10, acquired resistance against viruses is commonly expressed as a reduction in the size of lesions formed on plants which had previously expressed hypersensitive resistance to that virus, or had been treated with a variety of other inducers. In some cases, the number of lesions formed on plants with acquired resistance is also reduced. The evidence on whether virus multiplication is actually reduced in the smaller lesions formed on plants with acquired resistance is variable. The particular data presented in Chapter 10 show clear evidence for reduction: other reports suggest that the smaller lesions may be associated with amounts of virus multiplication similar to those on previously untreated plants (e.g. Fraser 1979; Coutts and Wagih, 1983; Pennazio *et al.*, 1983). It has also been reported that multiplication of systemically-spreading strains of virus, which did not induce the hypersensitive response when inoculated on leaves with acquired resistance as a result of previous inoculation of lower leaves with a necrotic strain, was uninhibited (Fraser, 1979). How can these different results - inhibition or no inhibition of virus multiplication in leaves with acquired resistance - be reconciled, and how does the phenomenon of acquired resistance relate to the mechanism of the 'mainline' hypersensitive response?

As explained in Chapters 1 and 10, there is ample evidence that salicylic acid is a key component of the signal transduction pathway leading to systemic acquired resistance and accumulation of the pathogenesis-related (PR) proteins. Recently, Chivasa *et al.* (1997) showed that pre-treatment of leaf disks of a TMV-susceptible tobacco genotype with salicylic acid strongly reduced levels of viral RNA and coat protein accumulation after inoculation, in confirmation of earlier work on intact plants by Van Loon and Antoniw (1982). Chivasa *et al.* (1997) found that salicylhydroxamic acid strongly antagonised the inhibition by salicylic acid of TMV accumulation in susceptible plants, and the induction of systemic acquired resistance by salicylic acid in resistant plants, but did not inhibit the induction by salicylic acid of PR-1 protein. They concluded that the induction of resistance to TMV by salicylic acid involved a different pathway to that involved in induction of resistance to bacterial and fungal pathogens.

It is clear from the work by Chivasa *et al.* (1997) and others (e.g. Hooft van Huijsduijnen *et al.*, 1986) that salicylic acid can inhibit the multiplication of a number of viruses in whole plants, leaf disks or protoplasts. The mechanism is not clear, but Chivasa *et al.* (1997) suggested that it might involve the mitochondrial alternative oxidase, which is inhibited by salicylhydroxamic acid. What is clear, and perhaps concerning, is that the concentrations of salicylic acid used by Chivasa *et al.* (1997) and Hooft van Huijsduijnen *et al.* (1986) to inhibit virus multiplication (1-5 mM) were two to three orders of magnitude greater than the concentrations observed in planta during induction of acquired resistance, although some of the experimentally-applied compound may have become sequestered in the vacuole. It may be that the variable effects of acquired resistance on virus multiplication discussed above could reflect differences in the amounts of salicylic acid produced in different situations, and its consequent effects against virus multiplication.

In summary, the case for acquired resistance as a meaningful form of resistance against viruses is in some doubt, especially in questions of consistency of effect, and the mechanisms involved. The latter do not appear to involve PR proteins.

IV.   Conclusion: Use of Resistance in Disease Control

Breeding for natural resistance has undoubtedly been very successful in a number of crops, but has been disappointing or a failure in others. Resistance to ToMV in tomato exemplifies some of the problems and successes. The first gene introduced, *Tm-1*, was quickly overcome by virulent isolates of the virus (Fletcher, 1992). The two alleles at the *Tm-2* locus were initially transferred to tomato cultivars with associated undesirable characters from the donor lines or wild species, and extensive further selection was necessary to present the resistance in acceptable forms. Even so, the *Tm-2* allele was soon overcome, and resistance now depends solely on *Tm-2^2*. Fortunately, this has proved extremely durable, but it is not satisfactory that resistance should be on such a narrow base. In many other crops, no natural resistance is available for particular viruses. For example, J. A. Tomlinson (personal communication) estimated that in the vegetable crop species grown by the UK horticultural industry, no suitable source of resistance was available to a total of 23 economically important viruses.

Non-host resistance is sometimes mentioned as a possible route to crop protection against viruses. This is based more on the perception that most species are resistant to most viruses, than on any understanding of the mechanisms and how they might be exploited. If non-host resistance involved the negative type of mechanism discussed in section II, i.e. the non-host lacked a particular factor required by the virus, it would be impossible to exploit in crop protection.

Of the resistance mechanisms imposed on susceptible plants, only cross protection and transgenics have found application in crop protection. Cross protection is comparatively simple technology, but has some disadvantages, and may not see much further expansion to other crops and viruses. Transgenic protection, on the other hand, has great potential. One of the most noteworthy features is the enormous range of types of transgene which can confer meaningful levels of resistance. Many of these could be, or are being, tested under field conditions, without the underlying mechanisms being fully understood. The transgene constructs vary in their likely level of acceptability to the public, for ethical, safety and environmental reasons. The strategy also depends on the availability of efficient transformation methods for each target crop. It is still too early to estimate how much of a problem will be caused by viruses developing virulence against transgenic resistances (e.g. Hellwald and Palukaitis, 1994), although the possibilities of combining different types of transgenic resistance, or combining transgenic resistance with natural resistance genes (e.g. Barker *et al.*, 1994; Derrick and Barker, 1997), may offer means to combat this.

References

Angell SM and Baulcombe DC (1997) Consistent gene silencing in transgenic plants expressing a replicating potato virus X RNA. EMBO J 16: 3675—3684

Antoniw JF, White RF and Zie W (1990) Cryptic viruses of beet and other plants. In: Fraser RSS (ed) Recognition and Response in Plant-Virus Interactions, pp 273—286. Springer Verlag, Heidelberg

Ares X, Calamante G, Cabral S, Lodge J, Hemenway P, Beachy RN and Mentaberry A (1998) Transgenic plants expressing potato virus X ORF2 protein (p24) are resistant to tobacco mosaic virus and Ob tobamoviruses. J Virol 72: 731—738

Atabekov JG and Taliansky ME (1990) Expression of a plant virus-coded transport function by different viral genomes. Adv Virus Res 38: 201—248

Bald JG and Tinsley TW (1967a) A quasi-genetic model for plant virus host ranges. I. Group reactions within taxonomic boundaries. Virology 31: 616—624

Bald JG and Tinsley TW (1967b) A quasi-genetic model for plant virus host ranges. II. Differentiation between host ranges. Virology 32: 321—327

Bald JG and Tinsley TW (1967c) A quasi-genetic model for plant virus host ranges. III. Congruence and relatedness. Virology 32: 328—336

Barker H, Webster KD, Jolly CA, Reavy B, Kumar A and Mayo MA (1994) Enhancement of resistance to potato leafroll virus multiplication in potato by combining the effects of host genes and transgenes. Mol Plant-Microbe Interact 7: 528—530

Barnett OW (1995) Plant virus disease - economic aspects. In: Webster RG and Grannoff A (eds) Encyclopedia of Virology, CD-ROM Edition. Academic Press, London

Baulcombe DC (1996) Mechanisms of pathogen-derived resistance to viruses in transgenic plants. Plant Cell 8: 1833—1844

Baulcombe DC, Saunders GR, Bevan MW, Mayo MA and Harrison BD (1986) Expression of biologically active viral satellite RNA from the nuclear genome of transformed plants. Nature 321: 446—449

Beck DL, Van Dolleweerd CJ, Lough TJ, Balmori E, Voot DM, Andersen MT, O'Brien IE, Forster RL (1994) Disruption of virus movement confers broad-spectrum resistance against systemic infection by plant viruses with a triple gene block. Proc Natl Acad Sci USA 91: 10310—10314

Beffa RS, Hofer RM, Thomas M and Meins F (1996) Decreased susceptibility to viral disease of β-1,3-glucanase-deficient plants generated by antisense transformation. Plant Cell 8: 1001—1011

Bendahmane A, Kohn BA, Dedi C and Baulcombe DC (1995) The coat protein of potato virus X is a strain-specific elicitor of *Rx1*-mediated virus resistance in potato. Plant J 8: 933—941

Bendahmane M, Fitchen JH, Zhang G and Beachy R (1997) Studies of coat protein-mediated resistance to tobacco mosaic tobamovirus: correlation betweeen assembly of mutant coat proteins and resistance. J Virol 71: 7942—7950

Berzal-Herranz A, De la Cruz A, Tenllado F, Diaz-Ruiz JR, Lopez L, Sanz AI, Vaquero C, Serra MT and Garcia-Luque I (1995) The *Capsicum L3* gene-mediated resistance against the tobamoviruses is elicited by the coat protein. Virology 209: 498—505

Bonness MS, Ready, MP, Irvine JD and Mabry TJ (1994) Pokeweed antiviral protein inactivates pokeweed ribosomes: implications for the antiviral mechanism. Plant J 5: 173—183

Brunt AA, Crabtree K, Dallwitz MJ, Gibbs AJ and Watson L (1996) Viruses of Plants. CAB International, Wallingford

Burnett PA and Plumb RT (1998) Present status of controlling barley yellow dwarf virus. In: Hadidi A, Khetarpal RK and Koganezawa H (eds) Plant Virus Disease Control, pp 448—458. American Phytopathological Society Press, St Paul

Büschges R, Hollricher K, Panstrunga R, Simons G, Wolter M, Frijters A, van Daelen R, van der Lee T, Diergaarde P, Groenendijk J, Topsch S, Vos P, Salamini F and Schulze-Lefert P (1997) The barley *Mlo* gene: a novel control element of plant pathogen resistance. Cell 88: 695—705

Calder VL and Palukaitis P (1992) Nucleotide sequence analysis of the movement protein genes of resistance-breaking strains of tomato mosaic virus. J Gen Virol 73: 165—168

Callaway A, Liu W, Andrianov V, Stenzler L, Zhao J, Wettlaufer S, Jayakumar P and Howell SH (1996) Characterization of cauliflower mosaic virus (CaMV) resistance in virus-resistant ecotypes of *Arabidopsis*. Mol Plant-Microbe Interact 9: 810—818

Caranta C, Palloix A, Gebre-Selassie K, Lefebvre V, Moury B and Daubèze AM (1996) A complementation of two genes originating from susceptible *Capsicum annuum* lines confers a new and complete resistance to pepper veinal mottle virus. Phytopathology 86: 739—743

Carr JP, Gal-On A, Palukaitis P and Zaitlin M (1994) Replicase mediated resistance to cucumber mosaic virus in transgenic plants involves suppression of both virus replication in the inoculated leaves and long-distance movement. Virology 199: 439—447

Carrington JC, Kasschau KD, Mahajan SK and Schaad MC (1996) Cell-to-cell and long distance transport of virus in plants. Plant Cell 8: 1669—1681

Carroll TW, Gossel PL and Hockett EA (1979) Inheritance of resistance to seed transmission of barley stripe mosaic virus in barley. Phytopathology 69: 431—433

Channon AG, Cheffins NJ, Hitchon GM and Barker J (1978) The effect of inoculation with an attenuated mutant strain of tobacco mosaic virus on the growth and yield of early tomato glasshouse crops. Ann Appl Biol 88: 121—129

Cheo PC and Garard JS (1971) Differences in virus-replicating capacity among species inoculated with tobacco mosaic virus. Phytopathology 61: 1010—1012

Chivasa S, Murphy AM, Naylor M and Carr JP (1997) Salicylic acid interferes with tobacco mosaic virus replication via a novel salicylhydroxamic acid-sensitive mechanism. Plant Cell 9: 547—557

Clark WG, Fitchen JH and Beachy RN (1995) Studies of coat protein-mediated resistance to TMV. 1. The PM2 assembly defective mutant confers resistance to TMV. Virology 208: 485—491

Cooper B, Lapidot M, Heick JA, Dodds JA and Beachy RN (1995) A defective movement protein of TMV in transgenic plants confers resistance to multiple viruses whereas the functional analog increases susceptibility. Virology 206: 307—313

Couch HB (1955) Studies on seed transmission of lettuce mosaic virus. Phytopathology 45:63—71

Coutts RHA and Wagih EE (1983) Induced resistance to viral infection and soluble protein alterations in cucumber and cowpea plants. Phytopathol Z 107: 57—69

Covey SN, Al-Kaff NS, Lángara A and Turner DS (1997) Plants combat infection by gene silencing. Nature 385: 781—782

Csillery G, Tobias I and Rusko J (1983) A new pepper strain of tomato mosaic virus. Acta Phytopathol Acad Sci Hung 18: 195—200

Culver JN and Dawson WO (1989) Point mutations in the coat protein gene of tobacco mosaic virus induce hypersensitivity in *Nicotiana sylvestris*. Mol Plant-Microbe Interact 2: 209—213

Dawson WO (1996) Gene silencing and virus resistance: a common mechanism. Trends Plant Sci 1: 107—108

Dawson WO and Hilf ME (1992) Host range determinants of plant viruses. Annu Rev Plant Physiol Plant Mol Biol 43: 527—555

Dawson WO, Bubrick P and Grantham GL (1988) Modifications of the tobacco mosaic virus coat protein gene affect replication, movement and symptomatology. Phytopathology 78: 783—789

Day KL (1984) Resistance to bean common mosaic virus in *Phaseolus vulgaris* L. PhD Thesis, University of Birmingham

De Feyter R, Young M, Schroeder K, Dennis ES and Gerlach W (1996) A ribozyme gene and an antisense gene are equally effective in conferring resistance to tobacco mosaic virus on transgenic tobacco. Mol Gen Genet 250: 329—338

De Haan P (1998) Mechanisms of RNA-mediated resistance to plant viruses. In: Foster GD and Taylor SC (eds) Plant Virology Protocols: from Virus Isolation to Transgenic Resistance, pp 533—546. Humana Press, Totowa

De la Cruz A, Lopez L, Tenllado F, Diaz-Ruiz JR, Sanz AI, Vaquero C, Serra MT and Garcia-Luque I (1997) The coat protein is required for the elicitation of the *Capsicum L2* gene-mediated resistance against the tobamoviruses. Mol Plant Microbe Interact 10: 107—113

De Zoeten GA and Fulton RW (1975). Understanding generates possibilities. Phytopathology 81: 585—586

Dempsey DA, Pathirana MS, Wobbe KK and Klessig DF (1997) Identification of an *Arabidopsis* locus required for resistance to turnip crinkle virus. Plant J 11: 301—311

Derrick PM and Barker H (1997) Short and long distance spread of potato leafroll luteovirus: effects of host genes and transgenes conferring resistance to virus accumulation in potato. J Gen Virol 78: 243—251

Ding B, Haudenshield JS, Hull RJ, Wolf S, Beachy RN and Lucas WJ (1992) Secondary plasmodesmata are specific sites of localization of the tobaccco mosaic virus movement protein in transgenic tobacco plants. Plant Cell 4: 915—928

Ding SW, Shi BJ, Li WX and Symons RH (1996) An interspecies hybrid RNA virus is significantly more virulent than either parental strain. Proc Natl Acad Sci USA 93: 7470—7474

Dougherty WG, Lindbo JA, Smith HA, Parks TD, Swaney S and Proebsting WM (1994) RNA-mediated virus resistance in transgenic plants: exploitation of a cellular pathway possibly involved in RNA degradation. Mol Plant-Microbe Interact 7: 544—552

Drijfhout E (1978) Genetic interaction between *Phaseolus vulgaris* and bean common mosaic virus with implications for strain identification and breeding for resistance. Agric Res Rep Wageningen 872: 1—98

Edington BV and Nelson RS (1992) Utilization of ribozymes in plants: plant viral resistance. In: Reickson RP and Izant JG (eds) Gene Regulation: Biology of antisense RNA and DNA, pp 209—221. Raven Press, New York

English JJ, Mueller E and Baulcombe DC (1996) Suppression of virus accumulation in transgenic plants exibiting silencing of nuclear genes. Plant Cell 8: 179—188

Fisher ML and Kyle MM (1994) Inheritance of resistance to potyviruses in *Phaseolus vulgaris* L. III. Cosegregation of phenotypically similar dominant resistance to nine potyviruses. Theor Appl Genet 89: 818—823

Fisher ML and Kyle MM (1996) Inheritance of resistance to potyviruses in *Phaseolus vulgaris* L. IV. Inheritance, linkage relations and environmental effects on systemic resistance to four potyviruses. Theor Appl Genet 92: 204—212

Fletcher JT (1992) Disease resistance in protected crops and mushrooms. Euphytica 63: 33—49

Flor HH (1956) The complementary genetic systems in flax and flax rust. Adv Genet 8: 29—54

Fraile A, Escriu F, Aranda MA, Malpica JM, Gibbs A and Garcia-Arenal F (1997). A century of tobamovirus evolution in an Australian population of *Nicotiana glauca*. J Virol 71: 8316—8320

Fraser RSS (1969) Effects of two TMV strains on the synthesis and stability of chloroplast ribosomal RNA in tobacco leaves. Mol Gen Genet 106: 73—79

Fraser RSS (1979) Systemic consequences of the local lesion reaction to tobacco mosaic virus in a tobacco variety lacking the *N* gene for hypersensitivity. Physiol Plant Pathol 14: 383—394

Fraser RSS (1985) Host range control and non-host immunity to viruses. In: Fraser RSS (ed) Mechanisms of Resistance to Plant Diseases, pp 13—28. Martinus Nijhoff/Dr W Junk, Dordrecht

Fraser RSS (1986a) Genes for resistance to plant viruses. Crit Rev Plant Sci 3: 257—294

Fraser RSS (1986b) Biochemistry of Virus-Infected Plants. Research Studies Press, Letchworth; John Wiley, New York

Fraser RSS (1990) The genetics of resistance to plant viruses. Annu Rev Phytopathol 28: 179—200

Fraser RSS (1992a) Plant viruses as agents to modify the phenotype for good or evil. In: Wilson TMA and Davies J (eds) Genetic Engineering with Plant Viruses, pp 1—23. CRC Press, Boca Raton

Fraser RSS (1992b) The genetics of plant-virus interactions: implications for plant breeding. Euphytica 63: 175—185

Fraser RSS (1993) Crop protection in UK horticulture: rationale, current practices and dependence on crop protection products. In: Tyson D (ed) Crop Protection: Crisis for UK Horticulture, pp 11—19. British Crop Protection Council, Farnham

Fraser RSS (1998) Introduction to classical cross protection. In: Foster GD and Taylor SC (eds) Plant Virology Protocols: from Virus Isolation to Transgenic Resistance, pp 13—24. Humana Press, Totowa

Fraser RSS and Gerwitz A (1987) The genetics of resistance and virulence in plant virus disease. In: Day PR and Jellis GJ (eds) Genetics and Plant Pathogenesis, pp 33—44. Blackwell Scientific, Oxford

Fraser RSS and Loughlin SAR (1980) Resistance to tobacco mosaic virus in tomato: effects of the *Tm-1* gene on virus multiplication. J Gen Virol 48: 87—96

Fraser RSS, Gerwitz A and Morris GEL (1986) Multiple regression analysis of the relationships between tobacco mosaic virus multiplication, the severity of mosaic symptoms, and the growth of tobacco and tomato. Physiol Mol Plant Pathol 29: 239—249

Frencel I and Pospieszny H (1979) Viruses in natural infections of yellow lupin (*Lupinus luteus* L.) in Poland. IV. Bean common mosaic virus (BCMV). Acta Phytopathol Acad Sci Hung 14: 279—284

Fuchs M, Ferrieira S and Gonsalves D (1997) Management of virus disease by classical and engineered protection. Mol Plant Pathol On-Line http://www.bspp.org.uk/mppol/1997/0116fuchs

Furusawa I and Okuno T (1978) Infection with BMV of protoplasts derived from five plant species. J Gen Virol 40: 489—491

Gallitelli D, Vovlas C, Martelli G, Montasser MS, Tousignant ME and Kaper JM (1991) Satellite mediated protection of tomato against cucumber mosaic virus. II. Field test under natural epidemic conditions in southern Italy. Plant Dis 75: 93—95

Gerber M and Sarkar S (1989) The coat protein of tobacco mosaic virus does not play a significant role for cross protection. J Phytopathol 124: 323—331

Gibson RW and Pickett JA (1983) Wild potato repels aphids by release of aphid alarm pheromone. Nature 302: 608—609

Godefroy-Colburn T, Schoumacher F, Erny C, Berna A, Moser O, Gagey MJ and Stussi-Garaud C (1990) The movement protein of some plant viruses. In: Fraser RSS (ed) Recognition and Response in Plant-Virus Interactions, pp 207—231. Springer Verlag, Heidelberg

Goulden MG and Baulcombe DC (1993) Functionally homologous host components recognize potato virus X in *Gomphrena globosa* and potato. Plant Cell 5: 921-930

Hall TJ (1980) Resistance at the *Tm-2* locus in the tomato to tomato mosaic virus. Euphytica 29: 189—197

Hamamoto H, Watanabe Y, Kamada H and Okada Y (1997) Amino acid changes in the putative replicase of tomato mosaic virus that overcomes resistance in *Tm-1* tomato. J Gen Virol 78: 461—464

Hammond J and Dienelt MM (1997) Encapsidation of potyviral RNA in various forms of transgene coat protein is not correlated with resistance in transgenic plants. Mol Plant-Microbe Interact 10: 1023—1027

Hartley MR, Chaddock JA and Bonness MS (1996) The structure and function of ribosome inactivating proteins. Trends Plant Sci 1: 254—256

Hayes RJ and Buck KW (1990) Complete replication of a eukaryotic virus RNA in vitro by a purified RNA-dependent RNA polymerase. Cell 63: 363—369

Hellwald KH and Palukaitis P (1994) Nucleotide sequence and infectivity of cucumber mosaic cucumovirus (strain K) RNA2 involved in breakage of replicase-mediated resistance in tobacco. J Gen Virol 75: 2121—2125

Hellwald KH and Palukaitis P (1995) Viral RNA as a potential target for two independent mechanisms of replicase-mediated resistance against cucumber mosaic virus. Cell 83: 937—946

Holmes FO (1938) Inheritance of resistance to tobacco mosaic virus in tobacco. Phytopathology 28: 553—561

Holmes FO (1955) Additive resistance to specific viral diseases in plants. Ann Appl Biol 42: 129—139

Hong Y, Saunders K, Hartley MR and Stanley J (1996) Resistance to geminivirus infection by virus-induced expression of dianthin in transgenic plants. Virology 220: 119—127

Hooft van Huijsduijnen RAM, Alblas SW, De Rijk RH and Bol JF (1986) Induction by salicylic acid of pathogenesis-related proteins and resistance to alfalfa mosaic virus infection in various plant species. J Gen Virol 67: 2135—2143

Hughes JD'A and Ollennu LAA (1994) Mild strain protection of cocoa in Ghana against cocoa swollen shoot virus: a review. Plant Pathol 43: 442—457

Hull R (1989) Movement of viruses within plants. Annu Rev Phytopathol 27: 213—240

Ikeda R, Watanabe E, Watanabe Y and Okada Y (1993) Nucleotide sequence of tobamovirus Ob which can spread systemically in *N* gene tobacco. J Gen Virol 73: 1939—1944

Ishikawa M, Diez J, Restrepo-Hartwig M and Ahlquist P (1997) Yeast mutations in multiple complementation groups inhibit brome mosaic virus RNA replication and transcription and perturb regulated expression of the viral polymerase-like gene. Proc Natl Acad Sci USA 94: 13810—13815

Jellis GJ (1992) Multiple resistance to pests and diseases in potato. Euphytica 63: 51—58

Jones AT and Catherall PL (1970) The relationship between growth rate and the expression of tolerance to barley yellow dwarf virus in barley. Ann Appl Biol 65: 137—145

Khetarpal RK, Maisonneuve B, Maury Y, Chalhoub B, Dinant S, Lecoq H and Varma A (1988) Breeding for resistance to plant viruses. In: Hadidi A, Khetarpal RK and Koganezawa H (eds) Plant Virus Disease Control, pp 14—32. American Phytopathological Society Press, St Paul

Köhm BA, Goulden MG, Gilbert JE, Kavanagh TA and Baulcombe DC (1993) A potato virus X resistance gene mediates an induced, nonspecific resistance in protoplasts. Plant Cell 5: 913—920

Kong Q, Wang J and Simon AE (1997) Satellite RNA-mediated resistance to turnip crinkle virus in *Arabidopsis* involves a reduction in virus movement. Plant Cell 9: 2051—2063

Lartey R, Ghoshroy S, Sheng J and Citovsky V (1997) Transport through plasmodesmata and nuclear pores: cell-to-cell movement of plant viruses and nuclear import of *Agrobacterium* T-DNA. In: McCrae MA, Saunders JR, Smyth CJ and Stow ND (eds) Molecular Aspects of Host-Pathogen Interactions, pp 253—280. Society for General Microbiology Symposium volume 55. Cambridge University Press, Cambridge

Lecoq H, Lemaire JM and Wipf-Scheibel C (1991) Control of zucchini yellow mosaic virus in squash by cross protection. Plant Dis 75: 208—211

Lee S, Stenger DC, Bisaro DM and Davis KR (1994) Identification of loci in *Arabidopsis* that confer resistance to geminivirus infection. Plant J 6: 525—535

Lindbo JA and Dougherty WG (1992a) Pathogen derived resistance to a potyvirus: immune and resistant phenotypes in transgenic tobacco expressing altered forms of a potyvirus coat protein nucleotide sequence. Mol Plant-Microbe Interact 4: 247—253

Lindbo JA and Dougherty WG (1992b) Untranslatable transcripts of the tobacco etch virus coat protein gene sequence can interfere with tobacco etch virus replication in transgenic plants and protoplasts. Virology 189:725—733

Lodge JK, Kaniewski WK and Tumer NE (1993) Broad spectrum virus resistance in transgenic plants expressing pokeweed antiviral protein. Proc Natl Acad Sci USA 90: 7089—7093

Lucas WJ, Ding B and van der Schoot C (1993) Plasmodesmata and the supracellular nature of plants. New Phytol 125: 435—476

Lucas WJ and Gilbertson RL (1994) Plasmodesmata in relation to viral movement within leaf tissues. Annu Rev Phytopathol 32: 387—411

Lugtenberg BJJ (1986) (ed) Recognition in Microbe-Plant Symbiotic and Pathogenic Interactions. Springer Verlag, Berlin

Macfarlane SA (1997) Natural recombination among plant virus genomes: evidence from tobraviruses. Seminars Virol 8: 25—31

McKinney HH (1929) Mosaic diseases in the Canary Islands, West Africa and Gibraltar. J Agric Res 39:557-578

Malyshenko SI, Kondakova OA, Nazarova JV, Kaplan IB, Taliansky ME and Atabekov JG (1993) Reduction of tobacco mosaic virus accumulation in transgenic plants producing non-functional viral transport proteins. J Gen Virol 74: 1149—1156

Matthews REF (1991) Plant Virology (Third Edition). Academic Press, San Diego

Mayo MA and Pringle CR (1998) Virus taxonomy - 1997. J Gen Virol 79:649—657

Miller ED and Hemenway C (1998) History of coat protein-mediated protection. In: Foster GD and Taylor SC (eds) Plant Virology Protocols: From Virus Isolation to Transgenic Resistance, pp 25—38. Humana Press, Totowa

Mitra A, Higgins DW, Langenberg WG, Nie H, Sengupta DN and Silverman RH (1996) A mammalian 2,5-A system functions as an antiviral pathway in transgenic plants. Proc Natl Acad Sci USA 93: 6780—6785

Montasser MS, Tousignant ME and Kaper JM (1991) Satellite mediated protection of tomato against cucumber mosaic virus: I. Greenhouse experiments and simulated epidemic conditions in the field. Plant Dis 75: 86—92.

Montesclaros L, Nicol N, Ubalijoro E, Leclerc-Potvin C, Ganivet L, Laliberté JF and Fortin MG (1997) Response to potyvirus infection and genetic mapping of resistance loci to potyvirus infection in *Lactuca*. Theor Appl Genet 94: 941—946

Motoyoshi F and Oshima N (1977) Expression of genetically controlled resistance to tobacco mosaic virus infection in isolated tomato leaf mesophyll protoplasts. J Gen Virol 34: 499—506

Müller GW and Costa AS (1977) Tristeza control in Brazil by preimmunization with mild strains. Proc Int Soc Citricult 3: 868—872

Murakishi HH and Carlson PS (1976) Regeneration of virus-free plants from dark green islands of tobacco mosaic virus-infected tobacco leaves. Phytopathology 66: 931—932

Namba S, Ling K, Gonsalves C, Slightom JL and Gonsalves D (1992) Protection of transgenic plants expressing the coat protein gene of watermelon mosaic virus II or zucchini yellow mosaic virus against six potyviruses. Phytopathology 82: 940—946

Niblett CL, Dickson E, Fernow KH, Horst RK and Zaitlin M (1978) Cross protection among four viroids. Virology 91: 198—203

Nicolas O, Pirone TP and Hellman GM (1996) Construction and analysis of infectious transcripts from a resistance-breaking strain of tobacco vein mottling potyvirus. Arch Virol 141: 1535—1552

Nicolas O, Dunnington SW, Gotow LF, Pirone TP and Hellmann GM (1997). Variations in the VPg protein allow a potyvirus to overcome *va* gene resistance in tobacco. Virology 237: 452—459

Nishiguchi N and Motoyoshi F (1987) Resistance mechanisms of tobacco mosaic virus strains in tomato and tobacco. In: Evered D and Harnett S (eds) Plant Resistance to Viruses, pp 38-56. Wiley, Chichester

Noris E, Accotto GP, Tavazza R, Brunetti A, Crespi S and Tavazza M (1996) Resistance to tomato yellow leaf curl geminivirus in *Nicotiana benthamiana* plants transformed with a truncated viral C1 gene. Virology 224: 130—138

Osman TAM and Buck KW (1997) The tobacco mosaic virus RNA polymerase complex contains a plant protein related to the RNA-binding subunit of yeast eIF-3. J Virol 71: 6075—6082

Padgett HS and Beachy RN (1993) Analysis of a tobacco mosaic virus strain capable of overcoming *N* gene-mediated resistance. Plant Cell 5: 577—586

Palukaitis P, Roosinck MJ, Dietzgen RG and Francki RIB (1992) Cucumber mosaic virus. Adv Virus Res 41: 281—348

Palukaitis P and Zaitlin M (1984) A model to explain the 'cross protection' phenomenon shown by plant viruses and viroids. In: Kosuge T and Nester EW (eds) Plant-Microbe Interactions. Molecular and Genetic Perspectives, pp 420-429. Macmillan, New York

Pelham J, Fletcher JT and Hawkins JH (1970) The establishment of a new strain of tobacco mosaic virus resulting from the use of resistant varieties of tomato. Ann Appl Biol 65: 293—297

Pennazio S, Roggero P and Lenzi R (1983) Some characteristics of the hypersensitive reaction of White

Burley tobacco to tobacco necrosis virus. Physiol Plant Pathol 22: 347—355

Pfitzner UM and Pfitzner AJ (1992) Expression of a viral avirulence gene in transgenic plants is sufficient to induce the hypersensitive defense reaction. Mol Plant-Microb Interact 6: 318—321

Ponz F, Glascock CB and Bruening G (1988) An inhibitor of polyprotein processing with the characteristics of a natural virus resistance factor. Mol Plant-Microbe Interact 1: 25—31

Powell-Abel PA, Nelson RS, De B, Hoffman N, Rogers SG, Fraley RT and Beachy RN (1986) Delay of disease development in transgenic plants that express the tobacco mosaic virus coat protein gene. Science 232: 738—743

Powell-Abel PA, Sanders PR, Tumer NE, Fraley RT and Beachy RN (1990) Protection against tobacco mosaic virus infection in transgenic plants requires accumulation of coat protein rather than coat protein RNA sequences. Virology 175: 124—130

Prins M, De Haan P, Luyten R, Van Veller M, Van Ginsven MQJM and Goldbach R (1995) Broad resistance to tospoviruses in transgenic tobacco plants expressing three tospoviral nucleoprotein gene sequences. Mol Plant-Microbe Interact 8: 85—91

Quadt R, Kao CC, Browning KS, Hershberger RP and Ahlquist P (1993) Characterisation of a host protein associated with brome mosaic virus RNA-dependent RNA polymerase. Proc Natl Acad Sci USA 90: 1498—1503

Ratcliffe F, Harrison BD and Baulcombe DC (1997) A similarity between viral defense and gene silencing in plants. Science 276: 1558—1560

Rast ATB (1975) Variability of tomato mosaic virus in relation to control of tomato mosaic in glasshouse crops by resistance breeding and cross protection. Agric Res Rep Wageningen 834: 1—76

Register JC and Beachy RNA (1988) Resistance to TMV in transgenic plants results from an interference with an early stage of infection. Virology 166: 524—532

Revers F, Yang SL, Walter J, Souche S, Lot H, Le Gall O, Candresse T and Dunez J (1997) Comparison of the complete nucleotide sequences of two isolates of lettuce mosaic virus differing in their biological properties. Virus Res 47: 167—177

Sanfaçon H, Cohen JV, Elder M, Rochon DM and French CJ (1993) Characterization of *Solanum dulcamara* yellow fleck Ob: a tobamovirus that overcomes the *N* resistance gene. Phytopathology 83: 400—404

Santa Cruz, S and Baulcombe DC (1995) Analysis of potato virus X coat protein genes in relation to resistance conferred by the genes *Nx*, *Nb* and *Rx1* of potato. J Gen Virol 76: 2057—2061

Schaad MC, Lellis AD and Carrington JC (1997) VPg of tobacco etch potyvirus is a host genotype-specific determinant for long-distance movement. J Virol 71: 8624—8631

Seppanen P, Puska R, Honkanen J, Tyulkina LG, Fedorkin O, Morozov SY and Atabekov JG (1997) Movement protein-derived resistance to triple gene block-containing plant viruses. J Gen Virol 78: 1241—1246

Séron K and Haenni A-L (1996) Vascular movement of plant viruses. Molec Plant-Microb Interact 9: 435—442

Sherwood JL (1987) Demonstration of the specific involvement of coat protein in tobacco mosaic virus (TMV) cross protection using a TMV coat protein mutant. J Phytopathol 118: 358—362

Shukla DD and Ward CW (1989) Identification and classification of potyviruses on the basis of coat protein sequence data and serology. Arch Virol 106: 171—200

Slack SA and Singh RP (1998). Control of virus affecting potatoes through seed potato certification programs. In: Hadidi A, Khetarpal RK and Koganezawa H (eds) Plant Virus Disease Control, pp 249—260. American Phytopathological Society Press, St Paul

Smirnov S, Shulaev V and Tumer NE (1997) Expression of pokeweed antiviral protein in transgenic plants induces virus resistance in grafted wild-type plants independently of salicylic acid accumulation and pathogenesis-related protein synthesis. Plant Physiol 114: 1113—1122

Smith HA, Swaney SL, Parks TD, Wernsman EA and Dougherty WG (1994) Transgenic plant virus resistance mediated by untranslatable sense RNAs: expression, regulation and fate of non-essential RNAs. Plant Cell 6: 1441—1453

Stace-Smith R (1990) Tissue culture. In: Mandahar CL (ed) Plant Viruses volume II, pp 295—320. CRC Press, Boca Raton

Sulzinski MA and Zaitlin M (1982) Tobacco mosaic virus replication in resistant and susceptible plants: in some resistant species virus is confined to a small number of initially infected cells. Virology 121: 12—19

Suzuki M, Masuta C, Takanami Y and Kuwata S (1996) Resistance against cucumber mosaic virus in plants expressing the viral replicon. FEBS Lett 379: 26—30

Taiwo MA, Gonsalves D, Provvidenti R and Thurston HD (1982) Partial characterisation and grouping of isolates of blackeye cowpea mosaic and cowpea aphid-borne mosaic viruses. Phytopathology 72: 590-596

Tang W and Leisner SM (1997). Cauliflower mosaic virus isolate NY8153 breaks resistance in *Arabidopsis* ecotype En-2. Phytopathology 87: 792—798

Tenllado F, Garcia-Luque I, Serra MT and Diaz-Ruiz JR (1996) Resistance to pepper mild mottle tabamovirus conferred by the 54-kDa gene sequence in transgenic plants does not require expression of the wild-type 54-kDa protein. Virology 219: 330—335

Thomas PE and Fulton RW (1968) Correlation of ectodesmata number with non-specific resistance to initial infection. Virology 34: 459-469

Tien P and Wu GS (1991) Satellite RNA for the biocontrol of plant disease. Adv Virus Res 39: 321—339

Tumer NE, Hwang, DJ and Bonness M (1997) C-terminal deletion mutant of pokeweek antiviral protein inhibits viral infection but does not depurinate host ribosomes. Proc Natl Acad Sci USA 94: 3866—3871

Tumer NE, Parikh BA, Li P and Dinman JD (1998) The pokeweed antiviral protein specifically inhibits Tyl-directed +1 frameshifting and retrotransposition in *Saccharomyces cerevisiae*. J Virol 72: 1036—1042

Urban LA, Sherwood JL, Rezende JAM and Melcher U (1990) Examination of mechanisms of cross protection with non-transgenic plants. In: Fraser RSS (ed) Recognition and Response in Plant Virus Interactions, pp 415—426. Springer Verlag, Heidelberg

Vanderplank JE (1984) Disease Resistance in Plants. Second Edition. Academic Press, New York

Van Dun CMP, Overduin B, Van Vloten-Doting L and Bol JF (1988) Transgenic tobacco expressing tobacco streak virus or mutated alfalfa mosaic virus coat protein does not cross protect against alfalfa mosaic virus infection. Virology 164, 383—389

Van Loon L C (1987) Disease induction by plant viruses. Adv Virus Res 33: 205—256

Van Loon LC and Antoniw JF (1982) Comparison of the effects of salicylic acid and ethephon with virus-induced hypersensitivity and acquired resistance in tobacco. Neth J Plant Pathol 88: 237—256

Voinnet O and Baulcombe CD (1997) Systemic signalling in gene silencing. Nature 389: 553

Walkey DGA (1992) Studies on the control of zucchini yellow mosaic virus in courgettes by mild strain cross protection. Plant Pathol 41: 762—771

Wang HL, Gonsalves D, Provvidenti R and Lecoq HL (1991) Effectiveness of cross protection by a mild strain of zucchini yellow mosaic virus in cucumber, melon and squash. Plant Dis 75: 203—207

Watanabe Y, Ogawa T, Takahashi H, Ishida I, Takeuchi Y, Yamamoto M and Okada Y (1995) Resistance against multiple plant viruses in plants mediated by a double-stranded-RNA specific ribonuclease. FEBS Lett 372: 165—168

Waterworth HE and Hadidi A (1998) Economic losses due to plant viruses. In: Hadidi A, Khetarpal RK and Koganezawa H (eds) Plant Virus Disease Control, pp 1—13. American Phytopathological Society Press, St Paul

Weber H, Schultze S and Pfitzner AJ (1993) Two amino acid substitutions in the tomato mosaic virus 30-kilodalton movement protein confer the ability to overcome the *Tm-2²* resistance gene in the tomato. J Virol 67: 6432—6438

Webster RG and Granoff A (1994) Encyclopedia of Virology Vols 1-3. Academic Press, London

Weiland JJ and Edwards MC (1994) Evidence that the αa gene of barley stripe mosaic virus encodes determinants of pathogenicity to oat (*Avena sativa*). Virology 201: 116—122

Whenham RJ (1989) Effect of systemic tobacco mosaic virus infection on endogenous cytokinin concentration in tobacco (*Nicotiana tabacum* L.) leaves: consequences for the control of resistance and symptom development. Physiol Mol Plant Pathol 35: 85—95

Whenham RJ, Fraser RSS, Brown LP and Payne JA (1986) Tobacco mosaic virus-induced increase in abscisic acid concentration in tobacco leaves: intracellular concentration in light and dark-green areas, and relationship to symptom development. Planta 168: 592—598

Whitham S, Dinesh-Kumar SP, Choi D, Hehl R, Corr C and Baker B (1994) The product of the tobacco mosaic virus resistance gene *N*: similarity to Toll and the interleukin-1 receptor. Cell 78: 1101—1115

Whitham S, McCormick S and Baker B (1996) The *N* gene of tobacco confers resistance to tobacco mosaic virus in transgenic tomato. Proc Natl Acad Sci USA 93: 8776—8781

Wilson TMA and Watkins AC (1986) Influence of exogenous viral coat protein on the cotranslational disassembly of tobacco mosaic virus (TMV) particles in vitro. Virology 149: 132—135

Wintermantel WM, Banerjee N, Oliver JC, Paolillo DJ and Zaitlin M (1997). Cucumber mosaic virus is restricted from entering minor veins in transgenic tobacco exhibiting replicase-mediated resistance. Virology 231: 248—257

Wyatt SD and Kuhn CW (1980) Derivation of a new strain of cowpea chlorotic mottle virus from resistant cowpeas. J Gen Virol 49:289—296

Yang X, Yie Y, Zhu F, Liu Y, Kang L, Wang X and Tien P (1997) Ribozyme-mediated high resistance against potato spindle tuber viroid in transgenic potatoes. Proc Natl Acad Sci USA 94: 4861—4865

Yeh SD, Gonsalves D, Wang, HL, Namba R and Chiu RJ (1988) Control of papaya ringspot virus by cross protection. Plant Dis 72: 375—380

Yu YG, Buss GR and Maroof MA (1996) Isolation of a superfamily of candidate disease resistance genes in soybean based on a conserved nucleotide-binding site. Proc Natl Acad Sci USA 93: 11751—11756

Zaitlin M (1976). Viral cross-protection: more understanding is needed. Phytopathology 66: 382—383

Zheng Y and Edwards MC (1990) Expression of resistance to barley stripe mosaic virus in barley and oat protoplasts. J Gen Virol 71: 1865—1868

CHAPTER 10

SYSTEMIC INDUCED RESISTANCE

L.C. VAN LOON

Faculty of Biology, Section Phytopathology
Utrecht University, Sorbonnelaan 16, 3584 CA Utrecht, The Netherlands
(L.C.Vanloon@bio.uu.nl)

Summary

Induced resistance is the phenomenon in which a plant, once appropriately stimulated, exhibits an enhanced resistance upon "challenge" inoculation with a pathogen. Induced resistance can be localized as well as systemic, and can be induced by limited pathogen infection, avirulent pathogens, certain non-pathogenic bacteria, and certain chemicals. Systemic induced resistance has been particularly well studied in tobacco, cucumber and *Arabidopsis*. Under the influence of the inducing stimulus, a mobile signal is generated and transported to other parts of the plant where it enhances the mechanisms normally functioning to limit infection, growth, multiplication and spread of fungi, bacteria and viruses. In the systemic acquired resistance (SAR) induced by pathogens causing hypersensitive necrosis, resistance mechanisms to challenge inoculation become operative earlier or operate at a higher level than in non-induced plants. The expression of SAR is dependent on the accumulation of salicylic acid (SA) and associated with the induction of pathogenesis-related proteins (PRs) with anti-pathogen activities. In contrast, rhizobacterially-mediated induced systemic resistance (ISR) does not require SA, can occur without production of PRs, and is dependent on ethylene and jasmonic acid (JA) signalling. Mutants in the signal-transduction pathways of SAR and ISR have been identified in *Arabidopsis* which point to both pathways converging at the final step, while additional antimicrobial peptides are independently regulated. A different type of systemic induced resistance, involving production of proteinase inhibitors against herbivores, also requires JA. Synthesis of JA is initiated on perception of the polypeptide systemin, which is released on wounding of tomato and potato and acts as the mobile signal. These various types of systemic induced resistance confer an enhanced defensive capacity on plants against a broad spectrum of attackers and offer great potential for exploitation in crop protection.

Abbreviations

ABA   abscisic acid
ACC   1-aminocyclopropane-1-carboxylic acid

A. Slusarenko, R.S.S. Fraser, and L.C. van Loon (eds), Mechanisms of Resistance to Plant Diseases, 521-574.
© 2000 *Kluwer Academic Publishers. Printed in the Netherlands.*

AMD	actinomycin D
BA	benzoic acid
Fod	*Fusarium oxysporum* f.sp. *dianthi*
For	*Fusarium oxysporum* f.sp. *raphani*
GUS	β-glucuronidase
INA	2,6-dichloroisonicotinic acid
JA	jasmonic acid
LAR	localized acquired resistance
LPS	lipopolysaccharide
MeJA	methyl jasmonate
MeSA	methyl salicylate
PAL	phenylalanine ammonia-lyase
PI	proteinase inhibitor(s)
PRs	pathogenesis-related proteins
Pst	*Pseudomonas syringae* pv. *tomato*
PVX	potato virus X
SA	salicylic acid
SAR	systemic acquired resistance
TMV	tobacco mosaic virus
TNV	tobacco necrosis virus
ToRSV	tomato ringspot virus
TRSV	tobacco ringspot virus
TuMV	turnip mosaic virus

I.   The Phenomenon of Induced Resistance

When a plant is infected by a pathogen, its metabolism is altered. Generally, respiration is increased, photosynthesis is reduced, water relations and transport processes are affected, hormonal balance is shifted, and stress metabolites are released or synthesized and may accumulate. A pathogen that subsequently encounters such a plant is confronted with a different host environment compared to one that attacks a healthy plant. Whether the former pathogen will incite disease more easily, or be prevented from doing so, depends on the pathogen as well as on the host plant and the nature of the primary infection. When a plant has become extensively damaged, even opportunistic pathogens and saprophytes, that are not normally harmful, may gain access to nutrients released from the cells and be able to colonize its tissues, completing the rotting and decay of the succumbing host. On the other hand, exhaustion of critical factors for pathogen activity as a result of prior infection, or expression of specific resistance responses, can impede or reduce the establishment of an effective feeding relationship for a subsequent attacker. The latter phenomenon is commonly referred to as induced resistance: after a previous encounter with a pathogen the host is a less suitable substrate for the next one.

A. TYPES OF INDUCED RESISTANCE

1. *Cross-Protection*

Induced resistance can take several forms. A specific form is so-called cross-protection between related viruses, described initially by McKinney (1929). Generally, preinfection of a plant with a virus strain causing mild or hardly observable symptoms protects against subsequent disease development due to strains of the same or closely related viruses that may cause severe symptoms. Multiplication of the second virus is prevented, though not always abolished. When strongly reduced, its symptoms are masked. Cross-protection can be a very effective form of induced resistance, but it is specific for the infecting virus. Moreover, it is dependent on the presence of the virus in the cells. If the second virus is able to spread in the plant beyond the zone infected with the first virus, it regains activity and does incite disease. Cross-protection, therefore, is a cellular phenomenon, depending on the interaction between the first and the second virus with the cells of the host.

Several hypotheses have been put forward to explain cross-protection, but the mechanism has not been elucidated (Palukaitis and Zaitlin, 1984). An elegant hypothesis, supported by viral coat protein-mediated resistance in transgenic plants, suggests that coat protein of the first virus interferes with co-translational disassembly or viral replication of the second virus (Register and Beachy, 1988). However, cross-protection has also been demonstrated between viral isolates lacking coat protein, and between viroids. Most likely, the first virus monopolizes critical cellular components needed for its multiplication, thereby excluding other strains and related viruses dependent on the same interactions. There is no evidence whatsoever for a type of immunity related to the system in animals, which depends on the circulation of specific antibodies with a memory function. Parts of the plant not infected by the first virus are fully susceptible to infection and disease caused by the second virus. This specific form of induced resistance has been described only for viruses and appears to be related to the close dependency of this type of pathogen on the cellular environment of its host for its replication. A more extended account of cross-protection is given in chapter 9.

2. *Localized Acquired Resistance*

A different form of acquired resistance was found by Ross (1961a) when searching for the mechanism of hypersensitive resistance in tobacco mosaic virus (TMV)-infected tobacco. At temperatures below 28°C tobacco species and cultivars containing the *N* gene (Holmes, 1938) react to infection with TMV with the formation of necrotic local lesions. Lesions become visible about 2 days after infection and enlarge for a further 2-5 days before becoming limited. Ross (1961a) argued that the limitation of lesion spread must be caused by a mechanism altering the cells in advance of the virus, such that the tissue surrounding an infection site would become highly resistant or immune. Indeed, he found that a high degree of resistance developed in a 1-2 mm zone around each TMV-induced lesion. Following a second inoculation with TMV, no lesions, or at most a few tiny ones, were seen to develop in this ring of tissue. The resistant zones increased in size and in degree of protection for about 6 days, coinciding with the period during which the original lesion was enlarging. The effect was not dependent on special environmental conditions, because the development of the resistant zone was not affected by relative humidity or by exclusion of light before or after the second inoculation. However, resistance was not detectable at temperatures near 30°C, where hypersensitivity breaks down, and was strongly suppressed in leaves that were also systemically infected by potato virus X (PVX) and in which TMV lesions were not limited but expanded at a substantially increased rate. Thus, this type of induced resistance was clearly dependent on the expression of an active virus-localizing mechanism.

Testing the virus specificity of protection, Ross (1961a) found that the zone around a TMV lesion was also resistant to other strains of TMV (tomato aucuba virus and the Holmes' ribgrass strain), tobacco necrosis virus (TNV), tobacco ringspot virus (TRSV) and tomato ringspot virus (ToRSV), but not to turnip mosaic virus (TuMV). Conversely, TNV, but not TuMV, induced localized resistance to TMV. Interestingly, areas around TMV or TNV lesions were resistant to TNV in leaves also infected with PVX, apparently because PVX did not affect the localizing mechanism of tobacco to TNV. Some degree of virus specificity was evident, because homologous combinations gave better protection than heterologous ones. Ross (1961a) concluded that the acquired resistance in localized infection may be due to the additive effects of at least two mechanisms: one that is specific for the virus inducing it, and another that is not. Aqueous extracts of heavily lesioned leaves when mixed with the virus suspensions had no effect on the rate of lesion enlargement after infection, indicating again a cellular phenomenon. This type of induced resistance was called localized acquired resistance (LAR).

3.   Systemic Acquired Resistance

When testing for LAR in tobacco, Ross (1961b) noted in some of the tests that the lesions induced by the second, "challenge" inoculation, were smaller and fewer in number than those induced by similar inoculation of previously noninoculated control plants. This was particularly evident in leaves in which the first, or "inducer" inoculation, had produced many lesions. Since the original inoculation led to restriction of the virus to the immediate vicinity of the lesions, its protective effect appeared to extend to considerable distances. Two types of tests confirmed this effect. In half-leaf tests either the apical or the right half of a tobacco leaf was inoculated with a high dose of TMV, such that it resulted in a high number of lesions. Seven days later the basal or the left half, respectively, were challenge inoculated with a lower dose of virus, giving rise to 50-75 lesions that could be easily counted and measured. In the multiple-leaf technique, plants were trimmed to 4 or 5 leaves. In the first inoculation, the lower 2 or 3 leaves were infected. After 7 days, challenge inoculation of the upper 1 or 2 leaves was carried out. The diameters of the lesions resulting from challenge inoculation of induced plants were consistently only one-fifth to one-third of those of lesions on plants that previously were either left uninoculated or had been inoculated with water or juice from healthy plants. Moreover, leaf halves of control plants inoculated with a 1:10 diluted juice from systemically infected tobacco plants completely collapsed, whereas similar inoculation of leaf halves of induced plants caused only tiny lesions and the total amount of necrosis was very small compared to the controls. Uninoculated half leaves or upper leaves of induced plants remained symptomless and proved to be free of virus. Thus, the protection had to be plant-mediated. Because induced resistance was not confined to the immediate vicinity of the primary lesions as in LAR, Ross (1961b) designated this phenomenon as systemic acquired resistance (SAR).

By varying the interval between the inducer and challenge inoculation, it was found that resistance became detectable in 2 to 3 days, once primary lesions had developed, and reached a maximum in 7-10 days. In both types of test, reductions in lesion size were much more consistent than those in lesion number, suggesting that primarily lesions in resistant areas increased in size at a much slower rate. The reduction in lesion spread was associated with a reduction in extractable virus infectivity (Fig. 1A).

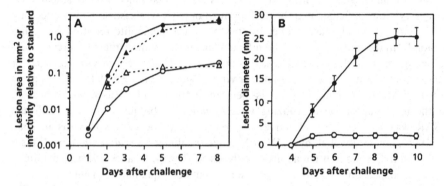

Figure 1. A. Systemic acquired resistance in upper leaves of Samsun NN tobacco challenge inoculated with TMV 7 days after lower leaves had been inoculated with, and developed necrotic lesions in response to, TMV (induced; open symbols), or had been rubbed with water plus Carborundum (non-induced; closed symbols): rate of TMV lesion enlargement (triangles) and increase in infectivity per unit area (circles) (after Ross, 1966); B. Rate of lesion enlargement on upper leaves of White Burley tobacco, excised 4 weeks after systemic resistance was induced by limited pathogen infection through injection of a sporangiospore suspension of *Peronospora tabacina* (open circles) or water (closed circles) into the stem base, and challenge inoculated with *P. tabacina* (after Stolle *et al.*, 1988).

The concentration of TMV in the inducer inoculum was not particularly critical, except that resistance was not induced to the maximum extent when inocula causing appreciably fewer than 100 lesions per leaf were used. However, even inocula causing about 16 lesions per lower leaf caused significant SAR in upper leaves. Increasing the ratio of the area first inoculated to that challenged did not greatly affect the extent of resistance attained, and subsequent inoculations of additional leaves to boost SAR attained after the primary infection, had no extra effect. In contrast, inocula causing rapid collapse of the primarily inoculated areas often induced less resistance than did less concentrated inocula, suggesting that the plant needs time to react in response to tissue necrotization in order to reach the resistant state. No resistance was induced by mechanical or chemical injury that caused rapid tissue necrosis.

As with LAR, the systemic resistance induced by TMV was not specific for this virus. Leaves with TMV-induced SAR were protected not only against TMV, but also against TNV, ToRSV, TRSV and TuMV. The resistance was not as effective against these viruses as against TMV. However, in each case lesion size was substantially reduced. Leaves highly resistant to TMV alone, upon challenge inoculation with a mixed inoculum of TMV and PVX, expressed only slight resistance, demonstrating a suppressive effect of PVX on SAR when TMV was the challenging pathogen. As in LAR, some degree of virus specificity was evident, even though SAR extended to all viruses tested that induced necrosis.

Resistance was induced by TMV infection in detached leaves, in leaves emerging from axillary buds on excised shoots, and in plants kept in the dark for 2 days before challenge inoculation. It was induced in lower leaves upon inoculation of upper leaves, as well as in upper leaves upon inoculation of lower leaves. Resistance developed somewhat more quickly in upper leaves than in adjacent half-leaves. Cutting the main vein of an upper leaf impeded development of resistance in the apical half. These observations indicated that the development of SAR was associated with necrotic lesion development and caused either by transport of a metabolite required for lesion expansion from the non-infected to the infected leaves, or by movement of a resistance-inducing signal from the infected areas to non-infected ones. Since viruses code for only a few proteins, and no viral particles or viral products have been detectable in areas expressing SAR, the moving factor must be plant-derived, and released or synthesized under the influence of the necrotizing viral infection. Subsequent studies etablished that SAR was induced very effectively by e.g southern bean mosaic virus in bean (Ross, 1966), but hardly at all by TMV in *N*-gene-containing *Nicotiana glutinosa* (Fraser *et al.*, 1979), indicating that the ability to express SAR upon localized virus infection is plant-species dependent in a strongly quantitative but not a qualitative manner.

Because the characteristics of LAR and SAR, as described by Ross (1961a,b, 1966) for tobacco, are qualitatively essentially similar, it might be inferred that LAR is an extreme form of SAR, i.e. viral lesion size is reduced to the extent that lesions are no longer macroscopically discernable. In fact, investigations of SAR reporting reductions in lesion numbers have often ignored the latter explanation, and microscopic examination has repeatedly shown no reduction in lesion numbers, only in size. As far as LAR is concerned, the matter has never been investigated systematically, but it is entirely possible that LAR is distinct from SAR in that a ring of tissue around a developing TMV lesion is indeed refractory to infection by a challenging virus. Other workers have independently observed similar phenomena in plants infected with pathogenic bacteria or fungi.

4.   *Localized Induced Resistance*

Introduction of living or heat-killed pathogenic or saprophytic bacteria into tobacco leaves has been reported to induce resistance against subsequent infection by compatible pathogenic bacteria, incompatible bacteria, and TMV (Sequeira, 1983). Localized necrosis ensued in the area of infiltration with living *Pseudomonas solanacearum*, which is avirulent on tobacco. When heat-killed cells or only the surface outer-membrane lipopolysaccharide (LPS) of the bacterium were injected into the leaves, no necrosis occurred. However, the injected areas were refractory to further infection by pathogenic bacteria. The LPS from the non-plant pathogens *Escherichia coli* and *Serratia marcescens* was also active (Graham *et al.*, 1977). This phenomenon, now designated localized induced resistance (LIR), has been little studied and, apart from the microbial determinant triggering LIR, nothing is known about the mechanism involved.

5.   *Systemic Induced Resistance*

Kuc (1982) coined the term "induced immunity" to refer to systemic protective effects against diseases caused by fungi, bacteria and viruses, resulting from restricted infection with fungi, bacteria and viruses. Because the protection has none of the characteristics of animal immunity, and plants do not become immune to infection but the effects of the infectious agents on the plants are reduced, the terms induced immunity, immunization, etc. are better abandoned in favor of the general term induced resistance (Van Loon, 1997). Kuc and co-workers (Hammerschmidt and Kuc, 1995) greatly extended the original observations by Ross and co-workers (Ross, 1961a,b, 1966; Bozarth and Ross, 1964) by analyzing induced resistance in green bean, cucurbits and tobacco.

Protection of green bean against anthracnose, caused by *Colletotrichum lindemuthianum*, was demonstrated upon inoculation of the hypocotyl with either cultivar-non-pathogenic races of the fungus or the avirulent pathogen *Colletotrichum lagenarium*. The treated plants developed minute lesions typical of a hypersensitive response and this reaction was sufficient to reduce subsequent infection with cultivar-pathogenic races of *C. lindemuthianum* both locally and systemically, allegedly to an extent that the outcome resembled a normal resistance reaction. Thus, resistance could be elicited in bean cultivars normally susceptible to *C. lindemuthianum*, protecting the plant from disease (Kuc, 1982). Induction of resistance in cucumber was demonstrated by primary infection with the fungi *C. lagenarium* or *Cladosporium cucumerinum*, the bacterium *Pseudomonas lachrymans*, or TNV (Box I). As in SAR, only living cells eliciting necrosis appeared to induce resistance. The resulting protection was manifest as a reduction, but hardly ever elimination, of symptoms. The induced resistance was non-specific with regard to both the inducer and the challenger, but the expression of resistance differed quantitatively and in details, depending on the inducing as well as the challenging organism. Further studies indicated that controlled infection with *C. lagenarium* or TNV protected cucumber, watermelon and muskmelon against diseases caused by a broad range of pathogens including biotrophic and necrotrophic fungi, local lesion- and systemic symptom-inducing viruses, and fungi and bacteria that caused wilt and those that caused limited or spreading lesions on foliage and fruits, and that this systemic induced resistance was effective against foliar as well as root pathogens (Table 1). However, this induced resistance was reported not to be effective against powdery mildew (Kuc, 1982).

In tobacco, restricted stem inoculation with the pathogen *Peronospora tabacina*, which causes blue mold, markedly reduced the lesioned area on foliage subsequently challenged with the fungus (Fig. 1B). Approximately 2 weeks were required for development of about 50% protection and 3 weeks for about 95% protection from necrosis caused by foliar inoculation with high inoculation doses of the fungus. There were fewer sporulating lesions and the sporulation per lesion was reduced in induced as compared to non-induced control plants (Stolle *et al.*, 1988).

Subsequent research by many workers using different plant species has amply confirmed that this systemic induced resistance is a general phenomenon and that a plant, once biotically induced, exhibits a higher level of resistance towards different types of pathogenic organisms, including fungi, bacteria, viruses and viroids and, in a few cases, also insects. On tobacco plants induced with TMV, the aphid *Myzus persicae* had a reduced lifespan and produced fewer offspring than on non-induced control plants (McIntyre *et al.*, 1981). However, upon induction of cucumber with *C. lagenarium*, no effect was found on two-spotted spider mite, fall armyworm or melon aphid, nor did previous feeding injury from spider mites or fall armyworms induce systemic resistance to *C. lagenarium* (Ajlan and Potter, 1991). Therefore, induced resistance certainly does not extend generally to insects (but see section IV).

Box I.   Characteristics of Systemic Induced Resistance in Cucumber (Caruso and Kuc, 1979)

A typical assay involved placing 30 drops of inoculum containing a limited concentration of the anthracnose-causing fungus *Colletotrichum lagenarium* on the surface of the first true leaf (leaf 1). Alternatively, plants were induced by swabbing suspensions of the bacterium *Pseudomonas lachrymans*, giving rise to angular leaf spot, on to the underside of leaf 1, or rubbing tobacco necrosis virus in the presence of the abrasive Carborundum on the leaf, resulting in necrotic lesions. Plants were challenge-inoculated on one or more upper leaves 7 days later. The number of lesions resulting from infection by the challenging pathogen, and lesion diameters reflecting infection progress, were used as parameters for assessing the extent of systemic resistance induced.

Primary infection with either *C. lagenarium* or *P. lachrymans* systemically protected susceptible cucumber cultivars against disease caused by subsequent challenge with either pathogen, and this induced resistance exhibited many characteristics of SAR. The concentration of inoculum used for induction and the number of lesions produced on the inducer leaf were directly related to the extent of resistance induced until a saturation dosage was attained. Thus, 10^4 cells ml^{-1} of the bacterium used as inducer on leaf 1 sufficed to significantly decrease the number of lesions of *C. lagenarium* applied as the challenger on leaf 2. Induction with 10^3 cells ml^{-1} already reduced *C. lagenarium* lesion size maximally to one half that on non-induced plants. A single lesion caused by *C. lagenarium*, or as few as 8 lesions caused by TNV, on the inducer leaf were reported to induce significant protection in upper leaves.

An interval of 72 to 96 h after the inducing inoculation was necessary for protection to be expressed. Protection was still evident 37 days after the initial infection, and leaves higher up the stem (leaf 8 and 10) also exhibited protection. The effect was apparent in different cultivars of cucumber, with *C. lagenarium* inducing a stronger resistance against *P. lachrymans* than the bacterium against itself. Conversely, the protection afforded by the bacterium was reduced when the concentration of conidia in the fungal inoculum used for challenge was increased. Killed cells or cell-free culture filtrate did not protect. Neither did mechanical injury or damage caused by dry ice, chemicals or fungal and plant extracts induce resistance. The bacteria *P. pisi*, *P. phaseolicola* and *P. angulata*, which caused local necrosis, also protected against both anthracnose and angular leaf spot, but the level of protection was significantly less than that elicited by *P. lachrymans*. Bacteria causing chlorosis or no symptoms were not protective. In protected leaves the number and diameter of the lesions of *C. lagenarium* were strongly reduced, whereas upon challenge with *P. lachrymans* the number of lesions as well as the rate of bacterial multiplication, but not mean lesion diameter, were decreased.

Thus, the systemic induced resistance was effective against both a fungal and a bacterial pathogen. Resistance was expressed as a reduction in disease severity but the level was dependent on the specific interactions between the pathogen and its host.

Table 1.   The biological spectrum of effectiveness of systemic resistance induced by foliar infection of cucumber with *Colletotrichum lagenarium* or tobacco necrosis virus (after Kuc, 1982 [1])

Disease	Pathogen
Fungi	
Anthracnose	*Colletotrichum lagenarium*
Scab	*Cladosporium cucumerinum*
Gummy stem blight	*Mycosphaerella melonis*
Fusarium wilt	*Fusarium oxysporum* f.sp. *cucumerinum*
Downy mildew	*Peronospora cubensis*
Local necrosis	*Phytophthora infestans*
Bacteria	
Angular leaf spot	*Pseudomonas lachrymans*
Bacterial wilt	*Erwinia tracheiphila*
Viruses	
Cucumber mosaic	Cucumber mosaic virus
Local necrosis	Tobacco necrosis virus

[1]   Bioscience 32: 856. Copyright 1982 American Institute of Biological Sciences

6.   Induced Systemic Resistance

Biotically-induced resistance has been associated almost invariably with limited necrotic infections. However, certain strains of non-pathogenic, plant growth-promoting rhizobacteria can induce resistance without inciting any visible symptoms in the host. Thus, colonization of roots of carnation by *Pseudomonas fluorescens* strain WCS417 reduced the number of plants developing wilting symptoms upon subsequent challenge inoculation in the stem with the pathogenic fungus *Fusarium oxysporum* f.sp. *dianthi* (Fod) (Van Peer *et al.*, 1991). No *Pseudomonas* bacteria were detectable in the stem, demonstrating effective separation between the inducing rhizobacteria and the challenging pathogen, and indicating that the effect must be plant-mediated. Since the effect of the rhizobacteria extended systemically, this phenomenon has been termed induced systemic resistance (ISR). Similar ISR has been found in radish, tobacco, cucumber and *Arabidopsis* against both soilborne and foliar fungi, as well as bacteria and viruses (Van Loon *et al.*, 1997, 1998). Induction of resistance is strain-specific. For instance, in the resistance induced in radish against *F. oxysporum* f.sp. *raphani* (For), of three rhizobacterial *Pseudomonas* spp. *P. fluorescens* strains WCS374 and WCS417 reduced the percentage of diseased plants, whereas *P. putida* WCS358 did not. OA$^-$ mutants of WCS374 and WCS417 deficient in the carbohydrate-containing O-antigenic side-chain of the bacterial outer membrane lipopolysaccharide (LPS) were inactive (Fig. 2). These results indicated that in the induction of systemic resistance in radish the bacterial LPS acted as the inducing determinant. Indeed, treatment with purified LPS from the wild type, but not from the OA$^-$ mutants, induced the same level of resistance as living bacteria (Leeman *et al.*, 1995b). About 2 days were required between induction and challenge for resistance to be detectable. A concentration of 10^5 cells g^{-1} root was necessary for induction, and higher concentrations only marginally increased the level of resistance (Leeman *et al.*, 1995a).

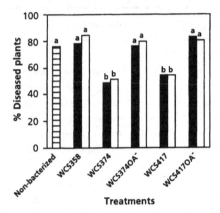

Figure 2.   Percentage of fusarium-wilted radish plants in an induced systemic resistance bioassay. The root tips were treated with *Pseudomonas putida* strain WCS358, *Pseudomonas fluorescens* strains WCS374 or WCS417, their O-antigen mutants 374OA⁻ or 417OA⁻ (closed bars), or their corresponding purified lipopolysaccharides (open bars) in talcum emulsion, 2 days before inoculation with the pathogen on the root base in peat. Bars with the same letter are not significantly different at p < 0.05 (after Leeman *et al.*, 1995b).

Rhizobacterially-mediated ISR was invariably expressed as a reduction in symptoms of the challenging pathogen and, therefore, appeared phenotypically similar to SAR and the systemic induced resistance described in sections I.A.3 and I.A.5, respectively. In their manifestations, virally-induced SAR, the non-specific systemic resistance induced by and active against fungi, bacteria and viruses, and ISR, all appear similar. However, a distinction can be made in that the former two seem to require necrosis for induction, and in that respect are more similar to each other than ISR, in which induction occurs without any symptom formation. On these grounds, the term SAR can be appropriately applied to the former two, retaining ISR for the latter type of induced resistance. Both SAR and ISR can be defined as a state in which the plant is to some extent protected against the effects of a challenging pathogen through previous biotic stimulation.

B.   CRITERIA FOR SYSTEMIC INDUCED RESISTANCE

Plants can also be protected against disease by direct antagonism between resident non-pathogenic micro-organisms and the attacking pathogen, without the plant being actively involved. To distinguish between such disease suppression and plant-mediated protection, criteria for the verification of induced resistance (Table 2) have been put forward by Steiner and Schönbeck (1995). Foremost is the establishment of the absence of toxic effects of the inducing agent on the pathogen. This criterion is easily fulfilled in the case of virally-induced SAR, but more difficult to ascertain for bacteria and fungi.

Bioassays are usually done such that the inducer is applied to a certain part of the plant, e.g. cucumber leaf 1, while the challenger is inoculated on a different part, e.g. leaf 2. However, the inducer may move into other plant parts or secrete toxic metabolites that are transported throughout the plant. Astonishingly little evidence has been reported to validate this point. However, in split-root experiments designed to test for induced resistance against soilborne *F. oxysporum*, the absence of the inducing fungus or bacterium has been verified (Kroon *et al.*, 1991; Liu *et al.*, 1995). Furthermore, inducing non-pathogenic rhizobacteria have been repeatedly shown not to be detectable at the site of challenge. Nevertheless, toxic compounds secreted by the inducer might still be present. Tests can be done to exclude such interactions in vitro, but this does not guarantee absence of toxic components in planta. Therefore, non-involvement of toxic compounds is difficult to prove when bacteria or fungi are used as inducers. It can be argued, however, that it is highly unlikely that an inducing micro-organism produces a toxin that is simultaneously active against fungi, bacteria and viruses, and the non-specificity of induced resistance appears to be the best indirect indication that the protection is plant-mediated.

Table 2.   Criteria for the verification of induced resistance (according to Steiner and Schönbeck, 1995)

- Absence of toxic effects of the inducing agent on the pathogens
- Suppression of the induced resistance by a previous application of specific inhibitors, such as actinomycin D, which affect gene expression of the plant
- Necessity of a time interval between application of the inducer and the onset of protection in the plant
- Absence of a typical dose-response correlation known for toxic compounds
- Non-specificity of protection
- Local as well as systemic protection
- Dependence on the plant genotype causing significant differences in level and type of protection in different cultivars

The non-specificity of protection is an inherent characteristic of induced resistance and suggests that a general defensive mechanism has become active.

Suppression of the induced resistance by a previous application of specific inhibitors, such as actinomycin D (AMD), which affect gene expression in the plant, is a criterion that can be applied meaningfully only in the case of viruses that replicate independently of DNA-dependent RNA synthesis. Indeed, development of TMV-induced SAR was found to be reduced in the presence of AMD (Loebenstein *et al.*, 1968), but AMD interferes with general plant metabolism and may affect symptom development directly, making interpretation of such experiments less than straightforward. Fungi and bacteria are themselves inhibited by AMD, as well as affected by other inhibitors that interfere with plant metabolism. However, some specific inhibitors do reduce the development of SAR, implicating certain plant metabolic pathways in the development of SAR (see section III).

The requirement for a time interval between application of the inducer and the onset of protection of the plant is one of the easiest to verify and is invariably fulfilled.

In conjunction with the previous criterion it suggests that plant metabolism has to operate before the resistant state is reached.

Absence of a typical dose-response correlation known for toxic compounds is evident from the common observations that even relatively low amounts of inducer inoculum can induce substantial protection. Although higher doses often lead to higher levels of induced resistance, the response quickly saturates. The state of induced resistance does seem to be easily reached once a threshold is crossed.

Local as well as systemic protection is common, but may not always be the case. As discussed in section I.A.3, it is entirely possible that over and above a mechanism that is active at a distance from the primary infection, one or more additional mechanisms are activated only in the immediate vicinity of the inducer site. LAR and LIR could be examples.

Dependence on the plant genotype causing significant differences in level and type of protection in different cultivars: this criterion presupposes that genetic variation exists within a plant species for either inducibility or expression of induced resistance. Apparently, SAR is expressed to different extents in different cultivars of cucumber (Jenns and Kuc, 1979) and barley (Steiner and Schönbeck, 1995). In carnation, resistance against Fod was induced by the rhizobacterial strain WCS417 in the moderately resistant cultivar Pallas, but less so and only occasionally in the susceptible cultivar Lena (Van Peer et al., 1991). In contrast, no differential inducibility was apparent in several radish cultivars differing in genetic resistance against For (Leeman et al., 1995a). Interestingly, Arabidopsis ecotypes RLD and Ws-0 were not inducible by rhizobacteria, whereas ecotypes Columbia (Col) and Landsberg erecta (Ler) were (Van Wees et al., 1997; Ton et al., 1999). In tobacco, TNV readily induced SAR in the cultivar Samsun NN, but not in the cultivar White Burley (Pennazio et al., 1983). Furthermore, inducibility can vary greatly between related plant species, such as the clear induction of SAR by TMV in Nicotiana tabacum as opposed to the slight induction in N. glutinosa (Fraser et al., 1979).

II.   Physiology and Biochemistry of Systemic Induced Resistance

A.   EXPRESSION OF INDUCED RESISTANCE

1.   *Necrosis is not a Prerequisite for Induction of Systemic Resistance*
Systemic resistance can be induced in all plant species investigated and is expressed invariably as an enhanced resistance against most pathogens used for challenge inoculation. Limited infection with pathogens, avirulent races of pathogens or non-pathogens can induce resistance, as long as the tissue reacts with localized necrosis, a hypersensitive reaction, or non-host pin-point lesions, respectively. Heat-killed pathogens, culture filtrates or pathogen-derived elicitors, not causing necrosis, do not commonly induce systemic protection, although they may induce powerful resistance reactions locally. Mechanical or chemical wounding, even when inflicting necrosis, hardly ever induce SAR. It would seem to be central, therefore, that some part of the plant be infected by a necrosis-inducing micro-organism. However, systemic resistance can be induced in the absence of necrosis. Thus, formation of localized symptomless starch lesions of TMV on cucumber induces SAR, as evidenced by a reduction

of TNV lesion diameter upon subsequent challenge inoculation (Roberts, 1982). A mutant of *Arabidopsis* that develops disease symptoms upon inoculation with an isolate of *Pseudomonas syringae* pv. *tomato* (Pst) that was avirulent on wild-type plants developed SAR, but exhibited a weaker response when challenged. This suggests that hypersensitive necrosis contributes to, but is not essential for, the induction of SAR (Cameron *et al.*, 1994). Moreover, non-pathogenic rhizobacteria eliciting ISR can induce a broad-spectrum resistance that may be as effective as SAR, without any necrosis being evident. For different inducer strains live bacteria are not required but can be replaced by the bacterial LPS or specific iron-regulated siderophores (Van Loon *et al.*, 1997). Thus, harmless microbial compounds can be as effective in inducing systemic resistance as live, necrosis-inducing pathogens (Table 3).

Table 3.   Treatments eliciting systemic induced resistance in plants

limited pathogen infection
avirulent forms of pathogens
cultivar-non-pathogenic races
non-pathogenic bacteria and microbial compounds
selected chemicals

Necrosis seems to be a contributing factor to the induction of SAR, but eliciting SAR or ISR appears to depend on (an)other mechanism(s), that can be triggered in diverse ways by different inducer molecules. When defense responses involving necrosis occur, an endogenous elicitor released from dying cells or synthesized in reaction in neighbouring cells, might act as the inducer of SAR. When no necrosis occurs, similar elicitors might still be set free, or molecules mimicking elicitors or components of the subsequent signal-transduction pathway could be involved. It has also been found that selected chemicals can induce resistance, either by activating the signalling pathway leading to SAR, or by functioning as intermediates in this same pathway (see section III).

2.   Characteristics of Induced Resistance in Various Plant-Pathogen Combinations

If systemic induced resistance involves a reduced ability of pathogens to incite disease, how is this state induced and maintained, how can it be expressed against different types of pathogens having distinctly different interactions with their hosts, and what type of signalling is responsible for the systemic induction of the effect? Several systems have been investigated to study these questions.

a. Tobacco - TMV.   Reductions in the number of diseased plants or in disease severity upon challenge inoculation of induced versus non-induced plants can result either from a reduced ability of the pathogen to use the plant as a host, or from an enhancement of available resistance mechanisms. For instance, in the case of *N*-gene-containing tobacco, the cellular environment could have become less suitable for TMV multiplication, resulting in less progeny virus being produced and transport to neighbouring cells being slowed. Thus, lesions would have expanded less by the time the resistance mechanism had become fully operative, and final lesion size would be reduced. Alternatively, resistance mechanisms might operate at the level of the transport of virus from cell to cell, preventing the virus from entering neighbouring cells more strongly without affecting viral multiplication per se. Again, by the time lesions become limited, the area affected would be less, resulting in reduced final lesion size. Without further knowledge about the interactions occurring between tobacco and TMV, it would be impossible to distinguish between these, and perhaps other possible mechanisms. In fact, so little is understood about pathogenesis and induction of resistance in TMV-infected tobacco, that no definitive answer can be given right now. From Fig. 1A it is seen that in induced leaves virus content was only half that in leaves from non-induced plants at the earliest time point measured, i.e. day 1, well before the appearance of necrotic lesions indicated the operation of N-gene-mediated resistance. This seems to imply reduced viral multiplication.

Inhibition of viral replication has been found early on in resistant plants (Otsuki et al., 1972), and similar activity was present in non-inoculated leaves from induced plants (Spiegel et al., 1989). Indeed, involvement of an inhibitor of viral replication has been advocated (Loebenstein and Gera, 1981), but the compound has never been characterized. On the other hand, due to the primary infection essential precursors for viral replication might have been withdrawn from the induced leaves. However, growth of induced plants is not affected by localized infection, nor have levels of nucleic acids, proteins, chlorophyll, free amino acids, etc. been found to be reduced in induced leaves (Bozarth and Ross, 1964). While this does not exclude the possibility of lack of a specific compound, it makes it less likely that an essential metabolite is missing. Induced resistance is operative not only against TMV, but also against other viruses and against fungi and bacteria, that have very different interactions with their host. Moreover, fertilization of plants, e.g. with nitrogen-containing compounds, did not consistently affect the level of induced resistance (Ross, 1961b; Bozarth and Ross, 1964).

Nevertheless, virus multiplication appears to be inhibited to a larger extent in induced plants than in non-induced ones. Occasionally, however, it has been inferred that viral multiplication was not reduced, and induced resistance involved masking of symptoms rather than reduced activity of the pathogen (Balasz et al., 1977; Fraser, 1979). It is difficult to dismiss such reports, but it must be taken into account that extraction of TMV from necrotizing tissue can hardly be expected to be quantitative, in view of the ready polymerization of proteins and phenolics in the reddish-brown to black pigments making up the necrotic lesion. Generally, however, less virus has been extracted from challenge-inoculated leaves from induced plants than from inoculated leaves on plants not previously induced (Van Loon, 1983a). As referred to above, careful inspection has shown no consistent reduction in lesion numbers on induced plants. Hence, smaller lesion size on induced leaves is associated with reduced virus content.

As seen from Fig. 1A, necrotic lesions became visible at the same time on non-induced and induced leaves. However, on induced leaves, they expanded at a lesser rate, and limitation appeared to occur at 5 days, as opposed to 8 days for non-induced plants. While reduced spread may be a consequence of reduced viral multiplication, it does not necessarily follow that viral cell-to-cell transport occurs unimpeded. N-gene-mediated resistance is also associated with reduced viral spread in the tissue (Takahashi, 1974) and, again, this mechanism might operate at a higher level in induced tissues. Finally, earlier lesion limitation on induced leaves is indicative of the resistance mechanism being more active in induced than in non-induced plants.

Whatever the exact mechanism(s) responsible for the expression of N-gene-mediated resistance against TMV in tobacco, these mechanisms clearly are operating more effectively in induced tissues. Whether these mechanisms are entirely the same, or additional mechanisms are becoming operative in induced tissues, cannot be deduced from these results alone, and require further insight into the physiological, biochemical and molecular mechanisms responsible. However, Ross (1961b, 1966) found that varying conditions always had the same effects on lesion size in non-induced and induced plants. Intermediate or low levels of resistance were detected in experiments in which the host-virus combinations or the environmental conditions favored continued lesion enlargement. The simplest explanation is that the mechanisms responsible for the expression of resistance are the same in both primarily inoculated and in induced and challenge-inoculated plants. However, in induced leaves it appears that the localizing mechanism becomes operative earlier than in non-induced leaves and/or operates at a higher level (Ross, 1966).

b. Tobacco - Peronospora tabacina. Induced resistance in tobacco against the blue mold fungus *P. tabacina* is expressed quite differently (Stolle *et al.*, 1988). When tobacco plants were inoculated by injecting 10^6 sporangiospores into stem tissue external to the xylem, necrosis along the cambium and external phloem developed. Further colonization of the tissue by the fungus was restricted and resistance was induced systemically, as evidenced by challenge through application of spore suspensions on the leaves. Upon inoculation of leaves of non-induced plants, spore germination and penetration, vesicle formation and growth of hyphae from vesicles led to continued tissue colonization for about 9 days. Necrotic lesions developed with mycelium extending approximately 1 mm beyond the lesion margins until lesion coalescence. More than 90% of the lesions sporulated. On leaves of induced plants, higher densities of sporangiophores at an inoculation site were required for symptoms of blue mold to appear, but no differences in fungal development were seen up to the stage of lesion development. However, tissue colonization by the fungus ceased by day 5, at which time lesions became visible, and lesion diameter did not subsequently increase (Fig. 1B). Mycelium was confined within the lesion area. Sporulation was not always observed on lesions on induced leaves and, if it was, the number of sporangiophores per cm^2 of colonized leaf area was reduced about 4-fold. Thus, there were fewer sporulating lesions and the sporulation was reduced in induced as compared to control plants. No effect of the induced state on development of the fungus was seen in the biotrophic phase of the fungus. Only when tissue necrosis ensued were associated defense responses activated to stop further fungal activity.

c. Cucumber - Colletotrichum lagenarium. In contrast, in cucumber infected with *C. lagenarium*, the induced resistant state in leaf 2 after prior induction of leaf 1 was manifested on challenge inoculation as a reduction of fungal development at the leaf surface, in the epidermis, and in the mesophyll (Kovats *et al.*, 1991a). Typical symptoms of anthracnose on non-induced plants consist of chlorosis, proceeding to necrosis 3 to 5 days after inoculation. On leaves of induced plants, challenge inoculation with a low dose of 5×10^4 spores ml^{-1} caused occasional chlorotic spots but no necrosis. Moreover, necrosis was strongly reduced at the higher concentration of 5×10^{-5} spores ml^{-1}. The formation of appressoria was only slightly reduced on induced resistant as compared to untreated plants. However, development of fungal infection hyphae from appressoria was strongly suppressed. Whereas in non-induced plants an average 9% of infections had progressed into the epidermis by 38 h and 22% by 72 h, these values were 0% and 1% respectively for induced plants. In leaves of non-induced cucumber plants, none of the infection droplets applied failed to lead to infection. In induced plants, however, most of the fungal penetration attempts aborted at the level of the outer cell wall, resulting in several infection droplets not progressing into infection (cf. Box I). This explains why, in contrast to TMV on tobacco, lesion number as well as lesion size can be used as a valid parameter for evaluating induced resistance against *C. lagenarium* on cucumber.

In order to avoid the influence of the epidermal layer on the infection process, Kovats *et al.* (1991a) inoculated cotyledons of induced and non-induced plants by injecting the spore suspension directly into the mesophyll. Necrotic areas developing were still reduced by over 90% in induced as compared to non-induced plants, indicating that induced defense reactions are also located in the mesophyll. Because in non-induced plants too some infections were stopped either at the epidermis or the mesophyll, it appears that cucumber possesses a certain level of basal resistance against *C. lagenarium*, which is breached by a sufficient dose of fungal inoculum. In induced tissues, these defenses are boosted and larger numbers of spores are required to initiate successful infection. A similar conclusion was reached for the induced resistance of tomato against the necrotrophic pathogen *Phytophthora infestans* (Kovats *et al.*, 1991b).

d. Radish - Fusarium oxysporum.   Expression of ISR against the soilborne fungus *F. oxysporum* has been studied in various plant species, e.g. cucumber (Liu *et al.*, 1995), tomato (Kroon *et al.*, 1991, carnation (Van Peer *et al.*, 1991), radish (Leeman *et al.*, 1995a) and *Arabidopsis* (Van Wees *et al.*, 1997). In radish, early symptoms are manifested as internal browning of the vascular strands, before external wilting occurs. Even in heavily infected soils, a small number of non-induced plants may escape infection, apparently because infection is a stochastic process and plants have a certain level of horizontal resistance that increases with age. Upon challenge inoculation of plants with induced resistance, a larger percentage of plants does not develop symptoms and appears to escape infection (Box II).

e. Arabidopsis - Peronospora tabacina and Pseudomonas syringae pv. tomato. Because of its excellent suitability for molecular-genetic analysis, *Arabidopsis* has lately received considerable attention for studies of systemic induced resistance. Two pathosystems have been used most: the fungus *Peronospora parasitica* and the bacterium *Pseudomonas syringae* pv. *tomato* strain DC3000 (Pst) (Uknes *et al.*, 1992).

Box II.   Rhizobacterially-mediated Induced Systemic Resistance in Radish against *Fusarium oxysporum* f.sp. *raphani* Increases Extant Defensive Mechanisms (Hoffland *et al.*, 1995)

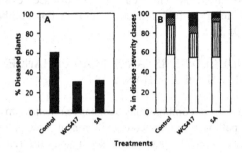

Part of the root system of radish seedlings was either control-treated (control; non-induced), or treated with the rhizobacterial inducer *Pseudomonas fluorescens* strain WCS417 or the chemical inducer salicylic acid (SA). Three days later a different part of the root system was challenged by inoculation with *Fusarium oxysporum* f.sp. *raphani* (For). Three weeks later the percentage of plants with symptoms was scored. As seen in A, both inducers were equally effective in reducing the percentage of diseased plants by about 50%.

In B only the diseased plants from the three treatments are considered and classified on the basis of disease severity. In radish, early symptoms are manifested as internal browning of the vascular strands in the root (☐), which thereupon extend into the tuber and shoot (▥). Later, external symptoms (▨) involving incipient wilt, leaf epinasty, yellowing and necrosis, progress into death of the shoot (■). Once infected, induced plants showed the same distribution of disease severity classes as non-induced plants. This means that in the interaction between radish and For systemic induced resistance must make plants escape disease, rather than reduce symptom severity. These observations are interpreted to mean that radish plants have a certain level of basal resistance against infection by For, but once the fungus has entered the vessels, it can no longer be stopped. In the induced plants, the basal resistance operating prior to vessel colonization by the fungus, is increased. This situation seems to apply to radish cultivars with different levels of resistance to For.

When untreated *Arabidopsis* was inoculated with a spore suspension of *P. parasitica*, a typical compatible downy mildew infection ensued. Intercellular hyphae proliferated with formation of intracellular haustoria and oospores, as well as a multitude of sporangiophores on the leaf surface. Necrosis was not evident until a late stage of the infection. In plants induced with the SAR-mimicking chemical INA asexual sporulation was significantly reduced. Growing intercellular hyphae mostly had haustoria reduced in size and encased in material presumably of host origin. Such encasements normally occurred in non-induced plants adjacent to older parts of the mycelium and never near the hyphal tips. Encased haustoria in induced plants were occasionally seen to become necrotic, which was never observed in non-induced plants. Apparently, defense reactions of the host, occurring late in infection of non-induced plants, were expressed earlier in induced plants.

Infection of control plants with Pst caused small grayish-brown lesions, surrounded by spreading chlorosis, often developing into necrotic flecks. Induced plants displayed fewer lesions and less chlorosis. This reduction in disease severity was commonly associated with an at least 10-fold reduction of bacterial numbers in the infected plants (Uknes *et al.*, 1992).

B.   SYSTEMIC INDUCED RESISTANCE IS AN ENHANCEMENT OF EXTANT RESISTANCE MECHANISMS

In any of these plant-pathogen combinations, resistance mechanisms are operating more effectively against the pathogen in induced than in non-induced plants. The mechanisms involved and the stages at which they are expressed appear different, as well as specific, for each combination. Because a single plant species, once induced, becomes more resistant to diverse types of pathogens, it is difficult to envisage an entirely novel defense mechanism that could be simultaneously active against fungi, bacteria and viruses, yet exert its effect at different stages of tissue colonization, depending on the specific pathogen involved. If a novel mechanism were so effective, it would also be surprising from an evolutionary point of view, that such a mechanism would not have been recruited as an inducible defense after primary infection. Rather, the enhanced resistance reactions seen upon challenge inoculation of induced tissues all seem to comprise defense responses occurring normally in non-induced plants. Apparently, upon primary inoculation the mechanisms involved are less effective or triggered too late during the infection to exert a maximal effect. In induced plants, defenses are boosted or become operative at an earlier stage of infection, thereby reducing disease severity up to a level resembling a resistant reaction (Kuc, 1982). At this stage, induced resistance may be re-defined as the phenomenon whereby a plant exhibits an enhanced level of resistance against pathogen infection after previous stimulation of its defensive mechanisms by the inducing inoculation. Because induced resistance is expressed only upon challenge inoculation, it can also be stated that induced resistance constitutes an enhanced defensive capacity of the plant (Van Loon, 1997).

Typical defense responses of plants to pathogens are synthesis of phytoalexins, accumulation of pathogenesis-related proteins (PRs) and reinforcement of cell walls.

1.   *Phytoalexins*
Phytoalexins are synthesized and accumulate locally around infection sites, notably in incompatible interactions (see chapter 7), but have never been found in systemically protected tissues before challenge inoculation. Hence, no evidence exists for accumulation of phytoalexins away from the site of induction in protected tissues. However, phytoalexins rapidly accumulate in induced tissue after challenge. Bean hypocotyls induced by local infection with the avirulent pathogen *C. lagenarium* accumulated phaseollin at sites challenged with the pathogen *C. lindemuthianum*, even though upon primary infection *C. lindemuthianum* induced high levels of phaseollineisoflavan and little or no phaseollin. Thus, prior infection with *C. lagenarium* primed the tissue to respond as if encountering an avirulent race of *C. lindemuthianum* (Kuc, 1982). In contrast, in tobacco induced by stem injection with *P. tabacina* and subsequently

challenged with the same fungus, reduced lesion development in the induced tissues was not associated with significant accumulation of phytoalexins. Accumulation of phytoalexins was also low on primary inoculation with the fungus and occurred only late in infection after the tissue had already been colonized. Thus, phytoalexins do not seem to play a role in restricting colonization of tobacco by *P. tabacina* either after primary or after challenge inoculation (Stolle *et al.*, 1988).

In carnation with ISR against Fod, induced by the rhizobacterial strain WCS417, an earlier and increased accumulation of dianthramide phytoalexins around the inoculation site occurred on challenge (Van Peer *et al.*, 1991). These phytoalexins accumulate in the tissue in response to infection with Fod, and supposedly function to restrict its spread in the plant. No phytoalexins were induced following induction by the rhizobacteria before challenge; only after challenge did a difference become noticeable (Fig. 3).

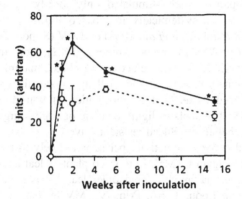

Figure 3. Accumulation of the phytoalexin methoxydianthramide S in stem segments of carnation cultivar Pallas after inoculation with *Fusarium oxysporum* f.sp. *dianthi* of plants without (open circles) or with (closed circles) induced systemic resistance as a result of treatment of the roots with water or with *Pseudomonas fluorescens* strain WCS417, respectively, 1 week prior to challenge inoculation. Asterisks indicate significant differences at p < 0.05 (after Van Peer *et al.*, 1991).

These observations indicate that in those host-pathogen combinations in which phytoalexin production is triggered by the pathogen, challenge inoculation of tissues with SAR or ISR leads to a (more) rapid synthesis and accumulation. Such enhanced production of phytoalexins may contribute to the increased resistance seen against fungi and probably bacteria. However, viruses are not affected by phytoalexins, and additional mechanisms must be responsible for the broad-spectrum protection observed.

2.   Cell Wall Reinforcement

Strengthening makes cell walls more difficult to penetrate and reduces intracellular spread of the pathogen (see chapter 5). Deposition of callose, lignin-like materials and suberin can also occur upon wounding, but during defense reactions to pathogenic attack, they are often highly stimulated in a zone surrounding the site of tissue penetration. Callose can block plasmodesmata and, thereby, inhibit cell-to-cell transport of viruses. Callose deposition has been found to be increased systemically in tobacco leaves upon local infection with TMV (Shimomura, 1979). It is conceivable that such induced callose formation contributes to early virus localization in conjunction with other mechanisms.

When *N*-gene-containing tobacco is infected with TMV and the temperature is raised to 30°C, the virus is no longer localized and systemic mosaic disease is initiated, provided lesion limitation had not yet stopped. Raising the temperature at different times after inoculation indicated that a barrier ring was being laid down around the developing lesion, from which the virus was increasingly unable to escape (Weststeijn, 1981). Lignin-like material was deposited in this ring, apparently due to polymerization of phenolic precursors synthesized after the stimulation of enzymes of aromatic biosynthesis and lignin-forming peroxidases (Legrand *et al.*, 1976). The key enzyme of phenylpropanoid metabolism, phenylalanine ammonia-lyase (PAL) is increased by both abiotic and biotic stresses, as is peroxidase. PAL and further enzymes of the phenylpropanoid pathway are stimulated only locally by resistance reactions, but peroxidase is increased systemically in tobacco following local infection with *Pseudomonas tabaci* (Lovrekovich *et al.*, 1968) or TMV (Simons and Ross, 1970; Van Loon and Geelen, 1971; Van Loon and Antoniw, 1982). Both the level of induced resistance and the increase in peroxidase activity reached a maximum around the same time and, thereafter, remained elevated. On challenge inoculation peroxidase increased more strongly in induced tissue than on inoculation of non-induced control plants. Increased peroxidation of phenols to lignin or toxic quinones might thus explain why TMV lesions remain smaller in induced resistant leaves.

However, increased peroxidase activity per se is unlikely to be responsible for the expression of resistance against TMV. In tobacco plants a leaf age-dependent gradient exists with young leaves having low, and older leaves having exponentially higher peroxidase activity (Van Loon, 1976). Primary TMV lesions are relatively large on young leaves but become progressively smaller on older leaves. Thus, an inverse correlation does exist between peroxidase activity and TMV lesion size in tobacco. But the reduction in size of TMV lesions developing upon challenge inoculation of systemically induced leaves is far greater than would be expected on the basis of the increased peroxidase activity of those leaves (Fig. 4). Moreover, abiotic treatments altering peroxidase activity did not alter TMV lesion size accordingly. Tobacco possesses several isoenzymes of peroxidase, but no association between any specific isoenzyme and leaf age or induced resistance was apparent (Van Loon, 1976, 1986). While increased peroxidase activity in systemically induced tissues constitutes an enhanced capacity for e.g. lignin formation, increased substrate availability is probably required for the enzyme to act. Indeed, greater depletion of phenolic precursors has been observed upon challenge inoculation of tobacco with TMV (Simons and Ross, 1971). This suggests that stimulation of phenylpropanoid metabolism might still be a limiting factor in the expression of induced resistance.

In cucumber plants, penetration of *C. lagenarium* beneath appressoria is reduced through deposition of papillae containing callose as a main structural component (Kovats *et al.*, 1991a). This material was identified by staining with aniline blue and digestion with β-1,3-glucanase. Enhanced formation of papillae was seen on challenge inoculation of induced tissues, correlated with reduced development of infection hyphae. This would allow the tissue more time to activate other defenses,

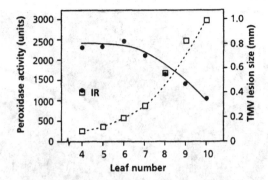

Figure 4. Relationship between peroxidase activity (open squares) of successive leaves of a Samsun NN tobacco plant and size of lesions (closed circles) developing on these leaves upon inoculation with TMV. IR denotes the situation in leaf 4 of a plant with SAR, three lower leaves of which had been induced by inoculation with TMV 14 days earlier. The reduction in lesion size is greater than would be expected on the basis of the increase in peroxidase activity if the two were causally connected (data taken from Van Loon, 1976).

such as increased lignification of entire cells, observed later during the infection (Dean and Kuc, 1987). As in TMV-induced tobacco, peroxidase activity was systemically increased in induced cucumber leaves, and peroxidase activity increased sooner in induced than in non-induced plants after challenge with *C. cucumerinum* or *C. lagenarium* (Hammerschmidt and Kuc, 1982; Hammerschmidt *et al.*, 1982), resulting in lignification of the fungus bound to wall components of the host.

In tobacco leaves infected with *P. tabacina*, some deposition of lignin occurred around old, degenerating haustoria (Ye *et al.*, 1992). In contrast, in TMV-induced plants, lignin-like material accumulated 2 days after challenge with *P. tabacina*. Some fungal hyphae were disorganized near the centre of infection sites. Adjacent host cells were plasmolyzed, and a few collapsed 3 days after challenge. Seventy-five percent of the cells showed wall-encased necrotic haustoria within the disintegrating host cells (Fig. 5). All infection sites in the induced plants were associated with necrotic cells 5-6 days after challenge. Upon primary inoculation, cell wall hydroxyproline was systemically increased and two peroxidases and four uncharacterized salt-soluble cell wall proteins were systemically induced. Increased peroxidase activity was positively correlated with induced resistance (Ye *et al.*, 1990). Both peroxidase activity and hydroxyproline content increased progressively from 2 to 6 days after challenge with *P. tabacina* in the induced plants, whereas following inoculation of the non-induced plants hydroxyproline content remained at almost the same level during the entire period of the experiment. Since hydroxyproline-rich glycoproteins seem to act as a matrix for lignification by peroxidases, and lignin-like materials accumulated upon challenge inoculation of induced tissues, early cell wall strengthening appears to be an important component in the expression of induced resistance.

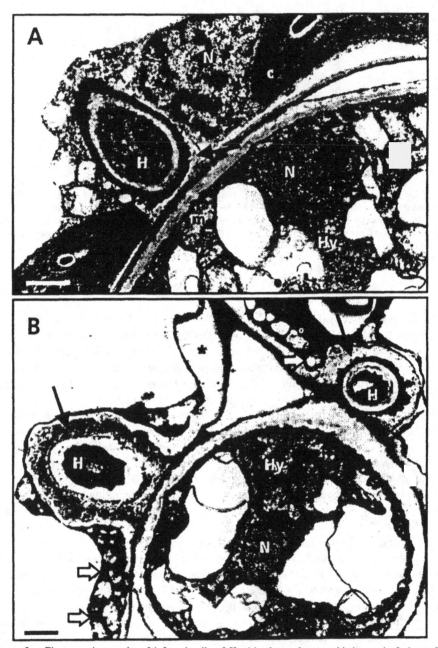

Figure 5. Electron micrographs of infected cells of Ky 14 tobacco leaves with haustoria 3 days after inoculation with *Peronospora tabacina* of (A) non-induced plants inoculated previously with water plus Carborundum, and (B) plants induced by inoculation with TMV 12 days prior to challenge inoculation. In A only slight wall apposition (arrow) occurred around the entry point, whereas in B a thick encasement (arrows) of haustoria (H), electron-dense cytoplasm of the host cells (open arrows), and an area where the host membrane is separating (*) from the cell wall are obvious. Hy, hypha; N, nucleus; c, chloroplast; m, mitochondrion. Bar = 1 μm (from Ye *et al.*, 1992).

3. *Pathogenesis-related Proteins*

When Van Loon and Van Kammen (1970) discovered PRs in hypersensitively reacting tobacco, they observed accumulation of these new protein components not only in the TMV-inoculated leaves, but also in non-infected upper leaves of the same plants. They suggested that these proteins could be responsible for SAR. Indeed, a correlation was found between the occurrence of PRs induced by different viruses, and induced resistance against both TMV and TNV in tobacco (Kassanis *et al.*, 1974; Van Loon, 1975). Since then, an association between the induction of PRs and SAR has been observed in various plant species in response to several types of treatments, both biotic and abiotic, suggestive of a causal relationship between the two (Van Loon, 1989; Ryals *et al.*, 1996; Schneider *et al.*, 1996). Further characterization of the types of PRs established that the acidic, extracellular proteins are most often induced systemically in response to necrotizing infections. Basic, vacuolar proteins are generally organ-specific and developmentally-regulated but, in addition, are strongly increased only locally during infections (Linthorst, 1991; Van Loon *et al.*, 1994).

Identification of some of these proteins as β-1,3-glucanases (PR-2) and chitinases (PR-3, -4, -8 and -11) immediately suggested that these PRs could function to degrade β-1,3-glucan- and chitin-containing fungal cell walls and, hence, constitute an inducible defense against fungal attack. Although glucanases and chitinases, particularly as mixtures, can restrict fungal growth in vitro, the evidence for their involvement in induced tissues in vivo is limited (see chapter 8). Cloning of their genes and plant transformation so far have not provided a single example in which an inducible acidic glucanase or chitinase, alone or in combination, enhances resistance against a pathogenic fungus. In contrast, overexpression of the basic proteins can reduce disease development, but not generally so. Thus, ironically, the glucanases and chitinases which are induced specifically by necrotizing infections, and which are also accumulating systemically in tissues exhibiting induced resistance, do not seem to act by restricting fungal infection. In tobacco the appearance of an additional glucanase has been associated with induced resistance against *P. tabacina* (Ye *et al.*, 1992), but direct proof is lacking. Some investigators have attempted to correlate total chitinase activity with the level of resistance in e.g. cucumber (Irving and Kuc, 1990) and tobacco (Pan *et al.*, 1992), but the relationships are not clear.

Oomycete fungi lack chitin in their cell walls and are not affected by chitinases. However, PRs of both the PR-1 and -5 families have been shown to have some inhibitory effect against at least some oomycetes (see chapter 8). These proteins could, therefore, contribute to induced resistance against these fungi. Endoproteinases of the PR-7 family could play an accessory role in attacking proteinaceous components of microbial cell walls. Chitinases of the PR-8 family also possess substantial lysozyme activity and might reduce bacterial activity. So far, no proof for this hypothesis is available. The role of the "ribonuclease-like" PR-10 family is unclear at present. The PR-6 proteinase inhibitors have a role against herbivorous insects (see section IV), and the PR-9 peroxidases could function in cell wall reinforcement through the formation of lignin-like deposits (cf. section II.B.2). Taking this evidence together, however, the contribution of PRs to systemic induced resistance appears to be minor (Van Loon, 1997). Moreover, most of the activities detected appear to be directed against fungi and none against viruses, whereas induced resistance is equally effective against all types of pathogens.

Involvement of PRs in ISR elicited by rhizobacteria has not been investigated in such detail. Induction of resistance by *P. fluorescens* strain CHAO in tobacco against TNV was associated with systemic induction of the full spectrum of acidic PRs (Maurhofer *et al.*, 1994). In contrast, induction of systemic resistance against several pathogens by strain WCS417 in both radish and *Arabidopsis* was not associated with accumulation of PRs (Hoffland *et al.*, 1995; Pieterse *et al.*, 1996). When compared to pathogen-induced SAR, rhizobacterially-mediated ISR was usually found to be somewhat less effective (Van Wees *et al.*, 1997). Such observations support the notion that accumulation of PRs is not a prerequisite for the induction of resistance, but that it is likely that these proteins contribute to the protective state (Van Loon, 1997). Even though PRs do not seem to have a central role in induced resistance, they nevertheless constitute good markers for SAR (Kessmann *et al.*, 1994).

PRs and associated "SAR genes" are coordinately induced at the mRNA level in both inoculated and non-infected leaves concomitant with induced resistance in e.g. tobacco (Ward *et al.*, 1991) (Fig. 6) and *Arabidopsis* (Uknes *et al.*, 1992). Moreover, maximal mRNA induction was apparent by the time induced resistance was also maximal. This suggests that, even though the two may not be causally connected, they share one or more steps in a common response pathway.

PRs have been shown to be induced by numerous conditions and chemical compounds, including senescence, callus culture, UV light, wounding, plasmolysis, polyacrylic acid, auxin, cytokinin, heavy metal salts, mannitol, amino acids, thiamine, arachidonic acid, ozone, hydrogen peroxide, etc. (Van Loon, 1983b; Kessmann *et al.*, 1994; Malamy *et al.*, 1996, Ryals *et al.*, 1996). Many of these conditions or treatments have been shown also to impart enhanced resistance on plants. This appears at variance with observations (see section II.A.1) that neither non-biotic conditions, nor mechanical or chemical wounding induce systemic resistance. However, in most cases plants were tested at sites previously exposed to the inducing condition or chemical, and local alterations reducing susceptibility to infection rather than genuine systemic induction may have played a role. In other cases, chemical toxicity may have given rise to a stressful situation, eliciting resistance induction even without, or prior to, visible necrosis becoming apparent. In the early experiments described in section I.A.3, cutting, abrasion or toxic chemicals inflicted damage rapidly and, apparently, few endogenous eliciting compounds were generated, because action was rapidly terminated. In contrast, persisting stressful conditions may be sufficiently damaging to initiate reactions that ultimately induce PRs, as well as SAR. Development of SAR seems to be invariably associated with slow, progressive necrosis, such as that occurring upon application of purified *Phytophthora* elicitins to tobacco leaves (Ricci *et al.*, 1989), or in the so-called *acd* (accelerated cell death; Greenberg *et al.*, 1994) and *lsd* (lesions simulating disease; Dietrich *et al.*, 1994) mutants of *Arabidopsis*, that form spontaneous lesions during development. Such systems have been advocated as suitable for the study of expression of induced resistance in the absence of pathogens. However, the associated induction of PRs is not a necessary component of the induced state.

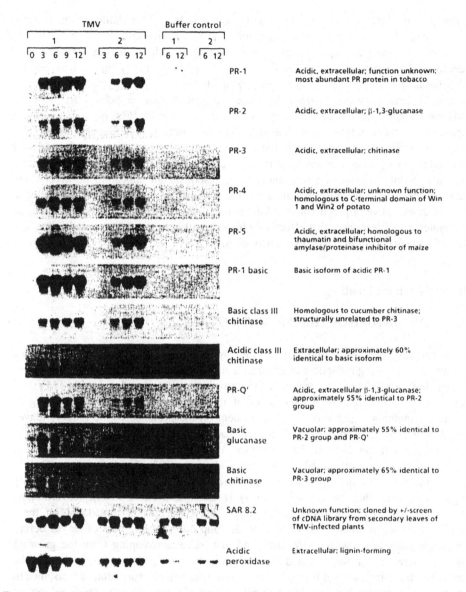

Figure 6. Northern blot analysis of the rate of SAR gene mRNA accumulation (in days) in (1°) lower leaves inoculated either with TMV or with buffer plus Carborundum, and in (2°) upper non-inoculated leaves of the same plants (from Ward *et al.*, 1991). Reprinted by permission of the publisher. Copyright American Society of Plant Physiologists.

Relatively little information is available on the rate and extent of further induction of PRs after challenge inoculation of induced tissues. On a tissue basis smaller increases in PR mRNAs and proteins seem to occur after challenge inoculation than on primary infection. This could be explained by the increased restriction of the pathogen and, consequently, the lesser amount of tissue engaged in defense after challenge inoculation. On a per cell basis, however, stimulation might well be increased.

In spite of many attempts, genes expressed solely in induced tissues have not been identified. Only some of the genes that are activated in primarily infected tissues are also induced systemically, as if in preparation for defense against further pathogen attack. On challenge inoculation, the same types of defense responses are induced. Whether considering phytoalexins, cell wall reinforcement, or PRs, in all cases where operation of these defensive mechanisms has been studied, none appears to be fully responsible for resistance. Rather, they all appear to be part of a coordinated and integrated set of responses that together are effective in limiting pathogen invasion. In plants exhibiting induced resistance, these responses all appear to be accelerated and enhanced on challenge inoculation. One may, therefore, further define induced resistance as a state of preparedness of the plant quickly and coordinately to express an integrated set of responses that effectively impede pathogen penetration, multiplication, spread and survival after challenge inoculation.

III.   Systemic Signalling

A.   THE NATURE OF THE "MOBILE SIGNAL"

Slow, progressive necrosis giving rise to PRs and SAR must either consume an essential metabolite repressing induction, or generate a signalling compound acting as an inducer. When three lower leaves of trimmed tobacco plants were inoculated with TMV and removed after different times, the level of resistance developing and persisting in the two remaining upper leaves was dependent on the presence of the inducer leaves for 3 to 4 days, i.e. once necrotic lesions had developed, the inducer leaves could be removed without loss of the resistance attained (Ross, 1961b). However, after induction of large, untrimmed plants, removal of the inducer leaves after 10 days caused those of the remaining leaves nearest the inducer leaves to develop substantially higher resistance than those farthest away (Bozarth and Ross, 1964). Nevertheless, development of induced resistance in upper leaves of plants inoculated on three lower leaves when young, and allowed to develop different numbers of leaves before challenge inoculation with TMV, indicated that leaves developing from the growing tip after removal of the infected lower leaves became resistant, even though the resistance-inducing infected leaves had been removed before expression of resistance in the tip leaves started (Box III). Mechanical girdling of stems inhibited the development of SAR in distal leaves (Ross, 1966), lending credence to the notion that a mobile signal is produced by the inducer leaves and transported through the phloem. When lesion development on primarily inoculated leaves was completed, or the inducer leaves were removed, the level of resistance induced tended to decline. This suggests that the signal is generated throughout the period of slow progressive necrosis. The signal tends to get "diluted", however, because more distal leaves develop quantitatively less resistance. On the other hand, induced resistance appears to persist for the remainder of the life of the leaf once attained. The induced state is not transmitted with the seed, although a single report by Roberts (1983) claims that plants grown from seed developing on a tobacco plant that had been induced by repeated inoculations with TMV, themselves exhibited typical SAR.

Box III.   Removal of the Site of Induction Reduces the Expression of Systemic Acquired Resistance in Distant Leaves (Bozarth and Ross, 1964; Dean and Kuc, 1986)

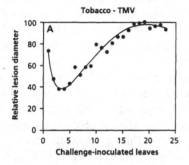
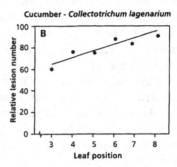

To study how far induced resistance extends systemically, young Samsun NN tobacco plants were inoculated with TMV on 3 leaves. Thereafter, they were allowed to develop 23 leaves before challenge inoculation of all upper leaves with TMV 49 days later. The primarily inoculated leaves had been removed 7 days after inoculation. As seen in frame A, depicting the ratio between mean lesion size on leaves of induced versus non-induced plants, the lowest remaining leaf (leaf 1) had only slight resistance, the next four were highly resistant, and successively higher leaves were progressively less resistant up to the 15th or 16th leaf. Thus, although primarily infected leaves were removed once systemic resistance had been induced, resistance was still expressed in leaves that developed subsequently and were at most mere initials at the time of primary inoculation.

Also in cucumber, leaves which emerged on growing plants after the inducer leaf 1 was removed developed progressively less systemic induced resistance. Frame B depicts the ratio of the number of lesions of *C. lagenarium* developing on leaves 3 to 8 on induced versus non-induced leaves. Leaf 1 was induced by inoculation with *C. lagenarium* and excised 7 days later, 10 days prior to challenge.

These results are difficult to interpret in terms of depletion of a factor repressing development of SAR. Rather, a diffusible signal seems to be generated in the primarily infected leaves, and to be distributed throughout the plant. In the absence of "source" leaves the signal tends to get "diluted", however, because more distal leaves develop progressively less resistance.

Similar results have been obtained in cucumber (Box III). Induced resistance of upper leaves was maintained when the first true leaf was induced and removed once symptoms had become apparent (Kuc and Richmond, 1977; Caruso and Kuc, 1979; Jenns and Kuc, 1980). Girdling the petioles of leaves to be challenged prevented induction of resistance in them (Guedes *et al.*, 1980), strongly suggesting that SAR was the result of a signal transported from the inducer leaf through the phloem.

The extent to which the signal can be transmitted was investigated by grafting experiments using plants induced by infecting leaf 1 with *C. lagenarium*. Protection was transmitted from the rootstock with the infected leaf to a grafted scion of an untreated plant. However, when the infected leaf on the rootstock was removed at the time of grafting, protection was not transmitted, even though leaf 2 present on the rootstock remained highly protected. Grafting a healthy scion from a previously infected plant onto an unprotected rootstock did not confer protection to the rootstock. These results demonstrated that the infected leaf is the source of the signal for SAR, and that the signal is not remobilized from or produced in systemically protected leaves (Dean and Kuc, 1986). The level of induced resistance attained was stronger for leaves above the inducer leaf than below it (Guedes *et al.*, 1980). However, roots can be induced through primary infection of leaves (Gessler and Kuc, 1982), or treatment with nonpathogenic rhizobacteria (Kloepper *et al.*, 1993).

Resistance induced in tobacco by limited infection with *P. tabacina* was reported to be transmitted to regenerants from tissue cultures from induced leaves (Tuzun and Kuc, 1987). However, in other experiments similar regenerants were found to be as susceptible to foliar challenge with the fungus as similar regenerants derived from unprotected control explants (Lucas *et al.*, 1985). This apparent discrepancy may be due to the choice of material taken for culturing, because the signal is not propagated in the absence of inducer tissues. Plants raised from seed of regenerants with induced resistance were no longer protected, confirming the earlier notion.

In cucumber, a single induction protected plants against various pathogens for 4-6 weeks. A booster inoculation with *C. lagenarium* or TNV 2-3 weeks after the first extended protection through the period of fruiting, allowing younger tissues to receive sufficient inducer signal. However, cucumbers could no longer be induced when they started to flower and set fruit. This seems to suggest that entering the reproductive phase turns off the mechanism of induction, without interference with the expression in previously induced tissues (Kuc and Richmond, 1977; Guedes *et al.*, 1980).

Grafting allowed not only the characterization of the signal as being mobile, but also its generality (Jenns and Kuc, 1979). Resistance induced by infection of leaf 1 of cucumber cultivar SMR-58 with *C. lagenarium* or TNV was transmitted to scions of watermelon and muskmelon. Likewise, using different cultivars of cucumber inoculated with *C. lagenarium* as rootstock induced resistance in cultivar SMR-58. However, pumpkin and squash rootstocks developing necrotic lesions in response to TNV did not induce resistance in SMR-58 cucumber scions. This might have been due to inefficient graft establishment and, hence, non-transmissibility of the signal. No controls with intact pumpkin or squash plants were conducted to establish whether induction of systemic resistance occurs upon inoculation of TNV. In view of the generality of the phenomenon, it seems unlikely that in these species systemic resistance would not be inducible. Judging from the results obtained with cucumber, watermelon and muskmelon, it must be concluded that the signal for systemic induced resistance is not cultivar-, genus-, or species-specific. However, the results with pumpkin and squash offer a note of caution that the signal may not be universal.

In tobacco graftings between different cultivars and species likewise provided no barrier to transmission of the signal between an induced rootstock and a non-induced scion (Gianinazzi and Ahl, 1983). Accompanying induction of SAR, PRs accumulated in the scions and these were characteristic of the cultivar or genus used as the scion rather than those expressed by the primarily inoculated rootstock. These experiments also demonstrated that the accumulation of PRs in systemically induced tissue is not caused by transport from the inoculated leaves, but that the proteins must be induced in the non-inoculated leaves as a result of the signal being transported into the scion. In view of the importance of unravelling the chemical nature of the signal(s), it is unfortunate that no further interspecific graftings seem to have been done to establish the specificity of the transported signal in different plant species.

B.   THE INVOLVEMENT OF ETHYLENE

In their early studies, Ross and Pritchard (1972) observed that when a single leaf of a healthy tobacco plant was enclosed in an atmosphere containing ethylene, other leaves on the same plant became more resistant. The role of ethylene appeared complex, however, because it was shown to promote leaf yellowing and senescence (Aharoni and Lieberman, 1979). Nevertheless, strongly increased ethylene production does occur early in hypersensitive reactions (Pritchard and Ross, 1975). In TMV-infected tobacco, the key enzyme of ethylene biosynthesis, 1-aminocyclopropane-1-carboxylic acid (ACC)-synthase, reaches a peak before the onset of necrosis. As a result ACC accumulates in the tissue until the last enzyme in the biosynthetic pathway, ACC oxidase, also increases concomitant with lesion development (De Laat and Van Loon, 1982). As a result, a peak of ethylene production occurs by the time lesions become visible. Reproducing the peak of ethylene production during lesion formation by pricking leaves of healthy tobacco plants with needles dipped in the ethylene-releasing compound ethephon, Van Loon (1977) observed localized necrosis resembling local lesions in size and appearance, both localized and systemically increased peroxidase activity and induction of PRs, as well as induced resistance. Wounding due to pricking with needles moistened with water or mineral acids did not provoke any of these effects, nor did treatment of leaves with liquid nitrogen, detergents, or extracts from infected plants (De Laat and Van Loon, 1983).

Taken together, these results suggested that the burst of ethylene occurring early in a hypersensitive reaction was responsible for initiating a signalling pathway leading to SAR and its associated biochemical alterations. Moreover, upon TMV infection, ACC oxidase activity was increased systemically (De Laat and Van Loon, 1983). Thus, on challenge inoculation the ACC generated by ACC synthase activity did not accumulate, but was immediately converted into ethylene, resulting in an earlier increase in ethylene production. Ethylene is a stress hormone that is produced at high levels during slow, progressive necrosis, and many of the conditions found to induce PRs in the absence of apparent necrosis, do increase ethylene production. Thus, the physiology of ethylene production appears consistent with a possible role for it in signalling (Van Loon, 1983a).

Ethylene itself does not appear to act as the mobile signal. It is produced by the cells surrounding the sites of infection (De Laat and Van Loon, 1983) and, as a gas, it diffuses in leaves over at most a few cell layers before escaping into the atmosphere. Neighbouring plants are not induced, probably because in the atmosphere the ethylene is diluted to such an extent that it is no longer active. ACC does not seem to act as a mobile signal either. Its accumulation is confined to the immediate vicinity of the lesions and no increase in ACC content or ethylene production was found systemically (De Laat and Van Loon, 1983).

However, several lines of evidence indicate that ethylene is not primarily responsible for the induction of SAR. Application of the precursor ACC is a poor inducer of resistance in tobacco and elicits primarily the basic rather than the acidic PRs that are associated with SAR (Brederode et al., 1991). However, applying inhibitors of ethylene synthesis such as aminoethoxyvinylglycine, or inhibitors of ethylene action such as norbornadiene, to TMV-infected leaves reduced the level of resistance attained in upper leaves (Van Loon and Pennings, 1993). Recently, the same result was obtained in tobacco plants transformed with a mutant etr1 gene from Arabidopsis, that confers dominant insensitivity to ethylene (Knoester, 1998). Whereas hypersensitivity in primarily inoculated leaves was unimpaired, upper leaves showed reduced SAR and did not express acidic PRs. These results strongly suggest that ethylene contributes to SAR, possibly by facilitating the release, synthesis or transport of the mobile signal.

C.   THE ROLE OF SALICYLIC ACID

1.   Requirement for Expression of SAR
In 1979, White accidentally observed that aspirin (acetylsalicylic acid) induces PRs and resistance in tobacco. Of all hydroxylated derivatives of benzoic acid (BA), only 2-hydroxybenzoic acid (salicylic acid; SA) and 2,6-dihydroxybenzoic acid acted as inducer (Van Loon, 1983b). Apparently, an ortho-hydroxylgroup was required for the effect, with further substitutions leading to loss of activity. The amounts of SA required were rather high, in the order of 1 mM, which is only slightly below the threshold for phytotoxicity. However, by using lack of increase in peroxidase activity as a marker for absence of toxic effects, Van Loon and Antoniw (1982) established that the effect of SA on inducing PRs and resistance is not systemic when SA is sprayed on or injected into individual leaves. In contrast, the effect was systemic when SA was applied as a soil drench. Under the latter conditions, SA appears to be taken up by the roots and transported into the shoot, where it acts locally. When SA is injected into a leaf, 99% is quickly conjugated to glucose and stored in the vacuole (Métraux et al., 1990; Ryals et al., 1996). Under these conditions, SA might be at sufficient concentration to act locally, but too little might be transported elsewhere to have an effect systemically. These considerations led Van Loon and Antoniw (1982) to speculate that ethylene, by stimulating phenylpropanoid metabolism, would lead to the synthesis of an aromatic compound resembling SA, that acts as the inducer of SAR.

In 1987, SA was identified as the signal compound responsible for thermogenesis in the inflorescences of some Araceae, indicating that SA has a hormonal role in plants (Raskin et al., 1987). Indeed, SA was found to be present in tobacco plants at nanogram concentrations and to be increased to about 1 μg g^{-1} fresh weight upon TMV inoculation (Fig. 7), associated with induction of PR-1 gene expression (Malamy et al., 1990). Moreover, SA increased similarly, but to only one tenth of this level, in the non-inoculated leaves. No increase in SA was detected when leaves were mechanically or chemically injured, indicating that the at least 20-fold increase in infected leaves and 5-fold increase in uninfected leaves after TMV inoculation are related to the state of induced resistance. Métraux *et al.* (1990) independently identified SA as appearing in the phloem sap of cucumber plants inoculated with either *C. lagenarium* or TNV, suggesting that SA itself could function as the endogenous signal in the transmission of SAR. Although the concentration range needed to elicit resistance by exogenous application was substantially higher than the endogenous concentrations measured, on feeding of 10 µg ml[-1] SA or inoculation with 1 µg TMV per leaf levels of unconjugated SA in the tissue were of the same magnitude, lending credence to the hypothesis that SA functions as the endogenous inducer of SAR (Yalpani *et al.*, 1991).

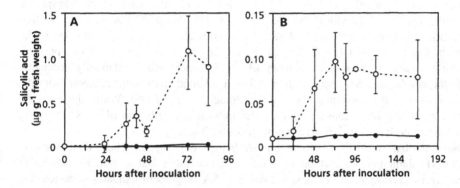

Figure 7.   Accumulation of salicylic acid in (A) lower leaves of Xanthi-nc tobacco inoculated with TMV (open circles) or buffer (closed circles), and (B) upper, non-inoculated leaves of the same plants. Necrotic lesions became apparent at 42 h. Adapted with permission from Malamy *et al.*, Science 250: 1003. Copyright 1990, American Association for the Advancement of Science.

Hybrids between *Nicotiana glutinosa* and *N. debneyi* were found to express the normally inducible PR-1f constitutively, and to be highly resistant against infection by TMV or TNV (Ahl and Gianinazzi, 1982). No further induction of resistance was possible in these hybrids. Thus, SAR was already turned on in these plants, and it was hypothesized that in these plants the signalling pathway was constitutively active. The finding that these plants contained elevated levels of SA (Yalpani *et al.*, 1993) conforms with this view and strengthens the association between SAR, SA and PRs (Malamy and Klessig, 1992). However, in tobacco SA does not induce PR mRNAs to the same level as TMV infection (Ward *et al.*, 1991), whereas ethephon induces much higher levels of the basic PRs than TMV. Wounding by razor blade cutting also induced basic PRs, but not resistance against TMV. This indicates that the basic PRs are not involved in resistance to TMV infection (Brederode *et al.*, 1991).

In tobacco leaves, SA is a product of aromatic biosynthesis, synthesized from benzoic acid (BA) by the enzyme BA-2-hydroxylase (Leon *et al.*, 1993). BA-2-hydroxylase activity was significantly increased following inoculation with TMV or application of BA, but the rate-limiting step may be the formation of BA from cinnamic acid. Stimulation of aromatic biosynthesis through increased PAL activity is instrumental in SA synthesis. Tobacco plants suppressed in PAL gene expression had SA levels approximately one fourth of those in control plants and did not develop SAR in response to TMV infection, suggesting that de novo synthesis of SA, and/or the presence or synthesis of other phenylpropanoids, is required for the expression of SAR (Pallas *et al.*, 1996). Using a specific inhibitor of PAL, the same conclusion was reached for *Arabidopsis* (Mauch-Mani and Slusarenko, 1996).

Knowledge of the biosynthetic pathway of SA was cleverly exploited to investigate whether indeed the SA found in non-inoculated tobacco leaves is transported there from inoculated inducer leaves. Inoculated leaves were enclosed in an atmosphere containing $^{18}O_2$, so that hydroxylation of benzoic acid would produce ^{18}O-labelled SA. Analysis of the upper, non-inoculated leaves indicated that at least half of the SA present was ^{18}O-labelled and, thus, had to have been synthesized in and transported from the TMV-inoculated leaf (Shulaev *et al.*, 1995). When ^{14}C-BA was injected into cucumber colyledons at the time of inoculation with TNV, ^{14}C-SA was detected in the phloem and in the first leaf 2 days after inoculation, before SAR became apparent. Because in leaf 1 the specific activity of ^{14}C-SA decreased substantially compared with the cotyledons, SA accumulation in leaf 1 resulted from both transport from the cotyledons and from synthesis in leaf 1 (Mölders *et al.*, 1996). While these results demonstrate that SA is transported, they also indicate that only part of the SA found in non-inoculated leaves is derived from the inoculated leaf.

Several observations cast doubt on SA being "the" signal. Thus, Rasmussen *et al.* (1991) demonstrated that on inoculation of cucumber leaf 1 with *Pseudomonas syringae* pv. *syringae*, the earliest increases in SA in phloem exudates occurred 8 h after inoculation in the petiole of leaf 1 and 12 h after inoculation in leaf 2. Detaching leaf 1 at intervals after inoculation demonstrated that it had to remain attached for only 4 h to result in the systemic accumulation of SA. Therefore, the induction of increased levels of SA seems to result from a different signal generated from leaf 1 within 4 h of inoculation.

SA is required for SAR to be expressed (Gaffney *et al.*, 1993). By generating transformed tobacco and *Arabidopsis* plants harboring the gene coding for the enzyme SA-hydroxylase (*NahG*) from *Pseudomonas putida*, plants were obtained in which any SA formed was rapidly converted into catechol (Vernooij *et al.*, 1994). Catechol was not active as an inducer of PRs or SAR. In these transformed plants SAR could no longer be induced, establishing SA as an essential signalling component (Fig. 8). Whether SA also functioned as the mobile signal was investigated by reciprocal grafts of *NahG* and untransformed tobacco (Box IV). These experiments revealed that SA is not itself the mobile signal, but its concentration is increased under the influence of the signal, and its accumulation is required for SAR to be expressed. Notably, *NahG*-containing plants are more susceptible to a variety of fungal, bacterial and viral pathogens (Delaney *et al.*, 1994), and in *NahG* plants TMV induces larger lesions. SA has been found to potentiate defense responses, such as elicitor-induced secretion of coumarin by suspension-cultured parsley cells (Kauss *et al.*, 1992), expression of asparagus PR-10 promoter-driven GUS expression in transgenic tobacco (Mur *et al.*, 1996) and induction of defense gene transcripts, H_2O_2 accumulation and hypersensitive cell death by an avirulent strain of *P. syringae* pv. *glycinea* in soybean (Shirasu *et al.*, 1997). Thus, SA seems to function as an agonist-dependent gain control mechanism for amplification of pathogen-derived signals.

Box IV.   SA is Not the Translocated Signal for Induction of Systemic Acquired Resistance in Tobacco (Vernooij *et al.*, 1994)

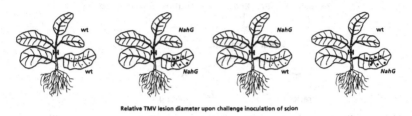

Relative TMV lesion diameter upon challenge inoculation of scion

| 41 | 99 | 104 | 40 |

To analyse whether salicylic acid (SA) functions as the mobile signal for inducing SAR in tobacco, reciprocal grafts were made between non-transformed Xanthi-nc tobacco (wt) and transformants containing the *NahG* gene (*NahG*). The latter gene codes for SA-hydroxylase that rapidly converts inducing SA in the plant to inactive catechol. Rootstock leaves of the grafted plants were either inoculated with water or with TMV, and scion leaves were challenge inoculated with TMV 7 days later. TMV lesions on *NahG* rootstocks were larger than lesions on wt rootstocks, indicating that lack of SA not only abolishes SAR but also reduces *N*-gene-mediated resistance to TMV. The figure illustrates the four combinations and indicates the relative lesion diameters resulting from challenge inoculation of the scions on induced versus non-induced rootstocks. In the combination wt/wt lesion diameter was reduced by 59%, whereas no reduction was seen in the combination *NahG/NahG*. No reduction was seen either in *NahG* scions on wt rootstocks. Wt rootstocks are expected to produce the mobile signal, but *NahG* scions did not express SAR, indicating that the signal requires the presence of SA in tissues distant from the infection site. Hence, SA is essential for resistance to be expressed under the influence of the translocated signal. *NahG* rootstocks are expected not to be able to provide wt scions with SA. Thus, if SA were the mobile signal, no reduction in lesion diameter should occur in this combination. Yet, wt scions on *NahG* rootstocks expressed the same level of resistance as wt/wt graftings. Thus, *NahG* rootstocks unable to accumulate SA were still fully capable of delivering the signal, rendering wt scions resistant to challenge. Concomitant with the expression of SAR, the scions showed an increase in endogenous SA and induction of PR mRNAs.

Similar results were obtained when, instead of TMV, the fungus *Cercospora nicotianae* was used as the challenging pathogen. This confirms that the same mechanism is operative against different types of pathogens. Thus, SA is not the translocated signal for induction of SAR in tobacco.

Chen et al. (1993) demonstrated that SA binds and inhibits tobacco catalase activity, which would lead to an increase in the endogenous level of H_2O_2. H_2O_2 in turn could act as a second messenger for the induction of defense responses. However, H_2O_2 did not induce PR gene expression in NahG plants at concentrations activating PR-1 gene expression in untransformed tobacco (Bi et al., 1995; Neuenschwander et al., 1995), indicating that H_2O_2 does not function downstream of SA in the induction of PRs. Moreover, it is questionable whether physiological concentrations of H_2O_2 in the tissue are sufficiently elevated to stimulate SA biosynthesis and induce SAR (Hunt *et al.*, 1996). Recently, a soluble, high-affinity SA-binding protein (SABP2) was identified, which reversibly binds biologically active, but not inactive analogs of SA in vitro (Du and

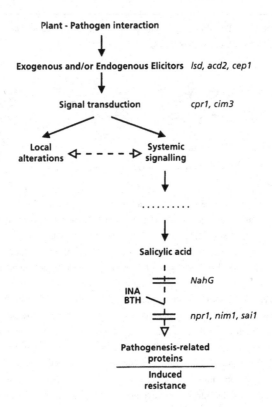

Figure 8.   Signal-transduction pathway leading to pathogen-induced systemic acquired resistance. Mutants identified in *Arabidopsis* at the different steps are indicated. INA, 2,6-dichloroisonicotinic acid; BTH, benzodiathiazole.

Klessig, 1997), but its role is not yet clear. SA has been shown to activate a 48 kDa mitogen-activated-type protein kinase (MAP kinase) in tobacco (Zhang and Klessig, 1997), as well as to mediate protein dephosphorylation (Conrath et al., 1997), implying both phosphorylation and dephosphorylation in the SA signal-transduction chain. Other workers have suggested that SA is part of a feed-back loop potentiating the action of other signalling compounds, perhaps H_2O_2. (Durner *et al.*, 1997). In cultured parsley cells, SA appeared to act by two mechanisms: it induced one group of defense genes, but only enhanced activation of a second group that was induced by application of the Pmg elicitor (Thulke and Conrath, 1998). Because induced resistance constitutes an enhancement of extant resistance mechanisms (see section II.B), both mechanisms can be involved in the induction of SAR by SA.

2.   Non-requirement for Expression of ISR

Surprisingly, *NahG*-transformed *Arabidopsis* plants were still inducible by non-pathogenic rhizobacteria, expressing ISR against both the fungal root pathogen For and the bacterial leaf pathogen Pst (Pieterse *et al.*, 1996; Van Wees *et al.*, 1997). Thus, induced resistance can operate in the absence of increased SA levels.

Rhizobacterially-mediated ISR was abolished, however, in mutants of *Arabidopsis* that were insensitive to ethylene (*etr1*) or unresponsive to jasmonic acid (JA) (*jar1*). In contrast, Pst-induced SAR was not impaired in these mutant plants (Pieterse *et al.*, 1998). Application of ACC or JA to wild type *Arabidopsis* induced resistance against Pst, implying that both ethylene and JA can act as inducers of resistance in *Arabidopsis*. These latter results seem at variance with those of Lawton *et al.* (1994, 1995) that SAR signal transduction in *Arabidopsis* is ethylene independent. The ethylene-releasing chemical ethephon induced SAR gene expression in both wild type and ethylene-insensitive mutants, whereas ethylene alone did not. These results suggested that acids liberated during the decomposition of ethephon, rather than ethylene itself, were the inducing compounds (Lawton *et al.*, 1994). Moreover, Pst induced SAR against *P. parasitica* equally in wild type plants and in ethylene-insensitive mutants, whereas the response was abolished in *NahG*-expressing plants (Lawton *et al.*, 1995). However, ethephon induction of PR-1 mRNA accumulation in *etr1* and *ein2* ethylene-insensitive mutants was less than observed in wild type plants. An SA dose-response experiment in the presence or absence of ethylene showed that exposure to ethylene immediately following SA application resulted in an increase in PR-1 mRNA at lower SA concentrations than observed with SA alone. These results indicated that ethylene could enhance the sensitivity of plants to low concentrations of SA by about 20-fold. Thus, SA appears to be an overriding signal in the induction of PRs and SAR, but in the absence of accompaning ethylene production, high levels are required.

During a hypersensitive reaction ethylene is generated in large amounts, consistent with lowering the threshold for SA action. Jasmonates may be another compound produced and sensitizing the tissue (Wasternack and Parthier, 1997). JA and its methyl ester (MeJA), when sprayed on potato or tomato leaves, were reported to express both local and systemic protection against *P. infestans* (Cohen *et al.*, 1993). Jasmonates also activated the coordinate expression of phenylpropanoid pathway genes in parsley cell cultures and transgenic tobacco (Ellard-Ivey and Douglas, 1996). In tobacco, MeJA induced the basic PR-5c, but not acidic PR-1a. However, in the presence of either SA or ethylene, induction of PR-5c was strongly increased and abundant PR-1a mRNA was induced (Xu *et al.*, 1994).

Since the induction of SAR in *Arabidopsis* by ethephon was abolished in *NahG* plants, ethylene seems to act before SA. However, ISR is still expressed in *NahG* plants, indicating that rhizobacterially-induced resistance bypasses the requirement for SA; yet sensitivity to ethylene and JA are required. Any of these requirements are lifted when plants are treated with 2,6-dichloroisonicotinic acid (INA). This chemical originated from a screening program for compounds inducing SAR, and appears to act after SA in the signalling pathway (Fig. 8).

D.   SIGNALLING PATHWAY MUTANTS IN *ARABIDOPSIS*

To further unravel the signalling pathway of SAR, mutants were sought in *Arabidopsis* that either exhibit SAR constitutively, or are impaired in its expression (Fig. 8). The *lsd* and *acd* mutants spontaneously develop necrotic lesions, were found to contain high levels of SA, and constitutively expressed PRs and SAR. As discussed in section

II.B.3, the expression of PRs and SAR in these mutants may be a consequence of the slow progressive necrosis occurring in these plants, rather than from a change directly affecting the SAR response. Most of these mutants are recessive, indicating that development of necrosis is under negative control. When the *lsd* mutants were crossed with *NahG* plants, lesion formation was suppressed in *lsd1*, *lsd6* and *lsd7*, but not in *lsd2* and *lsd4*, indicating that regulation of necrosis formation occurs both before and after the SA-requiring step (Weymann *et al.*, 1995; Ryals *et al.*, 1996). However, none of the *NahG*-transformed *lsd* mutants exhibited PR gene expression or SAR, again indicating the close relationship between SA production, accumulation of PRs and SAR.

The *cpr1* (constitutive expresser of PR genes) was identified in a screen for SAR mutants based on constitutive expression of these marker genes (Bowling *et al.*, 1994). Indeed, this mutant also expressed SAR constitutively. The *cim3* mutant (constitutive immunity) likewise showed constitutive expression of PRs, had elevated levels of both free and conjugated SA, and exhibited increased resistance to infection with the virulent bacterial pathogen Pst and the fungal pathogen *P. parasitica* (Ryals *et al.*, 1996). Neither *cpr1*, nor *cim3* develops spontaneous necrosis, and introduction of the *NahG* gene suppressed both PR gene expression and SAR. These mutants seem to be impaired in a step following necrotic lesion formation, but before SA accumulation. The *cpr1* mutation is recessive, whereas the *cim3* mutation is dominant (Durner *et al.*, 1997). The latter suggests that *cim3* may encode a gene product that is normally shutting down the signalling pathway leading up to SA production. Results from crosses between *lsd* and *cpr1* or *cim3* mutants have not been reported, but it would be most interesting to see whether the latter are epistatic over the former, as expected from the phenotypes observed, or if necrosis formation acts as an overriding determinant for the induction of PRs and SAR.

Mutants in which the SAR response is impaired were isolated on the basis of lack of PR gene expression: *npr1* (non-expresser of PR genes) (Cao *et al.*, 1994), or the inability to express SAR against *P. parasitica* following exogenous application of the inducers SA or INA: *nim1* (non-inducible immunity) (Delaney *et al.*, 1995). Both mutations have been shown to be allelic (Delaney, 1997), and to result from a mutation in a unique 60 kDa soluble protein containing ankyrin repeats (Cao *et al.*, 1997), and homologous to the mammalian transcription factor inhibitor IκB (Ryals *et al.*, 1997). Since ankyrin repeats are important for protein-protein interactions, the wild type gene probably functions by interacting with other proteins in the signalling pathway subsequent to SA. The mutants were no longer responsive to induction by SA or INA. The *nim1* plants accumulated both free and glucose-conjugated SA to levels in excess of those in wild type plants in response to pathogen infection, indicating that they are SA-insensitive. A similar SA-insensitive mutant, *sai*, was recently isolated on the basis of its inability to express the SA-inducible PR-1a promoter (Shah *et al.*, 1997). The phenotype of this mutant is similar to that of the *npr1* and *nim1* mutations, to which *sai* is allelic.

E.   COMPARISON OF SIGNALLING IN SAR AND ISR

As discussed in section III.C.2, rhizobacterially-induced ISR is not dependent on SA accumulation and is not associated with PR gene expression. Testing for ISR elicited

Systemic signalling pathways

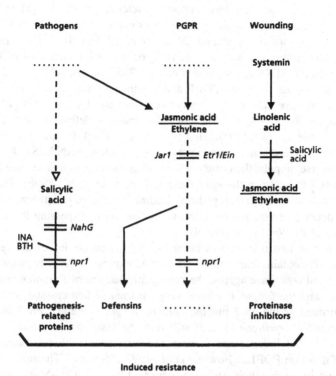

Figure 9. Systemic signalling pathways of resistances induced by pathogens (systemic acquired resistance), plant growth-promoting rhizobacteria (PGPR) (induced systemic resistance) and wounding (induced resistance against herbivores). Steps blocked by specific mutants or compounds are indicated by double crossed lines.

by WCS417 against Pst in *npr1* established that this mutant is unable to express not only SAR, but also ISR (Pieterse *et al.*, 1998). This result indicates that *npr1* must have a dual signalling function, e.g. on the one hand, induction of PR genes and, on the other hand, induced resistance, and that the two are separable, even though they are closely associated during the induction of SAR. Also, this result demonstrates that the signalling pathways of SAR, impaired in the *NahG* transformant but not in the *etr1* or *jar1* mutants, and of ISR, not impaired in the *NahG* transformant but absent in the *etr1* and *jar1* mutants, converge before or at the level of *npr1* (Fig. 9). An implication of this result is that expression of SAR and of ISR are both controlled by a similar gene late in the signalling pathway and therefore must be, at least partially, phenotypically similar. The latter is borne out by the similar enhanced defensive capacity of plants expressing SAR or ISR. The difference appears in the accumulation of PRs, which may enhance the level of resistance attained in SAR by their anti-pathogenic activities (Van Loon, 1997).

SA-independent induction of systemic resistance has been described in tobacco on treatment with cell wall-degrading enzymes of *Erwinia carotovora* subsp. *carotovora*. On challenge inoculation with this pathogen bacterial growth in planta and plant maceration were reduced. Systemic resistance was expressed similarly in *NahG* plants, indicating that it is not SA-mediated (Vidal *et al.*, 1998). In radish, induction of resistance against *Alternaria brassicicola* is associated with induction of the 5 kDa defensin proteins Rs-AFP3 and 4 (Terras *et al.*, 1995). In *Arabidopsis*, this pathogen induces the homologous defensins PDF1, and this induction is likewise independent of SA production, but dependent on sensitivity to ethylene and JA (Penninckx *et al.*, 1996). (see also chapter 8). MeJA, but not SA, also induces defense-related thionins in *Arabidopsis* (Epple *et al.*, 1995). However, ISR elicited by the rhizobacteria in *Arabidopsis* was not associated with expression of defensin mRNA (S.C.M. van Wees and C.M.J. Pieterse, unpublished observation), whereas its pathogen-induced induction was reported still to occur in the *npr1* mutant (Penninckx *et al.*, 1996). These results establish a second SA-independent pathway leading to induced resistance, which shares with ISR its dependence on JA and ethylene but differs in bypassing the requirement for the product of the *Npr1* gene (Fig. 9).

The existence of two pathways is borne out by the properties of the *cpr5* mutant which, like *cpr1*, contains elevated levels of SA, expresses PR-1 constitutively, and displays enhanced resistance against the pathogenic bacterium *Pseudomonas syringae* pv. *maculicola* and the fungus *Peronospora parasitica*. When crossed with either the *NahG* transformant or the *npr1* mutant, *cpr5* no longer expressed PR-1 and had lost resistance against *P. syringae*. Yet, it still retained resistance against *P. parasitica* in an SA- and NPR1-independent manner and associated with maintained elevated expression of defensin PDF1.2 (Bowling *et al.*, 1997). Recently, Thomma *et al.* (1998) established that in *Arabidopsis* pathogen-induced PR-1, -2 and -5 are dependent on SA perception and associated with induced resistance against the biotrophic fungus *P. parasitica*, whereas PDF1.2, basic PR-3 and PR-4-type hevein-like protein are activated by the ethylene/JA-dependent pathway and associated with induced resistance against the necrotrophic fungi *Fusarium oxysporum* f.sp. *matthiolae* and *Botrytis cinerea*. These data clearly iillustrate the occurrence of separate signal-transduction pathways in *Arabidopsis*, which lead to distinct sets of antimicrobial proteins and to enhanced resistance against different sets of pathogens.

IV.   Systemic Induced Resistance against Insect Attack

Jasmonate is also a signalling compound in an entirely different pathway, specific for the systemic induction of proteinase inhibitors (PI; PR-6), as well as other defensive proteins, in response to wounding in tomato and potato (Fig. 9). This response appears to be a reaction to, and to provide protection against, herbivores, particularly insects (Ryan, 1990). Trypsin and chymotrypsin inhibitors interfere with the breakdown of ingested proteins in the gut and impede nutrient utilization. Leaf damage by mechanical means, such as abrasion or razor blade cutting induces a proteinase inhibitor-inducing factor (PIIF), the nature of which has been identified (Pearce *et al.*, 1991). Initially, oligogalacturonides acting as endogenous elicitors were considered to be primary candidates as systemic signals in the wound response. However, pectin-degrading enzymes required to release oligogalacturonides from plant cell walls in response to wounding were not found in tomato leaves (Ryan, 1992), and labelled oligogalacturonide elicitors, when applied to wounds on tomato plants, did not move from the site of application. Using induction of PIs after absorption of fractionated extracts through the cut stem of young tomato plants as a bioassay, an endogenous polypeptide was isolated that was able to induce the synthesis of trypsin and chymotrypsin inhibitors at femtomole amounts per plant. This polypeptide contained 18 amino acids and, unlike the oligogalacturonides, was transported out of wounded to distal tissues. When ^{14}C-labelled polypeptide was placed on fresh wounds of tomato plants, the radioactivity moved out of the leaf within 30 min, and within 1-2 h was identified in the phloem exudate. Because of its mobility through the phloem, the inducing polypeptide was named "systemin" (Pearce *et al.*, 1991).

As in the case of SA, these results are suggestive, but do not prove, that systemin is the transported signal. Wildon *et al.* (1992) found that chilling the petiole of a damaged leaf inhibited phloem transport of ^{11}C-labelled photosynthate for at least 10 min, but still allowed passage of the signalling compound when the leaf was cut off after 5 min. The possibility of a hydraulic signal (1500 m s⁻¹) was discarded because the signal had not yet moved out of the treated leaf after 28 s. Electrical activity propagated through the plant at a speed of between 1 and 4 mm s⁻¹ and was not blocked by chilling the petiole. Indeed, mechanical wounding gave rise to an electrical pulse that was propagated systemically and was detectable in distant leaves whenever PI (*pin*) genes were activated. Propagated electrical impulses can also be induced by stimuli such as osmotic shock, heat shock, cold or touch, all of them stress responses, but those have not been tested for concomitant induction of PIs. Phloem transport can reach speeds of 0.5 to 4 mm s⁻¹, which might be sufficient for transport of systemin out of a damaged leaf within 28 s. The only argument to dismiss systemin as the signal would be the block in phloem transport at 4°C, but experiments showing that labelled systemin is not transported under these conditions have not been done.

Several lines of evidence indicate that release of systemin is essential for systemic induction of PIs. On wounding systemin is cleaved off from a larger protein, called prosystemin (McGurl *et al.*, 1992) (Fig. 10). In transgenic plants constitutively expressing systemin, PIs are also produced constitutively, whereas in plants transformed with an antisense systemin DNA-construct systemic induction of PIs in response to wounding was almost completely suppressed (Bergey *et al.*, 1996).

1	MGTPSYDIKNKGDDMQEEPKVKLHHEKGGDEKEKIIEKETPSQDINNKDT
51	ISSYVLRDDTQEIPKMEHEEGGYVKEKIVEKETISQYIIKIEGDDDAQEK
101	LKVEYEEEEYEKEKIVEKETPSQDINNKGDDAQEKPKVEHEEGDDKETPS
151	QDIIKMEGEGALEITKVVCEKIIVREDLAVQSKPPSKRDPPKMQTDNNKL

Figure 10. The amino acid sequence of prosystemin. Systemin is the 18-amino acid peptide underlined.

The plant hormone abscisic acid (ABA) has been shown to induce *pin* mRNA in tomato (Pena-Cortes *et al.*, 1989) and the wound-induced response was inhibited in ABA-deficient plants (Pena-Cortes *et al.*, 1996). This implicated ABA as a signal in the pathway leading to PI synthesis. However, reports are not consistent as to the role of ABA. Rather, JA and MeJA were demonstrated to be powerful inducers of PIs in tomato (Farmer and Ryan, 1992; Ryan, 1992). The JA precursors linolenic acid, 13(*S*)-hydroperoxylinolenic acid and 12-oxo-phytodienoic acid, when applied to the surfaces of tomato leaves, also acted as powerful inducers, whereas derivatives closely related in structure to the precursors but not intermediates in the JA biosynthetic pathway did not induce PI synthesis. These results suggested that a lipid-based signalling system was responsible for the induction of PI synthesis locally in response to endogenous elicitors and systemically as a result of the action of systemin. Systemin would be bound to a receptor in the plasma membrane of leaf cells, leading to activation of a lipase with consequent release of linolenic acid. Under the influence of lipoxygenase, linolenic acid would be converted, in several consecutive steps, to JA. JA would then be the direct inducer of PIs (Fig. 9). Pathogens secreting pectin-degrading enzymes could also induce PIs locally by liberating oligouronides from the plant cell walls. Presumably, these signals could feed in the signal-transduction pathway at the level of the plasma membrane lipase activation step. Interestingly, wound-induced ethylene production potentiates PI synthesis (O'Donnell *et al.*, 1996), whereas SA is inhibitory (Doares *et al.*, 1995). SA inhibits the activity of the last enzyme in the ethylene biosynthetic pathway, ACC oxidase, and appears to reduce *pin* mRNA expression both before and after the step requiring JA. Thus, the activation of the pathway leading to SAR will override and partly shut down the wounding response in tomato (Koiwa *et al.*, 1997).

Identification of the polypeptide systemin as a signal in systemic induction of PIs has prompted research into the possibility of peptide signal(s) responsible for SAR. However, so far no such a compound has been reported. Tobacco does not contain a prosystemin homologue (HJM Linthorst and JF Bol, unpublished observation), and systemic induction of PIs in response to pathogens is likely to be regulated by ethylene, together with the basic PRs in this species (Brederode *et al.*, 1991). However, PIs are also strongly induced in tobacco by JA and MeJA (Farmer and Ryan, 1990), but not in response to mechanical wounding. Thus, responses seem to be conserved in different plant species, but their activation in response to various stimuli appears to be different, as are the molecular signals involved.

When induced plants are enclosed in a chamber together with non-induced plants, the latter also become induced to synthesize PIs, because the volatile MeJA produced by the induced plants is perceived by the non-induced neighbours (Farmer and Ryan, 1990). In plants expressing SAR, part of the SA synthesized is converted into the volatile methyl-SA (MeSA). Similarly, non-induced plants start to express PRs and SAR when enclosed with plants producing MeSA, and it has even been suggested that MeSA acts as the endogenous "mobile signal" generating SAR (Shulaev *et al.*, 1997). It is questionable, however, whether such transfer from induced to non-induced plant parts or neighbouring plants occurs in nature. When plants were not enclosed, yet placed next to each other in a growth chamber or greenhouse, SAR was never transmitted from TMV-induced tobacco plants to neighbouring controls. Rather, the release of these volatiles appears to play a role in helping plants to attract predators of insect herbivores. For example, upon infestation by spider mites, bean plants produce mixtures of volatile allelochemicals, such as MeSA, (E)-β-ocimene, linalool and 4,8-dimethyl-1,3(E),7-nonatriene, that attract phytosciid mites that are predators of the spider mites. Predator-attracting "infochemicals" are produced systemically throughout the spider mite-infested plants (Dicke *et al.*, 1990). Similarly, corn plants that are infested by beet army worm larvae emit a blend of volatiles that attracts the parasitic wasp *Cotesia marginiventris*. The chemicals are only emitted upon herbivory and not or only in minor quantities following mechanical damage (Turlings *et al.*, 1990). Volicitin, a glutamic acid-conjugated fatty acid derivative present in the saliva of the larvae, appears to be responsible for the emission of volatile terpenoids from the maize seedlings (Alborn *et al.*, 1997). The blend of herbivore-induced chemicals seems to be rather specific: the natural enemies are able to discriminate between different plant-herbivore combinations. Downwind neighbours of spider mite-infested plants may themselves start producing infochemicals in anticipation of insect attack (Dicke *et al.*, 1990). Since these responses are far more specific than the types of induced resistance discussed earlier, and signalling is dependent on herbivore-specific elicitors, the signal-transduction pathway will again be different, although some components, e.g. for SA production, may be shared. Conceptually, however, this is a genuine systemic induced resistance phenomenon, in which the plant is stimulated to better defend itself.

V.   Relevance of SAR and ISR in Natural and Agricultural Situations

Since induced resistance is a general phenomenon, the question can be asked how far plants in natural and agricultural situations are induced over their lifetime (Tuzun and Kloepper, 1995). Induced resistance is not transmitted through seed and plants need to be induced to become resistant. *Pseudomonas* bacteria are naturally abundant in soils and can act as inducers of ISR (Van Loon *et al.*, 1997, 1998). Bacteria of some other genera and non-pathogenic soilborne fungi may also induce ISR. However, the ability of any one strain to induce ISR differs and soil microbial populations are exceedingly diverse. Although plant roots may contain up to 10^9 colony-forming units (cfu) of bacteria g^{-1} root, the level of a single bacterial strain is unlikely to reach 10^5 cfu g^{-1} root, the level minimally required for eliciting ISR in e.g. radish. Perhaps, different inducing strains might together reach the threshold. However, the rhizosphere is a highly dynamic environment and inducing strains can easily be antagonized and replaced by non-inducing bacteria. In bioassays, radish seedlings protected by application of 2×10^7 cfu *P. fluorescens* strain WCS374 to the young roots 5 days after germination developed substantial protection against infection by For, but infection was delayed rather than abolished, and this protection tended to break down by 4 weeks after inoculation (Table 4). Root colonization by the bacteria still amounted to 6×10^6 cfu g^{-1} root at this stage. Either build-up of the pathogen in the rhizosphere had overcome the induced resistance (Leeman *et al.*, 1995a), or plants were responsive to the bacterial signal only at the seedling stage, and the state of induced resistance was not maintained.

Thus, ISR can be induced, but may also be lost again. Under commercial greenhouse conditions, radish seed bacterization with WCS374 resulted in up to 50% more marketable yield (Leeman *et al.*, 1995c), indicating that in an agricultural situation treatment with an ISR-eliciting bacterial strain suppresses disease. Similar results have been seen in cucumber, where prior seed treatment or seed treatment plus a soil drench at transplanting with four rhizobacterial strains resulted in protection against challenge-inoculated angular leaf spot and naturally occurring anthracnose (Kloepper *et al.*, 1993; Wei *et al.*, 1996). Such results indicate that plants in agricultural situations are not usually already induced.

Table 4.   Expression of induced systemic resistance in radish infected with *Fusarium oxysporum* f.sp. *raphani* upon treatment with *Pseudomonas fluorescens* WCS374 2 days before inoculation [1]

Days after inoculation	Percentage of diseased plants	
	Nonbacterized	WCS374-treated
12	19	7
17	45	8
21	42	11
26	41	39

[1] H. Steijl, J. van den Heuvel and L.C. van Loon, unpublished data

Similarly, protection and yield increases have been observed by inducing SAR in cucumber and tobacco, and by treatment with resistance-inducing elicitors in small grains. In cucumber (Caruso and Kuc, 1977), plants were inoculated with *C. lagenarium* on leaf 1 at the seedling stage or left untreated, transplanted into the field, and challenge inoculated with the same fungus a few weeks later. After another 10 days both types of plants on average had the same number of leaves with lesions, indicating that protection had been insufficient or had not lasted long enough to reduce disease. In contrast, when the primarily inoculated leaves received a booster inoculation one week before challenge, the level of disease was drastically reduced, with fewer than half as many leaves per plant showing lesions. Moreover, the mean diameter of lesions and the area of lesions per leaf were significantly reduced. Since the inducer signal is progressively less effective in inducing resistance in newly developing leaves (cf. Box III), it is likely that the effect of the booster inoculation can be ascribed to the production of a renewed supply of the signal compound. Protection was also noted in field-grown watermelon as a reduction in the number and size of lesions, but best demonstrated in the survival of induced versus non-protected plants (Caruso and Kuc, 1977). Similarly, stem inoculation of tobacco at the seedling stage with *P. tabacina* has been employed to protect against blue mold under field conditions (Tuzun and Kuc, 1985).

There is a danger in these treatments in that induction is achieved using the pathogen proper. Occasionally, the intended limited inducer inoculation leads to extensive systemic infection and crop loss. Nevertheless, it is clear that the plants can be protected and, thus, are not naturally induced already at the seedling stage.

A more practical way to induce resistance has been employed by Schönbeck and colleagues by using bacterial culture filtrates. Metabolites from *Bacillus subtilis* or *Stachybotrys chartarum* reduced infection of *Erysiphe graminis* f.sp. *hordei* in barley, of *E. graminis* f.sp. *tritici* in wheat, and of brome mosaic virus in barley (Steiner and Schönbeck, 1995). When compared to a commercial fungicide application, the bacterial culture filtrate reduced mildew symptoms on barley only moderately. However, grain yield was increased as much as after fungicide treatment, and the combination of fungicide and inducer was not better than inducer alone (Kehlenbeck *et al.*, 1994). These effects were attributed to a stimulation of photosynthesis as a result of inducer application, pointing to complex effects on host physiology.

Induction of resistance in natural vegetation appears not to have been studied. Since PRs can be induced by small numbers of necrotic lesions, as may occur as a result of contact with many avirulent pathogens, or slow progressive necrosis, as can result from adverse environmental conditions, it is likely that plants may develop some level of SAR over their lifetime. Greenhouse-grown bean and tobacco exposed to polluted air and mature, field-grown tobacco have been shown to contain PRs (Wilson, 1983). Induction of PRs associated with senescence has also been observed. In tobacco PRs are induced spontaneously upon flowering (Fraser, 1981), when the whole plant starts dying off and older leaves develop age-related necrosis. Such natural induction of PRs may be associated with increased resistance of older plants against various pathogens.

If induced resistance is an effective mechanism to protect the plant against different pathogens, it may be asked why the response is not constitutive. The simplest answer would be that there must be a cost involved in maintaining the induced state. Certainly, the accumulation of PRs, often to high levels, causes a drain on nitrogen-containing amino acids and would be expected to reduce growth potential. *Arabidopsis cpr* and *cim* mutants, exhibiting SAR constitutively, develop more slowly and remain smaller than wild-type plants (Hoffland *et al.*, 1998). Primary inoculation of lower leaves of tobacco plants developing necrotic lesions in response to TMV, reduced the area of upper leaves associated with the development of SAR (Fraser, 1979). This inhibited growth of induced plants was attributable to an increase in the stress hormone ABA in the infected leaves and transport in sufficient quantities to the upper leaves to reduce growth (Whenham and Fraser, 1981). Likewise, repeated inoculations with an avirulent pathogen decreased yield of barley and lettuce (Schönbeck *et al.*, 1987), indicating that a permanent maintenance of post-infectional resistance mechanisms requires an increased consumption of photosynthetic products and energy. But growth reduction may be offset by increased protection against deleterious micro-organisms in the soil (Schippers *et al.*, 1987). In agricultural situations where high amounts of fertilizer are often applied, effects on plant growth or development are rarely obvious.

Induced resistance is an attractive means for plant disease control. Although it reduces severity rather than preventing disease, it is active against a wide range of pathogens and, thereby, increases the level of horizontal resistance (Fig. 11). By itself

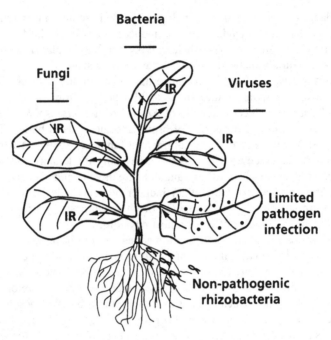

Figure 11. Diagrammatic representation of systemic resistance against fungi, bacteria and viruses induced locally by limited pathogen infection or root colonization by non-pathogenic rhizobacteria.

it could be sufficient to reduce diseases to economically acceptable levels, or postpone deleterious effects on the crop until after harvest. It could reduce pathogen build-up and, thereby, prolong durability of monogenic resistance genes. Also, it might be applied together with chemical crop protectants at lower doses than would be needed without induction of resistance.

Because of the risks incurred and the inpracticability of using pathogens for inducing resistance, several alternative strategies present themselves. Plants genetically modified to constitutively express a single or a combination of PRs have been generated and some show considerably improved resistance against major diseases (see chapter 8). Rhizobacterially-elicited ISR can be attractive to exploit, because the inducing bacteria are non-pathogenic and colonize the roots of many plant species. Moreover, the application of biochemical, molecular and genetic techniques has led to the identification of key components of the signalling pathways leading to defense responses. Genes such as *cim3*, that dominantly confer operation of the SAR pathway, might be used to engineer plants that are better able to defend themselves against a wide spectrum of pathogens. Constitutively enhanced levels of SA in crop plants may not be desirable, because high concentrations are phytotoxic and will not be nutritionally acceptable to the public. Another way is to develop compounds capable of inducing SAR beyond the SA-requiring step (Kessmann *et al.*, 1994; Sticher *et al.*, 1997). One such compound, INA, does not always reliably induce SAR, and also shows phytotoxicity. Recently, a compound with physiological properties similar to INA was identified, which is particularly active in inducing resistance in various crop plants (Görlach *et al.*, 1996. This benzothiadiazole has some structural similarity to INA, but is not phytotoxic. This "plant activator" has been found to protect cereals, tobacco, rice, vegetables and bananas effectively against a number of important diseases (Ciba, 1996; Kunz *et al.*, 1997). Identification of the key signalling components will undoubtedly be further exploited to activate the induced resistance pathways in plants and aid in environment-friendly plant protection.

References

Aharoni N and Lieberman M (1979) Ethylene as a regulator of senescence in tobacco leaf discs. Plant Physiol 64: 801—804

Ahl P and Gianinazzi S (1982) b-Protein as a constitutive component in highly (TMV) resistant interspecific hybrids of *Nicotiana glutinosa* x *Nicotiana debneyi*. Plant Sci Lett 26: 173—181

Ajlan AM and Potter DA (1991) Does immunization of cucumber against anthracnose by *Colletotrichum lagenarium* affect host suitability for arthropods? Entomol Exp Appl 58: 83—91

Alborn T, Turlings TCJ, Jones TH, Stenhagen G, Loughrin JH and Tumlinson JH (1997) An elicitor of plant volatiles from beet armyworm oral secretion. Science 276: 945—949

Balasz E, Sziraki I and Kiraly Z (1977) The role of cytokinins in the systemic acquired resistance of tobacco hypersensitive to tobacco mosaic virus. Physiol Plant Pathol 11: 29—37

Bergey DR, Howe GA and Ryan CA (1996) Polypeptide signaling for plant defensive genes exhibits analogies to defense signaling in animals. Proc Natl Acad Sci USA 93: 12053—12058

Bi YM, Kenton P, Mur L, Darby R and Draper J (1995) Hydrogen peroxide does not function downstream of salicylic acid in the induction of PR protein expression. Plant J 8: 235—245

Bowling SA, Guo A, Cao H, Gordon AS, Klessig DF and Dong X (1994) A mutation in *Arabidopsis* that leads to constitutive expression of systemic acquired resistance. Plant Cell 6: 1845—1857

Bowling SA, Clarke JD, Liu Y, Klessig DF and Dong X (1997) The *cpr5* mutant of *Arabidopsis* expresses both NPR1-dependent and NPR1-independent resistance. Plant Cell 9: 1573—1584

Bozarth RF and Ross AF (1964) Systemic resistance induced by localized virus infections: extent of changes in uninfected plant parts. Virology 24: 446—455

Brederode FT, Linthorst HJM and Bol JF (1991) Differential induction of acquired resistance and PR gene expression in tobacco by virus infection, ethephon treatment, UV light and wounding. Plant Mol Biol 17: 1117—1125

Cameron RK, Dixon R and Lamb C (1994) Biologically induced systemic acquired resistance in *Arabidopsis thaliana*. Plant J 5: 715—725

Cao H, Bowling SA, Gordon AS and Dong X (1994) Characterization of an *Arabidopsis* mutant that is nonresponsive to inducers of systemic acquired resistance. Plant Cell 6: 1583—1592

Cao H, Glazebrook J, Clarke JD, Volko S and Dong XN (1997) The Arabidopsis *NPR1* gene that controls systemic acquired resistance encodes a novel protein containing ankyrin repeats. Cell 88: 57—63

Caruso FL and Kuc J (1977) Field protection of cucumber, watermelon, and muskmelon against *Colletotrichum lagenarium* by *Colletotrichum lagenarium*. Phytopathology 67: 1290—1292

Caruso FL and Kuc J (1979) Induced resistance of cucumber to anthracnose and angular leaf spot by *Pseudomonas lachrymans* and *Colletotrichum lagenarium*. Physiol Plant Pathol 14: 191—201

Chen Z, Ricigliano JW and Klessig DF (1993) Purification and characterization of a soluble salicylic acid-binding protein from tobacco. Proc Natl Acad Sci USA 90: 9533—9537

Ciba (1996) Nature created the concept: the plant activator. Ciba-Geigy AG, Basle, Switzerland

Cohen Y, Gisi U and Niderman T (1993) Local and systemic protection against *Phytophthora infestans* induced in potato and tomato plants by jasmonic acid and jasmonic methyl ester. Phytopathology 83: 1054—1062

Conrath U, Silva H and Klessig DF (1997) Protein dephosphorylation mediates salicylic acid-induced expression of *PR-1* genes in tobacco. Plant J 11: 747—757

De Laat AMM and Van Loon LC (1982) Regulation of ethylene biosynthesis in virus-infected tobacco leaves II. Time course of levels of intermediates and in vivo conversion rates. Plant Physiol 69: 240—245

De Laat AMM and Van Loon LC (1983) The relationship between stimulated ethylene production and symptom expression in virus-infected tobacco leaves. Physiol Plant Pathol 22: 261—273

Dean RA and Kuc J (1986) Induced systemic protection in cucumbers: the source of the "signal". Physiol Plant Pathol 28: 227-233

Dean RA and Kuc J (1987) Rapid lignification in response to wounding and infection as a mechanism for induced systemic protection in cucumber. Physiol Mol Plant Pathol 31: 69—81

Delaney TP (1997) Genetic dissection of acquired resistance to disease. Plant Physiol 113: 5—12

Delaney TP, Uknes S, Vernooij B, Friedrich L, Weymann K, Negrotto D, Gaffney T, Gut-Rella M, Kessmann H, Ward E and Ryals J (1994) A central role of salicylic acid in plant disease resistance. Science 266: 1247—1250

Delaney T, Friedrich L and Ryals J (1995) *Arabidopsis* signal transduction mutant defective in chemically and biologically induced disease resistance. Proc Natl Acad Sci USA 92: 6602—6606

Dicke M, Sabelis MW, Takabayashi J, Bruin J and Posthumus MA (1990) Plant strategies of manipulating predator-prey interactions through allelochemicals: prospects for application in pest control. J Chem Ecol 16: 3091—3118

Dietrich RA, Delaney TP, Uknes SJ, Ward ER, Ryals JA and Dangl JL (1994) *Arabidopsis* mutants simulating disease resistance response. Cell 77: 565—577

Doares SH, Narvaez-Vasquez J, Conconi A and Ryan CA (1995) Salicylic acid inhibits synthesis of proteinase inhibitors in tomato leaves induced by systemin and jasmonic acid. Plant Physiol 108: 1741—1746

Du H and Klessig DF (1997) Identification of a soluble, high-affinity salicylic acid-binding protein in tobacco. Plant Physiol 113: 1319—1327

Durner J, Shah J and Klessig DF (1997) Salicylic acid and disease resistance in plants. Trends Plant Sci 2: 266—274

Ellard-Ivey M and Douglas CJ (1996) Role of jasmonates in the elicitor- and wound-inducible expression of defense genes in parsley and transgenic tobacco. Plant Physiol 112: 183—192

Epple P, Apel K and Bohlmann H (1995) An *Arabidopsis thaliana* thionin gene is inducible via a signal transduction pathway different from that for pathogenesis-related proteins. Plant Physiol 109: 813—820

Farmer EE and Ryan CA (1990) Interplant communication: airborne methyl jasmonate induces synthesis of proteinase inhibitors in plant leaves. Proc Natl Acad Sci USA 87: 7713—7716

Farmer EE and Ryan CA (1992) Octadecanoid precursors of jasmonic acid activate the synthesis of wound-inducible proteinase inhibitors. Plant Cell 4: 129—134

Fraser RSS (1979) Systemic consequences of the local lesion reaction to tobacco mosaic virus in a tobacco variety lacking the *N* gene for hypersensitivity. Physiol Plant Pathol 14: 383—394

Fraser RSS (1981) Evidence for the occurrence of the "pathogenesis-related" proteins in leaves of healthy tobacco plants during flowering. Physiol Plant Pathol 19: 69—76

Fraser RSS, Loughlin SAR and Whenham RJ (1979) Acquired systemic susceptibility to infection by tobacco mosaic virus in *Nicotiana glutinosa* L. J Gen Virol 43: 131—141

Gaffney T, Friedrich L, Vernooij B, Negrotto D, Nye G, Uknes S, Ward E, Kessmann H and Ryals J (1993) Requirement of salicylic acid for the induction of systemic acquired resistance. Science 261: 754—756

Gessler C and Kuc J (1982) Induction of resistance to Fusarium wilt in cucumber by root and foliar pathogens. Phytopathology 72: 1439—1441

Gianinazzi S and Ahl P (1983) The genetic and molecular basis of b-proteins in the genus *Nicotiana*. Neth J Plant Pathol 89: 275—281

Görlach J, Volrath S, Knauf-Beiter G, Hengy G, Beckhove U, Kogel KH, Oostendorp M, Staub T, Ward E, Kessmann H and Ryals J (1996) Benzothiadiazole, a novel class of inducers of systemic acquired resistance, activates gene expression and disease resistance in wheat. Plant Cell 8: 629—643

Graham TL, Sequeira L and Huang TSR (1977) Bacterial lipopolysaccharides as inducers of disease resistance in tobacco. Appl Environ Microbiol 34: 424—432

Greenberg JT, Guo A, Klessig DF and Ausubel FM (1994) Programmed cell death in plants: a pathogen-triggered response activated coordinately with multiple defense functions. Cell 77: 551—563

Guedes MEM, Richmond S and Kuc J (1980) Induced systemic resistance to anthracnose in cucumber as influenced by the location of the inducer inoculation with *Colletotrichum lagenarium* and the onset of flowering and fruiting. Physiol Plant Pathol 17: 229—233

Hammerschmidt R and Kuc J (1982) Lignification as a mechanism for induced systemic resistance in cucumber. Physiol Plant Pathol 20: 61—71

Hammerschmidt R and Kuc J (1995) Induced Resistance to Disease in Plants. Kluwer, Dordrecht

Hammerschmidt R, Nuckles EM and Kuc J (1982) Association of enhanced peroxidase activity with induced systemic resistance of cucumber to *Colletotrichum lagenarium*. Physiol Plant Pathol 20: 73—82

Hoffland E, Pieterse CMJ, Bik L and Van Pelt JA (1995) Induced systemic resistance in radish is not associated with accumulation of pathogenesis-related proteins. Physiol Mol Plant Pathol 46: 309—320

Hoffland E, Jeger MJ and Van Beusichem ML (1998) Is plant growth rate related to disease resistance? In: Lambers H, Poorter H and Van Vuuren MMI (eds) Inherent Variation in Plant Growth: Physiological Mechanisms and Ecological Consequences, pp 409—427. Backhuys, Leiden

Holmes FO (1938) Inheritance of resistance to tobacco mosaic disease in tobacco. Phytopathology 28: 553—561

Hunt MD, Neuenschwander UH, Delaney TP, Weymann KB, Friedrich LB, Lawton KA, Steiner HY and Ryals J (1996) Recent advances in systemic acquired resistance research - a review. Gene 179: 89—95

Irving HR and Kuc J (1990) Local and systemic induction of peroxidase, chitinase and resistance in cucumber plants by potassium phosphate monobasic. Physiol Mol Plant Pathol 37: 355—366

Jenns AE and Kuc J (1979) Graft transmission of systemic resistance of cucumbers to anthracnose induced by *Colletotrichum lagenarium* and tobacco necrosis virus. Phytopathology 69: 753—756

Jenns AE and Kuc J (1980) Characteristics of anthracnose resistance induced by localized infection of cucumber with tobacco necrosis virus. Physiol Plant Pathol 17: 81—91

Kassanis B, Gianinazzi S and White RF (1974) A possible explanation of the resistance of virus-infected tobacco to second infection. J Gen Virol 23: 11—16

Kauss H, Theisinger-Hinkel E, Mindermann R and Conrath U (1992) Dichloroisonicotinic and salicylic acid, inducers of systemic acquired resistance, enhance fungal elicitor responses in parsley cells. Plant J 2: 655—660

Kehlenbeck H, Krone C, Oerke EC and Schönbeck F (1994) The effectiveness of induced resistance on yield of mildewed barley. Z Pflanzenkrankh Pflanzenschutz 101: 11—21

Kessmann H, Staub T, Ligon J, Oostendorp M and Ryals J (1994) Activation of systemic acquired disease resistance in plants. Eur J Plant Pathol 100: 359—369

Kloepper JW, Tuzun S, Liu L and Wei G (1993) Plant growth-promoting rhizobacteria as inducers of systemic disease resistance. In: Lumsden RD and Vaughn JL (eds) Pest Management: Biologically Based Technologies, pp 156—165. American Chemical Society, Washington

Knoester M (1998) The Involvement of Ethylene in Plant Disease Resistance. Ph.D. Thesis Utrecht University

Koiwa H, Bressan RA and Hasegawa PM (1997) Regulation of protease inhibitors and plant defense. Trends Plant Sci 2: 379—384

Kovats K, Binder A and Hohl HR (1991a) Cytology of induced systemic resistance of cucumber to *Colletotrichum lagenarium*. Planta 183: 484—490

Kovats K, Binder A and Hohl HR (1991b) Cytology of induced systemic resistance of tomato to *Phytophthora infestans*. Planta 183: 491—496

Kroon BAM, Scheffer RJ and Elgersma DM (1991) Induced resistance in tomato plants against Fusarium wilt invoked by *Fusarium oxysporum* f.sp. *dianthi*. Neth J Plant Pathol 97: 401—408

Kuc J (1982) Induced immunity to plant disease. Bioscience 32: 854—860

Kuc J and Richmond S (1977) Aspects of the protection of cucumber against *Colletotrichum lagenarium* by *Colletotrichum lagenarium*. Phytopathology 67: 533—536

Kunz W, Schurter R and Maetzke T (1997) The chemistry of benzothiadiazole plant activators. Pestic Sci 50: 275—282

Lawton KA, Potter SL, Uknes S and Ryals J (1994) Acquired resistance signal transduction in *Arabidopsis* is ethylene independent. Plant Cell 6: 581—588

Lawton K, Weymann K, Friedrich L, Vernooij B, Uknes S and Ryals J (1995) Systemic acquired resistance in *Arabidopsis* requires salicylic acid but not ethylene. Mol Plant-Microbe Interact 8: 863—870

Leeman M, Van Pelt JA, Den Ouden FM, Heinsbroek M, Bakker PAHM and Schippers B (1995a) Induction of systemic resistance by *Pseudomonas fluorescens* in radish cultivars differing in susceptibility to fusarium wilt, using a novel bioassay. Eur J Plant Pathol 101: 655—664

Leeman M, Van Pelt JA, Den Ouden FM, Heinsbroek M, Bakker PAHM and Schippers B (1995b) Induction of systemic resistance against fusarium wilt of radish by lipopolysaccharides of *Pseudomonas fluorescens*. Phytopathology 85: 1021—1027

Leeman M, Van Pelt JA, Hendrickx MJ, Scheffer RJ, Bakker PAHM and Schippers B (1995c) Biocontrol of Fusarium wilt of radish in commercial greenhouse trials by seed treatment with *Pseudomonas fluorescens* WCS374. Phytopathology 85: 1301—1305

Legrand M, Fritig B and Hirth L (1976) Enzymes of the phenylpropanoid pathway and the necrotic reaction of hypersensitive tobacco to tobacco mosaic virus. Phytochemistry 15: 1353—1359

Leon J, Yalpani N, Raskin I and Lawton MA (1993) Induction of benzoic acid 2-hydroxylase in virus-inoculated tobacco. Plant Physiol 103: 323—328

Linthorst HJM (1991) Pathogenesis-related proteins of plants. Crit Rev Plant Sci 10: 123—150

Liu L, Kloepper JW and Tuzun S (1995) Induction of systemic resistance in cucumber against fusarium wilt by plant growth-promoting rhizobacteria. Phytopathology 85: 695—698

Loebenstein G and Gera A (1981) Inhibitor of virus replication released from tobacco mosaic virus infected protoplasts of a local lesion-responding cultivar. Virology 114: 132—139

Loebenstein G, Rabina S and Van Praagh T (1968) Sensitivity of induced localized acquired resistance to actinomycin D. Virology 34: 264—268

Lovrekovich L, Lovrekovich H and Stahmann MA (1968) The importance of peroxidase in the wildfire disease. Phytopathology 58: 193—198

Lucas JA, Dolan TE and Coffey MD (1985) Nontransmissibility to regenerants from protected tobacco explants of induced resistance to *Peronospora hyoscyami*. Phytopathology 75: 1222—1225

Malamy J and Klessig DF (1992) Salicylic acid and plant disease resistance. Plant J 2: 643—654

Malamy J, Carr JP, Klessig DF and Raskin I (1990) Salicylic acid: a likely endogenous signal in the resistance response of tobacco to viral infection. Science 250: 1002—1004

Malamy J, Sanchez-Casas P, Hennig J, Guo A and Klessig DF (1996) Dissection of the salicylic acid signaling pathway in tobacco. Mol Plant-Microbe Interact 9: 474—482

Mauch-Mani B and Slusarenko AJ (1996) Production of salicylic acid precursors is a major function of phenylalanine ammonia-lyase in the resistance of *Arabidopsis* to *Peronospora parasitica*. Plant Cell 8: 203—212

Maurhofer M, Hase C, Meuwly P, Métraux JP and Défago G (1994) Induction of systemic resistance of tobacco to tobacco necrosis virus by the root-colonizing *Pseudomonas fluorescens* strain CHA0: Influence of the *gacA* gene and of pyoverdine production. Phytopathology 84: 139—146

McGurl B, Pearce G, Orozco-Cardenas M and Ryan CA (1992) Structure, expression, and antisense inhibition of the systemin precursor gene. Science 255: 1570—1573

McIntyre JL, Dodds JA and Hare JD (1981) Effects of localized infections of *Nicotiana tabacum* by tobacco mosaic virus on systemic resistance against diverse pathogens and an insect. Phytopathology 71: 297—301

McKinney HH (1929) Mosaic diseases in the Canary Islands, West Africa and Gibraltar. J Agric Res 39: 557—578

Métraux JP, Signer H, Ryals J, Ward E, Wyss-Benz M, Gaudin J, Raschdorf K, Schmid E, Blum W and Inverardi B (1990) Increase in salicylic acid at the onset of systemic acquired resistance in cucumber. Science 250: 1004—1006

Mölders W, Buchala A and Métraux JP (1996) Transport of salicylic acid in tobacco necrosis virus-infected cucumber plants. Plant Physiol 112: 787—792

Mur LAJ, Naylor G, Warner SAJ, Sugars JM, White RF and Draper J (1996) Salicylic acid potentiates defence gene expression in tissue exhibiting acquired resistance to pathogen attack. Plant J 9: 559—571

Neuenschwander U, Vernooij B, Friedrich L, Uknes S, Kessmann H and Ryals J (1995) Is hydrogen peroxide a second messenger of salicylic acid in systemic acquired resistance? Plant J 8: 227—233

O'Donnell PJ, Calvert C, Atzorn R, Wasternack C, Leyser HMO and Bowles DJ (1996) Ethylene as a signal mediating the wound response of tomato plants. Science 274: 1914—1917

Otsuki Y, Shimomura T and Takebe I (1972) Tobacco mosaic virus multiplication and expression of the *N* gene in necrotic responding tobacco varieties. Virology 50: 45—50

Pallas JA, Paiva NL, Lamb C and Dixon RA (1996) Tobacco plants epigenetically suppressed in phenylalanine ammonia-lyase expression do not develop systemic acquired resistance in response to infection by tobacco mosaic virus. Plant J 10: 281—293

Palukaitis P and Zaitlin M (1984) A model to explain the "cross-protection" phenomenon shown by plant viruses and viroids. In: Kosuge T and Nester EW (eds) Plant-Microbe Interactions, Molecular and Genetic Perspectives, Vol 1, pp 420—429. MacMillan, New York

Pan SQ, Ye XS and Kuc J (1992) Induction of chitinases in tobacco plants systemically protected against blue mold by *Peronospora tabacina* or tobacco mosaic virus. Phytopathology 82: 119—123

Pearce G, Strydom D, Johnson S and Ryan CA (1991) A polypeptide from tomato leaves induces wound-inducible proteinase inhibitor proteins. Science 253: 895—898

Pena-Cortez H, Sanchez-Serrano JJ, Mertens R and Willmitzer L (1989) Abscisic acid is involved in the wound-induced expression of the proteinase inhibitor II gene in potato and tomato. Proc Natl Acad

Sci USA 86: 9851—9855

Pena-Cortez H, Prat S, Atzorn R, Wasternack C and Willmitzer L (1996) Abscisic acid-deficient plants do not accumulate proteinase inhibitor II following systemin treatment. Planta 198: 447—451

Pennazio S, Roggero P and Lenzi R (1983) Some characteristics of the hypersensitive reaction of White Burley tobacco to tobacco necrosis virus. Physiol Plant Pathol 22: 347—355

Penninckx IAMA, Eggermont K, Terras FRG, Thomma BPHJ, De Samblanx GW, Buchala A, Métraux JP, Manners JM and Broekaert WF (1996) Pathogen-induced systemic activation of a plant defensin gene in *Arabidopsis* follows a salicylic acid-independent pathway. Plant Cell 8: 2309—2323

Pieterse CMJ, Van Wees SCM, Hoffland E, Van Pelt JA and Van Loon LC (1996) Systemic resistance in *Arabidopsis* induced by biocontrol bacteria is independent of salicylic acid accumulation and pathogenesis-related gene expression. Plant Cell 8: 1225—1237

Pieterse CMJ, Van Wees SCM, Van Pelt JA, Knoester M, Laan R, Gerrits H, Weisbeek PJ and Van Loon LC (1998) A novel signaling pathway controlling induced disease resistance in plants. Plant Cell: in press

Pritchard DW and Ross AF (1975) The relationship of ethylene to formation of tobacco mosaic virus lesions in hypersensitive responding tobacco leaves with and without induced resistance. Virology 64: 295—307

Raskin I, Ehmann A, Melander WR and Meeuse BJD (1987) Salicylic acid: a natural inducer of heat production in Arum lilies. Science 237: 1601—1602

Rasmussen JB, Hammerschmidt R and Zook MN (1991) Systemic induction of salicylic acid accumulation in cucumber after inoculation with *Pseudomonas syringae* pv. *syringae*. Plant Physiol 97: 1342—1347

Register III JC and Beachy RN (1988) Resistance to TMV in transgenic plants results from interference with an early event in infection. Virology 166: 524—532

Ricci P, Bonnet P, Huet JC, Sallantin M, Beauvais-Cante F, Bruneteau M, Billard V, Michel G and Pernollet JC (1989) Structure and activity of proteins from pathogenic fungi *Phytophthora* eliciting necrosis and acquired resistance in tobacco. Eur J Biochem 183: 555—563

Roberts DA (1982) Systemic acquired resistance induced in hypersensitive plants by nonnecrotic localized viral infections. Virology 122: 207—209

Roberts DA (1983) Acquired resistance to tobacco mosaic virus transmitted to the progeny of hypersensitive tobacco. Virology 124: 161—163

Ross AF (1961a) Localized acquired resistance to plant virus infection in hypersensitive hosts. Virology 14: 329—339

Ross AF (1961b) Systemic acquired resistance induced by localized virus infections in plants. Virology 14: 340—358

Ross AF (1966) Systemic effects of local lesion formation. In: Beemster ABR and Dijkstra J (eds) Viruses of Plants, pp 127—150. North Holland, Amsterdam

Ross AF and Pritchard DW (1972) Local and systemic effects of ethylene on tobacco mosaic virus lesions in tobacco. Phytopathology 62: 786

Ryals JA, Neuenschwander UH, Willits MG, Molina A, Steiner HY and Hunt MD (1996) Systemic acquired resistance. Plant Cell 8: 1809—1819

Ryals J, Weymann K, Lawton K, Friedrich L, Ellis D, Steiner HY, Johnson J, Delaney TP, Jesse T, Vos P and Uknes S (1997) The *Arabidopsis NIM1* protein shows homology to the mammalian transcription factor inhibitor IκB. Plant Cell 9: 425—439

Ryan CA (1990) Proteinase inhibitors in plants: genes for improving defenses against insects and pathogens. Annu Rev Phytopathol 28: 425—449

Ryan CA (1992) The search for the proteinase inhibitor-inducing factor, PIIF. Plant Mol Biol 19: 123—133

Schippers B, Bakker AW and Bakker PAHM (1987) Interactions of deleterious and beneficial rhizosphere microorganisms and the effect of cropping practices. Annu Rev Phytopathol 25: 339—358

Schneider M, Schweizer P, Meuwly P and Métraux JP (1996) Systemic acquired resistance in plants. Int Rev Cytol 168: 303—340

Schönbeck F, Grunewaldt-Stöcker G and Wilde T (1987) Zur Auswirkung von Resistenzreaktionen auf Pflanzenwachstum und Ertrag in inkompatiblen Wirt-Parasit-Systemen. J Phytopathol 118: 32—49

Sequeira L (1983) Mechanisms of induced resistance in plants. Annu Rev Microbiol 37: 51—79

Shah J, Tsui F and Klessig DF (1997) Characterization of a salicylic acid-insensitive mutant (*sail*) of *Arabidopsis thaliana*, identified in a selective screen utilizing the SA-inducible expression of the *tms2* gene. Mol Plant-Microbe Interact 10: 69—78

Shimomura T (1979) Stimulation of callose synthesis in the leaves of Samsun NN tobacco showing systemic acquired resistance to tobacco mosaic virus. Ann Phytopathol Soc Japan 45: 299—404

Shirasu K, Nakajima H, Rajasekhar VK, Dixon RA and Lamb C (1997) Salicylic acid potentiates an agonist-dependent gain control that amplifies pathogen signals in the activation of defense mechanisms. Plant Cell 9: 261-270

Shulaev V, Leon J and Raskin I (1995) Is salicylic acid a transported signal of systemic acquired resistance in tobacco? Plant Cell 7: 1691—1701

Shulaev V, Silverman P and Raskin I (1997) Airborne signalling by methyl salicylate in plant pathogen resistance. Nature 385: 718—721

Simons TJ and Ross AF (1970) Enhanced peroxidase activity associated with induction of resistance to tobacco mosaic virus in hypersensitive tobacco. Phytopathology 60: 383—384

Simons TJ and Ross AF (1971) Changes in phenol metabolism associated with induced systemic resistance to tobacco mosaic virus in Samsun NN tobacco. Phytopathology 61: 1261-1265

Spiegel S, Gera A, Salomon R, Ahl P, Harlap S and Loebenstein G (1989) Recovery of an inhibitor of virus replication from the intercellular fluid of hypersensitive tobacco infected with tobacco mosaic virus and from uninfected induced-resistant tissue. Phytopathology 79: 258—262

Steiner U and Schönbeck F (1995) Induced disease resistance in monocots. In: Hammerschmidt R and Kuc J (eds) Induced Resistance to Disease in Plants, pp 86—110. Kluwer, Dordrecht

Sticher L, Mauch-Mani B and Métraux JP (1997) Systemic acquired resistance. Annu Rev Phytopathol 35: 235—270

Stolle K, Zook M, Shain L, Hebard F and Kuc J (1988) Restricted colonization by *Peronospora tabacina* and phytoalexin accumulation in immunized tobacco leaves. Phytopathology 78: 1193—1197

Takahashi T (1974) Studies on viral pathogenesis in host plants VI. The rate of primary lesion growth in the leaves of "Samsun NN" tobacco to tobacco mosaic virus. Phytopathol Z 79: 53—66

Terras FRG, Eggermont K, Kolaleva V, Raikhel NV, Osborn RW, Kester A, Rees SB, Torrekens S, Van Leuven F, Vanderleyden J, Cammue BPA and Broekaert WF (1995) Small cysteine-rich antifungal proteins from radish: their role in host defense. Plant Cell 7: 573—588

Thomma BPHJ, Eggermont K, Penninckx IAMA, Mauch-Mani B, Vogelsang R, Cammue BPA and Broekaert WF (1998) Separate jasmonate-dependent and salicylate-dependent defense response pathways in Arabidopsis are essential for resistance to distinct microbialpathogens. Proc Natl Acad Sci USA 95: 15107—15111

Thulke O and Conrath U (1998) Salicylic acid has a dual role in the activation of defence-related genes in parsley. Plant J 14: 35—42

Ton J, Pieterse CMJ and Van Loon LC (1999) Identification of a locus in Arabidopsis controlling both the expression of rhizobacteria-mediated induced systemic resistance (ISR) and basal resistance against *Pseudomonas syringae* pv. tomato. Mol Plant-Microbe Interact 12: 911—918

Turlings TCJ, Tumlinson JH and Lewis WJ (1990) Exploitation of herbivore-induced plant odors by host-seeking parasitic wasps. Science 250: 1251—1253

Tuzun S and Kloepper J (1995) Practical application and implementation of induced resistance. In: Hammerschmidt R and Kuc J (eds) Induced Resistance to Disease in Plants, pp 152-168. Kluwer, Dordrecht

Tuzun S and Kuc J (1985) A modified technique for inducing systemic resistance to blue mold and increasing growth in tobacco. Phytopathology 75: 1127—1129

Tuzun S and Kuc J (1987) Persistence of induced systemic resistance to blue mold in tobacco plants derived via tissue culture. Phytopathology 77: 1032—1035

Uknes S, Mauch-Mani B, Moyer M, Potter S, Williams S, Dincher S, Chandler D, Slusarenko A, Ward E and Ryals J (1992) Acquired resistance in *Arabidopsis*. Plant Cell 4: 645—656

Van Loon LC (1975) Polyacrylamide disc electrophoresis of the soluble leaf proteins from *Nicotiana tabacum* var. Samsun' and Samsun NN' IV. Similarity of qualitative changes of specific proteins after infection with different viruses and their relationship to acquired resistance. Virology 67: 566—575

Van Loon LC (1976) Systemic acquired resistance, peroxidase activity, and lesion size in tobacco reacting hypersensitively to tobacco mosaic virus. Physiol Plant Pathol 8: 231—242

Van Loon LC (1977) Induction by 2-chloroethylphosphonic acid of viral-like lesions, associated proteins, and systemic resistance in tobacco. Virology 80: 417—420

Van Loon LC (1983a) Mechanisms of resistance in virus-infected plants. In: Bailey JA and Deverall BJ (eds) The Dynamics of Host Defence, pp 123—190. Academic Press Australia, North Ryde

Van Loon LC (1983b) The induction of pathogenesis-related proteins by pathogens and specific chemicals. Neth J Plant Pathol 89: 265—273

Van Loon LC (1986) The significance of changes in peroxidase in diseased plants. In: Greppin H, Penel C and Gaspar T (eds) Molecular and Physiological Aspects of Plant Peroxidases, pp 405—418. Univ of Geneva, Geneva

Van Loon LC (1989) Stress proteins in infected plants. In: Kosuge T and Nester EW (eds) Plant-Microbe Interactions, Molecular and Genetic Perspectives, Vol 3, pp 198—237. MacGraw-Hill, New York

Van Loon LC (1997) Induced resistance in plants and the role of pathogenesis-related proteins. Eur J Plant Pathol 103: 753—765

Van Loon LC and Antoniw JF (1982) Comparison of the effects of salicylic acid and ethephon with virus-induced hypersensitivity and acquired resistance in tobacco. Neth J Plant Pathol 88: 237—256

Van Loon LC and Geelen JLMC (1971) The relation of polyphenoloxidase and peroxidase to symptom expression in tobacco var. "Samsun NN" after infection with tobacco mosaic virus. Acta Phytopathol Acad Sci Hung 6: 9—20

Van Loon LC and Pennings GGH (1993) Involvement of ethylene in the induction of systemic acquired resistance in tobacco. In: Fritig B and Legrand M (eds) Mechanisms of Plant Defense Responses, pp. 156—159. Kluwer, Dordrecht

Van Loon LC and Van Kammen A (1970) Polyacrylamide disc electrophoresis of the soluble leaf proteins from *Nicotiana tabacum* var. Samsun' and Samsun NN' II. Changes in protein constitution after infection with tobacco mosaic virus. Virology 40: 199—211

Van Loon LC, Pierpoint WS, Boller T and Conejero V (1994) Recommendations for naming plant pathogenesis-related proteins. Plant Mol Biol Reporter 12: 245—264

Van Loon LC, Bakker PAHM and Pieterse CMJ (1997) Mechanisms of PGPR-induced resistance against pathogens. In: Ogoshi A, Kobayashi K, Homma Y, Kodama F, Kondo N and Akino S (eds) Plant

Growth-Promoting Rhizobacteria - Present Status and Future Prospects, pp 50—57. Faculty of Agriculture, Hokkaido University, Sapporo

Van Loon LC, Bakker PAHM and Pieterse CMJ (1998) Systemic resistance induced by rhizosphere bacteria. Annu Rev Phytopathol 36: 453—483

Van Peer R, Niemann GJ and Schippers B (1991) Induced resistance and phytoalexin accumulation in biological control of fusarium wilt of carnation by *Pseudomonas* sp. strain WCS417r. Phytopathology 81: 728—734

Van Wees SCM, Pieterse CMJ, Trijssenaar A, Van 't Westende Y, Hartog F and Van Loon LC (1997) Differential induction of systemic resistance in *Arabidopsis* by biocontrol bacteria. Mol Plant-Microbe Interact 10: 716—724

Vernooij B, Friedrich L, Morse A, Reist R, Kolditz-Jawhar R, Ward E, Uknes S, Kessmann H and Ryals J (1994) Salicylic acid is not the translocated signal responsible for inducing systemic acquired resistance but is required in signal transduction. Plant Cell 6: 959—965

Vidal S, Eriksson ARB, Montesano M, Denecke J and Palva ET (1998) Cell wall-degrading enzymes from *Erwinia carotovora* cooperate in the salicylic acid-independent induction of a plant defense response. Mol Plant-Microbe Interact 11: 23—32

Ward ER, Uknes SJ, Williams SC, Dincher SS, Wiederhold DL, Alexander DC, Ahl-Goy P, Métraux JP and Ryals JA (1991) Coordinate gene activity in response to agents that induce systemic acquired resistance. Plant Cell 3: 1085—1094

Wasternack C and Parthier B (1997) Jasmonate-signalled plant gene expression. Trends Plant Sci 2: 302—307

Wei G, Kloepper JW and Tuzun S (1996) Induced systemic resistance to cucumber diseases and increased plant growth by plant growth-promoting rhizobacteria under field conditions. Phytopathology 86: 221—224

Weststeijn EA (1981) Lesion growth and virus localization in leaves of *Nicotiana tabacum* cv. Xanthi nc. after inoculation with tobacco mosaic virus and incubation alternately at 22°C and 32°C. Physiol Plant Pathol 18: 357—368

Weymann K, Hunt M, Uknes S, Neuenschwander U, Lawton K, Steiner HY and Ryals J (1995) Suppression and restoration of lesion formation in *Arabidopsis lsd* mutants. Plant Cell 7: 2013—2022

Whenham RJ and Fraser RSS (1981) Effect of systemic and local-lesion-forming strains of tobacco mosaic virus on abscisic acid concentration in tobacco leaves: consequences for the control of leaf growth. Physiol Plant Pathol 18: 267—278

White RF (1979) Acetylsalicylic acid (aspirin) induces resistance to tobacco mosaic virus in tobacco. Virology 99: 410—412

Wildon DC, Thain JF, Minchin PEH, Gubb IR, Reilly AJ, Skipper YD, Doherty HM, O'Donnell PJ and Bowles DJ (1992) Electrical signalling and systemic proteinase inhibitor induction in the wounded plant. Nature 360: 62—65

Wilson TMA (1983) Pathogenesis-related-protein synthesis in selected cultivars of beans and cowpeas following leaf damage by carborundum, treatment with aspirin, infection with tobacco mosaic virus, or with the bean or cowpea strain of southern bean mosaic virus. Neth J Plant Pathol 89: 313—317

Xu Y, Chang PFL, Liu D, Narasimhan ML, Raghothama KG, Hasegawa PM and Bressan RA (1994) Plant defense genes are synergistically induced by ethylene and methyl jasmonate. Plant Cell 6: 1077—1085

Yalpani N, Silverman P, Wilson TMA, Kleier DA and Raskin I (1991) Salicylic acid is a systemic signal and an inducer of pathogenesis-related proteins in virus-infected tobacco. Plant Cell 3: 809—818

Yalpani N, Leon J, Lawton M and Raskin I (1993) Pathway of salicylic acid biosynthesis in healthy and virus-inoculated tobacco. Plant Physiol 103: 315—321

Ye XS, Pan SQ and Kuc J (1990) Activity, isozyme pattern, and cellular localization of peroxidase as related to systemic resistance of tobacco to blue mold (*Peronospora tabacina*) and to tobacco mosaic virus. Phytopathology 80: 1295—1299

Ye XS, Järlfors U, Tuzun S, Pan SQ and Kuc J (1992) Biochemical changes in cell walls and cellular responses of tobacco leaves related to systemic resistance to blue mold (*Peronospora tabacina*) induced by tobacco mosaic virus. Can J Bot 70: 49—57

Zhang S and Klessig DF (1997) Salicylic acid activates a 48 kilodalton MAP kinase in tobacco. Plant Cell 9: 809—824

CHAPTER 11

TRANSGENIC APPROACHES TO CONTROL EPIDEMIC SPREAD OF DISEASES

BEN J.C. CORNELISSEN & ANDRÉ SCHRAM

Faculty of Biology, University of Amsterdam
Kruislaan 318, 1098 SM Amsterdam, the Netherlands
(cornelissen@bio.uva.nl aschram@bio.uva.nl)

Summary

To feed the growing world population in the future yield and quality of crops need to be enhanced drastically. One way to increase yield is to minimize losses due pathogen infections. Traditional approaches to control epidemic spread of diseases are no longer sufficient and hence the development of pathogen resistance traits has become an important target in plant biotechnology. Using molecular techniques various natural disease resistance genes have been isolated during the last five years. However, their use in molecular breeding programs is limited since they code for resistance to one specific race of a pathogen only. To engineer broad spectrum resistance traits very different strategies are being pursued. The first concept for virus resistance implied the constitutive expression of viral coat protein genes in transgenic plants. Engineered resistance based on this concept is documented very well and the first product (virus resistant squash) is about to enter the market in the USA. In the mean time it has become clear that expression of other viral sequences as well can bring about resistance. Published strategies for the engineering of bacterium resistance are limited in number and as yet not successful. The most wide-spread approach for fungus resistance is the expression of genes encoding proteins inhibiting fungus growth. Many of these proteins appear to act synergistically both in vitro and in planta. First results of field trials with plants expressing antifungal proteins indicate that levels of resistance are high enough to be commercially interesting. In the last few years strategies for fungus resistance have been explored based on the induction by pathogens of cell death at the site of infection. One of the concepts being pursued successfully to engineer nematode resistance implies the production in transgenic plants of compounds that directly affect nematode development. An alternative strategy aims at disruption of nematode feeding structures in the plant.

A. Slusarenko, R. Fraser, and L.C. van Loon (eds), Mechanisms of Resistance to Plant Diseases, 575-599.
© 2000 *Kluwer Academic Publishers. Printed in the Netherlands.*

I.   Introduction

One of the main challenges of the first half of the next millenium will be the production of food and feed in quantities large enough to feed the world population. This population continues to increase and could even double within the next decades. However, global agricultural acreage will not grow much further. To feed the entire world population in the future there is a clear need to improve yield and quality of crops. This may be accomplished by the development of cultivars of crops better adapted to the conditions they are grown in, and by improved control of epidemic spread of diseases. Plant diseases due to pathogen infections cause a yearly loss of up to 12% of agricultural production world wide (James *et al.*, 1991) and have negative effects on the quality of food as well. Losses occur despite the use of disease resistant cultivars of crops, large scale application of various husbandary techniques such as crop rotation, and application of increasing quantities of agrochemicals. As to the latter, without the use of pesticides the average current agricultural production would decrease by more than 50% (De Wit and Van Vloten-Doting, 1993). To give better control of epidemic spread of diseases, new pathogen resistant cultivars must be developed continously. Durable resistance traits are often polygenic and breeding for such traits is very time-consuming and difficult. Naturally occurring, monogenic resistance traits are abundantly known from conventional breeding programmes. Often they are race-specific and relatively easy to breed for. However, their use is limited by the specificity of the trait and by the relative ease with which they are overcome by the evolution of new races of the pathogen. In addition, for some diseases monogenic resistance traits are not available at all. As to the use of agrochmicals, many have been found to be harmful to environment and consumer, and people are increasingly becoming concerned about their continuing use. Therefore, new strategies are being developed to protect crops from diseases caused by pathogens. One of the new approaches currently being worked out relates to biological control (recently reviewed by Handelsman and Stabb, 1996). Another novel approach is the application in breeding programmes of molecular markers such as restriction length polymorphism (RFLP) markers, random amplified polymorphic DNA (RAPD) markers and amplified fragment length polymorphism (AFLP) markers. Using these markers linkage maps can be created, allowing breeders to select for a specific trait, either mono- or polygenic, in a fast and easy way. Yet another strategy has emanated from the development of transformation and regeneration technologies for important food crop species such as soybean, rice, wheat, maize and cassava. The state-of-the-art technology allows the transfer of traits into plants without altering their intrinsic properties. Here we will review emerging strategies using transgenic plants to control diseases caused by pathogens. These strategies include both the manipulation of natural disease resistance genes and the engineering of new traits to control epidemic spread of virus, bacterium, fungus and/or nematode diseases.

II. Manipulating Natural Disease Resistance Genes

The naturally occurring pathogen resistance genes most frequently employed in breeding programs are dominant and monogenic. Usually they protect plants against one or a few races of a pathogen species only. Often this so-called race-specific resistance is manifested in a hypersensitive response which is characterized by fast, localized necrosis at the site of infection. As a result the pathogen is contained within the region immediately surrounding the infection site and spread to non-infected parts of the plant is prevented. Genetically race-specific resistance is explained by the gene-for-gene hypothesis. This was put forward by Flor (1956, 1971) to explain the results of his genetic studies on the interaction between flax (*Linum usitatissimum*) and the rust fungus *Melampsora lini*. According to Flor the presence of a dominant resistance (*R*) gene in the plant together with a corresponding, dominant avirulence (*Avr*) gene in the invading pathogen results in resistance. To explain the gene-for-gene hypothesis biochemically and physiologically it has been proposed that activation of the signalling pathway leading to resistance is triggered by a specific recognition of a pathogen-derived ligand by a plant receptor. In this model the ligand and the receptor are encoded by a pathogen *Avr* gene and its corresponding *R* gene, respectively. Since Flor proposed his model many plant-pathogen interactions fitting the gene-for-gene model have been characterized genetically. However, to date only a few have been investigated at the biochemical or physiological levels.

To clone monogenic resistance genes various approaches have been, and are still being pursued, including map-based cloning and transposon tagging. Using a transposon tagging approach Johal and Briggs (1992) were the first to isolate a pathogen resistance gene, notably the genetically defined *Hm*1 locus in maize conferring resistance to race 1 of the fungus *Cochliobolus carbonum*. *Hm*1 encodes an NADPH-dependent reductase capable of reducing a small cyclic tetrapeptide, called HC-toxin. This pathogen-produced toxin mediates the specific pathogenicity of *C. carbonum* race 1 on maize. Since the isolation of *Hm*1 about 20 *R* genes have been cloned (Table 1). Except for *Mlo* they all are dominant and code for race-specific resistances. The recessive *Mlo* gene from barley codes for a non-race-specific resistance to powdery mildew (Büschges *et al.*, 1997). A general feature of the products of race-specific *R* genes, except Pto, is the presence of leucine-rich repeat (LRRs) motifs, which are believed to be involved in protein-protein interactions (Kobe and Deisenhofer, 1995); *Pto* confers resistance to *Pseudomonas syringae* pv *tomato* in tomato and encodes a protein kinase (Martin *et al.*, 1993). Computer analysis suggest that the products of some *R* genes (*Cf*-2,4,5 and 9, *Xa*-21 and *Hs*1[pro-1]) are transmembrane proteins with both an extra- and an intracellular domain, whereas the products of the others are cytoplasmatic proteins. The LRRs in the transmembrane proteins are located in the extracellular domains. This location coincides with the location of the corresponding *Avr* gene products. The predicted location of the bacterium race-specific resistance gene products is the cytoplasm into which the corresponding specific elicitors are delivered directly (reviewed by Bonas and Van den Ackerveken, 1997). All these observations are in line with the gene-for-gene model. However, a direct interaction between a specific elicitor encoded by a pathogen *Avr* gene, and the product of the corresponding *R* gene, has been shown for AvrPto and

Pto only (Tang *et al.*, 1996). For extensive reviews on *R* gene and their functions the reader is referred to Bent (1996), Jones and Jones (1996), Hammond-Kosack and Jones (1997) and Parker and Coleman (1997).

Table 1.   Cloned resistance genes

	gene	plant species	pathogen	references
virus	*N*	tobacco	tobacco mosaic virus	Whitham *et al.*, 1994
bacterium	*Pto*	tomato	*Pseudomonas syringae* pv *tomato*	Martin *et al.*, 1993
	Prf	tomato	*Pseudomonas syringae* pv *tomato*	Salmeron *et al.*, 1996
	RPS2	Arabidopsis	*Pseudomonas syringae* pathovars expressing AvrRpt2[1]	Bent *et al.*, 1994; Mindrinos *et al.*, 1994
	RPM1 (*RPS3*)[2]	Arabidopsis	*Pseudomonas syringae* pathovars expressing AvrRpm1 or AvrB[3]	Grant *et al.*, 1995
	Xa-21	rice	*Xanthomonas oryzae* pv *oryzae*	Song *et al.*, 1995
	Xa-1	rice	*Xanthomonas oryzae* pv *oryzae*	Yoshimura *et al.*, 1997
fungus	*Cf-2*	tomato	*Cladosporium fulvum*	Dixon *et al.*, 1996
	Cf-4	tomato	*Cladosporium fulvum*	In: Jones and Jones, 1996
	Cf-5	tomato	*Cladosporium fulvum*	In: Hammond-Kosack and Jones, 1997
	Cf-9	tomato	*Cladosporium fulvum*	Jones *et al.*, 1994
	I2	tomato	*Fusarium oxysporum* f.sp. *lycopersici*	Ori *et al.*, 1997; Simons *et al.*, 1998
	L6	flax	*Melampsora lini*	Lawrence *et al.*, 1995
	M	flax	*Melampsora lini*	Anderson *et al.*, 1997
	RPP5	Arabidopsis	*Perenospora parasitica*	Parker *et al.*, 1997
	*Hm*1	maize	*Cochliobolus carbonum*	Johal and Briggs, 1992
	Mlo	barley	*Erysiphe graminis* f.sp. *hordei*	Büschges *et al.*, 1997
nematode	*Hs1*[pro-1]	sugar beet	*Heterodera schachtii*	Cai *et al.*, 1997
	Cre3	wheat	*Heterodera avenae*	Lagudah *et al.*, 1997
	Mi	tomato	*Meloidogyne incognita*	Simons, 1997

[1] Originally *AvrRpt2* was isolated from *P. syringae* pv *tomato*; [2] *Rpm1* and *RPS3* are identical genes (Bisgrove *et al.*, 1994); [3] Originally *AvrRpm1* was isolated from *P. syringae* pv *maculicola and AvrB* from *P. syringae* pv *glycinea*; the products of the two genes are not similar; nevertheless *P. syringae* pathovars expressing either gene show incompatible interactions with *A. thaliana* strains containing resistance gene *RPM1*.

Given the availability of techniques for genetic transformation of crops, the transfer of cloned resistance genes to other cultivars of the same species lacking the trait is an obvious experiment. As a matter of fact, often a putative *R* gene from a given plant species is transferred to another cultivar of the same species to prove that an active *R* gene has been cloned rather than a homolog. One step further is the transfer of an *R* gene into another plant species. Provided that downstream signalling leading to resistance is possible, *R* genes may function in heterologous systems. Indeed, the transfer of *Pto* from tomato to tobacco results in resistance to *Pseudomonas syringae* pv *tabaci* expressing *avrPto* (Thilmony *et al.*, 1995). Also, *N*-gene mediated resistance to tobacco mosaic virus has been transferred successfully from tobacco to tomato (Whitham *et al.*, 1996). However, transfer of *Pto* from tomato to *Arabidopsis* and *RPS2* from *Arabidopsis* to tomato has not resulted in expanded resistance in the target plants (Bent, 1996). Apparently, for successful functioning of *R* genes in heterologous systems a close genetic relation of donor and acceptor species is required. Another drawback of transfer of naturally occurring *R* genes is the fact that the resulting transgenic plants will remain susceptible to races of the pathogen that do not contain the corresponding *Avr* gene. In addition, the wide scale growing of the newly bred pathogen-resistant cultivars is likely to lead to the evolution of new virulent races, and the cultivation of crops as monocultures will promote rapid epidemic outbreaks. Thus the use of resistance genes involved in race-specific interactions in the breeding for disease resistance is limited by the specificity of the traits and by the relative ease with which they are overcome by new races of fungal pathogens. However, with growing knowledge on the structure and function of *R* genes it may become feasible to design new resistance genes with lower specificities. Furthermore, naturally occurring *R* genes may be used in newly designed systems for resistance. De Wit (1992, 1995) proposed a two-component sensor system that provides transgenic crops with a potentially durable disease resistance directed towards a broad spectrum of pathogens. The key feature of this system is the presence of both an *R* gene and its corresponding *Avr* gene in one plant. Control of expression of either gene by a strictly pathogen-inducible promoter will result in the induction of *R* gene-mediated resistance reactions upon pathogen attack, provided the downstream signalling pathway required is present. Elaborating this system, Honée *et al.* (1997) introduced the gene encoding the AVR9 elicitor from *Cladosporium fulvum* under control of the *gst*1 promoter into tomato plants carrying the corresponding resistance gene (*Cf*-9). The *gst*1 promoter of potato directs the pathogen-inducible expression of a glutathione S-transferase gene (Strittmatter *et al.*, 1996). Transgenic plants were inoculated with races of *C. fulvum* lacking the *Avr*9 gene and were found to be resistant.

Studies such as those described above will rapidly lead to enhanced knowledge of the functioning of genes recognised by breeders as responsible for resistance. Identification of the products of *R* genes is a first step in the identification of components of the signal transduction cascade leading to defence reactions in the plant. Although used successfully in a number of cases it is questionable whether *R* genes as such will represent the preferred genetic tools in engineering resistance. Rather it is to be expected that insight into natural mechanisms of plant defence will allow the design of alternative strategies aimed at durable, broad spectrum pathogen resistance.

III.   Engineering Disease Resistance Traits

Disease resistance strategies involving the manipulation of naturally occurring resistance genes have emerged only recently. In contrast, the first reports on the engineering of disease resistance traits not based on natural resistance genes date back to the mid-eighties. At that time the first example of a transgenic approach obtaining virus resistant plants was published. Since then many papers have been published on the engineering of virus resistance as well as on how to enhance fungus resistance in plants. The numbers of papers on the engineering of bacterium and nematode resistance are still limited.

A.   APPROACHES TO ENGINEER VIRUS RESISTANCE

In plant virology cross protection is the phenomenon where systemic infection of a plant with a mild strain of a virus protects it from the effects of a superinfection with a severe strain of the same virus. The specificity of this resistance suggested that gene products encoded by the mild virus are responsible for the infection. This suggestion inspired Beachy and coworkers (Powell Abel *et al.*, 1986) to express the coat protein gene of tobacco mosaic virus (TMV) in transgenic tobacco. Upon challenge inoculation the transgenic plants were found to be resistant to TMV. Since then, many more examples have been published of virus resistance in plants brought about by the expression of coat protein genes (for recent reviews see Lomonossoff, 1995; Baulcombe, 1996). This so-called coat protein-mediated resistance is directed towards the virus for which the coat protein gene is expressed and to closely related viruses only. However, resistance can be broadened by the introduction into crops of different virus coat protein genes. In this way resistance has been engineered to mixed infections by two (Lawson *et al.*, 1990; Fuchs and Gonsalves, 1995) and three (Tricoli *et al.*, 1995) unrelated viruses. Application of coat protein-mediated resistance has led to the development of commercially interesting virus resistant potato and squash cultivars (see section IV).

Initially all reports on resistance brought about by the expression of viral coat protein genes were in line with a role of the protein itself, most likely through inhibition of virion disassembly in infected cells. However, in later years conflicting data were published. For example, in the cases of cucumber mosaic virus, potato virus X and potato leafroll virus (PLRV), RNA expression of coat protein genes in antisense orientation in transgenic plants was shown to result in protection against the virus as well (Cuozzo *et al.*, 1988; Hemenway *et al.*, 1998; Van der Wilk *et al.*, 1991; Kawchuk *et al.*, 1991). Also, resistance to PLRV was observed in coat protein gene transgenic potato plants in which the viral coat protein could not be detected, although the presence of its mRNA was demonstrated convincingly (Van der Wilk *et al.*, 1991; Kawchuk *et al.*, 1991). These observations were the first in a series accumulated over the years leading to the general conclusion that expression of viral sequences, either coding or non-coding, in transgenic plants often results in resistance to the virus from which the expressed sequences are derived. Although in many cases the mechanism underlying this so-called pathogen-derived resistance to viruses is still unclear, it seems as if in at least a number of cases resistance can be explained by gene silencing (reviewed by Baulcombe, 1996).

An alternative strategy to attenuate virus infection is the inactivation by antibodies of virally encoded proteins in the infected plant cell. Tavladoraki *et al.* (1993) engineered a gene encoding a cytoplasmic, single chain Fv antibody directed towards the coat protein of artichoke mottled crinkle virus. The constitutive expression of this gene in transgenic *Nicotiana benthamiana* plants reduced virus accumulation and delayed symptom development. After challenge inoculation with TMV similar results were obtained in transgenic tobacco plants expressing TMV-specific antibodies (Voss *et al.*, 1995). Although complete resistance has not yet been achieved using this approach alone, the strategy may be useful in combination with others.

The long list of publications reporting success in the engineering of virus-resistance by the introduction of viral sequences in the plant genome clearly indicates the strength of the technology. In addition, the broad applicability of the various forms of pathogen-derived resistance makes this technology a serious tool in the control of epidemic spread of virus diseases.

B.   APPROACHES TO ENGINEER BACTERIUM RESISTANCE

Thus far only a few papers have been published on the engineering of bacterium resistance in plants. In one of these the expression of a bacteriophage T4 lysozyme gene in transgenic potato was described (Düring *et al.*, 1993). Bacteriophage T4 lysozyme is the most active member of a class of bacteriolytic enzymes; similar enzymes occur in the plant kingdom (Tsugita, 1971). In the transgenic potato plants lysozyme was secreted into the intercellular spaces, the site of entry and spread of the bacterium *Erwinia carotovora*. Although expression levels of the transgene were found to be very low, the plants appeared to be less susceptible towards *E. carotovora atroseptica* infection than control plants.

Thionins are small, cystein-rich polypeptides occurring in various organs of a some plant species. They exhibit antimicrobial activities in vitro against bacteria and fungi (reviewed in Florack and Stiekema, 1994). The antimicrobial action is thought to be based on the ability of thionins to form pores in cell membranes, resulting in membrane disruption and cell death. In one paper, expression of a barley thionin gene in transgenic tobacco was reported to enhance resistance to phytopathogenic bacterium (Carmona *et al.*, 1993). However, this result could not be confirmed by others (Florack *et al.*, 1994).

Reactive oxygen species (ROS) such as as H_2O_2 and O_2^- are believed to play important roles in defence responses of plants to pathogens. They are produced during the oxidative burst, one of the earliest events in incompatible plant pathogen interactions (Wu *et al.*, 1995, and references therein). ROS are required for cross-linking of hydroxyproline-rich cell wall proteins and might be involved in some other processes related to the hypersensitive response. Glucose oxidase (GO) is an enzyme occurring in some bacteria and fungi and catalyses the oxidation of glucose, thereby producing H_2O_2 and glucuronic acid. Wu *et al.* (1995) introduced a GO gene from *Aspergillus niger* into potato plants. Transgenic tubers showed increased levels of H_2O_2 and were found to be less susceptible to *E. carotovora* subsp. *carotovora*, the causal agent of bacterial soft rot disease. Whether this non-specific resistance is due to the direct antimicrobial action of H_2O_2 or by possibly induced natural defence mechanisms is currently being investigated. Nevertheless, the result shows that production of H_2O_2-generating enzymes could be used to engineer transgenic plants resistant to bacteria.

Plants can also be engineered to produce antibodies to inactivate molecules necessary for bacteria to infect plants successfully. The possibility of producing functional plantibodies is being explored by different groups (reviewed in Van Engelen *et al.*, 1994).

In view of the low number of papers on the engineering of bacterium resistance the limited success scored thus far is not surprising. One can envisage that there is a clear need to increase the number of studies on bacterium-plant interactions for new leads to the development of bacterium resistance.

C.　　APPROACHES TO ENGINEER FUNGUS RESISTANCE

To engineer fungus resistant plants two types of approaches are being pursued. The first type focuses on the production in transgenic plants of natural compounds that directly affect the fungus. Such compounds include antifungal proteins and phytoalexins. The second type aims at the generation of plant responses leading to local cell death after infection.

1.　　*Antifungal Compounds*

Compounds that inhibit the growth of fungi in vitro or even kill them are abundant in nature. Whether they are involved in defence responses against fungi in vivo is not known. Nonetheless, genes encoding such compounds or enzymes involved in their synthesis can be used to render plants fungus resistant. Proteins that inhibit fungal growth include chitinases, β-1,3-glucanases, various other pathogenesis-related (PR) proteins, ribosome-inactivating proteins defensins and other small cystein-rich proteins such as thionins. Phytoalexins are low-molecular weight antimicrobial compounds that are produced by plants upon pathogen infection.

a. Chitinases and Glucanases.　　Chitinases and β-1,3-glucanases are probably the best studied antifungal proteins to date. They catalyze the hydrolysis of chitin and glucan, respectively, two major structural components in the cell wall of many fungi. Most chitinase genes described in the literature are of plant origin. However, chitinase genes have also been found in fungi, bacteria, insects and even in humans (Hakala *et al.*, 1993) and viruses (Hawtin *et al.*, 1995). Plant chitinase genes have been classified into four groups of families: Chi-a to Chi-d (Neuhaus *et al.*, 1996). Group Chi-a contains five families, group Chi-b and Chi-c contain one family each, and group Chi-d contains two families (Table 2). Within a group genes show high degrees of homology with each other, whereas virtually no homology is found between genes from different groups. Plant chitinase genes include various genes encoding PR proteins. In plants these proteins are induced upon pathogen attack or related situations, including wounding and application of elicitors or chemicals that induce host responses (Van Loon *et al.*, 1994). In some cases the mature product of a chitinase gene does not exhibit chitinase activity.

For example, Chi-a5 encodes the precursor of stinging nettle (*Urtica dioica*) agglutinin (UDA), a protein that consists of two very homologous domains with high affinity to chitin. From the UDA precursor the C-terminal chitinase domain is cleaved off to yield mature UDA (Lerner and Raikhel, 1992). Likewise, hevein, a protein of 43 amino acids from the rubber tree (*Hevea brasiliensis*), is encoded by a Chi-d gene. The C-terminal domain of the primary translation product of this gene is a chitinase which is not present in mature hevein (Lee *et al.*, 1991). The primary structure of hevein is very homologous to the chitin-binding domains of UDA and of some other plant lectins, and to the chitin-binding domain of many mature chitinases (Beintema, 1994).

Table 2. Classification of chitinase genes from plants

group	family	mature gene product	PR protein	characteristics of gene product
Chi-a	Chi-a1	class I chitinase	PR-3	chitin-binding domain present
	Chi-a2	class II chitinase	PR-3	no chitin-binding domain
	Chi-a4	class IV chitinase		
	Chi-a5	*Urtica dioca* agglutinin (UDA)		
	Chi-a6			long proline-rich domain present
Chi-b	Chi-b1	class III chitinase	PR-8	distantly related to PR-11
Chi-c	Chi-c1	class VI chitinase	PR-11	homologous to group A bacterial exo-chitinases
Chi-d	Chi-d1	class I PR-4 / hevein / *win* protein	PR-4	chitin-binding domain present
	Chi-d2	class II PR-4	PR-4	no chitin-binding domain

As far as tested most products of chitinase and glucanase genes, including UDA and hevein, display antifungal activity in vitro. However, some chitinases and glucanases do not show antifungal activity at all (reviewed by Cornelissen *et al.* 1996). Also the sensitivity of fungi to the purified proteins varies greatly. Interestingly, some chitinases have been found to inhibit fungal growth in vitro in a synergistic way when mixed with the products of other chitinase genes. For example, a Chi-a1 chitinase was found to act synergistically with UDA (Broekaert *et al.*, 1989) or with proteins encoded by Chi-d genes (Hejgaard *et al.*, 1992; Ponstein *et al.*, 1994). Strong synergy is often observed as well when chitinases are mixed with β-1,3-glucanases (reviewed by Cornelissen *et al.*, 1996). A chitinase from the mycoparasitic fungus *Trichoderma harzianum* was found to act synergistically in fungal growth inhibition when mixed with a chitobiosidase of the same origin (Lorito *et al.*, 1993).

Broglie *et al.* (1991) were the first to show enhanced fungal resistance in transgenic plants brought about by the expression of a single chitinase gene. A Chi-a1 gene from bean under the control of the cauliflower mosaic virus 35S promoter was introduced into both tobacco and canola (oil seed rape). Compared to control plants, transgenic plants were found to be more resistant to *R. solani*. Later, enhanced protection against *R. solani* was also observed in transgenic rice plants constitutively expressing a rice Chi-a1 chitinase gene (Lin *et al.*, 1995), in *Nicotiana sylvestris* plants transgenic for a tobacco Chi-a1 chitinase gene (Vierheilig *et al.*, 1993) and in tobacco plants constitutively expressing a ChiA exo-chitinase gene from the bacterium *Serratia marcescens* (Jach *et al.*, 1992; Howie *et al.*, 1994). Recently, the hevein gene has been expressed in transgenic tomato plants (Lee and Raikhel, 1995). Although processing of the preproprotein was poor, less severe disease symptoms appeared on slices of transgenic tomato fruit after inoculation with *Trichoderma hamatum* than on control slices; the susceptibility towards *Botrytis cinerea* and *Rhizoctonia solani* was not altered.

Synergy between chitinases and β-1,3-glucanases in inhibiting fungal growth has been observed in planta as well. The constitutive co-expression in tomato of a Chi-a1 chitinase gene and a class I β-1,3-glucanase, both from tobacco, resulted in increased resistance against *Fusarium oxysporum* (Fig. 1); plants expressing either one of the genes did not show enhanced resistance to this fungus (Van den Elzen *et al.*, 1993; Jongedijk *et al.*, 1995). Likewise, combined expression of a rice Chi-a1 chitinase gene and an alfalfa class I β-1,3-glucanase gene in transgenic tobacco resulted in significant levels of resistance against *Cercospora nicotianae,* whereas in plants transgenic for a single gene-construct resistance was hardly observed (Zhu *et al.*, 1994). Synergy in planta was found as well between a Chi-a2 chitinase and a class II β-1,3-glucanase, both from barley: after challenge-inoculation with *R. solani* up to 60% reduction in disease development was found in transgenic tobacco plants producing both enzymes (Jach *et al.*, 1995).

Figure 1.   Transgenic tomato showing enhanced resistance to *Fusarium oxysporum*. A fungus inoculated, transgenic tomato plant constitutively expressing a chitinase and a β-1,3-glucanase gene from tobacco (left), a fungus inoculated, non-transgenic plant (middle) and a non-inoculated, non-transgenic control plant (right). Photo: ZENECA MOGEN Leiden, the Netherlands.

b. PR-Proteins Other than Chitinases or β-1,3-Glucanases. Since PR proteins are induced during a hypersensitive response (HR), and concomitantly with the induction of systemically acquired resistance, they are believed to play a role in resistance (Van Loon *et al.*, 1994). This knowledge prompted scientists to test the usefulness of PR genes in the engineering of resistance. Besides genes encoding products with known enzymatic functions like chitinases or β-1,3-glucanases (see above) other genes for PR proteins of unknown or incompletely undertood function were tested as well. The enzymatic activity and biological function of PR1 proteins are unknown. Nonetheless, constitutive expression of the PR1a gene in tobacco results in enhanced resistance to *Peronospora tabacina* (Alexander *et al.*, 1993). Likewise, the precise function of PR5 proteins (or AP24, or thaumatin-like protein or osmotin) is still to be determined, although they have been speculated to permeabilise the fungal cell wall thereby causing lysis. Nonetheless, in vitro they show antimicrobial activity (Woloshuk *et al.*, 1991) and when overproduced in transgenic potato plants they delay symptom development upon infection with *Phytophthora infestans* (Liu *et al.*, 1994).

c. Plant Ribosome-Inactivating Proteins. Plant ribosome-inactivating proteins (RIPs) modify 28S RNAs of eukaryotic ribosomes, thereby inhibiting protein elongation (reviewed by Stirpe *et al.*, 1992). Plant RIPs inactivate ribosomes of distantly related species and of other eukaryotes including fungi. A RIP isolated from barley was shown to exhibit antifungal activity in vitro to a number of plant pathogenic fungi (Leah *et al.*, 1991). Tobacco plants expressing the barley gene under control of a wound-inducible promoter showed an increased resistance to *R. solani* (Logemann *et al.*, 1992; Jach *et al.*, 1995). The levels of resistance found in these plants were higher than reported earlier for transgenic tobacco plants constitutively expressing a bacterial ChiA exo-chitinase gene. However, the best results were obtained by combined expression of the barley RIP gene and a barley Chi-a2 chitinase gene in tobacco: up to 55% reduction in disease development was found after inoculation with *R. solani* (Jach *et al.*, 1995).

d. Small Cystein-Rich Proteins. In the literature a whole array of small cystein-rich proteins exhibiting in vitro fungal growth-inhibiting activity has been described. Small cystein-rich proteins from plants are divided into several groups, one of them being formed by hevein and UDA, encoded by chitinase genes Chi-d and Chi-a5, respectively (see under III.C.1.a). Another group consists of plant defensins, which are subdivided into at least three subgroups (Broekaert *et al.*, 1995). Members of two of these subgroups show in vitro antifungal activities and include Rs-AFP2, Mj-AMP2 and Ac-AMP2, peptides from *Raphanus sativus*, *Mirabilis jalapa* and *Amaranthus caudatus*, respectively. Transgenic tobacco producing Rs-AFP2 shows enhanced resistance to the fungus *Alternaria longipipes* (Broekaert *et al.*, 1995). In contrast, constitutive expression in transgenic tobacco of either Mj-AMP2 or Ac-AMP2 does not result in resistance, although the proteins purified from the transgenic tobacco plants still exert antifungal ativity in vitro (de Bolle *et al.*, 1996).

Yet another group of small cystein-rich proteins is formed by thionins. Although the published results on bacterium resistance brought about by the constitutive expression of thionin genes in transgenic plants are conflicting (see under III.B), the results on fungus resistance seem to be more positive. The constitutive expression of the *Arabidopsis* thionin Thi2.1 gene in transgenic *Arabidopsis* enhances resistance against the fungus *Fusarium oxysporum* f.sp. *matthiolae* (Epple *et al.*, 1997).

e. Phytoalexins. Phytoalexins are antimicrobial compounds produced in plants after pathogen attack and some abiotic stresses. They have long been implicated as playing an important role in plant defence (Kuc, 1995). In grapevine (*Vitis vinifera*) reduced susceptibilty to the fungus *Botrytis cinerea* has been associated with relatively high levels of the phytoalexin resveratrol, suggesting a role for this phytoalexin in resistance. A key enzyme in the biosynthesis of resveratrol is stilbene synthase. In tobacco the substrate for stilbene synthase is present, but a gene encoding stilbene synthase is lacking. In 1993 Hain *et al.* introduced a stilbene synthase gene from grapevine into tobacco plants. Expression of the gene resulted in production of resveratrol. More importantly, the plants showed enhanced resistance to *Botrytis cinerea*. However, resistance levels obtained thus far are too low to be interesting commercially. Nonetheless the results demonstrate that constitutive production of phytoalexins in heterologous plant systems may result in increased pathogen resistance.

2.   Fungus-Induced Cell Death

As discussed above, engineering fungus resistant plants by the introduction of genes encoding antifungal compounds has been shown to be successful in a number of cases. However, constitutive production of phytoalexins or antifungal proteins does not render plants fungus resistant to an absolute level, and in all cases resistance is observed towards a limited number of fungi only. Transgenic crops over-expressing a chitinase gene and found to be resistant to a number of fungal pathogens containing chitin in their cell walls, never appeared resistant to fungi lacking chitin. Moreover, fungi may become adapted to the changes in their hosts. For example, a fungus may modify its cell wall composition e.g. by the biosynthesis of more chitosan or glucan instead of chitin, and become insensitive to chitinases. Also fungal strains may evolve that can detoxify certain phytoalexins. Sexually reproducing fungi may overcome their sensitivity to phytoalexins or other fungal toxins even within a very short period of time. Furthermore, since plants are attacked by different microbial organisms during their life cycle, the possibility exists that other pathogens will benefit from the absence of the chitinase-sensitive pathogens. Therefore, strategies that might provide transgenic crops with a potentially more durable resistance directed towards a broader spectrum of pathogens are currently being investigated. These strategies involve local cell death induced by the attacking pathogen and are often based on general defence responses occurring in plants during incompatible plant-pathogen interactions.

As discussed under section II De Wit (1992, 1995) proposed a two-component sensor system that triggers a hypersensitive response and local cell death in plants upon infection with a pathogen. Strittmatter *et al.* (1995) developed a two-component system in which barnase, a cytotoxic protein with RNase activity, and its inactivator (barstar) are used. The genes encoding these proteins are both derived from *Bacillus amyloliquefaciens*. Since barnase activity is permitted at sites of infection only, the encoding gene was placed under control of the pathogen inducible *gst1* promoter from potato (see section II). To avoid cell death due to unwanted expression of the barnase gene, the barstar gene was constitutively expressed in all tissues. Cells are killed only if the barnase activity is higher than barstar activity. A number of transgenic potato lines expressing both genes showed severe necrosis of leaf tissue after inoculation with *Phytophthora infestans* spores. A strong reduction in symptom development could also be observed.

Expression of a gene encoding bacterio-opsin (bO), a light-driven proton pump from *Halobacterium halobium*, in transgenic tobacco plants resulted in a phenotype similar to lesion mimic mutants (Mittler *et al.*, 1995). Without pathogen infection, defence responses similar to those induced during systemic acquired resistance (SAR) occurred. Responses such as hypersensitive response-like lesions, accumulation of PR gene transcripts, phenylammonialyase and other compounds were observed. Induced ion fluxes are one of the early events in incompatible plant-pathogen interactions. Therefore, functional proton translocation by bO might induce the observed SAR-like defence responses. The process by which bO activates cell death is not clear, but might offer another tool to create broad spectrum resistance in plants.

To develop fungus resistant plants thus far most efforts have been put into studies on the constitutive production in transgenic plants of antifungal compounds, in particular proteins. Numerous antifungal polypeptides have been described and many of them seem to inhibit fungal growth in a synergistic way when mixed with a second antifungal protein, both *in vitro* and *in planta*. Fungus resistant transgenic plants have been described in a limited number of papers. However, in most cases the observed resistance was not absolute and restricted to a few fungi. For the antifungal compound strategy to be successful at the long term, levels of resistance in transgenic plants should be increased and the range of resistance should be broadened. This might be achieved by the application of either new genes encoding proteins with higher antifungal activity or by new combinations of genes. In the meantime concepts which aim at fungus-induced cell death at sites of infection, should be worked out further. Finally, through studies on the molecular basis of natural plant defence responses knowledge might be gained which can be used to develop novel systems to engineer fungus resistance.

D.   TRANSGENIC STRATEGIES TO CONTROL NEMATODE DISEASES

Economic losses due to sedentary nematodes, mainly cyst nematodes from the genera *Heterodera* and *Globodera*, and root-knot nematodes from the genus *Meloidogyne*, far exceed the negative economic effects of migratory nematodes. Hence research on the development of nematode-resistant crops has so far been focused on sedentary nematodes. Papers on the engineering of nematode resistance are still limited in number and often describe approaches that are also used to develop virus, fungus and, in particular, insect resistance (Burrows and De Waele, 1997). Basically two strategies are being pursued. As with the engineering of fungus resistance, the first one focuses on the production in transgenic plants of proteins that directly affect nematode development, and the second aims at induction of local cell death upon infection.

1.   Anti-Nematode Proteins

Most proteins speculated to have anti-nematode activity are products of genes shown to have fungicidal and/or insecticidal activity (Burrows and De Waele, 1997). They include enzymes (among others chitinases, collagenases and RIPs), enzyme inhibitors (e.g. proteinase inhibitors), lectins and toxins from *Bacillus thuringiensis* (Bt toxins). Many of these products are currently being evaluated as anti-nematode proteins. First successes have been reported for a proteinase inhibitor and a lectin. As yet no studies have appeared describing results of nematode-resistance assays on transgenic plants expressing Bt-genes, despite the fact that this particular approach has been found to be very successful in the control of insect pests. In addition to the above, it has been suggested that nematode resistance may be engineered by the expression of monoclonal antibodies in plants (so-called 'Plantibodies').

a. Proteinase inhibitors.   The enzymatic activity of proteinases may be inhibited by proteinase inhibitors (PI). PIs are ubiquitous in the plant kingdom and are often induced upon wounding and / or insect attack. Expression of PI genes in transgenic plants enhances resistance to insects (reviewed by Ryan, 1990). This observation inspired Hepher and Atkinson (1992) to assay transgenic potato plants expressing a serine proteinase inhibitor gene from cowpea for enhanced resistance to *Globodera pallida*. Despite the low abundance of serine proteinases as compared to cysteine proteinases in the intestine of *G. pallida* and other cyst nematodes (Lilley *et al.*, 1996), clear effects were observed: suppression of initial growth and alteration in sex ratio towards the production of males. Next, focusing on plant cysteine proteinase inhibitors, a cystatin gene from rice was expressed in a tomato hairy root system and shown to affect growth and development of *G. pallida* (Urwin *et al.*, 1995). An engineered version of the cystatin gene gave an even greater effect in the hairy root system (Urwin *et al.*, 1995). Finally, expression of the modified gene under control of the CaMV 35S promoter in transgenic *Arabidopsis* was found to result in resistance to both the beet-cyst nematode *Heterodera schachtii* and the root-knot nematode *Meloidegyne incognita*: no females of either species grew large enough to be able to produce eggs; the development of males of normal size was not prevented (Urwin *et al.* 1997).

b. Lectins.   Lectins are proteins with high and specific affinity to carbohydrates. Most lectins are toxic to mammals and birds, and some also to insects. *Galanthus nivalis* agglutinin (GNA) is a mannose binding lectin from snowdrop with toxic effects towards some insects, but with little effects on mammalian systems. Expression of the GNA gene in transgenic tobacco resulted in enhanced protection against insects (Hilder *et al.*, 1995). Although the first results on testing the use of the GNA gene in the engineering of nematode resistance in oilseed rape and potato are encouraging, the experiments need to be repeated on a larger scale (Burrows and De Waele, 1997).

c. Plantibodies.   Since the first report on the production of functional antibodies in transgenic plants (Hiatt *et al.*, 1989), there have been speculations on the engineering of disease resistance using genes encoding such so-called 'plantibodies' (Schots *et al.*, 1992). Indeed, as has been shown by Tavladoraki *et al.* (1993) and by Voss *et al.* (1995)

enhanced virus resistance has been achieved by expressing genes encoding specific antibodies. Functional plantibodies directed against secretions of nematodes have been produced in transgenic plants (Rosso *et al.*, 1996; Baum *et al.*, 1996). However, proof that these plants show enhanced resistance towards nematodes is still lacking.

2. Nematode-Induced Cell Death

Many plant parasitic nematodes depend on specific structures within their hosts for feeding. These structures are induced by juveniles and provide the nutritional requirements of the growing nematode, in particular of females. Disordering of those feeding structures will interfere directly with nematode development and may result in resistance. Disruption can be brought about by nematode-induced production of cytotoxic compounds such as DNases, RNases and proteases. It is clear that expression of genes encoding such cytotoxins should be restricted to the cell or cells that form the feeding structure to prevent unwanted side effects. The question has been raised whether promoters directing feeding structure-specific expression of genes do exist, since it is hard to see that plant genes could have evolved which solely serve a pathogen (Ohl *et al.*, 1997). However, nematode-responsive promoters with activity in other tissues as well may be modified to enhance feeding structure specificity. The validity of this idea has been demonstrated by Opperman *et al.* (1994). They found that a truncated promoter of a tobacco gene encoding a root-specific putative water channel protein had lost all its activity in healthy plants, but had retained its responsiveness to root-knot nematodes. This promoter may be used in combination with a cytotoxin gene to test the hypothesis that disruption of feeding structures results in nematode resistance.

To circumvent the necessity of using a promoter with very high feeding structure-specificity, Ohl *et al.* (1997) proposed the application of a two-component system consisting of a cytotoxin and a toxin-inactivating compound. In such a system leakage of the nematode-responsive promoter in tissues other than the feeding structure is to be neutralized by constitutive production of the anti-toxin in those tissues. Elaborating this idea an RNase gene under the direction of a nematode inducible promoter was introduced into *Arabidopsis* together with an RNase-inhibitor encoding gene driven by a promoter that is down-regulated in feeding structures. Both the RNase (barnase) and the inhibitor (barstar) gene were originally isolated from *Bacillus amyloliquefaciens*. The first results of testing transgenic *Arabidopsis* lines for resistance to *H. schachtii* indicate that the two-component approach may be successful, although full resistance has not yet been obtained (Ohl *et al.*, 1997).

The development of strategies to control parasitic nematodes is still in its infancy compared to virus and fungus resistance strategies. In part this is due to the obligate parasitic nature of this plant-root pathogen which makes it a difficult pathogen-host system to work with. In this context, the finding that *Arabidopsis* can be used as an experimental system to study interactions between both *M. incognita* and *H. schachtii,* and their hosts (Goddijn *et al.*, 1993), may be considered as an important breakthrough. Indeed, the first positive results on engineered nematode resistance have been reported in *Arabidopsis* (Urwin *et al.* 1997; Ohl *et al.*, 1997).

IV.   Field Trials and Market Introduction of Transgenic Crops

In the last decade the number of field trials carried out by research institutions and companies with transgenic crops has grown tremendously, and is still growing. The number of trials to test crops for improved yield and quality or for the production of specific proteins or oils, far exceeds the number of trials with crops with engineered protection. Futhermore, in the latter category most trials involved insect resistance brought about by the expression of various Bt-toxin genes, and herbicide resistance. Relatively few studies have been performed to assess the level and stability of virus and fungus resistance under field conditions. Thus far no (successful) studies have been carried to evaluate bacterium and nematode resistance in the field.

In 1988 the first field trial was carried out with two tomato lines expressing the coat protein gene of TMV (Nelson et al., 1988). Both lines were found to be less susceptible to infection by tomato mosaic virus. Yields from one line were equal to those of control plants, yields of the other were depressed. Although the scale of the trial was limited, the results clearly indicated the feasibility of engineering of resistance traits in agriculture. In 1992 results were published which were obtained from three consecutive years of field trials with transgenic cucumber expressing the coat protein gene of cucumber mosaic virus (Gonsalves et al., 1992). Levels of resistance were found to be comparable to the level conferred by a natural resistance gene present in a commercial cultivar. Furthermore, fruit yields and vegetative growth of transgenic lines were on average better then those of non-transgenic plants of the same cultivar.

Increased field resistance to potato viruses X (PVX) and Y (PVY) has been reported in Russet Burbank potato transgenic for the coat protein genes of the two viruses (Kaniewski et al., 1990). An extensive field evaluation of engineered virus resistance in potato has been published by Jongedijk et al. (1992). The PVX coat protein gene under the direction of the CaMV 35S promoter was introduced into potato cultivars Bintje and Escort. During three consecutive years trials were performed to assess levels of resistance against PVX and changes in intrinsic properties of the two cultivars. Improved resistance up to near immunity was observed and, very important for this vegetatively propagated crop, in many of the lines tested intrinsic cultivar properties were preserved. Despite this success, neither of the two engineered cultivars has yet entered the market. Thus far squash is the only crop engineered to disease resistance that has been cleared for market introduction in the USA. Asgrow Seed Company (Seminis Vegetable Seeds) developed various squash lines expressing the coat protein gene of either cucumber mosaic virus, watermelon mosaic virus 2 or zucchini yellow mosaic virus. In addition, lines were produced containing combinations of genes from two viruses and also lines with the coat protein gene of all three viruses (Tricoli et al., 1995; Fuchs and Gonsalves, 1995). Extensive field evaluations revealed several lines of commercial quality showing high levels of resistance to those viruses for which the coat protein gene was expressed. Horticultural performance of the multiple virus resistant squash lines was not negatively affected by the presence or expression of the coat protein genes. These results show that the spectrum of virus-derived resistance traits can be broadened by pyramiding genes of different viruses in a single crop.

Reports on enhanced protection against fungal diseases brought about by the constitutive expression of a gene or genes encoding antifungal proteins in field grown plants are still relatively scarce compared to the number of studies on antifungal proteins. Over two consecutive years Howie *et al.* (1994) have performed field trials to assess the effect of the constitutive expression of a *Serratia marcescens Chi*A chitinase gene in transgenic tobacco on infection by *R. solani*. Consistently, disease tolerance was observed in transgenic plants regardless of whether the chitinase accumulated intra- or extracellularly (Howie *et al.*, 1994). In field trials at two different locations transgenic canola constitutively expressing a tomato endo-chitinase gene was found to exhibit increased tolerance to three fungal pathogens (Grison *et al.*, 1996). However, horticultural performance of the transgenic canola has not yet been evaluated. MOGEN International nv (ZENECA MOGEN) has tested various genes and combinations of genes encoding PR proteins for their use in enhancing resistance to fungi in transgenic plants. In two consecutive years of field trials it was established that both the expression of a chitinase and a β-1,3-glucanase gene in combination (Fig. 2), and the expression of an AP24 gene, gave rise to broad spectrum fungal resistance in transgenic carrot plants (Stuiver *et al.*, 1996).

Figure 2. Field trial with transgenic carrots. Transgenic carrot plants constitutively expressing both a tobacco chitinase and a tobacco β-1,3-glucanase gene are resistant to *Alternaria* (left); non-transgenic carrot plants are severly affected by the fungus (right). Photo: ZENECA MOGEN Leiden, the Netherlands.

As mentioned earlier the number of field trials in this area (of which results have been published) is still limited compared to the number of papers on the engineering of disease resistance in plants. However, the results are promising and in one case (virus resistant squash) permission has been obtained for introduction into the market.

V.   Concluding remarks

Ever since its birth, the question of the possible contribution of plant genetic engineering to disease control has been asked. In the last fifteen years a lot of efforts have been put into studies of the introduction in transgenic plants of viral sequences, on the expression of transgenes encoding proteins affecting growth or development of pathogens and on pathogen-induced cell death. The results show that commercial levels of resistance to specific viruses can be obtained by the expression of viral sequences, and that the spectrum of resistance can be broadened by pyramiding viral sequences in the plant genome. However, one question that remains is the durability of the trait. It can be envisaged that by mutations at possibly specific sites in either coding or noncoding sequences of their genome, viruses could overcome resistance. In this context it is noteworthy that transgenic plants producing a mutant alfalfa mosaic virus coat protein with the N-terminal serine residue replaced by a glycine are resistant to the mutant virus, but susceptible to infection with wild type virus (Taschner *et al.*, 1994). If in a few years it becomes apparent that virus-derived resistance traits are broken relatively easy by mutant viruses, it may be reassuring that a new source for resistance, the genome of the mutant, is close at hand.

Numerous polypeptides that (potentially) interfere with pathogen growth or development have been described. Many antifungal proteins have been shown to inhibit fungal growth in a synergistic way when mixed. The results of various laboratory and greenhouse trials show that enhanced resistance to viruses (through the production of antibodies), fungi (antifungal proteins) and nematodes (proteinase inhibitor) can be achieved. However, in most cases the observed resistance was not absolute or restricted to a limited number of pathogen species. To be interesting commercially levels of resistance need to be increased in most cases and broadened in range. This might be achieved by the application of new genes or combinations of genes.

Honée *et al.* (1997) have reported a first positive result in the development of a resistance system based on pathogen-induced cell death using the two-component sensing sytem. It is expected that in the near future new concepts will be developed on broad-range pathogen resistance through locally induced cell death and that existing ones will worked out in more details. Furthermore, studies on the molecular basis of natural plant defence responses will generate knowledge which can be used in the development of new resistance systems: sytems that provide us with durable resistance to a broad spectrum of pathogens in a variety of economically important crops.

VI. References

Alexander D, Goodman RM, Gut-Rella M, Glascock C, Weymann K, Friedrich L, Maddox D, Ahl-Goy P, Luntz T, Ward E and Ryals J (1993) Increased tolerance to two oomycete pathogens in transgenic tobacco expressing pathogenesis-related protein 1a. Proc Natl Acad Sci USA 90 7327—7331

Anderson PA, Lawrence GJ, Morrish BC, Ayliffe MA, Finnegan EJ and Ellis JG (1997) Inactivation of the flax rust gene M associated with loss of a repeated unit within the leucine-rich repeat coding region. Plant Cell 9 641—651

Baulcombe DC (1996) Mechanisms of pathogen-derived resistance to viruses in transgenic plants. Plant Cell 8 1833—1844

Baum TJ, Hiatt A, Parrot H and Hussey RS (1996) Expression in tobacco of a functional monoclonal antibody specific to stylet secretions of the root-knot nematode. Mol Plant-Microbe Interact 9 382—387

Beintema JJ (1994) Structural features of plant chitinases and chitin-binding proteins. FEBS Lett 350 159—163

Bent AF (1996) Plant disease resistance genes: function meets structure. Plant Cell 8 1757—1771

Bent AF, Kunkel BN, Dahlbeck D, Brown, KL, Schmidt R, Giraudat J, Leung J and Staskawicz BJ (1994) Rps2 of Arabidopsis thaliana: a leucine-rich repeat class of plant disease resistance genes. Science 265 1856—1860

Bisgrove SR, Simonich MT, Smith NM, Sattler A and Innes RW (1994) A disease resistance gene in Arabidopsis with specificity for two different avirulence genes. Plant Cell 6 927—933

De Bolle MFC, Osborn RW, Goderis IJ, Noe L, Acland D, Hart CA, Torrekens S, van Leuven F and Broekaert WF (1996) Antimicrobial peptides from Mirabilis jalapa and Amaranthus caudatus: expression, processing, localization and biological activity in transgenic tobacco. Plant Mol Biol 31 993—1008

Bonas U and Van den Ackerveken G (1997) Recognition of bacterial avirulence proteins occurs inside the plant cell: a general phenomenon in resistance to bacterial diseases? Plant J 12: 1—7

Broekaert WF, Van Parijs J, Leyns F, Joos H and Peumans WJ (1989) A chitin-binding lectin from stinging nettle rhizomes with antifungal activity. Science 245 1100—1102

Broekaert WF, Terras FRG, Cammue BPA and Osborn RW (1995) Plant defensins: novel antimicrobial peptides as components of the host defence system. Plant Physiol 108 1353—1358

Broglie K, Chet I, Holliday M, Cressman R, Biddle P, Knowlton S, Mauvais CJ and Broglie R (1991) Transgenic plants with enhanced resistance to the fungal pathogen Rhizoctonia solani. Science 254 11941197

Burrows PR and De Waele D (1997) Engineering resistance against plant parasitic nematodes using anti-nematode genes. In: Fenoll C, Grundler FMW and Ohl SA (eds.) Cellular and Molecular Aspects of Plant-Nematode Interactions, pp 217—236, Kluwer Academic Publishers, Dordrecht

Büschges R, Hollricher K, Panstruga R, Somins G, Frijters A, Van Daelen R, Van der Lee T, Diergaarde P, Groenendijk J, Topsch S, Vos P, Salamini F and Schulze-Levert P (1997) The barley Mlo gene: a novel control element of plant pathogen resistance. Cell 88 695—705

Cai D, Kleine M, Kilfe S, Harloff HJ, Sandal NN, Marcker KA, Klein-Lankhorst RM, Salentijn EMJ, Lange W, Stiekema WJ, Wyss U, Grundler FMW and Jung C (1997) Positional cloning of a gene for nematode resistance in sugar beet. Science 275 832—834

Carmona MJ, Molina A, Fernandez JA, Lopez-Fando JJ and Garcia-Olmedo F (1993) Expression of the alfa-thionin gene from barley in tobacco confers enhanced resistance to bacterial pathogens. Plant J 3 457—462

Cornelissen BJC, Does M and Melchers LS (1996) Strategies for molecular resistance breeding (& transgenic plants). In: Sneh B, Jabaji-Hare S, Neate S and Deist G (eds.) Rhizoctonia Species: Taxonomy, Molecular Biology, Ecology, Pathology and Control, pp 529—536, Kluwer Academic Publishers, Dordrecht

Cuozzo M, O'Connell KM, Kaniewski WK, Fang R-X, Chua N-H and Tumer NE (1988) Viral protection in transgenic tobacco plants expressing the cucumber mosaic virus coat protein or its antisense RNA. Bio/Technology 6 549—557

Dixon MS, Jones DA, Keddie JS, Thomas CM, Harrison K and Jones JDG 1996 The tomato *Cf-2* disease resistance locus comprises two functional genes encoding leucine-rich repeats proteins. Cell 84 451—459

Düring K, Porsch P, Fladung M and Lörz H (1993) Transgenic potato plants resistant to the phytopathogenic bacterium *Erwinia carotovora*. Plant J 3 587—598

Van den Elzen PJM, Jongedijk E, Melchers LS and Cornelissen BJC (1993) Virus and fungal resistance: from laboratory to field. PhilosTrans R Soc Lond Biol Sci 342 835—838

Van Engelen FA, Schouten A, Molthoff JW, Roosien J, Salinas J, Dirkse WG, Schots A, Bakker J, Gommers FJ, Jongsma MA, Bosch D and Stiekema WJ (1994) Coordinate expression of antibody subunit genes yield high levels of functional antibodies in roots of transgenic tobacco. Plant Mol Biol 26 1701—1710

Epple P, Apel K and Bohlmann H (1997) Overexpression of an endogenous thionin enhances resistance of *Arabidopsis* against *Fusarium oxysporum*. Plant Cell 9 509—520

Flor HH (1956) The complementary genetic systems in flax and flax rust. Adv Genet 8 29—54

Flor HH (1971) The current status of the gene-for-gene concept. Annu Rev Phytopathol 9 275—296

Florack DEA and Stiekema WJ (1994) Thionins: properties, possible biological roles and mechanisms of action. Plant Mol Biol 26 25—37

Florack DEA, Dirkse WG, Visser B, Heidekamp F and Stiekema WJ (1994) Expression of biologically active hordothionin in tobacco. Effects of pre- and pro-sequences at the amino and carboxyl termini of the hordothionin precursor on mature protein expression and sorting. Plant Mol Biol 24 83—96

Fuchs M and Gonsalves D (1995) Resistance of transgenic hybrid squash ZW-20 expressing the coat protein genes of zucchini yellow mosaic virus and watermelon mosaic virus 2 to mixed infections by both potyviruses. Bio/Technology 13 1466—1473

Goddijn OJM, Lindsey K, Van der Lee, FM, Klap JC and Sijmons PC (1993) Differential gene expression in nematode-induced feeding structures of transgenic plants harbouring promoter-*gus*A fusion constructs. Plant J 4 863—873

Gonsalves D, Chee P, Provvidenti R, Seem R and Slightom JL (1992) Comparison of coat protein-mediated and genetically-derived resistance in cucumbers to infection by cucumber mosaic virus under field conditions with natural challenge inoculations by vectors. Bio/Technology 10 1562—1570

Grant MR, Godiard L, Straube E, Ashfield T, Lewald J, Sattler A, Innes RW and Dangl JL (1995) Structure of the *Arabidopsis Rmp*1 gene enabling dual specificity disease resistance. Science 269 843—846

Grison R, Grezes-Besset B, Schneider M, Lucante N, Olsen L, Leguay J-J and Toppan A (1996) Field tolerance to fungal pathogens of *Brassica napus* constitutively expressing a chimeric chitinase gene. Nature Biotechnol 4 643—646

Hain R, Reif H-J, Krause E, Langebartels R, Kindl H, Vornam B, Wiese W, Schmelzer E, Schreier PH, Stöcker RH and Stenzel K (1993) Disease resistance results from foreign phytoalexin expression in a novel plant. Nature 361 153—156

Hakala BE, White C and Recklies AD (1993) Human cartilage gp-39, a major secretory product of articular chondrocytes and synovial cells, is a mammalian member of a chitinase protein family. J Biol Chem 268 25802—25810

Hammond-Kosack KE and Jones JDG (1997) Plant disease resistance genes. Annu Rev Plant Physiol Plant Mol Biol 48 575—607

Handelsman J and Stabb EV (1996) Biocontrol of soilborne plant pathogens. Plant Cell 8 1855—1869

Hawtin RE, Arnold K, Ayers MD, Zanotto PM, Howard SC, Gooday GW, Chapell LH, Kitts PA, King LA and Possee RD (1995) Identification and preliminary characterization of a chitinase gene in the *Autographa californica* nuclear polyhedrosis virus genome. Virology 212 673—685

Hejgaard J, Jacobsen S, Bjorn SE and Kragh KM (1992) Antifungal activity of chitin-binding PR-4 type proteins from barley grain and stressed leaf. FEBS Lett 307 389-392

Hemenway C, Fang R-X, Kaniewski W, Chua N-H and Tumer NE (1988) Analysis of the mechanism of protection in transgenic plants expressing the potato virus X coat protein or its antisense RNA, EMBO J 7 1273—1280

Hepher A and Atkinson HJ (1992) Nematode control with proteinase inhibitors. European patent application number 9230 1890.7; publication number 0 492 730 A1

Hiatt A, Cafferkey R and Bowdish K (1989) Production of antibodies in transgenic plants. Nature 342 76—68

Hilder VA, Powell KS, Gatehouse AMR, Gatehouse JA, Gatehouse LN, Shi Y, Hamilton WDO, Merryweather A, Newell CA, Timans JC, Peumans WJ, Van Damme E and Boulter D (1995) Expression of snowdrop lectin in transgenic tobacco plants results in added protection against aphids. Transgenic Res 4 18—25

Honnée G, Stuiver M, Weide R, Tigelaar H, Melchers LS and De Wit PJGM (1997) Infection induced expression of the avirulence gene *Avr9* in transgenic CF9 tomato plants confers resistance to fungal pathogen attack. 5th International Congress of Plant Molecular Biology, Singapore, abstract 63

Howie W, Joe L, Newbigin E, Suslow T and Dunsmuir P (1994) Transgenic tobacco plants which express the *chi*A gene from *Serratia marcescens* have enhanced tolerance to *Rhizoctonia solani*. Transgenic Res 3 90-98

Jach G, Logemann S, Wolf G, Oppenheim A, Chet I, Schell J and Logemann J (1992) Expression of a bacterial chitinase leads to improved resistance of transgenic tobacco plants against fungal infection. Biopractice 1 33-40

Jach G, Grönhardt B, Mundy J, Logemann J, Pinsdorf E, Leah R, Echell J and Maas C (1995) Enhanced quantitative resistance against fungal disease by combinatorial expression of different barley antifungal proteins in transgenic tobacco. Plant J 8 97—109

James WC, Teng PS and Nutter FW (1991) Estimated losses of crops from plant pathogens. In: Boston PD (Ed) CRC Handbook of Pest Management in Agriculture 1, pp 15—51, CRC Press, Boca Raton

Johal GS and Briggs SP (1992) Reductase activity encoded by the *Hm*1 disease resistance gene in maize. Science 258 985—987

Jones DA and Jones JDG (1996) The role of leucine-rich repeat proteins in plant defences. Adv Bot Res 24 91—167

Jones DA , Thomas CM, Hammond-Kosack KE, Balint-Kurti PJ and Jones JDG (1994) Isolation of the *Cf-9* gene for resistance to *Cladosporium fulvum* by transposon tagging. Science 266 789—793

Jongedijk D, De Schutter AAJM, Stolte T, Van den Elzen, PJM and Cornelissen BJC (1992) Increased resistance to potato virus X and preservation of cultivar properties in transgenic potato under field conditions. Bio/Technology 10 422—429

Jongedijk E, Tigelaar H, Van Roekel JSC, Bres-Vloemans SA, Dekker I, Van den Elzen PJM, Cornelissen BJC and Melchers LS (1995) Synergistic activity of chitinases and β-1,3-glucanases enhances fungal resistance in transgenic tomato plants. Euphytica 85 173—180

Kaniewski W, Lawson C, Samsons B, Haley L, Hart J, Delannay X and Tumer N (1990) Field resistance of transgenic Russet Burbank potato to effects of infection by potato virus X and potato virus Y. Bio/Technology 8 750-754

Kawchuk LM, Martin RR and McPherson J (1991) Sense and antisense RNA-mediated resistance to potato leafroll virus in Russet Burbank potato plants. Mol Plant-Microbe Interact 4 247—253

Kobe B and Deisenhofer J (1995) A structural basis of the interactions between leucine-rich repeats and protein ligands. Nature 374 183—186

Kuc J. (1995) Phytoalexins, stress metabolism and disease resistance in plants. Annu Rev Phytopathol 33 275—297

Lawrence GJ, Finnegan EJ, Ayliffe MA and Ellis JG (1995) The *L6* gene for flax rust resistance is related to the *Arabidopsis* bacterial resistance gene *Rps2* and the tobacco viral resistance gene *N*. Plant Cell 7 1195—1206

Lawson C, Kaniewski W, Haley L, Rozman R, Newell C, Sanders P and Tumer NE (1990) Engineering resistance to mixed virus infection in a commercial potato cultivar: resistance to potato virus X and potato virus Y in transgenic Russet Burbank. Bio/Technology 8 127—134

Leah R, Tommerup H, Svendsen I and Mundy J (1991) Biochemical and molecular characterization of three barley seed proteins with antifungal properties. J Biol Chem 266 1464—1573

Lee H.-I and Raikhel NV (1995) Prohevein is poorly processed but shows enhanced resistance to a chitin-binding fungus in transgenic tomato plants. Braz J Med Biol Res 28: 743—750

Lee H, Broekaert WF and Raikhel NV (1991) Co- and post-translational processing of the hevein preproprotein of latex of the rubber tree (*Hevea brasiliensis*). J Biol Chem 266 15944—15948

Lerner DR and Raikhel NV (1992) The gene for stinging nettle lectin (*Urtica dioica* agglutinin) encodes both a lectin and a chitinase. J Biol Chem 267 11085—11091

Lilley CJ, Urwin PE, McPherson MJ and Atkinson HJ (1996) Characterisation of intestinally active proteases of cyst-nematodes. Parasitology 113 415—424

Lin W, Anuratha CS, Datta K, Potrykus I, Muthukrishnan S and Datta SK (1995) Genetic engineering of rice for resistance to sheath blight. Bio/Technology 13 686—691

Liu D, Raghothama KG, Hasegawa PM and Bressan RA (1994) Osmotin overexpression in potato delays development of disease symptoms. Proc Natl Acad Sci USA 91 1888—1892

Logemann J, Jach G, Tommerup H, Mundy J and Schell J (1992) Expression of a barley ribosome inactivating protein leads to increased fungal protection in transgenic tobacco plants. Bio/Technology 10 305—308

Lomonossoff GP (1995) Pathogen-derived resistance to plant viruses. Annu Rev Phytopathol 33 323—343

Van Loon LC, Pierpoint WS, Boller T and Conejero V (1994) Recommendations for naming plant pathogenisis-related proteins. Plant Mol Biol Rep 12 245—264

Lagudah ES, Moullet O and Appels R (1997) Map-based cloning of a gene encoding a nucleotide-binding domain and a leucine-rich region at the *Cre3* nematode resistance locus of wheat. Genome 40 659—665

Lorito M, Harman GE, Hayes CK, Broadway RM, Tronsmo A, Woo SL and Di Pietro A (1993) Chitinolytic enzymes produced by *Trichoderma harzianum*: antifungal activity of purified endochitinase and chitobiosidase. Phytopathology 83 302—307

Martin GB, Brommonschenkel SH, Chunwongse J, Frary A, Ganal MW, Spivey R, Wu T, Earle ED and Tanksley SD (1993) Map-based cloning of a protein kinase gene conferring disease resistance in tomato. Science 262 1432—1436

Mindrinos M, Katagiri F, Yu G-L and Ausubel FM (1994) The *A. thaliana* disease resistance gene *RPS2* encodes a protein containing a nucleotide binding site and leucine-rich repeats. Cell 78 1089—1099

Mittler R, Shulaev V and Lam E (1995) Coordinated activation of programmed cell death and defence mechanisms in transgenic tobacco plants expressing a bacterial proton pump. Plant Cell 7 29—42

Nelson RS, McCormick SM, Delannay X, Dube P, Layton J, Anderson EJ, Kaniewska M, Proksch RK, Horsch RB, Rogers SG, Fraley RT and Beachy RN (1988) Virus tolerance, plant growth, and field performance of transgenic tomato plants expressing coat proteins from tobacco mosaic virus. Bio/Technology 6 403—409

Neuhaus J-M, Fritig B, Linthorst HJM, Meins F, Mikkelsen JD and Ryals J (1996) A revised nomencalture for chitinase genes. Plant Mol Biol Rep 14 102—104

Ohl SA, van der Lee FM and Sijmons PC (1997) Anti-feeding structure approaches to nematode resistance. In: Fenoll C, Grundler FMW and Ohl SA (eds.) Cellular and Molecular Aspects of Plant-Nematode Interactions, pp 250—261, Kluwer Academic Publishers, Dordrecht

Opperman CH, Taylor CG and Conkling MA (1994) Root-knot nematode-directed expression of a plant root-specific gene. Science 263 221—223

Ori N, Eshed Y, Paran I, Presting G, Aviv D, Tanksley S, Zamir D and Fluhr R (1997) The *I2C* family from the wilt disease resistance locus *I2* belongs to the nucleotide binding, leucine-rich repeat superfamily of plant resistance genes. Plant Cell 9 521—532

Parker JE and Coleman MJ (1997) Molecular intimacy between proteins specifying plant-pathogen recognition. Trends Blochem Sci 22 291—296

Parker JE, Coleman MJ, Szabo V, Frost LN, Schmidt R, Van der Biezen EA, Moores T, Dean C, Daniels MJ and Jones JDG (1997) The *Arabidopsis* downy mildew resistance gene *Rpp5* shares similarity to the Toll and interleukin-1 receptors with *N* and *L6*. Plant Cell 9 879—894

Ponstein AS, Bres-Vloemans SA, Sela-Buurlage MB, Van den Elzen PJM, Melchers LS and Cornelissen BJC (1994) A novel pathogen- and wound-inducible tobacco (*Nicotiana tabacum*) protein with antifungal activity. Plant Physiol 104 109—118

Powell-Abel P, Nelson RS, De B, Hoffman N, Rogers SG, Fraley RT and Beachy RN (1986) Delay of disease development in transgenic plants that express the tobacco mosaic virus coat protein gene. Science 232 738—743

Rosso MN, Schouten A, Roosien J, Borstvrenssen T, Hussey RS, Gommers FJ, Bakker J, Schots A and Abad P (1996) Expression and functional characterisation of a single chain FV antibody directed against secretions involved in plant nematode infection process. Biochem Biophys Res Comm 220 255—263

Ryan CA (1990) Protease inhibitors in plants: genes for improving defences against insects and pathogens. Annu Rev Phytopathol 28 425—429

Salmeron JM, Oldroyd GED, Rommens CMT, Scofield SR, Kim HS, Lavelle DT, Dahlbeck D and Staskawicz BT (1996) Tomato *Prf* is a member of the leucine-rich repeat class of plant disease resistance genes and lies embedded within the *Pto* kinase gene cluster. Cell 86 123—133

Schots A, De Boer, J, Schouten A, Roosien J, Zilvertant JF, Pomp H, Bouwman-Smits L, Overmars H, Gommers FJ, Visser B, Stiekema WJ and Bakker J (1992) Plantibodies: a flexible approach to design resistance against pathogens. Neth J. Plant Pathol 98 183—191

Simons G (1997) Map based cloning of the *I2* and *Mi* genes from tomato, and the *Mlo* gene from barley. Abstract EMBO workshop: Plant diseases resistance gene function. Maratea, Italy, May 18-20

Simons G, Groenendijk J, Wijbrani J, Reijans M, Groenen J, Diergaarde P, van der Lee T, Bleeker M, Onstenk J, de Both M, Haring M, Mes J, Cornelissen B, Zabeau M and Vos P (1998) Dissection of the *Fusarium I2* gene cluster in tomato reveals six homologs and one active gene copy. Plant Cell 10 1055—1068

Song W-Y, Wang G-L, Chen L-L, Kim H-S, Pi L-Y, Holsten T, Gardner J, Wang B, Zhai W-X, Zhu L-H, Fauquet C and Ronald PC (1995) A receptor kinase-like protein encoded by the rice disease resistance gene *Xa21*. Science 270 1804—1806

Stirpe F, Barbieri L, Battelli LG, Soria M and Lappi DA (1992) Ribosome-inactivating proteins from plants: present status and future prospects. Bio/Technology 10 405—412

Strittmatter G, Janssens J, Opsomer C and Botterman J (1995) Inhibition of fungal disease development in plants by engineering controlled cell death. Bio/Technology 13 1085—1089

Strittmatter G, Gheysen G, Gianninazzi-Pearson V, Hahn K, Niebel A, Rohde W and Tacke E (1996) Infections with various types of organisms stimulate transcription from a short promoter fragment of the potato *gst*1 gene. Mol Plant-Microbe Interact 9 68—73

Stuiver MH, Tigelaar H, Molendijk L, Troost-van Deventer E, Sela-Buurlage MB, Storms J, Plooster L, Sijbolts F, Custers J, Apotheker-de Groot M and Melchers LS (1996) Broad spectrum resistance in transgenic carrot plants. In: Stacey G, Mullin B, and Gresshoff PM (eds) 8 th International Congress Molecular Plant-Microbe Interactions, Knoxville, TN, p. B-93

Taschner PE, Van Marle G, Brederode FT, Tumer NE and Bol JF (1994) Plants transformed with a mutant alfalfa mosaic virus coat protein gene are resistant to the mutant but not to wild-type virus. Virology 203 269—276

Tang X, Frederick RD, Zhou J, Halterman DA, Jia Y and Martin GB (1996) Initiation of plant disease resistance by physical interaction of AvrPto and Pto kinase. Science 274 2060—2963

Tavladoraki P, Benvenuto E, Trinca S, De Martinis D, Cattaneo A and Galeffi P (1993) Transgenic plants expressing a functional single-chain Fv antibody are specifically protected from virus attack. Nature 366 469—472

Thilmony RL, Chen Z, Bressan RA and Martin GB (1995) Expression of the tomato *Pto* gene in tobacco enhances resistance to *Pseudomonas syringae* pv *tabaci* expressing *avrPto*. Plant Cell 7 1529—1536

Tricoli DM, Carney KJ, Russell PF, McMaster JR, Groff DW, Hadden KC, Himmel PT, Hubbard JP, Boeshore ML and Quemada HD (1995) Field evaluation of transgenic squash containing single or multiple virus coat protein gene constructs for resistance to cucumber mosaic virus, watermelon mosaic virus 2 and zucchini yellow mosaic virus. Bio/Technology 13 1458—1465

Tsugita A (1971) Phage lysozyme and other lytic enzymes. In: Boyer PD (ed) The Enzymes, Vol 5, pp 344—411. Academic Press, New York

Urwin PE, Atkinson HJ, Waller DA and McPherson MJ (1995) Engineered oryzacystatin-I expressed in transgenic hairy roots confers resistance to *Globodera pallida*. Plant J 8 121—131

Urwin PE, Lilley CJ, McPherson MJ and Atkinson HJ (1997) Resistance to both cyst and root-knot nematodes conferred by transgenic *Arabidopsis* expressing a modified plant cystatin. Plant J 12 455—461

Vierheilig H, Alt M, Neuhaus JM, Boller T and Wiemke A (1993) Colonization of transgenic *Nicotiana sylvestris* plants expressing different forms of *Nicotiana tabacum* chitinase, by the root pathogen *Rhizoctonia solani* and by the mycorrhizal symbiont *Glomus mosseae*. Mol Plant Microbe Interact 6 261—264

Voss A, Niersbach M, Hain R, Hirsch HJ, Liao YC, Kreuzaler F and Fischer R (1995) Reduced virus infectivity in *N. tabacum* secreting TMV-specific full-size antibody. Mol Breeding 1 39—50

Whitham S, Dinesh-Kumar SP, Choi D, Hehl R, Corr C and Baker B (1994) The product of the tobacco mosaic virus resistance gene *N*: similarity to Toll and the interleukine-1 receptor. Cell 78 1101—1115

Whitham S, McCormick S and Baker B (1996) The *N* gene of tobacco confers resistance to tobacco mosaic virus in transgenic tomato. Proc Natl Acad Sci USA 93 8776—8781

Van der Wilk F, Postumus-Lutke Willink D, Huisman MJ, Huttinga H and Goldbach R (1991) Expression of the potato leafroll luteovirus coat protein gene in transgenic potato plants inhibits viral infection. Plant Mol Biol 17 431—439

De Wit PJGM (1992) Molecular characterization of gene-for-gene systems in plant fungus interactions and the application of avirulence genes in control of plant pathogens. Annu Rev Phytopathol 30 391—481

De Wit PJGM (1995) Fungal avirulence genes and plant resistance genes: unraveling the molecular basis of gene-for-gene interactions. In: Andrews JH and Tummerup IC (eds) Botanical Research, Vol 21, pp 147—185. Academic Press Limited, London

De Wit PJGM and Van Vloten-Doting L (1993) General introduction to biotechnology in plant breeding and crop protection. In: Vuijk DH, Dekkers JJ and Van der Plas HC (eds) Developing Agricultural Biotechnology in the Netherlands. pp. 19—23, Pudoc Scientific, Wageningen.

Woloshuk CP, Meulenhoff EJS, Sela-Buurlage M, Van den Elzen PJM and Cornelissen BJC (1991) Pathogen-induced proteins with inhibitory activity toward *Phytophthora infestans*. Plant Cell 3 619—628

Wu G, Shortt BJ, Lawrence EB, Levine EB, Fitzsimmons KC and Shah DM (1995) Disease resistance conferred by expression of a gene encoding H_2O_2 -generating glucose oxidase in transgenic potato plants. Plant Cell 7 1357—1368

Yoshimura S, Yamanouchi U, Katayose Y, Toki S, Wang Z-X, Kono I, Kurata N, Yano M and Sasaki T (1997) Map-based cloning of *Xa*-1, a bacterial blight resistance gene in rice. 5th International Congress of Plant Molecular Biology, Singapore, abstract 613.

Zhu Q, Maher EA, Masoud S, Dixon RA and Lamb CJ (1994) Enhanced protection against fungal attack by constitutive co-expression of chitinase and glucanase genes in transgenic tobacco. Bio/Technology 12 807—812

SUBJECT INDEX

abscisic acid (ABA) 381, 383, 501, 560
ACC oxidase 549, 560
ACC synthase 549
accelerated cell death (*acd*) 205, 302, 544
acd 301, 302, 405, 544, 555
acetophenone 241, 345, 346
acetylsalicylic acid, see aspirin
acquired resistance, see resistance
active oxygen species (AOS), see reactive
    oxygen intermediates
adhesion 250
ADP-ribose 293
aequorin 212
AFLP, see amplified fragment length
    polymorphism
aggressiveness 172 et seq., 180, 492
alarm pheromone 489
2S albumin 371, 373, 387, 430 et seq., 437,
    438
alkalinization 38, 65, 212, 213, 284, 290,
    437, 441
allelochemicals 561
allelomorphism 1, 9, 11
alternative oxidase 403, 510
amine oxidase 290
1-aminocyclopropane-1-carboxylic acid
    (ACC) 549, 550, 555
aminoethoxyvinylglycine (AVG) 550
amplified fragment length polymorphism
    (AFLP) 120, **121** et seq., 576
antagonism 430, 439, 530
antibacterial 300, 399, 426, 433, 439, 441
antifungal 142, 193, 194, 304, 326, 330, 332,
    335, 338 et seq., 342, 345, 358
antimicrobial compounds 325 et seq., 397,
    582, 586
antimicrobial protein 92, 371 et seq., 558
antisense 304, 371, 388
    - chitinase 401, 402
    - β-1,3-glucanase 15, 191, 401, 402
    - lipoxygenase 306
    - osmotin 402
    - systemin 559
    - viral RNAs 507, 580
antiviral 191, 375
    - chemical 481

- factor (AVF) 15
- protein 375, 398, 428, 442, 501
AOS, see ROI
apoptosis, see also PCD 191, 282, 292, 294,
    296 et seq., 306.
appressorial lobe 81
appressorium 81 et seq., 86, 87, 92, 93, 249,
    250, 252, 262,
arabinogalactan protein (AGP) 240, 261
arachidonic acid 197, 198, 544
arjunolic acid 341, 345, 346
artificial recombinants 11, 497
aspirin (acetylsalicylic acid) 550
ATPase
        H^+-ATPase 54, 65, 77, 283, 284, 286
        Na^+/K^+-ATPase 283
attenuation 502
autofluorescence 89, 346, 349
auxin, 544
avenacins 327, 331.- 334, 357, 358
avenacinase 333, 334
    - gene 333
avenacosides 327, 331, 332, 334
avirulence, see also virulence 10, 110, 124,
    144, 492
    - *Avr4* & *Avr9* regulation of 61
    - AVR4 & AVR9 peptides 62 et seq.
    - AVR4 & AVR9 plasmamembrane
    binding 67
    avrA 203
    avrB 203
    avrBs2 288
    avrBs3 203, 204, 288, 289
    avrC 203
    avrD, AVRD 200, 211
    avr/hrp box 203
    avrPph 203
    avrPto 198, 203, 209, 211, 299, 301,
    424, 577, 579
    avrRpm 203
    avrRps 203
    avrRpt 203
    - definition/explanation **55**, 103 et
    seq, 172
    - determinant/factor 5, 10 et seq., 54,
    55, 59, 103 et seq., 210, 213, 217,
    281, 282, 311, 352, 496, 498
    - exploitation 68, 577

- genes (*avr* genes) 42, 60 et seq., 80, 90, 190, 198 et seq., 199 (Table)
- proteins and type III secretion 286 et seq.
- proteins and targeting to the plant nucleus 204, 289
avoidance
- of defense responses 29, 62
- of disease (escape) 161, 167, 169, 170, 179, 180, 184
- programme 480

barnase 586, 587, 589
barstar 586, 587, 589
basal resistance, see resistance
bc-1, bc-2, bc-3, bc-u 495
Bdv1 118
benzoic acid 308, 340, 341, 344, 550 et seq.
benzothiadiazole (BTH) 554, 565
bioassay 26, 196, 338, 339, 372, 530, 531, 559, 561
bioluminescence 28
blight 23, 28, 29, 38
Bs2 42, 288
Bt toxins 588, 590

calcium (Ca^{2+}) 212, 213, 482
- antimicrobial proteins and 430, 433, 437, 439, 441, 444
- Ca^{2+} binding motif 290
- channel 212
- capsid dissembly and 6, 492
- cell wall 236, 238, 254, 255
- channel blocker 212, 285
- chelator 212, 260
- ionophore 212
- HR and 285, 291
- oxidative burst and 285, 290, 291
- papillae and 89, 259
- phytoalexin biosynthesis and 212
- programmed cell death and 297, 298
- XR and 284
callose 15, 53, 58, 60, 64, 77, 88 – 90, 191, 236, 239, 255, 260 – 263, 266, 401, 499, 539, 540
callose synthase 89
camalexin 300, 341, 345, 357

CaMV35S promoter 389, 396, 399, 402, 405, 588
caspase 296, 297, 301, 305, 311
- activated DNAse 297
catalase 303, 304, 307, 553
catechol 216, 327, 329, 552, 553
caulimovirus 482, 484
Ccn-D1 118
cell wall (bacterial) 420
cell wall (fungal – see also chitin) 58 et seq., 142, 257, 400
- elicitors 193, 196, 290, 352, 397, 420
- glucans 196
- PR-proteins and 375, 410, 420, 427, 543, 582, 585
cell wall (oomycete) 417, 543
cell wall (plant) 231 et seq.
- adherence/attachment to 57, 85
- appositions (see also papilla) 58, 78, 88, 90, 143, 205, 258, 259
- Ca^{2+} and 255
- composition 235 et seq.
- dicots 235
- monocots (grasses) 238
- degrading enzymes 29, 33, 36, 56, 242 et seq., 268
- disruption of 30, 31, 242, 353
- deposition of 261
- elicitors 189, 191 et seq.
- induced resistance and 541
- iron in 307
- penetration 80, 83, 84, 233, 234, 242
- by physical pressure 248, 249
- resistance to 249, 253 - 256
- proteins **40** et seq.
- reinforcement of 192, 213, 214, 234, 258, 289, 539
- sulphur in 347
- synthesis 142
- virus movement and 6, 15
- virus resistance and 14, 501
cell-to-cell spread 6, 15, 483, 486, 487, 491
cellulase 36, 247, 248
cellulose 30, 36, 235 et seq., 247, 248, 254, 260, 417, 442

Cf genes 53 et seq., 114, 121, 207, 295, 299, 577, 578
- AFLP markers and 121
- corresponding *Avr* genes 60 et seq.
- developmental regulation of 291
- engineered resistance and 579
- homologues of 68
- ligand-binding and 211
- modifying genes and 114
- structure of 65 et seq.
chalcone synthase 14, 91, 310, 345
chemotaxis (in pathogenic bacteria) 26, 35, 37, 43
chemotaxonomy 338
chitin
- binding domain 432
- oligomers 191, 193
- osmotin and 426
- similarity to Nod factors 218
chitinase (see also PR-3, -4, -8, -11) 59, 65, 91, 142, 310, 374 et seq., 385, 390 et seq., 402, 410 et seq., 543, 582
- antimicrobial activity 374 et seq.
- engineered resistance 582 et seq.
- synergism with glucanase 191, 419
chromosomes (dispensable) 355
cim 302, 556, 563, 564
cinnamyl alcohol dehydrogenase (CAD) 89, 270
- inhibitors 270
coat protein 3 et seq., 482, 483, 486, 500
- resistance gene 496, 498
coat protein-mediated resistance 505 et seq., 523, 580, 590, 592
coevolution 109, 149, 181, 183, 184, 232, 287, 399, 493
coi 382, 405
collar 84, 85
comovirus 482, 483, 500
compensation 162, 177 et seq., 184, 185
- of β-1,3-glucanase 401
conidiophores 54
conidium 54, 56, 58, 78 et seq., 136, 308, 310, 331, 346, 359, 528
co-suppression 112, 267
co-translational disassembly 6, 505, 523
coumarins 342, 345, 552
- secretion 341

cpr 300, 302, 404, 556, 558, 563
crop losses 8, 176, 179, 185
crop protection 2, 8, 149, 170, 184, 185, 481, 511, 521
cross links
- cell wall strengthening and 254, 259
- diphenolic 238
- in pectic polysaccharides 238, 240
- isodityosine 238, 254
cross protection 505
cryptic virus 480
cuticle 87, 233, 234, 241, 253, 267, 489
- adherence to 249
- appressorium formation and 252
- cutin 241, 242, 253
- hydrophobicity of 250
- penetration of 233, 243, 249, 268
- quiescent infections and 329
cutinase 81, 87, 233, **243**, 250, **268**
cyclic ADP-ribose 293
cyclic GMP 293
cysteine protease (see caspase)
cytogenetic manipulation 147
cytokine 12, 13, 57
cytokinins 502, 220
cytoplasmic aggregate 87, 256, 257
cytoplasmic male sterility (maize/*Helminthosporium maydis*) 144

dark green islands 501, 502, 507
defensin 373 et seq., **438** et seq., 558, 582, 585
detoxification 206, 271, 304
- avenacin 333
-phytoalexin 352 et seq.
2,6-dichloroisonicotinic acid (INA) 142, 537, 554 et seq., 564
dienes 306, 327, 329, 330
diffusible factors (DF) 35, 38, 43
diffusible signal factor (DSF) 38
diffusion potential, see membrane
disease control 24, 27, 330, **358** et seq., 510, 563, 592
disease resistance genes (see R genes)
disease terminology 127
dispersal 24, 111, **164** et seq., 168, 172, 182
Dm 110, 117
DNA laddering 296, 298, 306

DNA methylation 507
dnd 302, 308, 309
DPI, see NADPH oxidase
durable resistance 54, 64, 69, 111, 124, **143**, 149, 152, 497, 576, 586, 592

eds 298, 299
ein2 382, 555
electrolyte leakage 53, 60, 64, **283**, 311, 437
elicitins 196, 199, 216, 289, 291, 544
elicitor 40, 42, 189 et seq., 196, 256, 262, 281, 283, 285, 289, 301, 306, 309, 342, 380, 410, 552, 554, 582
    - endogenous 244, 247, 290, 310, 351, 533, 559, 560
    - general (see non-specific)
    - herbivore-specific 561
    - non-specific 58, 59, **60**, 64, 190, 345, 352
    - pathogen-derived 257, 352, 397, 420, 532
    - ROI and 290, 381
    - race-specific 53, 58, 59, **60** et seq., **64**, 67, 68, 91, 190, 291, 442, 577, 579
    - specific (see race-specific)
elongating secondary hypha 82, 86
endocytosis 257
endoglucanase 33, **36**, 38
endoplasmic reticulum
    - chaperone accumulation in 91
    - continuity with extrahaustorial membrane 289
    - phospholipids and 430
endoproteinases (see also PR-7) 543
engineered disease resistance 358, 390. 505, 506, **575** et seq.
enzyme secretion 33, **36**, 259, 333
epiphytic growth 25
epicuticular wax 26, 39, 235, 241
    - differentiation of appressoria and 252
    - germ tube growth and 251, 253
    - penetration of 243
    - sulphur accumulation in 346
ergosterol 198
escape 162, **167** et seq., 184, 536, 537
ethylene 64, 216, 294, **381** et seq.

    - biosynthesis 549, 560
    - inhibition of biosynthesis 270
    - insensitive mutants (see also named mutants) 555
    - receptor 405
    - response element binding protein 209, 301
    - signalling and 549, 558
    - wound-induced 560
etr1 382, 404, 550, 555, 557
exchange response, see XR
exocytosis 257
exoglucanase 36
extracellular
    - alkalinization 65, **284**, 290
    - antimicrobial proteins 384 et seq.
    - enzymes 33, 35, 36, 42, 43, 65, 290, 333
    -polysaccharide (EPS) **31**, 35, 42
        - biosynthesis 35
        - in pathogenesis or virulence 29, 30, 37
    - proteins (ECP) (see also cell wall proteins) 543
        - ECP1 56, 57
        - ECP2 56, 57
        - from *Phytophthora* spp. 196, 289
        - from yeast 197
    - suppressors 57
extrahaustorial matrix 82, **84** et seq., 261
extrahaustorial membrane 82, **84** et seq., 261, 289

fatty acid 191, 197, 306, 328, 329, 430 et seq.
    - epicuticular 329
    - hydroperoxides 306, 307
fibrillar material 25, 30
fibrillar matrix 31, 33
field trials 137, 400, 506, **590** et seq.
fitness 162, 176, **179** et seq., 211, 499
fungal proteins (in powdery mildew) 86
fungicides 78, 165, 176, 358, 387, 563
furanocoumarin (see coumarins)
furanoacetylenes 338, 352

G proteins 65, 214, 290

geminivirus 482, 508, 509
gene
    - disruption 334, 355
    - flow 164 et seq., 175, 181
    - for-gene relationship 9, 10, 33, 42,
    53 et seq., **102**et seq., 173, 174, 176,
    183, 190, 198, 204, 291, 308, 348,
    352, 492, 495, 577
    - mapping 10, 11, 80, 114, 116, 119,
    120, 122, 132, 134, 142, 493, 496,
    497
    - replacement 57
    - silencing 289, 502, 505, 507, 580
    -dosage dependence 487, 493, 494,
    500
genetic
    - manipulation 342, 358, 360
    - mapping 80, 116, 119, 131, 132,
    137, 148
    - resources 111, **145**, 146
genome organisation
    - TMV 5 et seq.
    - virus 482, 483 et seq., 491, 498,
    499, 506, 507
    - wheat 112, 113, 146
genotype x environment interaction 138, 496
germin, see oxalate oxidase
glucan 59, 191, **193** et seq.
β-1,3-glucanase (see also PR-2) 15, 40, 59,
    65, 91, 92, 142, 374 et seq., 386, **389**
    et seq.,**410** et seq., 540, 543, 582 et
    seq., 591
β-1,4-glucanase 247, 248, 410, 412
glucose oxidase (GO) 304, 581
glucosinolates 328, **334** et seq., 361
glucuronoarabinoxylan (GAX) 238, 240,
    247, 254
glutathione 64, 297, 303, 304
glutathione-S-transferase (GST) 292, 303,
    579, 587
glyceollin 282, 342, 344, 348, 360
glycine-rich protein (GRP) 240
glycopeptides 191, **194, 197**
glycoprotein 40, 60, 64, 66, 196, 215, 235,
    236, **238** et seq., 248, 352, 407, 409,
    417
golgi 262
guanidine 88, 90

guanylate cyclyse 293
guttation fluid 24, 26, 29, 31, 39, 43

halo 31, 88, 106, 258, 387
harpins 286, 287, **288**
haustorium 79, 82, **85** et seq., 90, 93, **260** et
    seq., 289, 298
heavy metals 544
hemicellulase 247
hemicellulose 30, 235, 238, 240, 254
hemolysin 202
herbivores 373, 557, 558, 561
hevein 373 et seq., 394, 396, 398, 413, 419,
    421, 422, **428** et seq., 432, 442, 558,
    583 et seq.
high affinity binding sites 53, 67, 69, **191** et
    seq., 441, 553, 583
histidine kinase 37
HM1 122, 141, 204, 577
H_2O_2, see hydrogen peroxide
hordothionin 399, 439
horizontal resistance (see resistance)
horseshoe crab (anti viral protein) 442, 443
host range
    - *Blumeria graminis*
    - PVX 64
    - TMV 7
    - viruses (general) **489** et seq.
    - *Xanthomonas campestris* 22 et seq.
host resistance (see resistance)
host selection (insects) 489
HR 3, 190, 256, **279** et seq., 360, 361, 493
    - *avr* genes and 281, **286** et seq.
    - *Blumeria graminis* 78, 80, 81, **90** et
    seq.
    - Ca^{2+} and 284, **285**, 290 et seq., 297,
    298
    - *Cladosporium fulvum* 55, 59 et seq.,
    67, 68, 337
    - collapse phase 281
    - *Colletotrichum lindemuthianum* 348
    et seq.
    - confluent necrosis 282, 283
    - host protein synthesis and 282
    - *hrp* genes and **286** et seq.
    - induction phase 281, 282, 296, 305,
    306
    - inoculum threshold and 282

- intrinsic plant genetic program 282, **294** et seq.
- ion fluxes 283 et seq., 291, **292**, 293, 311
- K$^+$ 284, 291
- K$^+$/H$^+$ exchange (XR) 284, 285
- latent phase 281, 282
- lipid peroxidation and 283, 284, 296, 304, 306, 307, 308
- low oxygen tension and 14
- MAP kinases and 292, 293
- morphology of 281, 282
- membrane changes and **283** et seq.
- nitric oxide (NO) and 215, 292, 293, 294, 305, 311
- oxidative burst and 14, 192, 214, 285, 289, **290** et seq.
- *Phytophthora infestans* 338
- *Pseudomonas syringae* and 200 et seq.
- salicylic acid (SA) and 292, 294, 299, 300, 302, 303, 307, **308**, 309, 311
- syringolide and 200
- TMV 1 et seq., 7, 9 et seq., 13
- *Verticillium dahliae* 346
- viruses (general) 496, 499, 509
- *Xanthomonas* 33 et seq., 39, 40, 348
hrc genes 286, 287
hrp genes (see also *hrc* genes) 31, **33** et seq., 201 et seq., 260, 282, **286** et seq.
    hrpA 35, 286, 287
    hrpB 34, 43
    hrp box 203
    hrpC 35
    hrpD 35
    hrpE 35
    hrpF 35
    hrp mutants 34, 39, 310
    hrpX 34, 43
    hrpXc 35, 43
    hrpXct 42
    hrpXv 34
- relationship to avr genes 201
Ht 117
Ht2 141
Htn1 141
hydathodes 22, 25, 26, 28, 39, 43, 282, 384

hydrogen peroxide (H$_2$O$_2$) 88, 90, 214, 258, 259, 290, 291, 307, 544, 581
hydrophobins 250
hydrophobicity 26, 233, 249 et seq.
hydroxyphaseollin 352
hydroxyproline-rich glycoprotein (HRGP) 41, 42, 238, 240, 261
hypersensitive reaction, see HR
hypersenstive response, see HR
hypersensitivity, see HR

IκB 300
immunisation 162, 170, 171, 184
immunity 299, 302, 491, 494, 523, 527, 556, 590
INA, see 2,6-dichloroisonicotinic acid
induced systemic resistance (ISR) **529**, 530, 533, 536, 537, 539, 544, 554 et seq., 561, 562, 564
infectivity assay 2
infochemicals 561
inositol-1,4,5-triphosphate 214
insect
    - α-amylase inhibitors 438
    - attractants 335, 489, 561
    - defensins 438, 439
    - development 13
    - gene-for-gene 107
    - resistance to 142, 373, 528, 558, 587, 590
    - toxic proteins and 430, 437 et seq., 442, 543, 588
    - as vectors 2, 167, 169, 172, 174, 232, 483, 484
interferon 15, 508
introgression 112, 122, 146, 147, 150
ion channel 62, 283, 285, 437
ion flux (see also HR, membrane) 212, 214, 256, **283** et seq., **291** et seq., 311, 437, 441, 587
ion pumps (see also HR, membrane) 283
isoflavones 327, 329
isoflavone reductase 345
isoflavonoids 338, 341, 342, 349, 352, 358, 360

jasmonic acid / methyl jasmonate
    (JA/MeJA) 215, 216, 304, **381** et
    seq., 555, 558, 560

K⁺/H⁺ exchange, see XR
2-keto-3-deoxyoctulosonate (KDO) 32
kievitone 344, 349, 358
kinase (see also individual kinases, his-,
    MAP-, protein kinase C, receptor-,
    ser/thr-) 115, 116, 206, 207, 209,
    210, 212, 213, 293, 294, 295, 299,
    577

L⁶ 115
lacinilenes 341, 346, 348, 349
LAR, see resistance
latent/quiescent infection 28, 329
leaf spot 22 et seq., 36, 107, 135, 138, 141,
    143, 302, 334, 528, 529, 562
lectin 41, 194, 413, 414, 421, 427 et seq.,
    583, 588
lesion mimic mutants 205, 206, 291, 587
leucine-rich repeat (LRR) **66** et seq., **206** et
    seq., 210, 295, 299, 301, 577
lignification 15, 64, 241, 248, 249, 254, 255,
    270, 362, 499, 541
lignin 30, 88, 90, 142, 213, 235, 236, **239** et
    seq., 253, **254**, 258, 342
    - biosynthesis 239, 259, 267
    - biosynthesis inhibitors 270, 271
    - degradation 248
    - induced resistance and 539 et seq.
    - wall appositions 260
ligno-peroxidase 248
linoleic acid 306
linolenic acid 215, 306, 560
lipid A 32, 40
lipid peroxidation 65, **283** et seq., 296, 304,
    306 et seq., 337
lipid transfer proteins 91, 373 et seq., 379,
    382 et seq., 395, 399, **430** et seq.,
    437, 438
lipopolysaccharide (LPS) 32, 33, 38, 42
    - biosynthetic mutants 32
    - HR and 39, 40
    - induced resistance and 526
lipoxygenase (LOX) 64, 198, 292, 304, **306**,
    330, 560

- antisense 306
Llsl 206, 301, **302**
localization of
    - AMPs 376, 387
    - β-1,3-glucanase
    - glucosinolate/myrosinase 335, 336
    - glycoproteins 40, 41
    - inhibitors
    - phytoalexin **345** et seq., 360
localized acquired resistance (LAR) see
    resistance
localized induced resistance (LIR) see
    resistance
LOD score 132, 138
LOX, see lipoxygenase
Lrl 117
Lr7 123
Lr-9 107, 117, 121, 123
Lr-10 116, 117
Lr-20 110
Lr-23 112
Lr24 117, 121, 123, **148**, **149**
Lr31 113
Lr34 118, 143, 144
LRR (see leucine-rich repeat)
lsd 291, 292, 301, **302**, 303, 308, 405, 544,
    555, 556
luteone 327
luteovirus 482, 484
lysozyme 33, 374, 410, 413, **415**, 416, 419,
    421, 423, 543, 581

macrophage 215, 285, 290, 407, 417
mannitol 56, 86, 220
MAP kinase 292, 293, 554
marketable quality 480
melanin 30, 249
membrane, see also HR
    - bacterial outer membrane 32, 286
        - proteins (see also *hrc*, *hrp*)
            23, 32, 285 et seq.
    - cytokine receptors 57
    - host plasma membrane 33
        - blebbing 295
        - coated pits/vesicles 257, 258
        - degradation 36, 64
        - depolarisation 212, 214, 256
        - diffusion potential

- electrogenic pump
- elicitor binding 67
- haustoria and 82, 84, 85
- H⁺-ATPase 65, 77, 283
- ion fluxes 291
- irreversible membrane damage 38, 40, 90, **281** et seq., **305** et seq., 542
- permeability changes 38, 282, 284
- peroxidation 283, 306
- pit- 266
- potential 212, 283, 297, 426
- proteins 67, 283, 299, 375
- receptors **192** et seq., 211, 257, 420, 560
- sterols 331 et seq.
- transport 12
- targets for AMPs **375** et seq., 424, 426, 441, 581
- virus 482
MeSA, see methylsalicylate
metalloprotease 36
metapopulation 166, 174, 175, 181 et seq.
methylsalicylate (MeSA) 381, 560, 561
7-methyl guanosine 483
Mi 143
microenvironment 35, 307, 326
microsatellites 120, 122
migration
    - nuclear 256, 295
    - pathogen 111, **164** et seq., 176
mild strains 504, 505, 580
mites 528, 561
mitogen-activated protein kinase, see MAP kinase
Ml-a **80**, 81, 114, 141
ml-o **80**, 117, 143, 205, 206, 260, 270, 295, 299, 307, 493, 577, 578
mobile signal 381, **546**, 549 et seq., 560
molecular markers 80, 93, 101, 114, 116, 117, 119, 122, 124, 131 et seq., 136, **137**, 150 et seq., 576
monoclonal antibodies 15, 23, 29, 32, 33, 40, **41**, 245, **263, 264**, 588
movement protein 6, 10, 15, 199, 483, 486, 487, 498, 500, 507, 508
MPS

multilines **150**, 168, 169, 171, 176, 182 et seq.
multipartite genome 483
mutagens/mutagenesis 6, 9, 11, 33, 67, 110, 114, 122, 180, 262, 268, 333, 411
mutants, see individual entries
mv 143

N-, N'-genes 7 et seq., **12**, 15, 115, 206, 293, 309, 496, 526, 533 et seq., 540, 553, 579
NADPH oxidase (plasma membrane) 65, **213** et seq., **259, 290**, 293, 294
    - inhibition by diphenylene iodonium (DPI) 290
nahG 64, 216, 304, 308, 381, 382, 404, 552, **553** et seq., **557**, 558
near-isogenic lines (NILs) 55, 56, 58, 60, 68, 69, 80, 83, **117**, 123, 336
neck band **82, 84**, 85
necrosis 11, 14, 15, 23, 26, 28 et seq., 38, 59, 60, 64, 118, 143, 196, 205, 206, 209, 216, 281, 282, **296**, 298, 304, 307, 309, 337, 350, 525 et seq., 530, 532 et seq., 544, 546, 549, 556, 563, 577, 587
NF-κB 300
NIM1/NPR1 292, 299, 300, 304, 404, 556 et seq.
nitric oxide (nitrogen II oxide), see NO
NO 215, 292, **293**, 294, 305, 311
NO synthase 215, **293**, 294
Nod factors 93, 218
non-host resistance, see resistance
non-race-specific resistance, see resistance
non-stringent surveillance 205
norbornadiene 550
nuclear localization signal (NLS) 204, **288** et seq., 304
nucleotide binding site (NBS) 12, 13, 115, 206, 207, 295, 299, 301, 496
null allele 9

O₂⁻, see superoxide anion
oligogenic resistance (see resistance)
oligogalacturonides 194, 559
O-methyltransferase (OMT) 91, 92

open reading frame (ORF) 4, 5, 11, 33, 62, 200, 204, 482, 483
osmotin 216, 394, 399, 402, 424 et seq., 585
ozone 383, 544
overexpression 267, 297, 299, 301, 303, 304, 388 et seq., 397 et seq., 403, 543
oxalate oxidase 91 et seq., 290
oxidative burst 14, 40, 64, 65, 214, 215, 256, 258, 259, 285, 289, **290**, 291, 293, 303, 307, 308, 311, 337, 581

PAL, see phenylalanine ammonia-lyase
PAD genes/mutants 299, 300, 311, 357
papilla 78, 80, 82 et seq., **88** et seq., 234, 256, 258 et seq., 265, 270, 307, 310, 360, 540
pathogenesis-related (PR) proteins 14, 15, 57, **59**, 63, 91, 92, 142, 200, 213, 216, **373** et seq. **381** et seq., **390** et seq., **403** et seq., 428, 510, 538, **543** et seq., 548, 549, 560, 563, 582
PR genes 65, 209, **299** et seq., 307 et seq., 372, 381 et seq.587
PR-1 91, 262, 265, 292, 293, **373** et seq., **377** et seq., 390, 400, **406** et seq., 510, 550, 554, 555, 585
PR-2 142, **373** et seq., **377** et seq., 390 et seq., 397, 400, 402, **410** et seq., 543
PR-3 373 et seq., **377** et seq., 390 et seq., 397, 400 et seq., **410** et seq., 429, 543, 583
PR-4 91, **373** et seq., **377** et seq., 394, 400, **421** et seq., 543, 583
PR-5 92, **373** et seq., **377** et seq., 394, 399, 400, 402, 423 et seq., 543, 585
PR-6 **373**, 543, **558**
PR-7 543
PR-8 373 et seq., **377** et seq., 391 et seq., **410** et seq., 543, 583
PR-9 543
PR-10 , 543,553
PR-11 **373** et seq., **377** et seq., **410** et seq., 543, 583
pathogenicity
  - definition **55**
pathogenic bottleneck 496
PCD, see programmed cell death, apoptosis

PcX 114
pectate lyase 194, 210, **246**
pectin 30, 235, 237, 238, 240, 241, 244 et seq., 254, 259, 260, 262, 266, 268, 269
pectin hydrolase **246**
pectin lyase **246**
pectin methyl esterase (PME) 30, 245
pectolytic enzymes 30, **244,** 268, 559, 560
penetration of the cell/leaf **233** et seq.
  - *Blumeria graminis hordei* 81, et seq., 87 et seq., 143, 259, 307
  - *Cladosporium fulvum* 337
  - *Colletotrichum lagenarium* and induced resistance 536
  - *Colletotrichum lindemuthianum* 310, 329
  - cutinase and 268
  - *Fusarium oxysporium* 261 et seq.
  - hydolytic enzymes and 242, 244, 247
  - endo- and exocytosis 257
  - intracellular events and 255 et seq.
  - *Magnoporthe grisea* 249
  - nuclear migration and 295
  -*Peronospora tabacina* and induced resistance 535
  - physical pressure and 248 et seq.
  - *Phytophthora infestans* 262
  - plant cell wall modefication and 258
  - *Plasmodiophora* 249
  - recognition and 192, 232
  - stomate recognition and 252
  - surface hydrophobicity and 250
permatins 424
peroxidase (see also ligno-peroxidase, PR-9) 40, 65, 88, 91, 142, 213, 258, 259, 267, 290, 303, 307, 342, 540 et seq., 549, 550
pesticides 102, 152, 346, 481, 504, 576
Pg3 117
Pg9 114
phage 484
  - lysozyme 413, 415, 581
  - plaques 333
  - type 23

phaseollin 337, 338, 341, 342, 349 et seq., 538

phaseollinisoflavan 349, 350, 538

phenolics (see also salicylic acid) 30, 58, 65, 88 et seq., 206, 234, 238, 240, 248, 254, 258 et seq., 267, 271, 302, 308, 329, 341, 492, 499, 534, 540

phenylalanine ammonia-lyase (PAL) 14, 89, 91, 92, 142, 214 et seq., 260, 267, 270, 293, 308, 310, 345, 540, 551, 552
- inhibitors 270, 271, 308

phenylpropanoids 77, 342, 344, 552

phospholipase 202, 214, 284, 285, 306

phylloplane 329

phytoalexin
- biosynthesis 309, 310, 342 et seq., 352, 354
- deficient mutants (see *PAD* genes/mutants)

phytoanticipins **325** et seq.

phytoreovirus 482, 483

PI, see protease-inhibitor

pigB 36, 38, 43

PIIF, see protease-inhibitor inducing-factor

pisatin 338, 354 et seq.

pisatin demethylase 354 et seq.

plant
- activator 565
- breeding 7, 9, 54, 62, 68, **101** et seq., 175 et seq., 336, 493, 497, 509, 510, 576
- growth regulators/hormones 14, 93, 218, 380, 382, 501, 549, 560, 563
- population 108, 117, 118, 123, 129, **132** et seq., **161** et seq., 204, 205, 489, 490, 492

plantibodies 506, 507, 582, **588**

plasmodesmata 6, 15, 486, 499, 502, 539

Pm1 110

Pm3 114, 117

pokeweed antiviral protein (PAP) 394, 398, 399, 501, 509

polyacrylic acid 544

polygalacturonase (PG) 36, 194, 244, 246, 268

polygalacturonase inhibiting protein (PGIP) 244

polygalacturonate lyase (PGL) 36

polyprotein 483, 500

potyvirus 482, 483, 486, 495 et seq., 504, 506, 507

predisposition 162, **170**, 171, 184

primary germ tube (PGT) 58, 78, 79, **81** et seq., 87 et seq.

programmed cell death (PCD) 14, 191, 270, 282, 292, 294, **296** et seq., 301 et seq., 305, 309, 311
- definition 296
- morphology 296
- phases 296
- role of mitochondria 297

proline-rich protein (PRP) 142, 238, 240, 431

protease/proteinase 33, 36, 61, 199, 248, 305, 431, 435, 483, 500, 543, 588, 589
- cysteine protease (see caspase)
- inhibitors (see also PR-6)62, 218, 373, 443, 500, 521, 543, 558, 588, 592
- inhibitor inducing-factor (PIIF) 558

protein kinase 116, 206, 208 et seq., **212** et seq., 293 et seq., 299, 301, 424, 554,577

protein kinase C 304

protein phosphorylation 212 et seq., 285

protocatechuic acid 327, 329

PR-protein, see pathogenesis related protein

Prf 12, 207, **209** et seq., 299, 578

Pti **209**

Pto 116, 122, 199, 203, 206 et seq., **209** et seq., 295, 299 et seq., 424, 577 et seq.

puroindolin 373, **430** et seq., 438

quantitative trait loci (QTL) 101, 119, 121, 122, 131, **132** et seq., 151, 152

quantitative resistance 102, 116, 122, **125** et seq., 172 et seq., 176, 185, 389

quiescent/latent infections 28, 329

quorum sensing 38

R-genes, see resistance

race-non-specific resistance, see resistance

race-specific resistance, see resistance

random amplified DNA fragment (RAPD) 120, **121** et seq., 576

RAPD, see random amplified DNA fragment

Rar1, Rar2 114, 299

Rcr-1, Rcr-2 114

reactive oxygen intermediates (ROI) 37, 38, 40, 192, 213, **214**, 215, 259, 279, **285**, 289, **290**, 291, **292** et seq., 296, 297, 302, 303, 305 et seq., 311, 337, 378, 381 et seq., 581

reactive oxygen species, see reactive oxygen intermediates

recA 35

receptor
- ethylene 382, 405
- elicitor 191, 193, 196, 197, 210, 217, 257, 380, 420
- FSH 210
- interleukin I/cytokine 12, 13, 57, 206
- kinase 116, 218, 293, 318, 424, 425
- in quorum-sensing 38
- R-gene domains 11, 59, 62, 104, 192, 193, 210, 212, 213, 217, 292, 577
- systemin 560
- Toll protein 13, 206
- TNF 57
- two component regulatory systems 37
- for viruses 484, 504

regulation of pathogenicity factors (*rpfA-F, rpfN*) 36 et seq., 43, 44

regulatory genes 33, 37

replicase 4 et seq., 10, 11, 199, 483, 486, 487, 491, 498, 500, 502, 507, 508

resistance
- basal 536, 537
- breeding 68, **101** et seq.
  - barley/powdery mildew 147
  - lettuce/downy mildew 110
  - rice/bacterial blight 122
  - sorghum/milo disease 145
  - tomato 146
  - wheat/stem rust 113, 146
- durable 8, 64, 69, 80, 101, 111, 124, 139, **143** et seq., 149, 151, 152, 175, 177, 497, 510, 576, 579, 586, 592

- genes (R-genes), see individual genes (*bc, Bdv, Bs, Ccn-D, Cf, Dm, HM, Ht, Htn, Lr, mi, Ml-a, ml-o, mv, N, N', PcX, Pg, Pm, Pto, Rpg, Rph, RPM, RPP, RPS, Rx, Sr, Tm, Xa, Yr*)
  - cereal rust (Table) **147**
  - flax rust R genes (*K,L,M,N,P*) 109
  - Tables 207, 295, 578
  - virus (Table) 498
- horizontal **124**, et seq., 139, 141, 143, 144, 492, 536, 563
- host (see specific)
- induced 39, 83, 170, **232**, 244, 250, **255** et seq., 265, 266, **269** et seq., 403, **521** et seq.
- localized acquired (LAR) **524** et seq., 532
- localized induced resistance (LIR) 526, 529, 533
- non-host **190**, 232, 335 et seq., 360, **489** et seq., 500, 511
- oligogenic
- qualitative
- quantitative
- race-non-specific (non-race-specific) 88, 260, 338, 577
- race-specific 190, 484
- specific, see race-specific
- systemic acquired (SAR) 170, 192, 216, 299, 300, 302, 304, 308, 309, 510, **524** et seq., 530 et seq., 537, 541, , 543, **546** et seq., 587
- terminology 127
- vertical **124**, 125, 139, 141

restriction fragment length polymorphism (RFLP) 23, 29, **119** et seq., 180, 496, 576

resveratrol 342 et seq., 358, 359, 586

retrovirus 484

reverse transcriptase 484

RFLP, see restriction fragment length polymorphism

rhabdovirus 482

rhizobacteria 529 et seq., 537, 539, 544, 548, 554, 557, 558, 562, 564

rhizosphere 329, 561

ribosome-inactivating protein (RIP) 373,
    375, 376, 378, 379, 384, 391, 395,
    398, 403, **427**, 428, 501, 509, 582,
    585, 588
ribozyme 508
ribonuclease (see also PR-10) 6, 507, 543
ribonuclease inhibitor 210
RIP, see ribosome inactivating protein
RNA polymerase 4, 482, 507
ROI, see reactive oxygen intermediates
root infection 26 et seq., 205, 211, 232, 244,
    249, 255, 258, 261 et seq., 327, 333,
    334, 385, 386, 389, 496, 527, 529 et
    seq., 554, 561, 564, 587 et seq.
Ror1, Ror2 299
ROS, see reactive oxygen intermediates
Rp1 204
Rp3,4 143
rpfA-H 37
rpfC 37
rpfF 36
rpfG 37
Rpg1 108, 115
Rpg4 200
Rph12 141
RPM1 206, 210
RPP5 207
RPP8 207
RPS2 115, 206, 210, 496, 579
Rx 496

SA, see salicylic acid
sai1 300, 556
sakuranetin 326, 327
salicylhydroxamic acid (SHAM) 510
salicylic acid (SA) 64, **216**, 271, 294, 299,
    311, 564
    - binding protein 229
    - biosynthesis 216, 551
    - hypersensitive response and 302,
    308, 403
    - hydroxylase, see *nahG*
    - induced resistance and 142, 300,
    403, 510, 537, **550** et seq.
    - oxidative burst and 14, 302, 303
    381, 554
    - PI synthesis and 560
    - radicals 307

- signalling and **216**, 292, 294, 382,
    383, 510, 551, 553, **554**, **557**
saponins 327, **331** et seq.
SAR, see resistance
satellite RNA 481, 502, 509
SCAR, see sequence-characterized amplified
    region
secondary metabolism 91, 334 et seq., 342,
    360, 362
seed certification 480
senescence 544, 549, 563
sequence analysis 12, 67, 286
sequence-characterized amplified region
    (SCAR) 121
sequence-tagged site (STS) 121, 123
serotypes 23
ser/thr protein kinase 116, 208, 299, 301,
    424
sesquiterpenes 58, 341, 348, 360
SHAM, see salicylhydroxamic acid
siderophores 533
signal transduction 13, 37, 38, 68, 69, 93,
    107, **209**, 211, **212** et seq., 257, 285,
    286, 297, 298, 300, 301, 305, 311,
    342, 357, 381, 382, 400, 487, 510,
    533 et seq., 558, 560, 561, 579
silicon 88, 90, 252, 255, 258 et seq.
simple sequence repeats (SSR), see
    microsatellites
site-directed mutagenesis 11, 268
size exclusion limit 6
SOD, see superoxide dismutase
Sr7a 113
Sr9b 114
Sr10 113
Sr11 114
Sr15 110
Sr26 111
Sr33 114
sterols 189, 191, 198, 331, 332, 426
stoma 24, 26, 39, 54, 59, 61, 89, 233, 243,
    244, **250** et seq., 282, 337, 383, 384
stringent surveillance 205
STS, see sequence-tagged site
stunting 2, 495
suberin 235 et seq., **253**, 255, 260, 266, 539
sub-genomic messenger RNAs 483
subliminal infection 491, 494, 500

superoxide anion (O$_2^-$) 213 et seq., 259,
 290, 307, 382, 383
superoxide dismutase (SOD) 38, 40, 87, 213,
 259, 291
suppressor 57, 112, 113, 197, 302, 360, 380
synteny **114**, 115, 120
syringolide 191,199, **200**, 201, 203, 211
systemic acquired resistance (SAR), see
 resistance
systemic spread 3, 8, 384, 524
systemic signal 192, 213, 216, 308, 507, 559
systemin 218, **559**, 560

termination codon 4, 483
thaumatin-like proteins, see PR5
thionin 373 et seq., 380, 383, 384, 387, 395,
 399, **434** et seq., 443, 558, 581, 582,
 585, 586
threonine-hydroxyproline-rich glycoprotein
 (THRGP) 240
Tm-1, Tm-2 495, 497
TNF, see tumor necrosis factor
TNFR, see tumor necrosis factor receptor
tobamovirus 7, 8, 482, 486, 492, 498, 508
tolerance (to disease) **127**, 131, 144, 151,
 162, **177** et seq., 184, 185, 591
tomatine 328, 331 et seq., 337
topography (leaf surface) 233, 251 et seq.
tospovirus 482
toxins 30, 56, 106, 145, 151, 200, 204, 205,
 255, 295, 298, 531, 586, 588, 590
transcription factors 12 et seq., 61, 209, 289,
 292 et seq., 299 et seq., 381, 556
transformation 11, 116, 122, 147, 203, 268,
 333, 334, 355, 389, 401, 505, 507,
 511, 543, 576, 579
transgenes 64, 65, 112, 211, 212, 216, 267,
 293, 303, **304**, 308, 359, 381, 382,
 388, 389, **390** et seq., **402** et seq.,
 405, 501, 503, 505 et seq., 509, 511,
 523, 552, 555, 559, **575** et seq., 581,
 592
transmission
 - bacterial 24
 - fungus 172, 174
 - virus 2, 481, 483, 484, **485**, 486,
 489, 501, 507, 509
transposon tagging 11, 204, 207, 577

tuliposides 328, 330, 331
tumor necrosis factor (TNF) 57
tumor necrosis factor receptor(TNFR) 57
T-urf13 145
turgor pressure 235, 242, 247, 249
two component gene cassette 68, 69, 579,
 586, 589, 592
two-component regulatory systems 37
tyloses 266, 346
tymovirus 482
type III secretion system 34, **201** et seq., **286**
 et seq., 292, 311

ultra violet radiation (U.V.) 43, 215, 377,
 378, 383, 544
 - fluorescence 88, 258, 357

variety mixtures **150**, 168, 171, 176, 179,
 185
vascular plugging 30
vascular wilt 345, 346
vectors 2, 172, 174, 176, 181, 232, 483, 484,
 486, 489, 501
vegetative propagation 481
vertical resistance, see resistance
viniferins 341 et seq.
viroids 481 et seq., 502, 505, 523, 528
virulence
 - definition **55**
 - factors 63, 64, 211
 - genes 55, 112, 182
 - dispersal 182
virus
 - definition 481
 - genetic map 4, **5**, 10, 496
 - 'lifestyle' 2
 - localization 15, 499, 540
 - multiplication 4, 6 et seq., 11, 13,
 14, 29, 30, 32, 37, 428, 481, 484, et
 seq., 491, 492, 494, 498, 500, 502,
 504 et seq., 509 et seq., 523, 534
 - particle structure
 - replication 5, 14, 191, 481, 483,
 484, **485**, 491, 492, 505, 523, 534
 - inhibitors 14
 - structure 3 et seq.
volicitin 561

wounds/wounding 2, 4, 26 et seq., 41, 42,
215, 218, 233, 236, 243, 244, 250,
255, 260, 266 et seq., 346, 377, 378,
383, 484, 486, 532, 539, 544, 549,
551, 557 et seq., 582, 588
wyerone derivatives 337 et seq., 346, 353,
354

Xa 117, 577
Xa21 122, 206
xanthan 31, 33, 35 et seq.
xanthomonadin 25, 35, 38, 43
XR (K$^+$/H$^+$ exchange response) 284, 285
X-ray microanalysis 346
xylanase, 194, 195, 216, 247
xyloglucan 235 et seq., 240, 247, 254, 262,
264

Yr18 118, 143

SPECIES INDEX

Aegilops ventricosa 119
African cassava mosaic virus (ACMV) 509
Agrobacterium 204, 288
Agropyron elongatum 111, 119, 147—148,
alfalfa 341—345, 360—393, 584
alfalfa mosaic virus (AMV) 389—390, 483,
    509, 592
Allium cepa: see onion
Alternaria alternata 298, 390, 397—398
Alternaria brassicicola 382, 434, 558
Alternaria cucumeris 393—394
Alternaria longipes 390, 396, 398
Alternaria radicini 393,
Amaranthus caudatus 396, 429, 432, 585
Ammophila arenaria 168
apple 106, 170, 178, 340, 341
Arabidopsis thaliana 12, 22, 31—32, 39,
    110, 115, 119, 142, 203, 206, 207,
    215—217, 271, 281, 288—295,
    298—311, 341, 357, 376—385,
    399—410, 415, 493, 496, 521, 529,
    532—538, 544, 550—563, 578—
    579, 586—589
Arachis hypogea: see groundnut
army worm 528, 561
Ascochyta pisi 135
Ascochyta rabiei 355
Aspergillus niger 581
Atriplex nummularia 429
Avena longiglumis 333
Avena sativa: see oat
Avena strigosa 333
avocado 327—330

Bacillus amyloliquefaciens 586, 589
Bacillus campestris 22—23
Bacillus thuringiensis 588
Bacterium campestris 23
banana 266, 329, 565
barley 77—100, 103, 106, 110, 114—119,
    126—147, 165—171, 177, 205—
    206, 253, 257—260, 270, 280—281,
    295—299, 307, 333, 374, 376—380,
    383—388, 391, 395, 399, 408, 410—
    416, 421—423, 428—429, 434

barley yellow dwarf virus (BYDV) 480
bean common mosaic virus (BCMV) 489,
    495
bean yellow mosaic virus (BYMV) 119
beet yellows virus (BYV) 489
blackcurrant 326
Blumeria graminis 77—100, 295—299, 307
Botrytis cinerea 242, 258, 339, 358—360,
    393—398, 402—405, 558, 584, 586
Botrytis fabae 326—327, 353—354
Botrytis tulipae 334, 337, 353
Brassica spp. 22, 26, 32, 40—41, 327, 336,
    376, 387
Brassica campestris: see turnip
Brassica juncea 335—336
Brassica napus: see oilseed rape
Brassica oleracea: see cabbage
Bremia lactucae 106, 110, 176
broad bean 177, 257, 262, 337—340, 346,
    353—354
Bromus spp. 280

cabbage 21—44, 142, 386, 424
cacao swollen shoot virus (CSSV) 481
Canavalia ensiformis (jack bean) 387, 415—
    416
Candida albicans 426
canola: see oilseed rape
Capsicum: see pepper
cassava 32, 43, 576
castor bean 427
cauliflower 22
caulifower mosaic virus (CaMV) 396, 402,
    583
Cercospora beticola 398
Cercospora nicotianae 267, 390—393, 397,
    402—404, 553, 584
cereals 114, 121, 126, 130, 235, 240, 255,
    427, 433—434, 480, 565
Chalara elegans 386
cherry 426, 504
citrus tristeza virus (CTV) 480, 503—504
Cladosporium cladosporioides 328
Cladosporium cucumerinum 527, 529, 541
Cladosporium fulvum 53—75, 106, 114,
    199, 207, 211, 291, 295, 299, 337,
    421, 442, 578—579

Cladosporium herbarum 339—340
Clavibacter michiganense 395, 433, 441
Claviceps purpurea 244—245
Clitoria ternatea 441
Cochliobolus carbonum 141, 204, 577—578
Cochliobolus heterostrophus 355
Cochliobolus sativus 136
cocoa 341, 345—347, 358
Colletotrichum spp. 107, 329
Colletotrichum circinans 258, 327
Colletotrichum dematium 145
Colletotrichum gloeosporoides 327
Colletotrichum lindemuthianum 290, 310, 348—352, 527, 538
Colletotrichum lagenarium 393—394, 527—529, 535—541, 547—551, 562
Cotesia marginiventris 561
cotton 215, 262—266, 283—284, 290, 341, 346—349
courgette 480, 503—504, 575
cowpea 257, 285, 295, 298, 306, 500, 588
cowpea chlorotic mottle virus (CCMV) 500
cowpea mosaic virus (CPMV) 500
Crambe abyssinica 435
crucifers 21—52, 249, 327, 334—335, 341, 345
cucumber 7, 281, 283, 307, 362, 382, 387, 392—397, 419, 503, 521, 527—536, 540—543, 547—552, 562—564, 590
cucumber mosaic virus (CMV) 394, 489, 496, 502, 508—509, 580, 590
Cylindrosporium concentricum 394, 400
cytomegalovirus 407

Datura spp. 32
Dothistroma pini 328
Drosophila 13, 417, 439, 440
Dutch elm disease: see *Ophiostoma ulmi*

elm 266
Entamoeba histolytica 433
Erwinia spp. 34, 286, 309, 437
Erwinia amylovora 201, 288
Erwinia carotovora 242, 304, 403—404, 558, 581
Erwinia chrysanthemi 202, 268
Erwinia tracheiphila 529

Erysiphe fischeri 174
Erysiphe graminis: see *Blumeria graminis*
Escherichia coli 34, 41—42, 200—203, 444, 526

flax 12, 103, 109, 115, 119, 207, 295, 426, 577, 578
frangipani 7
French bean 41—42, 106, 110, 121, 131, 135, 201, 252, 260, 281—301, 310, 341, 351, 358
Fulvia fulva, see *Cladosporium fulvum*
Fusarium spp. 143, 211
Fusarium avenacearum 334
Fusarium culmorum 334, 437, 441
Fusarium graminearum 334
Fusarium moniliforme 388
Fusarium oxysporum 207, 295, 328, 331, 380, 531, 536, 584
  f. sp. *cucumerinum* 529
  f. sp. *dianthi* 529, 539
  f. sp. *lycopersici* 261—262, 267, 386, 392, 398, 578
  f. sp *matthiolae* 395, 558, 586
  f. sp. *pisi* 261, 268
  f. sp. *radicis-lycopersici* 262, 267
  f. sp. *raphani* 529, 537, 562
  f. sp. *vasinfectum* 262-265
Fusarium solani 134, 268, 354

Gaeumannomyces graminis var. *avenae* 333—334
Gaeumannomyces graminis var. *tritici* 333—334, 357
Galanthus nivalis: see snowdrop
Galium aparine 163
garlic 481
Globodera pallida 588
Gloeocercospora sorghi 327
Glomus mosseae 400
Glycine canescens 174
Gomphrena globosa 496
Gossypium hirsutum: see cotton
grapevine 341—343, 359, 586
grasses 78, 81, 111, 146—149, 168, 235, 238, 240, 247, 251—255
groundnut 342

Halobacterium halobium 405, 587
Helminthosporium carbonam 327
Helminthosporium maydis 144
Helminthosporium victoriae 106
Heterodera glycines 140
Heterodera rostochiensis 106
Heterodera schachtii 578, 588—589
Heuchera sanguinea 441
Hevea brasiliensis: see rubber tree
hop 266
Hordeum chilense 253
Hordeum spontaneum 146—147
Hordeum vulgare: see barley
horse chestnut 441
horseshoe crab 442—443
human 15, 43, 235, 259, 289—293, 297,
        301, 407, 409, 417, 433, 438, 441,
        481, 492, 508—509, 582
human immunodeficiency virus (HIV) 484

Impatiens balsamina 443—444
Ipomoea purpurea 145

Kluyveromyces lactis 429, 432

Lactuca sativa: see lettuce
Leptosphaeria maculans 327, 336, 441
lettuce 103, 110, 117—121, 176, 281, 290,
        311, 341, 360, 486, 496, 563
lettuce mosaic virus (LMV) 486, 496
Linum marginale 174, 181
Linum usitatissimum: see flax
Lotus corniculatus 334
lupin 327, 329
Lycopersicon chilense 54
Lycopersicon esculentum: see tomato
Lycopersicon hirsutum 54, 146
Lycopersicon peruvianum 54, 143, 146
Lycopersicon pimpinellifolium 54, 146

Macadamia integrifolia 442, 444
Magnaporthe grisea 199, 249
Malus pumila: see apple
maize 11, 106, 108, 109, 114—145, 204,
        207, 260, 263, 295, 302—304, 355,
        388, 406, 408, 424—426, 431—433,
        442—444, 561, 576—578
mango 329

Medicago sativa: see alfalfa
Melampsora lini 12, 103, 106, 109, 174,
        181, 207, 295, 577—578
Meloidogyne incognita 143, 578, 588—589
Microbotryum violaceum 169
Micrococcus lysodeikticus 420
mildews 77—100, 103, 106, 110—115, 121,
        126, 130, 135—147, 165, 205, 242,
        257, 260, 270—271, 289, 298, 357,
        360, 496, 527, 529, 537, 563, 577
millet 135, 140
Mirabilis jalapa 396, 442, 444, 585
mistletoe 435
mouse 407, 417
Mycosphaerella zeae-maydis 144, 268
Mycosphaerella melonis 529
Myzus persicae 528

Nectria haematococca 328, 352, 354
Neurospora crassa 61, 333, 437, 441
Nicotiana alata 443
Nicotiana benthamiana 289, 392, 395, 508,
        581
Nicotiana debneyi 388, 551
Nicotiana digluta 7
Nicotiana glauca 492
Nicotiana glutinosa 1, 2, 7, 11, 388, 493,
        526, 532, 551
Nicotiana rustica 7
Nicotiana sylvestris 7, 9, 392, 397, 400—
        402, 583
Nicotiana tabacum: see tobacco
Nicotiana tomentosiformis 7, 9

oats 89, 106, 113—117, 280, 327, 330—334,
        357—358
Oidium graminis 79
oilseed rape 22, 134, 138, 142, 327, 336,
        391, 394, 400, 408, 410, 440, 588
onion 258, 327, 329, 341, 430, 433—434
Ophiostoma ulmi 266
orchid 7
Oryza sativa: see rice
Oudemansiella mucida 358

parsley 196, 215, 285, 290—291, 341—345,
        406, 552—555
papaya 268, 329, 415, 503

pa̱ ̱ ringspot virus (PRV) 503
p ̱ ̱18—119, 135, 214, 261—262, 268,
  281, 338, 354—357, 385, 414, 419
̱ea enation mosaic virus (PEMV) 119
pepper 7—8, 31, 32, 39—42, 135, 141, 203,
  204, 261, 288—289, 496, 498, 502,
  508
Periconia circinata 145, 200
Peronospora parasitica 207, 298, 308, 357,
  389, 404—405, 536, 537, 555—558,
  578
Peronospora tabacina 359, 383, 389—390,
  401—402, 525, 527, 535—548, 562,
  585
Petrostelium crispum: see parsley
Phakopsora pachyrhyzi 174
Phaseolus vulgaris: see French bean
Phoma lingam 394, 400
Phoma medicaginis 392, 393
Phomopsis subordinaria 172—174, 181
Phytolacca americana: see pokeweed
Phytomonas campestris 23
Phytophthora spp. 193, 196—199, 211, 332,
  496, 544
Phytophthora capsici 135, 141
Phytophthora infestans 126, 140—142, 166,
  175, 182, 197, 217, 262, 289, 306,
  309, 338, 384, 389, 394, 399, 402,
  404, 407, 529, 536, 585, 587
Phytophthora megasperma 193, 218, 392—
  393
Phytophothora parasitica 389—390, 394,
  397, 399, 401, 404
Pisum sativum: see pea
pine 328—329
Plantago spp. 7
Plantago lanceolata 172—173, 181
pokeweed 384, 394, 398, 428, 501
potato 69, 103, 106, 114—115, 119, 126,
  131, 140—142, 162, 166, 175, 181—
  182, 197, 207, 216—217, 262, 289,
  301, 304, 338, 341, 383—386, 394,
  398—402, 407, 421, 441—442, 480,
  506, 521, 555, 558, 579—581, 585—
  589
potato leafroll virus (PLRV) 580
potato spindle tuber viroid (PSTV) 482

potato virus X (PVX) 54, 199, 394—395,
  400, 496, 498, 508, 524, 590
potato virus Y (PVY) 389—390, 394—395,
  489, 496—497, 590
Pseudomonas campestris 23
Pseudomonas corrugata 215
Pseudomonas fluorescens 203, 310, 529—
  544, 561—562
Pseudomonas putida 216, 405, 529—530,
  552
Pseudomonas solanacearum 34, 142, 199,
  202, 286, 395, 433, 437, 441, 526
Pseudomonas syringae 12, 110, 116, 198,
  199—202, 207, 286—287, 298, 300,
  357, 399, 578
  pv. *glycinea* 107, 200—201, 215,
    282, 295, 552, 578
  pv. *maculicola* 207, 215, 288, 291—
    295, 299, 404, 405, 558, 578
  pv. *phaseolicola* 106, 201—202,
    281—290, 310
  pv. *pisi* 281, 283
  pv. *syringae* 202, 282, 288, 395, 552
  pv. *tabaci* 211, 283, 310, 390, 395,
    399, 404, 579
  pv. *tomato* 198, 200, 202, 207, 209,
    295, 299, 301, 380, 395, 399, 404—
    405, 424, 496, 533, 536, 577—578
Puccinia coronata 113
Puccinia dispersa 280
Puccinia graminis 103, 106, 180, 280
Puccinia hordei 177
Puccinia lagenophorae 170, 177
Puccinia sorghi 106, 108, 177, 204, 207
Puccinia striiformis 106
Puccinia recondita 103, 106, 163
Pyrenophora teres 136
Pyrularia pubera 438
Pythium spp. 332
Pythium aphanidermatum 391
Pythium sulvaticum 404

radish 22, 26, 374, 387, 396, 398, 433—441,
  529—537, 544, 558, 561—562
ragweed 429, 432
Ralstonia (Pseudomonas) solanacearum 34,
  286, 433, 437, 441
rapeseed: see oilseed rape

Raphanus sativus: see radish
Rhizobium 93, 218, 258, 262, 265
Rhizoctonia solani 389, 391—395, 420, 584
Rhynchosporium secalis 126
rice 34, 39, 106, 114—122, 131—142, 194,
    206—207, 240, 249, 295, 301, 326,
    329, 341, 390—393, 480, 506, 565,
    576, 578, 583, 584, 588
rubber tree 374, 394, 412, 415—416, 421—
    422, 428, 432, 583
Rumex crispus 170
rusts 102-130, 139—150, 163—165, 170,
    174—183, 205—208, 242, 248,
    251—262, 270, 280, 285, 289, 295,
    360, 577
rye 114, 130, 245, 280, 480

Saccharomyces cerevisiae: see yeast
Salmonella spp. 32, 201, 286
Sclerospora graminicola 140
Sclerotinia sclerotiorum 394, 400
Senecio vulgaris 170—177
Septoria avenae 327
Septoria lycopersici 334
Serratia marcescens 417, 526, 584, 591
Setosphaeria turcica 136, 141
Shigella 34, 201, 286
Shope fibroma virus 57
Silene alba 169
snowdrop 588
Solanum berthaultii 489
Solanum dulcamare 8
Solanum tuberosum: see potato
sorghum 327, 341, 346, 361—362, 387—
    388, 438, 440, 480
soybean 107, 110, 115, 131, 134, 138-141,
    179, 193, 196, 200—215, 257, 282—
    310, 348, 390, 397, 431, 433, 496,
    552, 576
spider mite 528, 561
squash: see courgette
Stachybotrys chartarum 563
Stagonospora nodorum 126, 128—131
Stemphylium alfalfae 392—393
Stemphylium loti 334
stinging nettle 413, 428—429, 583
Strobilurus tenacellus 358
sugarbeet 121, 207, 295, 506

sugarcane borer 135, 138, 142
sunflower 438, 440

Thaumatococcus danielli 374, 423, 425
Theobroma cacao: see cocoa
tobacco 1—20, 65, 69, 106—107, 115,
    141—143, 191—216, 267, 281—
    286, 291—295, 301, 304—308,
    358—359, 372, 376— 428, 438, 440,
    489, 493, 496—502, 506—509,
    523—553, 558—565, 578—592
tobacco etch virus (TEV) 390, 402
tobacco mild green mottle virus (TMGMV)
    492
tobacco mosaic virus (TMV) 1—20, 106—
    107, 141, 143, 191, 198—199, 207,
    211, 215, 267, 291—295, 304, 309,
    372, 380—382, 386—394, 401—
    404, 410, 413, 421, 423, 428, 483—
    486, 492, 496—510, 523—552, 561,
    563, 580—581, 590
tobacco vein mottling virus (TVMV) 390,
    402
tomato 7—12, 53—69, 106, 114—122, 137,
    142—146, 194—211, 218, 261—
    262, 266—267, 281, 289—291, 295,
    298—301, 328, 331—341, 374, 386,
    392—395, 398, 400, 406—408,
    419—421, 424, 431, 488—510, 521,
    536, 555, 558—560, 577—591
tomato mosaic virus (ToMV) 8, 488, 493—
    504, 510
tomato ringspot virus (ToRSV) 524, 525
tomato spotted wilt virus (TSWV) 489
Trichoderma hamatum 394, 584
Trichoderma harzianum 423, 583
Trichoderma longibrachatum 420
Trichoderma viride 194
Triticum aestivum: see wheat
Triticum monococcum 146
Triticum spelta 146
Triticum tauschii 112, 146
tulip 328—331, 337
turnip 22, 32, 36, 380
turnip crinkle virus (TCV) 380
turnip mosaic virus (TuMV) 395, 524

Uromyces appendiculatus 260
Uromyces rumicis 170
Uromyces vignae 199, 257, 261, 295, 306
Urtica dioica: see stinging nettle
Ustacystis waldsteiniae 258
Ustilago violacea 167—169
Ustilago nuda 169

Venturia inaequalis 106, 178
Verticillium albo-atrum 261
Verticillium dahliae 215, 267, 345—347
Vicia faba: see broad bean
Vigna sinensis 258
Vigna unguiculata: see cowpea
Viscaria vulgaris 167
Vitis vinifera: see grapevine

Waldsteinia geoides 258
wheat 78—79, 102—150, 162—164, 169,
        180—183, 207—208, 240, 252,
        279—280, 333—334, 358, 374, 395,
        405, 429—438, 480, 563, 576, 578
white rot fungi 248

Xanthomonas 21—52, 286,
Xanthomonas campestris
        pv. *armoraciae* 23—26
        pv. *campestris* 21—52, 110, 198—
        199, 207, 261, 283, 288
        pv. *glycineae* 307, 310
        pv. *malvacearum* 283, 348, 349
        pv. *manihotis* 32—33, 43
        pv. *raphani* 23, 39—41
        pv. *versicatoria* 395
Xanthomonas hyacinthi 25
Xanthomonas oryzae 34, 39, 106, 207, 295,
        578

yeast 115, 197, 203, 209—210, 289, 297—
        300, 345, 415—416, 426, 429, 437,
        491, 501, 509
Yersinia enterocolitica 34, 43, 201, 286

Zea mays: see maize
zucchini: see courgette
zucchini yellow mosaic virus (ZYMV) 480,
        503—504, 590